# Fundamentals of Technical Mathematics with Calculus

**Second Edition**

### Arthur D. Kramer

**Professor of Mathematics
New York City Technical College
City University of New York**

## GLENCOE
McGraw-Hill

New York, New York
Columbus, Ohio
Mission Hills, California
Peoria, Illinois

Production Services: York Production Services

**Dedicated to my dad, Joseph**

**Library of Congress Cataloging-in-Publication Data**

Kramer, Arthur D.
    Fundamentals of technical mathematics with calculus.

    Includes index.
    1. Mathematics—1961–    .  I. Title.
QA39.2.K687  1989b        510        88-13788
ISBN 0-07-035567-3

Send all inquiries to:
Glencoe/McGraw-Hill
936 Eastwind Drive
Westerville, Ohio 43081

4 5 6 7 8 9 10 11 12 13 14 15   RRC/PC   03 02 01 00 99 98 97 96 95

ISBN 0-07-035567-3

# Contents

# Preface

*Fundamentals of Technical Mathematics with Calculus* is primarily designed for students who are preparing for technical or scientific careers. This new edition has been substantially improved over the first edition. There is a more logical arrangement of chapters and a new chapter on matrices. Many new useful examples and exercises have been added, and more illustrations to clarify ideas have been included. Most important, the presentation and format for the text, the examples, and the exercises have been made more effective for the instructor to work with and for the student to understand.

While it is desirable for the student to have some background in elementary algebra, it is not essential. Chapter 2 includes a thorough treatment that can be used as a review or an introduction to algebra. The text provides comprehensive coverage of the mathematics necessary for the beginning student, starting with a review of arithmetic in Chapter 1, up to the calculus level student. The approach stresses a working knowledge of mathematics and the application of ideas to solving technical and practical problems.

Clearly, mathematics is best learned by doing many exercises and problems. The text now contains over 1100 worked-out examples and over 5500 exercises and problems. As much as possible, I have used meaningful applications taken from various scientific, technical, and practical areas, with an emphasis on electronics and computers. The applications do not require any prior knowledge of a specific subject and serve to develop an understanding of where and how mathematics is used in many fields. Many of the exercises relate back to the worked-out examples, which help the student to do the problems and reinforce the ideas in the text.

Each odd-numbered problem is similar to the following even-numbered problem, and the odd-numbered answers are given in Appendix C so they may be used for self-checking. Even-numbered answers are available in the *Instructor's Manual and Key*. Every chapter is followed by a series of review questions which not only cover the important chapter ideas but also strive to integrate several of them within some of the problems.

The chapters and chapter sections are arranged in a carefully thought-out, logical order so that ideas flow smoothly from one topic to another. Chapters 1, 2, 3, and 4 cover, respectively, basic material in arithmetic, algebra, variation, and geometry. Any topics in these chapters can be integrated into the course or used for review or reference. The metric system is presented in Chapter 4 and used equally throughout the text with the U.S. Customary system.

Chapters 5, 6, 7, 8, and 9 form a sequence of algebraic topics progressing from linear equations to logarithms. Chapters 10, 11, and 12 form a sequence of trigonometric topics from the basic trigonometric functions through graphs and identities. Chapters 13 to 16 contain further material in preparation for the calculus sequence, Chapters 19 to 25. Chapter 17 introduces inequalities and linear programming, topics of increasing importance in the technical world. Chapter 18, an introduction to computers and BASIC, is designed to augment other BASIC material; it presents applications to technology not usually found in introductory BASIC texts.

Many of the chapters can be studied independently, as there are cross-references to aid in their understanding. The chapter sequence is therefore flexible and can be arranged to meet various student needs. Calculator sections are presented at the end of those chapters which introduce calculator functions: Chapters 1, 2, 4, 9, and 10. These sections use examples in the chapter and are designed to reinforce understanding and strengthen the ability to estimate results and determine errors. They can be studied separately or integrated into the chapter material. In addition to these sections, many problems which lend themselves to calculator usage are marked with the symbol $\boxed{C}$.

I would like to thank the many users of the first edition whose suggestions have helped shape this second edition. Particular thanks to my colleagues Ken Stelzig of Chippewa Valley Technical College, Sandra Beken of Horry Georgetown Technical College, and Gary Simundza of Wentworth Institute of Technology for their thorough reviews of the manuscript and many useful comments. Last but not least, thanks again to my many students who are an inspiration to me and whose needs provide an invaluable guide in the writing of the text.

ARTHUR D. KRAMER

# C H A P T E R

# 1

# Review of Arithmetic

## 1-1 | *Laws of Arithmetic*

Two types of laws apply to the operations of addition and multiplication. The first type states that it makes no difference in what order you add or multiply two numbers. For example, $2 + 5 = 5 + 2$ and $3 \times 4 = 4 \times 3$. These are called the *commutative laws:*

$$A + B = B + A$$
$$\text{and} \quad (1\text{-}1)$$
$$A \times B = B \times A$$

**A** thorough understanding of arithmetic is essential for a good grasp of basic mathematics including algebra and geometry. Sections 1-1 to 1-4 are designed to reinforce your ability in arithmetic. Section 1-5 discusses arithmetic operations on the calculator and will help you estimate answers, interpret results, and identify errors. Section 1-6 explains precision and accuracy, which are important in technical applications. Chapters 2, 4, 9, and 10 introduce other calculator functions (sine, cosine, log, etc.) as the need arises. Finally, Section 1-7, Review Questions, contains many useful exercises which integrate the ideas in the chapter.

The second type states that if three numbers are to be added or multiplied, it makes no difference whether you start the operations with the first and second numbers or with the second and third. For example, in addition, $(2 + 3) + 5 = 2 + (3 + 5)$, or $5 + 5 = 2 + 8$. In multiplication, $(3 \times 4) \times 5 = 3 \times (4 \times 5)$, or $12 \times 5 = 3 \times 20$. These are called the *associative laws:*

$$(A + B) + C = A + (B + C)$$
$$\text{and} \quad (1\text{-}2)$$
$$(A \times B) \times C = A \times (B \times C)$$

When we apply these two types of laws together, it follows that three or more numbers can be added or multiplied in any order. For example, $2 + 3 + 4$ (or $2 \times 3 \times 4$) can be added (or multiplied) in six different ways:

$$2 + 3 + 4 \qquad 2 + 4 + 3 \qquad 3 + 2 + 4$$
$$3 + 4 + 2 \qquad 4 + 2 + 3 \qquad 4 + 3 + 2$$

Another important law of arithmetic which combines multiplication and addition is the *distributive law*. This law says that multiplication distributes over addition:

$$A \times (B + C) = A \times B + A \times C \qquad (1\text{-}3)$$

The distributive law is important in algebra and is applied in Chap. 2.

The *order of operations* in arithmetic, when there are no parentheses, is *multiplication or division first, addition or subtraction second*. Computers and most scientific calculators are programmed to perform the operations in this order. It is called *algebraic logic*, or the *algebraic operating system*.

### Example 1-1
Calculate the following:

$$5 \times 21 - 36 + 4 \div 2$$

**Solution**

Multiply and divide first:  $\qquad 105 - 36 + 2$

Then subtract and add:  $\qquad\qquad 69 + 2 = 71$

You can test if your calculator uses algebraic logic as follows. Enter this example exactly as it appears above and see if you get 71 when you press the equals key. See Sec. 1-5 for further discussion of calculator operation.

### Example 1-2
Calculate the following:  $\qquad 5 \times 21 - (36 + 4) \div 2$

**Solution**

Perform the operation in parentheses first:

$$5 \times 21 - 40 \div 2 = 105 - 20 = 85$$

See Example 1-20 for the calculator solution of Example 1-2.

### Example 1-3
Calculate the following:

$$\frac{5 \times 21 \times 12}{15 \times 7 \times 3}$$

**Solution**

You can multiply and divide in any order. One way is to multiply across the top and bottom and then to divide:

$$\frac{5 \times 21 \times 12}{15 \times 7 \times 3} = \frac{1260}{315} = 4$$

An easier way is to divide common factors in the top and bottom first and then to multiply:

$$\frac{\cancel{5} \times \cancel{21} \times \cancel{12}}{\cancel{15} \times \cancel{7} \times \cancel{3}} = \frac{4}{1} = 4$$

See Example 1-21 for the calculator solution of Example 1-3.

**Exercise 1-1**

In problems 1 to 16, test your understanding of arithmetic by mentally calculating the result. Check by doing the problem by hand. (You can check further with the calculator.)

**1.** $6 + 5 + 7 + 3 + 5 + 4$       **2.** $8 + 2 - 3 + 9 - 1$

**3.** $5 \times 2 \times 3 \times 4$       **4.** $12 \div 3 \div 2 \div 2$

**5.** $(800 + 20) \div 20$       **6.** $10 \div (5 + 40) \times 9$

**7.** $8 + 13 \times 2 - 4$       **8.** $7 - 6 \div 3 + 8 \div 4$

**9.** $5 + (8 - 1) \times 6 \div 2$       **10.** $(5 - 1) \div 2 + 3 \times 4$

**11.** $\dfrac{9 \times 4}{3 \times 6} + \dfrac{18}{6}$       **12.** $\dfrac{8 \times 9}{4} - \dfrac{15}{3}$

**13.** $\dfrac{12 \times 15}{5 \times 3 \times 2}$       **14.** $\dfrac{8 \times 7 \times 6}{4 \times 28}$

**15.** $\dfrac{(3 + 5) \times 2}{13 - 11}$       **16.** $\dfrac{6 + 8 \times (4 - 1)}{4 - 1 \times 2}$

In problems 17 to 24, solve each applied problem by hand. (You can check for accuracy with the calculator.)

**17.** One car travels 228 miles (mi) on 12 gallons (gal) of gasoline, and a second car travels 336 mi on 16 gal. How many more miles per gallon does the car with the better gas mileage get?

**18.** An experienced electronics technician earns $330 for a 40-hour week, and a civil engineer earns $287 for a 35-hour week. Who earns more per hour and how much more?

**19.** A Mariner space probe traveling at an average speed of 6000 miles per hour (mi/h) takes 400 days to reach Mars. What is the total distance traveled by the space probe?

20. A bus route is 22 kilometers (km) long. It takes the bus 50 minutes (min) to complete the route in one direction and 70 min to complete it in the other direction. What is the average rate of speed of the bus in kilometers per hour for the total trip back and forth? (Average rate = total distance/total time.)

21. The formula $N(N + 1)/2$ can be used to calculate the sum of the first $N$ numbers. Check the formula for the first 12 numbers by (a) adding 1 through 12 directly; (b) letting $N = 12$ in the formula and calculating the result.

22. A number is divisible by 9 if the sum of its digits is divisible by 9. Otherwise it is not. Test this with (a) the eight-digit serial number on a dollar or another bill and (b) your nine-digit social security number.

23. In BASIC and other computer languages, the following symbols are used for the arithmetic operations:

Addition +
Subtraction −
Multiplication *
Division /

Applying the order of operations, calculate the following example written in BASIC:

$$6 - 4/2 + 3 * 5$$

24. Do as in problem 23 for:

$$6 * 8/(3 + 1) - 4$$

## 1-2 | *Fractions*

Calculations with fractions, decimals, and percentages lead to mistakes because of a misunderstanding of the concepts involved. The calculator can prevent some of these mistakes, but it is not a substitute for clear understanding. The following examples review the basic arithmetic of fractions. Each example is designed to be done by hand. (You can check the results with the calculator.)

### Example 1-4
Simplify (reduce to lowest terms):

$$\frac{28}{42}$$

### Solution
Divide out any common factors (divisors) in the top and bottom:

$$\frac{28}{42} = \frac{\overset{1}{\cancel{2}} \times 2 \times \overset{1}{\cancel{7}}}{\underset{1}{\cancel{2}} \times 3 \times \underset{1}{\cancel{7}}} = \frac{2}{3}$$

It is not necessary to show all the factors. This is done to clearly illustrate the procedure. You can simply divide top and bottom by 14. The numbers $^{28}/_{42}$ and $^2/_3$ are called *equivalent fractions*. A fraction can be changed to an equivalent fraction by dividing out common factors or by multiplying top and bottom by the same factor. For example:

$$\frac{3}{4} = \frac{6}{8} = \frac{9}{12} = \frac{12}{16} \qquad \text{and so on}$$

## Example 1-5

Calculate each of the following:

1. $\dfrac{3}{8} \times \dfrac{2}{9}$

2. $\dfrac{5}{12} \div \dfrac{15}{16}$

3. $4 \times \dfrac{3}{14} \times \dfrac{5}{9}$

4. $\dfrac{5}{4} \times 8 \div \dfrac{1}{4}$

## Solution

1. $\dfrac{3}{8} \times \dfrac{2}{9}$

To *multiply fractions,* first divide out common factors that occur in any numerator and denominator. Then multiply the numerators and the denominators:

$$\frac{\overset{1}{\cancel{3}}}{\underset{4}{\cancel{8}}} \times \frac{\overset{1}{\cancel{2}}}{\underset{3}{\cancel{9}}} = \frac{1}{12}$$

2. $\dfrac{5}{12} \div \dfrac{15}{16}$

To *divide fractions,* invert the divisor, that is, the fraction after the division sign, and multiply:

$$\frac{\overset{1}{\cancel{3}}}{\underset{3}{\cancel{12}}} \times \frac{\overset{4}{\cancel{16}}}{\underset{3}{\cancel{15}}} = \frac{4}{9}$$

**3.** $4 \times \dfrac{3}{14} \times \dfrac{5}{9} = \dfrac{4}{1} \times \dfrac{\overset{2}{\cancel{4}}}{\underset{7}{\cancel{14}}} \times \dfrac{5}{\underset{3}{\cancel{9}}} = \dfrac{10}{21}$

**4.** $\dfrac{5}{4} \times 8 \div \dfrac{1}{4} = \dfrac{5}{\underset{1}{\cancel{4}}} \times \dfrac{8}{1} \times \dfrac{\overset{1}{\cancel{4}}}{1} = 40$

Notice in (3) and (4) that a whole number can be written with a denominator of 1. To multiply a fraction by a whole number, multiply the numerator by the whole number:

$$3 \times \frac{1}{4} = \frac{3}{4}$$

## Example 1-6
Combine the fractions:

$$\frac{2}{3} + \frac{5}{6}$$

## Solution
To combine fractions, that is, to add or subtract, first change each fraction to an equivalent fraction so that the denominators are the same. The easiest denominator to use is the lowest common denominator (lcd), which is the smallest number that each denominator divides into. Then combine the numerators over the lcd:

$$\frac{2(2)}{3(2)} + \frac{5}{6} = \frac{4}{6} + \frac{5}{6} = \frac{4+5}{6} = \frac{9}{6} = \frac{3}{2}$$

The lcd = 6, and only the first fraction needs to be changed. Note that the result % is reduced to ½.

## Example 1-7
Combine:

$$\frac{7}{15} + \frac{5}{12} - \frac{1}{6}$$

## Solution

The lcd is 60. This can be found by taking multiples of the largest denominator—15, 30, and so on—until each denominator divides into a multiple. Another way is to factor each denominator:

$$\frac{7}{(3)(5)} + \frac{5}{(2)(2)(3)} - \frac{1}{(2)(3)}$$

and to make up the lcd so that it contains all the factors appearing in each denominator: $(2)(2)(3)(5) = 60$. The solution is then:

$$\frac{7(4)}{15(4)} + \frac{5(5)}{12(5)} - \frac{1(10)}{6(10)} = \frac{28 + 25 - 10}{60} = \frac{43}{60}$$

Note that parentheses or a dot ($\cdot$) indicates multiplication. These signs are used in algebra to avoid confusing the multiplication sign ($\times$) with the letter x.

## Example 1-8

Calculate:

$$\frac{13}{8} - \frac{7}{5} \times \frac{15}{14} + 2 \div \frac{8}{15}$$

## Solution

Invert the last fraction, and change the operation of division to multiplication. Then divide common factors and multiply:

$$\frac{13}{8} - \frac{\overset{1}{\cancel{7}}}{\cancel{5}} \times \frac{\overset{3}{\cancel{15}}}{\underset{2}{\cancel{14}}} + \overset{1}{\cancel{2}} \times \frac{15}{\underset{4}{\cancel{8}}} = \frac{13}{8} - \frac{3}{2} + \frac{15}{4}$$

Now combine over the lcd, 8:

$$\frac{13}{8} - \frac{3(4)}{2(4)} + \frac{15(2)}{4(2)} = \frac{13 - 12 + 30}{8} = \frac{31}{8}$$

See Example 1-22 for the calculator solution of Example 1-8.

**Exercise 1-2**  In problems 1 to 6, simplify each fraction (reduce to lowest terms).

1. $\dfrac{6}{10}$    2. $\dfrac{12}{36}$

3. $\dfrac{28}{35}$    4. $\dfrac{27}{54}$

**5.** $\dfrac{39}{52}$                          **6.** $\dfrac{34}{51}$

In problems 7 to 32, calculate each by hand. (You can check with the calculator.)

**7.** $\dfrac{5}{9} \times \dfrac{6}{25}$                          **8.** $\dfrac{2}{21} \times \dfrac{7}{16}$

**9.** $\dfrac{8}{9} \div \dfrac{2}{3}$                          **10.** $\dfrac{3}{11} \div \dfrac{1}{22}$

**11.** $\dfrac{3}{5} \times \dfrac{15}{7} \times \dfrac{14}{9}$                          **12.** $6 \times \dfrac{4}{5} \div \dfrac{8}{15}$

**13.** $\dfrac{3}{17} \div \left( \dfrac{1}{34} \times \dfrac{1}{2} \right)$                          **14.** $\left( \dfrac{9}{8} \div \dfrac{3}{4} \right) \div \dfrac{3}{2}$

**15.** $\dfrac{3}{8} + \dfrac{1}{4}$                          **16.** $\dfrac{4}{15} + \dfrac{5}{6}$

**17.** $\dfrac{3}{4} - \dfrac{1}{2} + \dfrac{7}{10}$                          **18.** $\dfrac{1}{6} - \dfrac{2}{3} + \dfrac{11}{20}$

**19.** $2 + \dfrac{7}{8} + \dfrac{2}{3}$                          **20.** $\dfrac{5}{2} + \dfrac{5}{3} + \dfrac{5}{6}$

**21.** $\dfrac{16}{9} \times \dfrac{1}{2} + \dfrac{1}{4}$                          **22.** $\dfrac{1}{6} + \dfrac{3}{8} \div \dfrac{1}{4}$

**23.** $3 \times \dfrac{1}{6} + \dfrac{7}{2} - \dfrac{4}{5} \div 8$                          **24.** $\dfrac{3}{100} + \dfrac{7}{10} \times \dfrac{2}{35} - \dfrac{1}{50}$

**25.** $\left( \dfrac{1}{2} + \dfrac{1}{3} \right) \times \left( 8 \div \dfrac{4}{3} \right)$                          **26.** $\left( 1 + \dfrac{3}{8} \right) \div \left( 1 - \dfrac{3}{8} \right)$

**27.** A \$30,000 inheritance is distributed as follows: one-half to the spouse, two-thirds of what is left to the children, and the remainder to charity. How much money is given to charity?

**28.** A bookcase is to be 8 feet (ft) 3½ inches (in.) high and to contain six equally spaced shelves and a top, each ½ in. thick (seven pieces total). How many feet and inches apart should each shelf be?

**29.** Calculate the resistance of a series-parallel circuit given by:

$$R = \dfrac{1}{\dfrac{1}{12} + \dfrac{1}{4}} + 3$$

**30.** Calculate the focal length of a lens given by:

$$f = \dfrac{\dfrac{1}{10} \times \dfrac{3}{20}}{\left( \dfrac{3}{2} - 1 \right)\left( \dfrac{1}{10} + \dfrac{3}{20} \right)}$$

**31.** Calculate the following written in BASIC (See Exercise 1-1, problem 23):

$$(1/10) * 3/(3/2 - 1) * (20/3)$$

**32.** Do the same as problem 31 for:

$$(1/2 + 1/3)/(1/6) * 4$$

# 1-3 | *Decimals and Percentages*

## *Decimals*

Decimals represent fractions whose denominators are powers of 10: 10, 100, 1000, and so on. The number of decimal places equals the number of zeros in the denominator:

$$0.3 = \frac{3}{10} \qquad 0.21 = \frac{21}{100} \qquad 0.067 = \frac{67}{1000}$$

### Example 1-9
Calculate each of the following:

**1.** $6.23 + 17.87 + 0.15$     **2.** $1.3 \times 0.05$

**3.** $\dfrac{13.2}{0.12}$     **4.** $\dfrac{0.5 \times 0.02}{0.06 - 0.01}$

### Solution
**1.** $6.23 + 17.87 + 0.15$

*Add or subtract decimals* in the same way as whole numbers, lining up the decimal point and the columns:

$$
\begin{array}{r}
6.23 \\
17.87 \\
\underline{0.15} \\
24.25
\end{array}
$$

**2.** $1.3 \times 0.05$

To *multiply decimals,* multiply the numbers and add the decimal places in each number to determine the number of decimal places in the answer:

$$1.3 \times 0.05 = 0.065$$

(Decimal places: one + two = three)

3. $\dfrac{13.2}{0.12}$

To *divide decimals,* first move the decimal points in the numerator and denominator to the right as many places as there are in the denominator. Then divide the numbers:

$$\frac{13.2}{0.12} = \frac{1320.}{12.} = 110$$

Note that moving the decimal points in the numerator and denominator to the right is the same as multiplying the top and bottom by 10, 100, 1000, etc., and does not change the value of the fraction. Above, the top and bottom are multiplied by 100.

4. $\dfrac{0.5 \times 0.02}{0.06 - 0.01} = \dfrac{0.010}{0.05} = \dfrac{1.0}{5} = 0.2$

Study (4), which combines subtraction, multiplication, and division of decimals. Observe in the last step when 5 is divided into 1.0 the decimal point is kept in the same place so the number of decimal places remains the same.

## Percentages

Percentages are fractions with denominators of 100. *To change from a percentage to a decimal,* move the decimal point two places to the left, and vice versa.

### Example 1-10
Express each number as a fraction, decimal, and percentage.

1. $\dfrac{53}{100} = 0.53 = 53\%$

2. $\dfrac{1}{10} = 0.10 = 10\%$

3. $\dfrac{3}{50} = 0.06 = 6\%$

4. $\dfrac{27}{1000} = 0.027 = 2.7\%$

5. $\dfrac{7}{4} = 1.75 = 175\%$

6. $\dfrac{1}{3} = 0.333 \cdots = 33\dfrac{1}{3}\%$

## Solution

To change each fraction to a decimal, you must first insert the decimal point in the numerator and then add zeros before dividing. For example, in (3):

$$\frac{3}{50} = \frac{3.00}{50} = 0.06$$

## Example 1-11

The current in a circuit is 3.50 amperes (A). The voltage is increased by 4 percent, causing the current to increase by the same percentage.

1. What is the increase in current?
2. What is the new current?

## Solution

1. The increase in the current is:

$$3.50(4\%) = 3.50(0.04) = 0.14 \text{ A}$$

2. The new current is:

$$3.50 + 0.14 = 3.64 \text{ A}$$

## Example 1-12

The list price of a computer is $779. A discount store sells it for 20 percent off.

1. What is the discount price?
2. What is the final price, including a tax of 7½ percent?

## Solution

1. The discount price is:

$$779 - 779(20\%) = 779 - 779(0.20)$$
$$= 779 - 155.80$$
$$= \$623.20$$

Note that (1) can also be done by multiplying (779)(0.80).

2. The final price, including a tax of 7½ percent, is:

$$623.20 + 623.20(7\tfrac{1}{2}\%) = 623.20 + 623.20(0.075)$$
$$= 623.20 + 46.74$$
$$= \$699.94$$

Note that (2) can also be done by multiplying (623.20)(1.075).

### Example 1-13

If a total of 5.0 cubic meters (m³) of cement and sand produces 4.3 m³ of concrete, what is the percentage of shrinkage?

### Solution

The percentage change is given by:

$$\text{Percentage change (increase or decrease)} = \frac{\text{change} \times 100\%}{\text{original amount}} \qquad (1\text{-}4)$$

Therefore:

$$\text{Percentage of shrinkage} = \frac{(5.0 - 4.3)(100)}{5.0} = \frac{(0.7)(100)}{5.0} = 14\%$$

### Exercise 1-3

In problems 1 to 14, calculate each by hand. (You can check with the calculator.)

**1.** $1.05 + 8.98 + 0.06$

**2.** $15.64 + 4.36 - 19.09$

**3.** $8.1 \times 1.1$

**4.** $0.031 \times 0.20$

**5.** $\dfrac{5.1}{0.17}$

**6.** $\dfrac{3.60}{0.030}$

**7.** $3.2 \times \dfrac{0.10}{4.0}$

**8.** $\dfrac{0.72}{3.0 \times 0.3}$

**9.** $(1.3 + 2.8) \times (1.6 + 1.4)$

**10.** $(0.4 + 0.1) \times (0.08 - 0.04)$

**11.** $\dfrac{1.2 \times 1.0 \times 0.03}{0.0050 + 0.0040}$

**12.** $\dfrac{0.02 \times (3.0 + 5.0)}{0.004}$

**13.** $\dfrac{0.5 \times 7.0 + 0.6 \times 7.0}{2.3 - 1.2}$

**14.** $\dfrac{8.0 \times 0.6 \times 0.04}{40 \times 30 \times 0.02}$

In problems 15 to 26, express each number as a fraction, decimal, and percentage.

**15.** 20%

**16.** 86%

**17.** 0.17

**18.** 0.06

**19.** 5.6%

**20.** 8½%

**21.** 0.004

**22.** 2.00

**23.** $\dfrac{1}{4}$

**24.** $\dfrac{2}{3}$

**25.** $\dfrac{6}{5}$

**26.** $\dfrac{1}{1000}$

In problems 27 to 38, solve each applied problem by hand.

**27.** Four batteries connected in series have the following voltages: 13.01, 12.52, 6.21, and 8.38 volts (V). What is the total voltage? (*Note:* Total voltage = sum of series voltages.)

28. One discount technical supply store sells a micrometer (an instrument for measuring small distances) for one-third off its usual price of $59.40. Another discount store sells the same micrometer for 25 percent off its usual price of $56.00. Which micrometer is the best bargain and by how much?

29. About 0.003 percent of seawater is NaCl (sodium chloride, or table salt). How many grams of NaCl are contained in 1 million grams (g) (1 metric ton) of seawater?

30. A retailer pays the publisher 20 percent less than the list price for this text. The list price is $26.95. *(a)* How much does the retailer pay? *(b)* What is the percentage of markup that the retailer adds to sell at the list price? (The answer is not 20 percent.)

31. Suppose your calculator costs $19.00 today. *(a)* Approximately how much will it cost to replace it 1 year from now at an inflation rate of 15 percent per year? *(b)* Approximately how much will it cost to replace it 2 years from now?

32. The *specific gravity* of an object is its weight in air divided by the weight of an equal volume of water. A 50 percent solution of sulfuric acid ($H_2SO_4$) is made by adding 10 g of water to an equal volume of acid. If the solution weighs 28.31 g, what is the specific gravity of $H_2SO_4$? *(Hint:* Find the weight of the acid.)

33. The current in a circuit increases from 8.00 to 8.40 A when the voltage is increased from 12.0 V to a higher value. *(a)* What is the percentage increase in the current? *(b)* If the percentage increase in the voltage is the same as that in the current, what is the new voltage?

34. A computer cabinet whose volume is 1.5 $m^3$ expands to 1.6 $m^3$ when heated. What is the percentage increase in volume?

35. The earth's surface receives 317 British thermal units per hour per square foot [Btu/(h · $ft^2$)] of solar energy on a clear day. How many Btu per hour is converted by a solar panel 4 ft by 2 ft if 75 percent of it is covered with solar cells?

36. Arrange in order of increasing size: 1.19, ⅝%, 110%, 1⅙, $^{112}$⁄₁₀₀₀.

37. Calculate the following, written in BASIC:

$$(2.1 - 1.2) * 6.0/0.20$$

38. Do as in problem 37 for:

$$6.0 * 0.40 * 0.01/(10 * 3.0 * 0.02)$$

## 1-4    *Powers and Roots*

### *Powers*

A *power*, or *exponent*, defines repeated multiplication:

$$x^n = \underbrace{(x)(x)(x) \cdots (x)}_{n \text{ times}} \qquad n = \text{positive whole number} \qquad (1\text{-}5)$$

### Example 1-14
Evaluate each of the following:

**1.** $2^5 = (2)(2)(2)(2)(2) = 32$

**2.** $5^4 = (5)(5)(5)(5) = 625$

**3.** $\left(\dfrac{2}{3}\right)^3 = \left(\dfrac{2}{3}\right)\left(\dfrac{2}{3}\right)\left(\dfrac{2}{3}\right) = \dfrac{8}{27}$

**4.** $(0.01)^2 = (0.01)(0.01) = 0.0001$

In the order of operations, raising to a power (or taking a root) is done before multiplication or division as shown in the next example.

### Example 1-15
Evaluate each of the following:

**1.** $\dfrac{(2)^5(5)^2}{4^3} = \dfrac{(32)(25)}{64} = \dfrac{25}{2}$   or   $12.5$

**2.** $\dfrac{(0.3)^2 + (0.4)^2}{0.5} = \dfrac{0.09 + 0.16}{0.5} = \dfrac{0.25}{0.5} = \dfrac{1}{2}$   or   $0.5$

**3.** $\left(\dfrac{5}{2}\right)^2 - \dfrac{(2)^4(3)^2}{6^2} = \dfrac{25}{4} - \dfrac{(16)(9)}{36} = \dfrac{25}{4} - 4 = \dfrac{9}{4}$   or   $2.25$

See Example 1-25 for the calculator solution to Example 1-15(3).

## *Square Roots*

Taking a root is the *inverse* operation of raising to a power, just as division is the inverse of multiplication and subtraction is the inverse of addition. If a positive number squared (multiplied by itself) exactly equals a second number, then the first number is called the *square root* of the second number. This is indicated with the root, or radical, sign $\sqrt{\phantom{x}}$ .

### Example 1-16
Find each square root.

**1.** $\sqrt{9} = 3$      because      $3^2 = 9$

**2.** $\sqrt{625} = 25$      because      $25^2 = 625$

**3.** $\sqrt{\dfrac{16}{49}} = \dfrac{4}{7}$      because      $\left(\dfrac{4}{7}\right)^2 = \dfrac{16}{49}$

**4.** $\sqrt{1.21} = 1.1$      because      $(1.1)^2 = 1.21$

The numbers 9, 625, $^{16}/_{49}$, and 1.21 are called *perfect squares* because their square roots are whole numbers or fractions. (Note: Fractions include decimals.)

Whole numbers and fractions are called *rational numbers* because they can be expressed as a ratio of two whole numbers. Numbers that cannot be expressed as a ratio of two whole numbers are called *irrational numbers*. Square roots of numbers that are not perfect squares, such as $\sqrt{2}$ and $\sqrt{10}$, are irrational. Irrational numbers are infinite decimals with no pattern of repeating digits:

$$\sqrt{2} = 1.414213562373 \cdots$$
$$\sqrt{10} = 3.162277660167 \cdots$$

Because square roots cannot be expressed exactly as decimals, we often calculate with them in the form of radicals. Three basic rules of calculation for the square roots of two positive numbers $x$ and $y$ are:

$$\sqrt{x^2} = (\sqrt{x})^2 = (\sqrt{x})(\sqrt{x}) = x \qquad (1\text{-}6)$$
$$\sqrt{xy} = (\sqrt{x})(\sqrt{y}) \qquad (1\text{-}7)$$
$$\sqrt{\frac{x}{y}} = \frac{\sqrt{x}}{\sqrt{y}} \qquad (1\text{-}8)$$

In rule (1-6) all the expressions are equal, and they define the square root sign: *When a radical is squared or multiplied by itself, the radical sign is eliminated.* Rules (1-7) and (1-8) can be used either way to simplify radical expressions: Products or quotients under a radical sign can be separated into products or quotients of separate radicals, and vice versa.

### Example 1-17

Simplify each of the following by applying rules (1-6), (1-7), and (1-8):

**1.** $(\sqrt{3})^2 + \sqrt{5}^2 - (\sqrt{6})(\sqrt{6}) = 3 + 5 - 6 = 2$    by (1-6)

**2.** $\sqrt{12} = (\sqrt{4})(\sqrt{3}) = 2\sqrt{3}$    by (1-7)

**3.** $\sqrt{200} = (\sqrt{100})(\sqrt{2}) = 10\sqrt{2}$    by (1-7)

Notice in (2) and (3) that when you apply rule (1-7) and separate a radical into the product of two radicals, you look for factors (divisors) that are perfect squares (4, 9, 16, 25, . . . ). This reduces the number under the radical and simplifies the expression.

**4.** $(\sqrt{0.2})(\sqrt{1.8}) = \sqrt{0.36} = 0.6$    by (1-7)

**5.** $\dfrac{\sqrt{28}}{\sqrt{7}} = \sqrt{\dfrac{28}{7}} = \sqrt{4} = 2$    by (1-8)

**6.** $\sqrt{\dfrac{10}{9}} = \dfrac{\sqrt{10}}{\sqrt{9}} = \dfrac{\sqrt{10}}{3}$ or $\dfrac{1}{3}\sqrt{10}$    by (1-8)

Notice in (2), (3), and (6) that a number in front of a radical means multiplication ($2\sqrt{3} = 2 \times \sqrt{3}$). See Example 1-27 for the calculator verification of Example 1-17(2) and (6). Radicals are discussed in greater detail in Chap. 8.

### Example 1-18

Simplify each of the following by applying rules (1-7) and (1-8).

**1.** $\dfrac{\sqrt{24}}{(\sqrt{3})(\sqrt{2})} = \dfrac{\sqrt{24}}{\sqrt{6}} = \sqrt{4} = 2$

**2.** $\dfrac{(\sqrt{0.2})(\sqrt{2.5})}{\sqrt{2}} = \dfrac{\sqrt{0.50}}{\sqrt{2}} = \sqrt{0.25} = 0.5$

**3.** $\dfrac{3\sqrt{10}}{(\sqrt{3})(\sqrt{6})} = \dfrac{3\sqrt{10}}{\sqrt{18}} = \dfrac{3\sqrt{10}}{(\sqrt{9})(\sqrt{2})} = \dfrac{\cancel{3}\sqrt{10}}{\cancel{3}\sqrt{2}} = \sqrt{5}$

Note that there is more than one way to do each of the above. For example, another way you can do (1) is by first dividing:

$$\frac{\sqrt{24}}{(\sqrt{3})(\sqrt{2})} = \frac{\sqrt{12}}{\sqrt{3}} = \sqrt{4} = 2$$

## *Higher-Order Roots*

The properties of higher-order roots (cube roots, fourth roots, etc.) are similar to the properties of square roots. The $n$th root of a number, written $\sqrt[n]{\phantom{x}}$ , is the inverse of the $n$th power as follows:

### Example 1-19

Find the indicated roots.

**1.** $\sqrt[3]{1000} = 10$    because    $10^3 = 1000$

**2.** $\sqrt[4]{256} = 4$    because    $4^4 = 256$

**3.** $\sqrt[3]{\dfrac{8}{27}} = \dfrac{2}{3}$    because    $\left(\dfrac{2}{3}\right)^3 = \dfrac{8}{27}$

**4.** $\sqrt[3]{0.064} = 0.4$    because    $(0.4)^3 = 0.064$

The number in the crook of the radical is called the *index*. For a square root, the index is not written but is understood to be 2. Following are two rules for higher-order roots of positive numbers. The first rule is the definition and the extension of rule (1-6) for square roots:

$$\sqrt[n]{x^n} = (\sqrt[n]{x})^n = x \tag{1-9}$$

For example, $\sqrt[5]{2^5} = (\sqrt[5]{2})^5 = 2$. The second rule is useful in finding the $n$th root on the calculator and is applied in the next section:

$$\sqrt[n]{x} = x^{1/n} \tag{1-10}$$

Rule (1-10) says that taking the $n$th root of a number is the same as raising it to the $(1/n)$th power. For example, $\sqrt{3} = 3^{1/2}$, $\sqrt[3]{2} = 2^{1/3}$, etc. See Example 1-31 for the calculator application of rule (1-10). The reason for this rule is explained in Sec. 8-2.

**Exercise 1-4**

In problems 1 to 28, evaluate each by hand. (You can check for accuracy with your calculator.)

**1.** $3^4$

**2.** $4^3$

**3.** $\left(\dfrac{1}{2}\right)^2$

**4.** $\left(\dfrac{3}{5}\right)^4$

**5.** $(0.7)^3$

**6.** $(1.1)^2$

**7.** $\dfrac{4^4}{(8)^2(6)^2}$

**8.** $\dfrac{2^3 + 3^3}{5^2}$

**9.** $\left(\dfrac{1}{5}\right)^2 + \left(\dfrac{2}{5}\right)\left(\dfrac{1}{10}\right)^2$

**10.** $(6)\left(\dfrac{1}{4}\right)^2 - (4)\left(\dfrac{1}{6}\right)^2$

**11.** $\dfrac{(0.2)^3}{(0.5)^2 - (0.3)^2}$

**12.** $\dfrac{(0.7)^2(0.3)^3}{(2.1)^2} - (0.1)^3$

**13.** $\dfrac{2^5}{8^2} - \left(\dfrac{3}{4}\right)^3$

**14.** $\left(\dfrac{1}{2} + \dfrac{1}{3}\right)^2\left(\dfrac{1}{2} - \dfrac{1}{10}\right)^2$

Find each root.

**15.** $\sqrt{16}$

**16.** $\sqrt{121}$

**17.** $\sqrt{\dfrac{25}{4}}$

**18.** $\sqrt{\dfrac{9}{100}}$

**19.** $\sqrt{0.64}$

**20.** $\sqrt{1.44}$

**21.** $\sqrt[3]{64}$

**22.** $\sqrt[4]{81}$

**23.** $\sqrt[4]{5^4}$

**24.** $(\sqrt[3]{0.1})^3$

**25.** $\sqrt[4]{\dfrac{1}{16}}$

**26.** $\sqrt[3]{\dfrac{27}{1000}}$

**27.** $\sqrt[3]{0.008}$

**28.** $\sqrt[3]{0.125}$

In problems 29 to 46, simplify each expression (see Examples 1-17 and 1-18).

**29.** $\sqrt{2^2} + (\sqrt{5})^2$

**30.** $(\sqrt{7})(\sqrt{7}) - (\sqrt{3})^2$

**31.** $\sqrt{8}$

**32.** $\sqrt{32}$

**33.** $2\sqrt{50}$

**34.** $3\sqrt{27}$

**35.** $(\sqrt{8})(\sqrt{2}) + (\sqrt{10})^2$

**36.** $(\sqrt{2})(\sqrt{18}) - (\sqrt{12})(\sqrt{3})$

**37.** $\dfrac{\sqrt{12}}{\sqrt{3}}$

**38.** $\dfrac{\sqrt{3}}{\sqrt{75}}$

**39.** $\sqrt{\dfrac{7}{16}}$

**40.** $\sqrt{\dfrac{18}{25}}$

**41.** $\dfrac{(\sqrt{4})(\sqrt{12})}{\sqrt{3}}$        **42.** $\dfrac{\sqrt{20}}{(\sqrt{200})(\sqrt{10})}$

**43.** $\dfrac{\sqrt{0.3}}{(\sqrt{0.6})(\sqrt{0.18})}$        **44.** $\sqrt{\dfrac{(8)(0.1)^2}{0.02}}$

**45.** $\dfrac{(\sqrt{2})(\sqrt{24})}{\sqrt{6}}$        **46.** $\dfrac{\sqrt{200}}{\sqrt{18}} + \sqrt{\dfrac{4}{9}}$

**47.** The volume of a cube $V = s^3$, where $s =$ side. Find the volume of a cubic crystal of NaCl (sodium chloride) if $s = 1.2$ millimeters (mm).

**48.** At an inflation rate of 10 percent a year, the cost of a \$10,000 car will increase to approximately $10,000(1.1)^3$ in 3 years. How much will this approximate cost be?

**49.** The impedance $Z$ of an alternating-current (ac) circuit that has a resistance $R$ and a reactance $X$ is given by $Z = \sqrt{R^2 + X^2}$. Find $Z$ if $R = 0.90$ ohms ($\Omega$) and $X = 1.2 \, \Omega$.

**50.** The coefficient of restitution (measure of elasticity) for a material is given by $e = \sqrt{h'/h}$, where $h$ is the height of a small sphere of the material dropped onto a smooth rigid surface and $h'$ is the height of the rebound. Find $e$ for a semielastic material if $h = 20$ centimeters (cm) and $h' = 9.8$ cm.

**51.** In the BASIC computer language, an arrow ($\uparrow$) or caret ($\wedge$) denotes an exponent, and SQR($x$) denotes $\sqrt{x}$. For example, $\sqrt{81} - 2^3$ is written SQR(81) $- 2\wedge3$. Calculate the following example, written in BASIC:

$$4\wedge3/\text{SQR}(36) + 2/9$$

**52.** Do as in problem 51 for:

$$4\wedge2/\text{SQR}(81) + 2/9$$

# 1-5 | *Hand Calculator Operations*

We live in the age of electronic computation. Imagine a long-division problem that would take you 1 minute to do by hand. The hand calculator takes a few hundredths of a second to do such a problem. One of the world's fastest computers, the CRAY-1, can do 100,000,000 (one hundred million) of these problems in 1 second!

You need to master the calculator. It will allow you to spend less time doing tedious calculations and to spend more time getting a good grasp of arithmetic concepts. Remember, though, speed alone is of little value unless you can understand and apply these concepts. For scientific and technical use you need a "scientific" type of calculator. This type has the trigonometric keys $\boxed{\text{sin}}$, $\boxed{\text{cos}}$, $\boxed{\text{tan}}$; an exponential key $\boxed{y^x}$ or $\boxed{x^y}$; and the logarithmic keys $\boxed{\text{log}}$ and $\boxed{\text{ln}}$. This section is not designed to replace the instruction manual for your calculator, which you should read to understand the calculator's basic operations, but to help you master the calculator. For example, some calculators use what is called reverse Polish notation (RPN). These have an $\boxed{\text{ENT}}$ key instead of an $\boxed{=}$ key. In this text, all calculator operations are shown with an $\boxed{=}$ key, so if your calculator uses RPN, you will have to keep in mind the different procedure.

One of the important things to realize about your calculator is that *it cannot think!* You possess that unique ability. You must understand the problem, interpret the information, and key it into the calculator correctly. Furthermore, and most important, you must be able to judge if the answer makes sense. If it does not, and this can happen often, you must understand the mathematical concepts well enough to troubleshoot for the error. You should estimate or approximate answers, whenever possible, before calculating. This provides a check on the results and enables you to understand the concepts better. These are some of the skills that this section and others in the text are designed to help you develop.

## *Arithmetic Operations and Memory*

Most scientific calculators are programmed to perform the basic arithmetic operations $\boxed{+}$, $\boxed{-}$, $\boxed{\times}$, and $\boxed{\div}$ according to the order of operations discussed in Sec. 1-1. This is called algebraic logic or the *algebraic operating system* (AOS). Example 1-1 can be used to test whether your calculator employs AOS. When you key in the example exactly as it appears below, the display should show the number indicated by the arrow:

$$5 \boxed{\times} 21 \boxed{-} 36 \boxed{+} 4 \boxed{\div} 2 \boxed{=} \rightarrow 71$$

If your answer differs from 71, you may need to check your instruction manual for the correct procedure. Since most scientific calculators are programmed for AOS, the steps for these calculators are shown in the text.

**Example 1-20**

Calculate:     $5 \times 21 - (36 + 4) \div 2$.

**Solution**

This is Example 1-2. If your calculator has parentheses, the problem can be keyed in directly:

$$5 \boxed{\times} 21 \boxed{-} \boxed{(} 36 \boxed{+} 4 \boxed{)} \boxed{\div} 2 \boxed{=} \rightarrow 85$$

Another way to do the example without parentheses is to use the memory, which stores a result:

$$36 \boxed{+} 4 \boxed{=} \boxed{\text{M}_{\text{in}}} \quad 5 \boxed{\times} 21 \boxed{-} \boxed{\text{MR}} \boxed{\div} 2 \boxed{=} \rightarrow 85$$

$$\downarrow \qquad\qquad\qquad \uparrow$$

$$40 \qquad\qquad\qquad 40$$

The $\boxed{M_{in}}$ key stores the result, 40. The $\boxed{MR}$ key recalls the result and enters it into the operations being performed. Some calculators use $\boxed{STO}$ or $\boxed{x \rightarrow M}$ instead of $\boxed{M_{in}}$, and $\boxed{RCL}$ or $\boxed{RM}$ rather than $\boxed{MR}$. You can also write down the number on the calculator display instead of using $\boxed{M_{in}}$ and enter that number into the calculator instead of using $\boxed{MR}$. Some calculators have more than one memory, and it is necessary to key in the number of the memory after $\boxed{M_{in}}$ and $\boxed{MR}$, for example, $\boxed{M_{in}}$ 07 and $\boxed{MR}$ 07.

## Example 1-21
Calculate:

$$\frac{5 \times 21 \times 12}{15 \times 7 \times 3}$$

### Solution
This is Example 1-3. One way you can do it on the calculator is:

$$5 \boxed{\times} 21 \boxed{\times} 12 \boxed{\div} 15 \boxed{\div} 7 \boxed{\div} 3 \boxed{=} \rightarrow 4$$

Another way is:

$$5 \boxed{\div} 15 \boxed{\times} 21 \boxed{\div} 7 \boxed{\times} 12 \boxed{\div} 3 \boxed{=} \rightarrow 4$$

## Example 1-22
Calculate:

$$\frac{13}{8} - \frac{7}{5} \times \frac{15}{14} + 2 \div \frac{8}{15}$$

### Solution
This is Example 1-8. One calculator solution is:

$$13 \boxed{\div} 8 \boxed{-} 7 \boxed{\div} 5 \boxed{\times} 15 \boxed{\div} 14 \boxed{+} 2 \boxed{\times} 15 \boxed{\div} 8 \boxed{=} \rightarrow 3.875$$

When the arithmetic becomes more difficult, you depend more on the calculator. However, you need some way of checking that the calculator answer is approximately correct. You may be keying in the problem incorrectly or misinterpreting the problem or the data, or the batteries might be weak. The next two examples are first done approximately to provide a check on the calculator answer.

## Example 1-23
Approximate the answer by hand and then calculate:

$$\frac{(64 + 57) \times 320}{840 - 50 \times 8}$$

## Solution

Approximate the answer by rounding all the numbers to one figure and then doing the calculation by hand. If the second figure is 5 or greater in a number, round to the next larger figure as follows:

$$\frac{(64 + 57) \times 320}{840 - 50 \times 8} \approx \frac{(60 + 60) \times 300}{800 - 50 \times 8} = \frac{120 \times 300}{800 - 400} = \frac{36,0\cancel{0}\cancel{0}}{4\cancel{0}\cancel{0}} = 90$$

This gives you some idea of the size of the answer. One way the problem can then be done on the calculator is by using the memory:

$$840 \boxed{-} 50 \boxed{\times} 8 \boxed{=} \boxed{\text{M}_{\text{in}}} \ \ 64 \boxed{+} 57 \boxed{=} \boxed{\times} 320 \boxed{\div} \boxed{\text{MR}} \boxed{=} \rightarrow 88$$

$$\downarrow \qquad\qquad\qquad\qquad\qquad\qquad\qquad \uparrow$$

$$440 \qquad\qquad\qquad\qquad\qquad\qquad\qquad 440$$

or by using parentheses:

$$\boxed{(} \ 64 \boxed{+} 57 \boxed{)} \boxed{\times} 320 \boxed{\div} \boxed{(} \ 840 \boxed{-} 50 \boxed{\times} 8 \boxed{)} \boxed{=} \rightarrow 88$$

Suppose your calculator display shows an answer of 880 or 8.80. The approximate answer of 90 alerts you to the error, and you can check for it.

## Example 1-24

Approximate the answer by hand and then calculate:

$$\frac{(3.14 + 2.06) \times 0.0240}{1.20 \times 0.0650}$$

## Solution

Approximate the answer by rounding to one figure as follows:

$$\frac{(3 + 2) \times 0.02}{1 \times 0.07} = \frac{5 \times 0.02}{0.07} = \frac{0.10}{0.07} = 1.4 \qquad \text{(nearest tenth)}$$

One calculator solution is:

$$3.14 \boxed{+} 2.06 \boxed{=} \boxed{\times} 0.0240 \boxed{\div} 1.20 \boxed{\div} 0.0650 \boxed{=} \rightarrow 1.6$$

Note that the $\boxed{=}$ key can be used instead of parentheses. The approximate answer of 1.4 checks closely with the actual answer. When a lot of calculations are involved, the approximate answer can differ by a greater amount from the actual answer. However, it will rarely differ more than one place in the location of the decimal point.

## Powers and Roots

Powers (exponents) are done on the calculator by using the power key $\boxed{y^x}$ or $\boxed{x^y}$. For example, $2^5$ is done:

$$2 \; \boxed{y^x} \; 5 \; \boxed{=} \; \rightarrow \; 32$$

To square a number, just press $\boxed{x^2}$:

$$0.7 \; \boxed{x^2} \; \rightarrow \; 0.49$$

Calculators using algebraic logic are programmed to raise to a power or take a root before multiplying or dividing.

### Example 1-25
Calculate:

$$\left(\frac{5}{2}\right)^2 - \frac{(2)^4(3)^2}{6^2}$$

**Solution**
This is Example 1-15(3). One calculator solution is:

$$5 \; \boxed{\div} \; 2 \; \boxed{=} \; \boxed{x^2} \; \boxed{-} \; 2 \; \boxed{y^x} \; 4 \; \boxed{\times} \; 3 \; \boxed{x^2} \; \boxed{\div} \; 6 \; \boxed{x^2} \; \boxed{=} \; \rightarrow \; 2.25$$

If you are not sure of the order of operations of your calculator, use parentheses, or the memory, or write down intermediate results. Another solution that uses the memory is:

$$2 \; \boxed{y^x} \; 4 \; \boxed{\times} \; 3 \; \boxed{x^2} \; \boxed{\div} \; 6 \; \boxed{x^2} \; \boxed{=} \; \boxed{M_{in}} \; 5 \; \boxed{\div} \; 2 \; \boxed{=} \; \boxed{x^2} \; \boxed{-} \; \boxed{MR} \; \boxed{=} \; \rightarrow \; 2.25$$

$$\downarrow \qquad\qquad\qquad\qquad\qquad \uparrow$$
$$4 \qquad\qquad\qquad\qquad\qquad\quad 4$$

### Example 1-26
Approximate and then calculate:

$$\left(\frac{1}{0.25}\right)^2 + \frac{(4.5)^3}{(2.7)^2}$$

**Solution**
Approximate the answer by rounding to one figure:

$$\left(\frac{1}{0.3}\right)^2 - \frac{5^3}{3^2} = \frac{1}{0.09} + \frac{125}{9} = \frac{100}{9} + \frac{125}{9} = \frac{225}{9} = 25$$

The calculator solution can be done using the reciprocal key $\boxed{1/x}$:

$$0.25 \ \boxed{1/x} \ \boxed{x^2} \ \boxed{+} \ 4.5 \boxed{y^x} \ 3 \ \boxed{\div} \ 2.7 \ \boxed{x^2} \ \boxed{=} \rightarrow 28.5$$

The approximate answer of 25 agrees closely with the actual answer.

Square roots are done on the calculator by using the square root key $\boxed{\sqrt{x}}$ as shown in the next example.

### Example 1-27

With the calculator, verify rules (1-7) and (1-8) for radicals in each of the following.

**1.** $\sqrt{12} = (\sqrt{4})(\sqrt{3}) = 2\sqrt{3}$

**2.** $\sqrt{\dfrac{10}{9}} = \dfrac{\sqrt{10}}{\sqrt{9}} = \dfrac{\sqrt{10}}{3}$

### Solution

**1.** $\sqrt{12} = (\sqrt{4})(\sqrt{3}) = 2\sqrt{3}$

This is Example 1-17(2). To verify rule (1-7), calculate $\sqrt{12}$ and $2\sqrt{3}$ separately and show that they are the same (to four figures):

$$12 \ \boxed{\sqrt{x}} \ \rightarrow 3.464$$
$$2 \ \boxed{\times} \ 3 \ \boxed{\sqrt{x}} \ \boxed{=} \ \rightarrow 3.464$$

**2.** $\sqrt{\dfrac{10}{9}} = \dfrac{\sqrt{10}}{\sqrt{9}} = \dfrac{\sqrt{10}}{3}$

This is Example 1-17(6). To verify rule (1-8), calculate $\sqrt{10/9}$ and $\sqrt{10}/3$ separately and show that they are the same:

$$10 \ \boxed{\div} \ 9 \ \boxed{=} \ \boxed{\sqrt{x}} \rightarrow 1.054$$
$$10 \ \boxed{\sqrt{x}} \ \boxed{\div} \ \boxed{3} \ \boxed{=} \rightarrow 1.054$$

### Example 1-28

Approximate and then calculate:

$$\sqrt{\frac{(16.5)^2(2.20)}{(6.82)(3.10)}}$$

## Solution

Approximate the answer as follows, changing square roots to the nearest perfect square:

$$\sqrt{\frac{(20)^2(2)}{(7)(3)}} = \sqrt{\frac{800}{21}} \approx \sqrt{\frac{80\cancel{0}}{2\cancel{0}}} = \sqrt{40} \approx \sqrt{36} = 6$$

Notice you can round again to one figure as you approximate. This allows for quick approximation, with little loss in accuracy. One calculator solution is:

$$16.5 \; \boxed{x^2} \; \boxed{\times} \; 2.20 \; \boxed{\div} \; 6.82 \; \boxed{\div} \; 3.10 \; \boxed{=} \; \boxed{\sqrt{x}} \rightarrow 5.32 \qquad \text{(three figures)}$$

Since the original numbers are accurate to three figures, the answer is considered accurate to three figures. See the next section for further discussion on accuracy and approximate numbers.

## Example 1-29

Approximate and then calculate: $\qquad \sqrt{0.472} + \sqrt{0.0789}$

## Solution

To approximate the answer, change each decimal to the nearest perfect square:

$$\sqrt{0.472} + \sqrt{0.0789} \approx \sqrt{0.49} + \sqrt{0.09} = 0.7 + 0.3 = 1.0$$

The calculator solution is:

$$0.472 \; \boxed{\sqrt{x}} + 0.0789 \; \boxed{\sqrt{x}} \; \boxed{=} \rightarrow 0.968 \qquad \text{(three figures)}$$

## Example 1-30

Approximate and then calculate:

$$\frac{0.136\sqrt{0.0243}}{\sqrt{(0.791)(0.375)}}$$

## Solution

To approximate the answer, you can round to one figure and apply rule (1-8) as follows:

$$\frac{0.1\sqrt{0.02}}{\sqrt{(0.8)(0.4)}} = 0.1\sqrt{\frac{0.02}{(0.8)(0.4)}} = 0.1\sqrt{\frac{1}{16}} = 0.1\left(\frac{1}{4}\right) \approx 0.03$$

One calculator solution is:

$$0.136 \; \boxed{\times} \; 0.0243 \; \boxed{\sqrt{x}} \; \boxed{\div} \; \boxed{(} \; 0.791 \; \boxed{\times} \; 0.375 \; \boxed{)} \; \boxed{\sqrt{x}} \; \boxed{=} \rightarrow 0.0389$$

or applying rule (1-8) yields:

$$0.0243 \boxed{\div} 0.791 \boxed{\div} 0.375 \boxed{=} \boxed{\sqrt{x}} \boxed{\times} 0.136 \boxed{=} \rightarrow 0.0389$$

## Example 1-31
Calculate:
$$\sqrt[3]{4.86}$$

## Solution
Higher-order roots can be done in one or more than one way on your calculator, depending on which keys you have:

**1.** 4.86 $\boxed{\text{INV}}$ $\boxed{y^x}$ 3 $\boxed{=}$ $\rightarrow$ 1.69     (three figures)

**2.** 4.86 $\boxed{\sqrt[x]{y}}$ 3 $\boxed{=}$ $\rightarrow$ 1.69

**3.** 4.86 $\boxed{x^{1/y}}$ 3 $\boxed{=}$ $\rightarrow$ 1.69

**4.** 4.86 $\boxed{y^x}$ 3 $\boxed{1/x}$ $\boxed{=}$ $\rightarrow$ 1.69

Some calculators use $\boxed{\text{2ndF}}$ instead of $\boxed{\text{INV}}$, shown in (1). Solutions (3) and (4) apply rule (1-10), shown in the previous section.

   You can approximate a higher-order root by changing it to the closest perfect root:

$$\sqrt[3]{4.86} = \sqrt[3]{8} = 2$$

You can also check a calculator result by doing the inverse operation and seeing if you get the original number. To check Example 1-31, do not clear the answer on the display, but key in $\boxed{y^x}$ 3 $\boxed{=}$. This should give you 4.86 if the answer is correct.

## Example 1-32
Approximate and then calculate:

$$\sqrt[3]{\frac{(0.321)^2}{4.44}}$$

## Solution
The approximate solution is:

$$\sqrt[3]{\frac{(0.3)^2}{4}} = \sqrt[3]{\frac{0.09}{4}} \approx \sqrt[3]{0.022} \approx \sqrt[3]{0.027} = 0.3$$

One calculator solution is:

$$0.321 \boxed{x^2} \boxed{\div} 4.44 \boxed{=} \boxed{\text{INV}} \boxed{y^x} 3 \boxed{=} \rightarrow 0.285$$

More difficult calculations can be checked in stages by recording intermediate results, approximating answers, or doing inverse operations. In the following chapters use your calculator whenever it will simplify calculations. Exercises difficult to do without a calculator are marked with the symbol $\boxed{C}$.

Remember, your calculator cannot think. It cannot replace mathematical understanding and reasoning. *You should always check an answer and ask yourself if it makes sense.*

**Exercise 1-5**

In problems 1 to 12, calculate the approximate answer for each by hand by rounding each number to one figure. Choose the correct answer from one of the four choices and then check with the calculator.

**1.** $(82 + 68 - 86) \div 32$      (2, 12, 20, 22)

**2.** $930 \times 81 \div 9 - 90$      (82, 820, 828, 8280)

**3.** $\dfrac{72 + 52 - 20}{176 - 34 + 66}$      (0.1, 0.5, 0.9, 1.5)

**4.** $\dfrac{252 \times (86 + 61)}{(36 - 29) \times 36}$      (14.7, 47, 147, 1470)

**5.** $\dfrac{159 \times 91 \times 76}{247 \times 53 \times 35}$      (0.24, 2.4, 8.2, 24)

**6.** $\dfrac{31 \times 176 \times 90}{186 \times 120 \times 110}$      (0.20, 1.2, 2.0, 12)

**7.** $4.85 + 60.8 \div 3.20 - 2.75$      (0.211, 2.11, 21.1, 211)

**8.** $20.0 \times (1560 - 230) \div 28.0$      (9.5, 95, 650, 950)

**9.** $\dfrac{(0.0068)(0.48 - 0.31)}{0.0017}$      (0.068, 0.68, 6.8, 68)

**10.** $\dfrac{(0.0112)(0.0211)}{(0.0400)(0.700)}$      (0.00844, 0.0844, 0.344, 0.844)

**11.** $\dfrac{(0.301 + 0.271)(0.0504)}{(1.26)(1.60)}$      (14.3, 1.43, 0.0143, 0.0413)

**12.** $\dfrac{3.03}{2.02} - \dfrac{27.6 - 8.40}{30.0}$      (86.0, 8.60, 0.860, 0.0860)

In problems 13 to 34 do the same as in 1 to 12, but express answers to three figures.

**13.** $(3.20)(4.26)^2$      (5.81, 58.1, 88.0, 581)

**14.** $(0.913)^2(1.02)^3$      (0.185, 0.885, 8.85, 88.5)

**15.** $\left(\dfrac{0.112}{0.518}\right)^3$      (0.0101, 0.101, 0.110, 1.10)

**16.** $\dfrac{(9.87)^3}{(2.03)^4}$      (0.566, 5.66, 56.6, 566)

**17.** $\dfrac{(1.25)^4 - (0.831)^4}{3.00}$      (0.065, 0.655, 0.950, 0.955)

18. $\left(\dfrac{1}{8.51}\right)^2 + \left(\dfrac{1}{6.66}\right)^2$    (0.0364, 0.364, 3.64, 6.43)

19. $100\left(1.00 + \dfrac{0.180}{4.00}\right)^4$    (3.91, 11.9, 39.1, 119)

20. $13.5(0.281 + 0.591)^3$    (0.895, 6.63, 8.95, 66.3)

21. $\sqrt{0.0588}$    (0.242, 0.842, 2.42, 8.42)

22. $\sqrt[3]{25.6}$    (0.95, 2.95, 4.95, 6.95)

23. $(\sqrt{19.0})(\sqrt{171})$    (5.70, 27.0, 57.0, 570)

24. $(0.789)^2(\sqrt{0.105})$    (0.202, 0.388, 2.02, 3.88)

25. $\sqrt{93.2} + \sqrt{29.1}$    (1.50, 15.0, 150, 650)

26. $\sqrt[3]{8.31} - \sqrt[3]{1.28}$    (0.0940, 0.940, 9.40, 94.0)

27. $\dfrac{\sqrt{600}}{\sqrt{5.88}}$    (1.01, 10.1, 50.1, 101)

28. $\dfrac{\sqrt{(0.130)(22.4)}}{\sqrt{7.92}}$    (0.606, 6.06, 9.09, 60.6)

29. $\sqrt{(47.0)^2 + (33.0)^2}$    (1.74, 5.74, 57.4, 75.4)

30. $\sqrt[4]{0.00185}$    (0.0207, 0.207, 2.07, 20.7)

31. $\sqrt[3]{\dfrac{135}{(1.70)^2}}$    (0.360, 3.60, 8.60, 36.0)

32. $\dfrac{201}{\sqrt[3]{(3.14)^2}}$    (3.97, 9.37, 39.7, 93.7)

33. $\sqrt{\dfrac{(3.01)(5.05)^2}{10.2}}$    (1.47, 1.74, 2.74, 7.74)

34. $\dfrac{(\sqrt{65.6})(\sqrt{288})}{(24.0)(\sqrt{32.8})}$    (0.0100, 0.100, 1.00, 10.0)

In problems 35 to 46, use the calculator to solve each applied problem, to three figures.

35. The total resistance $R$ of three resistances $R_1$, $R_2$, and $R_3$ in parallel is given by:

$$R = \frac{R_1 R_2 R_3}{R_1 R_2 + R_2 R_3 + R_1 R_3}$$    (*Note:* $R_1 R_2 R_3$ means $R_1 \times R_2 \times R_3$.)

Calculate $R$ when $R_1 = 15.0\ \Omega$, $R_2 = 36.0\ \Omega$, and $R_3 = 48.0\ \Omega$.

36. What is the weight of a gold bar 6.50 in. long, 2.25 in. wide, and 0.560 in. thick if the density is 0.697 lb/in³? (*Note:* Weight = volume × density.)

37. The earth's crust contains 5.05 percent iron (Fe) by weight and 0.68 percent by volume. A typical rock weighing 456 kg has a volume of 0.255 m³. If the rock contains the above percentages of iron, how much iron does it contain by weight and volume?

38. A certain steel has a tensile strength of 60,000 lb/in². Approximately how much tensile force will a bar 2.62 in. wide and 1.38 in. thick withstand? (*Note:* Tensile force = area × tensile strength.)

**39.** In 1972, out of a total of 337,069 students enrolled in technology programs, 15,742 were electrical technology students, 64,440 were electronics students, 20,113 were civil technology students, and 26,395 were mechanical technology students. What percentage of the total was enrolled in each program?

**40.** The social security tax for 1987 was 12.3 percent of gross earnings up to \$43,800, with no tax on any amount above \$43,800. Carmen Velez earned \$44,850 in 1987. How much social security tax did she pay?

**41.** The heat energy $E$ radiated by a certain body is given by $E = k(T^4 - T_0^4)$, where $k = 0.0000488$, $T = 291$ K (Kelvin = absolute temperature of the body), and $T_0 = 273$ K. Calculate $E$.

**42.** The volume of a sphere $V = (4/3)\pi r^3$, where $r =$ radius. Find $V$ when $r = 3.28$ cm. Use $\pi = 3.142$ or the $\boxed{\pi}$ key on the calculator.

**43.** The velocity of sound $V$ in meters per second (m/s) is given by $V = V_0\sqrt{1 + t/273}$, where $V_0 =$ velocity of sound at 0 degrees Celsius (0°C) and $t =$ temperature in degrees Celsius. Find $V$ when $V_0 = 332$ m/s and $t = 15.0°C$.

**44.** The friction factor $f$ of flow in a smooth pipe is $f = 0.316/\sqrt[4]{R_e}$, where $R_e$ is a constant known as the *Reynolds number*. Calculate $f$ when $R_e = 3740$.

**45.** Your calculator cannot think, but it can display words. Calculate the following correctly and discover one of the artistic neighborhoods in New York by reading the display upside down:

$$\frac{82 \times 5 - 5}{16 + 24 \times 41}$$

**46.** Here's a good calculator trick: Tell somebody to *(a)* key in six 9s and divide by 7, *(b)* pick a number from 1 to 6 and multiply, *(c)* tell you the first digit in the display and *you will tell all the other digits*. Here is how it works: The number you get when you divide by 7 is a "cyclic" number. If you then multiply by any number from 1 to 6, you get the same order of six digits, starting with a different digit each time. Try it and see if you can memorize the cyclic number.

# 1-6 | *Precision and Accuracy of Numbers*

Consider the following problem. The weights of 11 players (in pounds) of a high school soccer team are 96.5, 98.0, 99.5, 101.5, 115, 127, 129, 133, 148, 161, and 162. What is the average weight? When you add the weights and divide by 11, you get $1370.5 \div 11 = 124.590909 \cdots$. How should you round the result? Before you can answer this question, you have to look at what is meant by precision and accuracy of numbers.

A number obtained by counting is an *exact* number. The number of players on a soccer team, 11, is an exact number. A number obtained by measurement or rounding is an *approximate* number. The weights of the 11 soccer players are *approximate* numbers. The *significant figures* of an approximate number are those that are determined by measurement. All figures are significant figures except zeros used as placeholders for the decimal point. The number 0.0201 has three significant figures. The first two zeros are placeholders, and the last is not.

*Precision* refers to the decimal place of the last significant figure. The number 11.5 is more precise than 115. *Accuracy* refers to the number of significant figures. The number 133 has three significant figures and is more accurate than 0.0033, which has two significant figures. Study Table 1-1, which compares the precision and accuracy of six numbers. Notice that 21,200 is shown as having three significant figures. This number can have more than three significant figures; however, you assume only three unless you have more information about the zeros.

The following rules apply to calculations with approximate numbers:

1. When you *add* or *subtract* approximate numbers, the result is only as precise as the *least* precise number.
2. When you *multiply* or *divide* approximate numbers, the result is only as accurate as the *least* accurate number.
3. The square root (or higher root) of an approximate number is only as accurate as the original number.
4. An exact number does not change the accuracy or the precision of a calculation.

**Table 1-1   COMPARISON OF THE PRECISION AND ACCURACY OF SIX NUMBERS**

| Number | Significant Figures | Explanation | Comment |
| --- | --- | --- | --- |
| 5718 | 4 | All nonzero figures are significant | Precise to nearest unit |
| 507.1 | 4 | Zero is not a placeholder | Precise to nearest tenth |
| 21,200 | 3 | Both zeros are assumed to be placeholders | Least precise |
| 0.0314 | 3 | Both zeros are placeholders | Most precise |
| 0.009 | 1 | All zeros are placeholders | Least accurate |
| 30.120 | 5 | Both zeros are not placeholders | Most accurate |

You can now answer the question about the average weight of the 11 soccer players. Some of the weights are precise to the nearest half pound and others only to the nearest pound. The sum of the weights, 1370.5, is therefore rounded to the nearest pound, 1371. Since 11 is an exact number, consider only 1371, which has four significant figures. The average weight is then accurate to four significant figures. Average weight = 1371/11 = $124.6363 \cdots \approx 124.6$ lb. Since the last digit of 1371 came from rounding, that digit is not totally reliable. The same applies to 124.6. The actual average lies somewhere between 124 and 125.

Most calculators can display 8 to 10 significant figures and store 2 or 3 more for rounding purposes. This degree of accuracy is far greater than is necessary for most calculations. For example, using $\pi$ accurate to nine decimal places (3.141592654), you can calculate the circumference ($C = 2\pi r$) of the earth to the nearest centimeter (1.00 cm = 0.40 in.)!

Study the next example, which applies the rules for approximate numbers.

### Example 1-33
Calculate each result and round the answer, applying the rules for approximate numbers.

**1.** $318.27 + 15.01 + 36.891 + 502.1 = 872.271 \approx 872.3$

The least precise number is 502.1. Therefore the result is precise to one decimal place.

**2.** $\dfrac{(813.27)(0.37)}{502.1} = 0.5993027 \approx 0.60$

The least accurate number is 0.37. Therefore the result is accurate to two significant figures.

**3.** $\dfrac{(0.065 + 30.1)(500.2 - 5.013)}{(6.35)(8.6)} = \dfrac{(30.2)(495.2)}{(6.35)(8.6)} = 273.8517 \approx 270$

The sum and difference are precise to one decimal place, and the result is accurate to two significant figures.

The square root (or higher-order root) of an approximate number is only as accurate as the original number. That is, the result has the same quantity of significant figures as the number whose root is being taken.

### Example 1-34
Calculate the following applying the rules for approximate numbers.

**1.** $\sqrt[3]{0.07070} = 0.4134977 \approx 0.4135$

**2.** $\sqrt{\dfrac{(212.1)^2}{38.7}} = 34.09457 \approx 34.1$

### Example 1-35

A bar of gold is 10.3 cm long, 2.3 cm wide, and 1.2 cm thick. If the density of gold is 19.296 grams per cubic centimeter ($g/cm^3$), what is the weight of the bar?

### Solution

$$Weight = (volume)(density) = (10.3)(2.3)(1.2)(19.296)$$
$$= 548.546688 \text{ g}$$

The last seven figures of the result have little meaning. The last figure of each measurement is approximate and has been estimated or rounded. The answer is 550 g and is accurate to only two significant figures.

Suppose one of the less accurate measurements, the width, was approximately measured as 2.4 rather than 2.3. Observe the effect on the result:

$$(10.3)(2.4)(1.2)(19.296) = 572.396544 \text{ g} \approx 570 \text{ g}$$

Only the first figure in the two answers is the same. The second figure in the answer 550 is not reliable and is subject to possible error.

Consider the following. Suppose 2.3 is closer to the true width than 2.4. The absolute error between the two results is then $570 - 550 = 20$ g. The *absolute error* is the difference between the true value and the approximate value of a number. The relative error or percentage of error shows more clearly the size of the mistake. The *relative error* is the *ratio* of the absolute error to the true value, expressed as a percentage:

$$\text{Relative error} = \frac{\text{absolute error}}{\text{true value}} = \frac{20}{550} = 0.036 = 3.6\%$$

**Exercise 1-6**

In problems 1 to 16, calculate the approximate answer by hand by rounding each number to one significant figure (see Sec. 1-5). Choose the correct answer from one of the four choices by applying the rules for approximate numbers, then check with the calculator.

1. $3.012 + 23.15 + 0.7156$      (2.69, 26.88, 26.9, 268.8)

2. $100.1 - 7.02 + (5.132)(6.02)$      (12.4, 124, 124.0, 1240)

3. $\dfrac{0.1014}{(1.501)(0.036)}$      (0.19, 1.88, 1.90, 1.9)

4. $\dfrac{(10.0)(5.0)(100.0)}{0.030}$      (16,700, 17,000, 167,000, 170,000)

5. $\dfrac{(6,700)(0.0780)(0.8900)}{51.20 + 5.150}$      (0.00825, 0.0825, 0.083, 0.8250)

**6.** $\dfrac{(5.6)(20.6)}{(1.02)(1.03)}$    (109, 109.8, 110, 110.0)

**7.** $(0.801 - 0.31)(0.080 - 0.04)$    (0.019, 0.020, 0.02, 0.20)

**8.** $(30,150 + 22,430)(0.542 + 0.777)$    (69,300, 69,353, 69,350, 69,400)

**9.** $\dfrac{0.101 + 0.202}{52.1} - \dfrac{47.3}{15,000}$    (0.0026, 0.00267, 0.026, 0.0267)

**10.** $\dfrac{333}{22.1 + 11.2} + \dfrac{5.99}{0.559}$    (2.07, 2.071, 20.7, 20.71)

**11.** $\sqrt{0.00488}$    (0.00699, 0.0699, 0.6986, 0.699)

**12.** $\sqrt[3]{25.06}$    (0.0293, 0.2926, 2.926, 2.9264)

**13.** $(0.062)^2(2.01)^3$    (0.03, 0.031, 0.0312, 0.312)

**14.** $(0.238)^2(\sqrt{110})$    (0.59, 0.594, 5.9, 5.94)

**15.** $\sqrt{\dfrac{1.00}{(2.2)^2}}$    (0.4, 0.45, 0.454, 0.4545)

**16.** $\dfrac{(\sqrt{33.0})(\sqrt{0.018})}{(7.107)^3}$    (0.002, 0.0021, 0.02, 0.021)

In problems 17 to 24, solve each by applying the rules for approximate numbers.

**17.** A student records the temperature of a steam line at four equal intervals during an 8-h period as follows: 220.5°, 221.5°, 225° and 227°F. He concludes the average temperature is 223.5°. Is this correct? If not, what is the correct answer?

**18.** A bridge column must support vertical forces of 1600, 2000, and 330 lb, all applied in the same direction. If the breaking strength of the column should be 5 times its total vertical load (safety factor of 5), *at least* how much vertical force should the column be designed to withstand?

**19.** A microprocessor chip measured with a ruler is found to be 5.3 cm long and 2.8 cm wide. The thickness, measured with a micrometer, is 0.0135 cm. *(a)* If the chip's density is 2.699 g/cm$^3$, what is its weight? *(b)* If the width is incorrectly measured as 2.7 cm, what are the absolute and relative errors? (See Example 1-35.)

**20.** The current in a circuit is measured by one ammeter as 12.2 A and an hour later by another ammeter as 13.15 A. *(a)* If the voltage has remained constant at 115 V, what is a reliable answer for the increase in power (watts)? *(Note:* Watts = volts $\times$ amperes.) *(b)* If the second ammeter is incorrect and the current has remained constant at 12.2 A, what are the absolute and relative errors of the incorrect current reading?

**21.** The earth's closest approach (perihelion) to the sun is 91,402,000 mi. If light travels at 186,000 mi/s, how many minutes does it take for sunlight to reach the earth at this distance?

**22.** The largest scientific building in the world is the Vehicle Assembly Building in Complex 39 at the John F. Kennedy Space Center, Cape Canaveral, Florida. It is 218 m (716 ft) long, 159 m (518 ft) wide, and 160 m (525 ft) high. Assuming the building has a rectangular shape, find the volume it occupies in cubic meters and in cubic feet.

**23.** The fastest base runner in baseball was Ernest Swanson, who took only 13.30 s to circle the bases, a distance of 360 ft, in 1932. What was his average speed in feet per second and in miles per hour (1 mi = 5280 ft)?

**24.** The radius of a sphere $r = \sqrt[3]{3V/(4\pi)}$, where $V$ is the volume of the sphere. Find $r$ if $V = 1126$ cm$^3$. Assume 3 and 4 are exact numbers, and use $\pi = 3.142$ or the $\boxed{\pi}$ key on the calculator.

# 1-7 | *Review Questions*

In problems 1 to 18, calculate each by hand. (You can check with the calculator.)

**1.** $8 \times 3 - 4 \div 2$

**2.** $(10 + 8) \times 3 \div 6$

**3.** $\dfrac{3 \times 15 \times 6}{2 \times 5}$

**4.** $\dfrac{5 + 4 \times (9 - 7)}{8 \times 3 + 2}$

**5.** $\dfrac{5}{6} \times \dfrac{12}{7} \div \dfrac{1}{21}$

**6.** $11 \times \dfrac{3}{22} \times \dfrac{4}{15}$

**7.** $\dfrac{2}{3} + \dfrac{3}{4} - \dfrac{5}{6}$

**8.** $5 + \dfrac{7}{10} \div \dfrac{21}{20}$

**9.** $3.2 + 2.1 \times 0.4$

**10.** $(0.10 - 0.09) \times (0.31 + 0.29)$

**11.** $\dfrac{1.2 + 3.0 \times 0.40}{0.02 \times 6.0}$

**12.** $\dfrac{0.03 \times 0.1}{6.0 \times 0.005}$

**13.** $\dfrac{6^2 + 8^2}{5^2}$

**14.** $(6)^2\left(\dfrac{1}{3}\right)^3 - (18)\left(\dfrac{1}{9}\right)^2$

**15.** $\sqrt{81}$

**16.** $\sqrt[3]{\dfrac{27}{64}}$

**17.** $\sqrt[3]{0.008}$

**18.** $\sqrt{6.25}$

In problems 19 to 22, simplify each expression.

**19.** $\sqrt{75}$

**20.** $(\sqrt{20})^2 - (\sqrt{5})(\sqrt{45})$

**21.** $\dfrac{(\sqrt{3})(\sqrt{32})}{6}$

**22.** $\dfrac{(0.06)(\sqrt{0.72})}{\sqrt{0.02}}$

In problems 23 to 30, approximate each answer by hand. Then choose the correct answer and check with the calculator.

**23.** $\dfrac{105 + 9 \times 25}{63 - 53}$  (3, 33, 60, 93)

**24.** $\dfrac{17 \times 2.0 \times 3.5}{2.5 \times 14}$  (3.4, 8.4, 34, 84)

**25.** $\dfrac{(0.08)(0.37 - 0.18)}{0.016}$  (0.19, 0.59, 0.95, 5.9)

**26.** $\dfrac{0.33}{5.5} - \dfrac{0.20}{28.3 + 71.7}$ $\quad$ (0.058, 0.098, 0.58, 0.98)

**27.** $(9.38)^2(\sqrt{9.91})$ $\quad$ (1.77, 2.77, 27.7, 277)

**28.** $(\sqrt[3]{28.3})(0.101)$ $\quad$ (0.308, 0.808, 3.08, 30.8)

**29.** $\dfrac{\sqrt{66.6}}{(\sqrt{50.0})(\sqrt{72.0})}$ $\quad$ (0.0136, 0.136, 0.936, 1.36)

**30.** $\sqrt{\dfrac{(1.21)(8.13)}{(0.331)^2}}$ $\quad$ (0.948, 1.98, 9.48, 19.8)

In problems 31 to 36, approximate each answer by hand. Choose the correct answer, applying the rules for approximate numbers, and check with the calculator.

**31.** $5.28 + 0.312 + 29.3$ $\quad$ (34.8, 34.89, 34.892, 34.9)

**32.** $(3.7)(8.12) + 1.002$ $\quad$ (31, 31.0, 31.04, 31.046)

**33.** $\dfrac{(39.03)(86.1)}{(10.0)(5.0)}$ $\quad$ (67, 67.2, 67.21, 670)

**34.** $\dfrac{0.707 + 0.808}{0.011}$ $\quad$ (137, 137.7, 138, 140)

**35.** $(\sqrt{0.4440})(\sqrt{6.126})$ $\quad$ (1.64, 1.649, 1.65, 1.66)

**36.** $(3.30)^2(\sqrt[3]{680})$ $\quad$ (95.8, 96, 100, 960)

In problems 37 to 44, use your calculator to solve each applied problem.

**37.** A cassette player sells for one-third off its usual price of $59.94. What is the total cost including an 8 percent sales tax?

**38.** The length of a wire is shortened from 15 to 12 cm, causing its resistance to decrease by the same percentage. *(a)* What is the percent decrease in the length and the resistance? *(b)* If the resistance of the original wire was 4.0 $\Omega$, what is the resistance of the shortened wire?

**39.** Under certain conditions, when the driver jams on the brakes of an automobile weighing 4000 lb and traveling at 60 mi/h, the car will skid a distance $d$ in feet given by:

$$d = \frac{4000 \times (13.9)^2}{1200 \times 1.22 \times 9.81}$$

Calculate this distance to three figures.

**40.** The velocity $v$ of a fluid flowing due to gravity in a pipeline is given by:

$$v = \sqrt{2g(h_2 - h_1)}$$

where $g$ = gravitational acceleration, $h_2$ = initial height, and $h_1$ = final height. Find $v$ if $g$ = 9.81 meters per second per second (m/s²), $h_2$ = 42.5 m, and $h_1$ = 37.5 m.

**41.** The thickness of a gear is measured by five students using a micrometer as 3.85, 3.9, 3.85, 4.0, and 3.95 mm. What is a reliable answer for the average thickness?

**42.** A lightning discharge from a cloud 2.5 mi high is measured at a speed of 550 mi/s for the downstroke and 10,000 mi/s for the powerful return stroke (ground to cloud). By applying the rules for approximate numbers, what is the total time required for both strokes?

**43.** Calculate the following, written in BASIC:

$$0.10 * 3.5/(2.7 + 4.3)$$

**44.** Calculate the following written in BASIC:

$$6 \wedge 2/SQR(2 + 1/4)$$

# 2

# Fundamentals of Algebra

## 2-1 | *Constants and Variables*

The symbols and equations used in algebra are not new to you. Years ago, when you saw your first mathematical formula, it was a type of algebraic equation. Perhaps it was the formula for the area of a rectangle:

$$\text{Area} = \text{length} \times \text{width} \qquad \text{or} \qquad A = lw$$

The multiplication sign $\times$ is not used in algebra because it can be confused with the letter x. Instead, letters are written next to each other to indicate that they are to be multiplied. The letters in the above formula assume different values as follows.

**The** fundamentals of algebra must be understood before you can learn the mathematics necessary for your technical field. This chapter forms the basis for understanding later chapters. The first half presents algebraic symbols and operations, the rules for exponents, and their application to scientific notation. The related calculator operations are shown in these sections and explained in detail in Sec. 2-10. The second half covers first degree equations and their application to verbal problems and formulas. Remember, algebra is of little practical use unless you can apply it to solving real verbal problems.

### Example 2-1
Find the area of each rectangle.

**1.** Length = 5 in., width = 2 in.

**2.** Length = 13 cm, width = 6 cm

**Solution**

**1.** Length = 5 in., width = 2 in.:

$$A = lw = (5)(2) = 10 \text{ in}^2$$

**2.** Length = 13 cm, width = 6 cm:

$$A = lw = (13)(6) = 78 \text{ cm}^2$$

Letters or literal symbols that assume different values are called *variables*. The letters $A$, $l$, and $w$ in the formula above are variables. Letters representing numbers that do not vary are called *constants*. For example, the velocity $v$ of a falling body is expressed by the formula:

$$v = v_0 + gt$$

For a particular body in motion, $v_0$ = initial (starting) velocity, $g$ = acceleration due to gravity, and $t$ = time elapsed. During the motion of the body, $v$ and $t$ assume different values and hence are variables; $v_0$ and $g$ assume only one value and so are constants. Consider the following examples.

### Example 2-2
A woman throws a ball straight down from the top of a cliff with a velocity of 50 cm/s. Assume $g = 980 \text{ cm/s}^2$.

**1.** Find the velocity after 2 s.

**2.** Find the velocity after 7 s.

### Solution
**1.** Find the velocity after 2 s.

The initial velocity is $v_0 = 50$ cm/s. The acceleration due to gravity $g = 980 \text{ cm/s}^2$. That is, the velocity increases 980 cm/s every second. By substituting into the formula above, the velocity after 2 s is:

$$v = v_0 + gt = 50 + (980)(2) = 50 + 1960 = 2010 \text{ cm/s}$$

**2.** Find the velocity after 7 s:

$$v = v_0 + gt = 50 + (980)(7) = 50 + 6860 = 6910 \text{ cm/s}$$

In each case the variables, $v$ and $t$, assume different values, and the constants, $v_0$ and $g$, do not change.

A number in a formula or an equation is also a constant since it does not vary, as shown in the next example.

### Example 2-3
The formula for converting temperature in degrees Celsius (°C) to degrees Fahrenheit (°F) is:

$$F = \frac{9}{5}C + 32$$

How many degrees Fahrenheit is 20°C?

**Solution**

In this formula $F$ and $C$ are variables, whereas 9/5 and 32 are constants:

$$F = \frac{9}{5}(20) + 32 = 36 + 32 = 68°F$$

Formulas and equations are studied in greater depth in the second part of this chapter.

**Exercise 2-1**

In problems 1 to 14, identify the variables and constants in each formula based on the given explanation.

1. $P = 2l + 2w$ (geometry). Formula for the perimeter $P$ of a rectangle: $l$ = length, $w$ = width.

2. $E = IR$ (electricity—Ohm's law). Given a circuit whose fixed resistance is $R$: $E$ = applied voltage, $I$ = resulting current.

3. $W = Fd$ (mechanics). Definition of work $W$: $F$ = force applied, $d$ = distance through which the force moves.

4. $F = ma$ (physics—Newton's second law). Given a body of fixed mass $m$: $F$ = force applied, $a$ = acceleration produced.

5. $P_0V_0 = PV$ (chemistry—Boyle's law). A gas has an initial pressure $P_0$ and an initial volume $V_0$. When the temperature is held constant, $P$ = final pressure, $V$ = final volume.

6. $v = C_v\sqrt{2gh}$ (fluid dynamics). Given a fluid in an open container whose coefficient of velocity is $C_v$: $v$ = discharge velocity resulting from a small opening $h$ units below the surface, $g$ = gravitational acceleration.

7. $C = 1/C_1 + 1/C_2$ (electricity). For any electric circuit: $C$ = total capacitance of two capacitances $C_1$ and $C_2$ connected in series.

8. $T = Fr$ (mechanics). Given a rotating body: $T$ = torque produced, $F$ = force applied, $r$ = distance of the force from the center of rotation.

9. $I = E/(R + r)$ (electricity). Given a battery having fixed voltage $E$ and internal resistance $r$: $R$ = resistance of a circuit, $I$ = resulting current in the circuit.

10. $C = 2\pi r$ (geometry). Formula for the circumference $C$ of any circle of radius $r$.

11. $V = 331.5 + 0.607t$ (physics). Formula for the velocity of sound $V$ (m/s) in air: $t$ = air temperature (°C).

12. $H = \pi nT/225{,}000$ (mechanics). Formula for the horsepower $H$ of a rotating shaft: $n$ = revolutions per minute, $T$ = torque produced.

13. $l = l_0 + al_0(T - T_0)$ (thermodynamics). Given a solid of fixed length $l_0$ at temperature $T_0$: $l$ = length at temperature $T$, $a$ = coefficient of linear expansion for the solid.

**14.** $s = v_0t + \frac{1}{2}gt^2$ (physics). Formula for the distance $s$ of a falling body: $v_0 = $ initial velocity, $t = $ time elapsed, $g = $ gravitational acceleration.

In problems 15 to 20, find the values of the indicated variables.

**15.** Using the formula in problem 1, find the perimeter of a rectangle where *(a)* $l = 6$ ft, $w = 4$ ft; *(b)* $l = 3$ m, $w = 5$ m.

**16.** Find the velocity in feet per second of a body that falls from rest after *(a)* 1 s, *(b)* 2 s, *(c)* 5 s. Use $g = 32$ ft/s$^2$. (*Hint:* See Example 2-2 and note that $v_0 = 0$.)

**17.** Using Ohm's law from problem 2, find the voltage in a circuit whose resistance $R = 6.2\ \Omega$ when *(a)* $I = 2.0$ A and *(b)* $I = 3.1$ A.

**18.** Using the formula in problem 7, find the total capacitance $C$ when $C_1 = 5$ microfarads ($\mu$F) and $C_2 = 10\ \mu$F.

**19.** The boiling temperature of ethyl alcohol is 78°C. How many degrees Fahrenheit is this? (See Example 2-3.)

**20.** Using the formula in problem 11, find the velocity of sound in meters per second when $t = 30.5$°C.

# 2-2 | *Signed Numbers and Zero*

One of the basic features of algebra is the use of positive and negative numbers. You are familiar with the use of signed numbers on a thermometer. In Fig. 2-1, the positive sign indicates the direction above zero, and the negative sign the direction below zero. Whenever there is a need to indicate direction on a scale of measurement or to indicate gain or loss, positive and negative numbers are used. Example 2-4 illustrates some uses of signed numbers.

### Example 2-4
List some examples of the use of signed numbers.

1. *Electric current:* Positive indicates current in one direction, and negative indicates current in the opposite direction.

2. *Motion:* Motion up, or to the right, is usually considered positive; down, or to the left, negative.

3. *Money:* Money owed to an account is designated negative (debit); money in an account is designated positive (credit).

4. *Angles:* Positive indicates counterclockwise rotation; negative indicates clockwise rotation.

5. *Sports:* In football a gain of yards is positive, and a loss negative.

The *integers* consist of the positive and negative whole numbers and zero:

$$\ldots -3, -2, -1, 0, 1, 2, 3, \ldots$$

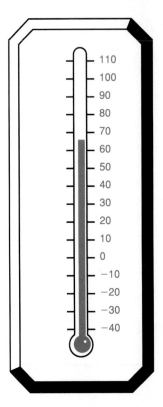

FIG. 2-1  *Thermometer*

A *rational number* is any number that can be expressed as a ratio of two integers. Rational numbers consist of the integers and fractions. Numbers that are not rational are called *irrational*. Irrational numbers include square roots of numbers that are not perfect squares, higher-order roots that are not perfect roots, the number $\pi$, and other numbers you will encounter in further study. Irrational numbers are infinite decimals with *no* pattern of repeating digits:

$$\sqrt{3} = 1.732050807568 \cdots$$
$$\sqrt[3]{7} = 1.91293118277 \cdots$$
$$\pi = 3.141592653590 \cdots$$

Rational numbers, when expressed as infinite decimals, show a pattern of repeating digits:

$$\frac{1}{3} = 0.33333 \cdots$$

$$\frac{9}{100} = 0.09000 \cdots$$

$$\frac{1}{7} = 0.142857142857 \cdots$$

The rational and irrational numbers together make up the set of *real numbers*. Study the following example.

### Example 2-5
Answer these three questions for each real number: Positive or negative? Integer? Rational or irrational?

### Solution

| Number | Positive or Negative | Integer | Rational (R) or Irrational (I) |
|---|---|---|---|
| 2 | + | Yes | R |
| −5 | − | Yes | R |
| $-\frac{3}{4}$ | − | No | R |
| $\frac{12}{3}$ | + | Yes | R |
| $-\sqrt{4}$ | − | Yes | R |
| 3.14 | + | No | R |
| $\sqrt{2}$ | + | No | I |
| $-\pi$ | − | No | I |
| $-\frac{13}{3}$ | − | No | R |
| 0.07 | + | No | R |
| $\sqrt{7}$ | + | No | I |
| 0 | Neither | Yes | R |

Notice that $-\sqrt{4} = -2$, which is an integer. Notice that 3.14 and 0.07 can both be written as fractions, 314/100 and 7/100, and are therefore rational. Zero is an integer. It is misleading to think of zero as representing "nothing." Zero is a real number positioned so that it divides the positive and negative numbers, and it can be an acceptable answer to a problem.

All the real numbers can be represented as points on a number line. Figure 2-2 shows the real number line and the location of the numbers in Example 2-5.

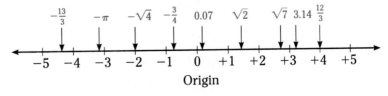

FIG. 2-2 *Real number line, Example 2-5*

Notice in Fig. 2-2 that $\sqrt{2}$ is a little less than halfway between 1 and 2 since $\sqrt{2} \approx 1.414$. Similarly $\sqrt{7} \approx 2.646$ and $-\pi \approx -3.142$. Inequality signs express which number is larger than another or to the right on the number line. The symbol $>$ means "greater than" and the symbol $<$ "less than." For example, $4 > 3$ and $-1 < 3$.

### Example 2-6

Insert the correct inequality sign or signs between the numbers in each set.

**1.** $2, -5$      $2 > -5$      or      $-5 < 2$

**2.** $-5, -\dfrac{3}{4}$      $-\dfrac{3}{4} > -5$      or      $-5 < -\dfrac{3}{4}$

**3.** $\sqrt{7}, \sqrt{2}$      $\sqrt{7} > \sqrt{2}$      or      $\sqrt{2} < \sqrt{7}$

**4.** $-\pi, -3$      $-3 > -\pi$      or      $-\pi < -3$

**5.** $0.07, 0, -\dfrac{13}{3}$      $0.07 > 0 > -\dfrac{13}{3}$      or      $-\dfrac{13}{3} < 0 < 0.07$

**6.** $1, \sqrt{2}, 2$      $2 > \sqrt{2} > 1$      or      $1 < \sqrt{2} < 2$

Compare these answers with the order of the numbers on the number line in Fig. 2-2. Notice that in (5) and (6) you can use the inequality sign twice to show the order of three numbers.

## Combining Signed Numbers

When no sign is shown in front of a quantity, the quantity is understood to be positive. Combining (adding) signed numbers uses the concept of *absolute value*. This is the value of a number without a sign, or the positive value. Absolute value is indicated by vertical lines:

$$|+4| = 4 \qquad |-4| = 4 \qquad |7| = 7 \qquad \left|-\dfrac{3}{4}\right| = \dfrac{3}{4}$$

The rules for combining (adding) signed numbers are as follows:

### Combining like signs

To combine two signed numbers with like signs, add their absolute values and attach the same sign to the result.

### Example 2-7

Combine each of the following.

**1.** $(+2) + (+7) = +9$      or      $2 + 7 = 9$

**2.** $(-2) + (-7) = -9$      or      $-2 + -7 = -9$

The expressions on the right show each example in a simpler form. The positive signs can be left out and are understood.

---

### Combining unlike signs
To combine two signed numbers with unlike signs, find the difference of their absolute values and attach the sign of the larger absolute value to the result.

---

### Example 2-8
Combine each of the following.

**1.** $(+7) + (-4) = +3 \qquad$ or $\qquad 7 - 4 = 3$

**2.** $(-7) + (+4) = -3 \qquad$ or $\qquad -7 + 4 = -3$

The following shows various examples of combining signed numbers:

### Example 2-9
Combine each of the following.

**1.** $18 - 2 = 16$

**2.** $-8 - 7 = -15$

**3.** $-10 + 10 = 0$

**4.** $10 - 15 = -5$

**5.** $-3.1 + 2.9 = -0.2$

**6.** $-\dfrac{1}{2} - \dfrac{3}{4} = -\dfrac{5}{4}$

**7.** $-3 + 2 - 1 = -4 + 2 = -2$

**8.** $5.3 - 1.1 - 4.0 = 5.3 - 5.1 = 0.2$

In (7) and (8) first like signs are combined and then unlike signs. They can also be done by combining from left to right:

**7.** $-3 + 2 - 1 = -1 - 1 = -2$

**8.** $5.3 - 1.1 - 4.0 = 4.2 - 4.0 = 0.2$

Subtraction in algebra is not a separate operation. It is the same as combining numbers with the sign changed.

---

### Subtracting signed numbers
To subtract two signed numbers, change the sign of the number that follows the negative sign and combine the numbers.

### Example 2-10
Subtract each of the following.

1. $18 - (+2) = 18 - 2 = 16$
2. $-8 - (+7) = -8 - 7 = -15$
3. $-10 - (-10) = -10 + 10 = 0$
4. $10 - (+15) = 10 - 15 = -5$
5. $-3.1 - (-2.9) = -3.1 + 2.9 = -0.2$
6. $-\dfrac{1}{2} - \left(+\dfrac{3}{4}\right) = -\dfrac{1}{2} - \dfrac{3}{4} = -\dfrac{5}{4}$

Compare these examples with (1) to (6) in Example 2-9. They are equivalent. A negative sign in front of a plus or minus sign simply changes it to the opposite sign. Study the next example.

### Example 2-11
Combine each of the following.

1. $-3 - (-2) - (1) = -3 + 2 - 1 = -4 + 2 = -2$
2. $3 + (-4) - (-3) + (2) = 3 - 4 + 3 + 2 = 8 - 4 = 4$
3. $1.01 - (8.09) - (-5.26) = 1.01 - 8.09 + 5.26 = 6.27 - 8.09 = -1.82$

When there are several subtractions in one problem, first change any sign that has a negative sign in front of it and then combine.

## Multiplication and Division of Signed Numbers

Besides parentheses, a dot is sometimes used to indicate multiplication: $(-3)(2)$ and $-3 \cdot 2$ both mean $-3$ times 2. The division sign $\div$ is rarely used in algebra. Division is indicated with a fraction line: $-\frac{3}{2}$ means $-3$ divided by 2. The expression $-3(\frac{2}{5})$ is the same as writing $-3 \cdot \frac{2}{5}$ and means $-3$ times 2 divided by 5.

> ### Multiplying or dividing signed numbers
> To multiply or divide two signed numbers, multiply or divide their absolute values. If the two signs are alike, the result is positive; if the two signs are unlike, the result is negative.

### Example 2-12
Calculate each of the following.

1. $-5 \cdot 3 = -15$          2. $-4/(-4) = 1$
3. $-56/7 = -8$               4. $(-8.2)(-7.1) = 58.22$
5. $1.5/(-3.0) = -0.5$        6. $(\frac{1}{2})(-\frac{3}{4}) = -\frac{3}{8}$

The laws of arithmetic stated in Sec. 1-1 apply equally to operations in algebra. The associative law states that several operations of multiplication and division can be done in any order. Some of the ways to do these operations are illustrated in the next example.

### Example 2-13
Calculate each of the following.

1. $-5(-3)(-2) = 15(-2) = -30$    or    $-5(6) = -30$

2. $\left(\dfrac{1}{2}\right)\left(-\dfrac{1}{4}\right)\left(-\dfrac{1}{2}\right)\left(\dfrac{1}{4}\right) = \left(-\dfrac{1}{8}\right)\left(-\dfrac{1}{8}\right) = \dfrac{1}{64}$

   or    $\left(-\dfrac{1}{4}\right)\left(-\dfrac{1}{16}\right) = \dfrac{1}{64}$

3. $\dfrac{5(-3)}{-15} = \dfrac{-15}{-15} = 1$    or    $\dfrac{5(1)}{5} = 1$    or    $\dfrac{1(-3)}{-3} = 1$

Notice that in (3) you can divide out the common factor of $-3$ or $5$ first and then multiply.

The order of operations stated in Sec. 1-1 (multiplication or division first, then addition or subtraction) also applies to operations in algebra, as the next example shows.

### Example 2-14
Perform the indicated operations.

1. $-8 \cdot 2 - 2 \cdot 3 = -16 - 6 = -22$

2. $-\dfrac{1}{2}(-2.2) - \dfrac{8(-3)}{-5} = 1.1 - \dfrac{-24}{-5} = 1.1 - 4.8 = -3.7$

3. $\dfrac{-0.8(0.2)}{0.2 - (+0.4)} = \dfrac{-0.16}{0.2 - 0.4} = \dfrac{-0.16}{-0.2} = 0.8$

4. $\dfrac{-8(0) - 5(0)}{3} = \dfrac{0 - 0}{3} = \dfrac{0}{3} = 0$

Observe in (3) that the fraction line serves the same purpose as parentheses. Any additions or subtractions in the top or bottom must be done before the division. Study (4) above. Any number times zero is zero. Any number divided *into* zero is also zero, with one exception. Zero divided by zero has no answer. Compare the two fractions:

$$\frac{10(0)}{5(0)} \quad \text{and} \quad \frac{6(0)}{2(0)}$$

Both simplify to % if you first multiply the top and the bottom of each. However, if you could divide out the zeros first, the first fraction would become $^{10}/_5 = 2$ and the second $^6/_2 = 3$. That cannot be, since two different answers would represent the same fraction, $^0/_0$. Because division by zero leads to contradictory results, it is not possible to define this operation. *Therefore, for any number a, $^a/_0$ has no answer.*

**Exercise 2-2**

1. Give an example of the use of signed numbers for: *(a)* measuring time and *(b)* plotting graphs.

2. Give an example of the use of signed numbers in your field of study.

In problems 3 to 20, answer the three questions for each real number given: Positive or negative? Integer? Rational or irrational?

| | | | |
|---|---|---|---|
| **3.** 2 | **4.** $-1$ | **5.** $-\dfrac{9}{3}$ | **6.** $\dfrac{5}{4}$ |
| **7.** $\sqrt{6}$ | **8.** $-\sqrt{5}$ | **9.** $2\sqrt{9}$ | **10.** $\sqrt[3]{27}$ |
| **11.** $-3.20$ | **12.** $0.001$ | **13.** $5\%$ | **14.** $8\frac{1}{2}\%$ |
| **15.** $10{,}000$ | **16.** $10{,}001$ | **17.** $2\pi$ | |
| **18.** $\pi + 1$ | **19.** $10^6$ | **20.** $1.1 \times 10^6$ | |

21. Sketch a number line like the one in Fig. 2-2, and locate the numbers in problems 3 to 7.

22. Do the same as in problem 21 for problems 8 to 12.

In problems 23 to 34, insert the correct inequality sign or signs between the numbers in each set.

| | | |
|---|---|---|
| **23.** 8, 4 | **24.** 7, $-2$ | **25.** $-5$, $-3$ |
| **26.** $-1$, $-\dfrac{3}{2}$ | **27.** $\sqrt{5}$, 3 | **28.** $\dfrac{16}{5}$, $\pi$ |
| **29.** $-1.01$, $-0.99$ | **30.** $\dfrac{7}{3}$, 2.30 | **31.** 400, $10^3$ |
| **32.** $10^3$, $2 \times 10^2$ | **33.** 8, 7.21, $-11$ | **34.** $-3.2$, $-\pi$, $-3.1$ |

In problems 35 to 84, perform the indicated operations.

| | |
|---|---|
| **35.** $5 - 2$ | **36.** $3 - 7$ |
| **37.** $-4 - 6$ | **38.** $-2 + 8$ |
| **39.** $3 - (-1)$ | **40.** $-9 - (-9)$ |
| **41.** $15 - (+10)$ | **42.** $-8 - (+5)$ |
| **43.** $\dfrac{1}{3} - \dfrac{1}{2}$ | **44.** $-\dfrac{1}{5} + \dfrac{1}{10}$ |
| **45.** $2.1 - (-3.2)$ | **46.** $-0.58 - 0.12$ |
| **47.** $4 - (-5) + 6$ | **48.** $8 - 6 - (-2) + 1$ |
| **49.** $2.3 + 0.0 - (-5.5)$ | **50.** $1 - \left(-\dfrac{1}{4}\right) - \dfrac{1}{8} - \dfrac{1}{2}$ |
| **51.** $-8 \cdot 5$ | **52.** $10 \cdot 11$ |

**53.** $(-6)\left(-\dfrac{1}{2}\right)$

**54.** $12(-13)$

**55.** $\dfrac{-63}{9}$

**56.** $\dfrac{-8}{-2}$

**57.** $\dfrac{-5}{-10}$

**58.** $\dfrac{0.2}{-0.1}$

**59.** $(8)(-1)(-6)$

**60.** $(-5)(5)(-3)$

**61.** $(-2)(0.1)(-0.3)$

**62.** $\left(\dfrac{2}{5}\right)\left(-\dfrac{15}{4}\right)(-2)$

**63.** $\dfrac{(6)(-3)}{-21}$

**64.** $\dfrac{(-2)(-2)}{18}$

**65.** $\dfrac{-1 - (-1)}{-1}$

**66.** $\dfrac{(8)(0)}{7 - 7}$

**67.** $\dfrac{(-6)(0.5)}{-0.1}$

**68.** $\dfrac{1.21}{(10)(-1.1)}$

**69.** $-2 \cdot 6 + 3 \cdot 8$

**70.** $-5 \cdot 5 - 4 \cdot 4$

**71.** $\dfrac{3 - 8}{(-1)(-2)}$

**72.** $\dfrac{(-4)(3)}{6 + 9(-2)}$

**73.** $\dfrac{6(0) - 6(-3)}{2(-5)}$

**74.** $\dfrac{2}{(5)(0) - (-1)(1)}$

**75.** $\dfrac{(-6)(5)(-8)}{(2)(-5) - 5}$

**76.** $\dfrac{2 - 8}{(5)(-1) + (-1)(-5)}$

**77.** $\dfrac{(-0.3)(-6.0)}{3.3 - 2.7}$

**78.** $\dfrac{1.1 + 1.3}{(8.0)(-0.3)}$

**79.** $5\left(\dfrac{-3}{-10}\right) - 4\left(\dfrac{7}{-12}\right)$

**80.** $\left(\dfrac{1}{2}\right)\left(\dfrac{1}{4}\right) - \left(\dfrac{1}{4}\right)\left(-\dfrac{1}{2}\right)$

**81.** $-6 - \dfrac{-16}{2} + 3(-2)$

**82.** $3(-7) - \dfrac{6}{-3} - (-3)$

**83.** $2(-0.5) + (-0.6)\left(\dfrac{4.0}{-8.0}\right)$

**84.** $\dfrac{8}{2(-0.2)} + \left(\dfrac{1}{2}\right)(-8.0)(5)$

**85.** The temperature changes from $-5$ to $-10°C$. What signed number represents this change?

**86.** The current in a circuit changes from $-10$ to $20$ A. What signed number represents this change?

**87.** Mount Everest, the highest mountain on earth, is 8848 m (29,028 ft) above sea level. The Mariana Trench, the greatest sea depth, is 11,034 m (36,201 ft) below sea level. Express these distances as signed numbers, and calculate the difference between the two numbers in meters and in feet.

**88.** What is the difference between the boiling point of oxygen, $-183.0°C$, and the melting point, $-218.4°C$?

89. A gear rotates 20° clockwise, 45° counterclockwise, then 35° clockwise. Express the three rotations as signed numbers, and combine them to calculate the net rotation.

90. A football team is pushed back 8 yards (yd) on a poor play. The team then completes a pass, gaining 12 yd, but incurs a penalty of 5 yd. Express these changes as signed numbers, and combine them to determine the team's gain or loss.

91. Using the formula $v = v_0 + gt$, calculate the velocity at $t = 3$ s if $v_0 = 5$ ft/s and $g = -32$ ft/s². (See Example 2-2.)

92. What temperature in degrees Fahrenheit is equal to $-10°C$? (See Example 2-3.)

93. What is the sign of the product of an even number of negative numbers?

94. What is the sign of the product of an odd number of negative numbers divided by a negative number?

# 2-3 | *Combining Algebraic Terms*

An *algebraic term* is a combination of letters, numbers, or both, joined by the operations of multiplication or division:

$$b, \quad -2xy^2, \quad \frac{E}{I}, \quad 1.8C, \quad \pi d, \quad Prt, \quad -\frac{mv^2}{2},$$

Each of the seven terms above has two parts: a *coefficient,* which is the number in front, and a *literal part:*

$$\text{Coefficient} \longleftarrow \left(-2\right) \quad \left(xy^2\right) \longrightarrow \text{literal part}$$

When there is no coefficient, it is understood to be 1: $b = (1)b$. *Like terms* have *exactly* the same literal part. Three pairs of like terms are:

$$2x, \, 3x \quad -3at^2, \, 2at^2 \quad -PV, \, PV$$

It is important to recognize like terms, since they represent amounts of the same quantity and may be combined. *Unlike terms* do not have the same literal part and *cannot* be combined. Three pairs of unlike terms are:

$$2, \, 3x \quad -3at^2, \, 2a^2t \quad PV, \, P_1V_1$$

Notice that the subscripts in $P_1$ and $V_1$ mean they are different quantities than $P$ and $V$.

> **Combining like terms**
> To combine like terms, combine *only* the coefficients; do not change the literal part.

### Example 2-15
Combine each of the following.

**1.** $2x + 3x = (2 + 3)x = 5x$

**2.** $2xy - 4xy - yx = 2xy - 4xy - xy = (2 - 4 - 1)xy = -3xy$

Notice in (2) that $yx$ is the same as $xy$ by the commutative law, (1-1). The above example uses the following variation of the distributive law, (1-3): $BA + CA = (B + C)A$.

An *algebraic expression* contains one or more algebraic terms connected by plus or minus signs. Study the following examples.

### Example 2-16
Simplify each algebraic expression. Remember, you can combine *only* like terms.

**1.** $10x^2 - 3x^2 + 2x + 1 - 5 = 7x^2 + 2x - 4$

**2.** $5.1at^2 - 2a^2t - 2.9at^2 + a^2t = (5.1 - 2.9)at^2 + (-2 + 1)a^2t$
$$= 2.2at^2 - a^2t$$

An algebraic expression consisting of one term is called a *monomial*, of two terms a *binomial*, and of three terms a *trinomial*. Binomals and trinomials are types of *polynomials*, which are expressions consisting of two or more terms. Example 2-16(1) simplifies to a trinomial, and Example 2-16(2) to a binomial.

### Example 2-17
Combine the two trinomials to form one polynomial:

$$(x^3 + 3x^2 + 2x) - (x^2 + x + 2)$$

**Solution**
Notice the negative sign is in front of the second trinomial. *A negative sign in front of a parenthesis changes the sign of every term within the parentheses:*

$$x^3 + 3x^2 + 2x - x^2 - x - 2 = x^3 + 2x^2 + x - 2$$

### Example 2-18
Simplify the algebraic expression:

$$2n - (3n - 5p) - 3p + (-5n + 4p)$$

**Solution**
First remove the parentheses:      $2n - 3n + 5p - 3p - 5n + 4p$

Then combine terms:     $2n - 8n + 9p - 3p = -6n + 6p$
$$= 6p - 6n$$

Notice that a positive sign in front of a parenthesis does not change the sign of any term within the parentheses.

When it is necessary to enclose a set of parentheses in a larger grouping, brackets [ ] are used and then braces { }, as shown in the next two examples.

### Example 2-19
Simplify the algebraic expression:

$$2x - [(a - x) - (2x - a)]$$

**Solution**
Work from the inside out.

First remove parentheses:          $2x - [a - x - 2x + a]$

Combine terms:                          $2x - [2a - 3x]$

Then remove brackets:                 $2x - 2a + 3x$

Combine terms:                          $5x - 2a$

One of the commonest mistakes in algebra is not paying careful attention to a negative sign in front of a parenthesis. Remember: *A negative sign is like a flashing red light—you must slow down and proceed cautiously, making sure to change the sign of every term within the parentheses.*

### Example 2-20
Simplify the algebraic expression:

$$3\pi - \{7\pi - [2r + \theta - (r - 2\theta)]\}$$

**Solution**
First remove parentheses and simplify:

$$3\pi - \{7\pi - [2r + \theta - r + 2\theta]\} = 3\pi - \{7\pi - [r + 3\theta]\}$$

Then remove the brackets and finally the braces:

$$3\pi - \{7\pi - r - 3\theta\} = 3\pi - 7\pi + r + 3\theta = r + 3\theta - 4\pi$$

**Exercise 2-3**   In problems 1 to 32, simplify each algebraic expression.

1. $9a - 11a$

2. $12x - 10x + 3x$

3. $11y + 2x - 8x$

4. $5a - 2b - 3b - 6a$

5. $5np + 14p - 3pn + 2p$

6. $abc + bca + cab$

**7.** $at^2 - a^2t + 3a^2 - at^2$

**8.** $2a^2y + 3ay - 6ay - ay^2$

**9.** $5.3R_1 - 2.8R_2 - 6.8R_1 + 0.5R_2$

**10.** $\dfrac{1}{2}v_1 + \dfrac{2}{5}v_2 - v_1 - \dfrac{1}{4}v_2$

**11.** $ax - (3 - 2ax)$

**12.** $(by - 4) - (2by - 3)$

**13.** $(\theta + 2r) - (\theta - 3r)$

**14.** $5E - (-2E + 4I) + 3I$

**15.** $-(x - 2y) + (2x - y)$

**16.** $3 + (5 + 8t) - (-9 - t)$

**17.** $(8x^2 + 2x + 1) - (3x^2 + 3x + 8)$

**18.** $(x^3 + 2x^2 - x) + (x^2 - 4x + 4)$

**19.** $(2u + 3v - 1) + (3v - 4w + 2)$

**20.** $(3a - 2b + 3c - d) - (2a + 5b + 6c - d)$

**21.** $\dfrac{1}{3}F_x - \dfrac{1}{5}F_y - \left(\dfrac{2}{3}F_x - \dfrac{3}{5}F_y\right)$

**22.** $-8.6\pi - (3r - 2.1\pi) + (6.8\pi + r)$

**23.** $3d - [(x - d) - (d - x)]$

**24.** $5 + [by - (by - 1)]$

**25.** $x^2y + [xy^2 - (2 - x^2y)]$

**26.** $-[(3ab + ac) - (2ab - ac)]$

**27.** $[n^2 + (2n + 1)] - [2n^2 - (n - 1)]$

**28.** $[3m - (2x + 1)] - (2m + x)$

**29.** $4 - \{[f - g) - [(f + g) - 3]\}$

**30.** $8H + \{3H - [2H + (H - 1)]\}$

**31.** $5x - \{2x + [(3y + 1] - (2y + 4)]\}$

**32.** $7 + \{[(5a - 2b) + 6] - (8a - b)\}$

**33.** Simplify the following expression from a problem involving the volume of a solid: $(\pi r^2 h + 2\pi r) - (\pi r - 2\pi r^2 h)$.

**34.** Simplify the following expression from a problem in mechanics: $3.26 - [\frac{1}{2}mv^2 + (mv^2 - 5.18)]$.

**35.** The work output of one computer is $W/3 - 5$, and that of another computer is $W/4 - 7$. How much more work, in terms of $W$, is done by the computer with the greater output?

**36.** Two forces are pulling on a body in the same direction. If $F_1 = 2d^2 - 3d + 1$ and $F_2 = 3d^2 + 4d - 5$, what is the resultant force $F_1 + F_2$ on the body?

**37.** The voltage drop across one resistance is given by $E_1 = 0.18IR - 0.15$ and the voltage drop across another resistance by $E_2 = 0.30IR + 0.09$. What is the total voltage drop $E_1 + E_2$ across both resistances?

**38.** The following expression arises from a problem concerning engine power: $P_1V_1 - P_2V_2 + P_1V_2 - P_2V_1$. What would the expression look like if the last two terms were enclosed in parentheses with a negative sign in front?

## 2-4 | *Rules for Exponents; Multiplying and Dividing Terms*

### Rules for Exponents

As shown in Sec. 1-4, an *exponent* defines repeated multiplication:

$$x^n = \underbrace{(x)(x)(x) \cdots (x)}_{n \text{ times}} \qquad (n = \text{positive integer}) \qquad (2\text{-}1)$$

The number $x$ is called the *base*. When the base is negative, the result is positive for an even exponent and negative for an odd exponent.

### Example 2-21
Calculate each of the following.

**1.** $(-3)^4 = (-3)(-3)(-3)(-3) = 81$

**2.** $-3^4 = -(3)(3)(3)(3) = -81$

**3.** $(-0.4)^3 = (-0.4)(-0.4)(-0.4) = -0.064$

Notice that (1) and (2) are *not* the same. The negative sign is affected by the exponent only if the sign is in the parentheses.

Exponents provide a shortcut way of multiplication as follows.

### Example 2-22
Simplify each of the following.

**1.** $(x^4)(x^3) = (x \cdot x \cdot x \cdot x)(x \cdot x \cdot x) = x^{4+3} = x^7$

**2.** $(2^2)(2^5) = (2 \cdot 2)(2 \cdot 2 \cdot 2 \cdot 2 \cdot 2) = 2^{2+5} = 2^7 = 128$

Example 2-22 leads to the *addition rule* for multiplying with exponents:

$$(x^m)(x^n) = x^{m+n} \qquad (2\text{-}2)$$

Look at (2) in Example 2-22. You might be tempted to add the exponents *and* multiply the 2s, to get $4^7$. This is incorrect because when you add the exponents, you *are* doing the multiplication. You do not change the base. Study the next example, which shows the shortcut method of division using exponents.

### Example 2-23
Simplify each of the following.

**1.** $\dfrac{x^4}{x^3} = \dfrac{x \cdot x \cdot x \cdot x}{x \cdot x \cdot x} = x^{4-3} = x^1 = x$

2. $\dfrac{2^5}{2^2} = \dfrac{2 \cdot 2 \cdot 2 \cdot 2 \cdot 2}{2 \cdot 2} = 2^{5-2} = 2^3$ or 8

Notice in the answer to (1) that if there is no exponent, the exponent is understood to be 1.

Example 2-23 leads to the *subtraction rule* for dividing with exponents:

$$\frac{x^n}{x^m} = x^{n-m} \qquad (x \neq 0) \tag{2-3}$$

In Example 2-23 the subtraction rule, (2-3), is applied when $n > m$. It also applies when $n = m$ and $n < m$ as follows.

## Example 2-24
Simplify each of the following.

1. $\dfrac{x^3}{x^5} = \dfrac{x \cdot x \cdot x}{x \cdot x \cdot x} = \dfrac{1}{1} = 1$ or by (2-3) $\qquad \dfrac{x^3}{x^3} = x^{3-3} = x^0$

2. $\dfrac{x^3}{x^5} = \dfrac{x \cdot x \cdot x}{x \cdot x \cdot x \cdot x \cdot x} = \dfrac{1}{x^2}$ or by (2-3) $\qquad \dfrac{x^3}{x^5} = x^{3-5} = x^{-2}$

Comparing the two sets of results in Example 2-24 leads to the definition of a *zero exponent* and a *negative exponent* ($x \neq 0$):

$$x^0 = 1 \tag{2-4}$$

$$x^{-n} = \left(\frac{1}{x}\right)^n = \frac{1}{x^n} \qquad \left(\frac{x}{y}\right)^{-n} = \left(\frac{y}{x}\right)^n \tag{2-5}$$

Definition (2-4) states that any number to the zero power, except 0, is 1. For example, $8^0 = 1$, $(-\frac{1}{2})^0 = 1$, $(ab)^0 = 1$, etc. (*Note:* $0^0$ is undefined although some calculators show it equal to 1.) Definition (2-5) states that a negative sign in front of an exponent means reciprocal, that is, invert and raise to the positive exponent. In particular, note that $x^{-1} = 1/x$. Also observe that *a negative exponent has nothing to do with the sign of the base*. Study the next example, which illustrates these definitions.

## Example 2-25
Simplify each of the following.

1. $4^{-2} = \left(\dfrac{1}{4}\right)^2 = \dfrac{1}{16}$ or 0.0625

2. $10^{-1} = \dfrac{1}{10}$ or 0.10

**3.** $\left(-\dfrac{2}{5}\right)^{-3} = \left(-\dfrac{5}{2}\right)^{3} = -\dfrac{125}{8}$ or $-15.625$

**4.** $(0.3)^{0}(0.2)^{-2} = (1)\left(\dfrac{1}{0.2}\right)^{2} = \dfrac{1}{0.04} = 25$

Two calculator solutions to (1) are:

$$4 \;\boxed{y^x}\; 2 \;\boxed{+/-}\;\boxed{=} \longrightarrow 0.0625$$

and

$$4 \;\boxed{1/x}\;\boxed{y^x}\; 2 \;\boxed{=} \longrightarrow 0.0625$$

See Example 2-63 for further calculator solutions to Example 2-25.

The next example illustrates the use of the addition and subtraction rules, (2-2) and (2-3), when the exponents $n$ or $m$ are negative integers.

### Example 2-26
Simplify each of the following.

**1.** $x^{-3} \cdot x^{6} = x^{-3+6} = x^{3}$

**2.** $10^{2} \cdot 10^{-4} = 10^{2-4} = 10^{-2}$     or     $\dfrac{1}{100} = 0.01$

**3.** $\dfrac{x^{-5}}{x^{2}} = x^{-5-2} = x^{-7}$     or     $\dfrac{1}{x^{7}}$

In the answer to (3), $1/x^{7}$ is usually preferred to $x^{-7}$.

**4.** $\dfrac{5^{-3}}{5^{-4}} = 5^{-3-(-4)} = 5^{-3+4} = 5$

Remember, the addition and subtraction rules for exponents apply only if the bases are the same. If you have $x^{2}y^{3}$ or $2^{3} \cdot 3^{2}$, you cannot simplify by adding the exponents. The expression $x^{2}y^{3}$ must be left as is: $2^{3} \cdot 3^{2}$ can be left as is or multiplied: $2^{3} \cdot 3^{2} = 8 \cdot 9 = 72$. The next example illustrates the multiplication rule for exponents.

### Example 2-27
Simplify each of the following.

**1.** $(x^{2})^{3} = (x \cdot x)(x \cdot x)(x \cdot x) = x^{2 \cdot 3} = x^{6}$
**2.** $(xy)^{3} = (x \cdot y)(x \cdot y)(x \cdot y) = x^{3}y^{3}$

Example 2-27(1) leads to the *multiplication rule* for exponents:

$$(x^{m})^{n} = x^{mn} \tag{2-6}$$

Example 2-27(2) leads to another useful rule for exponents:

$$(xy)^n = x^n y^n \qquad \left(\frac{x}{y}\right)^n = \frac{x^n}{y^n} \qquad\qquad (2\text{-}7)$$

Study the next example, which illustrates most of the rules (2-1) to (2-7).

### Example 2-28
Simplify each of the following.

**1.** $(x^2 \cdot x)^3 = (x^3)^3 = x^9$

**2.** $\left(\dfrac{2^2}{2^3}\right)^4 = (2^{2-3})^4 = (2^{-1})^4 = 2^{-4} \qquad$ or $\qquad \dfrac{1}{2^4} = \dfrac{1}{16}$

**3.** $\dfrac{(\pi r^2)^2}{2\pi r} = \dfrac{(\pi)^2(r^2)^2}{2\pi r} = \dfrac{\pi^2 r^4}{2\pi r} = \dfrac{\pi r^3}{2}$

**4.** $\dfrac{10^2 \cdot 10^{-3}}{10^{-4}} = \dfrac{10^{-1}}{10^{-4}} = 10^{-1-(-4)} = 10^3 \qquad$ or $\qquad 1000$

**5.** $(-2t)^3(t^{-1})^2 = (-2)^3(t)^3 t^{(-1)(2)} = -8t^3 t^{-2} = -8t$

Rules (2-1) to (2-7) have introduced the basic operations with exponents. Exponents and radicals are studied further in Chap. 8.

## Multiplying and Dividing Terms

**To multiply or divide algebraic terms**
Multiply or divide the coefficients and apply the rules for exponents to the literal parts.

### Example 2-29
Perform the operations for each of the following.

**1.** $(-5x^2)(3x^3) = (-5)(3)(x^{2+3}) = -15x^5$

**2.** $\dfrac{16V^5}{4V^3} = \left(\dfrac{16}{4}\right)V^{5-3} = 4V^2$

**3.** $\left(\dfrac{1}{2}mv^2\right)\left(\dfrac{1}{2}m^2v\right) = \dfrac{1}{4}m^3v^3$

**4.** $\dfrac{0.8x^2y^3}{0.2xy^2} = 4.0xy$

When one term is to be multiplied by each term in a polynomial, the polynomial is enclosed in parentheses as follows.

### Example 2-30
Perform the operations for each of the following.

1. $5x(x^2 + 2x - 1) = 5x(x^2) + 5x(2x) + 5x(-1) = 5x^3 + 10x^2 - 5x$
2. $at(t^2 - at - a^2) = at^3 - a^2t^2 - a^3t$
3. $-\dfrac{1}{2}(8IR - 6E) = -4IR + 3E \quad \text{or} \quad 3E - 4IR$
4. $0.3\pi rh(3.0r - h) = 0.9\pi r^2h - 0.3\pi rh^2$

Notice in (3) that the negative sign of the term $-\frac{1}{2}$ is multiplied by each term in the parentheses. Observe that a minus sign alone in front of a parenthesis is the same as multiplying by $-1$:

$$-1(a - b) = -(a - b) = -a + b \qquad \text{or} \qquad b - a$$

The above also shows that the negative of $a - b$ is $b - a$, a result that will be useful later.

When one term is to be divided into each term in a polynomial, a fraction line is used as follows.

### Example 2-31
Perform the operations for each of the following.

1. $\dfrac{5x^3 + 10x^2 - 5x}{5x} = \dfrac{5x^3}{5x} + \dfrac{10x^2}{5x} - \dfrac{5x}{5x} = x^2 + 2x - 1$
2. $\dfrac{at^3 - a^2t^2 - a^3t}{at} = \dfrac{at^3}{at} - \dfrac{a^2t^2}{at} - \dfrac{a^3t}{at} = t^2 - at - a^2$

Notice that (1) and (2) are the inverse of the operations in (1) and (2) from Example 2-30. When you compare them, you see that the *fraction line is like a set of parentheses*. Every term in the numerator is affected by the denominator or by any operation on the fraction.

### Example 2-32
The relation between the total resistance $R$ of an electric circuit and three resistances $R_1$, $R_2$, and $R_3$ in parallel is expressed by the formula:

$$\frac{1}{R} = \frac{R_2R_3 + R_1R_3 + R_1R_2}{R_1R_2R_3}$$

Express the formula another way by dividing the denominator into the numerator of the fraction.

## Solution

$$\frac{1}{R} = \frac{\frac{1}{\cancel{R_2}\cancel{R_3}}}{R_1\cancel{R_2}\cancel{R_3}} + \frac{\frac{1}{\cancel{R_1}\cancel{R_3}}}{\cancel{R_1}R_2\cancel{R_3}} + \frac{\frac{1}{\cancel{R_1}\cancel{R_2}}}{\cancel{R_1}\cancel{R_2}R_3} = \frac{1}{R_1} + \frac{1}{R_2} + \frac{1}{R_3}$$

**Exercise 2-4**

In problems 1 to 64, perform the indicated operations.

**1.** $(-2)^6$

**2.** $-2^6$

**3.** $-0.3^2$

**4.** $(-1.1)^2$

**5.** $x^2 \cdot x^4$

**6.** $y^3 \cdot y^5 \cdot y$

**7.** $\dfrac{p^7}{p^5}$

**8.** $\dfrac{q^2}{q^5}$

**9.** $a^2 \cdot \dfrac{a^4}{a^8}$

**10.** $\dfrac{n^5}{n^3 \cdot n}$

**11.** $\dfrac{(0.1)^5}{(0.1)^3(0.1)^2}$

**12.** $\dfrac{5^2 \cdot 5^3}{5^4}$

**13.** $(w^3)^3$

**14.** $(-v^2)^3$

**15.** $(3mn)^4$

**16.** $(x \cdot x^2)^3$

**17.** $\left(\dfrac{ab}{2x}\right)^3$

**18.** $\left(\dfrac{y^4}{y^2}\right)^3$

**19.** $8.8x^0$

**20.** $(8.8x)^0$

**21.** $5^{-2}$

**22.** $(-3)^{-3}$

**23.** $\left(-\dfrac{2}{9}\right)^{-1}$

**24.** $\dfrac{1}{7^{-2}}$

**25.** $C^2 \cdot C^{-3} \cdot C$

**26.** $10^2 \cdot 10^0 \cdot 10^{-4}$

**27.** $\dfrac{4}{4^{-2}}$

**28.** $\dfrac{d^{-1}}{d^2}$

**29.** $(-3t)^2(t^{-2})^2$

**30.** $(\pi r^2)(2\pi r)^2$

**31.** $\dfrac{(ab^3)^2}{a^2b^3}$

**32.** $\dfrac{-v^3}{(-2v)^2}$

**33.** $[(0.2)^2 v^3]^2$

**34.** $-Fd(2Fd^2)(Fd)^2$

**35.** $(3x^2)(-4x^3)$

**36.** $(2by^2)(12b^2y)$

**37.** $(5I)(2IR)$

**38.** $(-7mc)(8c)$

**39.** $(1.2a^3b)(-3.0ab^2)$

**40.** $(0.1rs)(0.3r)(0.2s)$

**41.** $\dfrac{18y^6}{6y^3}$

**42.** $\dfrac{28x^8}{7x^7}$

**43.** $\dfrac{2.8ax^2}{1.4ax}$

**44.** $\dfrac{100c^2d^3}{25cd}$

**45.** $\dfrac{(3h)(2ht)(5t^2)}{15ht}$

**46.** $\dfrac{-3ax(a^2x)}{2x(ax^2)}$

**47.** $2x(x^2 - x + 3)$

**48.** $-3n(2n^2 + 3n + 1)$

**49.** $-px(p^2 - x^2)$

**50.** $vt^2(v^2 + t^2)$

**51.** $ab^2c^3(a - b + c)$

**52.** $-3xy(x^2 - 3xy + y^2)$

**53.** $\dfrac{2}{3}xy(12ax - 21by)$

**54.** $0.6\pi r(2.0\pi - 3.0r)$

**55.** $x^2y(x + y) - xy^2(x - y)$

**56.** $(-5uv)(-3uv)(uv^2 + u^2v)$

**57.** $\dfrac{6x^2 + 12x}{2x}$

**58.** $\dfrac{9y^3 - 6y^2}{-3y^2}$

**59.** $\dfrac{4ab^3 - 6a^2b^2 + 4a^3b}{-2ab}$

**60.** $\dfrac{10k^3 - 5k^2 + 15k}{5k}$

**61.** $\dfrac{C_1^2 + C_1C_2}{C_1}$

**62.** $\dfrac{\pi r^2h + 2\pi rh}{\pi rh}$

**63.** $\dfrac{(xy)(x^2y - xy^2)}{x^2y^2}$

**64.** $\dfrac{-8(n^2p^2 - np)}{-4np}$

**65.** In calculating the final volume of gas $V_f$, the expression arises:

$$V_f = 3.2t(1.5 + 0.0037t) + 2.1t(0.0016t - 1.2)$$

Multiply and combine like terms.

**66.** The pressure of a gas after a temperature change from $T_1$ to $T_2$ is given by:

$$P\left(1 + \frac{T_2 - T_1}{T_1}\right)$$

Simplify this expression by first dividing out the fraction.

**67.** The force of a strong electric charge is $qr^2/r$, and the force of a weaker one is $(rq^2 - 2q)/q$. Write the expression that represents the difference between the two charges, and simplify the expression.

**68.** The total resistance $R$ of a series-parallel circuit is given by:

$$R = \left(\frac{1}{R_1} + \frac{1}{R_2}\right)^{-1} + R_3$$

Find $R$ when $R_1 = 0.80 \ \Omega$, $R_2 = 0.20 \ \Omega$, and $R_3 = 0.50 \ \Omega$.

**69.** Multiply the following BASIC expression and express the result in BASIC:

$$2 * X(X \wedge 2 - 3 * X + 5)$$

**70.** Divide the following BASIC expression and express the result in BASIC:

$$(2 * X \wedge 3 + 4 * X \wedge 2 - 6 * X)/(2 * X)$$

**71.** Using definition (2-5), show that $x^{a-b} = 1/x^{b-a}$.

**72.** If $a - b$ is the negative of $b - a$, simplify the fraction:

$$\frac{a - b}{b - a}$$

# 2-5 | *Scientific Notation*

Consider the following problem. The star nearest to earth (excluding the sun) is Proxima Centauri, which is 4.28 light-years away. This means it takes 4.28 years (yr) for light to travel from this star to earth. If light travels at 300,000 km/s, or 186,000 mi/s, how far away is Proxima Centauri? The distance is calculated as follows:

(300,000 km/s)(60 s/min)(60 min/h)(24 h/d)(365 d/yr)(4.28 yr) =

40,500,000,000,000 km    or    25,100,000,000,000 mi (three sig. fig.)

The answer is more than 40 trillion kilometers, or 25 trillion miles! Very large and very small numbers are often encountered in scientific and technical work. These numbers are usually expressed in *scientific notation,* which means as *a number between 1 and 10, times a power of 10.* The answer to the above problem in scientific notation is:

$$4.05 \times 10^{13} \text{ km} \quad \text{or} \quad 2.51 \times 10^{13} \text{ mi}$$

See Example 2-65 for the calculator solution to this problem.

The decimal point in scientific notation is placed *to the right of the first significant figure.* This is called the *zero position.* If the decimal point is originally in the zero position, it corresponds to the power of 10 equal to 0. For information on significant figures see Sec. 1-6.

> **Changing ordinary notation to scientific notation**
> Count the number of places the decimal point is to the right or left of the zero position. The number of places corresponds to the power of 10: positive if the decimal point is to the right and negative if it is to the left. Write the number, with the decimal point in the zero position, times the power of 10.

Study the following example. The arrow shows the zero position.

### Example 2-33
Change to scientific notation.

1. $5{,}830{,}000. = 5.83 \times 10^6$     $(10^6 = 1{,}000{,}000)$
2. $58{,}300. = 5.83 \times 10^4$     $(10^4 = 10{,}000)$
3. $58.3 = 5.83 \times 10^1$     $(10^1 = 10)$
4. $5.83 = 5.83 \times 10^0$     $(10^0 = 1)$
5. $0.0583 = 5.83 \times 10^{-2}$     $(10^{-2} = 0.01)$
6. $0.0000583 = 5.83 \times 10^{-5}$     $(10^{-5} = 0.00001)$

In each example, notice that the number of places between the arrow and the decimal point is equal to the exponent. In (4) the decimal point is in the zero position, and the power of 10 is therefore 0.

### Changing scientific notation to ordinary notation
Move the decimal point to the right if the exponent is positive and to the left if it is negative, the number of places indicated by the exponent.

### Example 2-34
Change to ordinary notation.

1. $3.86 \times 10^3 = 3860$
2. $8.10 \times 10^{-2} = 0.0810$
3. $5.25 \times 10^0 = 5.25$
4. $1.01 \times 10^{-4} = 0.000101$

### Example 2-35
Calculate by using scientific notation:

$$\frac{(328{,}000{,}000{,}000)(0.000850)}{2{,}380{,}000{,}000{,}000}$$

**Solution**
First write the numbers in scientific notation:

$$\frac{(3.28 \times 10^{11})(8.50 \times 10^{-4})}{2.38 \times 10^{12}}$$

You can calculate this by hand, applying the rules for exponents, or use the calculator. The calculator solution is shown here. First, as a check on the calculator, round each number to one significant figure for the approximate solution:

$$\frac{(3 \times 10^{11})(9 \times 10^{-4})}{2 \times 10^{12}} = \frac{27 \times 10^{11-4}}{2 \times 10^{12}} = \frac{27}{2} \times 10^{7-12}$$
$$\approx 14 \times 10^{-5}$$

Notice that the approximate answer is *not* in scientific notation although it is expressed as a power of 10. To change a number written as a power of 10 to scientific notation, move the decimal point to the zero position and balance the exponent by increasing it for every place you move the point to the left, and vice versa: $14 \times 10^{-5} = 1.4 \times 10^{-4}$. On the calculator, numbers are entered in scientific notation by using the EE, EEX, or EXP key, which enters the exponent. The calculator solution is, to three significant figures:

$$3.28 \boxed{EXP}\ 11 \boxed{\times}\ 8.50 \boxed{EXP}\ 4 \boxed{+/-} \boxed{\div}\ 2.38 \boxed{EXP}\ 12$$
$$\boxed{=} \rightarrow 1.17(-04)$$

The answer is $1.17 \times 10^{-4}$, or 0.000117, which agrees closely with the approximate answer.

Scientific notation allows you to clearly identify the significant figures when the last figures are zeros. For example, the number 380,000 could have two or more significant figures; however, if it is written $3.80 \times 10^3$, it clearly has three significant figures.

The size of numbers written in scientific notation can be misleading. The increase from $10^2$ to $10^3$ is only 900, but the increase from $10^6$ to $10^7$ is 9,000,000! Figure 2-3 gives you an idea of how powers of 10 relate to some of the world's important (and not so important) scientific and physical phenomena. For example, if you could add all the words ever spoken since people first started babbling, the number would still be less than $10^{17}$! Scientific notation and powers of 10 are important in the metric system, which is discussed in Chap. 4.

**Exercise 2-5**

In problems 1 to 12, change each number to scientific notation.

1. 42,600,000,000
2. 0.0000000930
3. 117,800,000,000
4. 0.00003001
5. 3.001
6. 0.62
7. $\dfrac{1}{1,000,000}$
8. $\dfrac{7}{10,000}$
9. $564 \times 10^4$
10. $36 \times 10^{-4}$
11. $\dfrac{1}{100} \times 10^6$
12. $\dfrac{1}{10} \times 10^{-4}$

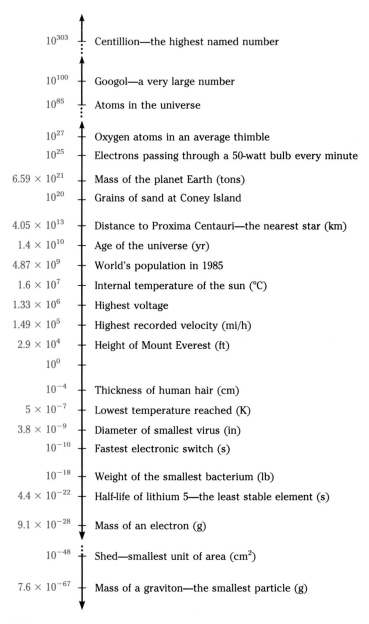

| | |
|---|---|
| $10^{303}$ | Centillion—the highest named number |
| $10^{100}$ | Googol—a very large number |
| $10^{85}$ | Atoms in the universe |
| $10^{27}$ | Oxygen atoms in an average thimble |
| $10^{25}$ | Electrons passing through a 50-watt bulb every minute |
| $6.59 \times 10^{21}$ | Mass of the planet Earth (tons) |
| $10^{20}$ | Grains of sand at Coney Island |
| $4.05 \times 10^{13}$ | Distance to Proxima Centauri—the nearest star (km) |
| $1.4 \times 10^{10}$ | Age of the universe (yr) |
| $4.87 \times 10^{9}$ | World's population in 1985 |
| $1.6 \times 10^{7}$ | Internal temperature of the sun (°C) |
| $1.33 \times 10^{6}$ | Highest voltage |
| $1.49 \times 10^{5}$ | Highest recorded velocity (mi/h) |
| $2.9 \times 10^{4}$ | Height of Mount Everest (ft) |
| $10^{0}$ | |
| $10^{-4}$ | Thickness of human hair (cm) |
| $5 \times 10^{-7}$ | Lowest temperature reached (K) |
| $3.8 \times 10^{-9}$ | Diameter of smallest virus (in) |
| $10^{-10}$ | Fastest electronic switch (s) |
| $10^{-18}$ | Weight of the smallest bacterium (lb) |
| $4.4 \times 10^{-22}$ | Half-life of lithium 5—the least stable element (s) |
| $9.1 \times 10^{-28}$ | Mass of an electron (g) |
| $10^{-48}$ | Shed—smallest unit of area (cm$^2$) |
| $7.6 \times 10^{-67}$ | Mass of a graviton—the smallest particle (g) |

FIG. 2-3 *Scientific notation*

In problems 13 to 20, change each number to ordinary notation.

**13.** $2.64 \times 10^{10}$      **14.** $3.90 \times 10^{-6}$

**15.** $9.306 \times 10^{-11}$      **16.** $1.115 \times 10^8$

**17.** $3.20 \times 10^{-5}$      **18.** $8.71 \times 10^0$

**19.** $4.44 \times 10^7$      **20.** $1.25 \times 10^{-7}$

In problems 21 to 30, choose the correct answer by approximating the result, using the rules for exponents. Check by hand or with the calculator, and express answers to three significant figures. See Example 2-35.

**21.** $(3.28 \times 10^7)(8.30 \times 10^5)$    $(2.72 \times 10^{12}, 2.72 \times 10^{13}, 2.72 \times 10^{14},$ $27.2 \times 10^{14})$

**22.** $(9.30 \times 10^{-8})(3.33 \times 10^{13})$    $(30.9 \times 10^4, 3.09 \times 10^5, 3.10 \times 10^6,$ $3.10 \times 10^7)$

**23.** $(0.00000000101)(9,330,000)$    $(0.000942, 0.00942, 0.0942, 0.942)$

**24.** $(32,200)(86,500)(28,300)$    $(7.88 \times 10^{10}, 7.89 \times 10^{11}, 7.88 \times 10^{12},$ $7.88 \times 10^{13})$

**25.** $\dfrac{8.89 \times 10^{15}}{9.89 \times 10^7}$    $(8.99 \times 10^7, 8.98 \times 10^8, 8.99 \times 10^8, 8.98 \times 10^9)$

**26.** $\dfrac{3.03 \times 10^{-3}}{6.06 \times 10^3}$    $(0.505 \times 10^{-6}, 5.00 \times 10^{-6}, 0.500 \times 10^{-7},$ $5.00 \times 10^{-7})$

**27.** $\dfrac{6.56 \times 10^{-8}}{1.23 \times 10^{-6}}$    $(0.00533, 0.0533, 0.533, 5.33)$

**28.** $\dfrac{(0.000210)(0.00349)}{451}$    $(1.62 \times 10^{-7}, 1.62 \times 10^{-8}, 1.63 \times 10^{-8}, 1.63 \times 10^{-9})$

**29.** $\dfrac{1.26 \times 10^{12}}{(3.50 \times 10^{-4})(9.00 \times 10^5)}$    $(4.00 \times 10^8, 0.400 \times 10^9, 4.00 \times 10^9, 40.0 \times 10^9)$

**30.** $\dfrac{(340)^2(450)^2}{(0.980)^3}$    $(24.9 \times 10^8, 2.49 \times 10^9, 2.49 \times 10^{10}, 24.9 \times 10^{10})$

**31.** General Motors made one of the greatest profits in 1978 ever achieved by an industrial company. Worldwide sales totaled \$63,221,100,000, with a net profit of \$3,508,000,000. Express these numbers in scientific notation, and calculate what percentage of worldwide sales the net profit was.

**32.** Parker Bros., Inc., manufacturer of the board game *Monopoly,* printed \$18,500,000,000,000 of toy money in 1979 for all its games. That is more than all the real money in circulation in the world. If all this "money" were distributed equally among the world's population, estimated at 4,321,000,000 in 1979, how much would each person get? Express the numbers in scientific notation, and do the calculation.

**33.** The shortest blip of light produced at the Center of Laser Studies in California lasts $0.20 \times 10^{-12}$ s [0.2 picosecond (ps)]. If the speed of light is $3.0 \times 10^8$ m/s, how many meters does the light travel in that time?

34. The world produced more than 6,818,000,000,000 kilowatt-hours (kWh) of electricity in 1978. If 1 kWh is equivalent to 0.076 gal of oil, how many gallons of oil would it take to produce that much electric energy?

35. One of the world's fastest computers, the CRAY-1, can do a simple addition in less than $5.5 \times 10^{-10}$ s, or 0.55 nanoseconds (ns). How many additions can this computer do in less than 1 min?

36. The resonant frequency $f$ in hertz (Hz) of a series resonance circuit is given by $f = 1/(2\pi\sqrt{LC})$. Calculate $f$ when the inductance $L = 1.50 \times 10^{-4}$ henry (H) and the capacitance $C = 2.94 \times 10^{-10}$ farad (F).

37. The lowest temperature ever achieved on this planet was $5.00 \times 10^{-7}$ K by Professor Abragam in France in 1969. The Kelvin scale starts at absolute zero ($-273.15°$C, or $-459.67°$F), the temperature at which a gas would theoretically show no pressure. Absolute temperatures $T$ are defined in terms of the ratio $273.15/T$. Compute this ratio for the lowest temperature ever achieved on earth.

38. The number of molecules in a substance whose mass in grams is equal numerically to its molecular mass (1 mole) is given by Avogadro's number: $6.02 \times 10^{23}$ molecules. If 1 mole of oxygen ($O_2$) is 32 g, that is, there are $6.02 \times 10^{23}$ molecules in 32 g of $O_2$, how many grams does one molecule weigh?

39. The world's most expensive pipeline is the Alaska pipeline, which is built to carry 2,000,000 barrels a day (bbl/day) of crude oil. If $6.65 \times 10^6$ bbl of oil (1 million tons) can generate $4.0 \times 10^9$ kWh of electricity, how many kilowatthours could be generated in a year from Alaska pipeline oil?

40. Engineering notation expresses numbers in powers of 10 that are multiples of 3. For example, $58,600,000 = 58.6 \times 10^6$ and $0.215 = 215 \times 10^{-3}$. Change each to engineering notation: *(a)* 81,200, *(b)* 0.0335, *(c)* $2.13 \times 10^7$, *(d)* $7.91 \times 10^{-5}$.

## 2-6 | *Multiplying Binomials and Polynomials*

Binomials occur in many algebraic problems, and you need to master the method for multiplying two binomials. Study the following examples.

### Example 2-36
Multiply the binomials:     $(x - 3)(2x + 5)$

### Solution
The distributive law, (1-3), says that each term in the first binomial must be multiplied by each term in the second binomial, which produces four products:

$$x(2x + 5) - 3(2x + 5) = x(2x) + x(5) - 3(2x) - 3(5)$$
$$= 2x^2 + 5x - 6x - 15$$
$$= 2x^2 - x - 15$$

A convenient way of doing this is the *FOIL method*. FOIL stands for the order in which you multiply the numbers: First, Outside, Inside, Last:

$$\text{First} \quad \text{Outside} \quad \text{Inside} \quad \text{Last}$$
$$(x - 3)(2x + 5) = x(2x) + \quad x(5) \quad - 3(2x) - 3(5)$$

Most problems involve similar binomials, that is, binomials whose first terms are similar and whose last terms are similar, as in Example 2-36. In this case the inside and outside products are similar and can be combined. With a little practice you can multiply similar binomials mentally.

### Example 2-37
Multiply the binomials mentally:

$$(x - 3)(2x + 5)$$

### Solution
Write the first product: $\quad (x - 3)(2x + 5) = 2x^2$

Write the last product: $\quad (x - 3)(2x + 5) = 2x^2 \quad - 15$

Mentally calculate the inside and outside products:

$$(x - 3)(2x + 5) = 2x^2 \quad \begin{array}{l} \rightarrow (5x) \\ \rightarrow (-6x) \end{array} - 15$$

Mentally combine and write down the middle term:

$$(x - 3)(2x + 5) = 2x^2 - x - 15$$

To help yourself learn binomial multiplication, see if you can mentally calculate the binomials in the next example.

### Example 2-38
Multiply each binomial pair.

1. $(x + 3)(x + 4) = x^2 + 7x + 12$

2. $(3x + 4)(2x - 5) = \overset{F}{6x^2} - \overset{O}{15x} + \overset{I}{8x} - \overset{L}{20} = 6x^2 - 7x - 20$

3. $(8n - 3)(n + 2) = 8n^2 + 13n - 6$

4. $(2a - 3b)(3a - 2b) = \overset{F}{6a^2} - \overset{O}{4ab} - \overset{I}{9ab} + \overset{L}{6b^2} = 6a^2 - 13ab + 6b^2$

5. $(r + 2t)(r - t) = r^2 + rt - 2t^2$

When two binomials are identical, their product is a *binomial square*, as follows.

### Example 2-39
Square each binomial.

1. $(x + y)^2 = (x + y)(x + y) = x^2 + xy + xy + y^2 = x^2 + 2xy + y^2$
2. $(2a + 3x)^2 = (2a + 3x)(2a + 3x) = 4a^2 + 2(2a)(3x) + 9x^2$
$$= 4a^2 + 12ax + 9x^2$$

Notice that the middle term of a binomial square is twice the product of the two terms in the binomial.

A useful binomial product is the sum of two terms multiplied by their difference. The result is called the *difference of two squares,* as follows.

### Example 2-40
Multiply each binomial pair.

1. $(A + B)(A - B) = A^2 - AB + AB - B^2 = A^2 - B^2$

Observe that in the difference of two squares the inside product cancels the outside product.

2. $(2p + 3q)(2p - 3q) = (2p)^2 - (3q)^2 = 4p^2 - 9q^2$

The difference of two squares is used in simplifying expressions with radicals. It is the only binomial product that has two terms, and it should not be confused with the binomial square, which results in three terms.

### Example 2-41
Multiply the binomial pair:

$$(\sqrt{x} + \sqrt{y})(\sqrt{x} - \sqrt{y}) = (\sqrt{x})^2 - (\sqrt{y})^2 = x - y$$

By rule (1-6), a radical and a square are inverse operations and cancel each other; therefore the result is free of radicals. This product is studied further in Chap. 7.

When the binomials are not similar, the inside and outside products are not similar and the result is a polynomial of four terms, as follows.

### Example 2-42
Multiply each binomial pair.

$$\begin{array}{cccc} \text{F} & \text{O} & \text{I} & \text{L} \end{array}$$
1. $(a + b)(c + d) = ac + ad + bc + bd$
2. $(2m - a)(m - 2b) = 2m^2 - 4mb - ma + 2ab$

When a binomial is multiplied by a polynomial of three or more terms, the example can be set up by grouping similar products as follows.

**Example 2-43**

Multiply:      $(x + 3)(x^2 + 2x - 2)$

**Solution**

$$x^2 + 2x - 2$$
$$x + 3$$

Multiply $x$ by $x^2 + 2x - 2$:      $x^3 + 2x^2 - 2x$
Multiply 3 by $x^2 + 2x - 2$:          $3x^2 + 6x - 6$
Combine similar terms:                $x^3 + 5x^2 + 4x - 6$

The procedure is similar to long multiplication in arithmetic. This example can also be done by writing the six products in a line and combining terms.

**Example 2-44**

Under certain load conditions, the vertical displacement of a beam of length $L$, at a distance $x$ from one end, is expressed by:

$$(3x - L)(2x - L)(x - L)$$

Multiply this expression.

**Solution**

Multiply the first two binomials:

$$(3x - L)(2x - L) = 6x^2 - 5xL + L^2$$

Then multiply the result by the last binomial, using the long method:

$$6x^2 - 5xL + L^2$$
$$x - L$$

Multiply $x$ by $6x^2 - 5xL + L^2$:      $6x^3 - 5x^2L + xL^2$
Multiply $-L$ by $6x^2 - 5xL + L^2$:       $- 6x^2L + 5xL^2 - L^3$
Combine similar terms:                 $6x^3 - 11x^2L + 6xL^2 - L^3$

Division of polynomials is discussed in Chap. 15, on polynomial functions (Sec. 15-2), where it is used in the solution of polynomial equations.

**Exercise 2-6**

In problems 1 to 46, multiply each of the products. Try to calculate the middle term mentally for two similar binomials.

**1.** $(x + 2)(x - 4)$          **2.** $(b + 3)(b + 5)$

**3.** $(3a - 1)(a - 3)$        **4.** $(y - 6)(2y + 4)$

**5.** $(t + 1)(5t + 2)$        **6.** $(1 - 4w)(2 - w)$

**7.** $(2d + 4)(3d - 5)$      **8.** $(3h + 4)(4h - 3)$

**9.** $(5m - n)(4m + n)$     **10.** $(7p + 2q)(3p - q)$

**11.** $(2x - 3y)(3x - 2y)$

**12.** $(4a + 3b)(2a + 5b)$

**13.** $(4ab + 7)(ab - 5)$

**14.** $(x^2 + 1)(2x^2 - 1)$

**15.** $(R + 0.4)(R + 0.6)$

**16.** $(0.5F - 1.0)(0.5F - 3.0)$

**17.** $(a + 3)^2$

**18.** $(4 - x)^2$

**19.** $(3x - 4y)^2$

**20.** $(2ax - b)^2$

**21.** $3(vt + a)^2$

**22.** $2(a^2 - x)^2$

**23.** $(y + 3)(y - 3)$

**24.** $(2r - s)(2r + s)$

**25.** $(5a + 2x)(5a - 2x)$

**26.** $(mp + 7)(mp - 7)$

**27.** $4\pi(r - h)(r - 2h)$

**28.** $2x(L + x)(L - x)$

**29.** $(\sqrt{x} + 1)(\sqrt{x} - 1)$

**30.** $(\sqrt{y} + \sqrt{2})(\sqrt{y} - \sqrt{2})$

**31.** $(\sqrt{a} + \sqrt{b})(\sqrt{a} - \sqrt{b})$

**32.** $(\sqrt{ab} + 3)(\sqrt{ab} - 3)$

**33.** $(x + a)(x + b)$

**34.** $(r - h)(r + 2)$

**35.** $(x^2 - 2)(x + 3)$

**36.** $(3l - 2h)(2w - h)$

**37.** $(x + 2)(x^2 - 3x + 1)$

**38.** $(x - 1)(2x^2 + 2x - 3)$

**39.** $(a + b)(a^2 - ab + b^2)$

**40.** $(a - b)(a^2 + ab + b^2)$

**41.** $(x + 2)^3$

**42.** $(y - 3)^3$

**43.** $(1 + j)^2(1 - 2j)$

**44.** $(1 - e)(1 + e)(2 + e)$

**45.** $(v_1 + v_2)^2(v_1 + v_2)$

**46.** $(x^2 + 1)^2(x - 2)$

**47.** A variation of Boyle's law leads to the expression:

$$\left(p + \frac{a}{v^2}\right)(v - b)$$

Multiply these binomials.

**48.** The side of a cube is $x$. Each side is lengthened by 3, increasing the volume. Write the expression for the new volume, and multiply out the expression.

**49.** The relation between the voltage $E$ and the temperature $t$ of a transmission line is given by:

$$E^2 = 350R(1 + 0.003t)(t - 24)$$

where $R$ = resistance of the line. Multiply this expression.

**50.** The work done by a force is found to be:

$$W = (r + d)(r - 2d)(2r - d)$$

Multiply these binomials.

## 2-7 | *First-Degree Equations*

Algebraic equations occur in all scientific and technical areas. The solution of equations is one of the most important processes in algebra. A *first-degree equation,* or *linear equation,* is a statement of equality that contains only variables to the first power. This section covers the solution of first-degree equations in one variable. Linear equations containing two or three variables are discussed in Chap. 5. To solve for the unknown variable in an equation, the following basic principle is applied:

Adding, multiplying, or dividing each side of an equation by the same quantity does not change the equality. (Division by zero is not allowed.)

$$(2\text{-}8)$$

Study the following examples, which show how to apply principle (2-8).

### Example 2-45

Solve and check the linear equation $3x + 5 = 11$ for the unknown variable $x$.

### Solution

To solve the equation, apply principle (2-8) to isolate the unknown on one side of the equation:

$$3x + 5 = 11$$

Add $-5$ to both sides: $\qquad 3x + 5 + (-5) = 11 + (-5)$

Simplify: $\qquad\qquad\qquad\qquad 3x = 6$

Divide both sides by 3: $\qquad\qquad \dfrac{\cancel{3}x}{\cancel{3}} = \dfrac{6}{3}$

Simplify: $\qquad\qquad\qquad\qquad x = 2$

To check the solution, substitute $x = 2$ into the original equation:

$$3(2) + 5 = 11?$$
$$6 + 5 = 11 \checkmark$$

### Example 2-46

Solve and check the linear equation:

$$4x - 3 = 5 + 2x$$

### Solution

| | |
|---|---|
| Add 3 to both sides: | $4x - 3 + (3) = 5 + 2x + (3)$ |
| Simplify: | $4x = 8 + 2x$ |
| Add $-2x$ to both sides: | $4x + (-2x) = 8 + 2x + (-2x)$ |
| Simplify: | $2x = 8$ |
| Divide both sides by 2: | $\dfrac{\cancel{2}x}{\cancel{2}} = \dfrac{8}{2}$ |
| Simplify: | $x = 4$ |

The solution check for $x = 4$ is:

$$4(4) - 3 = 5 + 2(4)?$$
$$16 - 3 = 5 + 8?$$
$$13 = 13 \checkmark$$

The process of solving an equation can be simplified by noticing that adding the same quantity to both sides of an equation is equivalent to *transposing*, which means moving that quantity to the other side of the equation and *changing its sign*.

### Example 2-47

Solve Example 2-46, using the method of transposing.

### Solution

| | |
|---|---|
| Transpose $-3$ and $2x$: | $4x - 3 = 5 + 2x$ |
| | $4x - 2x = 5 + 3$ |
| Combine similar terms: | $2x = 8$ |
| Divide by 2: | $x = 4$ |

Notice that in one step you transpose all the terms containing the unknown to one side and all the constant terms to the other side.

You should simplify your equation solving by using the method of transposing whenever possible. Study the following example closely until you understand every step.

### Example 2-48

Solve and check the equation:     $0.5w - 0.7 = 1.5(w - 1.2)$

### Solution

| | |
|---|---|
| First remove parentheses: | $0.5w - 0.7 = 1.5w - 1.8$ |
| Transpose $-0.7$ and $1.5w$: | $0.5w - 1.5w = -1.8 + 0.7$ |

Combine similar terms:                          $-1.0w = -1.1$

                                                or $-w = -1.1$

Multiply both sides by $-1$:                    $w = 1.1$

Look at the next-to-last step. You might be tempted to stop here, thinking you have the solution. The number $-1.1$, however, represents the negative value of the unknown. You must go one step further and multiply both sides by $-1$. This changes the sign of both sides and gives the correct answer: the positive value of the unknown.

The solution check for $w = 1.1$ is:

$$0.5(1.1) - 0.7 = 1.5(1.1 - 1.2)?$$

$$0.55 - 0.7 = 1.5(-0.1)?$$

$$-0.15 = -0.15 \checkmark$$

*Note:* Each side of an equation must be worked separately when you are checking the equation.

Observe that Example 2-48 can also be solved by first multiplying by 10 to clear decimals, resulting in the step:

$$5w - 7 = 15(w - 12)$$

**Example 2-49**

Solve and check:

$$\frac{A}{4} + \pi = 2\pi$$

**Solution**

Transpose $\pi$:                                $\dfrac{A}{4} = 2\pi - \pi$

Combine terms:                                  $\dfrac{A}{4} = \pi$

Multiply both sides by 4:         $(4)\dfrac{A}{4} = 4\pi$   or   $12.6$

The solution check for $A = 4\pi$ is:

$$\frac{4\pi}{4} + \pi = 2\pi?$$

$$\pi + \pi = 2\pi \checkmark$$

### Example 2-50

The following equation results from changing 27°C to degrees Fahrenheit. Solve and check:

$$27 = \frac{5}{9}(F - 32)$$

### Solution

Remove parentheses:

$$27 = \frac{5}{9}F - \frac{160}{9}$$

Transpose $\dfrac{-160}{9}$:

$$27 + \frac{160}{9} = \frac{5}{9}F$$

Rewrite and combine terms:

$$\frac{5}{9}F = \frac{243 + 160}{9} = \frac{403}{9}$$

Multiply by 9:

$$(\cancel{9})\frac{5F}{\cancel{9}} = \frac{403}{\cancel{9}}(\cancel{9})$$

Divide by 5:

$$\frac{\cancel{5}F}{\cancel{5}} = \frac{403}{5}$$

$$F = 80.6$$

Notice in the third step that the equation is rewritten with the unknown term on the left for uniformity.

   Example 2-50 can also be solved by first multiplying both sides by 9 and getting rid of the fraction as follows.

### Example 2-51

Solve Example 2-50 by first eliminating the fraction.

### Solution

Multiply by 9:

$$(9)27 = (\cancel{9})\frac{5}{9}(F - 32)$$

Remove parentheses:                    $243 = 5F - 160$

Transpose $-160$ and combine terms:    $403 = 5F$

Divide by 5:                           $F = 80.6$

   Eliminating fractions as soon as possible in an equation can make the solution easier. Eliminating fractions is very helpful when equations contain several fractions with different denominators. The solution of equations with fractions is discussed in detail in Chap. 7.

   Here are the steps for solving linear equations in one unknown:

1.  Simplify each side of the equation as much as possible by removing parentheses and eliminating fractions.
2.  Combine similar terms on each side.
3.  Transpose all terms containing the unknown to one side and the remaining terms to the other side.
4.  Combine similar terms, and divide or multiply both sides of the equation to make the coefficient of the unknown +1.

**Exercise 2-7**  In problems 1 to 32, solve and check each equation.

**1.** $x + 2 = 7$

**2.** $x - 3 = 9$

**3.** $3x + 2 = 5$

**4.** $4y - 8 = 12$

**5.** $4 + 5y = y - 12$

**6.** $6x - 10 = x - 15$

**7.** $3 - 3t = 6$

**8.** $16 = 8 - 4w$

**9.** $\dfrac{h}{8} = \dfrac{1}{2}$

**10.** $\dfrac{L}{2} = 1.5$

**11.** $5a + 9 = 6a + 3$

**12.** $3 - 7b = 3b - 2$

**13.** $K = \dfrac{1}{2}(4K + 2)$

**14.** $2R - \dfrac{1}{3} = 1 + \dfrac{2}{3}$

**15.** $0.2p + 3.0 = 1.4$

**16.** $1.2m + 1.3m = 5.0$

**17.** $0.7 = 2.0n - 2.3$

**18.** $0.06 + 0.05q = 0.03q$

**19.** $2(x - 4) = x - 5$

**20.** $6 + 2x = 2(2x + 3)$

**21.** $5 - (y + 2) = 5y$

**22.** $4 - (y - 4) = 2y$

**23.** $x(x + 2) = x(x - 4) + 12$

**24.** $x^2 + 4 = x(x - 6)$

**25.** $0.26 = 0.20(v - 1.0)$

**26.** $2.0(w - 1.0) = 6.5 + 3.0w$

**27.** $0.7x - 0.2x = 1.2(x + 1.4)$

**28.** $0.3a = 0.2(0.3a + 1.2)$

**29.** $3Q - \dfrac{5}{9} = \dfrac{2}{9}Q$

**30.** $\dfrac{9}{5}C + 32 = 4C - 1$

**31.** $\dfrac{4\pi(r + 2)}{2} = 3\pi$

**32.** $\dfrac{2}{3}(6x + 1) = \dfrac{1}{2}(4x - 1)$

**33.** The following equation results when 15°C is changed to degrees Fahrenheit: $15 = \dfrac{5}{9}(F - 32)$. Solve and check the equation.

**34.** The formula relating degrees Celsius to degrees Fahrenheit is:

$$F = \frac{9}{5}C + 32$$

Find $C$ when $F = -10°F$.

**35.** The voltage in a circuit is given by:

$$V = I_1R_1 + I_2R_2$$

Find $I_2$ when $V = 12.2$ V, $I_1 = 1.0$ A, $R_1 = 3.4\ \Omega$, and $R_2 = 4.2\ \Omega$.

**36.** In a series-parallel circuit, the following equation results from solving for an unknown resistance $R_1$:

$$0.40R_1 + 0.76(R_1 + 0.40) = 1.0(R_1 + 0.40)$$

Solve and check the equation.

**37.** Remove the parentheses and explain why the equation has *no* solution for $x$:

$$4(x - 1) = 3x - (1 - x)$$

**38.** Remove the parentheses and explain why *any* value of $x$ is a solution to the equation $4(x - 1) = 3x - (4 - x)$. This equation is called an *identity*.

## 2-8 | *Solving Verbal Problems*

No doubt you are aware that translating verbal problems to mathematics is not always easy. However, people communicate ideas to each other in Spanish, English, and some other verbal language. Therefore you must be able to translate verbal problems to mathematics to apply the skills you are learning. Verbal problems come in many different types. Most require that an equation be solved to get an answer. In general, you have to be able to formulate the necessary equation. That requires a certain skill. The best way to develop this skill is by *repeated practice*. Verbal problems are included in almost every chapter of this book to help you. Do not be discouraged. Try them, and you will soon develop the ability to handle these problems. Furthermore, you will sharpen your thought processes and find working with mathematics more enjoyable. Study the following examples carefully. Some you will find useful in your day-to-day experiences.

### Example 2-52
Adam paid $1200 for a home computer. The salesperson said that the present price is actually one-quarter less than last year's price. What was last year's price?

### Solution
Observe the order of steps used in solving the problem. First, the unknown can usually be set equal to what you are asked to find:

Let $x =$ last year's price

Then ask yourself: What should the equation say? Here is the "verbal equation":

Last year's price − ¼(last year's price) = present price

This is what you should be formulating in your mind. The algebraic equation is then:

$$x - \frac{1}{4}(x) = 1200$$

The solution is, first multiplying the equation by 4:

$$(4)x - (4)\frac{1}{4}(x) = (4)1200$$

$$4x - x = 4800$$

$$3x = 4800$$

$$x = 1600$$

It is important to check the answer in the original equation:

$$1600 - \frac{1}{4}(1600) = 1200?$$

$$1600 - 400 = 1200 \checkmark$$

### Example 2-53
Suppose the grade of your first test in this course is 76. What grade do you need to get on the second test if you want to raise your average for the two tests to 80?

### Solution
The unknown can be set equal to what you are asked to find:

Let $t$ = grade needed on the second test

What should the equation say? Here is the "verbal" equation:

$$\text{Average of two tests} = \frac{\text{first test} + \text{second test}}{2} = 80$$

You now need an expression for the average of the two tests in terms of $t$:

$$\text{Average} = \frac{76 + t}{2}$$

The algebraic equation is then:

$$\frac{76 + t}{2} = 80$$

The solution is:

$$(\cancel{2})\frac{76 + t}{\cancel{2}} = 80(2)$$

$$76 + t = 160$$

$$t = 160 - 76 = 84$$

Check the answer in the original equation:

$$\frac{76 + 84}{2} = \frac{160}{2} = 80 \ \checkmark$$

Even more important, check whether the answer makes sense logically. This last step is essential. Very often you can quickly discover a wrong answer to a verbal problem by seeing that it does not fit the facts. For example, if you got an answer of 70, you would know right away that something was wrong. If you want to raise your test average above 76, you need a grade larger than 76.

The above examples illustrate the basic steps in problem solving:

1. Study the problem to determine what is given and what you are to find.
2. Choose a meaningful letter for the unknown, and write down clearly what it represents.
3. Determine what the equation should say in words, that is, the verbal equation.
4. Write down other quantities needed for the equation in terms of the unknown.
5. Write and solve the algebraic equation.
6. Check that the answer not only satisfies the equation but also logically satisfies the conditions of the problem.

Here is another useful verbal problem. Your calculator may come in handy for this one.

### Example 2-54
Yesterday you bought a new pair of jogging sneakers for $34.50, which included 7 percent sales tax. Today, while traveling in another state, you discover that you can buy an identical pair of sneakers and pay only 3 percent sales tax.

How much could you have saved if you had bought the sneakers at the 3 percent tax rate?

## Solution
You need to find the price of the sneakers before tax and then determine the saving. Let $p$ = price of the sneakers before tax. The verbal equation is:

$$\text{Cost} = \text{price} + \text{tax} = \$34.50$$

The tax is 7 percent of the price = $0.07p$. The algebraic equation is:

$$p + 0.07p = 34.50$$

The solution is:

$$1.00p + 0.07p = 34.50$$
$$1.07p = 34.50$$
$$p = \frac{34.50}{1.07} = 32.242 = \$32.24$$

This answer seems about right and checks:

$$32.24 + 0.07(32.24) = 32.24 + 2.26 = 34.50$$

The cost of the sneakers at the 3 percent tax rate will be:

$$32.24 + 0.03(32.24) = 33.21$$

Therefore you could have saved $\$34.50 - \$33.21 = \$1.29$. This saving is probably not worth traveling to another state. However, if you were buying a more expensive item or were making the trip anyway for another reason, then it might be worthwhile.

The next two examples illustrate technical applications of verbal problems.

### Example 2-55
A copper wire 30.0 cm (11.8 in.) long is to be cut in such a way that the electric resistance of one piece is two-thirds the electric resistance of the other piece. How long should each piece be in centimeters and in inches (1.00 cm = 0.3937 in.)?

### Solution
Electric resistance depends directly on length. The length of the short piece should be two-thirds the length of the long piece. Since the length of the short

piece is given in terms of the length of the long piece, let $l$ = length of the long piece. The verbal equation is:

$$\text{Long piece} + \text{short piece} = 30.0 \text{ cm}$$

The short piece = $\frac{2}{3}l$. Therefore the algebraic equation is:

$$l + \frac{2}{3}l = 30.0$$

The solution is:

$$(3)l + (3)\frac{2}{3}l = (3)30.0$$

$$3l + 2l = 90.0$$

$$5l = 90.0$$

$$l = 18.0 \text{ cm}$$

and

$$\text{Short piece} = \frac{2}{3}(18.0) = 12.0 \text{ cm}$$

The solution check is:

$$18.0 + \frac{2}{3}(18.0) = 30.0?$$

$$18.0 + 12.0 = 30.0 \checkmark$$

The answers in inches are approximately:

$$\text{Long piece} = (18.0 \text{ cm})(0.3937 \text{ in.}/\text{cm}) = 7.09 \text{ in.}$$
$$\text{Short piece} = (12.0 \text{ cm})(0.3937 \text{ in.}/\text{cm}) = 4.72 \text{ in.}$$

The last example is a ''mixture'' type of problem. This type of problem has many applications.

### Example 2-56
A 10-quart (qt) cooling system contains a solution of 20 percent ethylene glycol (antifreeze) and 80 percent water. How much solution must be replaced with pure antifreeze to raise the strength to 50 percent antifreeze?

### Solution
Let $S$ = quarts of solution to be replaced with antifreeze. In a mixture problem, you express the equation in terms of one of the components in the mixture. You

can therefore solve the problem by using an equation in terms of antifreeze (solution 1 below) or in terms of water (solution 2 below) (Fig. 2-4).

**1.** The verbal equation in terms of antifreeze is:

$$\text{Antifreeze remaining} + \text{antifreeze added} = (50\%)(10 \text{ qt})$$

You remove $S$ qt of solution, so you have $10 - S$ qt remaining, 20 percent of which is antifreeze. Therefore:

$$\text{Antifreeze remaining} = 0.20(10 - S)$$

and

$$\text{Antifreeze added} = S$$

The algebraic equation and solution are then:

$$0.20(10 - S) + S = 0.50(10)$$
$$2.0 - 0.20S + S = 5.0$$
$$2.0 + 0.80S = 5.0$$
$$0.80S = 5.0 - 2.0 = 3.0$$
$$S = \frac{3.0}{0.80} = 3.75 \text{ qt}$$

*Check:*

$$0.20(10 - 3.75) + 3.75 = 0.50(10)?$$
$$1.25 + 3.75 = 5.0 \checkmark$$

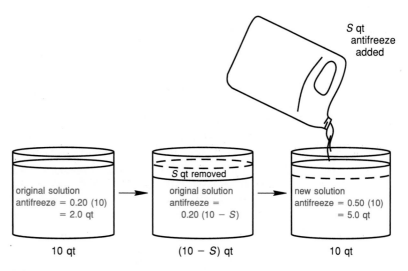

FIG. 2-4 *Mixture problem, Example 2-56*

**2.** In terms of water, the equation is:

$$0.80(10 - S) = 0.50(10)$$

Try to understand what the equation says and how to set it up. Solve the equation, and check that you get the same answer as in (1).

**Exercise 2-8**  For each problem *(a)* set up an algebraic equation and solve the problem, and *(b)* check that the answer makes logical sense and that it satisfies the equation.

1. Yolanda and Raphael together earn $32,000 a year. If Raphael earns $5000 more than Yolanda, what are their salaries?

2. The cost of renting a computer and a laser printer is $500 a month. If the cost for the computer is 3 times that for the printer, what is the monthly cost for each?

3. Your boss is raising your salary to $19,800 and indicates that this is a 10 percent raise. Based on that figure, what is your present salary?

4. An insurance company gives you $3300 to replace a stolen car and informs you that the car is worth only 30 percent of its original value. What is the insurance company's estimate of the original value?

5. An automobile costs 1.5 times as much as it did 5 years ago. If it costs $9000 today, what was the cost 5 years ago?

6. A discount sporting goods store gives one-third off list prices. If Erica paid $30 for a tennis racket before tax, what is the list price?

☐ 7. Mary Ellen paid $34.50 for a digital watch, a price that included 8 percent sales tax. How much could she have saved if she bought the watch in another state at a sales tax rate of 3 percent?

8. A calculator costs $20.00 with rechargeable batteries. There is a 15 percent price increase with these batteries. What is the price without rechargeable batteries?

9. The average of all your class tests is 75. The instructor indicates that your final average equals two-thirds your class test average plus one-third of your final examination grade. What do you need on the final examination to raise your average to 80?

10. To be eligible for a certain government job, you must average at least 75 on three tests. If your first two scores are 65 and 70, what do you need to get on the third test to raise your average to 75?

11. The total current in three parallel resistors is 26 A. The current in the second resistor is 1.5 times the current in the first and the current in the third is 0.75 times the current in the first. How much current flows through each resistor?

12. The total output of three "identical" steam-turbine generators in a power plant is 20,000 megawatts (MW, 1 MW = $10^6$ W). An electromechanical technician finds that two of the generators have the same power output but that the third generator puts out 20 percent more power than one of the other two. What is the power output of the third generator?

13. A new high-speed computer printer operates 3.5 times faster than an older printer. Together they print 1350 lines per minute. How many lines per minute does the new machine print?

14. In a machine shop, one jig completes a job in 20 minutes less than another jig. If the total operating time of both jigs is 3 hours, how long did the faster jig take to complete the job?

15. Two square solar panels, one 4 cm by 4 cm and the other 10 cm by 10 cm, receive a total of 174 calories per minute (cal/min) of solar energy on a clear day. How many calories per minute fall on 1 cm² of these panels on a clear day?

C 16. The surface area of the earth covered by water is approximately $3.617 \times 10^8$ km². If this represents 70.92 percent of the earth's total surface area, what is the total surface area of the earth (four figures)?

17. An aluminum wire 21.0 cm (8.27 in.) long is to be cut so that the electric resistance of one piece will be 75 percent of the resistance of the other piece. How long should each piece be in centimeters and in inches (0.3937 in. = 1.000 cm)? (See Example 2-55.)

C 18. A copper wire 0.914 m (1 yd) long is to be cut in three pieces in such a way that the second piece has twice the electric resistance of the first piece and the third piece has twice the resistance of the second piece. What should the length of the longest piece be in meters and in feet (1 m = 3.281 ft)? (See Example 2-55.)

19. An 18-qt cooling system contains 10 percent ethylene glycol (antifreeze). How many quarts should be replaced with pure antifreeze to increase the strength to 20 percent? (See Example 2-56.)

20. A container is filled with 60 milliliters (mL) of a 12 percent solution of sulfuric acid. How many milliliters of acid must be added to make a 25 percent acid solution? (See Example 2-56.)

21. Karolyn pays $1.50 a gallon for gasoline. Her car gets 20 mi/gal, but if she had the engine overhauled at a cost of $300, the car would get 30 mi/gal. How many miles would she have to drive to recover the cost of the overhaul? (*Hint:* First determine how many gallons she must save.)

22. The fuel of a two-cycle gasoline engine requires 1 part oil to 24 parts gasoline. How much oil should be added to 8 L of a gas-oil mixture that contains 1 percent oil to make the required mixture? (See Example 2-56.)

# 2-9 | *Formulas and Literal Equations*

Some verbal problems involving formulas were introduced earlier in this chapter (Section 2-1). This section discusses verbal problems involving technical applications and formulas that lead to linear equations.

### Example 2-57

The resistance of a circuit is 80.0 $\Omega$. If the voltage changes from 120 to 220 V, what is the change in the current $I_2 - I_1$?

### Solution

Ohm's law (Exercise 2-1) states that $E = IR$, where $E$ = voltage, $I$ = current, and $R$ = resistance. When $E = 120$ V substitute into the formula to obtain:

$$120 = 80.0I_1$$

Then
$$I_1 = \frac{120}{80.0} = 1.50 \text{ A}$$

When $E = 220$ V:

$$220 = 80.0I_2$$

and
$$I_2 = \frac{220}{80.0} = 2.75 \text{ A}$$

Therefore:

$$\text{Current change} = I_2 - I_1 = 2.75 - 1.50 = 1.25 \text{ A}$$

You can do this problem in a more direct way as follows.

### Example 2-58
Do Example 2-57 by first solving Ohm's law for $I$.

### Solution
$$E = IR$$

Divide both sides by $R$:
$$\frac{E}{(R)} = \frac{I\cancel{R}}{(\cancel{R})}$$

Therefore:
$$I = \frac{E}{R}$$

Now substitute the given information and directly calculate the two currents and their difference in one series of steps:

$$\text{Current change} = I_2 - I_1 = \frac{220}{80.0} - \frac{120}{80.0}$$

$$= \frac{220 - 120}{80.0} = \frac{100}{80.0} = 1.25 \text{ A}$$

Solving a formula for one of the variables is known as *solving a literal equation,* which is useful in the solution of problems. The next example shows how to solve some literal equations.

### Example 2-59
Solve each formula or literal equation for the indicated letter.

**1.** $P = \dfrac{F}{A}$; $A$    **2.** $v = v_0 + at$; $a$    **3.** $A = \pi r(2h + r)$; $h$

## Solution

**1.** $P = \dfrac{F}{A}$; $A$ (fluid mechanics: $P$ = pressure)

Multiply by $A$:         $(A)P = \dfrac{F}{\cancel{A}}(\cancel{A})$

Divide by $P$:         $\dfrac{A\cancel{P}}{(\cancel{P})} = \dfrac{F}{(P)}$

therefore         $A = \dfrac{F}{P}$

**2.** $v = v_0 + at$; $a$ (physics: $v$ = velocity)

Transpose $v_0$:         $at = v - v_0$

Divide by $t$:         $a = \dfrac{v - v_0}{t}$

**3.** $A = \pi r(2h + r)$; $h$ (geometry: $A$ = surface area)

Clear parentheses:         $A = 2\pi rh + \pi r^2$

Transpose $\pi r^2$:         $2\pi rh = A - \pi r^2$

Divide by $2\pi r$:         $h = \dfrac{A - \pi r^2}{2\pi r}$

The next two examples further illustrate the usefulness of solving literal equations.

### Example 2-60

The voltage in a circuit is given by:

$$V = V_0 + IR$$

where $V_0$ = initial voltage, $I$ = current, and $R$ = resistance. If $V_0 = 110$ V and $V = 230$ V, what is the decrease in resistance when the current changes from 2 to 3 A?

### Solution

First solve the formula for $R$:

$$V = V_0 + IR$$
$$V - V_0 = IR$$
$$R = \dfrac{V - V_0}{I}$$

Then substitute the given values and calculate the decrease in resistance:

$$\frac{230 - 110}{2} - \frac{230 - 110}{3} = \frac{120}{2} - \frac{120}{3} = 60 - 40 = 20 \ \Omega$$

### Example 2-61

Given the formula for the expansion of a gas under constant pressure $V_t = V_0(1 + at)$, where $t$ = temperature in degrees Celsius, $V_t$ = volume at $t°C$, $V_0$ = volume at $0°C$, and $a$ = coefficient of expansion.

1. Solve the formula for $t$.

2. Find the temperature in degrees Celsius and in degrees Fahrenheit if the gas doubles in volume from $V_0$ to $2V_0$ and $a = 0.00367$.

### Solution

1. Solve the formula for $t$:

$$V_t = V_0(1 + at)$$
$$V_t = V_0 + aV_0t$$
$$V_t - V_0 = aV_0t$$
$$t = \frac{V_t - V_0}{aV_0}$$

2. Find the temperature in degrees Celsius and in degrees Fahrenheit if the gas doubles in volume from $V_0$ to $2V_0$ and $a = 0.00367$. Let $V_t = 2V_0$ and $a = 0.00367$:

$$t = \frac{2V_0 - V_0}{0.00367V_0}$$

$$= \frac{\cancel{V_0}}{0.00367\cancel{V_0}}$$

$$= \frac{1}{0.00367} = 272°C$$

To find the temperature in degrees Fahrenheit, use the formula:

$$F = \frac{9}{5}C + 32 = \frac{9}{5}(272) + 32 = 490 + 32 = 522°F$$

As mentioned previously, skill in problem solving comes mainly through constant practice. As with baseball, it is not difficult to understand the rules, but only through constant practice can you play well!

The next example shows how to use a formula to set up an equation.

### Example 2-62

A sailboat can sail 5 knots (kn), or nautical miles per hour, when there is no current [1 nautical mile (nmi) = 1.15 statute or land miles]. If the boat sails upstream in 4 h and takes 3 h to return downstream the same distance, how fast is the current in nautical and in statute miles per hour?

### Solution

This is a "motion" problem, and you need to apply the procedure for problem solving outlined in Sec. 2-8:

Let $c$ = speed of the current in knots. What can you use to set up an equation? Since the distance upstream is the same as the distance downstream, the verbal equation is:

$$\text{Distance upstream} = \text{distance downstream}$$

The formula for distance is $D = RT$, where $D$ = distance, $R$ = average rate of speed, and $T$ = time. Then:

$$\text{Rate upstream} = 5 - c$$
$$\text{Rate downstream} = 5 + c$$
$$\text{Distance upstream} = RT = (5 - c)(4)$$
$$\text{Distance downstream} = (5 + c)(3)$$

The algebraic equation and solution are then:

$$(5 - c)(4) = (5 + c)(3)$$
$$20 - 4c = 15 + 3c$$
$$5 = 7c$$
$$c = \frac{5}{7} = 0.714 \text{ kn}$$

or $\qquad c = (0.714 \text{ nmi/h})(1.15 \text{ mi/nmi}) = 0.822 \text{ mi/h}$

You can carry the connection between baseball and mathematics further. You do not develop much skill in playing baseball by watching a lot of games. Likewise, studying a lot of text examples will increase your skill in problem solving only a little. Mathematics is not a spectator sport! You must participate and *solve as many problems yourself as possible.*

**Exercise 2-9**

In problems 1 to 8, solve each formula for the indicated letter.

**1.** $F = ma$; $a$ (mechanics: $F$ = force)

**2.** $\theta = \dfrac{s}{r}$; $r$ (geometry: $\theta$ = angle)

3. $x = \dfrac{y - b}{m}$; $y$ (geometry: equation of line)

4. $T = \dfrac{R - R_0}{aR_0}$; $R$ (electricity: $T$ = temperature)

5. $E = I(R + r)$; $R$ (electricity: $E$ = voltage)

6. $P = 2(I + w)$; $w$ (geometry: $P$ = perimeter)

7. $s = s_0 + v_0 t - 4.9t^2$; $v_0$ (physics: $s$ = distance)

8. $V = B - \dfrac{V_0}{A}$; $V_0$ (electronics: $V$ = transistor voltage)

9. The current in a circuit is 2.4 A, and the resistance is 100 $\Omega$. If the resistance increases to 150 $\Omega$, how much should the voltage be increased to maintain a current of 2.4 A? (See Example 2-58.)

10. The formula for power in watts is $P = EI$. (a) Solve this formula for $I$. The voltage of an electronic calculator is 9.0 V. (b) What is the current increase if the power increases from 3.6 to 3.9 W?

11. The boiling point $T(°F)$ of water for a height $h$ ft above sea level is given approximately by the formula $T = 212 - h/550$. (a) Solve the formula for $h$. (b) What is the increase in height if $T$ changes from 211 to 209°F?

12. (a) Solve the temperature conversion formula $F = (9/5)C + 32$ for $C$. (b) What is the increase in degrees Celsius if the temperature rises from 100 to 199°F?

13. In Example 2-60, find $R_1 - R_2$ when the current changes from 1.5 to 2.0 A.

14. In Example 2-60, (a) solve the formula for $I$, and (b) find the increase in current when the resistance changes from 50 to 40 $\Omega$.

☐ 15. Using the gas expansion formula in Example 2-61, find the temperature in degrees Celsius and Fahrenheit if the volume of the gas increases from $V_0$ to $1.5V_0$.

16. Find the temperature when degrees Celsius and Fahrenheit have the same value. (*Hint:* Let $F = C$ in the formula.)

17. You are given the formula for the velocity of a falling body: $v = v_0 + gt$. (a) Solve the formula for $t$. (b) If $v_0 = 7.06$ m/s, how long will it take for this velocity to triple? Use $g = 9.80$ m/s$^2$.

☐ 18. A certain steel rod of length $l_0$ at 0°C will increase in length to $l_t$ at $t$°C according to the formula $l_t = l_0(1 + 0.0000109t)$. (a) Solve the formula for $t$. (b) At what temperature will the steel rod have increased in length by 1 percent? (*Hint:* $l_t = 1.01l_0$.)

19. A boat sails 6 kn with no current. If the sailboat takes 3 h to sail upstream and 2 h to sail downstream, how fast is the current in knots and in miles per hour? (See Example 2-62.)

20. A train leaves New York City averaging 30 mi/h. A car leaves 20 min later averaging 50 mi/h. After how many miles will the car catch up to the train? (*Hint:* Distances are equal. See Example 2-62.)

21. The pitch diameter $D$ of a gear is related to the outside diameter $D_0$ and the number of teeth $N$ by the formula $D(N + 2) = D_0 N$. (a) Solve the formula for $D_0$. (b) If $D = 9$ cm, what is the change in $D_0$ when the number of teeth is increased from 10 to 15?

22. The rate of flow $R$ in an oil pipeline is given approximately by the formula $R = VA$, where $V$ = velocity of the oil and $A$ = cross-sectional area. If the rate of flow increases by 50 percent and the area increases by 25 percent, by what percentage does the velocity change?

23. A commercial jet flying between New York and Chicago, a distance of 714 statute miles, encounters favorable winds both ways. The trip west takes 2 h with a tailwind of 20 mi/h, and the trip east takes 1 h 45 min flying with a jetstream of 80 mi/h. *(a)* Was the plane flying faster relative to the air while going west or east? Find the airspeed each way. *(b)* How much faster was the faster airspeed? (See Example 2-62.)

$\boxed{C}$  24. A boat explosion is heard 7 s sooner through the water than through the air. The speed of sound is 1.44 km/s through the water and 0.344 km/s through the air. *(a)* How long did the sound take to travel through the air? *(b)* How far away was the explosion?

# 2-10 | *Hand Calculator Operations*

Most calculators will combine, multiply, and divide positive and negative numbers, applying the rules for signed numbers. They also will square a positive or a negative number by using the key $\boxed{x^2}$. However, some calculators are not programmed to raise a negative base to a power. This is due to the general concept of an exponent, which includes the roots of numbers. For example, by rule (1-10), $4^{1/2} = \sqrt{4} = 2$, whereas $(-4)^{1/2} = \sqrt{-4}$, which is an imaginary number. Imaginary numbers are not displayed on the calculator. The general concept of an exponent is explained further in Chap. 8, and imaginary numbers are discussed in Chap. 13. A negative base can be handled as in Example 2-63(3) below if your calculator is not programmed to perform this operation.

## Negative and Zero Exponents

To raise a number to a negative power, you can use the keys $\boxed{+/-}$ or $\boxed{1/x}$ as follows.

### Example 2-63
Calculate each of the following. (This is Example 2-25.)

**1.** $4^{-2}$    **2.** $10^{-1}$    **3.** $\left(\dfrac{-2}{5}\right)^{-3}$    **4.** $(0.3)^0(0.2)^{-2}$

### Solution
**1.** $4^{-2}$

Three calculator solutions are:

$$4 \boxed{y^x} 2 \boxed{+/-} \boxed{=} \rightarrow 0.0625$$
$$4 \boxed{1/x} \boxed{y^x} 2 \boxed{=} \rightarrow 0.0625$$
$$4 \boxed{1/x} \boxed{x^2} \rightarrow 0.0625$$

The first two methods apply to any power. The last method applies only to the $-2$ power.

2. $10^{-1}$

Three calculator solutions are:

$$10 \boxed{y^x} 1 \boxed{+/-} \boxed{=} \rightarrow 0.1$$
$$10 \boxed{1/x} \boxed{y^x} 1 \boxed{=} \rightarrow 0.1$$
$$10 \boxed{1/x} \rightarrow 0.1$$

The last method is the most direct but applies only to the $-1$ power since it is the same as taking the reciprocal.

3. $\left(\dfrac{-2}{5}\right)^{-3}$

To raise a negative base to a power, make the base positive if your calculator is not programmed to perform this operation, and attach the correct sign to the result:

$$2 \boxed{\div} 5 \boxed{=} \boxed{y^x} 3 \boxed{+/-} \boxed{=} \rightarrow 15.625$$

The correct solution is then $-15.625$. If your calculator is programmed to perform this operation, insert $\boxed{+/-}$ after the first $\boxed{=}$ above.

4. $(0.3)^0(0.2)^{-2}$

One calculator solution is:

$$0.3 \boxed{y^x} 0.0 \boxed{\times} 0.2 \boxed{1/x} \boxed{x^2} \boxed{=} \rightarrow 25$$

A more direct way is to key in 1 for any term to the zero power or to ignore the term if it is a factor, as above.

### Example 2-64

Calculate:     $\left(\dfrac{1.18}{5.29}\right)^{-2} + (2.47)^{-3}(3.55)^2$

### Solution

As a check on the calculator, the approximate solution is:

$$\left(\frac{1}{5}\right)^{-2} + (2)^{-3}(4)^2 = 25 + \left(\frac{1}{8}\right)(16) = 27$$

One calculator solution is:

1.18 $\boxed{\div}$ 5.29 $\boxed{=}$ $\boxed{y^x}$ 2 $\boxed{+/-}$ $\boxed{+}$ 2.47 $\boxed{y^x}$ 3 $\boxed{+/-}$ $\boxed{\times}$ 3.55 $\boxed{x^2}$ $\boxed{=}$

$\rightarrow$ 20.9     (three figures)

## Scientific Notation

To enter a number in scientific notation, (1) enter the significant figures with the decimal point to the right of the first figure; (2) press $\boxed{EE}$ $\boxed{EEX}$, or $\boxed{EXP}$; and (3) enter the exponent. For example, to enter $8.97 \times 10^{-3}$, press 8.97 $\boxed{EXP}$ 3 $\boxed{+/-}$. The exponent appears to the right on the display:

$$\boxed{8.97 \qquad -03}$$

When a calculation in ordinary notation produces a result with too many places for the display, the calculator automatically switches to scientific notation.

### Example 2-65
Calculate the distance to the nearest star, Proxima Centauri, which is 4.28 light-years away.

### Solution
This is the problem discussed at the beginning of Sec. 2-5. The calculator solution is to three figures:

300,000 $\boxed{\times}$ 60 $\boxed{\times}$ 60 $\boxed{\times}$ 24 $\boxed{\times}$ 365 $\boxed{\times}$ 4.28 $\rightarrow$ 4.05   (13)

Notice that the display switches to scientific notation as you calculate.

### Example 2-66
Calculate: $\dfrac{(7.07)^3(3.81 \times 10^8)}{6.60 \times 10^6}$

### Solution
Applying the rules for exponents, we find the approximate solution:

$$\frac{(7)^3(4 \times 10^8)}{7 \times 10^6} = (7)^2(4 \times 10^2) \approx 200 \times 10^2 = 2 \times 10^4$$

For the calculator solution it is not necessary to enter every number into the calculator in scientific notation. The calculator automatically changes ordinary notation to scientific notation when necessary:

7.07 $\boxed{y^x}$ 3 $\boxed{\times}$ 3.81 $\boxed{EXP}$ 8 $\boxed{\div}$ 6.60 $\boxed{EXP}$ 6 $\boxed{=}$ $\rightarrow$ 2.04   (04)

**Exercise 2-10**

In problems 1 to 8, calculate by hand and check with the calculator.

1. $(-4)^4$

2. $3^{-2}$

3. $\left(\dfrac{5}{3}\right)^{-3}$

4. $(-0.2)^5$

5. $(6)^{-1}(-2)^3$

6. $\left(\dfrac{9}{10}\right)^2\left(\dfrac{3}{4}\right)^{-3}$

7. $(1.2)^{-2}(-0.4)^3$

8. $(-0.8)^{-1}(2)^4$

In problems 9 to 20, choose the correct answer by approximating the result. Check with the calculator, and express answers to three figures.

9. $(2.82)^2(5.13)^{-1}$    (0.155, 1.55, 15.5, 155)

10. $(0.470)^{-2}(6.31)^{-3}$    (0.0180, 0.180, 1.80, 8.10)

11. $\dfrac{(10.0)^{-1}}{(0.238)^2}$    (0.176, 0.761, 1.76, 7.61)

12. $\dfrac{(6.10)^{-2}}{(9.83)^{-3}}$    (0.552, 2.55, 5.52, 25.5)

13. $(560^{-1} + 390^{-1} + 670^{-1})^{-1}$    (1.71, 17.1, 171, 1710)

14. $\dfrac{(3.74)^3(1.58)^{-5}}{(2.72)^2}$    (0.0718, 0.178, 0.718, 1.78)

15. $(5.86 \times 10^4)(9.19 \times 10^{-3})$    (5.39, 53.9, 539, 5390)

16. $(86,600)(94,100)(5310)$    ($4.33 \times 10^{13}$, $43.3 \times 10^{13}$, $4.33 \times 10^{14}$, $43.3 \times 10^{14}$)

17. $\dfrac{55,000,000,000}{220}$    ($25.0 \times 10^6$, $2.50 \times 10^7$, $2.50 \times 10^8$, $25.0 \times 10^8$)

18. $\dfrac{6.67 \times 10^3}{2.11 \times 10^{-6}}$    ($3.16 \times 10^6$, $3.16 \times 10^7$, $3.16 \times 10^8$, $3.16 \times 10^9$)

19. $\dfrac{1.01 \times 10^7}{(7.11 \times 10^7)(0.813)^2}$    (0.215, 0.512, 2.15, 5.12)

20. $\dfrac{(2.5 \times 10^2)^2}{(5.00)^4}$    (1.00, 10.0, 100, 1000)

## 2-11 | *Review Questions*

In problems 1 to 34, perform the indicated operations.

1. $10 - (-15) + (-1)$

2. $-(-2) - 3 + 4$

3. $(-5)(6) + (-2)(-7)$

4. $(-0.50)(6.0) + (-10)(-0.10)$

5. $\dfrac{(-3)(3) + 2(9)}{-(6)(5)}$

6. $\dfrac{(-7)(-3)(-1)}{(4)(6) - 3}$

**7.** $\left(\dfrac{5}{8}\right)\left(\dfrac{-2}{5}\right) - 4\left(\dfrac{-3}{8}\right)$

**8.** $(6)(-2) + 5 - \left(\dfrac{-14}{2}\right)$

**9.** $x^2 + 2x - 3x^2 + x$

**10.** $a^2b - ab^2 + 3a^2b + ab^2$

**11.** $r - 3t - (3t - r)$

**12.** $5 - (8 - 3px) + (2xp - 1)$

**13.** $(2a + b) - [(2b - a) + 2a]$

**14.** $1 - \{k + [2 - (3 - k)]\}$

**15.** $(2x^2 \cdot x)^2$

**16.** $(ab^2)(a^2b)(ab)$

**17.** $\dfrac{3n^6 \cdot n^{-2}}{6n^3}$

**18.** $\dfrac{(5^2)^2}{(5)(5^2)}$

**19.** $(5y^3)(-3y)^2(2y^2)$

**20.** $(0.4t^0)(2.0t^2)(5.0t^2)$

**21.** $\dfrac{7.5m^2v^2}{2.5mv}$

**22.** $\dfrac{3hx}{(5x^2)(15h)}$

**23.** $-b^2y^2(b^2 - y^2)$

**24.** $(2x)(3x)(x^2 - 2x + 1)$

**25.** $\dfrac{8x^3y + 12x^2y^2 - 4xy^2}{4xy}$

**26.** $\dfrac{-7(x^3y - xy^3)}{21xy}$

**27.** $(x - 4)(x + 2)$

**28.** $(2y - 3x)(3y + 2x)$

**29.** $(5a + b)(3a - b)$

**30.** $(3 + 2k)^2$

**31.** $(2px + 1)(2px - 1)$

**32.** $3(c + d)(c - d)$

**33.** $(x - 2)(x^2 + 2x - 1)$

**34.** $(H + 1)^2(H - 2)$

In problems 35 to 40, solve and check each equation.

**35.** $2x - 2 = 10$

**36.** $5y + 11 = 6y - 6$

**37.** $4(w + 2) = 5 - w$

**38.** $3.5 - 2.5m = 0.5 - m$

**39.** $\dfrac{5}{2}x + \dfrac{3}{2}x = 2$

**40.** $2x - (1 - x) = 2(x + 1)$

In problems 41 to 44, solve for the indicated letter.

**41.** $T = \dfrac{D}{R}$; $R$ (motion: $T$ = time)

**42.** $E = \dfrac{mv^2}{2}$; $m$ (physics: $E$ = energy)

**43.** $A = P + Prt$; $r$ (business: $A$ = amount)

**44.** $V = R(I_1 + I_2)$; $I_1$ (electricity: $V$ = voltage)

In problems 45 to 50, choose the correct answer by approximating the result. Check by hand or with the calculator.

**45.** $(5.21 \times 10^5)(8.11 \times 10^{-3})$    (4.23, 42.3, 423, 4230)

**46.** $(28,100)(32,500)(100,000)$    ($9.13 \times 10^{12}$, $9.13 \times 10^{13}$, $91.3 \times 10^{13}$, $9.13 \times 10^{14}$)

**47.** $\dfrac{101,000,000,000}{50,500,000,000}$    (0.200, 2.00, 20.0, 200)

**48.** $\dfrac{3.33 \times 10^5}{7.89 \times 10^{-4}}$    ($4.22 \times 10^8$, $42.2 \times 10^8$, $4.22 \times 10^9$, $42.2 \times 10^9$)

**49.** $(0.115)^{-2}(0.511)^3$     $(0.101,\ 1.01,\ 10.1,\ 101)$

**50.** $(5.00)^4(50.0)^3(100)^{-1}$     $(7.81 \times 10^3,\ 78.1 \times 10^3,\ 7.81 \times 10^4,$
$7.81 \times 10^5)$

**51.** The greatest temperature change recorded in a day is from 44 to $-56°F$ at Browning, Montana, on January 23, 1916. Express the change as a signed number.

**52.** The power output of one generator is given by $R(I_1^2 + 5I_2^2)$ and the power output of another generator by $R(3I_1^2 - I_2^2)$. What formula gives the total power output of both generators?

**53.** Simplify the expression which arises from a problem in structural design: $1.2Lx(x + 10) - \frac{2}{3}[9Lx + 1.2x(Lx - 5L)]$.

**54.** The area of the sector of a ring formed by two circles of radii $r_1$ and $r_2$ whose angle is $\theta$ is given by $\theta/2(r_1 + r_2)(r_1 - r_2)$. Multiply this expression.

**55.** The world's largest store is R. H. Macy & Co., Inc., in New York City. The total sales in 1978 were $1,834,000,000. Express this number in scientific notation.

**56.** The thickness of a human hair is approximately $3.94 \times 10^{-5}$ in. Express this number in ordinary notation.

**57.** A laser beam at the Lawrence Livermore Laboratory in California concentrated $2.6 \times 10^{10}$ kW of power onto a pinhead-size target for $9.5 \times 10^{-11}$ s on May 18, 1978. How many kilowatthours of energy does this represent?

**58.** The inductance $L$ in henrys of a resonant circuit is given by $L = 1/(4\pi^2 f^2 C)$. Calculate $L$ when $f = 1.7 \times 10^3$ Hz and $C = 3.5 \times 10^{-6}$ F.

In problems 59 to 66, *(a)* set up an algebraic equation and solve the problem, and *(b)* check that the answer makes logical sense and satisfies the equation.

**59.** A high-volume discount store makes only 4 percent profit on sales. A shoplifter walks out with a $3.00 pen. How much does the store have to sell to make up the loss?

**60.** An old-model calculator is selling for $9.35, a discount of 15 percent from the original price. What was the original price?

**61.** The average of three instrument readings is 38.5. If two of the readings are 40.0 and 35.0, what is the third reading?

**62.** Sixty kilograms of a cement-sand mixture contains 50 percent cement. How much sand must be added to decrease the mixture's cement content to 40 percent?

**63.** The resistance in a circuit is 7.5 $\Omega$. If the voltage changes from 12.3 to 12.0 V, what is the change in the current?

**64.** The formula for power is $P = I^2R$, where $I$ = current and $R$ = resistance. *(a)* Solve this formula for $R$. *(b)* If $I = 2$ A, find the change in the resistance when the power increases from 100 to 200 W.

**65.** The volume of a cylinder is given by $V = \pi r^2 h$. If $V = 50$ cm$^3$, what is the difference in the height $h$ when the radius $r$ doubles from 2 to 4 cm?

**66.** Two trains 180 mi apart leave at the same time and travel toward each other. If the average rate of one train is 40 mi/h and that of the other is 50 mi/h, how long will it take for the trains to meet?

# 3

# Ratio, Proportion, and Variation

## 3-1 | *Ratio and Proportion*

A *ratio* is a comparison of two quantities written with either a fraction line, such as 3/2, or a ratio sign, such as 3:2 (read ''3 to 2''). Study the following.

**E**xamples of ratio and proportion can be found in almost every chapter of this text. The ideas of ratio and proportion are basic to many natural and scientific relationships. This chapter focuses on these ideas and on many of their applications.

**Example 3-1**

List some examples of ratios.

**1.** Ratio of units for conversion:

$$\frac{2.54 \text{ cm}}{1.00 \text{ in.}}$$

**2.** Electric resistance = ratio of the voltage $V$ to the current $I$:

$$R = \frac{V}{I}$$

**3.** Rate of speed = ratio of distance to time:

$$\frac{100 \text{ mi}}{2 \text{ h}} = \frac{50 \text{ mi}}{1 \text{ h}} \quad \text{or} \quad 50 \text{ mi/h}$$

**4.** Fuel efficiency = ratio of distance traveled to fuel consumed:

$$\frac{257 \text{ mi}}{11.3 \text{ gal}} = 22.7 \text{ mi/gal}$$

5. Density = ratio of weight to volume:

$$\text{Density of kerosene} = 0.82 \text{ g/cm}^3 \quad (51 \text{ lb/ft}^3)$$

6. Specific gravity = ratio of the density of a substance to the density of water:

$$\text{Specific gravity of kerosene} = \frac{0.82 \text{ g/cm}^3}{1.00 \text{ g/cm}^3} = 0.82$$

7. Linear coefficient of thermal expansion = ratio of the change in unit length per 1° change in temperature:

$$\text{Coefficient for lead at } 20°C = 29.4 \times 10^{-6}/°C$$

8. Magnification of a microscope = ratio of object size to image size:

$$\frac{15.0 \text{ cm}}{0.500 \text{ mm}} = \frac{150 \text{ mm}}{0.500 \text{ mm}} = \frac{300}{1} \quad \text{or} \quad 300:1$$

Notice in (6) and (8) that when the units are the same, they are not written by the ratio.

### Example 3-2
A wire 30 cm long is to be cut into three pieces in such a way that the lengths are in the ratio 5:4:3. How long should each piece be?

### Solution
You can represent the lengths as $5x$, $4x$, and $3x$, where $x$ is the common factor. Then:

$$5x + 4x + 3x = 30$$
$$12x = 30$$
$$x = 2.5$$

The lengths are therefore $5x = 5(2.5) = 12.5$ cm, $4x = 10$ cm, and $3x = 7.5$ cm.

A *proportion* is an equation stating that two ratios are equal. The following examples show how to solve a proportion.

## Example 3-3

Solve the proportion for $x$:

$$\frac{32}{x} = \frac{65}{26}$$

The proportion can also be written $32:x = 65:26$.

### Solution

The unknown is in the denominator. To clear fractions, multiply both sides by the lowest common denominator (lcd), which in this case is the product of the two denominators $26x$:

$$(26)(\cancel{x})\frac{32}{\cancel{x}} = \frac{65}{\cancel{26}}\cancel{(26)}(x)$$

$$(26)(32) = (65)(x)$$

$$65x = 832$$

$$x = 12.8$$

Examine the first two steps above. Multiplying by the product of the two denominators is the same as *cross multiplication:*

$$\frac{32}{x} \diagtimes \frac{65}{26}$$

Cross multiplication is a direct way to solve a proportion when the unknown is in the denominator.

## Example 3-4

The current $I$ in a circuit is 5.3 A. If the resistance is constant and the voltage $V$ is increased from 12 to 15 V, what is the new value of the current?

### Solution

The ratio $V/I$ equals the resistance and is constant. If $I$ = new value of the current, you can therefore write the proportion:

$$\frac{12}{5.3} = \frac{15}{I}$$

Cross multiply and solve for $I$:

$$(12)(I) = (15)(5.3) = 79.5$$

$$I = \frac{79.5}{12} = 6.6 \text{ A}$$

Conversion of units can be done by setting up a proportion, as the next example shows.

## Example 3-5

If 0.609 m = 2.00 ft, how many meters are in 5.00 ft?

## Solution

Let $x$ = number of meters in 5.00 ft. You can write the proportion:

$$\frac{0.609 \text{ m}}{2.00 \text{ ft}} = \frac{x}{5.00 \text{ ft}}$$

Then
$$x = \frac{(5.00)(0.609)}{2.00} = 1.52 \text{ m}$$

Notice that when the unknown is in the numerator, you do not need to cross multiply.

## Exercise 3-1

In problems 1 to 8, simplify each ratio. Change to the same units if possible.

1. $\dfrac{60 \text{ mi}}{1.5 \text{ h}}$
2. $\dfrac{51 \text{ ft}}{1.7 \text{ s}}$

3. $\dfrac{5.0 \text{ kg}}{5.0 \text{ g}}$
4. $\dfrac{33 \text{ cm}}{1.1 \text{ m}}$

5. $\dfrac{60°}{180°}$
6. $\dfrac{22.5°}{7.5°}$

7. $\dfrac{5.7 \times 10^4}{1.9 \times 10^2}$
8. $\dfrac{3.0 \times 10^{-2}}{1.2 \times 10^{-5}}$

In problems 9 to 14, solve each proportion.

9. $\dfrac{x}{7} = \dfrac{3}{5}$
10. $\dfrac{10}{11} = \dfrac{8}{y}$

11. $\dfrac{3.21}{R} = \dfrac{7.85}{31.4}$
12. $\dfrac{1.01}{2.02} = \dfrac{d}{0.212}$

13. $\dfrac{2.9 \times 10^{-6}}{4.0 \times 10^{-6}} = \dfrac{h}{8.2 \times 10^{-6}}$
14. $\dfrac{F}{5.2 \times 10^3} = \dfrac{7.8 \times 10^3}{2.6 \times 10^3}$

15. Douglas measures the circumference of a circle, whose diameter is 12 cm, to be 37 cm. Virginia measures the circumference to be 38 cm. Which measurement is more accurate? Compute the experimental ratio $\pi = C/d$ for each.

16. The *Mach number* is the ratio of the velocity of an object to the velocity of sound. If the velocity of sound is 330 m/s, what is the Mach number of a rocket whose velocity is 1000 m/s?

**17.** The percentage by volume of nitrogen ($N_2$) in the atmosphere is 78.0 percent, and that of oxygen ($O_2$) is 21.0 percent. What is the ratio of the volume of $N_2$ to the volume of $O_2$?

**18.** A power plant produces 1000 kW, or $3.6 \times 10^4$ joules per hour (J/h), of electric energy for every $1.2 \times 10^5$ J/h of energy input. What is the efficiency (the ratio of output to input) of the power plant?

**19.** In a bronze bearing weighing 250 g, the ratio of copper to tin to zinc is $40:8:2$. How many grams of each metal does the bearing contain? (See Example 3-2.)

**20.** Mr. Chin's will states that $1 million of his estate is to be divided among his three children in the ratio $7:4:3$. Approximately how much does each child inherit? (See Example 3-2.)

**21.** The specific gravity of gasoline is 0.67. What is the ratio of the weight of gasoline to the weight of kerosene? [See Example 3-1(6).]

**22.** The brightness of a star is measured in magnitudes: A first-magnitude star is $\sqrt[5]{100}$ times brighter than a second-magnitude star, which is $\sqrt[5]{100}$ times brighter than a third-magnitude star, and so on. What is the ratio of brightness of a first-magnitude star to a sixth-magnitude star?

In problems 23 to 26, set up and solve the necessary proportion.

**23.** If the interest on $600 for 2 years is $90, what will be the interest on $2000 for 2 years?

**24.** On a scale model of a building 3 in. = 10 ft. If the model is 2 ft 3 in. high, how high will the building be?

**25.** If a 0.5 pint (pt) of oil is mixed with 2 gal of fuel for a 2-cycle gasoline engine, how much oil should be mixed with 1.5 gal?

**26.** The current in a circuit increases from 7.8 to 8.3 A when a battery producing 12.0 V is replaced with a new one. If the resistance is constant, what is the voltage of the new battery?

**27.** If 5.0 nmi equals 9.3 km, how many kilometers equal 3.5 nmi?

C **28.** The acceleration due to gravity $g$ is 978.039 cm/s², or 32.0878 ft/s², at 0° latitude (equator). If $g = 983.217$ cm/s² at 90° latitude (north pole), what is its value in feet per second squared at the pole?

**29.** If a 12-V solar cell can produce 36 ampere-hours (Ah) of electricity in 15 h, how many hours will it take to produce 350 Ah?

**30.** What will be the energy output of the power plant in problem 18 if the input is $1.5 \times 10^5$ J/h?

**31.** If resistances $R_1$, $R_2$, $R_3$, and $R_x$ form the arms of a Wheatstone bridge (an instrument used for measuring resistance), the following proportion results when the bridge is balanced:

$$\frac{R_1}{R_x} = \frac{R_2}{R_3}$$

If $R_1 = 0.112$ Ω, $R_2 = 0.200$ Ω, and $R_3 = 0.303$ Ω, find $R_x$.

C **32.** For a hydraulic lift the ratio of the applied force $F_1$ to the resulting force $F_2$ is equal to the ratio of the squares of the diameters of the pistons:

$$\frac{F_1}{F_2} = \frac{D_1{}^2}{D_2{}^2}$$

What applied force is necessary to lift a 2300-kg automobile if $D_1 = 9.5$ cm and $D_2 = 75$ cm?

C **33.** The principle of photometry states that if two light sources produce the same illumination on a surface,

$$\frac{I_1}{I_2} = \frac{d_1{}^2}{d_2{}^2}$$

where $I_1$ and $I_2$ are the intensities of the light sources and $d_1$ and $d_2$ are their distances from the surface. Find $I_2$ when $I_1 = 20.5$ candelas (cd), $d_1 = 38.2$ m, and $d_2 = 23.3$ m.

**34.** For two pulleys connected by a belt,

$$\frac{r_1}{r_2} = \frac{\omega_2}{\omega_1}$$

where $r_1$ and $r_2$ are the radii and $\omega_1$ and $\omega_2$ are the angular velocities in radians per second. If $r_1 = 2.3$ and $r_2 = 5.8$ in., how fast will the large pulley turn if the angular velocity of the small one is 6.3 rad/s?

**35.** A current of 15 A flowing through a conductor produces $8.22 \times 10^8$ J of heat energy in 2 h. How many joules of heat energy are produced in 5 h?

**36.** The effect of a moving load on the stress in a beam results in the influence diagram in Fig. 3-1. If $BC = 10.5$ m, find $AB$ and $CD$. (*Hint:* The three right triangles are similar, so corresponding sides are in proportion. See Sec. 4-4.)

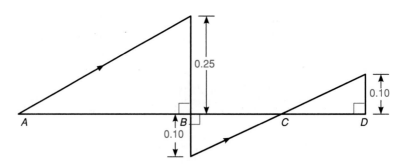

FIG. 3-1 *Influence diagram, problem 36*

## 3-2 | *Direct Variation*

When the ratio of two quantities $y/x$ is constant, $y$ is *directly proportional* (or proportional) to $x$, or $y$ *varies directly* as $x$. The constant value $k$ is called the *constant of proportionality* or the *constant of variation*:

$$\frac{y}{x} = k \quad \text{or} \quad y = kx \quad (3\text{-}1)$$

For any two pairs of values $x_1$, $y_1$ and $x_2$, $y_2$, you can write the proportion:

$$\frac{y_1}{x_1} = \frac{y_2}{x_2} \quad \text{or} \quad x_1 y_2 = x_2 y_1 \quad (3\text{-}2)$$

Section 3-1 contains many relations that involve direct variation. Some of these are included in the next example.

### Example 3-6
List some examples of direct variation.

1. $C = 2\pi r$. The circumference $C$ of a circle is directly proportional to the radius $r$, and $2\pi$ ($\approx 6.28$) is the constant of proportionality.

2. $F = ma$. Newton's second law of motion. The force $F$ is directly proportional to the acceleration $a$. The mass $m$ is the constant of proportionality.

3. $E = IR$. Ohm's law. When the resistance $R$ is constant, the voltage $E$ is directly proportional to the current $I$. Here $R$ is the constant of proportionality.

4. $V = kT$. Charles' law. Under constant pressure the volume $V$ of a gas is directly proportional to its absolute temperature $T$ in kelvin. The value of the constant of proportionality $k$ depends on the original volume of the gas.

5. $P = I^2 R$. Electric power. When the resistance $R$ is constant, the power $P$ is directly proportional to the square of the current $I$; and $R$ is the constant of proportionality.

6. $l = kT^2$. Pendulum. The length $l$ of a pendulum is directly proportional to the square of its period $T$. The constant of proportionality $k = g/(4\pi^2)$, where $g$ = gravitational acceleration.

7. $R = \sigma T^4$. Stefan-Boltzmann law. The rate of radiation $R$ of a perfect radiator is proportional to the fourth power of its absolute temperature $T$ in kelvin. Sigma ($\sigma$) is called the *Stefan-Boltzmann constant*.

## Example 3-7

A balloon filled with 10 m$^3$ of hydrogen at 20°C experiences a temperature decrease to $-20$°C, under constant pressure.

1. Using Charles' law, $V = kT$, find the value of the constant of proportionality.
2. Find the reduced volume.

### Solution

1. Using Charles' law, $V = kT$, find the value of the constant of proportionality.

   Express the temperature in kelvin, using the formula $K = C + 273$ and substitute to find $k$:

$$10 = k(20 + 273) = k(293)$$

$$k = \frac{10}{293} = 0.034 \text{ m}^3/\text{K}$$

2. Find the reduced volume.

   Using the value of $k$ from (1), substitute to find $V$:

$$V = (0.034)(-20 + 273) = 8.6 \text{ m}^3$$

   You can also find the reduced volume by using the proportion:

$$\frac{10}{293} = \frac{V}{253}$$

The examples of direct variation in Example 3-6 are examples of linear functions which are studied in Sec. 5-1 (see Example 5-2). *A first-degree equation of direct variation $y = kx$ expresses $y$ as a linear function of $x$.* This means that the graph of $y$ vs. $x$ is a straight line, as is shown in the next example.

## Example 3-8

Hooke's law states that stress is proportional to strain for an elastic body. When it is applied to a spring, the law means that the force $F$ varies directly as the distance $x$ the spring stretches. A force of 8.7 lb stretches a spring 0.75 in.

1. Find the constant of variation.
2. Draw the graph of $F$ vs. $x$ for $F = 0.0, 5.0, 10, 15,$ and 20 lb.

### Solution

1. Find the constant of variation.

Since $F$ varies directly as $x$, $F = kx$. Then:

$$8.7 = k(0.75)$$

$$k = \frac{8.7}{0.75} = 11.6 \text{ in./lb}$$

2. Draw the graph of $F$ vs. $x$ for $F = 0.0$, 0.5, 10, 15, and 20 lb.

Substitute each value of $F$ into the formula $F = 11.6x$, and solve for the corresponding value of $x$. You then obtain the table

| $F$ | 0.0 | 5.0 | 8.7 | 10 | 15 | 20 |
|---|---|---|---|---|---|---|
| $x$ | 0.0 | 0.43 | 0.75 | 0.86 | 1.3 | 1.7 |

Use different scales for $F$ and $x$ to draw the graph (Fig. 3-2).

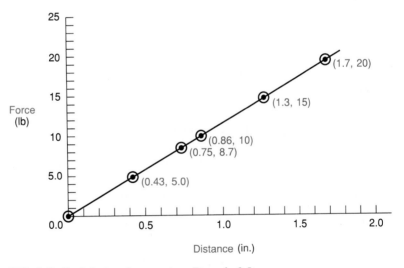

FIG. 3-2 *Hooke's law for a spring, Example 3-8*

In general, for any linear equation in $x$ and $y$ written $y = mx + b$, the *change* in $y$ is proportional to the *change* in $x$. The ratio of the change in $y$ to the change in $x$ is called the *slope m* of the line. (Slope is studied in Chap. 14.) The next example illustrates a proportional relation that is not a linear equation.

### Example 3-9

The visible distance $D$ of the horizon at sea varies directly as the square root of the height $h$ of an observer's eye. Alicia, whose eye height is 15 m (49 ft) above sea level, can see 8.0 nmi to the horizon. How much farther will she be able to see if she climbs a mast which is 23 m (75 ft) above sea level?

## Solution

The equation of variation is:

$$D = k\sqrt{h}$$

Substitute and solve for $k$:

$$8.0 = k\sqrt{15} = k(3.9)$$

$$k = \frac{8.0}{3.9} = 2.1$$

Then:

$$D = 2.1\sqrt{23} = 2.1(4.8) = 10 \text{ nmi}$$

Therefore the observer will be able to see 2 nmi farther from the mast. You can also find $D$ by using the proportion:

$$\frac{8.0}{\sqrt{15}} = \frac{D}{\sqrt{23}}$$

**Exercise 3-2**

In problems 1 to 4, (a) express each as an equation of direct variation and (b) find the indicated quantities.

1. The distance $D$ is directly proportional to the time $T$ at a constant rate of speed $R$. Find $R$ if $D = 304$ mi and $T = 5$ h 20 min.

2. The resistance $R$ of a wire is directly proportional to the length $l$. Determine the constant of proportionality $k$ if $R = 5.32 \ \Omega$ when $l = 10.7$ cm.

3. The volume $V$ of a sphere varies directly as the cube of the radius $r$. The constant of proportionality is $4\pi/3$. Find the ratio of the new volume to the old volume if $r$ doubles to $2r$.

4. The pressure $P$ of a column of liquid varies directly as the height $h$ (the pressure head). The constant of proportionality is the density $\gamma$ (gamma). Find the pressure exerted by a column of mercury 760 cm high (1 atmosphere of pressure) if the density of mercury $\gamma = 13.5$ g/cm³.

For problems 5 to 22, solve each using a proportion when helpful.

5. A force of 150 N produces an acceleration of 10 m/s² on a mass $m$. (a) Find $m$ in kilograms, using Newton's law $F = ma$. (b) What force is required to produce an acceleration of 15 m/s²?

6. A balloon filled with 15 m³ of air at 15°C is heated under constant pressure to 40°C. (a) Using Charles' law, $V = kT$, find the value of the constant of proportionality, and (b) find the increased volume. (See Example 3-7.)

7. The cost $C$ of computer time on a certain computer varies directly as the amount of time $t$ used. Suppose that 10 min costs $2000. (a) Find the value of the constant of proportionality and tell what it represents. (b) How much does 45 min cost?

8. The hull speed of a boat (maximum speed without planing) varies directly as the square root of the waterline length $L$. If a boat whose waterline length $= 32$ ft has a hull speed of 7.1 kn (nmi/h), find the constant of variation.

[C] 9. If a pendulum of length $l = 0.883$ m has a period $T = 1.89$ s, find the constant of proportionality. [See Example 3-6(6).]

[C] 10. Find the Stefan-Boltzmann constant $\sigma$ if $T = 600$ K and $R = 6330$ kcal/(m$^2 \cdot$ h) [see Example 3-6(7)].

11. A force $F$ of 8.2 kg stretches a spring a distance $x = 3.3$ cm. (*a*) Find the constant of proportionality. (*b*) Draw a graph of $F$ vs. $x$ for $F = 0.0, 5.0, 10, 15,$ and 20 kg. (See Example 3-8.)

12. A voltage $E$ of 115 V produces a current $I$ of 10.5 A. (*a*) Find the resistance $R$. (*b*) Draw a graph of $E$ vs. $I$ for $E = 100, 105, 110, 115,$ and 120 V. [See Example 3-6(3) and 3-8.]

13. The wavelength $\lambda$ (lambda) of a vibrating string is directly proportional to its length $l$. If $\lambda = 20$ cm when $l = 40$ cm, (*a*) find the constant of proportionality; (*b*) draw a graph of $\lambda$ vs. $l$ for $l = 0.0, 10, 20, 30,$ and 40 cm.

14. The change in area $\Delta A$ (delta $A$) of a square plate of lead with unit area at 0°C varies directly as the change in temperature $\Delta T$. When $\Delta T = 100$°C, $\Delta A = 0.0058$ m$^2$. (*a*) Find the constant of proportionality. (*b*) Draw a graph of $\Delta A$ vs. $\Delta T$ for $\Delta T = 0, 50, 100, 150,$ and 200°C.

15. Under certain conditions the stopping distance of a car is proportional to the square of the speed. If a car traveling 20 mi/h stops in 35 ft, how many feet will it take to stop at 55 mi/h?

16. The volume of a cylinder varies directly as the square of the radius. The volume of a cylindrical tank is 9.8 ft$^3$. What will be the volume if the radius is tripled?

[C] 17. The velocity $v$ of a fluid flowing out of a discharge pipe at the bottom of an open tank varies directly as the square root of the height $h$ of the fluid's surface. When the tank is full, $h = 12.5$ m and $v = 14.9$ m/s. What will be the velocity of the fluid when the tank is half full?

[C] 18. The value of $g$ in San Francisco has been experimentally determined to be 979.965 cm/s$^2$. What should be the period of a pendulum 10.20 m long in the "city by the bay"? [See Example 3-6(6).]

19. The change in resistance of a substance is proportional to the temperature change. If the resistance of an aluminum wire increases 0.046 $\Omega$ from $T = 0$ to $T = 60$°C, what is the increase in resistance from $T = 0$ to $T = 100$°C?

20. The energy produced by a solar panel is proportional to the cosine of the angle $\theta$ that the sun's rays make with a line perpendicular to the surface of the panel. If $\theta$ changes from 0 to 45°, by what percentage is the panel's energy output reduced? (See Sec. 4-8 on Trigonometry.)

[C] 21. The flux density $B$ of a magnetic field is proportional to the field intensity $H$. The constant of proportionality $\mu$ (mu) is called the *permeability*. Find $H$ in amperes per meter if $B = 3.50 \times 10^{-5}$ tesla (T) and $\mu = 4\pi \times 10^{-7}$ T $\cdot$ m/A.

[C] 22. The kinetic energy $E$ of a molecule of gas is proportional to the thermodynamic temperature $T$ in kelvin: $E = (3/2) kT$, where Boltzmann's constant $k = 1.38 \times 10^{-23}$ J/K. Find $E$ when $T = 360$ K.

# 3-3 | *Inverse Variation and Joint Variation*

Many formulas involve more than direct variation. Two other types of variation are inverse variation and joint variation.

## Inverse Variation

When the product of two quantities $x$ and $y$ is constant, $y$ is *inversely proportional* to $x$, or $y$ *varies inversely* as $x$:

$$ xy = k \qquad \text{or} \qquad y = \frac{k}{x} \tag{3-3} $$

For any two pairs of values $x_1$, $y_1$ and $x_2$, $y_2$, you can write the proportion:

$$ \frac{x_1}{x_2} = \frac{y_2}{y_1} \qquad \text{or} \qquad x_1 y_1 = x_2 y_2 \tag{3-4} $$

## Joint Variation

When one quantity $y$ varies directly as the product of two (or more) quantities $x$ and $z$, $y$ *varies jointly* as $x$ and $z$, or $y$ is *proportional* to $x$ and $z$:

$$ y = kxz \tag{3-5} $$

### Example 3-10
List examples of inverse variation and joint variation and examples combining more than one type of variation.

1. $PV = k$. Boyle's law. At a constant temperature, the pressure $P$ of a gas is inversely proportional to the volume $V$.
2. $Fx = Wd$. Lever. The force $F$ needed to lift a weight $W$ a distance $d$ from a fulcrum is inversely proportional to the force's distance $x$ from the fulcrum. The constant of variation $Wd$ is called the *moment arm* (Fig. 3-3).
3. $E = IR$. Ohm's law. When the voltage $E$ is constant, the current $I$ is inversely proportional to the resistance $R$.
4. $H = mvr$. Angular momentum. The angular momentum $H$ of a particle varies jointly as the linear velocity $v$ and the radius of rotation $r$. The constant of variation is the mass $m$.
5. $P = \gamma Ah$. Fluid force. The total force $P$ exerted on the bottom of a container of fluid varies jointly as the height $h$ of the fluid and the area of the bottom $A$. The density of the fluid $\gamma$ (gamma) is the constant of variation.

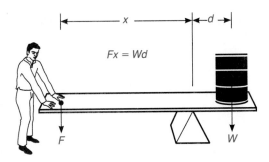

FIG. 3-3  *Principle of the lever, Example 3-10(2)*

**6.** $R = \rho l/A$. Resistance. The resistance $R$ of a conductor varies directly as the length $l$ and inversely as the cross-sectional area $A$. The resistivity $\rho$ (rho) is the constant of variation.

**7.** $F = kQ_1Q_2/r^2$. Coulomb's law. The electric force $F$ between two point charges $Q_1$ and $Q_2$ varies jointly as $Q_1$ and $Q_2$ and inversely as the square of the distance $r$ between them.

**8.** $S_{max} = 2\tau/(\pi R^3)$. Torsional stress. The maximum torsional stress $S_{max}$ on a circular shaft varies directly as the torque $\tau$ (tau) and inversely as the cube of the radius $R$.

### Example 3-11

In a steam engine the expansion of steam in the cylinder is an isothermal (constant-temperature) process that satisfies Boyle's law, $PV = k$. At the beginning of the expansion stroke, the pressure $P = 1080$ kilopascals (kPa), and the volume $V = 1.25$ m³.

**1.** Find the constant of variation, and express the relation of inverse variation between $P$ and $V$.

**2.** Find $P$ when $V = 3.75$ m³.

### Solution

**1.** Find the constant of variation, and express the relation of inverse variation between $P$ and $V$.

Substitute to find $k$:

$$PV = (1080)(1.25) = 1350 \text{ kPa} \cdot \text{m}^3 = k$$

The relation is then:

$$PV = 1350 \qquad \text{or} \qquad P = \frac{1350}{V}$$

**2.** Find $P$ when $V = 3.75$ m$^3$. Substitute to find $P$:

$$P = \frac{1350}{3.75} = 360 \text{ kPa}$$

You can also find $P$ by solving the proportion:

$$\frac{1.25}{3.75} = \frac{P}{1080}$$

The next example illustrates how to set up and solve a problem involving more than one type of variation.

### Example 3-12

The number $y$ varies jointly as $x$ and $z$ and inversely as $r^2$. It is given that $y = 15$ when $x = 60$, $z = 18$, and $r = 12$.

**1.** Find the constant of variation, and express the equation of variation.
**2.** Find $y$ when $x = 25$, $z = 15$, and $r = 14$.

### Solution

**1.** Find the constant of variation, and express the equation of variation.

The equation of variation takes the form:

$$y = \frac{kxz}{r^2}$$

Substitute to find $k$:

$$15 = \frac{k(60)(18)}{(12)^2} = 7.5k$$

$$k = \frac{15}{7.5} = 2.0$$

The equation of variation is then:

$$y = \frac{2.0xz}{r^2}$$

**2.** Find $y$ when $x = 25$, $z = 15$, and $r = 14$.

Substitute to find $y$:

$$y = \frac{2.0(25)(15)}{(14)^2} = 3.8 \quad \text{(two figures)}$$

## Example 3-13

Two equal and oppositely charged parallel plates, each of area $A$ and separated by a distance $d$, are the components of a parallel-plate capacitor. The capacitance $C$ of a parallel-plate capacitor varies directly as $A$ and inversely as $d$. If $C = 1.2 \times 10^{-9}$ F when $A = 0.35$ m$^2$ and $d = 0.015$ m, find $C$ when $A = 0.55$ m$^2$ and $d = 0.020$ m.

### Solution

First find $k$, using the equation of variation:

$$C = \frac{kA}{d}$$

You can solve for $k$ and substitute:

$$k = \frac{Cd}{A}$$

$$k = \frac{(1.2 \times 10^{-9})(0.015)}{0.35} = 5.1 \times 10^{-11} \text{ F/m}$$

Then find $C$, substituting the value for $k$:

$$C = \frac{(5.1 \times 10^{-11})(0.55)}{0.020} = 140 \times 10^{-11}$$

$$= 1.4 \times 10^{-9} \text{ F} \quad \text{or} \quad 1.4 \text{ nanofarad (nF, 1 nF} = 10^{-9} \text{ F)}$$

The constant $k$ is called the *permittivity*. Its value depends on the insulating material between the plates.

## Example 3-14

The maximum force $F$ exerted on the vane of a wind generator (when the vane is perpendicular to the wind direction) varies jointly as the area $A$ of the vane and the square of the wind velocity $v$. If the wind velocity increases by 50 percent, by what factor should the area be reduced to produce the same maximum force?

### Solution

The equation of variation is $F = kAv^2$. Let $A'$ equal the reduced area that produces the same maximum force $F$ when $v$ increases by 50 percent to $1.5\,v$. Then:

$$F = kAv^2 = kA'(1.5v)^2$$

$$kAv^2 = 2.25kA'v^2$$

$$A' = \frac{kAv^2}{2.25kv^2} = \frac{A}{2.25} = 0.44A$$

Therefore you need to reduce the area by 56 percent to produce the same maximum force on the vane.

**Exercise 3-3**

In problems 1 to 6, find the constant of variation and express the equation of variation.

**1.** $y$ varies inversely as $x$; $y = 15$ when $x = 3$.

**2.** $y$ varies inversely as the square root of $x$; $y = 3$ when $x = 196$.

**3.** $F$ varies jointly as $m_1$ and $m_2$; $F = 10$ when $m_1 = 3.2$ and $m_2 = 8.6$.

**4.** $E$ varies jointly as $m$ and $v^2$; $E = 120$ when $m = 5.36$ and $v = 4.00$.

**5.** $S$ varies directly as $w^3$ and inversely as $d$; $S = 18$ when $w = 2.2$ and $d = 5.3$.

**6.** $R$ varies directly as $l$ and inversely as $A$; $R = 0.35$ when $l = 5.0$ and $A = 0.085$.

In problems 7 to 12, find the unknown, using the equation of variation.

**7.** $y$ varies inversely as $x$. If $y = 75$ when $x = 10$, find $y$ when $x = 15$.

**8.** $P$ varies jointly as $R$ and $I^2$. If $P = 1200$ when $R = 30$ and $I = 5$, find $P$ when $R = 25$ and $I = 6$.

**9.** $V$ varies directly as the square root of $h$ and inversely as $r$. If $V = 100$ when $h = 225$ and $r = 3.00$, find $V$ when $h = 289$ and $r = 5.00$.

**10.** $T$ varies directly as $x$ and inversely as the square of $f$. If $T = 1.3$ when $x = 9.6$ and $f = 3.0$, find $T$ when $x = 6.9$ and $f = 4.0$.

☐ **11.** $d$ varies jointly as $a$ and $b$ and inversely as the square root of $c$. If $d = 2.80$ when $a = 4.15$, $b = 3.79$, and $c = 6.25$, find $d$ when $a = 5.14$, $b = 2.97$, and $c = 7.84$.

☐ **12.** $p$ varies directly as $q$ and inversely as $r$ and the square of $s$. If $p = 2.7$ when $q = 0.31$, $r = 5.6$, and $s = 0.18$, find $p$ when $q = 0.56$, $r = 1.3$, and $s = 0.17$.

**13.** A 1000-lb drum to be lifted by a lever is placed 5 ft from the fulcrum. (*a*) What is the moment arm (the constant of variation)? [See Example 3-10(2).] (*b*) What force exerted 20 ft from the fulcrum on the other side of the lever will just lift the drum?

**14.** In Example 3-11, the pressure at the *end* of the expansion stroke is 180 kPa. The volume is at its maximum at this point. What is the maximum volume?

☐ **15.** Using the equation of variation for the resistance of a conductor in Example 3-10(6), find the resistivity $\rho$ of silver if a wire 15.0 cm long whose diameter $d = 0.150$ cm has a resistance $R = 0.00136 \ \Omega$. (*Note:* Cross-sectional area $A = \pi d^2/4$.)

**16.** Find the capacitance in nanofarads of the parallel-plate capacitor in Example 3-13 if $A = 0.45 \ \text{m}^2$ and $d = 0.018$ m.

**17.** To sell all the building lots in a land development, the price of each lot is set to vary jointly as the area and the distance from a nuclear power plant. If a 0.5-acre lot 5 mi from the plant sells for $5000, for how much will a 1-acre lot 3 mi from the plant sell?

**18.** The odometer of a car is designed to work with 26-in.-diameter tires. If the tires are changed to 28-in.-diameter, how many miles on the odometer will be equivalent to a 250-mi trip? (*Hint:* Use inverse variation.)

19. The gravitational potential energy $E_g$ varies jointly as the mass $m$ of a body and its height $h$ above the ground. The constant of variation $g = 9.8$ m/s$^2$. (*a*) Express the equation of variation. (*b*) Find $E_g$ (in m $\cdot$ kg) when $m = 1000$ kg (1 metric ton) and $h = 25$ m.

20. Newton's law of cooling states that heat loss from a body is proportional to the area of the surface $A$ and to the difference in temperature between the body and the surroundings $T_2 - T_1$. A 50-cm$^2$ metal plate at a temperature $T_2$ of 120°C loses 40 kcal/h to the surrounding air, whose temperature $T_1 = 20$°C. Find the proportionality constant $h$, which is called the *convection coefficient*.

21. The maximum force that a rectangular cantilever beam (a beam supported at one end) can withstand at its free end varies directly as the beam's width $w$ and the square of its height $h$ and inversely as its length $l$. If $F = 5500$ lb when $w = 0.45$ ft, $h = 0.70$ ft, and $l = 15$ ft, find $F$ when $w = 0.55$ ft, $h = 0.60$ ft, and $l = 20$ ft.

[C] 22. The compressive load $F$ that a cylindrical column can withstand is directly proportional to the fourth power of its diameter $d$ and inversely proportional to the square of its length $l$. If $F = 15$ metric tons (t) when $d = 0.75$ m and $l = 7.5$ m, find $F$ when $d = 0.70$ m and $l = 9.0$ m.

23. For an ideal gas, the general gas law states that the pressure $P$ varies directly as the thermodynamic temperature $T$ in kelvin and inversely as the volume $V$. At 293 K (20°C) and 101 kPa of pressure (1 atm), 2.3 m$^3$ of an ideal gas is compressed to 1.0 m$^3$. If the temperature increases by 50 K, what is the pressure exerted by the gas?

24. Given a wire conductor like the one in problem 15, by what factor does the resistance change if the diameter is doubled? (See Example 3-14.)

25. The frequency of vibration $f$ for a given string is directly proportional to the square root of the tension $T$ and inversely proportional to the length $l$ and to the diameter $d$. If $l$ and $d$ are both doubled, by what factor should $T$ be increased to produce the same frequency (pitch)? (See Example 3-14.)

26. The force $F$ per unit length between two parallel wires carrying a current $I$ is directly proportional to $I^2$ and inversely proportional to the distance $d$ between the wires. This is expressed by $F = \mu I^2/(2\pi d)$, where $\mu$ is an experimental constant called the *permeability*. If $I$ increases from 10 to 15 A and $d$ remains constant, by what factor does $F$ increase? (See Example 3-14.)

## 3-4 | *Review Questions*

In problems 1 to 16, solve each proportion.

1. $\dfrac{x}{15} = \dfrac{42}{25}$

2. $\dfrac{3.5}{8.8} = \dfrac{2.2}{a}$

3. A turbojet engine operates at temperature $T_1 = 875$ K while the exhaust temperature $T_2 = 500$ K. *Efficiency* is defined as the ratio of the difference of these two temperatures to the exhaust temperature. What is the efficiency of the engine?

4. A log 85 ft long is cut so that the lengths are in the ratio 2:3:5. How long is each piece?

5. A transformer contains a primary wire coil and a secondary wire coil. The voltages of the coils are related by the equation:

$$\frac{V_1}{V_2} = \frac{N_1}{N_2}$$

where $V_1$ = primary voltage, $V_2$ = secondary voltage, $N_1$ = primary turns, and $N_2$ = secondary turns. What voltage is produced in the secondary coil if $V_1 = 4000$ V, $N_1 = 10,000$, and $N_2 = 600$?

6. Tincture of iodine is made by adding 70 g of iodine ($I_2$) and 50 g of potassium iodide (KI) to 50 mL of water and diluting with 1 L of alcohol. How many grams of KI should be added if 100 g of $I_2$ is used in the mixture?

7. Weight varies directly as gravity $g$ where mass is the constant of variation. (*a*) Express this idea as an equation of variation. (*b*) Find $m$ in kilograms when $W = 813$ N and $g = 9.81$ m/s$^2$ (1 N = 1 kg · m/s$^2$).

8. The force $F$ required to move a weight $W$ at a uniform rate along a horizontal surface varies directly as the weight. If 93 lb of force will move a 2500-lb automobile, (*a*) what is the constant of variation (the coefficient of friction) and (*b*) what force is required to move a 3500-lb automobile?

9. The heat developed in a conductor is proportional to the square of the current. If a current of 5.00 A produces $1.44 \times 10^8$ J/h, how much heat is produced by a current of 18.0 A?

10. Wind pressure varies directly as the square of the velocity. A fresh breeze of 20 mi/h produces a pressure of 1.5 lb/ft$^2$. (*a*) Find the constant of variation. (*b*) Find the pressure produced by a gale force wind of 50 mi/h.

11. The number $x$ varies inversely as $y^2$. If $x = 7$ when $y = 4$, what is the constant of variation?

12. $A$ varies jointly as $h$ and $r^2$. If $A = 44$ when $h = 3.5$ and $r = 2.0$, find $A$ when $h = 4.5$ and $r = 3.0$

13. A variation of Coulomb's law [Example 3-10(7)] applied to magnetism states that the force of attraction $F$ between two magnets varies jointly as the pole strengths $p_1$ and $p_2$ and inversely as the square of the distance $r$ between them. (*a*) Express this relation as an equation of variation. (*b*) Find the constant of variation in newtons per ampere squared if $F = 1.0$ N when $p_1 = 10^2$ A · m, $p_2 = 10$ A · m, and $r = 0.010$ m.

14. For a given concave mirror, the size of the image is directly proportional to the size of the object and inversely proportional to the distance of the object. An object 5.2 cm long at a distance of 15 cm from the mirror will produce the same size image as an object 2.2 cm long placed at what distance from the mirror? (*Hint:* Set up a proportion.)

**15.** The volume of a rectangular box with a square base varies jointly as the height and the square of the base. If the side of the base is decreased by 25 percent, by what percentage should the height be increased to produce the same volume? (See Example 3-14.)

**16.** A force on one end of a lever just balances a weight on the other end that is 5 times greater than the force. If the force is 9.0 ft from the fulcrum, how far is the weight from the fulcrum?

# 4

# Measurement: Basic Geometry and Trigonometry

## 4-1 | *Metric System (SI)*

The United States is moving "inch by 2.54 centimeters" toward using the metric system. In 1960 the metric system, first established in France in 1790, became the more complete International System (abbreviated SI, which stands for *Système International*). Since then almost every country has officially converted to SI except the United States. The U.S. Customary System, which evolved from the old English system, is gradually being replaced by the metric system.

Many scientific and technical fields already use SI units: computer science, medicine, electronics, optics, physics, chemistry. The transitional period is not easy for scientific and technical personnel (not to mention everyone else) because a good working knowledge of both systems is necessary at this time. Therefore, both systems are used throughout this text, with an emphasis on SI. Whatever you are accustomed to and understand well is generally easier than something new. However, as you become familiar with the metric system, you will see that it is not difficult to work with and why it is preferred to the U.S. system.

**T**his chapter begins with the metric system and its relation to the U.S. Customary System. Since measurement is an important part of geometry, these ideas lead to a treatment of basic geometry. Some of the geometry may be familiar to you, and it can be used as review or reference material. Radian measure, introduced in Sec. 4-3, the pythagorean theorem in Sec. 4-7, and the trigonometry covered in Sec. 4-8 are important and are needed for work in later chapters. The use of trigonometric functions on the calculator is shown in Sec. 4-8 and discussed further in Sec. 4-9.

The metric system is based on units of 10, as is our number system (the decimal system). The U.S. system is based on such numbers as 12 and 16, which were the basis of the old English system. In part, these numbers were used because they had several divisors and were easy to work with as fractions. This is no longer important in the age of electronic calculation.

There are seven basic SI units, and the other units are derived in terms of these base units (Table 4-1). For example, the base unit of mass, the kilogram, is the mass of a certain cylinder of platinum-iridium, called the *international prototype kilogram*. It is kept in a vault in Sèvres, France, by the International Bureau of Weights and Measures. The unit of force, the newton, is a derived unit. It is the force that produces an increase in velocity of 1 meter per second when applied to a 1-kilogram mass for 1 second ($kg \cdot m/s^2$). More tables for the SI system are given in App. A.

Prefixes based on powers of 10 are used to denote quantities of metric units. (Table 4-2). For example, 1 kilometer (km) = 1000 m, 1 milligram (mg) = 0.001 g, and 1 megavolt (MV) = 1,000,000 V. Liters, which are used to measure fluid (gas or liquid) capacity, are an exception—the only prefix used with

## Table 4-1    SI BASE UNITS AND OTHER COMMON SI UNITS

|  | Quantity | Unit | Symbol |
|---|---|---|---|
| Base units | Length | meter | m |
| | Time | second | s |
| | Mass | kilogram | kg |
| | Electric current | ampere | A |
| | Thermodynamic temperature | kelvin | K |
| | Molecular substance | mole | mol |
| | Light intensity | candela | cd |
| | Force | newton | N |
| | Temperature | degree Celsius | °C |
| | Capacity | liter | L |
| | Angle | radian | rad |
| | Area | square meter | $m^2$ |
| | Volume | cubic meter | $m^3$ |
| | Velocity | meters per second | m/s |
| | Energy | joule | J |
| | Frequency | hertz | Hz |
| | Pressure | kilopascal | kPa |
| | Power | watt | W |
| | Electric voltage | volt | V |
| | Electric resistance | ohm | Ω |

## Table 4-2   COMMON SI PREFIXES

| Power of 10 | Prefix | Symbol | Factor |
|---|---|---|---|
| $10^{-6}$ | micro- | $\mu$ (mu) | millionth (0.000001) |
| $10^{-3}$ | milli- | m | thousandth (0.001) |
| $10^{-2}$ | centi- | c | hundredth (0.01) |
| $10^{3}$ | kilo- | k | thousand (1000) |
| $10^{6}$ | mega- | M | million (1,000,000) |

liter is *milli*. Note that *1 milliliter (mL) = 1 cubic centimeter (cm³)* exactly and L = 1000 cm³. Concerning temperature, the kelvin is the base unit for thermo-dynamic or absolute temperature, but degrees Celsius is more commonly used. The conversion formula is $K = C + 273.16$. For time, s is used for seconds, min for minutes, h for hours, and d for days. Changes within the metric system are much easier to make than within the U.S. system because of the powers of 10. Usually all that is necessary is to move the decimal point or to change the exponent.

Study the next example.

### Example 4-1
Change each of the following units.

**1.** 1.28 km to meters       **2.** 50.5 mg to grams

**3.** 8 MW to kilowatts       **4.** 0.034 m³ to cubic centimeters

### Solution
**1.** 1.28 km to meters

Since 1000 m = 1 km, you can divide this equation by 1 km and express it as a unit ratio:

$$\frac{1000 \text{ m}}{1 \text{ km}} = \frac{1 \text{ km}}{1 \text{ km}} = 1$$

Then multiply 1.28 km by this unit ratio, which does not change the value of a quantity:

$$1.28 \text{ km} \left( \frac{1000 \text{ m}}{1 \text{ km}} \right) = 1280 \text{ m}$$

Observe that you divide out units in the same way as factors. This helps you to determine the correct units for the answer. Notice that the change amounts to moving the decimal point three places to the right.

**2.** 50.5 mg to grams

Since 1 mg = 0.001 g:

$$\frac{0.001 \text{ g}}{1 \text{ mg}} = 1$$

and:

$$50.5 \text{ mg} \left( \frac{0.001 \text{ g}}{1 \text{ mg}} \right) = 0.0505 \text{ g}$$

**3.** 8 MW to kilowatts

Since 1 MW = 1,000,000 W and 1 kW = 1000 W, it follows that 1 MW = 1000 kW. Then:

$$8 \text{ MW} \left( \frac{1000 \text{ kW}}{1 \text{ MW}} \right) = 8000 \text{ kW}$$

**4.** 0.034 m³ to cubic centimeters

Since 1 m = 100 cm, or $10^2$ cm, it follows that $(1 \text{ m})^3 = 1 \text{ m}^3 = (10^2 \text{ cm})^3 = 10^6 \text{ cm}^3$. Then:

$$0.034 \text{ m}^3 \left( \frac{10^6 \text{ cm}^3}{1 \text{ m}^3} \right) = 0.034 \times 10^6 \text{ cm}^3$$

This number can be written in scientific notation as $3.4 \times 10^4$ cm³ or in ordinary notation as 34,000 cm³.

## Mass, Force, and Weight

In everyday life, the terms *mass* and *weight* are used interchangeably. However, mass and weight are not the same. In science and technology, one must make the distinction. *Weight* is the *force* exerted on a body by gravity. Weight decreases as a body moves farther away from the center of the earth, becoming zero in outer space. *Mass* is a measure of the *resistance to motion* (inertia) that a body possesses. The mass of a body *never* changes. The greater the force necessary to produce a given velocity of a body, the greater the mass of the body. On the surface of the earth, the gravitational force is nearly constant. If the standard value of $g$, 980.665 cm/s² (32.174 ft/s²), is used, mass and weight are equivalent. When you say, "That person weighs 50 kg," you really mean, "If the standard value of $g$ is used, that person weighs 50 kg and so that person's mass is 50 kg." The force of gravity on a 1-kg mass is 9.8 N. In technology, therefore, the unit 1 kilogram-force (kgf) = 9.8 N is sometimes used. A person weighing 50 kg exerts a force of 50 kgf, or (50)(9.8) = 490 N, on the earth. The term *weight* must be used carefully in technical and scientific applications.

**Exercise 4-1**     In problems 1 to 12, change each of the following units.

1. 2.3 kg to grams
2. 310 mm to meters
3. 600 kV to megavolts
4. 0.038 MHz to kilohertz
5. 600 mm² to square centimeters
6. 0.074 m³ to cubic centimeters
7. 5.7 L to milliliters
8. 100 L to cubic meters (1 mL = 1 cm³)
9. $3.1 \times 10^{-3}$ MΩ to ohms
10. $2.0 \times 10^{5}$ μA to amperes
11. 100°C to kelvin
12. 3.14 rad to degrees (1 rad = 57.3°)

13. The mass of the earth is $5.98 \times 10^{24}$ kg. How many metric tons is this equal to? [1 metric ton(t) = $10^3$ kg.]

14. The density of sulfuric acid is 1.8 kg/L. What is the mass in grams of 35 mL of the acid?

15. The U.S. system unit of electric energy, the kilowatthour, will eventually be replaced by joules; 1 kWh = 3.6 MJ. The Three Mile Island nuclear power station has a total capacity of 1725 MW. How many megajoules per hour is this equal to?

16. One ampere of current is equal to an electric charge of one coulomb per second. If the charge of one electron is $1.60 \times 10^{-19}$ coulomb, how many electron charges per second equal one ampere?

17. The formula for work is $W = Fd$. When the force $F$ is expressed in newtons and the distance $d$ in meters, $W$ is given in joules: 1 J = 1 newton-meter (N · m). Find $W$ in joules if $F = 98.0$ N and $d = 55.0$ cm.

18. An old unit of pressure being phased out is the bar: 1 bar = 100 kPa. One standard atmosphere (the pressure exerted by 76 cm of mercury) is equal to 1010 millibars (mbar). How many kilopascals is 1 atm?

19. The wavelength $\lambda$ of a radio wave in meters is given by $\lambda = 3.00 \times 10^8/f$, where $3.00 \times 10^8$ is the velocity of the wave in meters per second and $f$ is the frequency in hertz. Find the wavelength of a radio wave if $f = 1010$ kHz.

20. Using Ohm's law, $E = IR$, find $E$ in millivolts when $R = 5.2$ kΩ and $I = 25$ μA.

21. The men's world record for the 20-km run is held by Jos Hermens of the Netherlands, who ran 20 km in 57 min, 24.2 s. How many meters per second did he average?

22. Of the chemical elements, helium has the lowest melting point: −271.72°C. What temperature in kelvin is that equal to?

**23.** One pound of force (weight) is equivalent to 454 g of mass on the earth's surface (standard gravity). *(a)* Use your weight in pounds to calculate your mass in kilograms. *(b)* Find the gravitational force exerted on your body in newtons (force on 1 kg = 9.8 N).

## 4-2 | *Relations: U.S. Customary System to the Metric System*

It is more than just a coincidence that 12 in. is close to the size of the average human foot. This is how it all began, with "royal" feet, arms, hands, and so on forming the standards for the base units of the old English system. People have been struggling with these units ever since. Now you must learn them more thoroughly to better understand the conversion to metric during the coming years.

Some of the major differences and conversion factors between units of the U.S. Customary System and the International System are shown in Table 4-3. The base units for length and mass are different in both systems, but the other

**Table 4-3**   MAJOR DIFFERENCES AND CONVERSIONS
BETWEEN THE U.S. CUSTOMARY SYSTEM AND SI

| Quantity | USCS Unit | Symbol | SI Unit and Conversion Factor |
|---|---|---|---|
| Length | foot | ft | 1 m = 3.281 ft |
| Mass | pound (avoirdupois) | lb | 1 kg = 2.205 lb |
| Force | pound (force) | lb | 1 N = 0.2248 lb |
| Temperature | degree Fahrenheit | °F | $C = \dfrac{5}{9}(F - 32)$ |
| Capacity | gallon | gal | 1 L = 0.2642 gal |
| Area | square foot | ft$^2$ | 1 m$^2$ = 10.76 ft$^2$ |
| Volume | cubic foot | ft$^3$ | 1 m$^3$ = 35.32 ft$^3$ |
| Velocity | feet per second | ft/s | 1 m/s = 3.281 ft/s |
| Pressure | pounds per square foot | lb/ft$^2$ | 1 kPa = 20.89 lb/ft$^2$ |
| Work energy | foot-pound | ft · lb | 1 J = 0.7376 ft · lb |
| Heat energy | British thermal unit | Btu | 1 J = 9.485 × 10$^{-4}$ Btu |
| Mechanical power | horsepower | hp | 1 W = 1.341 × 10$^{-3}$ hp |

five base units (not shown) are essentially the same. In SI the unit for energy, the joule, is used for work, heat, and electric energy. The unit for power, the watt, is used for mechanical and electric power. In the U.S. system the unit for electric energy is the joule, and the unit for electric power is the watt. The other three units, work and heat energy and mechanical power, are different, as shown at the bottom of Table 4-3. More conversion factors for both systems are given in App. A. "Thinking metric," notice in Table 4-3 that

1. A meter is a little longer than a yard.
2. A kilogram is a little heavier than 2 lb.
3. A newton is about one-quarter of a pound force.
4. A liter is slightly more than a quart.

### Example 4-2
Convert each of the following.

1. 28 ft to meters    2. 1.5 lb to grams
3. 37 hp to kilowatts    4. 14.7 lb/in² to kilopascals

### Solution
1. 28 ft to meters

   Conversions can be done by multiplying by unit ratios and dividing out units, as in Example 4-1. Using Table 4-3 you obtain:

$$28 \text{ ft} \left( \frac{1 \text{ m}}{3.281 \text{ ft}} \right) = 8.534 \text{ m}$$

2. 1.5 lb to grams

   Using the conversion factor 1 lb = 453.6 g from App. A you have:

$$1.5 \text{ lb} \left( \frac{453.6 \text{ g}}{1 \text{ lb}} \right) = 680 \text{ g}$$

3. 37 hp to kilowatts:

$$37 \text{ hp} \left( \frac{1 \text{ W}}{1.341 \times 10^{-3} \text{ hp}} \right) \left( \frac{1 \text{ kW}}{1000 \text{ W}} \right) = 27.6 \text{ kW}$$

4. 14.7 lb/in² to kilopascals

   By using Table 4-3, one solution is:

$$14.7 \text{ lb/in}^2 \left( \frac{144 \text{ in}^2}{1 \text{ ft}^2} \right) \frac{1 \text{ kPa}}{20.89 \text{ lb/ft}^2} = 101 \text{ kPa}$$

Notice in (4) that normal atmospheric pressure is 14.7 lb/in². This makes the kilopascal a convenient measure for a barometer since normal atmospheric pressure is approximately 100 kPa.

The next example illustrates a practical situation you may be faced with someday.

### Example 4-3
On a trip through Canada, you stop for gas. The pump shows that it took 48.6 L to fill the tank. If your odometer indicates you traveled 220 mi since you last filled up, how many miles per gallon and kilometers per liter did your car average?

### Solution
When the conversion factor 1 mi = 1.609 km from App. A is used, one solution is:

$$\text{Gas mileage} = \frac{220 \text{ mi}}{48.6 \text{ L}} \left( \frac{1.609 \text{ km}}{1 \text{ mi}} \right) = 7.28 \text{ km/L}$$

By using Table 4-3, the other solution is:

$$\text{Gas mileage} = \frac{220 \text{ mi}}{48.6 \text{ L}} \left( \frac{1 \text{ L}}{0.2642 \text{ gal}} \right) = 17.1 \text{ mi/gal}$$

**Exercise 4-2**    In problems 1 to 16, convert each of the units (use App. A if necessary).

1. 120 lb to kilograms
2. 50 ft² to square meters
3. 5.3 in. to centimeters
4. 135 W to horsepower
5. 10 MJ to British thermal units
6. ¼ hp to watts
7. 9.2 yd³ to cubic meters
8. 60 mi/h to meters per second
9. 2090 lb/ft² to kilopascals
10. 1.0 L to cubic inches
11. 6 ft 3 in. to centimeters
12. 1 lb 4 oz to grams
13. 95°F to degrees Celsius
14. 185 lb to newtons
15. 10 ft/s to kilometers per hour
16. 100 mi² to square kilometers

17. One of the world's thinnest calculators is manufactured by Sharp Electronics Corporation. It is 1.6 mm thick. How many inches thick is this calculator?

18. The tallest office building in the world is the Sears Tower in Chicago, which was completed in 1974 and has 110 stories that rise to a total of 1454 ft. Its gross area is 4,400,000 ft². *(a)* What is the building's height in meters? *(b)* What is the gross area in square meters?

19. The speed of sound in water is 1440 m/s. How many miles per hour is this velocity?

20. If 1 nmi = 1.15 statute miles (mi), how many kilometers equal 1 nmi?

C 21. The resistance of a no. 8 copper wire whose cross section is 8.366 mm² is 2.061 $\Omega$/km. *(a)* What is the cross section in square inches? *(b)* What is the resistance in ohms per 1000 ft?

22. The tensile strength of a specially treated nickel steel is $2.50 \times 10^5$ lb/in². What is the strength in megapascals?

23. Mercury, the closest planet to the sun, completes its orbit of $2.28 \times 10^8$ mi every 88 d. How fast does Mercury move in kilometers per second?

24. The velocity of seismic waves at a depth of 45 km is 6.75 km/s. *(a)* What is the depth in miles? *(b)* What is the velocity in miles per hour?

25. The density of ethyl alcohol is 0.80 g/cm³. Which bottle of ethyl alcohol is the best buy, 16 oz (1 pt) for $1.29, 500 mL for $1.39, or 400 g for $1.35?

26. The average solar energy falling on 1 cm² of the earth is 1.93 cal/min. If 1 kW = 239 cal/s, how many kilowatts per square meter fall on the earth?

27. Using Einstein's formula for atomic energy, $E = mc^2$, calculate how many joules of atomic energy are contained in 1 mg of mass. (*Note:* 1 J = 1 kg · m²/s² and $c$ = speed of light = 300,000 km/s.)

28. The coefficient $\alpha$ of linear expansion of a substance is the increase in length per unit length per degree of temperature. If $\alpha = 1.84 \times 10^{-5}$/°C for a certain grade of aluminum, what is $\alpha$ per degree Fahrenheit? [*Note:* 1°C = (9/5)°F.]

## 4-3 | *Lines and Angles: Radian Measure*

The words *metric, meter,* and *geometry* come from the ancient word *mete,* which means to measure. The prefix *geo* means earth, and therefore *geometry* means earth measurement. Geometry began in ancient Egypt with the need to measure people's property and other areas. Every year the Nile River overflowed, washing away the boundaries between farms. It was necessary to measure the land carefully to reestablish boundaries.

The presentation of the geometry in this chapter assumes you are familiar with certain elementary ideas. The chapter centers on important concepts that are necessary for further mathematical study. Lines and angles are the basic elements in geometric figures. *Lines* show length and direction. An *angle* measures rotation in a circle between two lines (radii). The point where the radii meet is the center of the circle, which is called the *vertex* of the angle. *One*

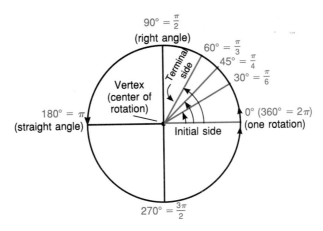

FIG. 4-1 *Basic angles*

*rotation* is defined as 360° or $2\pi$ ($\approx$6.284) *radians* (rad). When there is no degree (°) symbol after the angle, it is understood to be in radian measure.

Figure 4-1 shows the basic angles: 30, 45, 60, 90 (right angle), 180 (straight angle), and 360°. Rotation is counterclockwise, starting from the right at 0°. For an angle that is a divisor of 360°, the radian measure is generally given in terms of $\pi$: 360° = $2\pi$, 180° = $\pi$, 90° = $\pi/2$, 60° = $\pi/3$, 45° = $\pi/4$, 30° = $\pi/6$, and so on. Using the equation $\pi$ rad = 180°, you can divide both sides by $\pi$ to obtain the number of degrees in one radian:

$$1 \text{ rad} = \frac{180°}{\pi} \approx \frac{180°}{3.142} \approx 57.3° \qquad (4\text{-}1)$$

Therefore *to change from radians to degrees,* multiply by the conversion factor 180°/$\pi$, or 57.3°/1 rad. *To change from degrees to radians,* multiply by $\pi$/180°, or 0.0175 rad/1°.

### Example 4-4
Change each angle in degrees to radians, and vice versa.

**1.** $10° = 10° \left(\dfrac{\pi}{180°}\right) = \dfrac{\pi}{18}$ or 0.175 rad

**2.** $40.2° = 40.2° \left(\dfrac{\pi}{180°}\right) = 0.704$ rad

**3.** $\dfrac{3\pi}{4} = \dfrac{3\pi}{4} \left(\dfrac{180°}{\pi}\right) = 135°$

**4.** $1.73 \text{ rad} = 1.73 \text{ rad} \left(\dfrac{180°}{\pi}\right) = 99.1°$

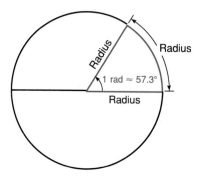

FIG. 4-2  *Radian measure*

See Example 4-26 for the use of the calculator in changing to degrees and radians.

    The name *radian* was chosen because an angle of 1 rad cuts off an arc of one radius in length on a circle (Fig. 4-2). In Sec. 4-5, the usefulness of radian measure is explained further.

    The following are explanations of basic terms used to describe lines and angles (Fig. 4-3).

*Perpendicular* lines (denoted ⊥) meet at right angles. Lines $L_2$ and $L_3$ are ⊥.
    The right angle is indicated by a little square at the vertex.
An *acute* angle is greater than zero but smaller than a right angle. Angle 1 (∡1)
    is an acute angle.
An *obtuse* angle is greater than a right angle but smaller than a straight angle.
    ∡5 is an obtuse angle.

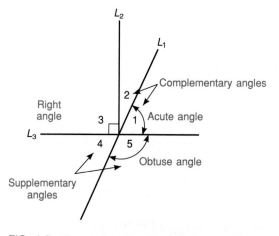

FIG. 4-3  *Lines and Angles, Example 4-5*

*Supplementary* angles are two angles whose sum is 180°. $\angle 4$ and $\angle 5$ are supplementary.

*Complementary* angles are two angles whose sum is 90°. $\angle 1$ and $\angle 2$ are complementary.

*Vertical angles* are opposite angles formed by intersecting lines and are equal to each other. $\angle 1$ and $\angle 4$ are vertical angles, and $\angle 1 = \angle 4$.

## Example 4-5

In Fig. 4-3, given $\angle 5 = 110°$, find $\angle 1$, $\angle 2$, and $\angle 4$.

## Solution

Now $\angle 4$ is the supplement of $\angle 5$, and $\angle 1 = \angle 4$; therefore $\angle 1 = \angle 4 = 180° - 110° = 70°$. And $\angle 2$ is the complement of $\angle 1$; so $\angle 2 = 90° - 70° = 20°$.

In degree measure $1° = 60'$ (minutes) and $1' = 60''$ (seconds). When an angle is given in degrees and minutes, you must change it to decimal degrees to enter it into the calculator (see Sec. 4-9).

## Example 4-6

Find the complement of 30°40′ in (1) degrees and minutes, (2) decimal degrees, and (3) radians.

## Solution

1. $90° - 30°40' = 89°60' - 30°40' = 59°20'$
2. $59°20' = 59° + (20/60)° = 59.3°$
3. $59.3° = 59.3°(\pi/180°) = 1.03$ rad

Notice in (1) that you change 90° to 89°60′ and subtract the degrees and minutes separately.

Degrees, minutes, and seconds are units that come from the original use of degrees to measure time. They are gradually being replaced with decimal degrees in most scientific and technical fields. In surveying, navigation, astronomy, and cartography (mapmaking), minutes and seconds are still used.

A *transversal* is a line that crosses two or more parallel lines. Consider the two parallel lines (indicated by arrows) and the transversal (not perpendicular) in Fig. 4-4. Eight angles are formed, but there are only two different sizes of angle. All the acute angles are equal: $\angle 1 = \angle 3 = \angle 5 = \angle 7$. All the obtuse angles are equal: $\angle 2 = \angle 4 = \angle 6 = \angle 8$. Furthermore, any of the acute angles is supplementary to any of the obtuse angles. The equal pairs $\angle 3 = \angle 5$ and $\angle 4 = \angle 6$ are called *alternate interior* angles. The equal pairs $\angle 1 = \angle 5$, $\angle 2 = \angle 6$, $\angle 3 = \angle 7$, and $\angle 4 = \angle 8$ are called *corresponding* angles. If the transversal is perpendicular to the parallel lines, all the angles are right angles and are all equal.

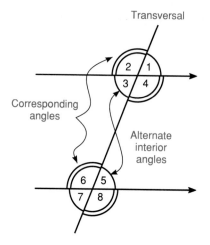

FIG. 4-4  *Parallel lines (identical marks on angles mean they are equal)*

### Example 4-7

In Fig. 4-4, given $\angle 7 = 45°$, find $\angle 2$ and $\angle 3$.

### Solution

Since $\angle 3 = \angle 7$, $\angle 3 = 45°$. Since $\angle 2$ is supplementary to $\angle 3$:

$$\angle 2 = 180° - 45° = 135°$$

**Exercise 4-3**   In problems 1 to 10, change each angle to radians.

   **1.** 60°          **2.** 90°
   **3.** 20°          **4.** 80°
   **5.** 15°          **6.** 150°
   **7.** 40° 30′       **8.** 110° 50′
   **9.** 67.3°         **10.** 114.6°

In problems 11 to 20, change each angle to degrees.

   **11.** $\pi/6$        **12.** $\pi/2$
   **13.** $\pi/9$        **14.** $2\pi/5$
   **15.** $2\pi/3$       **16.** $5\pi/6$
   **17.** 1.17 rad      **18.** 0.0195 rad
   **19.** $0.28\pi$      **20.** $\pi/7$

   **21.** Find the supplement of 160°20′ in *(a)* degrees and minutes, *(b)* decimal degrees, and *(c)* radians.
   **22.** Find the complement of 25°30′ in *(a)* degrees and minutes, *(b)* decimal degrees, and *(c)* radians.

In problems 23 to 30, find all numbered angles. Express answers in the angle measure given.

**23.**

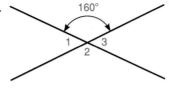

**24.**

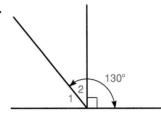

**25.**

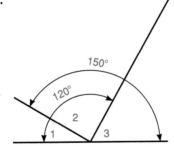

**26.**

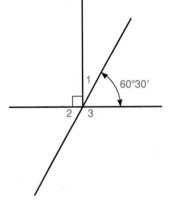

**27.**

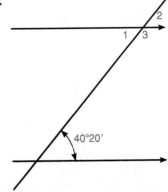

**28.**

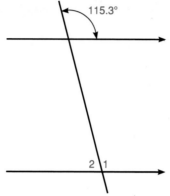

**29.**

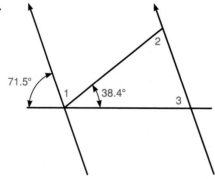

**30.**

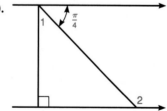

In problems 31 to 34, set up an equation and solve for $x$.

**31.**

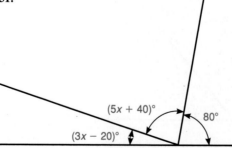

**32.**

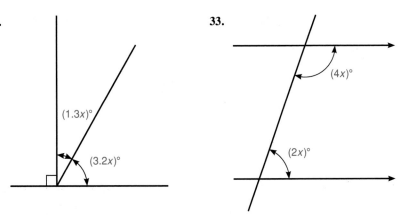

**33.**

**34.**

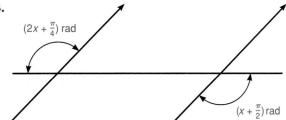

**35.** Longitude is measured from 0° (the longitude of London on the Greenwich meridian) traveling west parallel to the equator. There are twenty-four 1-h time zones. *(a)* How many degrees are in each time zone? *(b)* If the sun rises exactly 5 h later in Philadelphia than in London, what is the longitude of Philadelphia?

**36.** If it takes you 15 min to do this problem, how many degrees and radians do the *(a)* minute hand and *(b)* hour hand of a clock move in that time?

**37.** On a plane or a ship's compass angles are measured clockwise, beginning with 0° at north (N) (Fig. 4-5). A ship traveling halfway between north and west (NW) changes course to northeast (NE) and then to east (E). What is the total number of degrees the ship has turned?

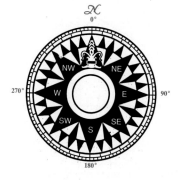

FIG. 4-5 *Ship's compass, problem 37*

38. While it is rounding a curve, one front wheel of a trailer truck rotates 286°40′ and a second front wheel rotates 5.00 rad. Which wheel is closer to the inside of the curve? (*Hint:* Inside wheels rotate less.)

39. The angular frequency $\omega$ of alternating current in radians per second is given by $\omega = 2\pi f$, where $f$ is the frequency. If $f = 60$ Hz for ordinary house current, what is the angular frequency in (*a*) radians per second and (*b*) degrees per second?

40. For a simple beam acted on by a concentrated load, the angle of deflection in radians is given by $\theta = Pl^2/(16EI)$. Find $\theta$ in (*a*) radians and (*b*) degrees when $P = 46{,}000$ N, $l = 800$ cm, and $EI = 10^{10}$ N · cm².

## 4-4 | *Triangles: Congruence and Similarity*

Triangles are the building blocks of all *polygons* (plane figures bounded by straight lines) and therefore form the basis for a thorough understanding of geometry. One of the most important geometric theorems states that *the sum of the angles of any triangle equals 180°*. The most useful triangle is the *right triangle,* which contains a right angle and two complementary angles (Fig. 4-6). The *hypotenuse* is the longest side, opposite the right angle. An *equilat-*

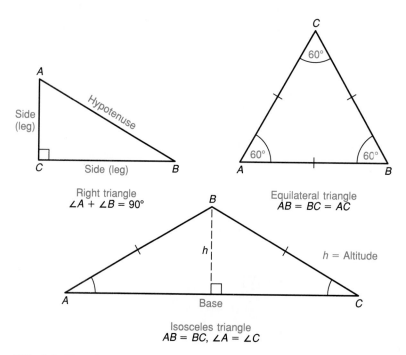

FIG. 4-6 *Three basic triangles (identical marks mean equal lines)*

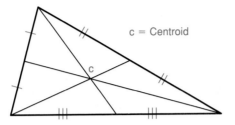

FIG. 4-7 *Medians of a triangle (identical marks mean equal lines)*

*eral triangle* has three equal sides and three equal angles of 60°. An *isosceles triangle* has two equal sides and two equal angles (opposite the equal sides) called the *base angles*.

The three *medians* of a triangle are the lines drawn from each vertex to the midpoint of the opposite side, and they meet in a point called the *centroid* of the triangle (Fig. 4-7). The centroid is the center of gravity, or "balance point," of the triangle and is important in structural applications. The *altitude,* or *height,* of a triangle is a line drawn from the vertex of one angle perpendicular to the opposite side or base. In an *acute triangle,* which has all acute angles, the three altitudes lie inside the triangle. In an *obtuse triangle,* which contains one obtuse angle, one or more altitudes lie outside the triangle (Fig. 4-8). Altitudes are used in finding areas of figures, which is discussed in the next section.

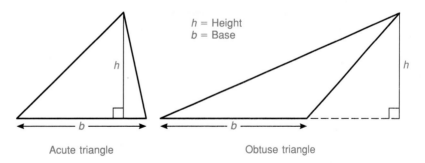

FIG. 4-8 *Altitudes of triangles*

### Example 4-8
In Fig. 4-9 $\angle 4 = 110°$ and $AB = BC$. Find $\angle 1$, $\angle 2$, and $\angle 3$.

### Solution
Angle 2 can be found first:

$$\angle 2 = 180° - \angle 4 = 180° - 110° = 70°$$

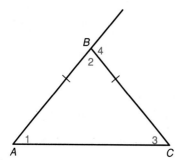

FIG. 4-9 *Example 4-8*

The sum of the angles of triangle *ABC* equals 180°:

$$\angle 1 + \angle 2 + \angle 3 = 180°$$

Since triangle *ABC* is isosceles, the base angles are equal: $\angle 1 = \angle 3$. Let $x = \angle 1 = \angle 3$. Then substituting in the above equation gives:

$$x + 70° + x = 180°$$
$$2x = 110°$$
$$x = 55°$$

Therefore $\angle 1 = 55°$, $\angle 2 = 70°$, and $\angle 3 = 55°$.

*Congruent triangles* are triangles that are identical except for their position. The corresponding sides and corresponding angles of each are equal. Congruent triangles have the same size and shape (Fig. 4-10).

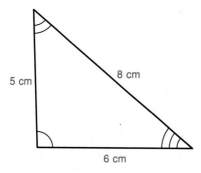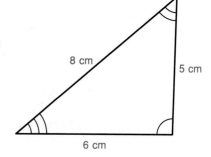

FIG. 4-10 *Congruent triangles*

### Example 4-9
The two right triangles in Fig. 4-11 are congruent (written $\triangle ABC \cong \triangle FDE$), with the angles shown equal. Find $x$, $y$, $\angle D$, $\angle E$, $DF$, and $EF$.

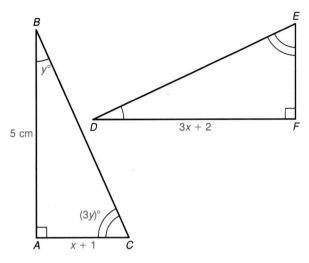

FIG. 4-11 *Congruent Triangles, Example 4-9*

## Solution

First identify the corresponding equal sides and equal angles:

$$AB = DF \qquad BC = DE \qquad AC = EF$$

and
$$\angle A = \angle F \qquad \angle B = \angle D \qquad \angle C = \angle E$$

Notice that the equal angles are opposite the equal sides. To find $x$ and $y$, you then write and solve the two equations:

$$3x + 2 = 5 \qquad \text{and} \qquad 3y + y = 90°$$

$$x = 1 \qquad\qquad y = \frac{90°}{4} = 22.5°$$

Then $\angle D = \angle B = y = 22.5°$, $\angle E = \angle C = 3y = 67.5°$, $DF = AB = 5$ cm, and $EF = x + 1 = 2$ cm.

*Similar triangles* (or similar figures) have the *same shape* but not necessarily the same size. If the corresponding angles of two triangles are equal, the triangles are similar. In Fig. 4-12 the large right triangle is similar to the small one since the corresponding angles are equal. *Corresponding sides of similar triangles are in proportion.* The large triangle is proportionately twice the size of the small triangle. This means that the ratios of the sides of the large triangle to the corresponding sides of the small triangle are equal: $^{10}/_5 = ^8/_4 = ^6/_3 = ^2/_1$. This proportion can also be written going from the small triangle to the large triangle: $^5/_{10} = ^4/_8 = ^3/_6 = ^1/_2$. Proportions are discussed in greater detail in Chap. 3.

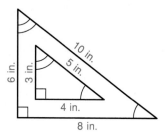

FIG. 4-12 *Similar triangles*

You encounter applications of similar figures in many scientific and technical problems. An engineering drawing or plan is drafted similar to the original. The corresponding angles are equal, and the corresponding sides are in proportion. Maps and photographs, images in a microscope, telescope, or camera, and scale models of buildings, cars, equipment, etc., are all examples of similarity, where the dimensions of a figure are proportional to those of the original. The following example illustrates a useful application of similar triangles.

### Example 4-10

The lowest noon elevation of the sun (angle above the horizon) occurs on the day of the winter solstice (around December 21). Suppose at that time, when shadows are long, the shadow of one of the World Trade Center towers is measured as 844 m, and the shadow of a 1.80-m pole as 3.70 m. What is the height of the tower in meters and in feet? See Fig. 4-13.

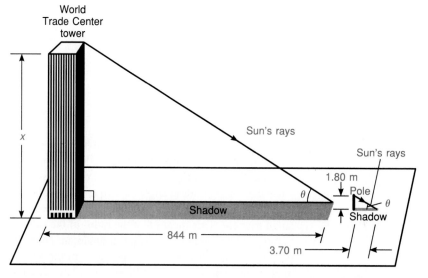

FIG. 4-13 *Example 4-10*

### Solution

The two right triangles in Fig. 4-13 are similar for the following reasons. The right angle and the sun's elevation $\angle \theta$ are correspondingly equal in both triangles. The third angles are also equal since the angles of any triangle must add to 180°. It is therefore true that *two triangles are similar if only two pairs of corresponding angles are equal*. The proportion is written going from the large triangle to the small triangle:

$$\frac{x}{1.80} = \frac{844}{3.70}$$

The solution for the height of the tower is:

$$x = \frac{844}{3.70}(1.80) = 411 \text{ m} \quad \text{(three figures)}$$

The height in feet is:

$$411 \text{ m} \left( \frac{3.281 \text{ ft}}{1 \text{ m}} \right) = 1350 \text{ ft}$$

**Exercise 4-4**

In problems 1 to 10, find the unknown variables. Express angles in the angle measure given. Identical marks on lines or angles indicate they are equal.

**1.**

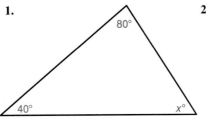

**2.**

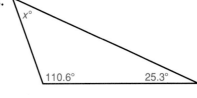

**3.**

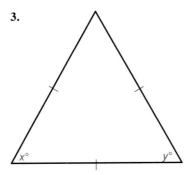

**4.**

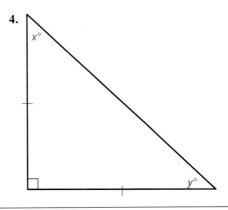

**5.**

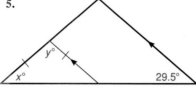

**6.**

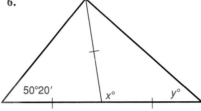

**7.**

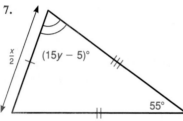

**8.**

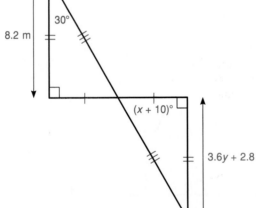

**9.**

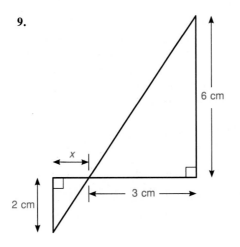

**10.**

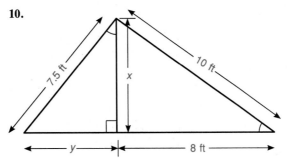

11. In Fig. 4-9 (Example 4-8) find $\angle 1$ and $\angle 4$ if $\angle 2 = 100°$.

12. In Fig. 4-11 (Example 4-9), given that $\angle B = y$, $\angle C = 2y$, $AC = 1$ in., $BC = x + 1$, and $EF = 2x - 1$, find $\angle D$ and $DE$.

13. The shadow of a building is 35 m. At the same time a meter stick casts a shadow of 70 cm. How high is the building? (See Example 4-10.)

14. In Fig. 4-13 (Example 4-10) find the length of the shadow of the World Trade Tower in meters and feet if the pole's shadow is 8 ft 6 in. (assume the height of the tower = 1350 ft).

15. The effect of a moving load on the stress in a certain beam can result in a diagram like Fig. 4-14. It is called an *influence diagram*. If $AB = 20$ ft, find $BC$ and $CD$ in the diagram.

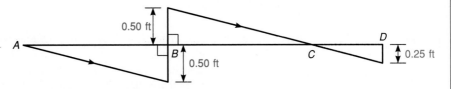

FIG. 4-14 *Influence diagram, problem 15*

16. The shadow of a woman standing 11 m from a streetlight is 1.5 m long. If the woman is 180 cm tall, how high is the streetlight in meters and in feet?

17. The image of a microprocessor circuit chip under a microscope is a rectangle 3.5 cm by 4.9 cm. If the length of the actual chip is 0.70 mm, what is the width? (*Hint:* The image is proportional to the actual chip.)

18. Derek is building a model to scale of a ketch, which is a type of two-masted sailboat. The masts of the model are 50 and 30 cm high. If the mainmast (the taller mast) of the ketch is 16 m high, how high is the mizzenmast (the shorter mast)? (*Hint:* The model is proportional to the original.)

# 4-5 | *Polygons and the Circle: Perimeter and Area*

## *Polygons*

Squares and rectangles abound in architecture, parallelograms and trapezoids are common in bridge design, and snow crystallizes in beautiful structures shaped like hexagons, which are also found in honeycombs. Common examples of these basic polygons are shown in Fig. 4-15.

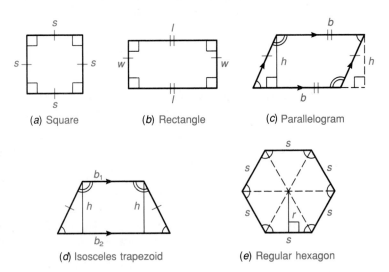

(*a*) Square  (*b*) Rectangle  (*c*) Parallelogram

(*d*) Isosceles trapezoid  (*e*) Regular hexagon

FIG. 4-15  *Basic polygons*

A *rectangle* is a quadrilateral, or four-sided polygon, with four right angles and with opposite sides equal. A *square* is a rectangle with all sides equal. A *parallelogram* has opposite sides parallel and equal and opposite angles equal. A *trapezoid* has only two parallel sides (bases); the other two sides are not parallel. An isosceles trapezoid, the most common type, has the two nonparal-

Table 4-4  ## AREAS OF BASIC POLYGONS

| Figure | Area $A$ | |
|---|---|---|
| Triangle | $A = \dfrac{1}{2}\, bh$ | (Fig. 4-8) |
| Square | $A = s^2$ | (Fig. 4-15a) |
| Rectangle | $A = lw$ | (Fig. 4-15b) |
| Parallelogram | $A = bh$ | (Fig. 4-15c) |
| Trapezoid | $A = \dfrac{1}{2}\, h(b_1 + b_2)$ | (Fig. 4-15d) |
| Regular hexagon | $A = \dfrac{3s^2\,\sqrt{3}}{2}$ or $2r^2\,\sqrt{3}$ | (Fig. 4-15e) |

lel sides equal and the corresponding angles opposite these sides equal. The angles of any quadrilateral add to 360°, since a quadrilateral can be divided into two triangles. A *hexagon* is a six-sided polygon. A regular hexagon, the most common type, has all sides and angles equal.

The formulas for the areas of these basic polygons are given in Table 4-4. Notice in Fig. 4-15c, in the parallelogram, that if the triangle on the left is moved to the dashed position on the right, the resulting rectangle has the same area as the parallelogram: $bh$. The formula for the area of the trapezoid applies to any trapezoid, not just the isosceles one shown in Fig. 4-15d. The area of the regular hexagon is the sum of the areas of the six congruent equilateral triangles that make it up. By using the pythagorean theorem, it can be shown that $r = s\sqrt{3}/2$ (see Sec. 4-7, problem 26). From this relation the two formulas for the area of a regular hexagon in Table 4-4 follow.

### Example 4-11
Find the area and perimeter (the distance around) of the wall and the gable in Fig. 4-16 (a gable is a triangular wall at the end of a ridged roof). Express answers in meters and in feet.

### Solution

$$\text{Perimeter} = 2(7.7) + 2(3.8) + 15 = 38 \text{ m}$$

or

$$38 \cancel{\text{ m}} \left( \frac{3.281 \text{ ft}}{1 \cancel{\text{ m}}} \right) = 125 \text{ ft}$$

$$\text{Area of gable (an isosceles triangle)} = \frac{1}{2}\,(13)(3.5) = 22.8 \text{ m}^2$$

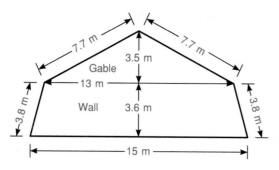

FIG. 4-16  *Wall and gable, Example 4-11*

Area of wall (an isosceles trapezoid) $= \dfrac{1}{2}\,(3.6)(13 + 15) = 50.4\ \text{m}^2$

The total area is:

$$22.8 + 50.4 = 73.2\ \text{m}^2 \quad \text{or} \quad 73.2\ \cancel{\text{m}^2}\left(\frac{10.76\ \text{ft}^2}{1\ \cancel{\text{m}^2}}\right) = 788\ \text{ft}^2$$

## The Circle and $\pi$

The formulas for the circumference $C$ and the area $A$ of a circle give rise to the irrational number $\pi$:

$$C = \pi D = 2\pi r \quad (D = \text{diameter},\ r = \text{radius}) \tag{4-2}$$

$$A = \pi r^2 \tag{4-3}$$

Now $\pi$ is a constant that tells us that the circumference of a circle is proportional to the radius and that the area is proportional to the square of the radius. For example, if the radius doubles, the circumference doubles, and the area increases by 4 times ($2^2 = 4$). Civilizations struggled for many centuries to find the exact value of $\pi$. You can almost trace the development of mathematics by how many decimal places of $\pi$ were known at a given time. One of the great number of decimal places to which $\pi$ has been calculated is $10^6$. This calculation was achieved by two French mathematicians on May 24, 1973, on a CDC 7600 computer working continuously for almost 2 days! The value obtained is $\pi = 3.14159265358979 \cdots 58151$ (omitting a "few" decimal places). The number fills a not very exciting 400-page book. You need $\pi$ accurate to only 3.142 for most scientific and technical work.

### Example 4-12

The circumference of the circular cross section of an electric conduit is 8.0 cm. Find (1) the radius and (2) cross-sectional area in centimeters and in inches.

### Solution

**1.** Solve Eq. (4-2) for $r$ and substitute.

$$r = \frac{C}{2\pi} = \frac{8.0}{2\pi} = \frac{8.0}{6.28} = 1.3 \text{ cm} \quad \text{or} \quad 1.3 \text{ cm} \left( \frac{0.394 \text{ in.}}{1 \text{ cm}} \right) = 0.52 \text{ in.}$$

**2.** Use Eq. (4-3).

$$A = \pi r^2 = 3.14(1.3)^2 = 5.3 \text{ cm}^2 \quad \text{or} \quad 5.3 \text{ cm}^2 \left( \frac{0.155 \text{ in.}^2}{1 \text{ cm}^2} \right) = 0.82 \text{ in.}^2$$

## Radian Measure and Arc Length

The formula for circumference $C = 2\pi r$ tells you that the radius of a circle fits around the circumference $2\pi$ times, or approximately 6.28 times. The definition of an angle of 360° as being equal to $2\pi$ rad directly relates the measure of the angle to the arc length it cuts off on the circle. An angle of $2\pi$ rad cuts off an arc equal to $2\pi$ radii. An angle of 2 rad cuts off an arc equal to 2 radii; 1 rad cuts off an arc equal to 1 radius, etc. (see Fig. 4-2). This leads to the formula for arc length $s$:

$$s = r\theta \quad (\theta \text{ in radians}) \tag{4-4}$$

The formula states that the angle $\theta$ and the arc length $s$ are proportional (Fig. 4-17a). Similarly, the area $A$ of a sector of a circle (Fig. 4-17b) comes from the formula for the area of the entire circle:

$$A = \frac{1}{2} \theta r^2 \quad (\theta \text{ in radians}) \tag{4-5}$$

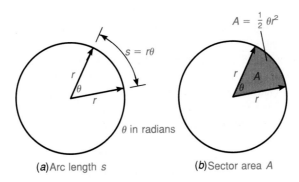

(a) Arc length $s$          (b) Sector area $A$

FIG. 4-17  *(a) Arc length s; (b) Sector area A.*

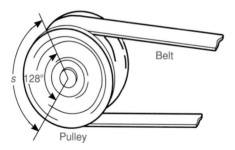

FIG. 4-18  *Belt and pulley, Example 4-13*

### Example 4-13

The diameter of a pulley is 27 in. How many feet does a belt on the pulley move when it turns 128°?

### Solution

The belt moves a distance equal to the arc length *s* cut off by the angle of 128° (Fig. 4-18). To use the formula for arc length, you need to change 128° to radians. You can do this within the calculation for *s*:

$$s = r\theta = \left(\frac{27}{2}\right)(128°)\left(\frac{\pi}{180}\right) = 30.2 \text{ in.}$$

or
$$30.2 \text{ in.} \left(\frac{1 \text{ ft}}{12 \text{ in.}}\right) = 2.52 \text{ ft}$$

**Exercise 4-5**  **1.** Find the area and perimeter of the isosceles triangle in Fig. 4-19.

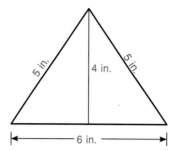

FIG. 4-19  *Problems 1 and 21*

**2.** Find the area of the obtuse triangle in Fig. 4-20.

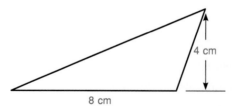

8 cm          FIG. 4-20  *Problem 2*

**3.** Find angle $x$ and the perimeter in feet and in meters of the parallelogram in Fig. 4-21.

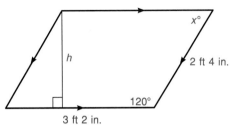

FIG. 4-21  *Problems 3 and 4*

**4.** Find the area of the parallelogram in Fig. 4-21 in square feet and in square meters if the altitude $h = 2.0$ ft.

**5.** Find the area and circumference of an automobile tire whose radius is 43 cm.

**6.** Find the cross-sectional area of a tree if the circumference is 4 ft 3 in.

**7.** Find the area and perimeter of the Norman doorway in Fig. 4-22. It is a rectangle with a semicircle on top. Express answers in meters and in feet.

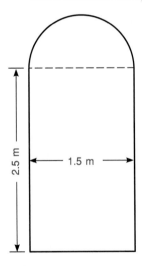

FIG. 4-22  *Norman doorway, problem 7*

**8.** Find the area and perimeter of the railroad trestle in the shape of the isosceles trapezoid shown in Fig. 4-23. Express answers in yards and in meters.

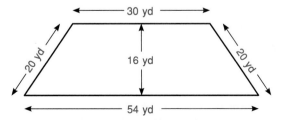

FIG. 4-23  *Railroad trestle, problem 8*

9. Given a gear of radius 4.0 cm, find the arc length and area of the sector cut off by a central angle of 40°.

10. A sheet of copper in the shape of a sector of a circle (Fig. 4-24) is to be rolled up to form a cone. Find the length of the arc and the area of the sheet.

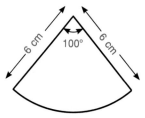

FIG. 4-24 *Problem 10*

11. The bridge section in Fig. 4-25 is a parallelogram. Find the height $h$ drawn to the base which measures 4.4 ft. (*Note:* The area formula for the parallelogram applies to either of the bases and the height drawn to that base.)

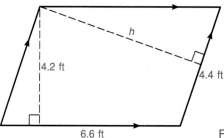

FIG. 4-25 *Bridge section, problem 11*

12. Find the cross-sectional area in square feet of the I beam in Fig. 4-26.

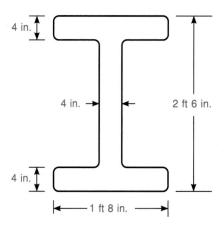

FIG. 4-26 *I beam, problem 12*

13. A circle is to be cut out of a square piece of steel 46.0 mm on a side, as shown in Fig. 4-27. What is the least amount of metal that can be wasted? Express the answer in square centimeters.

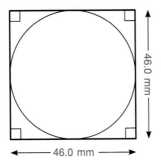

FIG. 4-27  *Problem 13*

C **14.** The cross-sectional area of a wire is proportional to its current-carrying capacity (amperage rating). The diameter of a no. 6 gauge wire is 0.2031 in., and the diameter of a no. 8 gauge wire is 0.1719 in. Find the ratio of the current capacity of the no. 6 wire to the current capacity of the no. 8 wire by finding the ratio of the cross-sectional areas.

**15.** The minute hand of a town clock is 52 cm long. *(a)* How far does a sleeping ladybug on the tip of the minute hand travel in 5 min? *(b)* What area is swept out by the minute hand in that time?

**16.** The price of a small-size 1.5-ft-diameter pizza is $5.00. The price of a large-size 2.0-ft-diameter pizza is $8.00. Which is a better buy?

C **17.** Find the area of the hexagonal bolt head in Fig. 4-28. Express answers in square inches and in square millimeters.

**18.** The largest ground area covered by any office building in the world is covered by the Pentagon, in Arlington, Virginia. The ground area is a regular pentagon, shown in Fig. 4-29, and each side is 281 m (921 ft). If this pentagon can be divided into five congruent isosceles triangles of height 193 m (633 ft), find the ground area in square meters and in square feet.

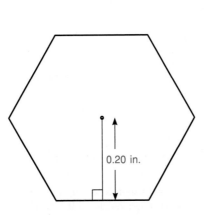

FIG. 4-28  *Bolt head, problem 17*

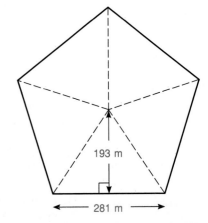

FIG. 4-29  *Pentagon building, problem 18*

**19.** A building lot is shaped like the trapezoid in Fig. 4-30. The area of an expanse of land or water is measured in hectares (ha), where 1 ha = 10,000 m². Find the area of the lot in hectares.

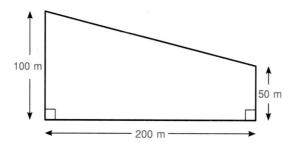

FIG. 4-30 *Building lot, problem 19*

**20.** A branch of a river basin is in the shape of an isosceles trapezoid with a semicircle on one base, as shown in Fig. 4-31. Find the surface area of the water contained in this branch of the river in hectares (see problem 19).

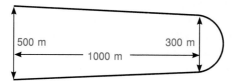

FIG. 4-31 *River basin, problem 20*

**21.** Heron of Alexandria, who lived in about A.D. 100, derived the following formula for the area of a triangle, which is used in surveying: $A = \sqrt{s(s - a)(s - b)(s - c)}$. The sides of the triangle are $a$, $b$, $c$, and the semiperimeter $s = (a + b + c)/2$. Find the area of the isosceles triangle in Fig. 4-19, using this formula.

**22.** How much will it cost to clean and maintain a baseball infield that is a square measuring 90 ft between the bases if a company charges $2.00 per square yard?

**23.** In Example 4-13, find how many meters the belt moves when the pulley turns 200°.

**24.** The orbit of Venus is very nearly a circle of radius $6.7 \times 10^7$ mi. If the orbital velocity of Venus is 21.8 mi/s, how many earth days are there in a Venusian year?

**25.** The resistance of a wire 7.62 mm in diameter is to be reduced by 50 percent by replacing it with a wire that has twice the cross-sectional area. What should be the diameter of the new wire?

**26.** The normal stress $s_n$ in a bar under tension is given by $s_n = P/A$, where $P$ is the tensile force and $A$ is the area of the cross section. Find the stress on a circular bar 8.0 cm in diameter acted on by a tensile force of 1500 kgf.

**27.** Cheyenne, Wyoming, and Colorado Springs, Colorado, lie on the same circle of longitude whose radius is the radius of the earth. The latitude of Cheyenne is 41°08′N, and that of Colorado Springs is 38°50′N. The central angle of the arc of longitude connecting the two cities is the difference of their latitudes: $2°18'$. Taking the radius of the earth to be 3960 mi, calculate how far Cheyenne is from Colorado Springs by finding the length of the arc of longitude.

**28.** Two old water mains 4.0 and 3.0 ft in diameter are to be replaced by one main with the same water-carrying capacity (cross section). What should be the diameter of the new main?

## 4-6 | *Geometric Solids: Surface Area and Volume*

We live in a three-dimensional world, surrounded by geometric solids. This text is the shape of a rectangular solid (a box). Sodium chloride (table salt) crystallizes in cubes. Pipes, wheels, gears, tanks and wire are cylindrical in shape. Steeples, funnels, and pulleys are made in conical shapes (not to mention ice cream cones). The sphere is very common because of its compact shape. It contains a given volume within the least amount of surface area compared to any other geometric solid. That is one of the reasons why stars, planets, most heavenly bodies, and the human brain assume spherical shapes. The atom is also envisioned to be spherical because of this property. Examples of basic geometric solids are shown in Fig. 4-32.

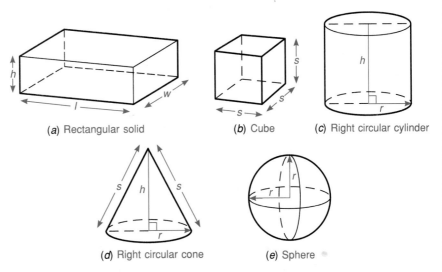

(*a*) Rectangular solid    (*b*) Cube    (*c*) Right circular cylinder

(*d*) Right circular cone    (*e*) Sphere

FIG. 4-32 *Basic geometric solids*

The six faces of a *rectangular solid* are rectangles that are perpendicular to one another. A *cube* is a rectangular solid whose faces are all squares. The sides of a *right circular cylinder* are perpendicular to the base, which is a circle. The

**Table 4-5**    AREAS AND VOLUMES OF SOLIDS (see Fig. 4-32)

| Solid | Surface Area $A$, Lateral Area $L$ | Volume |
|---|---|---|
| Rectangular solid | $A = 2lw + 2lh + 2hw$ | $V = lwh$ |
| Cube | $A = 6s^2$ | $V = s^3$ |
| Right circular cylinder | $L = 2\pi rh$ | $V = \pi r^2 h$ |
| Right circular cone | $L = \pi rs$ | $V = \frac{1}{3}\pi r^2 h$ |
| Sphere | $A = 4\pi r^2$ | $V = \frac{4}{3}\pi r^3$ |

cross section is therefore a rectangle. In a *right circular cone* a perpendicular line (height) drawn from the vertex (top point) meets the circular base at the center. The cross section that cuts the vertex and the center of the base is an isosceles triangle whose equal sides are the slant height $s$ of the cone. The formulas for the surface area and volume of each of the solids in Fig. 4-32 are given in Table 4-5.

The *lateral surface area $L$* of the cylinder and the cone means the area of the sides. The lateral area of a right circular cylinder can be thought of as that of a rectangle wrapped around two circular bases. The length of the rectangle is $2\pi r$, and the width is the height $h$ of the cylinder. The lateral area of a cone can be thought of as that of a sector of a circle wrapped around a circular base. The radius of the circle is the slant height $s$ of the cone.

### Example 4-14
A missile is to have a cylindrical shaft with an open bottom and a conical top as shown in Fig. 4-33. It is to be constructed of a special steel costing $2.00 per square foot. Find the cost of the steel and the volume of the missile in cubic feet and in cubic meters.

### Solution
The total area $A$ to be covered by the steel consists of the lateral area of the cylinder plus the lateral area of the cone:

$$A = 2\pi(10.0)(50.0) + \pi(10.0)(30.0)$$
$$= 4080 \text{ ft}^2 \quad \text{(three figures)}$$

The cost $C$ is:

$$C = (2.00)(4080) = \$8160$$

The volume $V$ is equal to the volume of the cylinder plus the volume of the cone:

30.0 ft

28.3 ft

50.0 ft

10.0 ft

FIG. 4-33  *Missile, Example 4-14*

$$V = \pi(10.0)^2(50.0) + \frac{1}{3}\pi(10.0)^2(28.3) = 18{,}700 \text{ ft}^3$$

or
$$18{,}700 \text{ ft}^3 \left( \frac{1 \text{ m}^3}{35.31 \text{ ft}^3} \right) = 529 \text{ m}^3$$

In the next example the principles of problem solving from Sec. 2-9 are applied to a geometric problem.

### Example 4-15

Two empty rectangular chemical tanks, one having a square base 10 m on a side and the other a square base 13 m on a side, are connected by a small open pipe at their bases (Fig. 4-34). If $1.8 \times 10^6$ L of chlorine is added to the large tank, to what height will the chlorine eventually level off in both tanks?

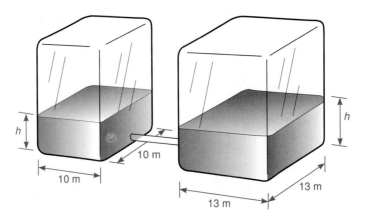

FIG. 4-34 *Chemical tanks, Example 4-15*

### Solution

Let $h$ = the common height. Neglecting the amount of chlorine in the pipe, the verbal equation is:

$$\begin{matrix} \text{Volume of chlorine} \\ \text{in large tank} \end{matrix} + \begin{matrix} \text{volume of chlorine} \\ \text{in small tank} \end{matrix} = \begin{matrix} \text{volume of} \\ \text{chlorine added} \end{matrix}$$

The volume added in cubic meters is:

$$(1.8 \times 10^6 \ \cancel{L}) \left( \frac{1 \ \text{m}^3}{10^3 \ \cancel{L}} \right) = 1800 \ \text{m}^3$$

The algebraic equation is then:

$$(13)(13)(h) + (10)(10)(h) = 1800$$
$$169h + 100h = 1800$$
$$h = \frac{1800}{269} = 6.7 \ \text{m}$$

**Exercise 4-6**

1. Find the surface area and the volume of a railroad container car of length 30.0 ft, width 10.0 ft, and height 12.0 ft, neglecting the thickness of the car. Express answers in feet and in meters.

2. Find the lateral area and the volume of a cylindrical gas tank whose radius is 15.0 m and height is 42.0 m.

3. *(a)* Find the surface area and the volume of a spherical tank of radius 1.00 m. *(b)* Find the surface area of a cube with the same volume as the sphere, and show that this area is greater than the surface area of the sphere. (*Hint:* First find the length of a side of the cube.)

**4.** Find the amount of canvas needed to make a conical tent in which the slant height $s$ is 6.5 ft and the base is 4.0 ft in diameter.

**5.** A church steeple consists of a cylinder with a conical top as shown in Fig. 4-35. One gallon of paint costs $20.00 and will cover 25 m². If the paint must be bought by the gallon, how much will it cost to paint the steeple?

**6.** On top of a building, a cylindrical tank 22.0 ft high with a radius of 6.00 ft is filled with water. If water weighs 62.4 lb/ft³, what is the weight of the water in the tank in pounds and in kilograms?

**7.** A rectangular piece of copper 8.00 cm by 12.0 cm is made into a box by cutting a square out of each corner and folding up the sides as shown in Fig. 4-36. Calculate the volume when *(a) s* = 2.00 cm and *(b) s* = 1.50 cm.

**8.** A jet's rectangular fuel tank measures 3.3 m by 2.6 m by 1.5 m. The jet is being fueled in midair at a rate of 100 L/min. If the tank was one-third full at the start, how long will it take to refuel completely?

**9.** If the tanks in Example 4-15 have square bases 9.0 m on a side and 12 m on a side, respectively, to what height will the chlorine level off?

**10.** If the tanks in Example 4-15 are cylinders of radii 5.0 and 7.0 m, respectively, to what height will the chlorine level off?

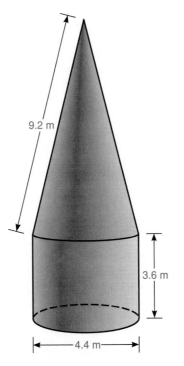

9.2 m

3.6 m

4.4 m

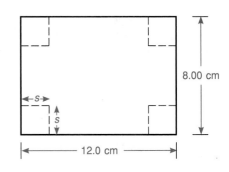

8.00 cm

12.0 cm

FIG. 4-35  *Church steeple, problem 5*      FIG. 4-36  *Problem 7*

[C] **11.** A cylindrical capacitor consists of two cylindrical plates, one inside the other. If the capacitor is 13 mm high and the radii of the plates are 3.2 and 5.6 mm, find the volume of the space between the plates in cubic centimeters.

[C] **12.** A coil of 0.102-in-diameter copper wire (no. 10 gauge) used for a solenoid weighs 1.20 lb. Find the length of the wire in the coil if the copper weighs 530 lb/ft$^3$.

[C] **13.** The scientific building with the largest cubic capacity in the world is the Vehicle Assembly Building at the JFK Space Center, the launch site for the Apollo moon craft. It measures 716 ft in length, 518 ft in width, and 525 ft in height. Assuming that the building is rectangular, what is the volume in cubic feet and in cubic meters?

[C] **14.** If the price of gold is 25 times that of silver, which would you rather have, a pure gold coin 0.850 mm thick and 10.5 mm in diameter or a pure silver coin 2.30 mm thick and 36.5 mm in diameter? Find the volume of each coin and compare.

[C] **15.** If the diameter of the earth is $7.9 \times 10^2$ mi and the diameter of the sun is $8.6 \times 10^5$ mi, what is the ratio of the volume of the sun to the volume of the earth?

[C] **16.** A sphere of diameter 5.0 in. is carved from a wood cube whose side is 5.0 in. How much volume has been removed from the cube?

**17.** Two cylindrical beakers have the same diameter, 10.0 cm, and together they have a total volume of 2.0 L. If the large beaker is 3 times the height of the small beaker, find the height of the small beaker.

**18.** A conical cistern (a water tank) 30 ft high has a top radius of 15 ft and is full of water. All the water is poured into an empty cylindrical boiler of radius 12 ft and height 20 ft. If the boiler is upright on its circular base, what is the height of the water in the boiler?

[C] **19.** Charles' law states that when the pressure is constant, the volume of a gas increases proportionally with the absolute temperature in kelvin. The diameter of a weather balloon is 80 cm. If the absolute temperature increases by 20 percent and the pressure remains constant, to what will the diameter of the balloon increase? (*Note:* Find the original volume and the new volume first.)

[C] **20.** Find the mass in kilograms of a hollow steel ball used as a buoy if the outside diameter is 2.30 m, the wall thickness is 10.0 mm, and the density of the steel is 8.00 g/cm$^3$.

# 4-7 *Pythagorean Theorem*

One of the most important and useful theorems in all of mathematics, especially geometry, is the pythagorean theorem. Although Babylonian and Egyptian surveyors used the theorem over 3000 years ago, the earliest record of its formal proof was left by the Greek Pythagoras around 520 B.C. The *pythagorean theorem* states:

*In a right triangle, the sum of the squares of the legs equals the square of the hypotenuse.*

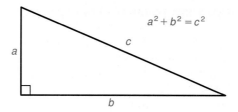

FIG. 4-37 *Pythagorean theorem*

This leads to the formula:

$$a^2 + b^2 = c^2 \qquad (4\text{-}6)$$

where $a$ and $b$ are the legs, or sides, of a right triangle and $c$ is the hypotenuse (Fig. 4-37).

### Example 4-16

A boat travels 5.00 km due south and then 12.0 km due east. How far is the boat from its starting point? See Fig. 4-38.

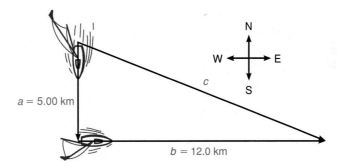

FIG. 4-38 *Boat trip, Example 4-16*

### Solution

In Fig. 4-38 you need to solve for $c$, the hypotenuse of the right triangle:

$$a^2 + b^2 = c^2$$
$$(5.00)^2 + (12.0)^2 = c^2$$
$$25.0 + 144 = c^2$$
$$c^2 = 169$$
$$c = \sqrt{169} = 13.0 \text{ km}$$

## Proof of the Pythagorean Theorem

There are over 400 proofs of the pythagorean theorem, including one by a president of the United States, James Garfield. The following geometric proof is probably similar to the one first used by Pythagoras.

Figure 4-39 translates into the following verbal equation:

Area of large square = area of small square + area of four right triangles

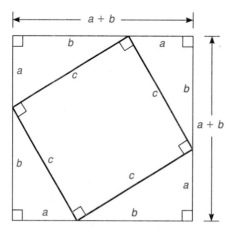

FIG. 4-39  *Proof of the pythagorean theorem*

This leads to:

$$(a + b)^2 = c^2 + 4 \left( \frac{1}{2} ab \right)$$

The area of one right triangle = ½ab since one leg is the base and the other is the height. Multiplying on each side gives:

$$a^2 + 2ab + b^2 = c^2 + 2ab$$

Canceling 2ab, you get:

$$a^2 + b^2 = c^2$$

### Example 4-17

Find the volume $V$ of a cone that has a radius $r = 4.0$ ft and slant height $s = 5.0$ ft (Fig. 4-32 *d*).

## Solution

You need to find the height $h$ because the volume $V = \frac{1}{3}\pi r^2 h$. By the pythagorean theorem $h^2 + r^2 = s^2$; therefore:

$$h^2 + (4.0)^2 = (5.0)^2$$
$$h^2 = 25 - 16 = 9$$
$$h = \sqrt{9} = 3.0$$

and
$$V = \frac{1}{3}\pi(4.0)^2(3.0) = 50 \text{ ft}^3 \quad \text{(two figures)}$$

The pythagorean theorem is basic for measuring distance in two dimensions. The next example shows how the theorem can be extended for measuring distance in three dimensions.

### Example 4-18

A helicopter ascends vertically to an altitude of 1000 ft, flies due north for 2000 ft, turns and flies due east 3000 ft. How far is the helicopter from its launching site? See Fig. 4-40.

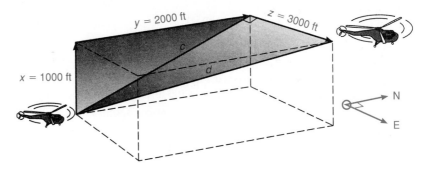

FIG. 4-40　*Helicopter flight, Example 4-18*

## Solution

In Fig. 4-40 you need to find the distance $d$, which is the diagonal of a rectangular box whose dimensions are $x = 1000$, $y = 2000$, and $z = 3000$ (assuming the earth is flat over short distances). First derive the formula for $d$ in terms of $x$, $y$, and $z$. From the two right triangles in Fig. 4-40 you can write the two equations:

$$x^2 + y^2 = c^2 \qquad z^2 + c^2 = d^2$$

Combining (adding) the equations gives:

$$x^2 + y^2 + z^2 + c^2 = c^2 + d^2$$

and canceling $c^2$ yields:

$$x^2 + y^2 + z^2 = d^2 \tag{4-7}$$

Notice the similarity between (4-7) and (4-6). Formula (4-7) is the extension of the pythagorean theorem to three dimensions. Now substitute and solve for the distance $d$:

$$1000^2 + 2000^2 + 3000^2 = d^2$$
$$1{,}000{,}000 + 4{,}000{,}000 + 9{,}000{,}000 = d^2$$
$$d^2 = 14{,}000{,}000$$
$$d = \sqrt{14{,}000{,}000}$$
$$= 3740 \text{ ft} \quad \text{(three figures)}$$

**Exercise 4-7**

In problems 1 to 10, find the missing side of each right triangle ($c$ = hypotenuse).

1. $a = 3$, $b = 4$, $c = ?$        2. $a = 5$, $b = 12$, $c = ?$
3. $a = ?$, $b = 8$, $c = 10$       4. $a = 8$, $b = ?$, $c = 17$
5. $a = 1$, $b = 2$, $c = ?$        6. $a = 3$, $b = 3$, $c = ?$
7. $a = 0.25$, $b = 0.60$, $c = ?$  8. $a = ?$, $b = 16.5$, $c = 18.7$
9. $a = \frac{2}{3}$, $b = \frac{1}{2}$, $c = ?$   10. $a = \frac{1}{2}$, $b = ?$, $c = 1$

11. Two high-speed commuter trains leave a station at exactly 6 p.m. One train travels north averaging 80 mi/h, and the other travels west averaging 60 mi/h. How far apart are they at 6:30 p.m.?

12. A ladder 8.0 m long is placed 4.8 m from a wall. How far up the wall does the ladder reach?

13. Find the amount of canvas needed to make a conical tent 9.0 ft in diameter and 6.0 ft high. (See Example 4-17.)

14. A gable (a triangular wall) is in the shape of an isosceles right triangle. If the hypotenuse is 10 m long, how long is each side? (*Hint:* Let each side = $x$.)

[C] 15. Two right triangles have the same angles and so are similar. The legs of the first triangle are 3.20 and 4.80 cm long. The hypotenuse of the second triangle is 11.2 cm long. Find the length of each leg of the second triangle.

16. A baseball diamond is a square whose area is 8100 ft². Find the distance from home plate to second base.

17. Two guy wires for a 60-m-high radio antenna are to be attached parallel to each other as shown in Fig. 4-41. At least how long should each guy wire be?

18. Roxanne cycles 8.0 km south, 9.0 km east, and 4.0 km farther south. How far is she from her starting point? (*Hint:* Draw a rectangle.)

19. In an ac circuit containing a resistance $R$ and an inductive reactance $X_L$, the effect of the resistance is at 90° to the effect of the reactance. The vector diagram is shown in Fig. 4-42. Using the pythagorean theorem, find the impedance $Z$ if $R = 4.5$ Ω and $X_L = 6.0$ Ω.

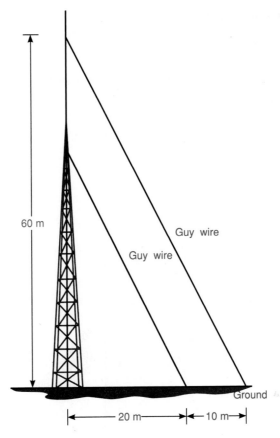

FIG. 4-41  *Radio antenna, problem 17*

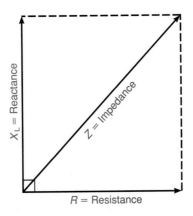

FIG. 4-42  *AC circuit, problems 19 and 20*

**20.** In Fig. 4-42 find $R$ if $Z = 19.5 \ \Omega$ and $X_L = 7.50 \ \Omega$.

**21.** A rectangular doorway is 4 ft 3 in. wide and 8 ft 2 in. high. Can a circular tabletop 2.70 m in diameter fit through the doorway? (*Hint:* Find the diagonal of the doorway.)

**22.** A boat traveling at 6.2 kn experiences a 2.1-kn current abeam (at right angles) to its heading. Find the resultant velocity $R$ shown in the vector diagram, Fig. 4-43.

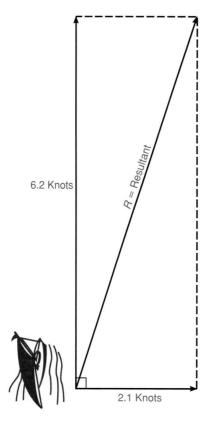

FIG. 4-43 *Velocity-current diagram, problem 22*

**23.** A hallway is 24.0 ft long, 6.0 ft wide, and 8.0 ft high. How far does a fly in one corner of the hallway have to travel in a straight line to reach the corner diagonally opposite? (See Example 4-18.)

**24.** After takeoff Carlos ascends vertically in his helicopter to an altitude of 1.0 km, flies due west 3.0 km, and then flies due north 2.5 km. How far is he from the takeoff site? (See Example 4-18.)

**25.** A cofferdam is a watertight structure placed in a body of water such as a river and pumped dry so that a bridge foundation or other structure can be built. How long should a cross brace be for a cofferdam in the shape of a rectangular box 22 m by 15 m by 44 m? (See Example 4-18.)

**26.** Show that the altitude of an equilateral triangle $h$ is equal to $s\ \sqrt{3}/2$, where $s =$ side. (*Note:* The altitude bisects the side.)

# 4-8 | *Trigonometry of the Right Triangle*

The examples and exercises in the last section show how often you encounter right triangles in problems. Trigonometry, or the geometry of triangles, focuses on right triangles for they are the key to solving all triangles. Two right triangles that have one acute angle equal have all angles equal and so are similar. Similar triangles have equal ratios for any two corresponding sides. It is therefore very useful to define these trigonometric ratios, or functions, for a right triangle.

The *three basic trigonometric functions* for angle $A$ in Fig. 4-44 are:

$$\text{sine } A \ (\sin A) = \frac{\text{opposite side}}{\text{hypotenuse}} = \frac{a}{c}$$

$$\text{cosine } A \ (\cos A) = \frac{\text{adjacent side}}{\text{hypotenuse}} = \frac{b}{c} \tag{4-8}$$

$$\text{tangent } A \ (\tan A) = \frac{\text{opposite side}}{\text{adjacent side}} = \frac{a}{b}$$

An easy way to remember these definitions is by the mnemonic "s-oh c-ah t-oa" (sin − opp/hyp, cos − adj/hyp, tan − opp/adj).

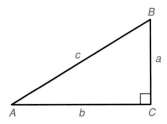

FIG. 4-44 *Trigonometric functions, Example 4-19*

## Example 4-19
Find the sine, cosine, and tangent of $\angle A$ and $\angle B$ in Fig. 4-44 if $a = 6$ and $b = 8$.

### Solution

First calculate $c$, using the pythagorean theorem:

$$c^2 = 6^2 + 8^2$$
$$c^2 = 100$$
$$c = \sqrt{100} = 10$$

Then:

$$\sin A = \frac{6}{10} = \frac{3}{5} \qquad\qquad \sin B = \frac{4}{5}$$

$$\cos A = \frac{8}{10} = \frac{4}{5} \qquad\qquad \cos B = \frac{3}{5}$$

$$\tan A = \frac{6}{8} = \frac{3}{4} \qquad\qquad \tan B = \frac{4}{3}$$

Notice that $\sin A = \cos B$ and $\cos A = \sin B$. $A$ and $B$ are complementary angles and it follows from the definitions (4-8) that the sine of an angle equals the cosine of its complement. For example, $\sin 60° = \cos 30°$, $\sin 20° = \cos 70°$, etc. Hence the cosine is called the *cofunction* of the sine.

The *three reciprocal trigonometric functions* for angle $A$ in Fig. 4-44:

$$\text{secant } A \text{ (sec } A) = \frac{\textbf{hypotenuse}}{\textbf{adjacent side}} = \frac{c}{b} = \frac{1}{\cos A}$$

$$\text{cosecant } A \text{ (csc } A) = \frac{\textbf{hypotenuse}}{\textbf{opposite side}} = \frac{c}{a} = \frac{1}{\sin A} \qquad (4\text{-}9)$$

$$\text{cotangent } A \text{ (cot } A) = \frac{\textbf{adjacent side}}{\textbf{opposite side}} = \frac{b}{a} = \frac{1}{\tan A}$$

Tangent and cotangent are cofunctions, and so are secant and cosecant. The reciprocal functions are introduced here for completeness. They are discussed further in Chap. 10.

### Example 4-20

Find the value of each trigonometric function.

**1.** $\sin 40.2°$     **2.** $\cos 70°10'$     **3.** $\tan 0.2967$

### Solution

**1.** $\sin 40.2°$

The values of the trigonometric functions are obtained by using a calculator or a table of values like those in App. B. Both methods are shown below.

| Degrees | Sin | Cos | Tan | Cot |
|---------|-----|-----|-----|-----|
| 40.0 | 0.6428 | 0.7660 | 0.8391 | 1.192 |
| 0.1 | 0.6441 | 0.7649 | 0.8421 | 1.188 |
| ▶ 0.2 | (0.6455) | 0.7638 | 0.8451 | 1.183 |
| 0.3 | 0.6488 | 0.7627 | 0.8481 | 1.179 |

FIG. 4-45 *Decimal degree table, Example 4-20*

*Calculator.* When the angle is in decimal degrees, set the calculator for the degree mode with the $\boxed{\text{DRG}}$ key or $\boxed{\text{MODE}}$ key. Most calculators are in the degree mode when they are turned on, and it is not necessary to press this key. Enter the angle, and press the trigonometric function key:

$$40.2 \;\boxed{\text{sin}} \rightarrow 0.6455$$

*Table.* For angles less than 45°, use the left-hand side of the table and the top headings. Figure 4-45 shows the portion of the decimal degree table used (App. B-1). If the angle falls between two values in the table, for example, 40.25°, take the closest value (40.3°). Greater accuracy can be obtained by interpolation.

**2.** cos 70°10′

*Calculator.* Change degrees and minutes to decimal degrees, and press the function key:

$$70 \;\boxed{+}\; 10 \;\boxed{\div}\; 60 \;\boxed{=}\; \boxed{\text{cos}} \rightarrow 0.3393$$

*Table.* For angles greater than 45°, use the right-hand side of the table, reading from the bottom up with the bottom headings. Figure 4-46 shows the portion of the degree-minute table used (App. B-2). The tables are arranged with top and bottom headings because of the cofunction property of sine and cosine, tangent and cotangent.

| | | | | |
|---|---|---|---|---|
| 0.3365 | 0.9417 | 0.3574 | 2.798 | 20′ |
| (0.3393) | 0.9407 | 0.3607 | 2.773 | 10′ ◀ |
| 0.3420 | 0.9397 | 0.3640 | 2.747 | 70°00′ |
| 0.3448 | 0.9387 | 0.3673 | 2.723 | 50′ |
| **Cos** | **Sin** | **Cot** | **Tan** | **Degrees** |

FIG. 4-46 *Degree and minute table, Example 4-20*

**3.** tan 0.2967

*Calculator.* The angle 0.2967 is in radians, because there is no degree mark. Enter the angle, set the calculator for radians, using the $\boxed{\text{DRG}}$ key or $\boxed{\text{MODE}}$ key, and press the function key:

$$0.2967 \;\boxed{\text{DRG}}\; \boxed{\text{tan}} \rightarrow 0.3057$$

*Table.* The decimal degree table (App. B-1) gives the angle in radians and is used as described in (1) above.

The trigonometric functions enable you to solve for distances by measuring angles, as the next example shows.

### Example 4-21

To find the height of a cliff, Nuria, a surveyor, chooses a point 100 m from its base and sets up her theodolite—a telescope that can be rotated to measure horizontal and vertical angles. She then measures the *angle of elevation* (the angle above the horizontal) of the top of the cliff to be 33.8° (Fig. 4-47). How high is the cliff? (Neglect the height of the theodolite.)

### Solution

Using the given angle, set up an equation, choosing the basic trigonometric function that involves the side known (100 m is *adjacent* to 33.8°) and the side

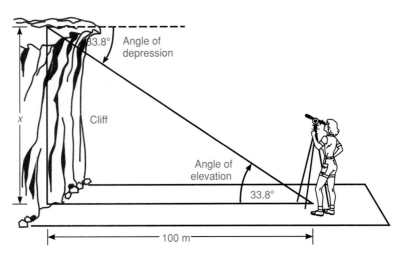

FIG. 4-47 *Surveyor, Example 4-21*

you want to find (the height $x$ is *opposite* to 33.8°). Therefore you choose the tangent:

$$\tan 33.8° = \frac{\text{opposite}}{\text{adjacent}} = \frac{x}{100}$$

Using a calculator or table, find tan 33.8° and solve for $x$:

$$\tan 33.8° = 0.6694 = \frac{x}{100}$$

$$x = 0.6694(100) = 66.9 \text{ m} \quad \text{(three figures)}$$

For an observer on top of the cliff looking down at the surveyor, the *angle of depression* (the angle below the horizontal) is equal to the angle of elevation of the surveyor. This is because the horizontal lines in Fig. 4-47 are parallel and the two acute angles are alternate interior angles.

If you know any side and an acute angle, or any two sides of a right triangle, you can find any other side or acute angle by using trigonometric functions. The next example illustrates how to find an angle when you know a trigonometric function of the angle.

### Example 4-22

Find angle $A$ in Example 4-19 given $\sin A = \dfrac{3}{5} = 0.6000$.

### Solution

*Calculator.* Enter the value and press INV sin or 2nd F sin :

$$0.6000 \boxed{\text{INV}} \boxed{\text{sin}} \rightarrow 36.87°$$

The answer means that sin 36.87° = 0.6000. (See Sec. 4-9 for further calculator information.) The inverse trigonometric functions are discussed further in Chap. 10.

*Table.* Use the table backward by finding the closest value to 0.6000 in the sine column and reading the corresponding angle in the degree column: 36.9°, or 36°50′. Figure 4-48 shows the portion of the decimal degree table used. Greater accuracy can be obtained by interpolation.

Study the next two examples, which illustrate the use of both trigonometric and inverse trigonometric functions.

| Degrees | Sin |
|---|---|
| 36.0 | 0.5878 |
| $\vdots$ | $\vdots$ |
| 0.7 | 0.5976 |
| 0.8 | 0.5990 |
| (0.9) | 0.6004 ◀ |
| 37.0 | 0.6018 |

FIG. 4-48  *Decimal degree table, Example 4-22*

### Example 4-23

A rocket is supported by a cable attached at a point 200 ft high and to the ground 100 ft from the base of the rocket (Fig. 4-49).

**1.** Assuming the cable is taut, what angle $\theta$ does it make with the ground?

**2.** How long is the cable?

### Solution

**1.** Assuming the cable is taut, what angle $\theta$ does it make with the ground?

To find $\theta$, use the inverse of the tangent, since you know the opposite side (200 ft) and the adjacent side (100 ft).

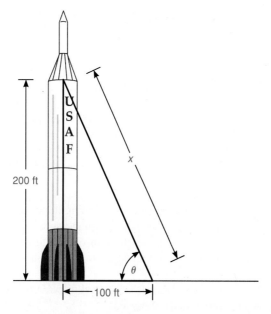

FIG. 4-49  *Rocket, Example 4-23*

*Calculator.*  200 $\boxed{\div}$ 100 $\boxed{=}$ $\boxed{\text{INV}}$ $\boxed{\text{tan}}$ $\rightarrow$ 63.4°

*Table.* Find the closest value to tan $\theta$ = 200/100 = 2.00 in the tangent column (bottom heading), and read the corresponding angle 63.4° or 63°30′ (right-hand side).

**2.** How long is the cable?

The length $x$ of the cable can be found by using the sine or cosine (or by using the pythagorean theorem). When the sine (or cosine) is used, the unknown appears in the denominator:

$$\sin 63.4° = \frac{200}{x}$$

You can solve for $x$ first and then use a calculator or table:

$$x \sin 63.4° = 200$$

$$x = \frac{200}{\sin 63.4°} = \frac{200}{0.8942} = 224 \text{ ft}$$

Observe that the trigonometric function and the unknown just switch places in this equation.

The next example shows an application of trigonometry to navigation, a field in which trigonometry found early application and in which it is still used extensively. Compass bearings or headings are measured clockwise starting from north, which is 0°.

### Example 4-24

An underwater missile is fired with a velocity of 50.0 km/h at a target whose bearing is 0° (north) and whose distance is 10.0 km. Six minutes later the bearing of the target changes to 15.0°, and then the target remains stationary. What should be the degree correction in the heading of the missile? (See Fig. 4-50.)

### Solution

After 6 min the missile travels (6/60)(50.0) = 5.00 km. You need to find $\theta$, the degree correction in the missile heading. First find the distance $x$ the target has moved, using the large right triangle in the figure:

$$\tan 15.0° = \frac{x}{10.0}$$

$$x = 10.0(\tan 15.0°) = 10.0(0.2679)$$
$$= 2.68 \text{ km}$$

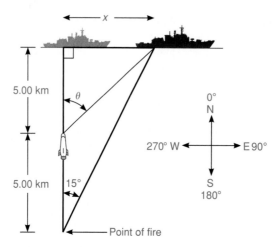

FIG. 4-50 *Underwater missile, Example 4-24*

Then using the small right triangle, find tan $\theta$ and then $\theta$:

$$\tan\theta = \frac{x}{5.00} = \frac{2.68}{5.00} = 0.536$$

and $\theta = 28.2°$, using the inverse of the tangent.

**Exercise 4-8**   In problems 1 to 10, find sin $A$, cos $A$, and tan $A$ for each right triangle ($c = $ hypotenuse).

**1.** $a = 8$, $b = 15$, $c = 17$      **2.** $a = 4$, $b = 3$, $c = 5$

**3.** $a = 10$, $b = 24$, $c = 26$      **4.** $a = 7$, $b = 24$, $c = 25$

**5.** $a = 12$, $b = 9$      **6.** $a = 12$, $c = 13$

**7.** $a = 0.60$, $c = 1.0$      **8.** $a = 1.0$, $b = 1.0$

**9.** $b = x$, $c = 2x$      **10.** $a = 3x$, $b = 4x$

In problems 11 to 18, find the value of each trigonometric function, using a calculator or table.

**11.** sin 80.6°      **12.** cos 39.3°

**13.** tan 39°50′      **14.** sin 45°20′

**15.** cos 1.3875      **16.** tan 0.4066

**17.** cot 87.6°      **18.** cot 25°10′

In problems 19 to 24, find angle $A$ in each equation, using a calculator or table.

**19.** sin $A = 0.3173$      **20.** cos $A = 0.0958$

**21.** tan $A = 2.135$      **22.** sin $A = 0.9820$

**23.** $\cos A = \dfrac{2}{3}$      **24.** $\tan A = \dfrac{1}{5}$

In problems 25 to 32, find all missing sides and acute angles of the following right triangles. Express angles in the form in which they are given.

**25.** $A = 50°$, $c = 3.0$          **26.** $B = 6.8°$, $c = 8.2$

**27.** $a = 5.0$, $c = 8.0$          **28.** $a = 2.5$, $b = 6.0$

**29.** $B = 73°$, $a = 2.0$          **30.** $A = 27°30'$, $a = 3.3$

**31.** $\tan A = 1.0$, $a = 2.0$          **32.** $\sin B = 0.60$, $c = 10$

**33.** The angle of elevation of a building is 60° at a point 50 m from its base. How high is the building?

**34.** The angle of depression of a sailboat from the bridge of an aircraft carrier 300 ft above sea level is 22°. How far is the sailboat from the carrier?

**35.** To find the distance across a river, Gordon, a surveying student, chooses a point $C$ on the riverbank directly opposite a tree $B$ on the other riverbank (Fig. 4-51). He measures 100 m from $C$ to a point $A$ on the same riverbank and measures angle $CAB$ to be 50°. Assuming angle $ACB$ is a right angle, how long is $BC$?

**36.** From a Rocky Mountain peak 9300 ft above sea level the angle of elevation of an adjacent peak is 15°30′. What is the *air* distance between the peaks if the second peak is 11,200 ft above sea level? (Neglect the earth's curvature.)

**37.** After takeoff the altimeter in a passenger jet indicates a rise of 2000 ft/min at a constant velocity of 300 mi/h. Assuming the angle of ascent to be constant, find its value.

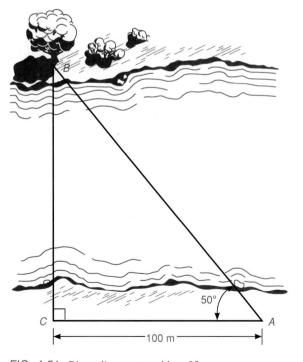

FIG. 4-51 *River distance, problem 35*

38. Joseph wants to find the height of the television antenna on top of the Eiffel Tower. From a point on the ground he measures the angle of elevation of the tower, which he knows is 301 m high, to be 71.6°. From the same point he then measures the angle of elevation of the antenna to be 72.7°. How high is the top of the antenna above the ground?

39. The force $F$ acting on a body at an angle $\theta$ from the horizontal is equal to the vector addition of the horizontal component $F_x$ and the vertical component $F_y$. Figure 4-52 shows the force rectangle. If $F_x = 28$ kg and $F_y = 36$ kg, find $F$ and $\theta$.

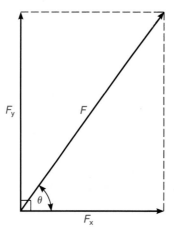

FIG. 4-52 *Force components, problem 39*

40. In the ac circuit of Fig. 4-42 (problem 19 of Sec. 4-7), the phase angle $\phi$ is the angle between the impedance $Z$ and the resistance $R$. Find $\phi$ if $R = 5.3\ \Omega$ and $X_L = 3.5\ \Omega$.

41. The rocket in Fig. 4-49 (Example 4-23) is to be supported by another cable attached to it at the same point as the first cable but attached to the ground 50 ft from the base. *(a)* What angle will the second cable make with the ground? *(b)* How long should this cable be?

42. Find the correction in the heading of the missile in Example 4-24 if 6 min after launching the bearing of the target changes to 20.5° and remains the same.

43. A ship heads due west for 30 min at 12 kn, then changes course and travels due south for 10 min. At this point the ship receives a distress call and must return to its original position. *(a)* What should the ship's heading be? *(b)* How many minutes will the return take at a maximum speed of 20 kn? (See Fig. 4-50 for compass headings.)

44. Enrique is traveling parallel to a straight coastline in his motorboat and wants to find his distance from the shore. Using a compass, he takes the bearing of the stack of a power plant and measures the angle between his heading and the stack to be 52° (Fig. 4-53). After he travels 0.30 mi, the angle between his heading and the stack is 90°. How far is he from the coast?

45. The *index of refraction* $\mu$ of a medium is defined as the ratio of the speed of light $c$ in air to the speed of light $c_m$ in the medium. It is equal to:

$$\mu = \frac{c}{c_m} = \frac{\sin \phi}{\sin \phi_m}$$

where $\phi$ is the angle of incidence and $\phi_m$ is the angle of refraction of a light ray (Fig. 4-54). If a light ray passes from air into water, where $\mu = 1.33$ and $\phi = 0.254$ rad, find the angle of refraction $\phi_m$ in radians.

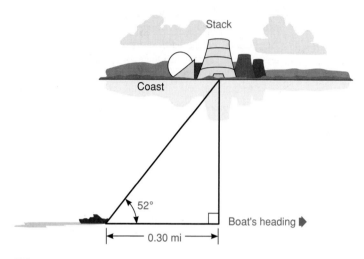

FIG. 4-53 *Shore distance, problem 44*

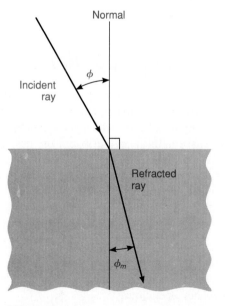

FIG. 4-54 *Light refraction, problem 45*

☐ **46.** The safe angle $\phi$ for the bank of a highway curve is given by $\tan \phi = v^2/(gR)$, where $v$ = velocity, $R$ = radius of curve, and $g$ = gravitational acceleration, 32.2 ft/s². Find $\phi$ if $v$ = 80.7 ft/s (55 mi/h) and $R$ = 950 ft.

**47.** The instantaneous voltage of alternating current is defined by the equation $V = V_m \sin \omega t$, where $V_m$ is the maximum voltage, $\omega$ is the angular velocity in radians per second, and $t$ is the time in seconds. Find $V$ if $V_m$ = 220 V and $\omega t = \pi/4$ rad.

**48.** A rocket fired at an angle $\phi = 70.3°$ with the horizontal is traveling at a velocity $v$ = 9560 km/h. Assuming a straight flight path, find the rocket's vertical component of velocity $v = v \sin \phi$.

☐ **49.** An observer on the earth measures with a telescope the angle that the moon subtends (the angle between the ends of the diameter) to be $0°31'$. If the distance to the moon is 239,000 mi, how long is its diameter? (*Hint:* Divide the triangle into two right triangles.)

**50.** Find the area of a right-triangular piece of land with an acute angle of 37.4° and a hypotenuse of 112 m. (*Hint:* First find the length of each leg of the triangle.)

**51.** Given $\sin A$ = 0.500, find the values of the other five trigonometric functions: $\cos A$, $\tan A$, $\sec A$, $\csc A$, and $\cot A$.

**52.** Given $\cot B$ = 1.00, find the values of the other five trigonometric functions: $\sin B$, $\cos B$, $\tan B$, $\sec B$, and $\csc B$.

**53.** Prove that $\sin 30° = \cos 60° = 0.5000$, using the fact that the altitude of an equilateral triangle divides it into two congruent right triangles. Let the side of the equilateral triangle be exactly equal to 2.

**54.** The ancient Greeks believed a rectangle with a length equal to $1 + \sqrt{5}$ and a width equal to 2 had perfect proportions. They called it a *golden rectangle*. What angle does the diagonal of this rectangle make with the longer side?

# 4-9 | *Hand Calculator Operations*

There are calculators programmed to do metric conversions, and some have a key that converts from radians to degrees. However, in both cases, the calculator is just multiplying by a factor, which you can do directly on any calculator. Most calculators have a ⎡DRG⎤ key or ⎡MODE⎤ key used for degree, radian, and grad measure. Grad measure is used in Europe and in road construction; 100 grad = 90°. A grad of 10 means an increase of 10 percent in the angle of rise from horizontal to vertical.

## *Angles*

On the calculator angles are expressed in decimal degrees or radians. An angle in degrees and minutes must first be changed to decimal degrees before it can be used in calculator operations.

## Example 4-25
Change each angle to decimal degrees.

**1.** 23°14'      **2.** 42°10.4'      **3.** 2°10'45"

## Solution
**1.** 23°14'

To change from decimal degrees, divide the minutes by 60:

$$23 \boxed{+} 14 \boxed{\div} 60 \boxed{=} \rightarrow 23.23°$$

**2.** 42°10.4'

$$42 \boxed{+} 10.4 \boxed{\div} 60 \boxed{=} \rightarrow 42.173°$$

**3.** 2°10'45"

To change seconds to decimal degrees, divide by 60 × 60, or 3600:

$$2 \boxed{+} 10 \boxed{\div} 60 \boxed{+} 45 \boxed{\div} 3600 \boxed{=} \rightarrow 2.1792°$$

To change from decimal degrees back to degrees and minutes, multiply the decimal part by 60.

After an angle has been changed to decimal degrees, it can then be changed to radians if necessary.

## Example 4-26
**1.** Change 23.23° to radians.      **2.** Change 2.027 rad to degrees.

## Solution
**1.** Change 23.23° to radians.

To change from decimal degrees to radians, multiply by $\pi/180$, or 0.01745:

$$23.23 \boxed{\times} \boxed{\pi} \boxed{\div} 180 \boxed{=} \rightarrow 0.4054$$

or:

$$23.23 \boxed{\times} 0.01745 \boxed{=} \rightarrow 0.4054$$

**2.** Change 2.027 rad to degrees.

To change from radians to decimal degrees, multiply by $180/\pi$, or 57.30:

$$2.027 \boxed{\times} 180 \boxed{\div} \boxed{\pi} \boxed{=} \rightarrow 116.1°$$

## *Trigonometric Functions*

Finding values of the three basic trigonometric functions by using the $\boxed{\text{sin}}$, $\boxed{\text{cos}}$, or $\boxed{\text{tan}}$ keys is explained in Example 4-20. Values of the reciprocal trigonometric functions can be obtained by using the reciprocal key $\boxed{1/x}$.

### Example 4-27

Find each of the following.

**1.** sin 42°22'　　　**2.** cot 1.30

### Solution

**1.** sin 42°22'

Change first to decimal degrees and then find the sine:

$$42 \;\boxed{+}\; 22 \;\boxed{÷}\; 60 \;\boxed{=}\; \boxed{\text{sin}} \to 0.6739$$

**2.** cot 1.30

Set the calculator for radians, press the reciprocal-function key, then press the reciprocal key:

$$1.30 \;\boxed{\text{DRG}}\; \boxed{\text{tan}}\; \boxed{1/x} \to 0.2776$$

The reciprocal functions are discussed further in Chap. 10.

To find an angle when given a trigonometric function of it, use the inverse trigonometric function keys $\boxed{\text{INV}}$ $\boxed{\text{sin}}$ or $\boxed{\text{2nd F}}$ $\boxed{\text{sin}}$ and so on. This is shown in Examples 4-22 and 4-23. These keys should be used cautiously. The sine or cosine of an angle cannot be greater than 1. If a number greater than 1 is entered and $\boxed{\text{INV}}$ $\boxed{\text{sin}}$ or $\boxed{\text{INV}}$ $\boxed{\text{cos}}$ is pressed, the display will flash or show error.

### Example 4-28

Find angle $A$ in each equation.

**1.** cos $A$ = 0.4512　　　**2.** sin $A$ = 1.312

### Solution

**1.** cos $A$ = 0.4512

$$0.4512 \;\boxed{\text{INV}}\; \boxed{\text{cos}} \to 63.18°$$

**2.** $\sin A = 1.312$

This equation has no solution, since the sine cannot be greater than 1. The inverse trigonometric functions are discussed further in Chap. 10.

**Exercise 4-9**

In problems 1 to 6, change each angle to decimal degrees.

**1.** $85°18'$       **2.** $19°09'$

**3.** $50°0.8'$      **4.** $100°10.7'$

**5.** $6°06'20''$     **6.** $1°10'06''$

In problems 7 to 10, change each angle to radians.

**7.** $7.08°$        **8.** $95.55°$

**9.** $66°06.7'$     **10.** $19°20'50''$

In problems 11 to 14, change each angle to degrees.

**11.** 0.1952 rad     **12.** 1.571 rad

**13.** 2.356 rad      **14.** $0.350\pi$

In problems 15 to 20, find the value of each trigonometric function.

**15.** $\cos 18.33°$     **16.** $\sin 57.04°$

**17.** $\tan 33°19'$     **18.** $\cot 50°10'$

**19.** $\cot 0.95$       **20.** $\cos 1.23$

In problems 21 to 26, find angle $A$ in each equation.

**21.** $\tan A = 3.412$     **22.** $\sin A = 0.5050$

**23.** $\cos A = 0.0110$     **24.** $\cot A = 1.707$

**25.** $\sin A = 2.215$     **26.** $\cos A = 1.500$

# 4-10 | *Review Questions*

In problems 1 to 6 convert each of the units (use App. A if necessary).

**1.** $0.51 \text{ cm}^2$ to square millimeters     **2.** 120 V to kilovolts

**3.** 23 kg to pounds              **4.** $70 \text{ ft}^3$ to cubic meters

**5.** 66.0 mi/h to kilometers per hour     **6.** 11.0 gal to liters

In problems 7 to 10, change degrees to radians and vice versa.

**7.** $50.2°$                    **8.** $81°30'$

**9.** $\dfrac{\pi}{10}$                **10.** 0.350 rad

In problems 11 to 14, find the unknown variables. Express angles in the angle measure given.

**11.**

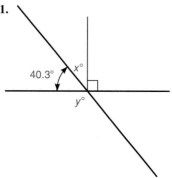

**12.**

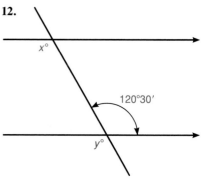

**13.**

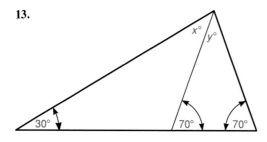

**14.**

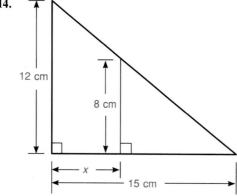

In problems 15 to 18, find sin $A$, cos $A$, and tan $A$ for each right triangle.

**15.** $a = 8$, $b = 6$, $c = 10$                 **16.** $a = 16$, $b = 30$, $c = 34$

**17.** $a = 10$, $b = 24$                              **18.** $a = 2.1$, $c = 7.5$

In problems 19 to 22, find the value of each trigonometric function.

**19.** tan $35.6°$                                           **20.** sin $7°30'$

**21.** cos $0.7243$                                        **22.** cot $30.0°$

In problems 23 to 26, find angle $A$.

**23.** $\cos A = 0.9890$     **24.** $\sin A = 0.8480$

**25.** $\tan A = 2.807$     **26.** $\cot A = 2.000$

**27.** The mean radius of the earth is $6.37 \times 10^6$ m. What is this value in kilometers?

**28.** A "22-caliber" rifle means the bore is 0.22 in. What is the caliber of a rifle whose bore is 7.62 mm?

**29.** The maximum density of water, which occurs at 4.0°C, is 1.0 g/mL. What is this density in pounds per cubic foot?

**30.** The fastest speed ever recorded for a wheeled vehicle occurred in 1970. The liquid-gas rocket-propelled vehicle covered a measured kilometer in 3.543 s. What was the average speed in miles per hour?

**31.** The shadow of a power plant's 100-m-high stack is 40 m long. At the same time an antenna casts a shadow 16 m long. How high is the antenna?

**32.** Find the area and perimeter of an equilateral triangle with a side measuring 20 ft and an altitude of 17 ft.

**33.** Find the circumference and area of a cam in the shape of a circle whose radius is 60.0 mm. Express answers in centimeters.

**34.** Find the area of the isosceles trapezoid in Fig. 4-55. [*Hint:* Use the pythagorean theorem (4-6) to find the altitude.]

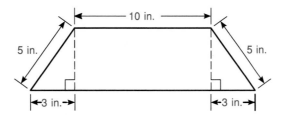

FIG. 4-55  *Problem 34*

**35.** The diameter of a wheel is 30 cm. How far does the wheel roll if it turns through an angle of 80°? (See Example 4-13.)

**36.** Two electric cables, each of diameter 15 in., are to be packed in a conduit that has a rectangular cross section. What is the smallest rectangular cross-sectional area that can accommodate the cables?

**37.** A rectangular fuel tank measures 30 cm by 50 cm by 80 cm. How many liters does the tank hold?

**38.** A spherical capacitor consists of two concentric spheres (having the same center). Find the volume of the space between the spherical plates if the radii are 6.83 and 5.12 mm.

**39.** Two planes leave an airport at the same time. One plane travels east averaging 300 mi/h, and the other travels south averaging 350 mi/h. How far apart are they after 2 h?

**40.** A cube-shaped container has a side equal to 5.0 in. How long is the distance from one corner through the cube to the diagonally opposite corner (the main diagonal)?

**41.** The angle of elevation of the top of a wind generator is 75° at a point 20 m from the base of the tower. How high is the generator?

**42.** From a helicopter 1000 ft in the air, the angles of depression of two landing platforms, both due south of the helicopter, are measured to be 21° and 16°. How far apart are the landing platforms?

**43.** A rectangular circuit chip measures 2.30 mm by 4.80 mm. What angle does the diagonal make with the longer side?

**44.** Find the angle $\theta$ in the design problem of Figure 4-56.

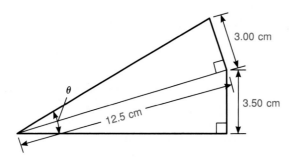

FIG. 4-56 *Design problem, problem 44*

**45.** A road has a 6 percent grade, which means a 6-ft rise for every 100 ft of horizontal distance. What is the angle of rise?

**46.** In Chicago, Fernando measures the altitude of the sun (the angle of elevation) with a sextant to be 43°50′. How high is the world's tallest office building, the Sears Tower in Chicago, if its shadow at this time is 461 m? Express the answer in meters and in feet.

# 5

# Linear Systems

## 5-1 | *Linear Functions and Graphs*

Suppose you want to compare the Celsius temperature scale to the Fahrenheit one. The Celsius scale was devised by the Swedish astronomer Anders Celsius in 1742. The Fahrenheit scale was developed by the German physicist Gabriel Fahrenheit in 1714. Consider the familiar conversion formula $F = (9/5)C + 32$, or $F = 1.8C + 32$. This is a *first-degree, or linear, equation in two variables, F and C*. In Example 5-1 this equation is used to compare the two scales for the range $-30$ to $40°C$. This is the range of temperatures most people experience in temperate climates during the course of a year.

> **I**n Chap. 2 first-degree equations are also called *linear equations.* This is because the graph of a first-degree equation is a straight line. This chapter begins with graphs and the relationship between linear equations and straight lines. It then studies the algebraic and graphical solutions of two linear equations in two unknowns. The chapter concludes with the algebraic solution of linear systems containing three unknowns.

### Example 5-1

Make a table of values in degrees Fahrenheit for the following temperatures in degrees Celsius: $-30, -20, -10, 0, 10, 20, 30$, and $40°$, and draw a graph of degrees Fahrenheit vs. degrees Celsius.

### Solution

For each value of $C$ calculate the corresponding value of $F$. For example, when $C = -20$, $F = 1.8(-20) + 32 = -36 + 32 = -4$. The table of values is (the degree mark being understood):

| C | $-30$ | $-20$ | $-10$ | 0 | 10 | 20 | 30 | 40 |
|---|---|---|---|---|---|---|---|---|
| F | $-22$ | $-4$ | 14 | 32 | 50 | 68 | 86 | 104 |

The table helps you compare certain temperatures in degrees Celsius with those in degrees Fahrenheit. The graph of the table will show how the two scales compare for *all values* within the range of the table. To construct the graph use a horizontal axis, or *x axis,* for Celsius values and a vertical axis, or *y axis,* for Fahrenheit values (Fig. 5-1).

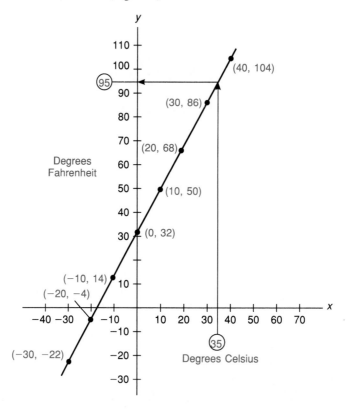

FIG. 5-1 *Fahrenheit vs. Celsius, Example 5-1*

The point where the axes cross is called the *origin*. It represents the point where $x = 0$ and $y = 0$. Up and to the right is positive, down and to the left is negative. Such a graph is called a *rectangular coordinate system*. Whenever possible, the same scales are used for $x$ and $y$. From the graph you can determine any value of $F$ for any given value of $C$. For example, when $C = 35$, the arrows on the graph show that $F = 95$. The most important thing to notice about the graph is that all the points lie on the same straight line. This is more than a coincidence. *The graph of any first-degree equation in two variables is always a straight line*. For this reason first-degree equations are called linear equations.

## Linear Function

Example 5-1 leads to the concept of a function, an important idea that you will see examples of throughout the text. Note that *each value of C determines a single value of F.* When that relation exists between two variables, the second variable (*F* in this case) is said to be a *function* of the first variable (*C*):

A variable *y* is said to be a function of a variable *x* if, given a set of values for *x*, there corresponds one and only one value of *y* for each value of *x*.                                                                      (5-1)

Also observe in Example 5-1 that since the graph of $F = 1.8C + 32$ is a straight line, *F is a linear function of C.* A first-degree equation in two variables expresses one of the variables as a linear function of the other variable. Study the following example.

### Example 5-2
List some examples of linear functions.

1. $C = 2\pi r$. The circumference *C* of a circle is a linear function of its radius *r*.
2. $F = ma$. Newton's second law of motion. The force *F* on a body is a linear function of the acceleration *a* (*m* = mass).
3. $E = IR$. Ohm's law. When the resistance *R* of a circuit is constant, the voltage *E* is a linear function of the current *I*.
4. $V = 0.607t + 332$. The velocity of sound (in meters per second) is a linear function of the temperature *t* (in degrees Celsius).
5. $v = v_0 + at$. The velocity *v* of a body with a constant acceleration *a* is a linear function of the time *t* ($v_0$ = velocity at $t = 0$).
6. $M = Kit$. When the current *i* is constant, the mass *M* deposited on an electrode in a solution is a linear function of the time *t* (*K* = constant).
7. $l = at + l_0$. The length *l* of a metal rod is a linear function of the temperature *t* ($l_0$ = length at 0°C; *a* = constant).

Observe that examples 1, 2, 3 and 6 are also equations of direct proportion (Sec. 3-2). Later chapters introduce other types of functions whose graphs are not straight lines, such as quadratic functions, trigonometric functions, exponential functions, and logarithmic functions.

## *Rectangular Coordinate System*

The rectangular coordinate system and graphs of linear functions will now be studied in more detail. Notice in Fig. 5-1 that each point is identified by two coordinates. The first coordinate is the *x* coordinate, and the second is the *y* coordinate.

### Example 5-3
Draw a rectangular coordinate system, and plot the following points: $A(6, 5)$, $B(-3, 2)$, $C(0, 4)$, $D(4, 0)$, $E(-5, -6)$, $F(-5, 0)$, $G(2, -3)$, and $H(0, -5)$.

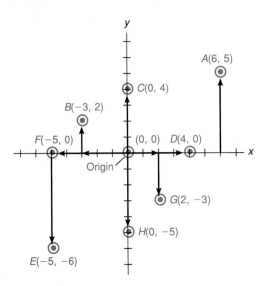

FIG. 5-2  *Rectangular coordinate system, Example 5-3*

## Solution

First draw the $x$ and $y$ axes. This divides the graph into four quadrants (Fig. 5-2). The scales on the $x$ and $y$ axes are left out when each box is 1 unit. To locate each point, measure the distance of the first coordinate along the $x$ axis to the right (+) or left (−) and the move up (+) or down (−) the distance of the second coordinate.

Note that the origin has the coordinates (0, 0). Every point on the $x$ axis has a $y$ coordinate of zero, and every point on the $y$ axis has an $x$ coordinate of zero.

## Example 5-4

Draw the graph of the linear equation $2x - y = -1$.

## Solution

This equation is in *standard form* $Ax + By = C$, where $A$, $B$, and $C$ are constants. To show more clearly that $2x - y = -1$ expresses $y$ as a linear function of $x$, solve for $y$:

$$2x - y = -1$$
$$2x + 1 = y$$
$$y = 2x + 1$$

In this form of the equation $x$ is called the *independent variable,* and $y$ is the *dependent variable*. To graph the equation, you can now assign any value to $x$ and calculate the corresponding $y$ value. Two points are all that is necessary to

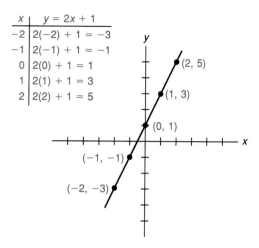

| $x$ | $y = 2x + 1$ |
|---|---|
| $-2$ | $2(-2) + 1 = -3$ |
| $-1$ | $2(-1) + 1 = -1$ |
| $0$ | $2(0) + 1 = 1$ |
| $1$ | $2(1) + 1 = 3$ |
| $2$ | $2(2) + 1 = 5$ |

FIG. 5-3  *Linear function, Example 5-4*

graph a straight line or linear equation. However, if one point is incorrect, the entire line is wrong. You should plot three points or more. Choose simple values for $x$ such as those chosen for the five points in Fig. 5-3.

Example 5-4 can also be done by leaving the equation in standard form, $2x - y = -1$, and graphing the intercepts as shown in the next example.

## Example 5-5
Draw the graph of $2x - y = -1$, using the intercept method.

### Solution
You can assign any values to $x$ or $y$. The values $x = 0$ and $y = 0$ give the points where the line intercepts each axis:

$$
\begin{array}{ll}
y \text{ intercept:} & x \text{ intercept:} \\
\text{Let } x = 0 & \text{Let } y = 0 \\
0 - y = -1 & 2x - 0 = -1 \\
y = 1 & x = -\dfrac{1}{2}
\end{array}
$$

The point $(0, 1)$ is where the line crosses the $y$ axis and is the $y$ intercept. The point $(-\frac{1}{2}, 0)$ is the $x$ intercept. Since the $x$ intercept falls between two units on the graph, estimate its position (Fig. 5-4). Choose one more simple value for $x$ as a checkpoint, such as $x = 1$, and solve for $y$ as done above. Then plot the three points and draw the line.

Linear functions and straight lines are studied further in Chap. 14.

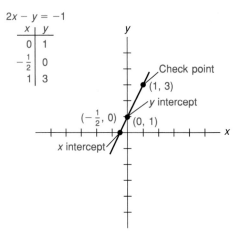

$2x - y = -1$

| x | y |
|---|---|
| 0 | 1 |
| $-\frac{1}{2}$ | 0 |
| 1 | 3 |

FIG. 5-4 *Intercept method, Example 5-5*

**Exercise 5-1**

In problems 1 to 10, plot the points on a rectangular coordinate system:

**1.** $(3, 1)$     **2.** $(1, 4)$     **3.** $(0, 5)$

**4.** $(-5, 0)$     **5.** $(-3, -2)$     **6.** $(-3, 1)$

**7.** $(-5, -5)$     **8.** $(0, 0)$     **9.** $\left(\frac{7}{2}, -\frac{1}{2}\right)$

**10.** $(1.5, -2.5)$

In problems 11 to 18, draw the graphs of the linear functions. First solve for $y$ if the equation is not in that form.

**11.** $y = x$     **12.** $y = 3x$     **13.** $y = 4x - 3$

**14.** $y = \frac{1}{2}x + 1$     **15.** $2x + y - 4 = 0$     **16.** $x + 2y + 2 = 0$

**17.** $x - 4y = 4$     **18.** $5x - 5y = 8$

In problems 19 to 24, draw the graphs, using the intercept method.

**19.** $x - y = 3$     **20.** $x + y = -5$     **21.** $4x + 2y = 8$

**22.** $3x - 4y = 12$     **23.** $3x - 4y = 6$     **24.** $3x + 6y = 2$

**25.** The force $F$ (in pounds) exerted on a spring is a linear function of the distance $x$ the spring stretches: $F = kx$ ($k = $ constant). If $k = 1.5$ lb/ft for a particular spring, draw the graph of $F$ vs. $x$ from $x = 0$ to $x = 6$ ft.

**26.** The resistance $R$ of a circuit is 2.0 $\Omega$. Assuming the resistance is constant, that is, it does not change with a change in the voltage, graph $E$ vs. $I$ from $I = -3.0$ to $I = 3.0$ A. [See Example 5-2(3).]

**27.** Given the linear function $v = v_0 + at$, where $v_0 = 100$ cm/s and $a = 50$ cm/s, graph $v$ vs. $t$ from $t = -3$ to $t = 5$ s. Use different scales for $v$ and $t$.

28. Draw the graph of the velocity of sound $V$ as a linear function of the temperature $t$ from $t = -20$ to $t = 40°C$. Use different scales for $V$ and $t$, beginning $V$ at 310 m/s. [See Example 5-2(4).]

29. Extend the graph in Fig. 5-1, using values of $C < -30$, and find the point where $F = C$.

30. Assume the reaction time required to apply the brakes of an automobile in an emergency is 0.40 s. Graph the distance traveled in feet during that time vs. speed from 20 to 80 mi/h.

📟 31. The resistance $R$ of a wire is a linear function of its length $l$. This relation is given by $R = \rho l/A$, where rho ($\rho$) is a constant called the resistivity and $A$ equals the cross-sectional area. Graph $R$ vs. $l$ for a copper wire when $\rho = 1.75 \times 10^{-6}\ \Omega \cdot m$ and the diameter $d = 10$ mm. Use different scales for $R$ and $l$, and plot values from $l = 1.0$ to $l = 5.0$ m.

📟 32. The length $L$ of a pendulum is a linear function of the square of its period $T^2$. From the following experimental data for $L$ and $T$, calculate $T^2$ and draw the graph of $L$ vs. $T^2$. Plot $T^2$ on the horizontal axis, and use different scales for $L$ and $T^2$:

| $L$ (cm) | 20 | 40 | 60 | 80 | 100 |
|---|---|---|---|---|---|
| $T$ (s) | 0.894 | 1.27 | 1.55 | 1.79 | 2.00 |
| $T^2$ | | | | | |

# 5-2 | *Solution of Two Linear Equations*

Many problems involve two or more unknown quantities that are related by linear equations. This section introduces two equations in two unknowns and the methods for solving them together, or simultaneously.

### Example 5-6
Two forces acting in the same direction on an object exert a total force of 4 lb. When the two forces act in opposite directions, the result is a force of 2 lb (Fig. 5-5). How much is each force?

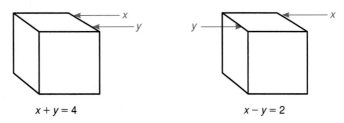

$$x + y = 4 \qquad\qquad x - y = 2$$

FIG. 5-5 *Forces on an object, Example 5-6*

## Solution

Let $x$ = the greater force and $y$ = the lesser force. You can then write the two simultaneous equations:

$$x + y = 4$$
$$x - y = 2$$

These equations are solved for $x$ algebraically by combining (adding) equations and eliminating $y$:

$$
\begin{aligned}
x + \cancel{y} &= 4 \\
\underline{x - \cancel{y} = 2} \\
2x \qquad &= 6 \\
x \qquad &= 3 \text{ lb}
\end{aligned}
$$

Solve for $y$ by substituting $x = 3$ into one of the original equations:

$$(3) + y = 4$$
$$y = 1 \text{ lb}$$

The simultaneous equations above can also be solved *graphically* by drawing the two lines on the same graph and finding the point of intersection as follows.

## Example 5-7

Solve the simultaneous equations $x + y = 4$ and $x - y = 2$ graphically.

## Solution

For each equation find the $x$ and $y$ intercepts and one checkpoint. Then draw both lines on the same graph (Fig. 5-6). The point of intersection, $(3, 1)$, is the solution of the two equations. It represents the values of $x$ and $y$ that satisfy both equations simultaneously.

Graphical methods of solving problems are generally not as precise as algebraic methods. However, graphs provide a better understanding of a problem and a check on the algebra. In many technical and applied problems, graphical methods of solution are more direct and more effective. Sometimes they are the only way to solve a problem.

## Example 5-8

Solve the linear system of two equations algebraically and graphically:

$$2x - y = -5$$
$$3x + 2y = 3$$

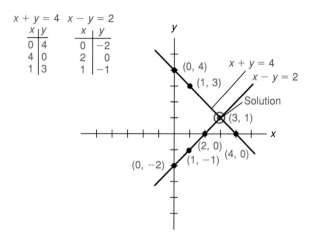

FIG. 5-6 *Simultaneous equations, Example 5-7*

## Solution

*Algebraic Solution.* Solve the system algebraically and then check whether the graph gives the same solution. To eliminate $x$, you need to multiply both equations. You can eliminate $y$ more easily by multiplying the first equation by 2 and combining equations:

$$2[2x - y = -5] \longrightarrow 4x - 2y = -10$$
$$\frac{3x + 2y = \quad 3}{7x \quad = -7}$$
$$x = -1$$

To find $y$, substitute $x = -1$ into the first equation:

$$2(-1) - y = -5$$
$$-y = -3$$
$$y = 3$$

The solution is therefore $(-1, 3)$.

*Graphical Solution.* Find the intercepts and one checkpoint for each equation. Then draw both lines on the same coordinate system (Fig. 5-7).

Simultaneous equations may also be solved algebraically by the *substitution method* as follows.

## Example 5-9

Solve Example 5-8 by substitution.

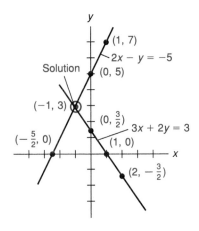

$$2x - y = -5 \qquad 3x + 2y = 3$$

| x | y |
|---|---|
| 0 | 5 |
| $-\frac{5}{2}$ | 0 |
| 1 | 7 |

| x | y |
|---|---|
| 0 | $\frac{3}{2}$ |
| 1 | 0 |
| 2 | $-\frac{3}{2}$ |

FIG. 5-7  *Example 5-8*

## Solution

To solve by substitution, solve one of the equations for one of the variables and substitute this expression into the other equation. The first equation is solved for $y$ and substituted into the second equation:

$$y = \boxed{2x + 5}$$
$$3x + 2y = 3$$
$$3x + 2(2x + 5) = 3$$
$$3x + 4x + 10 = 3$$
$$7x = -7$$
$$x = -1$$

Then find $y$, as described in Example 5-8. The substitution method is helpful when variables cannot be eliminated by combining equations. It is used in Chap. 14 for a linear quadratic system of two equations.

## Example 5-10

Solve the linear system algebraically and graphically to the nearest tenth:

$$5A + 2B = 4$$
$$3A + 5B = 6$$

## Solution

*Algebraic Solution.*  It does not make too much difference whether you eliminate $A$ or $B$. Suppose you choose $A$. Multiply the first equation by 3 and the second equation by $-5$ (you could also use $-3$ and 5). This pair of numbers comes from switching the coefficients of $A$:

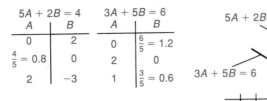

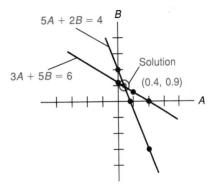

FIG. 5-8 *Example 5-10*

$$3[5A + 2B = 4] \longrightarrow \cancel{15A} + 6B = \phantom{-}12$$
$$-5[3A + 5B = 6] \longrightarrow -\cancel{15A} - 25B = -30$$
$$-19B = -18$$
$$B = \frac{-18}{-19} = 0.94 \approx 0.9$$

Substitute into the first equation to find $A$:

$$5A + 2(0.94) = 4$$
$$5A = 2.12$$
$$A = 0.42 \approx 0.4$$

The solution is therefore (0.4, 0.9), to the nearest tenth. Notice that both answers are first expressed to two decimal places and then rounded.

*Graphical Solution.* Plot $A$ on the $x$ axis and $B$ on the $y$ axis. Find the intercepts and one checkpoint for each equation. When plotting points, estimate decimal values (Fig. 5-8).

Notice that for $5A + 2B = 4$, $A = 1$ is not chosen as a checkpoint. It is too close to the intercept $A = 0.8$ to be a good check. The coordinates of the point of intersection are estimated on the graph. If desired you can make the graphical solution more accurate by using more boxes for each unit.

**Exercise 5-2**

In problems 1 to 12, solve each linear system algebraically and graphically, and show that the answers agree.

1. $2x - y = 4$
   $x + y = 5$

2. $3x + y = 6$
   $x - 2y = 2$

3. $x + 2y = 4$
   $2x + y = 5$

4. $3x - y = 7$
   $x - 2y = 9$

**5.** $3A + 2B = -1$
$6A - 4B = -6$

**6.** $5f - 2g = 20$
$f + 4g = 15$

**7.** $2x - 3y - 3 = 0$
$3x - 2y + 3 = 0$

**8.** $x + 2y - 3 = 0$
$3x + 4y - 5 = 0$

**9.** $3s = 4 - 8t$
$s = 4t$

**10.** $u = 1 - v$
$v = 7u + 7$

**11.** $3a - d = 2.1$
$a + 2d = 1.4$

**12.** $c/2 + m = 3$
$c - m = 0$

In problems 13 to 20, solve the linear systems by any method (or as directed by the instructor).

**13.** $3x - y = 1$
$x + 2y = 12$

**14.** $x - 2y = 7$
$2x + 3y = 0$

**15.** $y = x - 4$
$y = 2x - 7$

**16.** $y = x + 6$
$x = 2 - y$

**17.** $2p - 3q = 4$
$3p - 2q = -2$

**18.** $5s + 2t = -4$
$3s - 5t = 6$

**19.** $\dfrac{1}{2}x + y = 4$

$x - \dfrac{1}{2}y = -5$

**20.** $1.5x - 2.0y = -1.0$
$3.0x + 2.5y = 3.2$

**21.** The following equations result from a problem concerning electric circuits:

$$-0.2I_1 + 0.5I_2 = 0.4$$
$$0.6I_1 - 0.2I_2 = 1.4$$

Solve for the two currents $I_1$ and $I_2$ in amperes.

**22.** The following equations result from a problem involving the forces on a structure:

$$\frac{2}{3}F_x + \frac{1}{3}F_y = 3$$

$$\frac{1}{3}F_x - \frac{2}{3}F_y = 4$$

Solve for the two forces $F_x$ and $F_y$ in newtons.

**23.** Solve graphically and explain why there are *many solutions* for $x$ and $y$:

$$2x + 3y = 6$$
$$4x + 6y = 12$$

This is called a *dependent* system.

**24.** Solve graphically and explain why there is *no solution* for *x* and *y*:

$$2x + 3y = 6$$
$$2x + 3y = 3$$

This is called an *inconsistent* system.

# 5-3 | *Verbal Problems with Applications*

The algebraic solutions of the following verbal problems illustrate the application of the techniques discussed in Secs. 2-8 and 2-9 to two unknowns.

### Example 5-11

Adam bought 160 ft of fencing to enclose the front and two sides of his house within a rectangular fence (Fig. 5-9). He decides the front section should be 3 times as long as each side. What should the dimensions of the rectangle be?

### Solution

Let $l$ = length and $w$ = width. From the first sentence you have the verbal equation:

$$\text{Length} + \text{width} + \text{width} = 160 \text{ ft}$$

One algebraic equation is then:

$$l + w + w = 160$$

FIG. 5-9  *House fence, Example 5-11*

The second sentence gives the verbal equation:

$$\text{Length} = 3(\text{width})$$

The other algebraic equation is then:

$$l = 3w$$

Put both equations in standard form $(Ax + By = C)$:

$$l + 2w = 160$$
$$l - 3w = 0$$

Multiplying the second equation by $-1$ gives the solution:

$$\cancel{l} + 2w = 160$$
$$-1[l - 3w = 0] \longrightarrow \underline{-\cancel{l} + 3w = 0}$$
$$5w = 160$$
$$w = 32 \text{ ft}$$

and
$$l = 3w = 3(32) = 96 \text{ ft}$$

Check by substituting the values for $w$ and $l$ into the first equation:

$$96 + 2(32) = 96 + 64 = 160 \checkmark$$

This example can also be solved by the substitution method.

A problem can often be set up more easily by using two unknowns than by using one. The next example shows the solution of the mixture problem of Example 2-56 using two equations in two unknowns. Compare the two solutions.

### Example 5-12
A 10-qt cooling system contains a solution of 20 percent antifreeze and 80 percent water. How much solution must be replaced with pure antifreeze to raise the strength to 50 percent antifreeze?

### Solution
The problem can be thought of as follows. You want to make 10 qt of a 50 percent solution of antifreeze by mixing pure antifreeze with a 20 percent solution (Fig. 5-10). Let $A$ = quarts of pure antifreeze and $B$ = quarts of 20 percent solution.

The first algebraic equation is then:

$$A + B = 10$$

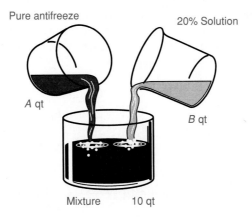

Pure antifreeze

20% Solution

A qt

B qt

Mixture    10 qt

FIG. 5-10  *Mixture problem, Example 5-12*

The verbal equation in terms of antifreeze is:

$$\text{Pure antifreeze} + 20\% \text{ (solution)} = 50\%(10 \text{ qt})$$

The second algebraic equation is then:

$$A + 0.20B = 0.50(10) = 5$$

Multiplying the second equation by $-1$ gives the solution:

$$-1[A + 0.20B = 5] \longrightarrow \begin{array}{r} \cancel{A} + B \phantom{0.20B} = 10 \\ \underline{-\cancel{A} - 0.20B = -5} \\ 0.80B = 5 \\ B = 6.25 \text{ qt} \end{array}$$

and by using the first equation:

$$A = 10 - 6.25 = 3.75 \text{ qt}$$

The next example illustrates an application of the linear function $v = v_0 + at$ to a motion problem. [See Example 5-2(5).]

## Example 5-13

General Motors is conducting tests on a new electric car. Starting from rest, the car accelerates at a constant rate for 10 s. Starting at 20 mi/h, the car accelerates at the same constant rate and takes 6 s to reach the same final velocity. What are the final velocity in miles per hour and the acceleration in miles per hour per second?

## Solution

Let $v$ = final velocity and $a$ = constant acceleration. In the first trial $v_0 = 0$ mi/h, and in the second trial $v_0 = 20$ mi/h. Use of the formula $v = v_0 + at$ gives the two equations:

$$v = 0 + a(10)$$
$$v = 20 + a(6)$$

Writing the equations in standard form and multiplying the first equation by $-1$ yield the solution:

$$-1[v - 10a = 0] \longrightarrow -v + 10a = 0$$
$$\underline{v - 6a = 20}$$
$$4a = 20$$
$$a = 5 \text{ mi/(h/s)}$$

and by using the first equation:     $v = (5)10 = 50$ mi/h

The next example shows an application to electric circuits using Ohm's law, $E = IR$ [see Example 5-2(3)], and Kirchhoff's voltage law (see below).

## Example 5-14

Rachel wants to find the voltage of a battery and a resistance. She connects the battery in series with the unknown resistance and a 2.4-$\Omega$ resistor. An ammeter in the circuit reads 2.5 A (Fig. 5-11). She repeats the experiment, substituting a 10-$\Omega$ resistor for the 2.4-$\Omega$ resistor, and the ammeter reads 1.0 A. What are the battery voltage and the unknown resistance?

## Solution

Let $E$ = battery voltage and $R$ = resistance. Kirchhoff's voltage law states:

The sum of the voltages around any closed loop equals zero.     (5-2)

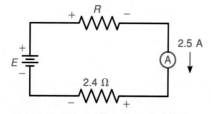

FIG. 5-11 *Kirchhoff's voltage law, Example 5-14*

In going clockwise around the circuit, the verbal equation for both experiments is:

Battery voltage − voltage across $R$ − voltage across resistor = 0

By Ohm's law, $E = IR$, the first equation is:

$$E - (2.5)(R) - (2.5)(2.4) = 0$$
$$E - 2.5R = 6.0$$

The second equation is:

$$E - (1.0)(R) - (1.0)(10) = 0$$
$$E - R = 10$$

Multiplying the first equation by −1 yields the solution:

$$-1[E - 2.5R = 6.0] \longrightarrow -E + 2.5R = -6.0$$
$$\underline{E - R \quad\; = 10}$$
$$1.5R = 4.0$$
$$R = 2.7\ \Omega$$

and by using the second equation:

$$E = 10 + 2.7 = 12.7\ \text{V}$$

**Exercise 5-3**   In problems 1 to 18, solve algebraically by setting up two equations in two unknowns.

1. The total area to be occupied by a building and a parking lot is 4000 m². If the area occupied by the building is to be 500 m² more than that covered by the parking lot, what is the area occupied by each?

2. The total number of fire alarms in New York City on a certain day is 105. If twice the number of false alarms equals 5 times the number of real alarms, how many of each kind were there?

3. Two men push in the same direction on an object with a total force of 180 lb. A woman joins the men and pushes with a force equal to one of them. This results in a total force of 260 lb. What is the force exerted by each person?

4. A copper wire 1 m long is to be cut into two pieces so that the resistance of the shorter piece is one-third the resistance of the longer piece. How long should each piece be, in centimeters? *Note:* Length is proportional to resistance.

5. Pamela bought 35.0 m (115 ft) of fencing to enclose a vegetable garden in her backyard. The garden is to be a rectangle, twice as long as it is wide, with a fence across the middle parallel to the width. What should the length of her garden be in meters and in feet?

6. A powerful computer takes 35 ns (1 ns = $10^{-9}$ s) to perform 5 operations of one type and 7 operations of a second type. It takes 37 ns for the computer to perform 7 operations of the first type and 5 operations of the second type. How many nanoseconds are required for each operation?

7. Solder made of 40 percent tin is to be melted with solder made of 10 percent tin to produce 60 g of solder containing 30 percent tin. How many grams of each solder should be used? (See Example 5-12.)

8. A mixture contains 60 percent sand and 40 percent cement. Some of the mixture is to be combined with pure cement to make 100 lb of a mixture that is 40 percent sand and 60 percent cement. How many pounds of each substance should be combined? (See Example 5-12.)

9. Two skiers are on a smoothly sloped hill, one skier above the other. The lower skier heads straight down the hill from rest while the upper skier pushes off with a velocity of 10 mi/h. The first skier takes 7.0 s to reach a certain velocity; the skier higher up on the hill takes 5.0 s to reach the same velocity. Assuming the acceleration is constant and the same for both skiers, find their acceleration (in miles per hour per second) and the velocity they both reach. (See Example 5-13.)

10. Two identical rockets are traveling toward space. Both must reach the same escape velocity to go into orbit. The first rocket, traveling at 7.0 km/s, fires its booster and reaches escape velocity in 20 s. The second rocket, traveling at 6.5 km/s, fires an identical booster and reaches escape velocity in 30 s. Assuming the acceleration is constant and the same for both rockets, what are the escape velocity and the acceleration for the rockets? (See Example 5-13.)

11. A battery of unknown voltage is connected in series to an unknown resistance and a resistor of 4.0 Ω. An ammeter in the circuit reads 4.0 A. When the 4.0-Ω resistor is replaced by a 2.4-Ω resistor, the ammeter reads 5.0 A. What are the battery voltage and the unknown resistance? (See Example 5-14.)

12. A battery is connected to a resistance of 2.7 Ω. When a 12-V battery and a resistance of 2.4 Ω are added in series to the circuit, the current remains the same. What are the voltage of the original battery and the current? (See Example 5-14.)

13. Bill Chan lives 5.0 mi from his job. One morning he decides to get some exercise by riding his bike part of the way and jogging the rest. He estimates he can average 18 mi/h on the bike and 6.0 mi/h jogging. If he has to get to work in 30 minutes, how long should he ride and how long should he jog?

14. A tug pushes a barge 10 mi upriver in 2.5 h and downriver the same distance in only 1 h. What are the speed of the current and the speed of the tug relative to the water? (See Example 2-62.)

15. Greg and his wife together had an unpaid balance of $500 on their credit cards for 1 month. The company that issued Greg's credit card charges interest of 2 percent a month on the unpaid balance; his wife's charges are 1½ percent a month. If the total finance charge for both was $9.00, how much was each unpaid balance?

16. Police officer Sabrena fires a rifle at a practice target and hears the bullet strike 1.5 s later. The muzzle velocity of the rifle is 1000 m/s (3280 ft/s), and the speed of sound is 332 m/s (1090 ft/s). If both velocities are constant, how long did it take for the bullet to hit the target, and what is the distance of the target in meters and in feet?

☐ **17.** The volume $V_t$ of a metal bar at $t°C$ is given by $V_t = V_0(1 + bt)$, where $V_0$ is the volume at $0°C$ and $b$ is the volume coefficient of expansion. For a silver bar $V_0 = 10.30$ cm$^3$ and $b = 5.831 \times 10^{-5}$. For a tin bar $V_0 = 10.20$ cm$^3$ and $b = 6.889 \times 10^{-5}$. At what temperature will the volume of the bars be the same?

**18.** The total resistance $R$ of a circuit containing two resistances $R_1$ and $R_2$ in parallel is given by $1/R = 1/R_1 + 1/R_2$. In such a circuit $R$ changes from 2.0 to 3.0 $\Omega$ when $R_1$ is doubled. Find $R_1$ and $R_2$. Do not clear fractions in the equations. Treat $1/R_1$ and $1/R_2$ as the unknowns, and note that $1/(2R_1) = (\frac{1}{2})(1/R_1)$.

# 5-4 | *Solution of Three Linear Equations*

The previous section contains many types of problems that lead to two linear equations in two unknowns for their solution. Similar problems involving three or more unknowns lead to three or more linear equations. This section shows the algebraic method of solving three linear equations in three unknowns. It is similar to the method of elimination used for two linear equations but has more steps.

## Example 5-15
Solve the system of three linear equations (I–III):

  **I.** $x + y - 3z = 6$

  **II.** $2x - y + 2z = -2$

**III.** $3x + 2y + 5z = 2$

## Solution
By combining two different pairs of equations and eliminating the *same* variable each time, you reduce the system to two equations in two unknowns. You can eliminate any variable first; however, certain variables are easier to work with because of their coefficients. Choose $y$ for this reason and combine Eqs. (I) and (II):

$$\begin{array}{rl} x + \cancel{y} - 3z &= 6 \\ \underline{2x - \cancel{y} + 2z} &= \underline{-2} \\ 3x \qquad\quad -z &= 4 \end{array}$$

Now eliminate the same variable, $y$, from a different pair of equations. This will give you a second equation in the variables $x$ and $z$. Multiply (II) by 2 and combine with (III):

$$2[2x - y + 2z = -2] \longrightarrow \begin{array}{rl} 4x - \cancel{2y} + 4z &= -4 \\ \underline{3x + \cancel{2y} + 5z} &= \underline{2} \\ 7x \qquad\quad +9z &= -2 \end{array}$$

Next solve the two equations in $x$ and $z$, eliminating $z$:

$$9[3x - z = 4] \longrightarrow 27x - 9z = 36$$
$$\frac{7x + 9z = -2}{34x \qquad = 34}$$
$$x = 1$$

Substitute $x$ into one of the two equations in two unknowns and find $z$:

$$3x - z = 4$$
$$3(1) - z = 4$$
$$z = -1$$

Substitute $x$ and $z$ into any one of the original equations to find $y$. By using Eq. (I):

$$x + y - 3z = 6$$
$$(1) + y - 3(-1) = 6$$
$$y = 2$$

The solution is therefore $x = 1$, $y = 2$, $z = -1$.

To represent three linear equations in three unknowns graphically, you must draw them on a three-dimensional coordinate system. Each point in space has three coordinates $(x, y, z)$, and each equation produces a plane (a flat surface) in space. The point where the three planes from the three equations come together, if it exists, represents the solution to the three equations. In the above example the solution can be written $(1, 2, -1)$, which represents the point in space where the three planes intersect (Fig. 5-12).

The next two examples illustrate important applications of three linear equations in three unknowns. The first example is a mixture problem.

### Example 5-16

Three iron ores A, B, and C are to be mixed to produce 1000 kg (one metric ton) of an ore that contains 60 percent iron, 30 percent other minerals, and 10 percent ash. The following table lists the contents of each ore:

|   | Iron, % | Other, % | Ash, % |
|---|---------|----------|--------|
| A | 70      | 10       | 20     |
| B | 60      | 40       | 0      |
| C | 50      | 40       | 10     |

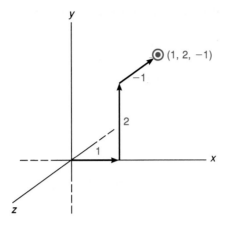

FIG. 5-12 *Three-dimensional coordinate system, Example 5-15*

How many kilograms of each ore should be mixed together?

**Solution**
First write down the unknowns. Let $x$ = kilograms of A, $y$ = kilograms of B, and $z$ = kilograms of C. The first equation expresses the sum of the three amounts. Verbal equation:

Kilograms of A + kilograms of B + kilograms of C = 1000 kg

Algebraic equation:

$$x + y + z = 1000$$

Each of the other two equations must equate one of the components, iron, other, or ash, in each ore. You could choose any two. Ash is easy to work with because there is 0 percent of it in ore B. Suppose you choose iron and ash. You then have
Verbal equation:

Iron in A + iron in B + iron in C = iron in mix

Algebraic equation:

$$0.70x + 0.60y + 0.50z = 0.60(1000) = 600$$

Verbal equation:

Ash in A + ash in B + ash in C = ash in mix

Algebraic equation:

$$0.20x + 0.0y + 0.10z = 0.10(1000) = 100$$

Multiplying the last two equations by 10 eliminates the decimals, giving the system:

**I.** $x + y + z = 1000$

**II.** $7x + 6y + 5z = 6000$

**III.** $2x + z = 1000$

It is only necessary to eliminate $y$ once by using Eqs. (I) and (II) to reduce the system to two equations in two unknowns:

$$-6[x + y + z = 1000] \longrightarrow -6x - 6y - 6z = -6000$$
$$\underline{7x + 6y + 5z = 6000}$$
$$x \qquad\quad -z = 0$$

Combine this equation with Eq. (III):

$$2x + z = 1000$$
$$\underline{x - z = 0}$$
$$3x \quad\;\; = 1000$$
$$x \quad\;\; = 333 \text{ kg} \quad \text{(three figures)}$$

Substitute back to find $y$ and $z$:

$$y = 333 \text{ kg} \qquad z = 333 \text{ kg}$$

The answers do not check exactly in each equation because they have been rounded. Note that even if you do not know the total number of kilograms in the mixture, you can still solve the problem in terms of percentages. Use 100 percent or 100 instead of 1000; use 60 instead of 600, and 10 instead of 100.

## Kirchhoff's Laws

Example 5-17 below illustrates an application of Kirchhoff's laws to the series-parallel circuit in Fig. 5-13. Kirchhoff's law for current states:

$$\text{The sum of the currents into any volume equals zero.} \qquad (5\text{-}3)$$

Applying this law to the junction marked by the circle in Fig. 5-13 produces the equation:

$$I_1 + I_2 - I_3 = 0 \qquad\qquad (5\text{-}4)$$

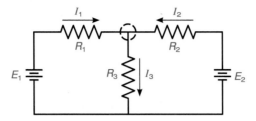

FIG. 5-13  *Kirchhoff's laws, Example 5-17*

Applying Kirchhoff's voltage law (Example 5-14) to each loop produces the equations:

$$E_1 - I_1 R_1 - I_3 R_3 = 0$$
$$E_2 - I_2 R_2 - I_3 R_3 = 0$$

or

$$I_1 R_1 + I_3 R_3 = E_1 \qquad (5\text{-}5)$$
$$I_2 R_2 + I_3 R_3 = E_2$$

When the voltages $E_1$, and $E_2$ and the resistances $R_1$, $R_2$, and $R_3$ are known, these equations represent three linear equations in the three unknowns $I_1$, $I_2$, and $I_3$.

### Example 5-17

Given the series-parallel circuit in Fig. 5-13, with $E_1 = 14.0$ V, $E_2 = 12.6$ V, $R_1 = 6.00\ \Omega$, $R_2 = 5.50\ \Omega$, and $R_3 = 4.00\ \Omega$, find $I_1$, $I_2$, and $I_3$.

### Solution

By substituting into the three equations above, (5-4) and (5-5) you obtain:

$$I_1 + I_2 - I_3 = 0$$
$$6.00 I_1 + 4.00 I_3 = 14.0$$
$$5.50 I_2 + 4.00 I_3 = 12.6$$

The solution of these three equations is $I_1 = 1.05$ A, $I_2 = 0.886$ A, and $I_3 = 1.93$ A. The solution is shown in Sec. 6-1 (Example 6-6) using determinants.

The solution of higher-order linear systems of four or more equations can be done by using the same elimination process for two and three equations. This process is known as *gaussian elimination,* and it can be effectively programmed on a computer.

**Exercise 5-4**   In problems 1 to 12, solve each system of linear equations.

1.  $x + y + z = 6$
    $x - y + 2z = 5$
    $2x - y - z = -3$

2.  $3x + y - z = 2$
    $x - 2y - 2z = 6$
    $4x + y + 2z = 7$

3.  $2x + y - 3z = -2$
    $x + 3y - 2z = -5$
    $3x + 2y - z = 7$

4.  $2x + 3y + z = 4$
    $x + 5y - 2z = -1$
    $3x - 2y + 4z = 3$

5.  $8u + v + w = 1$
    $7u - 2v + 9w = -3$
    $4u - 6v + 8w = -5$

6.  $2a + 3b + 6c = 3$
    $3a + 4b - 5c = 2$
    $4a - 2b - 2c = 1$

7.  $2d + e + 3f = -2$
    $5d + 2e = 5$
    $2e - 3f = -7$

8.  $r = s - 1$
    $s = t + 2$
    $t = 2r - 4$

9.  $2.0x - 3.0y = -1$
    $3.0x + 5.0z = 4.5$
    $x + 6.0y - 10z = -1.5$

10. $0.2x + 0.1y + 0.2z = 0.8$
    $0.4x + 0.2y - 0.4z = 0.4$
    $0.4x - 0.3y + 0.6z = 1.4$

11. $x - y + z + t = 4$
    $x - y + 2z = 1$
    $x + 2y + t = -1$
    $2x - z - t = 1$

12. $4a + b = 3$
    $3b - 2c = 4$
    $a + c + 2d = 2$
    $b + 4c - d = -2$

13. In analyzing the forces on a beam, the following equations result. Solve for $F_x$, $F_y$, and $F_z$:

$$0.20F_x + 0.10F_y = 30$$
$$0.20F_y - 0.10F_z = 25$$
$$0.10F_x + 0.20F_z = 35$$

14. A computer analysis produces the following equations. Solve for $x_1$, $x_2$, and $x_3$:

$$2x_1 - x_2 + 4x_3 = 1.9$$
$$x_1 + 5x_2 - x_3 = -1.6$$
$$x_1 - 2x_2 - x_3 = 0.50$$

In problems 15 to 22, solve by setting up three equations in three unknowns.

15. Stacey wants to determine the volume in liters of three oddly shaped bottles. All she has is a 1-gal (3.8-L) container filled with water. After much experimenting she finds the following. The gallon exactly fills all three containers. Twice the volume of the second container exactly fills the first and the third. Three times the volume of the third exactly fills the first. What are the volumes of the three bottles in gallons and in liters?

16. Three pumps can fill a jet fuel tank that holds 144,000 L in 8 h. When the first and second pumps are operating, it takes 12 h to fill the tank. All three pumps are turned on. However, it is found that the first pump is clogged and is only pumping at one-half its normal rate. With the defective pump, it takes 9 h to fill the tank. What

is the normal pumping rate in liters per hour for each pump? [*Note:* (Rate)(time) = volume pumped.]

17. Three crude oils are to be mixed and loaded aboard a 500,000-ton supertanker. The crudes contain the following percentages of light-, medium-, and heavyweight oils:

|  | Light, % | Medium, % | Heavy, % |
|---|---|---|---|
| I | 0 | 20 | 80 |
| II | 10 | 40 | 50 |
| III | 60 | 0 | 40 |

How much of each crude *in thousands of tons* should be mixed so that the resulting crude contains 30 percent light-, 20 percent medium-, and 50 percent heavyweight oils? (See Example 5-16.)

18. Solve Example 5-16 by using an equation involving the *other* minerals in the ores instead of the iron. Check that you get the same results.

19. Applying Equations (5-4) and (5-5) to the series-parallel circuit in Fig. 5-13, find $I_1$, $I_2$, and $I_3$ when $E_1 = 10$ V, $E_2 = 12$ V, $R_1 = 2.0$ Ω, $R_2 = 6.0$ Ω, and $R_3 = 3.0$ Ω.

20. Do the same as in problem 19 when $E_1 = 30$ V, $E_2 = 24$ V, $R_1 = 2.5$ Ω, $R_2 = 4.0$ Ω, and $R_3 = 4.0$ Ω.

21. A section of the road going through the Wilkerson Pass near Pike's Peak, in Colorado, is 18 km long. A bus going from east to west on this section of the road climbs for 6 min at one speed, levels off for 6 min at a second speed, and descends for 6 min at a third speed. Going from west to east on the same section of the road, the bus climbs for 12 min, levels off for 6 min, and descends for 3 min. A passenger records the time for the first half of the trip going from west to east as 13 min. What are the speeds of the bus climbing, on the level, and descending in kilometers per *hour?*

22. When Kirchhoff's laws are applied to the circuit in Fig. 5-14, the following equations result:

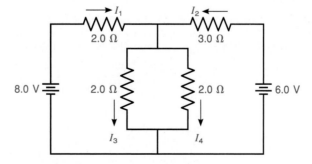

FIG. 5-14 *Kirchhoff's laws, problem 22*

$$I_1 + I_2 - I_3 - I_4 = 0$$
$$2.0I_1 + 2.0I_3 = 8.0$$
$$3.0I_2 + 2.0I_4 = 6.0$$
$$2.0I_3 - 2.0I_4 = 0.0$$

Solve this system for the currents $I_1$, $I_2$, $I_3$, and $I_4$.

## 5-5 | *Review Questions*

In problems 1 to 4, draw the graphs of the linear functions.

**1.** $y = 2x - 3$          **2.** $y = -3x + 1$

**3.** $2x + 4y = 8$          **4.** $x - 3y + 6 = 0$

In problems 5 to 14, solve each linear system.

**5.** $3x - y = -5$          **6.** $x - 2y = 0$
    $x + 4y = 7$              $3x + 6y = 3$

**7.** $2b - y = -0.1$        **8.** $k = n + 2$
    $3b + 2y = 3.0$          $n = 1 - 3k$

**9.** $2a + 3t = 9$          **10.** $2.4h - 3.0x = 2.4$
    $3a - 4t = 5$              $2.0h + 1.5x = 6.8$

**11.** $x - y + 2z = 6$      **12.** $2x + 2y - 3z = 5$
     $2x + y - z = 4$            $4x - 3y + 2z = -4$
     $x + 3y + 3z = 3$          $6x + 5y - 4z = 13$

**13.** $2a - b + c = 1$      **14.** $1.0u + 2.0v + 3.0w = 1.1$
      $7a - 4b = 4$               $1.2u - 1.5v - 2.0w = 2.4$
      $5c + 3 = 5b$              $0.8u + 0.5v + 1.0w = 1.2$

**15.** The "normal" weight of a person is a linear function of height. The weight is given approximately by the formula $W = 0.97H - 100$ when $W$ is in kilograms and $H$ is in centimeters. Graph $W$ vs. $H$ from $H = 150$ cm (4 ft 11 in.) to $H = 200$ cm (6 ft 7 in.).

[C] **16.** The thermoelectric power of copper with respect to lead is a linear function of the temperature $t$ given by the equation $Q = 2.76 + 1.22t$, where $Q$ = power in microvolts per degree Celsius. Graph $Q$ vs. $t$ from $t = -10$ to $t = 10°C$.

**17.** Alcohol weighing 0.80 g/mL is mixed with water weighing 1.0 g/mL to produce 1 L of solution. If the solution weighs 900 g, how many milliliters of alcohol and water are in it? (See Example 5-12.)

**18.** Jack, a jolly car salesman, offers some specials on a new Ford: standard car with stereo and air conditioner, $21,600, and standard car with just stereo, $19,200. If the air conditioner has 5 times the value of the stereo, how much is the standard car alone worth? (*Hint:* Let $x$ = standard car, $y$ = stereo.)

**19.** Two cars accelerate at the same constant rate to the same final velocity. The first car takes 10 s, starting at an initial velocity of 0 mi/h. The second car takes 15 s, starting at an initial velocity of 10 mi/h. What are the acceleration in miles per hour per second and the final velocity of the cars? (See Example 5-13.)

20. A sailboat leaves a harbor at 0600 (6 a.m.) and motors under power with no wind. At 0900 (9 a.m.) the wind picks up, and the skipper shuts off the boat's engine and sails till 1200 (noon), covering 18 mi. At this time the skipper decides to use both power and sail to make more progress. The boat arrives in port at 1800 (6 p.m.) after covering a total of 54 mi. If the speed under power and sail equals the speed under power plus the speed under sail, how fast did the boat travel under power, under sail, and under power and sail?

21. Three alloys A, B, and C contain the following percentages of gold, silver, and copper:

| | Copper, % | Silver, % | Gold, % |
|---|---|---|---|
| A | 60 | 30 | 10 |
| B | 40 | 30 | 30 |
| C | 30 | 70 | 0 |

How many grams of each alloy should be mixed to produce 1 kg of an alloy that is 44 percent copper, 38 percent silver, and 18 percent gold? (See Example 5-16.)

22. Using Eqs. (5-4) and (5-5) for the series-parallel circuit in Fig. 5-13, find $I_1$, $I_2$, and $I_3$ when $E_1 = 6.0$ V, $E_2 = 10$ V, $R_1 = 5.0$ $\Omega$, $R_2 = 3.0$ $\Omega$, and $R_3 = 2.0$ $\Omega$.

# CHAPTER

# 6

# Determinants and Matrices

## 6-1 | *Determinants and Linear Systems*

A *matrix* is a rectangular array of numbers, called *elements,* arranged in rows and columns. A matrix is enclosed in parentheses, and its size is given by: number of rows $\times$ number of columns. Here are some examples. The size is written next to each matrix:

$$(1 \quad 2 \quad 3)_{1 \times 3}$$

$$\begin{pmatrix} 5 & -8 & 1 \\ 3 & 2 & -3 \end{pmatrix}_{2 \times 3}$$

$$\begin{pmatrix} 3 \\ 5 \\ -4 \\ 0 \end{pmatrix}_{4 \times 1}$$

The following matrices have the same number of rows as columns and are called *square matrices:*

$$\begin{pmatrix} -6 & -3 & -2 \\ -5 & 4 & 2 \\ -8 & 3 & 0 \end{pmatrix}_{3 \times 3} \qquad \begin{pmatrix} 2 & 4 \\ -2 & 3 \end{pmatrix}_{2 \times 2}$$

Every *square matrix* has a number attached to it called its *determinant*. Determinants of square matrices provide a key to solving linear systems. The determinant of a square matrix is represented by parallel lines adjacent to the elements as shown below. The determinant of the $2 \times 2$ matrix given above is calculated as follows:

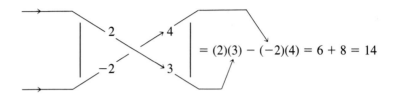

$$= (2)(3) - (-2)(4) = 6 + 8 = 14$$

Note that *the second product has a negative sign in front*.
    The formula for a $2 \times 2$ determinant is then:

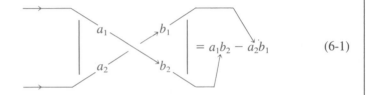

$$= a_1 b_2 - a_2 b_1 \qquad (6\text{-}1)$$

## Example 6-1
Evaluate each $2 \times 2$ determinant:

**1.** $\begin{vmatrix} 3 & 1 \\ 1 & 2 \end{vmatrix} = (3)(2) - (1)(1) = 6 - 1 = 5$

**2.** $\begin{vmatrix} 8 & -5 \\ -3 & 0 \end{vmatrix} = (8)(0) - (-3)(-5) = 0 - 15 = -15$

**3.** $\begin{vmatrix} \frac{1}{2} & -\frac{1}{4} \\ \frac{1}{4} & \frac{1}{4} \end{vmatrix} = (\frac{1}{2})(\frac{1}{4}) - (\frac{1}{4})(-\frac{1}{4}) = \frac{1}{8} + \frac{1}{16} = \frac{3}{16}$

## *Solution of Two Linear Equations*

The solution of two linear equations using determinants is shown in the following example. The general formula for any two equations is then given.

## Example 6-2
Solve the linear system in Example 5-8, using determinants:

$$2x - y = -5$$
$$3x + 2y = 3$$

## Solution

From Example 5-8 you know that the solution is $(-1, 3)$. The method of combining equations to eliminate one of the unknowns is equivalent to the following solution for $x$ and $y$ with the use of determinants. Study the solution carefully:

Constants

$$x = \frac{\begin{vmatrix} -5 & -1 \\ 3 & 2 \end{vmatrix}}{\begin{vmatrix} 2 & -1 \\ 3 & 2 \end{vmatrix}} = \frac{(-5)(2) - (3)(-1)}{(2)(2) - (3)(-1)} = \frac{-10 + 3}{4 + 3} = \frac{-7}{7} = -1$$

Coefficient matrix

Constants

$$y = \frac{\begin{vmatrix} 2 & -5 \\ 3 & 3 \end{vmatrix}}{\begin{vmatrix} 2 & -1 \\ 3 & 2 \end{vmatrix}} = \frac{(2)(3) - (3)(-5)}{(2)(2) - (3)(-1)} = \frac{6 + 15}{4 + 3} = \frac{21}{7} = 3$$

Coefficient matrix

Look closely at the above expressions. Here is how the determinants are set up to solve two linear equations.

The *denominator* for $x$ and $y$ is the *same*. It is called the *coefficient matrix* and is made up of the four coefficients of the unknowns arranged exactly as they appear in the equations in standard form. The *numerator for $x$* is obtained from the coefficient matrix by *replacing the column of $x$ coefficients by the constants* as they appear on the right side of the equations in standard form. The *numerator for $y$* is obtained from the coefficient matrix by *replacing the column of $y$ coefficients by the constants*. These procedures are expressed algebraically by *Cramer's rule* (no relation to the author):

Given a linear system in *standard form:*

$$a_1x + b_1y = k_1$$

$$k_1, \ k_2 = \text{constants}$$

$$a_2x + b_2y = k_2$$

the solutions for $x$ and $y$ are:

$$x = \frac{\begin{vmatrix} k_1 & b_1 \\ k_2 & b_2 \end{vmatrix}}{\begin{vmatrix} a_1 & b_1 \\ a_2 & b_2 \end{vmatrix}} \qquad y = \frac{\begin{vmatrix} a_1 & k_1 \\ a_2 & k_2 \end{vmatrix}}{\begin{vmatrix} a_1 & b_1 \\ a_2 & b_2 \end{vmatrix}} \qquad (6\text{-}2)$$

Note how the constants $k_1$ and $k_2$ replace the coefficients in each numerator. Remember that the equations *must be written in standard form* when the determinants are constructed.

### Example 6-3

The resistance of 1000 ft of no. 14 gauge copper wire at 20°C is 2.6 $\Omega$; at 75°C it is 3.1 $\Omega$. The resistance $R$ is a linear function of the temperature $t$ given by $R = R_0 + at$, where $R_0$ is the resistance at 0°C and $a$ is a constant. From the given data find $R_0$ and $a$ for the wire.

### Solution

You can write two equations, substituting the given data into the formula:

$$2.6 = R_0 + a(20)$$
$$3.1 = R_0 + a(75)$$

Write the equations in standard form and solve, using Cramer's rule:

$$20a + R_0 = 2.6$$
$$75a + R_0 = 3.1$$

$$a = \frac{\begin{vmatrix} 2.6 & 1 \\ 3.1 & 1 \end{vmatrix}}{\begin{vmatrix} 20 & 1 \\ 75 & 1 \end{vmatrix}} = \frac{(2.6)(1) - (3.1)(1)}{(20)(1) - (75)(1)} = \frac{-0.50}{-55} = 0.0091$$

$$R = \frac{\begin{vmatrix} 20 & 2.6 \\ 75 & 3.1 \end{vmatrix}}{\begin{vmatrix} 20 & 1 \\ 75 & 1 \end{vmatrix}} = \frac{(20)(3.1) - (75)(2.6)}{(20)(1) - (75)(1)} = \frac{-133}{-55} = 2.4 \ \Omega$$

## *Solution of Three Linear Equations*

Three linear equations can be solved with $3 \times 3$ determinants by using a procedure similar to the one for solving two linear equations. The value of a $3 \times 3$ determinant can be expanded in terms of the *elements in the first row and $2 \times 2$ determinants* as follows:

$$\begin{vmatrix} a_1 & b_1 & c_1 \\ a_2 & b_2 & c_2 \\ a_3 & b_3 & c_3 \end{vmatrix} = a_1 \begin{vmatrix} b_2 & c_2 \\ b_3 & c_3 \end{vmatrix} - b_1 \begin{vmatrix} a_2 & c_2 \\ a_3 & c_3 \end{vmatrix} + c_1 \begin{vmatrix} a_2 & b_2 \\ a_3 & b_3 \end{vmatrix} \qquad (6\text{-}3)$$

Each of the 2 × 2 determinants is called the *minor* of the element it is multiplied by. A minor of an element is the determinant formed by all the elements *not* in the same row and *not* in the same column as the given element. Look at the three 2 × 2 minors above. If you cross out the elements in the same row and column, first for $a_1$, then $b_1$, and then $c_1$, you form each of the three minors:

$$\begin{vmatrix} a_1 & b_1 & c_1 \\ a_2 & b_2 & c_2 \\ a_3 & b_3 & c_3 \end{vmatrix} \qquad \begin{vmatrix} a_1 & b_1 & c_1 \\ a_2 & b_2 & c_2 \\ a_3 & b_3 & c_3 \end{vmatrix} \qquad \begin{vmatrix} a_1 & b_1 & c_1 \\ a_2 & b_2 & c_2 \\ a_3 & b_3 & c_3 \end{vmatrix}$$

$$a_1 \text{ minor} \qquad\qquad b_1 \text{ minor} \qquad\qquad c_1 \text{ minor}$$

### Example 6-4

Evaluate the 3 × 3 determinant, applying formula (6-3), using the elements in the first row and expanding by minors:

$$\begin{vmatrix} 3 & 1 & -2 \\ 4 & 2 & 1 \\ 2 & 3 & -2 \end{vmatrix} = 3 \begin{vmatrix} 2 & 1 \\ 3 & -2 \end{vmatrix} - 1 \begin{vmatrix} 4 & 1 \\ 2 & -2 \end{vmatrix} + (-2) \begin{vmatrix} 4 & 2 \\ 2 & 3 \end{vmatrix}$$

$$= 3[(2)(-2) - (3)(1)] - 1[(4)(-2) - (2)(1)] - 2[(4)(3) - (2)(2)]$$

$$= 3(-4 - 3) - (-8 - 2) - 2(12 - 4) = 3(-7) + 10 - 2(8)$$

$$= -21 + 10 - 16 = -27$$

Observe that the sign of the second minor in the expansion is *negative* when the elements in the first row are used. A 3 × 3 determinant can be expanded in terms of the elements in any row (or column) and their respective minors. The sign of each minor depends on the location of its element. Only the first row expansion (6-3) is presented here.

Cramer's rule also applies to the solution of three equations in three unknowns. Given a linear system of three equations in *standard form:*

$$a_1 x + b_1 y + c_1 z = k_1$$
$$a_2 x + b_2 y + c_2 z = k_2 \qquad k_1, k_2, k_3 = \text{constants}$$
$$a_3 x + b_3 y + c_3 z = k_3$$

the solutions for $x$, $y$, and $z$ are:

$$x = \frac{\begin{vmatrix} k_1 & b_1 & c_1 \\ k_2 & b_2 & c_2 \\ k_3 & b_3 & c_3 \end{vmatrix}}{\begin{vmatrix} a_1 & b_1 & c_1 \\ a_2 & b_2 & c_2 \\ a_3 & b_3 & c_3 \end{vmatrix}} \qquad y = \frac{\begin{vmatrix} a_1 & k_1 & c_1 \\ a_2 & k_2 & c_2 \\ a_3 & k_3 & c_3 \end{vmatrix}}{\begin{vmatrix} a_1 & b_1 & c_1 \\ a_2 & b_2 & c_2 \\ a_3 & b_3 & c_3 \end{vmatrix}} \qquad z = \frac{\begin{vmatrix} a_1 & b_1 & k_1 \\ a_2 & b_2 & k_2 \\ a_3 & b_3 & k_3 \end{vmatrix}}{\begin{vmatrix} a_1 & b_1 & c_1 \\ a_2 & b_2 & c_2 \\ a_3 & b_3 & c_3 \end{vmatrix}} \qquad (6\text{-}4)$$

The denominator for $x$, $y$, and $z$ is again the coefficient matrix. The numerator for each unknown is the coefficient matrix with the coefficients of the unknown replaced by the constants.

### Example 6-5
Using determinants, solve the linear system:

$$2x + 3y + z = 5$$
$$3x - 2y + 4z = -3$$
$$5x + y + 2z = -1$$

### Solution
Apply Cramer's rule (6-4) to obtain the solutions for $x$, $y$, and $z$:

$$x = \frac{\begin{vmatrix} 5 & 3 & 1 \\ -3 & -2 & 4 \\ -1 & 1 & 2 \end{vmatrix}}{\begin{vmatrix} 2 & 3 & 1 \\ 3 & -2 & 4 \\ 5 & 1 & 2 \end{vmatrix}}$$

$$= \frac{5\begin{vmatrix} -2 & 4 \\ 1 & 2 \end{vmatrix} - 3\begin{vmatrix} -3 & 4 \\ -1 & 2 \end{vmatrix} + 1\begin{vmatrix} -3 & -2 \\ -1 & 1 \end{vmatrix}}{2\begin{vmatrix} -2 & 4 \\ 1 & 2 \end{vmatrix} - 3\begin{vmatrix} 3 & 4 \\ 5 & 2 \end{vmatrix} + 1\begin{vmatrix} 3 & -2 \\ 5 & 1 \end{vmatrix}}$$

$$= \frac{5(-4 - 4) - 3(-6 + 4) + (-3 - 2)}{2(-4 - 4) - 3(6 - 20) + (3 + 10)} = \frac{-40 + 6 - 5}{-16 + 42 + 13}$$

$$= \frac{-39}{39} = -1$$

Once you have the value of the denominator, you do not have to evaluate it again for $y$ and $z$:

$$y = \frac{\begin{vmatrix} 2 & 5 & 1 \\ 3 & -3 & 4 \\ 5 & -1 & 2 \end{vmatrix}}{39} = \frac{2\begin{vmatrix} -3 & 4 \\ -1 & 2 \end{vmatrix} - 5\begin{vmatrix} 3 & 4 \\ 5 & 2 \end{vmatrix} + 1\begin{vmatrix} 3 & -3 \\ 5 & -1 \end{vmatrix}}{39}$$

$$= \frac{2(-6+4) - 5(6-20) + (-3+15)}{39} = \frac{78}{39} = 2$$

$$z = \frac{\begin{vmatrix} 2 & 3 & 5 \\ 3 & -2 & -3 \\ 5 & 1 & -1 \end{vmatrix}}{39} = \frac{2\begin{vmatrix} -2 & -3 \\ 1 & -1 \end{vmatrix} - 3\begin{vmatrix} 3 & -3 \\ 5 & -1 \end{vmatrix} + 5\begin{vmatrix} 3 & -2 \\ 5 & 1 \end{vmatrix}}{39}$$

$$= \frac{2(2+3) - 3(-3+15) + 5(3+10)}{39} = \frac{39}{39} = 1$$

The solution is thus $(-1, 2, 1)$. It can be checked by substitution into the original equations.

### Example 6-6
Solve Example 5-17 (Kirchhoff's laws), using determinants.

### Solution
The equations in standard form are:

$$1.00I_1 + 1.00I_2 - 1.00I_3 = 0$$
$$6.00I_1 \qquad\qquad + 4.00I_3 = 14.0$$
$$5.50I_2 + 4.00I_3 = 12.6$$

Apply Cramer's rule (6-4) to obtain the solution for $I_1$:

$$I_1 = \frac{\begin{vmatrix} 0 & 1.00 & -1.00 \\ 14.0 & 0 & 4.00 \\ 12.6 & 5.50 & 4.00 \end{vmatrix}}{\begin{vmatrix} 1.00 & 1.00 & -1.00 \\ 6.00 & 0 & 4.00 \\ 0 & 5.50 & 4.00 \end{vmatrix}} = \frac{0(-22.0) - 1(56 - 50.4) + (-1)(77)}{1(-22.0) - 1(24) + (-1)(33)}$$

$$= \frac{-82.6}{-79} = 1.05 \text{ A}$$

Similarly:

$$I_2 = \frac{-70}{-79} = 0.886 \text{ A} \qquad I_3 = \frac{-152.6}{-79} = 1.93 \text{ A}$$

You can see from the preceding examples how the solution of linear systems with the use of determinants can be readily programmed on a computer. For larger linear systems, matrix solutions are more desirable. These are shown in the following sections.

**Exercise 6-1**

In problems 1 to 6, evaluate each determinant.

1. $\begin{vmatrix} 2 & 3 \\ 1 & 4 \end{vmatrix}$ 

2. $\begin{vmatrix} -1 & 5 \\ 3 & 1 \end{vmatrix}$

3. $\begin{vmatrix} 0 & 7 \\ -3 & 5 \end{vmatrix}$ 

4. $\begin{vmatrix} \frac{1}{2} & -\frac{1}{2} \\ -\frac{1}{2} & -\frac{1}{2} \end{vmatrix}$

5. $\begin{vmatrix} 3.0 & 2.5 \\ 1.0 & -1.5 \end{vmatrix}$ 

6. $\begin{vmatrix} 0.4 & 0.2 \\ 0.1 & 0.3 \end{vmatrix}$

In problems 7 to 12, evaluate each determinant. (See Example 6-4.)

7. $\begin{vmatrix} 2 & 1 & -2 \\ 3 & 2 & -1 \\ 1 & 1 & 3 \end{vmatrix}$ 

8. $\begin{vmatrix} 1 & -2 & 4 \\ 3 & 2 & -1 \\ 4 & -1 & -2 \end{vmatrix}$

9. $\begin{vmatrix} 4 & 0 & -3 \\ -2 & 1 & 0 \\ 5 & -3 & 1 \end{vmatrix}$ 

10. $\begin{vmatrix} -2 & 1 & 0 \\ 0 & 8 & -3 \\ 4 & -6 & 5 \end{vmatrix}$

11. $\begin{vmatrix} 1.0 & 2.0 & 0.8 \\ 0.5 & -1.0 & 3.0 \\ 2.0 & -2.0 & 0.4 \end{vmatrix}$ 

12. $\begin{vmatrix} 2.3 & 1.9 & 4.3 \\ 0.0 & 3.8 & 5.2 \\ 1.1 & 4.4 & 0.0 \end{vmatrix}$

In problems 13 to 20, solve the indicated problem from Exercise 5-2, using determinants.

13. Problem 1          14. Problem 2

15. Problem 5          16. Problem 6

17. Problem 7          18. Problem 10

19. Problem 11         20. Problem 20

In problems 21 to 28, solve the indicated problem from Exercise 5-4, using determinants.

21. Problem 1          22. Problem 2

23. Problem 5          24. Problem 6

25. Problem 7          26. Problem 8

27. Problem 9          28. Problem 10

In problems 29 to 32, solve each linear system using determinants.

**29.** $3x - 5y = 5$
    $4x + 2y = 10$

**30.** $\dfrac{1}{5}x - \dfrac{1}{2}y = \dfrac{3}{2}$
    $\dfrac{1}{4}x + \dfrac{2}{5}y = \dfrac{1}{3}$

**31.** $2.0h + 0.2r = 1.4$
    $2.0h + 0.7r = -3.2$

**32.** $0.01e_1 + 0.03e_2 = 0.13$
    $0.02e_1 - 0.01e_2 = 0.08$

In problems 33 to 36, solve each linear system by using determinants.

**33.** $3x + \phantom{2}y - 2z = 7$
    $2x + 3y - \phantom{2}z = 21$
    $\phantom{2}x + 2y - 3z = 4$

**34.** $2x + 3y + 4z = 14$
    $3x + 2y - 3z = -14$
    $5x + 5y + \phantom{3}z = 40$

**35.** $0.2i + 0.1j + 0.2k = 0.06$
    $0.1i - 0.3j - 0.2k = 0.01$
    $0.1i - 0.2j - 0.1k = 0.02$

**36.** $3.0F_a + 3.0F_c = 2.0F_b$
    $2.0F_a \phantom{+ 3.0F_c} = 1.0F_b$
    $1.0F_a + 1.0F_b = 1.1 - 1.0F_c$

In problems 37 to 48, solve each applied problem by using determinants.

**37.** It is found that the values of loads $W_1$ and $W_2$, in metric tons (1 t = 1000 kg), at two supports of a bridge satisfy the following equations:

$$1.58W_1 - 0.732W_2 = 1.30$$
$$0.643W_1 + 0.766W_2 = 8.40$$

Find $W_1$ and $W_2$.

**38.** The resistance of 1000 ft of no. 21 aluminum wire at 50°C is 23.3 Ω; at 100°C it is 27.1 Ω. Using the linear function from Example 6-3, $R = R_0 + at$, find $R_0$ and $a$ for this wire.

**39.** Solve problem 16, Sec. 5-3, using determinants.

**40.** Solve problem 17, Sec. 5-3, using determinants.

**41.** Explain what happens when you try to solve the dependent system of problem 23, Sec. 5-2, using determinants.

**42.** Explain what happens when you try to solve the inconsistent system of problem 24, Sec. 5-2.

**43.** A total of $13,000 is invested in three money market funds. At the end of 1 year the first fund yields 9.20 percent interest, the second 8.40 percent, and the third 10.26 percent. The total income from the three investments is $1232. If the income from the third fund equals the income of the other two funds combined, how much was invested in each fund?

**44.** Solve problem 13, Sec. 5-4.

**45.** Max the jeweler wants to make an 18-carat gold watch containing 75 percent gold, 15 percent silver, and 10 percent copper out of the following three gold alloys:

| | Gold, % | Silver, % | Copper, % |
|---|---|---|---|
| 14-carat | 60 | 25 | 15 |
| 20-carat | 85 | 10 | 5 |
| Gold coinage | 90 | 0 | 10 |

What percentage of each alloy should be used? (Represent the total amount by 100. See Example 5-16.)

**46.** Solve problem 17, Sec. 5-4.

**47.** Using Eqs. (5-4) and (5-5) for the series-parallel circuit in Fig. 5-13, find $I_1$, $I_2$, and $I_3$ when $E_1 = 9.0$ V, $E_2 = 6.0$ V, $R_1 = 4.5 \ \Omega$, $R_2 = 4.0 \ \Omega$, and $R_3 = 3.0 \ \Omega$.

**48.** Solve problem 20, Sec. 5-4.

# 6-2 | *Matrices and Matrix Addition*

As shown at the beginning of Sec. 6-1, a matrix is an ordered rectangular array of numbers, called elements. Its size is given by $m \times n$, where $m$ = number of rows and $n$ = number of columns. The elements are indicated in general by letters with two subscripts, as shown in the $2 \times 3$ matrix:

$$\begin{pmatrix} a_{11} & a_{12} & a_{13} \\ a_{21} & a_{22} & a_{23} \end{pmatrix}$$

The first subscript is the row, and the second subscript is the column. For example, $a_{23}$ is the element in the second row and the third column. *Two matrices A and B are equal, A = B, if they have the same number of rows and the same number of columns and if corresponding elements are equal.*

Two matrices *can be added only if they have the same number of rows and the same number of columns* as follows:

To *add two matrices A and B*, add the corresponding elements

in each row or column to produce $A + B$. (6-5)

Study the next example, which shows how to add two matrices.

### Example 6-7
Add each pair of matrices.

**1.**
$$\begin{pmatrix} 2 & 4 \\ 3 & 1 \\ -5 & 2 \end{pmatrix} + \begin{pmatrix} 0 & 8 \\ -1 & 7 \\ 3 & 8 \end{pmatrix} = \begin{pmatrix} 2+0 & 4+8 \\ 3-1 & 1+7 \\ -5+3 & 2+8 \end{pmatrix} = \begin{pmatrix} 2 & 12 \\ 2 & 8 \\ -2 & 10 \end{pmatrix}$$

**2.** $\begin{pmatrix} 3 & 2 \\ 1 & -3 \end{pmatrix} + \begin{pmatrix} 5 & 1 & 3 & -2 \\ 7 & -2 & 0 & 5 \end{pmatrix}$

The matrices in (2) cannot be added because they are not the same size. The first matrix is a $2 \times 2$ square matrix, and the second is a $2 \times 4$ matrix.

A matrix may be multiplied by a number $k$, called a *scalar*, as follows:

To *multiply a matrix A by a scalar k,* multiply each element by
the scalar to produce $kA$.                                                   (6-6)

### Example 6-8
Multiply the given matrix by the scalar.

**1.** $2\begin{pmatrix} 3 & 8 & 0 \\ 1 & -2 & 6 \end{pmatrix} = \begin{pmatrix} 2(3) & 2(8) & 2(0) \\ 2(1) & 2(-2) & 2(6) \end{pmatrix} = \begin{pmatrix} 6 & 16 & 0 \\ 2 & -4 & 12 \end{pmatrix}$

**2.** $-1\begin{pmatrix} 8 & 6 \\ 2 & -3 \end{pmatrix} = \begin{pmatrix} -1(8) & -1(6) \\ -1(2) & -1(-3) \end{pmatrix} = \begin{pmatrix} -8 & -6 \\ -2 & 3 \end{pmatrix}$

In matrix (2), note that multiplication by $-1$ changes the sign of every element in a matrix. That is, matrix $(-1)A = -A$ is the *negative* of matrix $A$. Also, as in algebra, $A + (-B) = A - B$ for two matrices $A$ and $B$. The *zero matrix,* denoted 0, is a matrix where all elements are equal to zero.

As a result of the preceding definitions [(6-5) and (6-6)], the following algebraic laws hold for matrices $A$ and $B$ and a scalar $k$:

$(A + B) + C = A + (B + C)$      associative law                           (6-7)

$A + B \qquad = B + A$             commutative law                            (6-8)

$k(A + B) \qquad = kA + kB$                                                   (6-9)

$A + (-A) = 0 \qquad$ and $\qquad A + 0 = A$                                  (6-10)

These laws are also valid for real numbers. Matrix multiplication, however, does not obey the multiplication laws for real numbers, as is shown in Sec. 6-3.

### Exercise 6-2

**1.** State the row and column that each of the matrix elements are in: (a) $a_{32}$, (b) $a_{41}$, (c) $b_{55}$, (d) $b_{ij}$.

**2.** Fill in the missing subscripts of the $4 \times 4$ square matrix:

$$\begin{pmatrix} a & a & a_{13} & a \\ a & a & a & a \\ a & a & a & a \\ a & a_{24} & a & a \end{pmatrix}$$

In problems 3 to 6, add each pair of matrices.

**3.** $\begin{pmatrix} 3 & 8 & 2 \\ -6 & 7 & 1 \\ 2 & 9 & 7 \end{pmatrix} + \begin{pmatrix} 2 & -1 & 5 \\ 0 & 2 & 9 \\ 1 & 3 & 4 \end{pmatrix}$

**4.** $\begin{pmatrix} 3 & 8 \\ -7 & 6 \\ 5 & -2 \end{pmatrix} + \begin{pmatrix} 3 & 9 \\ 7 & -3 \\ 2 & 0 \end{pmatrix}$

**5.** $\begin{pmatrix} \frac{1}{2} & \frac{1}{4} \\ -1 & -\frac{1}{2} \end{pmatrix} + \begin{pmatrix} \frac{1}{4} & \frac{1}{4} \\ 1 & -\frac{1}{2} \end{pmatrix}$

**6.** $\begin{pmatrix} 1.0 & 2.0 & 1.5 \\ -0.5 & -1.0 & 2.5 \end{pmatrix} + \begin{pmatrix} -2.0 & 0.0 & 0.5 \\ 1.0 & 0.5 & -1.5 \end{pmatrix}$

In problems 7 to 12, perform the indicated operations on the given matrices.

$$A = \begin{pmatrix} 3 & 1 & -5 \\ 0 & 6 & 10 \end{pmatrix} \qquad B = \begin{pmatrix} 2 & 9 & 7 \\ -3 & 0 & -1 \end{pmatrix} \qquad C = \begin{pmatrix} 3 & 7 \\ 9 & -2 \end{pmatrix}$$

**7.** $2A + B$   **8.** $B + 3A$

**9.** $A - B$   **10.** $B - 2A$

**11.** $A + C$   **12.** $C + B$

**13.** For matrices $A$ and $B$ above, show that (6-9) is true when $k = 2$.

**14.** Compute:

$$3\begin{pmatrix} 5 & -1 \\ 8 & 2 \end{pmatrix} - 1\begin{pmatrix} 3 & 8 \\ 2 & 0 \end{pmatrix} + 4\begin{pmatrix} -2 & 1 \\ 1 & -1 \end{pmatrix}$$

# 6-3 | *Matrix Multiplication and Inverse*

## *Matrix Multiplication*

Matrix multiplication is defined in a way which lends to the solution of linear systems. Two matrices can be multiplied *only if the number of columns of the first equals the number of rows of the second*. Study the following example, which illustrates the multiplication algorithm before it is formally stated.

### Example 6-9
Find the matrix product:

$$AB = \begin{pmatrix} 3 & 2 \\ 1 & -1 \end{pmatrix} \begin{pmatrix} 1 & 3 & 4 \\ -2 & 1 & 3 \end{pmatrix}$$

## Solution

Observe that the first matrix $A$ has two columns and the second matrix $B$ has two rows. Therefore, since the number of columns of the first equals the number of rows of the second, the matrices can be multiplied. Calculate the element in the *first row* and *first column* of the product as follows: Multiply the elements in the *first row* of $A$ by the corresponding elements in the *first column* of $B$ and add the products:

$$3(1) + 2(-2) = 3 - 4 = -1$$

Calculate the element in the *first row* and *second column* of the product as follows. Multiply the elements in the *first row* of $A$ by the corresponding elements in the *second column* of $B$ and add the products:

$$3(3) + 2(1) = 9 + 2 = 11$$

In general, calculate the element in the *ith row* and *jth column* of the product as follows. Multiply the elements in the *ith row* of $A$ by the corresponding elements in the *jth column* of $B$ and add the products. The product $AB$ is then:

$$\begin{pmatrix} 3 & 2 \\ 1 & -1 \end{pmatrix} \begin{pmatrix} 1 & 3 & 4 \\ -2 & 1 & 3 \end{pmatrix} =$$

$$\begin{pmatrix} 3(1) + 2(-2) & 3(3) + 2(1) & 3(4) + 2(3) \\ 1(1) + (-1)(-2) & 1(3) + (-1)(1) & 1(4) + (-1)(3) \end{pmatrix} = \begin{pmatrix} -1 & 11 & 18 \\ 3 & 2 & 1 \end{pmatrix}$$

Note that the product matrix has two rows and three columns. The product matrix will have the same number of rows as the first matrix and the same number of columns as the second matrix.

In general, given two matrices $A$ and $B$, such that the number of columns of $A$ is equal to $n$, and this equals the number of rows of $B$, the element $c_{ij}$ in the *ith row* and the *jth column* of the *product matrix* $AB = C$ is calculated as follows:

$$a_{i1}b_{1j} + a_{i2}b_{2j} + a_{i3}b_{3j} + \cdots + a_{in}b_{nj} = c_{ij} \tag{6-11}$$

Multiply the elements in the *ith row* of $A$ by the corresponding elements in the *jth column* of $B$ and add the products.

Observe that *matrix multiplication is not commutative,* that is, $AB \neq BA$. For instance, in Example 6-9 the product $BA$ cannot be calculated because the number of columns of $B$ (three) does not equal the number of rows of $A$ (two).

## Example 6-10

Find the matrix product:

$$AB = \begin{pmatrix} 3 & 0 & -2 \\ 1 & 3 & 5 \\ 2 & -2 & 4 \end{pmatrix} \begin{pmatrix} 5 & 0 \\ 1 & -1 \\ 4 & 3 \end{pmatrix}$$

**Solution**

Apply formula (6-11):

$$\begin{pmatrix} 3 & 0 & -2 \\ 1 & 3 & 5 \\ 2 & -2 & 4 \end{pmatrix} \begin{pmatrix} 5 & 0 \\ 1 & -1 \\ 4 & 3 \end{pmatrix} =$$

$$\begin{pmatrix} 3(5) + 0(1) + (-2)(4) & 3(0) + 0(-1) + (-2)(3) \\ 1(5) + 3(1) + 5(4) & 1(0) + 3(-1) + 5(3) \\ 2(5) + (-2)(1) + 4(4) & 2(0) + (-2)(-1) + 4(3) \end{pmatrix} = \begin{pmatrix} 7 & -6 \\ 28 & 12 \\ 24 & 14 \end{pmatrix}$$

Exercise 6-3, problems 23 and 24, shows the relationship between matrix multiplication and linear systems; this relationship is studied more thoroughly in Sec. 6-4.

## Inverse of a Matrix

Note that the first matrix in Example 6-10 is a $3 \times 3$ square matrix. Square matrices are important in the solution of linear systems. Every $n \times n$ square matrix $A$ has an $n \times n$ *identity matrix* $I$ associated with it. The $I$ matrix has all 1s on the main diagonal, upper left to lower right, and 0s elsewhere. For example, the $3 \times 3$ identity matrix is:

$$I = \begin{pmatrix} 1 & 0 & 0 \\ 0 & 1 & 0 \\ 0 & 0 & 1 \end{pmatrix}$$

When a square matrix $A$ is multiplied by the identity matrix $I$, the result is the original matrix $A$. For example, consider:

$$AI = \begin{pmatrix} 3 & 0 & 2 \\ 1 & 3 & 5 \\ 2 & -2 & 4 \end{pmatrix} \begin{pmatrix} 1 & 0 & 0 \\ 0 & 1 & 0 \\ 0 & 0 & 1 \end{pmatrix} = \begin{pmatrix} 3+0+0 & 0+0+0 & 0+0+2 \\ 1+0+0 & 0+3+0 & 0+0+5 \\ 2+0+0 & 0-2+0 & 0+0+4 \end{pmatrix}$$

$$= \begin{pmatrix} 3 & 0 & 2 \\ 1 & 3 & 5 \\ 2 & -2 & 4 \end{pmatrix}$$

For a square matrix $A$, the following is always true:

$$AI = IA = A \qquad (6\text{-}12)$$

If the product of two square matrices equals the identity matrix, one matrix is the *inverse* of the other matrix:

$$AA^{-1} = A^{-1}A = I \tag{6-13}$$

where $A^{-1}$ is the inverse of $A$. There are several methods for finding the inverse of a matrix. The method shown below is similar to the gaussian elimination method for solving a linear system (Chap. 5) and can be readily programmed on a computer. The following two processes, which derive from gaussian elimination, can be used on a matrix to find its inverse:

A row may be multiplied by any number except zero. (6-14*a*)

A row may be multiplied by any number except zero and

added to another row to replace either row. (6-14*b*)

Study the next example carefully. It shows how to find the inverse of a matrix.

### Example 6-11

Find the inverse of the matrix:

$$A = \begin{pmatrix} 2 & 1 \\ 3 & -2 \end{pmatrix}$$

### Solution

The method for finding the inverse of a matrix can be considered in terms of the matrix equation $AA^{-1} = I$. By applying rules (6-14*a*) and (6-14*b*) repeatedly to matrix $A$ on the left side of the equation to transform it to the identity matrix, and simultaneously performing the same steps to the matrix $I$ on the right side, you will obtain the equation $IA^{-1} = A^{-1}$. This transforms the identity matrix $I$ to $A^{-1}$. Begin by placing the given matrix $A$ alongside the identity matrix $I$:

$$\overset{A}{\begin{pmatrix} 2 & 1 \\ 3 & -2 \end{pmatrix}} \,\bigg|\, \overset{I}{\begin{pmatrix} 1 & 0 \\ 0 & 1 \end{pmatrix}}$$

Transform matrix $A$ to $I$ by starting with the *first column* and changing the 2 to 1. Apply (6-14*a*) and multiply row 1 by ½. Do the same to $I$ simultaneously:

$$\begin{pmatrix} (\tfrac{1}{2})2 & (\tfrac{1}{2})1 \\ 3 & -2 \end{pmatrix} \,\bigg|\, \begin{pmatrix} (\tfrac{1}{2})1 & (\tfrac{1}{2})0 \\ 0 & 1 \end{pmatrix} \longrightarrow \begin{pmatrix} 1 & \tfrac{1}{2} \\ 3 & -2 \end{pmatrix} \,\bigg|\, \begin{pmatrix} \tfrac{1}{2} & 0 \\ 0 & 1 \end{pmatrix}$$

Then change 3 to 0 by applying (6-14*b*). Multiply row 1 by $-3$, and add it to row 2 to replace row 2. Do the same to $I$:

$$\begin{pmatrix} 1 & \frac{1}{2} \\ 3+(-3)1 & -2+(-3)\frac{1}{2} \end{pmatrix} \Bigg| \begin{pmatrix} \frac{1}{2} & 0 \\ 0+(-3)\frac{1}{2} & 1+(-3)0 \end{pmatrix} \longrightarrow \begin{pmatrix} 1 & \frac{1}{2} \\ 0 & -\frac{7}{2} \end{pmatrix} \Bigg| \begin{pmatrix} \frac{1}{2} & 0 \\ -\frac{3}{2} & 1 \end{pmatrix}$$

Next work on column 2 and first change $-\frac{7}{2}$ to 1. Multiply the second row by $-\frac{2}{7}$:

$$\begin{pmatrix} 1 & \frac{1}{2} \\ (-\frac{2}{7})0 & (-\frac{2}{7})(-\frac{7}{2}) \end{pmatrix} \Bigg| \begin{pmatrix} \frac{1}{2} & 0 \\ (-\frac{2}{7})(-\frac{3}{2}) & (-\frac{2}{7})1 \end{pmatrix} \longrightarrow \begin{pmatrix} 1 & \frac{1}{2} \\ 0 & 1 \end{pmatrix} \Bigg| \begin{pmatrix} \frac{1}{2} & 0 \\ \frac{3}{7} & -\frac{2}{7} \end{pmatrix}$$

Finally change $\frac{1}{2}$ to zero. Multiply row 2 by $-\frac{1}{2}$ and add row 1 to replace row 1:

$$\begin{pmatrix} 1+(-\frac{1}{2})0 & \frac{1}{2}+(-\frac{1}{2})1 \\ 0 & 1 \end{pmatrix} \Bigg| \begin{pmatrix} \frac{1}{2}+(-\frac{1}{2})\frac{3}{7} & 0+(-\frac{1}{2})(-\frac{2}{7}) \\ \frac{3}{7} & -\frac{2}{7} \end{pmatrix} \longrightarrow \begin{pmatrix} 1 & 0 \\ 0 & 1 \end{pmatrix} \Bigg| \begin{pmatrix} \frac{2}{7} & \frac{1}{7} \\ \frac{3}{7} & -\frac{2}{7} \end{pmatrix}$$

The required inverse matrix is then given by:

$$A^{-1} = \begin{pmatrix} \frac{2}{7} & \frac{1}{7} \\ \frac{3}{7} & -\frac{2}{7} \end{pmatrix}$$

To show that this is correct, multiply $AA^{-1}$ to obtain $I$:

$$AA^{-1} = \begin{pmatrix} 2 & 1 \\ 3 & -2 \end{pmatrix} \begin{pmatrix} \frac{2}{7} & \frac{1}{7} \\ \frac{3}{7} & -\frac{2}{7} \end{pmatrix} = \begin{pmatrix} 2(\frac{2}{7})+1(\frac{3}{7}) & 2(\frac{1}{7})+1(-\frac{2}{7}) \\ 3(\frac{2}{7})-2(\frac{3}{7}) & 3(\frac{1}{7})-2(-\frac{2}{7}) \end{pmatrix} = \begin{pmatrix} 1 & 0 \\ 0 & 1 \end{pmatrix}$$

The sequence of steps shown above is important. You should work from left to right, obtaining the 1 in each column first and then the 0s. Study the next example, performing each calculation yourself and checking the result.

### Example 6-12
Find the inverse of the matrix:

$$A = \begin{pmatrix} 1 & 3 & 1 \\ -2 & -5 & 0 \\ 2 & 4 & 0 \end{pmatrix}$$

### Solution
Work from left to right, obtaining the 1 in each column first (if necessary) and then the 0s. The steps are as follows, starting with column 1:

$$\begin{array}{cc} A & I \end{array}$$

Start with column 1: $\begin{pmatrix} 1 & 3 & 1 \\ -2 & -5 & 0 \\ 2 & 4 & 0 \end{pmatrix} \Bigg| \begin{pmatrix} 1 & 0 & 0 \\ 0 & 1 & 0 \\ 0 & 0 & 1 \end{pmatrix}$

Multiply row 1 by 2 and add to row 2 $\longrightarrow$ $\begin{pmatrix} 1 & 3 & 1 \\ 0 & 1 & 2 \\ 2 & 4 & 0 \end{pmatrix} \left| \begin{pmatrix} 1 & 0 & 0 \\ 2 & 1 & 0 \\ 0 & 0 & 1 \end{pmatrix} \right.$

Multiply row 1 by $-2$ and add to row 3 $\longrightarrow$ $\begin{pmatrix} 1 & 3 & 1 \\ 0 & 1 & 2 \\ 0 & -2 & -2 \end{pmatrix} \left| \begin{pmatrix} 1 & 0 & 0 \\ 2 & 1 & 0 \\ -2 & 0 & 1 \end{pmatrix} \right.$

Now work on column 2:

Multiply row 2 by $-3$ and add to row 1 $\longrightarrow$ $\begin{pmatrix} 1 & 0 & -5 \\ 0 & 1 & 2 \\ 0 & -2 & -2 \end{pmatrix} \left| \begin{pmatrix} -5 & -3 & 0 \\ 2 & 1 & 0 \\ -2 & 0 & 1 \end{pmatrix} \right.$

Multiply row 2 by 2 and add to row 3 $\longrightarrow$ $\begin{pmatrix} 1 & 0 & -5 \\ 0 & 1 & 2 \\ 0 & 0 & 2 \end{pmatrix} \left| \begin{pmatrix} -5 & -3 & 0 \\ 2 & 1 & 0 \\ 2 & 2 & 1 \end{pmatrix} \right.$

Finally, work on column 3:

Multiply row 3 by $\frac{1}{2}$ $\longrightarrow$ $\begin{pmatrix} 1 & 0 & -5 \\ 0 & 1 & 2 \\ 0 & 0 & 1 \end{pmatrix} \left| \begin{pmatrix} -5 & -3 & 0 \\ 2 & 1 & 0 \\ 1 & 1 & \frac{1}{2} \end{pmatrix} \right.$

Multiply row 3 by $-2$ and add to row 2 $\longrightarrow$ $\begin{pmatrix} 1 & 0 & -5 \\ 0 & 1 & 0 \\ 0 & 0 & 1 \end{pmatrix} \left| \begin{pmatrix} -5 & -3 & 0 \\ 0 & -1 & -1 \\ 1 & 1 & \frac{1}{2} \end{pmatrix} \right.$

Multiply row 3 by 5 and add to row 1 $\longrightarrow$ $\begin{pmatrix} 1 & 0 & 0 \\ 0 & 1 & 0 \\ 0 & 0 & 1 \end{pmatrix} \left| \begin{pmatrix} 0 & 2 & \frac{5}{2} \\ 0 & -1 & -1 \\ 1 & 1 & \frac{1}{2} \end{pmatrix} \right.$

The required inverse matrix is then:

$$A^{-1} = \begin{pmatrix} 0 & 2 & \frac{5}{2} \\ 0 & -1 & -1 \\ 1 & 1 & \frac{1}{2} \end{pmatrix}$$

The check is left as problem 10 in Exercise 6-3.

**Exercise 6-3**    In problems 1 to 8, find the product of the two matrices.

1. $\begin{pmatrix} 1 & -1 \\ 0 & 2 \end{pmatrix} \begin{pmatrix} 2 \\ 1 \end{pmatrix}$

2. $(3 \quad 2 \quad 1) \begin{pmatrix} 3 & 1 \\ 4 & 0 \\ -1 & 2 \end{pmatrix}$

**3.** $\begin{pmatrix} 8 & 1 \\ 5 & 3 \\ 1 & 2 \end{pmatrix} \begin{pmatrix} 3 & -2 \\ 4 & 5 \end{pmatrix}$

**4.** $\begin{pmatrix} 3 & 1 \\ 2 & 5 \end{pmatrix} \begin{pmatrix} 4 & -1 \\ -1 & 0 \end{pmatrix}$

**5.** $\begin{pmatrix} 2 & 1 \\ 2 & -1 \end{pmatrix} \begin{pmatrix} 1 & 2 & -3 \\ 4 & 0 & 1 \end{pmatrix}$

**6.** $\begin{pmatrix} 3 & 5 & -2 \\ 1 & 0 & 1 \end{pmatrix} \begin{pmatrix} 1 & 2 \\ 1 & 4 \\ 3 & 2 \end{pmatrix}$

**7.** $\begin{pmatrix} 1 & -1 & 4 \\ 3 & 0 & 2 \\ 2 & 2 & 1 \end{pmatrix} \begin{pmatrix} 3 & 1 & 4 \\ 3 & 2 & 1 \\ 0 & 0 & 3 \end{pmatrix}$

**8.** $\begin{pmatrix} 0 & 1 & -2 & 2 \\ -1 & 0 & 3 & 1 \end{pmatrix} \begin{pmatrix} 1 & 3 & -1 & 2 \\ 6 & 3 & 1 & 4 \\ 0 & 2 & 2 & 1 \\ 1 & 4 & 1 & -4 \end{pmatrix}$

**9.** In problem 6, show that the product is not commutative by reversing the matrices and comparing the products.

**10.** In Example 6-12, check that the solution is correct by multiplying $AA^{-1}$ and obtaining $I$.

In problems 11 to 20, find the inverse of the given matrix and check that it is correct by multiplication. (See Examples 6-11 and 6-12.)

**11.** $\begin{pmatrix} 1 & 3 \\ 1 & 2 \end{pmatrix}$

**12.** $\begin{pmatrix} -1 & -2 \\ 3 & 5 \end{pmatrix}$

**13.** $\begin{pmatrix} 2 & 6 \\ 0 & -2 \end{pmatrix}$

**14.** $\begin{pmatrix} 1 & -1 \\ 3 & -2 \end{pmatrix}$

**15.** $\begin{pmatrix} 3 & 1 \\ 9 & 4 \end{pmatrix}$

**16.** $\begin{pmatrix} 2 & 3 \\ 4 & 5 \end{pmatrix}$

**17.** $\begin{pmatrix} 3 & -4 & -1 \\ -4 & 5 & 2 \\ 2 & -3 & -1 \end{pmatrix}$

**18.** $\begin{pmatrix} 1 & 2 & -1 \\ 3 & 7 & -5 \\ -1 & -2 & 0 \end{pmatrix}$

**19.** $\begin{pmatrix} 2 & 3 & 1 \\ -2 & -5 & -1 \\ 0 & 4 & 2 \end{pmatrix}$

**20.** $\begin{pmatrix} 1 & 3 & -1 \\ 2 & 4 & 2 \\ 0 & 3 & -2 \end{pmatrix}$

**21.** The BMI company produces three types of computers, X, Y, and Z. The number sold per week, production cost per unit, and income per unit are as follows:

|             | X      | Y      | Z      |
|-------------|--------|--------|--------|
| Number/week | 20     | 10     | 10     |
| Cost/unit   | $1000  | $2000  | $3000  |
| Income/unit | $2000  | $4000  | $5000  |

When the number produced is represented by the matrix $A = (20 \quad 10 \quad 10)$.

and the cost and price by the matrix:

$$B = \begin{pmatrix} 1000 & 2000 \\ 2000 & 4000 \\ 3000 & 5000 \end{pmatrix}$$

then the product matrix $AB$ compares the total cost and total income per week in a $1 \times 2$ matrix. Compute the product matrix $AB$.

22. A telecommunications company produces two types of computer phone modules, I and II, in two factories, A and B. The production hours per unit and the number of units produced per week in each factory are as follows:

| A | I | II |
|---|---|---|
| Hours/unit | 20 | 15 |
| Number of units/week | 19 | 36 |

| B | I | II |
|---|---|---|
| Hours/unit | 25 | 12 |
| Number of units/week | 15 | 45 |

The total number of production hours per week for factory A is given by the matrix product:

$$(20 \quad 15) \begin{pmatrix} 19 \\ 36 \end{pmatrix}$$

Compute the total production hours per week for factories A and B, and determine which is less.

23. The linear system:

$$2x + 3y = 8$$
$$3x + 4y = 11$$

can be represented by the matrix equation:

$$\begin{pmatrix} 2 & 3 \\ 3 & 4 \end{pmatrix} \begin{pmatrix} x \\ y \end{pmatrix} = \begin{pmatrix} 8 \\ 11 \end{pmatrix}$$

Find the values of $x$ and $y$ by solving the system, and then check the matrix equation by computing the matrix product.

24. Do as in problem 23 for the linear system:

$$x - 2y + z = 7$$
$$2x + y - z = 3$$
$$3x + y - 2z = 4$$

and the matrix equation:

$$\begin{pmatrix} 1 & -2 & 1 \\ 2 & 1 & -1 \\ 3 & 1 & -2 \end{pmatrix} \begin{pmatrix} x \\ y \\ z \end{pmatrix} = \begin{pmatrix} 7 \\ 3 \\ 4 \end{pmatrix}$$

## 6-4 | *Matrix Applications*

### *Computer Variables*

As mentioned at the beginning of the chapter, matrices are used in an increasing number of computer applications. A variable in a computer program that does not have subscripts is a *scalar variable* such as X, NAME, RF, and so on.

A *subscripted variable,* such as X(2,3), SSNO(5), SALARY(28), and so on, is considered part of an array where the number of subscripts is the *dimension* of the array. A one-dimensional array is called a *linear* or *vector array,* and a two-dimensional array is called a *matrix array.* In a matrix array, the first subscript is the row, and the second is the column, the same as in a matrix. The following example illustrates the use of arrays in a computer program.

### Example 6-13

An electronic circuit board contains 10 identical integrated circuits (ICs), each containing an array of 25 components. To identify each component in a computer design program, a linear array is used for the circuits: CIRCUIT (K), where $K$ takes on values from 1 to 10, and a matrix array is used for the components: X(I,J), where $I$ and $J$ take on values from 1 to 5.

### *Linear Systems*

A linear system:

$$\begin{aligned} a_{11}x_1 + a_{12}x_2 &= c_1 \\ a_{21}x_1 + a_{22}x_2 &= c_2 \end{aligned} \tag{6-15}$$

can be written as the matrix equation:

$$\begin{pmatrix} a_{11} & a_{12} \\ a_{21} & a_{22} \end{pmatrix} \begin{pmatrix} x_1 \\ x_2 \end{pmatrix} = \begin{pmatrix} c_1 \\ c_2 \end{pmatrix} \tag{6-16}$$

or simply:

$$AX = C \tag{6-17}$$

where:

$$A = \begin{pmatrix} a_{11} & a_{12} \\ a_{21} & a_{22} \end{pmatrix} \quad X = \begin{pmatrix} x_1 \\ x_2 \end{pmatrix} \quad \text{and} \quad C = \begin{pmatrix} c_1 \\ c_2 \end{pmatrix}$$

The matrix $A$ is called the *coefficient matrix* of the system (6-15). Matrix $X$ is the *solution matrix,* and matrix $C$ is the *constant matrix.* In a computer one stores a combination of matrices $A$ and $C$ called the *augmented matrix* of the system, which is given by:

$$\begin{pmatrix} a_{11} & a_{12} & c_1 \\ a_{21} & a_{22} & c_2 \end{pmatrix} \tag{6-18}$$

Note that the system is completely determined by its augmented matrix. If you multiply both sides of matrix equation (6-17) by the inverse of $A$, you obtain:

$$(A^{-1})AX = (A^{-1})C$$
$$IX = A^{-1}C$$
or
$$X = A^{-1}C \tag{6-19}$$

since $(A^{-1})A = I$ and $IX = X$. Equation (6-19) enables you to solve a linear system by multiplying the constant matrix by the inverse of the coefficient matrix. Study the next two examples, which illustrate this procedure.

### Example 6-14
Solve the linear system, using matrices:

$$x_1 - x_2 = 3$$
$$2x_1 + 3x_2 = 1$$

### Solution
The coefficient matrix is:

$$A = \begin{pmatrix} 1 & -1 \\ 2 & 3 \end{pmatrix}$$

Use the method shown in Example 6-11 to find $A^{-1}$:

$$\begin{pmatrix} 1 & -1 \\ 2 & 3 \end{pmatrix} \; \middle| \; \begin{pmatrix} 1 & 0 \\ 0 & 1 \end{pmatrix}$$

Since $a_{11} = 1$ in the coefficient matrix, the first step is to change $a_{21} = 2$ to zero:

Multiply row 1 by $-2$ and add to row 2 $\longrightarrow \begin{pmatrix} 1 & -1 \\ 0 & 5 \end{pmatrix} \bigg| \begin{pmatrix} 1 & 0 \\ -2 & 1 \end{pmatrix}$

Multiply row 2 by $\frac{1}{5}$ $\qquad \longrightarrow \begin{pmatrix} 1 & -1 \\ 0 & 1 \end{pmatrix} \bigg| \begin{pmatrix} 1 & 0 \\ -\frac{2}{5} & \frac{1}{5} \end{pmatrix}$

Add row 2 to row 1 $\qquad \longrightarrow \begin{pmatrix} 1 & 0 \\ 0 & 1 \end{pmatrix} \bigg| \begin{pmatrix} \frac{3}{5} & \frac{1}{5} \\ -\frac{2}{5} & \frac{1}{5} \end{pmatrix}$

Then $\qquad\qquad\qquad\qquad A^{-1} = \begin{pmatrix} \frac{3}{5} & \frac{1}{5} \\ -\frac{2}{5} & \frac{1}{5} \end{pmatrix}$

Apply Eq. (6-19) to find the solution matrix:

$$X = A^{-1}C = \begin{pmatrix} \frac{3}{5} & \frac{1}{5} \\ -\frac{2}{5} & \frac{1}{5} \end{pmatrix} \begin{pmatrix} 3 \\ 1 \end{pmatrix} = \begin{pmatrix} \frac{3}{5}(3) + \frac{1}{5}(1) \\ -\frac{2}{5}(3) + \frac{1}{5}(1) \end{pmatrix} = \begin{pmatrix} 2 \\ -1 \end{pmatrix}$$

Therefore, $x_1 = 2$ and $x_2 = -1$.

### Example 6-15
Solve the linear system, using matrices:

$$\begin{aligned} 3x - 4y - z &= 5 \\ -4x + 5y + 2z &= -8 \\ 2x - 3y - z &= 4 \end{aligned}$$

### Solution
The procedure shown in Example 6-14 and Eq. (6-19) can be extended to solve any linear system. The coefficient matrix is:

$$A = \begin{pmatrix} 3 & -4 & -1 \\ -4 & 5 & 2 \\ 2 & -3 & -1 \end{pmatrix}$$

It is necessary to find $A^{-1}$. Apply the method shown in Example 6-12 (the steps are omitted for clarity):

$$A^{-1} = \begin{pmatrix} 1 & -1 & -3 \\ 0 & -1 & -2 \\ 2 & 1 & -1 \end{pmatrix}$$

The solution matrix is then given by:

$$X = A^{-1}C = \begin{pmatrix} 1 & -1 & -3 \\ 0 & -1 & -2 \\ 2 & 1 & -1 \end{pmatrix} \begin{pmatrix} 5 \\ -8 \\ 4 \end{pmatrix}$$

$$= \begin{pmatrix} 1(5) - 1(-8) - 3(4) \\ 0(5) - 1(-8) - 2(4) \\ 2(5) + 1(-8) - 1(4) \end{pmatrix} = \begin{pmatrix} 1 \\ 0 \\ -2 \end{pmatrix}$$

The solutions are thus $x = 1$, $y = 0$, and $z = -2$. These solutions can be easily checked in the original system.

The matrix methods of solution described above may seem tedious and require much computation. However, the procedure lends to an algorithmic approach and hence can be readily programmed on a computer. The number of calculations is not a factor since the computer does them so rapidly. The method is what is important since it consists of clearly defined steps, which is necessary for computer processing.

**Exercise 6-4**    In problems 1 to 10, solve each linear system, using matrices.

1. $3x_1 + x_2 = 1$
   $2x_1 + x_2 = 3$

2. $2x_1 + 4x_2 = 12$
   $x_1 + 3x_2 = 8$

3. $x_1 - x_2 = 4$
   $3x_1 - 2x_2 = 11$

4. $5x_1 + 2x_2 = 6$
   $3x_1 + x_2 = 3$

5. $2x + y = 2$
   $4x - 3y = -1$

6. $2x + 3y = 5$
   $4x + 5y = 8$

7. $x + 2y - z = 2$
   $3x + 7y - 5z = 6$
   $x + 2y = 3$

8. $x + 3y - 2z = 2$
   $2x - 7y - 3z = 16$
   $x - y - 3z = 7$

9. $x_1 + 2x_2 + x_3 = 8$
   $x_1 + 3x_2 + x_3 = 10$
   $2x_1 + x_2 - x_3 = 1$

10. $x_1 + 3x_2 + x_3 = 5$
    $2x_1 + 5x_2 + x_3 = 7$
    $4x_2 + x_3 = 9$

11. Given the series-parallel circuit in Example 5-17 (Fig. 5-13), where $E_1 = 16$ V, $E_2 = 16$ V, $R_1 = 2.0$ $\Omega$, $R_2 = 2.0$ $\Omega$, and $R_3 = 3.0$ $\Omega$, solve the linear system for $I_1$, $I_2$, and $I_3$, using matrices.

12. Application of Kirchhoff's laws to the circuit in Fig. 6-1 results in the following equations for the three currents $I_1$, $I_2$, and $I_3$:

$$I_1 - I_2 + I_3 = 0$$
$$I_1 + I_2 = 5$$
$$I_2 + 2I_3 = 5$$

Solve this linear system, using matrices.

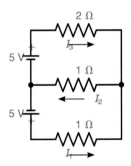

FIG. 6-1 *Series-parallel circuit, Exercise 6-4, problem 12*

# 6-5 | *Review Questions*

In problems 1 to 4, evaluate each determinant.

**1.** $\begin{vmatrix} 1 & 3 \\ 2 & 4 \end{vmatrix}$
**2.** $\begin{vmatrix} -3 & 1 \\ 5 & 6 \end{vmatrix}$

**3.** $\begin{vmatrix} 1 & 0 & 3 \\ 5 & 1 & -2 \\ 2 & 6 & -1 \end{vmatrix}$
**4.** $\begin{vmatrix} 2 & 2 & -3 \\ 4 & 0 & 2 \\ 6 & 5 & -4 \end{vmatrix}$

In problems 5 to 14, perform the indicated operations for the given matrices.

$$A = \begin{pmatrix} 3 & 1 \\ -2 & -3 \end{pmatrix} \qquad B = \begin{pmatrix} 3 & 0 \\ 2 & -2 \end{pmatrix}$$

$$C = \begin{pmatrix} 1 & 4 & 1 \\ 2 & -1 & 0 \end{pmatrix} \qquad D = \begin{pmatrix} 1 & 1 & -1 \\ 0 & -2 & 2 \\ 2 & 0 & 1 \end{pmatrix}$$

**5.** $A + B$
**6.** $A + B + C$

**7.** $3A - B$
**8.** $AB$

**9.** $CD$
**10.** $BA + A$

**11.** $B^{-1}$
**12.** $D^{-1}$

**13.** $CD^{-1}$
**14.** $B^{-1}C$

In problems 15 to 20, solve the linear systems, using *(a)* determinants and *(b)* matrices.

**15.** $x - y = 1$
$2x + y = 11$

**16.** $2x - 5y = 5$
$x + 3y = 8$

**17.** $2x_1 + 3x_2 = 10$
$3x_1 - 2x_2 = 2$

**18.** $2x_1 - 3x_2 = 3$
$4x_1 - 5x_2 = 4$

**19.**  $x + y + 2z = 9$
$\quad\ 2x + y - z = 2$
$\quad\ \ x + y - 3z = -6$

**20.**  $x_1 - x_2 - x_3 = 1$
$\quad\ x_1 - \quad 3x_3 = 4$
$\quad\ x_1 + 2x_2 \quad = 3$

Solve each applied problem, using determinants or matrices.

**21.** An impact printer prints 1 page using regular type and 3 pages using bold type in 23 s. If the printer prints 2 pages using regular type and 4 pages using bold type in 34 s, how long does it take to print 1 page for each typeface?

**22.** Two planes accelerate at the same rate to the same takeoff velocity. The first plane takes 30 s, starting at an initial velocity of 50 mi/h. The second plane takes 25 s, starting at an initial velocity of 75 mi/h. What are the acceleration in miles per hour per second and the takeoff velocity of the planes? (See Example 5-13.)

**23.** Application of Kirchhoff's laws to the series-parallel circuit in Fig. 6-2 results in the following equations for the three currents $I_1$, $I_2$, and $I_3$:

$$I_1 + I_2 - I_3 = 0$$
$$I_1 - I_2 = 0$$
$$I_2 + 2I_3 = 5$$

Solve this system for $I_1$, $I_2$, and $I_3$.

**24.** A sailboat leaves a harbor at 0600 (6 a.m.) and motors under power with no wind. At 0800 (8 a.m.) the wind picks up, and the skipper shuts off the boat's engine and sails until 1200 (noon), covering 22 mi. At this time the skipper decides to use both power and sail to make more progress. The boat arrives in port at 1800 (6 p.m.) after covering a total of 64 mi. If the speed under power and sail equals the speed under power plus the speed under sail, how fast did the boat travel under power, under sail, and under both power and sail?

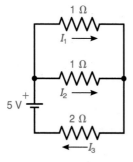

FIG. 6-2  *Series-parallel circuit, problem 23*

# C H A P T E R

# 7

# Factoring, Fractions, and the Quadratic Equation

## 7-1 | *Monomial Factors*

The key word in this chapter is *factor*. To factor means to separate a number into divisors. Factoring is the reverse of multiplication, as follows.

**T**his chapter builds on the algebra foundation introduced in Chap. 2. Simplification of polynomials and operations with fractions are studied first. This will enable you to handle more involved algebraic expressions, to solve equations with fractions, and to solve second-degree or quadratic equations. These topics are therefore studied in the second half of this chapter. Quadratic equations are very common in scientific and technical applications, and many applications of these equations are shown in this chapter.

### Example 7-1
Factor each number into the smallest divisors.

**1.** $6 = 2 \cdot 3$
**2.** $8 = 2 \cdot 4 = 2 \cdot 2 \cdot 2$ or $2^3$
**3.** $18 = 2 \cdot 9 = 2 \cdot 3 \cdot 3$ or $2 \cdot 3^2$
**4.** $28 = 2 \cdot 2 \cdot 7$ or $2^2 \cdot 7$
**5.** $33 = 3 \cdot 11$
**6.** $210 = 2 \cdot 3 \cdot 5 \cdot 7$

Observe that the divisors 2, 3, 5, 7, 11, . . . have no factors, except the number itself and 1. Such a number is called a *prime*. To factor completely means to separate a number into its prime factors. Study the next example.

### Example 7-2
Show the prime factors in each monomial and tell how many each has.

| Monomial | Prime Factors | Number of Prime Factors |
|---|---|---|
| $6x^2y$ | $2 \cdot 3 \cdot x \cdot x \cdot y$ | 5 |
| $-12pqr$ | $-2 \cdot 2 \cdot 3 \cdot p \cdot q \cdot r$ | 6 |
| $21a^2b^3$ | $3 \cdot 7 \cdot a \cdot a \cdot b \cdot b \cdot b$ | 7 |
| $40I^2R_2$ | $2 \cdot 2 \cdot 2 \cdot 5 \cdot I \cdot I \cdot R_2$ | 7 |
| $-51\pi\theta$ | $-3 \cdot 17 \cdot \pi \cdot \theta$ | 4 |

Prime factors allow you to simplify expressions and perform algebraic operations more easily. They are important in combining fractions (Sec. 7-5).

When you factor a polynomial, the first step is to *separate any common monomial factors from each term*. If a polynomial has no factors, it is called prime also.

### Example 7-3
Separate the *highest* common factor from each polynomial.

**1.** $2xy + 6x = 2x(y + 3)$

**2.** $30rst + 42qrs = 6rs(5t + 7q)$

**3.** $1.2a^2b - 1.5ab^2 = 0.3ab(4a - 5b)$

**4.** $I^2R_1 + I^2R_2 = I^2(R_1 + R_2)$

**5.** $2\pi r + 2\pi rh - 2\pi r^2 = 2\pi r(1 + h - r)$

Notice in (2) that you could also factor the expression as follows: $30rst + 42qrs = 3rs(10t + 14q)$. However, the factor $10t + 14q$ is not prime; it contains a common factor of 2. Notice in (5) that you must place a 1 within the parentheses if the entire term is the common factor. When you multiply back the common factor, you should get the original expression. Example 7-3 is the reverse procedure of the multiplication shown in Example 2-30. Separating common factors in polynomials is important in solving literal equations, as follows.

### Example 7-4
Solve the literal equation for $I$:

$$IR_1 + IR_2 = E$$

### Solution
The variable $I$ occurs in two terms, so you need to factor out $I$ first:

$$I(R_1 + R_2) = E$$

Then divide by the other factor to obtain the solution:

$$\frac{I(\cancel{R_1 + R_2})}{(\cancel{R_1 + R_2})} = \frac{E}{R_1 + R_2}$$

**Exercise 7-1** In problems 1 to 6, show all the prime factors for each term.

**1.** 36            **2.** 150

**3.** $98xy^2$            **4.** $-68a^2b^3$

**5.** $-39F_x F_y$         **6.** $80I^2R_1$

In problems 7 to 22, factor each expression as completely as possible by separating the highest common factor.

**7.** $6y + 3y^2$           **8.** $42uv - 28vw$

**9.** $10ab - 20b^2$       **10.** $100pqr + 200qrs$

**11.** $0.6kT_1 + 0.2kT_2$     **12.** $3.3I^2R_1 + 9.9I^2R_2$

**13.** $10x^2 + 20x - 15$      **14.** $2am^2 - 4am + 6a$

**15.** $x^3yz - x^2yz + xyz$    **16.** $\pi r^2h + \pi rh - \pi rh^2$

**17.** $m_0x - 5m^2x$        **18.** $21fd + 18Fd$

**19.** $\frac{1}{2}M\theta_1 - \frac{1}{4}M\theta_2$     **20.** $\frac{5}{6}mv_1^2 + \frac{1}{6}mv_2^2$

**21.** $36a^4b^2 + 16a^3b^3 - 12a^2b^4$    **22.** $2b^4h - 3b^3h^2 + 7b^2h^3$

In problems 23 to 26, solve the literal equation for the given letter. (See Example 7-4.)

**23.** $2ax + 4x = b;\quad x$       **24.** $EI_1 + EI_2 = P;\quad E$

**25.** $CT_1 - CT_2 = 10Q;\quad C$    **26.** $2Fs = mv - 3m;\quad m$

**27.** The electric power is $5.6I_1^2R$ in one circuit and $4.2I_2^2R$ in a second circuit. Factor the expression that represents the total power in the two circuits.

**28.** The volume change of a gas under constant pressure caused by a change in temperature from $T_1$ to $T_2$ is given by $(VT_2/T_1) - V$, where $V$ is the original volume. Factor this expression.

**29.** Factor the expression $\pi rh^2 - \pi h^3/3$, which represents the volume of a segment of a sphere of radius $r$ and height $h$.

**30.** Write the expression for the sum of the volume of a sphere of radius $r$ and the volume of a right circular cylinder of radius $r$ and height $h = \frac{2}{3}$. Factor this expression (see Table 4-5).

# 7-2 | *Binomial Factors*

To multiply binomials recall that the FOIL method shown in Sec. 2-6 is used. The reverse of that multiplication method is used to factor a polynomial into two binomials. Consider one of the easier cases first.

## Difference of Two Squares

The basic formula for factoring the difference of two squares is:

$$x^2 - y^2 = (x + y)(x - y) \qquad \text{(7-1)}$$

When the squares of two numbers are separated by a minus sign, the factors are the sum $x + y$ and the difference $x - y$ of the two numbers. Example 2-40 illustrates the multiplication of two binomials of this type.

### Example 7-5

Factor each difference of two squares completely.

1. $4a^2 - 49 = (2a)^2 - (7)^2 = (2a + 7)(2a - 7)$
2. $25v_1^2 - 36v_2^2 = (5v_1)^2 - (6v_2)^2 = (5v_1 + 6v_2)(5v_1 - 6v_2)$
3. $0.01 - 0.09I^2 = (0.1)^2 - (0.3I)^2 = (0.1 + 0.3I)(0.1 - 0.3I)$
4. $T^4 - 16 = (T^2 + 4)(T^2 - 4) = (T^2 + 4)(T + 2)(T - 2)$
5. $3xr^2 - 3xh^2 = 3x(r^2 - h^2) = 3x(r + h)(r - h)$      (Common factor separated first)

Observe in (4) that the factor $T^2 - 4$ is not prime, it is also the difference of two squares. You can factor again to obtain *three* prime factors. The factor $T^2 + 4$ *cannot be factored*. You may think $T^2 + 4$ is factored as $(T + 2)^2$, but remember: $(T + 2)^2 = (T + 2)(T + 2) = T^2 + 4T + 4$. *The sum of two squares $x^2 + y^2$ cannot be factored into two binomials*. Notice in (5) that the common factor is separated first, revealing the difference of two squares and producing four prime factors.

## Trinomial Factoring

The FOIL method is not essential for factoring the difference of two squares. However, to factor a *trinomial* into two binomials, you must understand the FOIL method well enough to multiply binomials mentally, as shown in Sec. 2-6. Examine Example 2-37 to refresh your memory. A clear understanding of this procedure and an awareness of the basic sign relations illustrated below will enable you to factor trinomials. Study the next example carefully. It shows all the possible sign combinations, and binomial factors, of the trinomial whose terms are $x^2$, $5x$, and $6$.

### Example 7-6

Factor each trinomial completely.

1. $x^2 + 5x \oplus 6 = (x + 3)(x + 2)$ ⎫
2. $x^2 - 5x \oplus 6 = (x - 3)(x - 2)$ ⎬ like binomial signs

**3.** $x^2 + 5x \ominus 6 = (x + 6)(x - 1)$ ⎫
**4.** $x^2 - 5x \ominus 6 = (x - 6)(x + 1)$ ⎬ unlike binomial signs

When the sign of the last term is positive, the binomials have like signs. When the last term is negative, the binomials have unlike signs. You can check the factors by multiplying back mentally:

When you factor a trinomial, it is necessary to try different pairs of binomials until you have the correct combination of factors. Make sure the first product produces the first term of the trinomial and the last product produces the last term of the trinomial. Then, using the proper signs, *check that the outside and the inside products combine to give the middle term of the trinomial*. The following example illustrates the procedure step by step.

## Example 7-7
Factor completely the trinomial $a^2 + 4a - 12$.

## Solution
The only factors of $a^2$ are $a \cdot a$. The factors of 12 are $1 \cdot 12$, $2 \cdot 6$, and $3 \cdot 4$. Since the last term is negative, the binomial factors have the form $(a + )(a - )$. Try different possible pairs until you arrive at the correct pair:

$$(a + 1)(a - 12) = a^2 - 11a - 12$$
$$(a + 12)(a - 1) = a^2 + 11a - 12$$
$$(a + 2)(a - 6) = a^2 - 4a - 12$$
$$(a + 6)(a - 2) = a^2 + 4a - 12 \qquad \text{correct}$$

With a little practice much of the work of factoring can be done mentally and you can arrive at the answer faster.

The trinomials factored so far all have 1 as the first coefficient. When the first coefficient is not 1, there are generally more possibilities to consider as follows.

## Example 7-8
Factor completely:

$$6y^2 - 19y + 8$$

**Solution**

The last term is positive, and the middle term is negative, so the answer has the form $(-)(-)$. You have to consider binomials whose first terms are factors of $6y^2$ and whose last terms are factors of 8. The factors of $6y^2$ are $6y \cdot y$ and $2y \cdot 3y$. The factors of 8 are $1 \cdot 8$ and $2 \cdot 4$. There are eight possible combinations of binomials that have the above form and these factors. See if you can list the combinations. After trying some of them, you will arrive at the correct answer: $6y^2 - 19y + 8 = (2y - 1)(3y - 8)$.

### Example 7-9

Factor completely:

$$m^2 + 2m + 3$$

**Solution**

The only factors that could possibly work do not: $(m + 1)(m + 3) = m^2 + 4m + 3$. This trinomial is therefore *prime and not factorable*.

### Example 7-10

Factor completely:

$$4a^2 - 20ab + 25b^2$$

**Solution**

The factors are the same:

$$4a^2 - 20ab + 25b^2 = (2a - 5b)(2a - 5b) = (2a - 5b)^2$$

This trinomial is a *perfect square*. (See Example 2-39.)

### Example 7-11

Factor completely:

$$4\pi r^2 - 6\pi r + 2\pi$$

**Solution**

This trinomial has a common factor, which should always be factored out first:

$$4\pi r^2 - 6\pi r + 2\pi = 2\pi(2r^2 - 3r + 1) = 2\pi(2r - 1)(r - 1)$$

## *Sum and Difference of Two Cubes*

These two special products are useful and should be memorized:

$$a^3 + b^3 = (a + b)(a^2 - ab + b^2) \qquad\qquad (7\text{-}2)$$
$$a^3 - b^3 = (a - b)(a^2 + ab + b^2) \qquad\qquad (7\text{-}3)$$

**Exercise 7-2**    In problems 1 to 52, factor each expression completely. Separate common factors first.
(*Note:* Some expressions are prime.)

1. $a^2 - 4b^2$

2. $16x^2 - 25y^2$

3. $100m^2 - 121m_0^2$

4. $225F_1^2 - 144F_2^2$

5. $1 - K^4$

6. $I^4 - 81$

7. $4b^2 + 9a^2$

8. $36f^2 + 25g^2$

9. $p^2 - 0.09$

10. $0.04c^2 - 0.01$

11. $0.81x^2 - 0.16y^2$

12. $0.25 - n^2$

13. $y^2 + 6y + 5$

14. $x^2 + 3x + 2$

15. $w^2 - 4w + 3$

16. $v^2 - 8v + 12$

17. $k^2 + 3k - 10$

18. $q^2 + q - 30$

19. $a^2 + 2a + 3$

20. $b^2 - 2b + 3$

21. $m^2 + 10m + 25$

22. $E^2 - 16E + 64$

23. $3x^2 + 4x + 1$

24. $6y^2 - 7y - 3$

25. $5a^2 + ab - 6b^2$

26. $8x^2 - 2xy - 15y^2$

27. $9v^2 - 15v + 4$

28. $8v_0^2 + 31v_0 - 4$

29. $a^2 + a + 1/4$

30. $0.36b^2 - 0.60b + 0.25$

31. $2.0p^2 - 1.5p + 0.18$

32. $3t^2 - (1/2)t - 1$

33. $12e^2 - e - 1$

34. $3R^2 + 4R - 4$

35. $6h^2 - 24$

36. $\pi r^2 - \pi R^2$

37. $8x^2 + 28x - 16$

38. $18t^2 + 60t + 50$

39. $10ax^2 + 15ax - 25a$

40. $5by^2 + 30by + 25b$

41. $x^4 + 2x^2 + 1$

42. $y^4 - 6y^2 + 9$

43. $f^2 + (2 \times 10^3)f + 10^6$

44. $c^2 - (2 \times 10^4)c + 10^8$

45. $32x^8 - 2$

46. $3Q^4 - 48$

47. $2a^4 - 5a^2 - 3$

48. $6d^4 + 13d^2 + 6$

Use formulas (7-2) and (7-3) for problems 49 to 52.

49. $x^3 + y^3$

50. $h^3 - x^3$

51. $8r^3 - R^3$

52. $t^3 + 0.027$

53. The volume $V$ of a box in terms of the width $w$ is given by $V = 2w^3 - 3w^2 - 2w$. Factor this expression.

54. The electric power in one circuit is $0.18I_1^2R$; it is $0.32I_2^2R$ in a second circuit. Factor the expression that represents the difference in power between the circuits.

55. The energy $E$ radiated by a blackbody is given by $E = kT^4 - kT_0^4$, where $k =$ constant, $T =$ absolute temperature of the body, and $T_0 =$ absolute temperature of the surroundings. Factor this expression.

56. The deflection $y$ of a beam of length $L$ at a distance $x$ from one end is given by $y = 2L^2x^2 + 4Lx - 16$. Factor this expression.

**57.** The following expression arises from a problem in heat transfer: $Q^4H + QH^4$. Factor this expression. [See formula (7-2).]

**58.** Factor the following expression from a problem in optics:

$$\frac{1}{p^2} + \frac{2}{pq} + \frac{1}{q^2}$$

# 7-3 | *The Quadratic Equation: Solution by Factoring*

Consider this problem: A printed rectangular circuit board is to have an area of 48 cm². If the length is to be 2 cm longer than the width, what should the dimensions of the board be?

The verbal equation is:

$$(\text{Length})(\text{width}) = \text{area} = 48$$

Let $w$ = width. Then $w + 2$ = length, and the algebraic equation is:

$$w(w + 2) = 48$$

or $\qquad\qquad w^2 + 2w - 48 = 0 \qquad$ (moving terms to one side)

This is an example of a quadratic equation. A *quadratic equation,* or *second-degree equation,* contains variables to the second power but to no higher power. The *general quadratic equation* in $x$ is:

$$ax^2 + bx + c = 0 \qquad a, b, c = \text{constants}, a \neq 0 \qquad (7\text{-}4)$$

You can readily solve a quadratic equation in the general form shown in (7-4) if the left side of the equation can be factored. To solve the above equation $w^2 + 2w - 48 = 0$, factor the left side:

$$(w + 8)(w - 6) = 0$$

In factored form the equation states: The product of two numbers, $w + 8$ and $w - 6$, equals zero. A product of two numbers is zero only if *one (or both) of the numbers equals zero.* Therefore you set each factor equal to zero and solve each linear equation for $w$:

$$w + 8 = 0 \qquad\qquad w - 6 = 0$$
$$w = -8 \qquad \text{and} \qquad w = 6$$

Both answers satisfy the original equation $w(w + 2) = 48$:

$$w = -8 \longrightarrow -8(-8 + 2) = -8(-6) = 48$$
$$w = 6 \longrightarrow 6(6 + 2) = 6(8) = 48$$

However, since the width must be positive, the only acceptable answer is $w = 6$. The length is then $w + 2 = 8$.

Note that the roots, or solutions, of a quadratic equation can be equal if the factors are the same. This is called a *double root*.

### Example 7-12
Solve the quadratic equation $\qquad 2x^2 + 5x - 3 = 4x + 12$

### Solution
To solve a quadratic equation by factoring, first put the equation in the general quadratic form (7-4). Move all the terms to the left side and simplify:

$$2x^2 + 5x - 3 - 4x - 12 = 0$$
$$2x^2 + x - 15 = 0$$

Then factor and set each factor equal to zero:

$$(2x - 5)(x + 3) = 0$$

$$2x - 5 = 0 \qquad\qquad x + 3 = 0$$
$$x = 5/2 \qquad \text{and} \qquad x = -3$$

Observe that you could have factored the left side of the original equation: $2x^2 + 5x - 3 = (2x - 1)(x + 3)$. However, that would not help, since you must have a zero on the right side to be able to set each factor equal to zero.

When the constant term $c$ in the general quadratic equation (7-4) equals zero, the equation can always be solved by factoring as follows.

### Example 7-13
Solve the quadratic equation $\qquad 3.5m^2 = 7.3m$

### Solution

$$3.5m^2 - 7.3m = 0$$
$$(m)(3.5m - 7.3) = 0$$
$$m = 0 \qquad \text{and} \qquad 3.5m - 7.3 = 0$$

$$m = \frac{7.3}{3.5} = 2.1 \text{ (two figures)}$$

When the first-degree term is missing ($b = 0$), the general quadratic equation can be solved by factoring if it is the difference of two squares as follows.

### Example 7-14
Solve the quadratic equation $2x^2 - 18 = 0$.

### Solution

Common factor separated first
$$2(x^2 - 9) = 0$$

Common factor divided out
$$\frac{\cancel{2}}{\cancel{2}}(x^2 - 9) = \frac{0}{2} = 0$$

$$(x + 3)(x - 3) = 0$$

$$x = -3 \qquad \text{and} \qquad x = 3$$

A common factor in a quadratic equation (or in any equation) should first be divided out of every term to reduce the terms and make solving the equation easier. Example 7-14 can also be done by solving for $x^2$ and taking the square root of each side of the equation. (See Example 7-34.)

### Example 7-15
Solve the literal quadratic equation for the variable $I$:

$$RI^2 = 2VI$$

### Solution
Put the equation in the general form, and separate the common factor $I$:

$$RI^2 - 2VI = 0$$
$$I(RI - 2V) = 0$$

Set each factor equal to zero and solve each for $I$:

$$I = 0 \qquad RI - 2V = 0$$

$$I = \frac{2V}{R}$$

**Exercise 7-3**

In problems 1 to 26, solve each quadratic equation by factoring.

1. $x^2 - 4 = 0$
2. $2K^2 = 32$
3. $q^2 - 3q - 10 = 0$
4. $y^2 + 6y + 8 = 0$
5. $3t^2 - 8t + 4 = 0$
6. $5a^2 + 2a = 7$

**7.** $p^2 - 0.09 = 0$                    **8.** $3r^2 - 0.75 = 0$

**9.** $56x^2 = 8x$                    **10.** $10y^2 + 18y = 0$

**11.** $7.28H^2 + 3.15H = 0$                    **12.** $0.38v_1^2 = 0.59v_1$

**13.** $10x^2 + 5x = 5$                    **14.** $6n^2 - 20n + 16 = 0$

**15.** $F_x^2 + 1.2F_x + 0.36 = 0$                    **16.** $F_y^2 - 1.4F_y + 0.49 = 0$

**17.** $3x^2 + x = x^2 - 4x + 3$                    **18.** $5y^2 - 10 = y^2 - 3y$

**19.** $64L^2 + 16L = 16L^2 + 32$                    **20.** $2.0R^2 + 1.0 = 2.5R + 0.50$

**21.** $x(x + 3) = 40$                    **22.** $(y - 2)(y + 3) = 6$

**23.** $(3h - 1)^2 - h^2 = 0$                    **24.** $2m^2 + 2 = (m + 1)^2$

**25.** $2.0T^2 - 0.5T = 1.0 - 1.0T^2$                    **26.** $4.2w + 0.20 = 2.0w^2 + 2.0$

In problems 27 to 32, solve the literal equation for the given variable by factoring. (See Example 7-15.)

**27.** $\pi r^2 - \pi rs = 0$;   $r$                    **28.** $VI - RI^2 = 0$;   $I$

**29.** $16Q^2 - 8Qh + h^2 = 0$;   $Q$                    **30.** $2d_1^2 - d_1 d_2 - d_2^2 = 0$;   $d_1$

**31.** $v_0 t = gt^2/2$;   $t$                    **32.** $2ax + bx^2 = cx$;   $x$

Solve each applied problem by setting up and solving a quadratic equation.

**33.** The distance $s$ above the ground of a rocket fired vertically is given by $s = v_0 t - \frac{1}{2}gt^2$, where $v_0 =$ initial velocity and $g =$ gravitational acceleration, $9.8$ m/s$^2$. Find how many seconds it takes a rocket fired at $v_0 = 196$ m/s to return to the ground. Let $s = 0$ and solve for $t$.

**34.** The rocket in problem 33 reaches a maximum height $s = 1960$ m. How long does the rocket take to reach this height?

**35.** A rectangular solar panel has an area of $1.0$ m$^2$. The length is 150 cm more than the width. Find the dimensions of the panel.

**36.** The short leg of a right triangular brace is 9.0 ft shorter than the hypotenuse and 7.0 ft shorter than the other leg. What are the lengths of the three sides of the brace? [*Hint:* Use the pythagorean theorem (4-6).]

**37.** The power in a circuit having two resistances $R_1$ and $R_2$ is expressed by the following equation: $VI = R_1 I^2 + R_2 I^2$. Solve for $I$.

**38.** A beam of length $L$ has a displacement of zero at a distance $x$ from one end. This is expressed by the following equation: $2x^2 - 3xL + L^2 = 0$. Solve the equation for $x$.

**39.** In an ac circuit the resistance $R$ is $1.8\ \Omega$ less than the impedance $Z$, and the reactance $X$ is $0.40\ \Omega$ less than $Z$. Find $R$, $X$, and $Z$. (*Note:* $R^2 + X^2 = Z^2$.)

**40.** When the radius of a pipeline is increased by 1.0 ft, the cross-sectional area increases by 125 percent. What is the original radius?

## 7-4 | *Reducing, Multiplying, and Dividing Fractions*

### Reducing Fractions

When you are working with algebraic fractions, factors play an important role. The more readily you can factor expressions, the more easily you will be able to perform operations with fractions. A common factor in the numerator and the denominator of a fraction can be divided out to reduce the fraction to lowest terms. For example:

$$\frac{12x^2}{28x} = \frac{\overset{3x}{\cancel{12x^2}}}{\underset{7}{\cancel{28x}}} = \frac{3x}{7} \qquad (4x \text{ divided out})$$

$$\frac{15x^2 + 5x}{5x} = \frac{\overset{1}{\cancel{5x}}(3x + 1)}{\underset{1}{\cancel{5x}}} = 3x + 1$$

When fractions have one term in the denominator, as the fractions above do, you can perform much of the factoring and dividing mentally. However, when fractions have more than one term in the denominator, you must *separate the factors before dividing*. Study the next example closely.

### Example 7-16
Reduce each fraction to lowest terms.

1. $\dfrac{12x^2 + 4x}{2x^2 - 2x} = \dfrac{\overset{2}{\cancel{4x}}(3x + 1)}{\underset{1}{\cancel{2x}}(x - 1)} = \dfrac{2(3x + 1)}{x - 1}$

2. $\dfrac{R^2 - 4T^2}{R - 2T} = \dfrac{(R + 2T)\,\overset{1}{\cancel{(R - 2T)}}}{\underset{1}{\cancel{R - 2T}}} = \dfrac{R + 2T}{1} = R + 2T$

3. $\dfrac{a^2 + 2ab + b^2}{a^2 - ab - 2b^2} = \dfrac{\overset{1}{\cancel{(a + b)}}\,(a + b)}{\underset{1}{\cancel{(a + b)}}\,(a - 2b)} = \dfrac{a + b}{a - 2b}$

When reducing fractions, you may feel a strong desire to cancel terms that are not factors. Consider (1) above. You may be tempted to divide $2x^2$ into $12x^2$ and $-2x$ into $4x$. The result would be $6 - 2 = 4$, which is very different from the correct answer. In fact, you probably will make this mistake at least once in your life. Try not to make it more than once! *When you cancel factors, you are doing division,* which is the inverse of multiplication. Therefore, when you cancel terms, you must make sure they are separated from the rest of the expression by multiplication signs, not by plus or minus signs. Also, since canceling factors is a division process, the canceled factor does not disappear but is replaced by 1:

$$\frac{\cancel{x}}{\cancel{x}} = 1$$

Study the next two examples, which illustrate special cases of reducing fractions.

### Example 7-17
Reduce to lowest terms:

$$\frac{u^2 + v^2}{u^2 - v^2} = \frac{u^2 + v^2}{(u + v)(u - v)} \qquad \text{no common factors}$$

You may be tempted to cancel the $u$'s and the $v$'s, but do not do it. The sum of two squares cannot be factored. This fraction has no common factors and *cannot be reduced.*

### Example 7-18
Reduce to lowest terms:

$$\frac{a - b}{b - a} = \frac{(-1)(-a + b)}{b - a} = \frac{-1(\cancel{b - a})}{\cancel{b - a}} = -1$$

Notice that $a - b$ is the negative of $b - a$, just as $-7$ is the negative of $+7$. When the order of two terms that have a negative sign in between them is switched, such as $a - b$ and $b - a$, it is the same as multiplying by a factor of $-1$.

## *Multiplying and Dividing Fractions*

The rules for multiplying and dividing algebraic fractions are the same as those in arithmetic:

Multiplication: Multiply numerators and denominators:

$$\frac{A}{B} \cdot \frac{C}{D} = \frac{AC}{BD} \qquad (7\text{-}5)$$

Division: Invert and multiply:

$$\frac{A}{B} \div \frac{C}{D} = \frac{A}{B} \cdot \frac{D}{C} = \frac{AD}{BC} \tag{7-6}$$

However, as in arithmetic, you first divide out common factors that appear in *any* numerator and *any* denominator before multiplying as follows.

### Example 7-19
Perform the indicated operations.

**1.** $\dfrac{12y^3}{7x^2} \cdot \dfrac{21x}{24y^2} = \dfrac{\overset{y}{\cancel{12y^3}}}{\underset{x}{\cancel{7x^2}}} \cdot \dfrac{\overset{3}{\cancel{21x}}}{\underset{2}{\cancel{24y^2}}} = \dfrac{3y}{2x}$     Divide out $7x$ and $12y^2$.

**2.** $\dfrac{a^2 - b^2}{4a} \cdot \dfrac{8a^2}{2a + 2b} = \dfrac{\overset{1}{(\cancel{a+b})}\,(a-b)}{\underset{1}{\cancel{4a}}} \cdot \dfrac{\overset{a}{\cancel{8a^2}}}{\underset{1}{2(\cancel{a+b})}}$

$$= \frac{a(a-b)}{1} = a(a-b)$$

**3.** $\dfrac{mc^2 + 4mc + 4m}{3c} \div \dfrac{2c^2 + c - 6}{6c - 9}$

$$= \frac{\overset{m(c^2 + 4c + 4)}{\cancel{mc^2 + 4mc + 4m}}}{3c} \cdot \frac{6c - 9}{2c^2 + c - 6}$$

$$= \frac{m(c + 2)\cancel{^2}}{\cancel{3}c} \cdot \frac{\overset{1}{\cancel{3}(\cancel{2c-3})}}{(\cancel{2c-3})\,(\cancel{c+2})} = \frac{m(c + 2)}{c}$$

Notice in (2) and (3) that you must first factor as much as possible before dividing factors. Study the next example, which illustrates some special cases of multiplying and dividing fractions.

### Example 7-20
Perform the indicated operations.

**1.** $\dfrac{R^2 - r^2}{\pi R - \pi r} \cdot 2\pi r = \dfrac{(\cancel{R-r})(R + r)}{\underset{1}{\pi(\cancel{R-r})}} \cdot \dfrac{2\pi r}{1} = 2r(R + r)$

**2.** $\dfrac{3m^2 + 5mn + 2n^2}{m + n} \div (3m + 2n)$

$$= \dfrac{\overset{1}{\cancel{(3m + 2n)}}\ \overset{}{\cancel{(m + n)}}}{\underset{1}{\cancel{m + n}}} \cdot \dfrac{1}{\underset{1}{\cancel{3m + 2n}}} = \dfrac{1}{1} = 1$$

Notice in (2) that when all the factors divide out, the answer is 1, not 0.

**3.** $\dfrac{\dfrac{4 - x^2}{6}}{\dfrac{x - 2}{3}} = \dfrac{4 - x^2}{6} \cdot \dfrac{3}{x - 2} = \dfrac{\overset{1}{(2 - x)}(2 + x)}{\underset{2}{\cancel{6}}} \cdot \dfrac{\overset{1}{\cancel{3}}}{-1\underset{1}{\cancel{(2 - x)}}} = \dfrac{2 + x}{-2}$

In (3) division of fractions is expressed by a *complex fraction*. Notice that $x - 2$ is written $-1(2 - x)$ so that it can divide out with $2 - x$. The answer to (3) can be written in three ways:

$$\dfrac{2 + x}{-2} = \dfrac{-(2 + x)}{2} = -\dfrac{2 + x}{2}$$

A minus sign in front of a fraction is equivalent to multiplying the numerator or the denominator (but not both) by $-1$.

**Exercise 7-4**

In problems 1 to 20, reduce each fraction to lowest terms. (*Note:* Some fractions cannot be reduced.)

**1.** $\dfrac{210y^3}{270y}$

**2.** $\dfrac{66b^2}{165b^3}$

**3.** $\dfrac{ac}{ab + ad}$

**4.** $\dfrac{qx - qy}{tx - ty}$

**5.** $\dfrac{2c^2 - 2d^2}{10c + 10d}$

**6.** $\dfrac{5.1\pi a + 5.1\pi b}{1.7a + 1.7b}$

**7.** $\dfrac{a^2 + 6ab + 9b^2}{3a^2 + 8ab - 3b^2}$

**8.** $\dfrac{4x^2 - 6xy - 4y^2}{4x - 8y}$

**9.** $\dfrac{2h^2 + 2x^2}{5h^2 - 5x^2}$

**10.** $\dfrac{m^2 + 8mn + 16n^2}{m^2 + 4mn + 4n^2}$

**11.** $\dfrac{7t_1^2 + 13t_1t_2 - 2t_2^2}{9t_1^2 + 14t_1t_2 - 8t_2^2}$

**12.** $\dfrac{16v^2 - 16v_0^2}{v_0^2 - vv_0 - 2v^2}$

**13.** $\dfrac{2x - 2y}{y - x}$

**14.** $\dfrac{p^2 - m^2}{m^2 - p^2}$

**15.** $\dfrac{5.5f^2N - 5.5F^2N}{1.1f^2 - 1.1F^2}$    **16.** $\dfrac{6V + 15R}{6V^2 + 9VR - 15R^2}$

Use formulas (7-2) and (7-3) for problems 17 to 20.

**17.** $\dfrac{a^3 + b^3}{a^2 - b^2}$    **18.** $\dfrac{x^3 - y^3}{x^2 - y^2}$

**19.** $\dfrac{8r^3 + R^3}{2r + R}$    **20.** $\dfrac{x_2^3 - x_1^3}{x_2 - x_1}$

In problems 21 to 48, perform the indicated operations. Divide out **factors first.**

**21.** $\dfrac{9x}{2} \cdot \dfrac{4}{3x^2}$    **22.** $\dfrac{5b^2y}{8by^3} \cdot \dfrac{4y}{10b}$

**23.** $\dfrac{5.6}{0.80t} \cdot \dfrac{1.0t^2}{1.4}$    **24.** $\dfrac{0.09E}{0.6I} \div \dfrac{0.3E}{I}$

**25.** $\dfrac{21u - 7v}{9u^2 - v^2} \cdot \dfrac{3u + v}{uv}$    **26.** $\dfrac{8a^2 - 8b^2}{16b} \cdot \dfrac{5b}{7a + 7b}$

**27.** $\dfrac{t^2 - 2th + h^2}{t} \div (t - h)$    **28.** $\dfrac{x + h}{x - h}(x^2 - h^2)$

**29.** $\dfrac{x^2 + 5x + 4}{2x + 2} \cdot \dfrac{4x + 4}{3x + 12}$    **30.** $\dfrac{6y^2 - 13y + 6}{4y + 1} \cdot \dfrac{16y^2 - 1}{6y - 4}$

**31.** $(2d^2 - 5d + 2) \div \dfrac{d^2 - 4}{5d + 10}$

**32.** $\dfrac{4m^2 - 12mn + 9n^2}{2n} \div (2m - 3n)$

**33.** $\dfrac{Q_h^2 - Q_h}{Q_h^2} \cdot \dfrac{Q_h + 1}{Q_h - 1}$

**34.** $\dfrac{T_1 - T_2}{T_1} \div \dfrac{T_1^2 - T_2^2}{T_2}$

**35.** $\dfrac{3a^2 - 12ax + 12x^2}{2a^2 + 2a - 12} \div \dfrac{6a^2 - 24x^2}{8a + 24}$

**36.** $\dfrac{3w^2 - 27}{6w^2 - 6} \cdot \dfrac{6w^2 - 12w + 6}{15w + 45}$

**37.** $\dfrac{(9 \times 10^9)(R - 1)}{R} \div \dfrac{(18 \times 10^9)(R + 1)}{2R}$

**38.** $\dfrac{(6 \times 10^{-4})F}{F^2 + Fd} \cdot \dfrac{F + d}{3 \times 10^{-3}d}$

**39.** $\dfrac{3.5x}{2.5y} \cdot \dfrac{0.5x^2y}{2.1xy^2} \div \dfrac{0.1x^2}{0.3y^2}$

**40.** $\dfrac{1.2c^2d}{0.4cd^2} \div \left(\dfrac{0.6c}{0.3d} \cdot \dfrac{1.8c}{1.2d}\right)$

**41.** $\dfrac{3x - 3}{2x^2 - x - 1} \cdot \dfrac{2x + 1}{x + 1} \cdot \dfrac{2x^2 + 3x + 1}{6x + 3}$

**42.** $\dfrac{2b + y}{2b^2 + 3by + y^2} \cdot \dfrac{2b - y}{4b + 8y} \div \dfrac{1}{4b + 4y}$

**43.** $\dfrac{a^4 - 1}{a^2 + 1} \cdot \dfrac{1 + a^2}{1 - a^2}$

**44.** $\dfrac{kT^4 - kT_0^4}{T^2 + T_0^2} \cdot \dfrac{kT + kT_0}{kT - kT_0}$

**45.** $\dfrac{(\pi R - \pi r)/(\pi R)}{(\pi r - \pi R)/(\pi r)}$

**46.** $\dfrac{(x - y)/(2a)}{(y - x)/(3a)}$

Use formulas (7-2) and (7-3) for problems 47 and 48.

**47.** $\dfrac{p^3 + q^3}{p^2 - pq + q^2} \cdot \dfrac{1}{p - q}$

**48.** $\dfrac{x^3 - y^3}{x^3 + y^3} \div \dfrac{x - y}{x + y}$

**49.** The input force of a machine is expressed by the formula $1.5F_x - 1.5F_y$ and the output force by $3F_x^2 - 3F_y^2$. The mechanical advantage = output force/input force. Write and simplify the expression for the mechanical advantage.

**50.** The height of a box is $6H + 2h$, and the volume is $24H^3 - 16H^2h - 8Hh^2$. Find and simplify the expression for the area of the base. (*Note:* Base area = volume/height.)

**51.** In a circuit, the resistance $R$ as a function of time is given by $R = 0.5t^2 + 1.0t + 0.5$ and the current $I$ by $t/(t + 1)$. Express the power $P = I^2R$ as a function of $t$, and simplify the expression.

**52.** The Doppler effect is the apparent change in frequency of a sound when there is movement between the source of the sound and the observer. If the source is moving toward the observer, the frequency heard by the observer is given by the equation $f_t = f_s v/(v - v_s)$, where $f_s$ = frequency of the source, $v$ = velocity of sound, and $v_s$ = velocity of the source with respect to the observer. If the source is moving away from the observer, the frequency heard is given by $f_a = f_s v/(v + v_s)$. Express the ratio of $f_t$ to $f_a$ and simplify the expression.

**53.** The volume strain of a beam is given by:

$$\dfrac{(a^3 - a'^3)/a^3}{(a^2 - a'^2)/a^2}$$

Simplify the expression.

**54.** Using the definitions (4-8) of $\sin A$, $\cos A$, and $\tan A$ from Sec. 4-8, show that $\tan A = \sin A/\cos A$.

## 7-5 | *Combining Fractions*

Combining (adding or subtracting) algebraic fractions requires a clear under-standing of the process of combining arithmetic fractions and a good grasp of factoring. The following discussion reviews only the process of adding arith-metic fractions, based on the assumption that by now you can factor well (true?). The ideas needed for combining algebraic fractions are emphasized.

How do you add $5/24$ and $7/30$? You look for the *lowest common denominator* (lcd) that both denominators divide into. This is usually done by trying multi-ples of the larger denominator—30, 60, 90, and so on—until you find a multi-ple that both denominators divide into. Since the lcd is the smallest quantity that contains all the factors of each denominator, a more direct way to find it is to factor each denominator into primes:

$$\frac{5}{\cancel{24}} \quad + \quad \frac{7}{\cancel{30}}$$
$$2 \cdot 2 \cdot 2 \cdot 3 \quad\quad 2 \cdot 3 \cdot 5$$

The lcd is then the product $2 \cdot 2 \cdot 2 \cdot 3 \cdot 5 = 2^3 \cdot 3 \cdot 5 = 120$. You *obtain the lcd by selecting the highest power of each prime factor*. All the denominators divide into the lcd since it contains the prime factors of each:

$$\frac{120}{24} = \frac{\cancel{2} \cdot \cancel{2} \cdot \cancel{2} \cdot \cancel{3} \cdot 5}{\cancel{2} \cdot \cancel{2} \cdot \cancel{2} \cdot \cancel{3}} = 5 \qquad \frac{120}{30} = \frac{2 \cdot 2 \cdot \cancel{2} \cdot \cancel{3} \cdot \cancel{5}}{\cancel{2} \cdot \cancel{3} \cdot \cancel{5}} = 4$$

This is the same method used to find the lcd for algebraic fractions. After you find the lcd, change each denominator to the lcd by applying the *fundamental rule of fractions*:

> Multiplying or dividing the numerator and the denominator of a fraction by the same number (other than zero) does not change the value of the fraction.

Multiply the top and the bottom of each fraction by the factor obtained when the denominator is divided into the lcd, as shown above. This changes the denomi-nator of each fraction into the lcd so that you can combine the fractions:

$$\frac{5\,(5)}{24\,(5)} + \frac{7\,(4)}{30\,(4)} = \frac{25}{120} + \frac{28}{120} = \frac{25 + 28}{120} = \frac{53}{120}$$

The following example now applies the above ideas to algebraic fractions.

### Example 7-21

Combine: $\dfrac{3}{6x^2y} + \dfrac{7}{18xy}$

### Solution

The prime factors of each denominator are:

$$6x^2y = 2 \cdot 3 \cdot x \cdot x \cdot y = 2 \cdot 3 \cdot x^2 \cdot y$$

and

$$18xy = 2 \cdot 3 \cdot 3 \cdot x \cdot y = 2 \cdot 3^2 \cdot x \cdot y$$

Taking the highest power of each prime factor, the lcd is:

$$2 \cdot 3^2 \cdot x^2 \cdot y = 18x^2y$$

For each fraction divide the denominator into the lcd and multiply the top and bottom by the result. This changes the denominator of each fraction to the lcd so that you can combine:

$$\frac{3\,(3)}{6x^2y\,(3)} + \frac{7\,(x)}{18xy\,(x)} = \frac{9}{18x^2y} + \frac{7x}{18x^2y} = \frac{9 + 7x}{18x^2y}$$

This example can be done in a more convenient way by combining and multiplying in one step, since you know all the denominators will be the lcd:

$$\frac{3}{6x^2y} + \frac{7}{18xy} = \frac{3(3) + 7(x)}{18x^2y} = \frac{9 + 7x}{18x^2y}$$

### Example 7-22

Combine: $\dfrac{5}{a} + \dfrac{a + 1}{a - 1}$

### Solution

When the denominator contains more than one term, you must carefully identify prime factors. The denominator $a - 1$ represents only one factor. The lcd contains two prime factors: $a$ and $a - 1$. Change each fraction to the lcd, combine, and simplify:

$$\frac{5\,(a - 1)}{a\,(a - 1)} + \frac{(a + 1)\,(a)}{(a - 1)\,(a)} = \frac{5(a - 1) + (a + 1)(a)}{\underbrace{(a)(a - 1)}_{\text{lcd}}}$$

$$= \frac{5a - 5 + a^2 + a}{(a)(a - 1)} = \frac{a^2 + 6a - 5}{(a)(a - 1)}$$

Remember that *the fraction line has the same effect as parentheses*. If there is more than one term in the numerator, enclose it in parentheses, as $a + 1$ is above, to avoid an error in multiplication.

### Example 7-23

Combine:     $3.0 - \dfrac{0.80t}{v} + \dfrac{1.0v + 0.60t}{t}$

### Solution

$$\underbrace{\dfrac{3.0(vt) - 0.80t(t) + (1.0v + 0.60t)(v)}{vt}}_{\text{lcd}}$$

$$= \dfrac{3.0vt - 0.80t^2 + 1.0v^2 + 0.60vt}{vt}$$

$$= \dfrac{1.0v^2 + 3.6vt - 0.80t^2}{vt}$$

Notice that 3.0 is treated as 3.0/1 and is multiplied top and bottom by the lcd.

### Example 7-24

Combine:     $\dfrac{2}{T + 1} - \dfrac{T - 3}{T^2 - 1}$

### Solution
Factor the second denominator into prime factors to find the lcd:

$$\dfrac{2}{T + 1} - \dfrac{T - 3}{\underset{(T + 1)(T - 1)}{\cancel{T^2 - 1}}} = \underbrace{\dfrac{2(T - 1) - (T - 3)}{(T + 1)(T - 1)}}_{\text{lcd}}$$

$$= \dfrac{2T - 2 - T + 3}{(T + 1)(T - 1)} = \dfrac{T + 1}{(T + 1)(T - 1)}$$

$$= \dfrac{\overset{1}{\cancel{T + 1}}}{\cancel{(T + 1)}(T - 1)} = \dfrac{1}{T - 1}$$

Observe that the denominator of the second fraction does not need to be changed but that every term in the numerator must be multiplied by the minus sign. Notice also that the answer can be simplified (reduced) by dividing out the common factor of $T + 1$. Answers should be simplified whenever possible.

## Example 7-25

Combine: $\dfrac{r + 2}{8r + 4} + \dfrac{1 - r}{2r^2 - 3r - 2}$

### Solution

Factor each denominator first:

$$\underset{\underset{4(2r + 1)}{}}{\dfrac{r + 2}{\cancel{8r + 4}}} + \underset{\underset{(2r + 1)(r - 2)}{}}{\dfrac{1 - r}{\cancel{2r^2 - 3r - 2}}} = \dfrac{(r + 2)(r - 2) + (1 - r)(4)}{\underbrace{4(2r + 1)(r - 2)}_{\text{lcd}}}$$

$$= \dfrac{r^2 - 4 + 4 - 4r}{4(2r + 1)(r - 2)} = \dfrac{r^2 - 4r}{4(2r + 1)(r - 2)}$$

The numerator of the result can be factored as $r(r - 4)$, but the fraction cannot be simplified.

Study the next example closely. It involves several steps that need to be done carefully.

## Example 7-26

Combine: $\dfrac{x^2 + y^2}{x^2 - y^2} - \dfrac{x + y}{2x - 2y}$

### Solution

$$\underset{\underset{(x + y)(x - y)}{}}{\dfrac{x^2 + y^2}{\cancel{x^2 - y^2}}} - \underset{\underset{2(x - y)}{}}{\dfrac{x + y}{\cancel{2x - 2y}}}$$

$$= \dfrac{(x^2 + y^2)(2) - (x + y)(x + y)}{\underbrace{2(x + y)(x - y)}_{\text{lcd}}}$$

$$= \dfrac{2x^2 + 2y^2 - (x^2 + 2xy + y^2)}{2(x + y)(x - y)}$$

$$= \dfrac{2x^2 + 2y^2 - x^2 - 2xy - y^2}{2(x + y)(x - y)}$$

$$= \dfrac{x^2 - 2xy + y^2}{2(x + y)(x - y)} = \dfrac{(x - y)^{\cancel{2}}}{2(x + y)\cancel{(x - y)}}$$

$$= \dfrac{x - y}{2(x + y)}$$

Note that the numerator of the second fraction is multiplied by $x + y$ and by $-1$. This is shown in two steps for clarity. The answer can be reduced as shown.

### Example 7-27

Combine:     $\dfrac{1}{R_1} + \dfrac{3}{R_2} - \dfrac{5}{R_1 + R_2}$

### Solution

$$\underbrace{\frac{1(R_2)(R_1 + R_2) + 3(R_1)(R_1 + R_2) - 5(R_1)(R_2)}{(R_1)(R_2)(R_1 + R_2)}}_{\text{lcd}}$$

$$= \frac{R_1 R_2 + R_2^2 + 3R_1^2 + 3R_1 R_2 - 5R_1 R_2}{(R_1)(R_2)(R_1 + R_2)} = \frac{3R_1^2 - R_1 R_2 + R_2^2}{R_1 R_2 (R_1 + R_2)}$$

Parentheses are used to emphasize the factors $R_1$ and $R_2$ but are not necessary in the answer.

    The next example shows how to simplify a complex fraction. A *complex fraction* is a fraction whose numerator or denominator contains a fraction.

### Example 7-28

The radius $R$ of a spherical mirror is given by the equation $R = 2pq/(p + q)$, where $p$ = object distance and $q$ = image distance from the mirror. If $q = pf/(p - f)$, where $f$ = focal length, find $R$ in terms of $p$ and $f$.

### Solution

Substitute the expression for $q$ in the formula for $R$ to obtain:

$$R = \frac{2p\left(\dfrac{pf}{p - f}\right)}{p + \left(\dfrac{pf}{p - f}\right)}$$

This *complex fraction* can be simplified two ways. The *first method* is by combining the two terms in the denominator, inverting, and multiplying:

$$\frac{2p\left(\dfrac{pf}{p - f}\right)}{\dfrac{p(p - f) + pf}{p - f}} = \frac{\dfrac{2p^2 f}{p - f}}{\dfrac{p^2 - pf + pf}{p - f}} = \frac{2p^2 f}{p - f} \cdot \frac{p - f}{p^2} = 2f$$

The *second method* is by multiplying the top and bottom of the complex fraction by the lcd of the fractions in the numerator and denominator, which is $p - f$:

$$\frac{2p\left(\dfrac{pf}{p-f}\right)(p-f)}{p + \dfrac{pf}{p-f}(p-f)} = \frac{\dfrac{2p^2f}{\cancel{p-f}}(\cancel{p-f})}{p(p-f) + \dfrac{pf}{\cancel{p-f}}(\cancel{p-f})}$$

$$= \frac{2p^2f}{p^2 - \cancel{pf} + \cancel{pf}} = \frac{2p^2f}{p^2} = 2f$$

Notice that both terms in the denominator must be multiplied by $p - f$.

**Exercise 7-5**    In problems 1 to 36, combine and simplify.

1. $\dfrac{3}{8} + \dfrac{7}{12}$

2. $\dfrac{4}{9} + \dfrac{11}{15}$

3. $\dfrac{7x}{18} + \dfrac{13x}{30}$

4. $\dfrac{11y}{90} - \dfrac{7y}{60}$

5. $\dfrac{1}{6 \times 10^3} + \dfrac{8}{3 \times 10^4}$

6. $\dfrac{5}{10^6} - \dfrac{3}{6 \times 10^5}$

7. $\dfrac{3x}{y} + \dfrac{2y}{x}$

8. $\dfrac{t}{s} + \dfrac{3t}{2s}$

9. $\dfrac{6}{5ab^2} - \dfrac{8}{15ab}$

10. $\dfrac{x}{at^2} + \dfrac{y}{a^2t}$

11. $1 + \dfrac{3c}{d} - \dfrac{5c^2}{2d^2}$

12. $\dfrac{3w}{x} + \dfrac{x}{w} - 2$

13. $\dfrac{4}{a^2 - 1} - \dfrac{3}{a + 1}$

14. $\dfrac{\pi}{r + 1} - \dfrac{\pi}{r - 1}$

15. $\dfrac{2.3}{T} + \dfrac{5.8}{T + kT}$

16. $\dfrac{0.31}{p - ap} - \dfrac{0.46}{p}$

17. $\dfrac{4}{3} - \dfrac{D + 1}{D} + \dfrac{2D + 6}{6D}$

18. $\dfrac{x - 1}{x} + \dfrac{x + 2}{2x} - \dfrac{1}{2}$

19. $\dfrac{1}{2x} - \dfrac{7}{6x} + \dfrac{4x - 1}{6x^2 - 6x}$

20. $\dfrac{8y - 2}{5y + 15} + \dfrac{1}{y + 3} - \dfrac{11}{10}$

21. $\dfrac{3y - 5}{y^2 - 1} - \dfrac{4}{y + 1}$

22. $\dfrac{2m^2}{k^2 - m^2} - \dfrac{k}{k + m}$

23. $\dfrac{a^2 + b^2}{a^2 - b^2} - \dfrac{a + b}{2a - 2b}$

24. $\dfrac{v_x + v_y}{v_x - v_y} + \dfrac{v_x v_y}{v_x^2 - v_y^2}$

25. $\dfrac{x - 1}{3x^2 + 4x + 1} + \dfrac{x}{3x + 3}$

26. $\dfrac{2b + 2}{b^2 + 6b + 9} + \dfrac{3b}{2b + 6}$

**27.** $\dfrac{3}{R^2 + 3Rx + 2x^2} - \dfrac{5}{R^2 - x^2}$

**28.** $\dfrac{x}{x^2 - y^2} - \dfrac{y}{x^2 + 2xy + y^2}$

**29.** $\dfrac{3}{I_1 + I_2} - \dfrac{1}{I_1} + \dfrac{7}{I_2}$

**30.** $\dfrac{5}{E_1} - \dfrac{1}{E_2 - E_1} + \dfrac{8}{E_2}$

**31.** $\dfrac{3}{x^2 - 1} + \dfrac{4x}{(x - 1)^2} + \dfrac{2}{x + 1}$

**32.** $\dfrac{12t}{t - 4} + \dfrac{14t}{t + 4} - \dfrac{15}{t^2 - 16}$

**33.** $\dfrac{5a^2}{a^4 - 1} + \dfrac{2}{a^2 - 1}$

**34.** $\dfrac{16}{x^2 + 1} - \dfrac{9}{x^4 - 1}$

Use formulas (7-2) and (7-3) for numbers 35 and 36.

**35.** $\dfrac{3xy}{x^3 + y^3} + \dfrac{1}{x + y}$

**36.** $\dfrac{2}{p - q} - \dfrac{p^2 + q^2}{p^3 - q^3}$

In problems 37 to 42, perform the indicated operations and simplify.

**37.** $\left(\dfrac{x}{x^2 - 1} \cdot \dfrac{x - 1}{x + 1}\right) + \dfrac{1}{(x + 1)^2}$

**38.** $\dfrac{f}{f - 1} - \left[\dfrac{f^2 - 1}{2} \cdot \dfrac{f}{(f - 1)^2}\right]$

**39.** $\dfrac{(1/x) - x}{(1/x) + 1}$

**40.** $\dfrac{(a^2/b) - b}{(a/b) + 1}$

**41.** $\dfrac{k[(2m + k)/m]}{(k^2/2) + km}$

**42.** $\dfrac{x + [xy/(x + y)]}{x[xy/(x + y)]}$

**43.** The average rate $R$ for a round trip whose one-way distance is $D$ is given by:

$$R = \dfrac{2D}{(D/r_1) + (D/r_2)}$$

where $r_1$ = rate going and $r_2$ = rate coming. (a) Simplify the complex fraction. (b) Find $R$ when $r_1 = 20$ mi/h and $r_2 = 30$ mi/h. (Would you have guessed this answer?)

**44.** Using the pythagorean theorem (4-6) (Sec. 4-7) and the definitions of sin $A$ and cos $A$ (4-8) (Sec. 4-8), show that $\sin^2 A + \cos^2 A = 1$.

**45.** The total resistance $R$ of an electric circuit containing three resistors $R_1$, $R_2$, and $R_3$ in parallel is given by:

$$\dfrac{1}{R} = \dfrac{1}{R_1} + \dfrac{1}{R_2} + \dfrac{1}{R_3}$$

Find the expression for $R$ by combining the three fractions and inverting.

**46.** The following expression arises in a problem involving microprocessors:

$$\left(\dfrac{3}{i + k} - \dfrac{4}{i}\right)v + \dfrac{v}{i}$$

Simplify this expression.

**47.** In a dc circuit two parallel resistances $R_1$ and $R_2$ have the equivalent resistance $R = R_1 R_2 / (R_1 + R_2)$. Show that the equivalent conductance $G$ is equal to $G_1 + G_2$, where $R = 1/G$, $R_1 = 1/G_1$, and $R_2 = 1/G_2$. (*Note:* Substitute, simplify, and then invert.)

**48.** When an elastic body of mass $m_0$ collides with an elastic body of mass $m$, both bodies moving at a velocity $v$ toward each other before impact, the rebound velocity of $m_0$ is given by:

$$\left(\frac{m}{m + m_0} - \frac{m_0}{m + m_0}\right)v - \frac{2vm_0}{m + m_0}$$

Simplify the expression.

**49.** The maximum deflection $\delta$ (delta) of a simple beam of length $l$ acted on by a uniform load $f$ and a central concentrated load $F$ is:

$$\delta = \frac{fl^4}{384EI} + \frac{Fl^3}{48EI}$$

where $E$ = modulus of elasticity and $I$ = moment of inertia. Simplify the expression.

**50.** The volume change of a gas under constant pressure gives rise to the expression:

$$V_1\left(1 + \frac{T_2 - T_1}{T_1}\right) - V_2\left(\frac{T_2 - T_1}{T_2} - 1\right)$$

Simplify the expression.

**51.** In Example 7-28 show that the same expression for $R$ results when $qf/(q - f)$ is substituted for $p$.

**52.** If $f_1$ and $f_2$ are the focal lengths of two thin lenses separated by a distance $d$, the focal length $F$ of the system is expressed by:

$$\frac{1}{F} = \frac{1}{f_1} + \frac{1}{f_2} - \frac{d}{f_1 f_2}$$

Find the expression for $F$ by combining the three fractions and inverting.

# 7-6 | *Equations with Fractions*

Chapter 2 contains the solutions of some simple equations with fractions. This section shows how to solve more involved equations with fractions.

## Example 7-29
Solve and check:

$$\frac{2x}{3} + \frac{3x}{2} = 26$$

## Solution

You could combine the two fractions; however, you would still have to eliminate the denominator. It is easier to *eliminate the denominators first by multiplying every term in the equation by the lcd.* Therefore multiply every term by 6:

$$(6)\frac{2x}{3} + (6)\frac{3x}{2} = (6)\,26$$

$$(\cancel{6})\frac{2}{\phantom{}}\frac{2x}{\cancel{3}} + (\cancel{6})\frac{3}{\phantom{}}\frac{3x}{\cancel{2}} = 156$$

$$4x + 9x = 156$$

$$13x = 156$$

$$x = \frac{156}{13} = 12$$

*Check:*

$$\frac{2(12)}{3} + \frac{3(12)}{2} = 26?$$

$$8 + 18 = 26 \ \checkmark$$

Observe that when each term is multiplied by the lcd, each denominator divides out. Do not confuse this procedure with the one of combining fractions done in the previous section. To *combine fractions, you multiply the top and the bottom* of each fraction. To *clear fractions in an equation, you multiply only the top* of each fraction.

## Example 7-30

The total resistance $R$ of an electric circuit containing two resistors $R_1$ and $R_2$ in parallel is given by:

$$\frac{1}{R} = \frac{1}{R_1} + \frac{1}{R_2}$$

If $R_1 = 9.0\ \Omega$ and $R_2 = 6.0\ \Omega$, what is the value of $R$?

## Solution

Substitute the given values and eliminate fractions by multiplying each term by the lcd = $18R$:

$$\frac{1}{R} = \frac{1}{9.0} + \frac{1}{6.0}$$

$$(\cancel{18R})\frac{1}{\cancel{R}}^{18} = (\cancel{18R})\frac{1}{\cancel{9.0}}^{2R} + (\cancel{18R})\frac{1}{\cancel{6.0}}^{3R}$$

$$18 = 2R + 3R$$

$$5R = 18$$

$$R = \frac{18}{5} = 3.6 \ \Omega$$

## Example 7-31

Solve and check:

$$\frac{2}{w-2} - \frac{9}{w} = \frac{1}{w}$$

## Solution

Eliminate the denominators first by multiplying each term by the lcd, $(w)(w-2)$:

$$(w)(\cancel{w-2})\frac{2}{\cancel{w-2}} - (\cancel{w})(w-2)\frac{9}{\cancel{w}} = (\cancel{w})(w-2)\frac{1}{\cancel{w}}$$

$$2w - 9(w-2) = w - 2$$

$$2w - 9w + 18 = w - 2$$

$$w = \frac{-20}{-8} = \frac{5}{2}$$

Notice that the minus sign in front of the second fraction must be multiplied by each term.

*Check:*

$$\frac{2}{5\!/\!2 - 2} - \frac{9}{5\!/\!2} = \frac{1}{5\!/\!2} \ ?$$

$$2\left(\frac{2}{1}\right) - 9\left(\frac{2}{5}\right) = \frac{2}{5} \ ?$$

$$\frac{20}{5} - \frac{18}{5} = \frac{2}{5} \ \checkmark$$

The next example shows how to solve a literal equation containing fractions.

## Example 7-32

In analyzing a system of forces, the following equation arises:

$$\frac{F_1 d_1}{F} + \frac{F_2 d_2}{4.5F} - \frac{3.6}{F^2} = 0$$

Solve this literal equation for $F$.

### Solution

First multiply by the lcd of $4.5F^2$:

$$(4.5F^2)\frac{F_1 d_1}{F} + (4.5F^2)\frac{F_2 d_2}{4.5F} - (4.5F^2)\frac{3.6}{F^2} = 0$$

$$4.5FF_1 d_1 + FF_2 d_2 - 16.2 = 0$$

Isolate the terms containing $F$ on one side, factor out $F$, and divide by the other factor:

$$4.5FF_1 d_1 + FF_2 d_2 = 16.2$$

$$F(4.5F_1 d_1 + F_2 d_2) = 16.2$$

$$\frac{F(4.5F_1 d_1 + F_2 d_2)}{4.5F_1 d_1 + F_2 d_2} = \frac{16.2}{4.5F_1 d_1 + F_2 d_2}$$

The last example shows an application where equations with fractions are used. It involves different work rates to complete a job.

## Example 7-33

Under certain conditions a solar cell can charge a dead storage battery in 14 h, whereas a wind generator takes 30 h to charge the same battery. How long will it take to charge the battery if both devices are charging simultaneously?

### Solution

The key to solving this combined rate problem is to consider the fraction of the work done by each input device in 1 h (or in another unit of time). If $t =$ number of hours to charge the battery, the verbal equation is:

$$(t)\left(\begin{array}{c}\text{fraction charged by}\\ \text{solar cell in 1 h}\end{array}\right) + (t)\left(\begin{array}{c}\text{fraction charged by}\\ \text{wind in 1 h}\end{array}\right)$$

$$= 1 \text{ ``whole'' battery}$$

The solar cell charges one-fourteenth and the wind generator charges one-thirtieth of the battery in 1 h. The algebraic equation is then:

$$t\left(\frac{1}{14}\right) + t\left(\frac{1}{30}\right) = 1$$

Solve for $t$:

$$(\overset{15}{\cancel{210}})\frac{t}{14} + (\overset{7}{\cancel{210}})\frac{t}{\cancel{30}} = (210)1$$

$$15t + 7t = 210$$

$$t = \frac{210}{22} = 9.5 \text{ h} \quad \text{(two figures)}$$

**Exercise 7-6**   In problems 1 to 28, solve and check each equation.

**1.** $\dfrac{x}{3} + \dfrac{1}{3} = 2$          **2.** $y - \dfrac{y}{2} = \dfrac{3}{4}$

**3.** $\dfrac{3a}{4} + \dfrac{a}{6} = 11$        **4.** $13 - \dfrac{2b}{3} = \dfrac{5b}{12}$

**5.** $\dfrac{q-1}{2} = \dfrac{3q}{8}$        **6.** $\dfrac{2t+1}{3} = \dfrac{3t+2}{9}$

**7.** $\dfrac{3}{5} - \dfrac{4-15y}{10} = 2y$     **8.** $\dfrac{2x}{3} - \dfrac{3x-1}{8} = 1$

**9.** $\dfrac{3.2m_0}{15} + \dfrac{8.8m_0}{45} = \dfrac{9.2}{5.0}$    **10.** $\dfrac{0.15V_r}{10} - \dfrac{0.16}{6.0} = \dfrac{0.25V_r}{30}$

**11.** $\dfrac{3}{2d} - \dfrac{9}{d^2} = \dfrac{6}{5d}$      **12.** $\dfrac{5}{4c} - \dfrac{7}{10c} = \dfrac{11}{2c^2}$

**13.** $\dfrac{5}{x+3} - \dfrac{1}{x} = \dfrac{3}{2x}$     **14.** $\dfrac{10}{2n} + \dfrac{3}{n-4} = \dfrac{20}{n}$

**15.** $\dfrac{2w-1}{w+3} + \dfrac{1}{w} = 2$    **16.** $\dfrac{3v}{v-2} - 3 = \dfrac{5}{v}$

**17.** $\dfrac{6.6}{F-1} = \dfrac{8.8}{F+1}$      **18.** $\dfrac{0.03}{2R-1} = \dfrac{0.05}{3R+1}$

**19.** $\dfrac{2}{p-3} - \dfrac{5}{2p-6} = \dfrac{1}{5}$   **20.** $\dfrac{4m}{m-2} - \dfrac{13}{3m-6} = \dfrac{1}{3}$

**21.** $\dfrac{4}{5y+5} - \dfrac{1}{20} = \dfrac{7}{10y+10}$   **22.** $\dfrac{2}{4y+12} + \dfrac{1}{3} = \dfrac{13}{2y+6} - 1$

**23.** $\dfrac{6}{3x^2-2x} + \dfrac{1}{x} = \dfrac{1}{3x-2}$   **24.** $\dfrac{1}{4x+6} - \dfrac{2}{2x^2+3x} = \dfrac{5}{4x}$

**25.** $\dfrac{8}{9-q^2} = \dfrac{3}{3+q} + \dfrac{1}{3-q}$   **26.** $\dfrac{2}{T-1} = \dfrac{5}{T^2-1} + \dfrac{3}{2T+2}$

27. $\dfrac{2}{x^2 + 2x + 1} - \dfrac{1}{x + 1}$

$= \dfrac{3}{2x + 2}$

28. $\dfrac{1}{2x^2 - 3x + 1} - \dfrac{6}{x - 1}$

$+ \dfrac{2}{2x - 1} = 0$

In problems 29 to 32, solve each linear system. Clear fractions first.

29. $\dfrac{x}{2} - \dfrac{y}{6} = \dfrac{1}{3}$

$\dfrac{x}{6} + \dfrac{y}{3} = \dfrac{1}{2}$

30. $\dfrac{x}{5} - \dfrac{y}{10} = -\dfrac{1}{2}$

$\dfrac{x}{4} + \dfrac{y}{6} = \dfrac{1}{4}$

31. $\dfrac{2F_x}{3} + \dfrac{3F_y}{5} = \dfrac{8}{15}$

$\dfrac{4F_x}{5} - \dfrac{3F_y}{10} = \dfrac{3}{10}$

32. $\dfrac{I_1}{7} + \dfrac{I_2}{2} = \dfrac{1}{2}$

$\dfrac{I_1}{6} + \dfrac{I_2}{2} = \dfrac{2}{3}$

In problems 33 to 36, solve each literal equation for the given letter. (See Example 7-32.)

33. $\dfrac{5}{x^2} - \dfrac{3}{ax} = \dfrac{1}{bx}$;   $x$

34. $\dfrac{5}{2k} - \dfrac{4}{3ky} + \dfrac{3}{4} = 0$;   $y$

35. $\dfrac{3t}{u} + \dfrac{t}{3u - 1} = \dfrac{5}{3u^2 - u}$;   $u$

36. $\dfrac{3}{2n} - \dfrac{np}{n - p} = \dfrac{2}{n}$;   $p$

37. The capacitance $C$ of a spherical capacitor is given by the formula:

$$\dfrac{k}{(9 \times 10^9)C} = \dfrac{1}{R_2} - \dfrac{1}{R_1}$$

where $R_2$ = outside radius, $R_1$ = inside radius, and $k$ = dielectric constant. Solve this formula for $C$.

38. The following equation is from a problem in celestial mechanics:

$$\dfrac{m_1}{G + 1} + \dfrac{m_2}{G - 1} = \dfrac{m_1 m_2}{G^2 - 1}$$

Solve for $G$ in terms of $m_1$ and $m_2$.

39. The following equation is from a problem which involves rotating machinery:

$$\dfrac{9}{2\pi rh} - \dfrac{6}{\pi r^2} + \dfrac{3}{4rh} = 0$$

Solve for $r$ in terms of $\pi$ and $h$.

40. A formula for the increase in resistance caused by a rise in temperature is:

$$\dfrac{R_2}{R_1} = \dfrac{M + T_2}{M + T_1}$$

Solve for $M$.

41. The focal length $f$ of a lens is given by the lensmaker's equation:

$$\frac{1}{f} = (\mu_m - 1)\left(\frac{1}{R_1} + \frac{1}{R_2}\right)$$

where $R_1$ and $R_2$ = radii of curvature and $\mu_m$ = index of refraction. Solve for $f$.

**42.** The formula for the efficiency $E$ of the diesel engine cycle is given by:

$$E = 1 - \frac{T_4 - T_1}{a(T_3 - T_2)}$$

where $a$ is a constant. If $E = 40$ percent, solve for $T_4$.

Solve each applied problem. See Example 7-33 for problems 45 to 48.

**43.** Find the total resistance $R$ in Example 7-30, given $R_1 = 12\ \Omega$ and $R_2 = 8\ \Omega$.

**44.** Find the total resistance $R$ in Example 7-30, given $R_1 = 2R$ and $R_2 = 7\ \Omega$.

**45.** Under certain conditions one tug can push a series of barges up a river in 6 h; another tug can do the job in 4 h. Under these conditions, how long will it take to push the barges up the river if both tugs work together?

**46.** One pump takes 20 h to unload an oil tanker, while a second pump takes 25 h. *(a)* If the pumps unload the tanker together, how long will it take? *(b)* If the second pump breaks down after 8 h, how long will the unloading take?

**47.** One microprocessor can process a set of inputs in 3 ms, while another microprocessor takes 6 ms. If they process the inputs together, how many seconds should it take?

**48.** A portable gasoline-powered generator can fully charge the dead storage battery in Example 7-33 in 10 h. *(a)* How long will it take if the wind generator and the gas-powered generator are charging the battery simultaneously? *(b)* How long will it take if all three devices—solar, wind, and gas—are charging simultaneously?

**49.** The first side of a triangular bridge structure is three-fifths the length of a second side. The third side is four-fifths the length of the first side. The perimeter of the structure is 260 m. What is the length of each side?

**50.** An alloy of gold contains copper and silver. The amount of copper is three fifths that of silver, and the amount of silver is one half that of gold. How many grams of each metal are in 144 g of the alloy?

**51.** The capacitance $C$ of a circuit containing two capacitances $C_1$ and $C_2$ in series is given by:

$$C = \frac{1}{1/C_1 + 1/C_2}$$

*(a)* Solve for $C_1$. *(b)* Find $C_1$ when $C = 2.5\ \mu F$ and $C_2 = 5.0\ \mu F$.

**52.** The total resistance $R$ of the series-parallel circuit in Fig. 7-1 is given by the expression:

$$R = R_1 + \frac{1}{1/R_2 + 1/R_3}$$

Find $R_2$ and $R_3$ when $R = 14\ \Omega$, $R_1 = 8.0\ \Omega$, and $R_3 = 2R_2$.

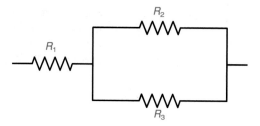

FIG. 7-1  *Series-parallel circuit, problem 52*

## 7-7 | *The Quadratic Equation: Solution by Formula*

Section 7-3 introduces the *general quadratic equation* in $x$:

$$ax^2 + bx + c = 0$$

where $a$, $b$, and $c$ are constants and $a \neq 0$.

Quadratic equations first appeared thousands of years ago in geometric problems involving areas of squares and other figures. The prefix *quad* and "$x$ squared" for $x^2$ come from their early use with squares.

Whenever a formula or equation in a problem contains a second-degree term (but no higher term), the problem usually leads to solving a quadratic equation. Applications of the pythagorean theorem in Sec. 4-7 result in simple quadratic equations that can be solved by taking square roots. Exercise 7-3 contains applied problems that lead to quadratic equations which can be solved by factoring.

The first three examples in this section show special cases of quadratic equations that can be solved by taking square roots.

An incomplete quadratic equation of the form $ax^2 + c = 0$, that is, $b = 0$, can be solved by taking square roots, as follows.

### Example 7-34
Solve by taking square roots: $2x^2 = 18$

### Solution
This is Example 7-14, which is solved by factoring in Sec. 7-3. First solve the equation for $x^2$, and then take the square root of both sides:

$$x^2 = \frac{18}{2} = 9$$

$$x = \pm\sqrt{9} = \pm 3$$

$$\text{or} \quad x = 3 \quad \text{and} \quad x = -3$$

In the applications of the pythagorean theorem (4-6) in Sec. 4-7, negative square roots do not apply. However, in general, when you take the square root of both sides of an equation, you must consider both positive and negative square roots, expressed as $\pm\sqrt{\phantom{x}}$ . In a practical problem, if an answer does not apply, you reject it.

### Example 7-35
Solve by taking square roots: $3.8I^2 - 0.75 = 0$

### Solution

$$3.8I^2 = 0.75$$

$$I^2 = \frac{0.75}{3.8} = 0.197$$

$$I = \pm\sqrt{0.197} = \pm0.44 \quad \text{(two figures)}$$

If a quadratic equation consists of a trinomial that is a perfect square on one side of the equal sign and a constant term on the other side, the equation can be solved by taking square roots as follows.

### Example 7-36
Solve by taking square roots: $4x^2 + 4x + 1 = 16$

### Solution
Factor the left side and take the square root of both sides:

$$(2x + 1)^2 = 16$$
$$\sqrt{(2x + 1)^2} = \pm\sqrt{16}$$

By applying rule (1-6), Sec. 1-4, the square and the square root cancel:

$$2x + 1 = \pm4$$

Solve for $x$:

$$x = \frac{-1 \pm 4}{2}$$

The two solutions are then:

$$x = \frac{-1 + 4}{2} = \frac{3}{2} \quad \text{and} \quad x = \frac{-1 - 4}{2} = -\frac{5}{2}$$

These roots both check in the original equation:

$$4\left(\frac{3}{2}\right)^2 + 4\left(\frac{3}{2}\right) + 1 = 16? \qquad 4\left(-\frac{5}{2}\right)^2 + 4\left(-\frac{5}{2}\right) + 1 = 16?$$

$$9 + 6 + 1 = 16 \checkmark \qquad\qquad 25 - 10 + 1 = 16 \checkmark$$

You may now be asking yourself, How do I solve a quadratic equation that cannot be solved by factoring or by taking square roots as in previous examples? Perhaps you know the answer: by applying the *quadratic formula*.

## The Quadratic Formula

Given the general quadratic equation $ax^2 + bx + c = 0$, the solution for $x$ is:

$$x = \frac{-b \pm \sqrt{b^2 - 4ac}}{2a} \tag{7-7}$$

Note that the numerator contains both $-b$ and the radical. This basic and important formula comes from the general quadratic equation by changing it to a perfect square (completing the square) and then solving for $x$ by taking square roots as shown in Example 7-36. Study the proof:

Given: $\qquad\qquad\qquad ax^2 + bx + c = 0$

Isolate $x$: $\qquad\qquad\quad ax^2 + bx = -c$

Divide by $a$: $\qquad\qquad x^2 + \frac{b}{a}x = -\frac{c}{a}$

Complete the square * by adding $b^2/(4a^2)$ to both sides:

$$x^2 + \frac{b}{a}x + \frac{b^2}{4a^2} = \frac{b^2}{4a^2} - \frac{c}{a}$$

Before factoring, multiply by $4a^2$ to clear fractions:

$$4a^2x^2 + 4abx + b^2 = b^2 - 4ac$$

Now factor the left side, which is a perfect square:

$$(2ax + b)^2 = b^2 - 4ac$$

Take square roots: $\qquad 2ax + b = \pm\sqrt{b^2 - 4ac}$

Solve for $x$: $\qquad\qquad\; 2ax = -b \pm \sqrt{b^2 - 4ac}$

$$x = \frac{-b \pm \sqrt{b^2 - 4ac}}{2a}$$

∗ Chapter 14 (Sec. 14-3, Example 14-17) explains the process of completing the square further. Study the following examples, which show how to use the quadratic formula.

### Example 7-37

Solve $5x^2 + 8x = 6x + 3$ by the quadratic formula, and check your answer by factoring.

### Solution

First put the equation in the general form:

$$5x^2 + 8x - 6x - 3 = 0$$
$$5x^2 + 2x - 3 = 0$$

Identify $a$, $b$, and $c$:

$$a = 5 \qquad b = 2 \qquad c = -3$$

Substitute in the formula:

$$x = \frac{-b \pm \sqrt{b^2 - 4ac}}{2a}$$

$$= \frac{-2 \pm \sqrt{(2)^2 - 4(5)(-3)}}{2(5)} = \frac{-2 \pm \sqrt{4 + 60}}{10}$$

Separate the two solutions:

$$x = \frac{-2 \pm \sqrt{64}}{10} \quad \nearrow \quad x = \frac{-2 + 8}{10} = \frac{6}{10} = \frac{3}{5} \quad \text{or} \quad 0.6$$

$$\searrow \quad x = \frac{-2 - 8}{10} = \frac{-10}{10} = -1$$

Notice that under the radical you multiply the minus sign in front of $-4(5)(-3)$ before combining that term with $2^2$. The roots (solutions) are *rational* (integers or fractions). This occurs when $b^2 - 4ac$ *is a perfect square*. An equation where $b^2 - 4ac$ is a perfect square can also be solved by factoring:

*Check:*

$$5x^2 + 2x - 3 = 0$$

$$(5x - 3)(x + 1) = 0$$

$$5x - 3 = 0 \qquad x + 1 = 0$$

$$x = \frac{3}{5} \qquad\qquad x = -1$$

The roots agree with those from the formula solution and serve as a check in this example. The next example cannot be solved by factoring.

### Example 7-38

Solve $3x^2 - 2x - 2 = 0$ by the quadratic formula. Express roots in *simplest radical form* and as *decimals to the nearest hundredth*. Check each decimal root.

### Solution

This equation is in the general form with $a = 3$, $b = -2$, and $c = -2$:

$$x = \frac{-b \pm \sqrt{b^2 - 4ac}}{2a}$$

$$= \frac{-(-2) \pm \sqrt{(-2)^2 - 4(3)(-2)}}{2(3)}$$

$$= \frac{2 \pm \sqrt{4 + 24}}{6} = \frac{2 \pm \sqrt{28}}{6}$$

Simplify the radical by applying rule (1-7), Sec. 1-4:

$$x = \frac{2 \pm \sqrt{4}\sqrt{7}}{6} = \frac{2 \pm 2\sqrt{7}}{6}$$

Then reduce the fraction by dividing out the common factor:

$$x = \frac{2(1 \pm \sqrt{7})}{6} = \frac{\overset{1}{\cancel{2}}(1 \pm \sqrt{7})}{\underset{3}{\cancel{6}}} = \frac{1 \pm \sqrt{7}}{3}$$

These are the roots in simplest radical form. Notice that when $b$ is negative, the minus sign in the formula makes the first term in the numerator positive.

Observe that the common factor of 2 is divided out to simplify the radical answer. The roots are *irrational*. This occurs when $b^2 - 4ac$ *is positive and not a perfect square*. An equation where $b^2 - 4ac$ is not a perfect square *cannot* be solved by factoring.

The roots in decimal form are:

$$x = \frac{1 + \sqrt{7}}{3} = \frac{1 + 2.65}{3} = 1.22$$

and

$$x = \frac{1 - \sqrt{7}}{3} = \frac{1 - 2.65}{3} = -0.55$$

The decimal roots can also be calculated without simplifying the radical. Which form of the roots is used depends on the application of the equation. To check the decimal roots approximately, you can use the calculator:

$$3(1.22)^2 - 2(1.22) - 2 = 4.47 - 2.44 - 2 = 0.03$$
$$3(-0.55)^2 - 2(-0.55) - 2 = 0.91 + 1.10 - 2 = 0.01$$

The results are close enough to zero to verify the answers.

### Example 7-39
Solve by the quadratic formula: $x^2 + 2 = x$

### Solution
Put the equation in the general form:

$$x^2 - x + 2 = 0$$

It cannot be factored. Use the formula, substituting $a = 1$, $b = -1$, and $c = 2$:

$$x = \frac{1 \pm \sqrt{(-1)^2 - 4(1)(2)}}{2(1)} = \frac{1 \pm \sqrt{-7}}{2}$$

The answers are *complex roots*. This occurs when $b^2 - 4ac$ is *negative*. They contain a real number combined with the square root of a negative number, or *imaginary number*. Complex numbers are discussed in detail in Chap. 13. At this point we note that such answers exist and that they can be left in the above form.

Table 7-1 summarizes the types of roots of a quadratic equation which are determined by the discriminant $b^2 - 4ac$.

The next example applies the methods shown in Sec. 7-6.

Table 7-1    TYPES OF ROOTS OF A QUADRATIC EQUATION

| Discriminant $b^2 - 4ac$ | Types of Roots |
|---|---|
| $b^2 - 4ac > 0$ perfect square | Real, rational, and unequal |
| $b^2 - 4ac > 0$ not perfect square | Real, irrational, and unequal |
| $b^2 - 4ac = 0$ | Real, rational, and equal |
| $b^2 - 4ac < 0$ | Complex (real + imaginary) |

**Example 7-40**

Solve by any method:    $\dfrac{5}{2x - 2} + \dfrac{3x}{4} = \dfrac{4}{x - 1}$

**Solution**

This is an equation with fractions which may not appear to be quadratic, but when you multiply by the lcd, $4(x - 1)$, you see that it is:

$$\overset{2}{\cancel{4}}(x - 1)\frac{5}{\cancel{2}(x - 1)} + \cancel{4}(x - 1)\frac{3x}{\cancel{4}} = 4\cancel{(x - 1)}\frac{4}{\cancel{x - 1}}$$

$$10 + 3x^2 - 3x = 16$$

$$3x^2 - 3x - 6 = 0$$

$$x^2 - x - 2 = 0 \qquad \text{(divide out common factor)}$$

Try factoring. This is usually easier than using the quadratic formula:

$$(x - 2)(x + 1) = 0$$

$$x = 2 \qquad \text{and} \qquad x = -1$$

The next example shows the solution of a literal quadratic equation.

**Example 7-41**

Solve the literal equation for $x$: $p^2x^2 + px = 1$

**Solution**

$$p^2x^2 + px - 1 = 0$$

$$a = p^2 \qquad b = p \qquad c = -1$$

$$x = \frac{-p \pm \sqrt{p^2 - 4(p^2)(-1)}}{2(p^2)}$$

$$= \frac{-p \pm \sqrt{p^2 + 4p^2}}{2p^2} = \frac{-p \pm \sqrt{5p^2}}{2p^2}$$

$$= \frac{-p \pm \sqrt{p^2}\sqrt{5}}{2p^2} = \frac{-p \pm p\sqrt{5}}{2p^2}$$

$$= \frac{\overset{1}{\cancel{p}}(-1 \pm \sqrt{5})}{2p^{\cancel{2}}} = \frac{-1 \pm \sqrt{5}}{2p}$$

Notice how the square root is simplified by separating $\sqrt{p^2} = p$. Radicals are discussed further in Chap. 8.

The next example illustrates an applied problem in which it is necessary to interpret the answers. Study it carefully.

### Example 7-42

A computer company produces 60 computers per month at a cost of $2000 per computer. Because of a decrease in sales, the company must cut back production or close. An analysis determines that for each computer less than 60 computers per month produced, the cost per computer rises $50. The company agrees to increase its production cost from $120,000 to $124,000 per month, but it must produce at least 50 computers per month to stay in operation. How many computers less than 60 per month should the company produce?

### Solution

Let $n$ = no. of computers less than 60 per month that the company should produce. The verbal equation is

(No. of computers/month)(cost per computer) = cost/month = $124,000

Then:

$$60 - n = \text{no. of computers/month}$$
$$2000 + 50n = \text{cost per computer}$$

The algebraic equation is:

$$(60 - n)(2000 + 50n) = \$124,000$$
$$120,000 + 1000n - 50n^2 = \$124,000$$

Move the terms to the right side to make the first term positive, and divide out the common factor of 50:

$$0 = 50n^2 - 1000n + 4000$$
$$n^2 - 20n + 80 = 0$$

The unknown $n$ takes the place of $x$ in the quadratic formula, in which $a = 1$, $b = -20$, and $c = 80$:

$$n = \frac{20 \pm \sqrt{400 - 4(1)(80)}}{2} = \frac{20 \pm \sqrt{80}}{2} = \frac{20 \pm 4\sqrt{5}}{2}$$

$$= 10 \pm 2\sqrt{5} = \begin{array}{l} \nearrow n = 14.5 \\ \rightarrow n = 5.5 \end{array}$$

Since the company must produce at least 50 computers per month, $n$ must be 10 or less. The only acceptable answer is $n = 5.5$. The nearest integral answer is 6; however, this makes the cost per month greater than \$124,000:

$n = 6$:   Cost/month $= (54)(2300) = \$124,200$

$n = 5$:   Cost/month $= (55)(2250) = \$123,750$

Therefore choose $n = 5$ as the best practical answer, and the company should cut production to 55 computers per month.

**Exercise 7-7**

In problems 1 to 10, solve each quadratic equation by taking square roots.

**1.** $3x^2 = 12$

**2.** $5.0I^2 - 3.2 = 0$

**3.** $4.2x^2 - 23 = 0$

**4.** $8.6 - 9.3r^2 = 0$

**5.** $x^2 - 4x + 4 = 49$

**6.** $9x^2 + 6x + 1 = 9$

**7.** $x^2 + 2x + 1 = 5$

**8.** $16y^2 - 16y + 4 = 8$

**9.** $(2y - 3)^2 - 4 = 0$

**10.** $16 - (x + 1)^2 = 0$

In problems 11 to 30, solve each quadratic equation by using the quadratic formula. Check rational roots by factoring. Express irrational roots in simplest radical form (or as decimals to the nearest hundredth if directed to do so by the instructor).

**11.** $x^2 + 4x + 3 = 0$

**12.** $x^2 + 2x - 8 = 0$

**13.** $3x^2 - x - 10 = 0$

**14.** $2x^2 - 7x + 3 = 0$

**15.** $2y^2 + 4 = 9y$

**16.** $3y^2 - y = 2$

**17.** $m^2 - 2m - 1 = 0$

**18.** $2p^2 - 1 = 2p$

**19.** $5v^2 = 5v + 1$

**20.** $6w - w^2 = 7$

**21.** $2x^2 - 6 = x^2 + 2x$

**22.** $4x^2 + 4x = 2 - 2x$

**23.** $2F^2 - 3F + \dfrac{1}{2} = 0$

**24.** $\dfrac{h^2}{3} - h = \dfrac{5}{6}$

**25.** $1.5t_0^2 + 1.0t_0 = 0.50$

**26.** $0.25R^2 + 1.0R + 0.25 = 0$

**27.** $x^2 + 2x + 2 = 0$

**28.** $3x^2 - x + 2 = 0$

**29.** $2y^2 = y - 1$

**30.** $3a - 3 = a^2$

In problems 31 to 44, solve by any method.

**31.** $x^2 + 3x = 0$

**32.** $2x^2 = x$

**33.** $12y^2 + 7y = 12$  **34.** $24y^2 + 5y = 36$

**35.** $\dfrac{4}{9}p^2 - 1 = 0$  **36.** $8n^2 = \dfrac{2}{9}$

**37.** $\dfrac{2}{k} - \dfrac{2}{2k+3} = 1$  **38.** $\dfrac{1}{u-1} + \dfrac{2}{3} = \dfrac{1}{u^2-1}$

ⓒ **39.** $1.1d^2 - 1.2d + 0.30 = 0$  ⓒ **40.** $L^2 = 5.2L - 3.8$

**41.** $6x^2 = 3x^2 - 3$  **42.** $x^2 + 21 = 9x$

**43.** $(x+2)^2 = 4(2x+3)$  **44.** $(3x-1)(2x+3)$
$= x^2 + 2(x-1)$

In problems 45 to 48, solve each literal equation for $x$.

**45.** $x^2 - mx - m^2 = 0$  **46.** $k^2x^2 + 2x = 3k$

**47.** $q^2x^2 - 2qx = 2$  **48.** $x^2 - 2ax = 2a^2$

**49.** The height $s$ of a rocket launched vertically from the ground at an initial velocity $v_0$ is given by $s = v_0t + \frac{1}{2}gt^2$. Solve this formula for the time $t$.

**50.** The deflection $y$ of a beam is given by $y = x^2 + Lx - 1.5L^2$, where $L =$ length and $x =$ distance from one end. Solve for the positive value of $x$ in terms of $L$ where the deflection is zero.

In problems 51 to 62, solve each applied problem by setting up and solving a quadratic equation.

**51.** One leg of a right triangular gable roof is 3 m longer than the other leg. If the area of the triangular roof is 35 m², find the length of the three sides.

**52.** A solar-cell collector on the roof of a house is to have the shape shown in Fig. 7-2. If the area of the collector must be 30 m² to collect enough energy to meet the needs of the house, what should be the dimension $x$? (*Note: x cannot be greater than 10.*)

**53.** The power plant of a small independent island community has 1000 customers who each pay a base rate of $20 per month for power. Plant officials estimate that for each increase of $1 per month in the base rate, the plant will lose 10 customers, who will change to wind, solar, or another kind of power. The company wants to in-

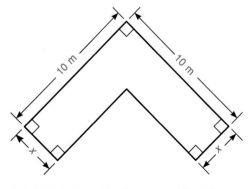

FIG. 7-2 *Solar cell collector, problem 52*

crease its monthly income to at least $30,000 but to lose as few customers as possible. How much should it increase the base rate for service? (See Example 7-42.)

**54.** Solve Example 7-42, assuming that the cost per computer rises $60 for each computer less than 60 per month produced.

**55.** A chemical tank was filled by two pumps in 2.5 h. The smaller pump alone takes 3.0 h more to fill the tank than the larger pump alone. How long does it take each pump to fill the tank alone? (See Example 7-33.)

**56.** At noon a ship sails due south from New York at an average speed of 6.0 nmi/h. One hour later another ship sails due east from New York at the same average speed. At what time will the ships be 30 nmi apart? (*Hint:* Use the pythagorean theorem.)

**57.** In a dc circuit, when a resistance $R$, an inductance $L$, and a capacitance $C$ are connected in series, the following equation must be solved to determine the instantaneous value of the current: $m^2 + (R/L)m + 1/(LC) = 0$. Find $m$ when $R = 30.0$ Ω, $L = 1.00$ H, and $C = 0.0100$ F.

**58.** The power output $P$ of a generator is given by $P = VI - RI^2$, where $V =$ generator voltage, $R =$ internal resistance, and $I =$ current. Find $I$ when $P = 40$ W, $V = 12$ V, and $R = 0.5$ Ω.

**59.** Ms. Hernandez, the owner of a fully rented 120-unit apartment house, charges rent of $300 per month for each apartment. She estimates that for each $10 increase in monthly rent she will rent one less apartment. What is the least amount she should increase the rent to bring her monthly income to $44,000? (See Example 7-42.)

**60.** Ms. Hernandez in problem 59 calculates that the maximum monthly income she can earn is $56,250. How much should she charge for monthly rent to increase her income to the maximum?

C **61.** The bending moment $M$ of a beam at a distance $x$ from a support is given by $M = 50x - 16(x - 3) - 1.5x^2$. Find the positive value of $x$ where $M = 0$.

**62.** A square machine base that is $s$ meters on a side is to be doubled in area to reduce the pressure on the floor. (*a*) If $x$ is the length to be added to each side, write the quadratic equation that relates $s$ and $x$. (*b*) Solve this equation for $x$ in terms of $s$, and reject the answer that does not apply.

## 7-8 | *Review Questions*

In problems 1 to 14, factor each expression completely.

**1.** $8x^2y - 12xy^2$     **2.** $1.5hk + 4.5hk^2$

**3.** $21a^3 + 15a^2 - 27a$     **4.** $\pi R^2h - \pi r^2h + 2\pi rh$

**5.** $9x^2 - y^2$     **6.** $2a^4 - 32$

**7.** $b^2 + 4b - 21$     **8.** $5x^3 + 20x^2 + 10x$

**9.** $3k^2 + 10k - 8$     **10.** $6x^2 - 17xy + 12y^2$

**11.** $6t^2 + 24t + 24$     **12.** $3.0p^2 - 3.0p + 0.75$

**13.** $2Q^4 + Q^2 - 1$                    **14.** $x^3 - x_1^3$

In problems 15 to 18, reduce each fraction to lowest terms.

**15.** $\dfrac{3x^2 + 7xy + 2y^2}{3x^2y + xy^2}$                    **16.** $\dfrac{mv_1^2 - mv_2^2}{v_1 + v_2}$

**17.** $\dfrac{a^3 - ab^2}{a^2 - 2ab + b^2}$                    **18.** $\dfrac{2n^2 - np - p^2}{2p^2 - pn - n^2}$

In problems 19 to 24, perform the indicated operations.

**19.** $\dfrac{6ax^2}{35} \cdot \dfrac{21a}{4a^2x}$                    **20.** $\dfrac{3x - 3y}{28y} \cdot \dfrac{7x + 7y}{6x^2 - 6y^2}$

**21.** $\dfrac{c^2 - d^2}{c^2 - 2cd + d^2} \div (c + d)$                    **22.** $\dfrac{2n^2 + n}{2n^2 + 9n - 5} \div \dfrac{2n + 1}{n^2 + 3n - 10}$

**23.** $\dfrac{x^2 + 1}{x^4 + 2x^2 + 1} \cdot \dfrac{x^4 - 1}{x^2 - 1}$                    **24.** $\dfrac{2}{3} \cdot \dfrac{3E - 3}{I + 1} \cdot \dfrac{5R + 5IR}{2E - 2}$

In problems 25 to 32, combine and simplify.

**25.** $\dfrac{5x}{6} + \dfrac{3x}{8}$                    **26.** $\dfrac{a}{3xy} - \dfrac{b}{5x^2}$

**27.** $\dfrac{4}{d - 1} - \dfrac{5}{d^2 - 1}$                    **28.** $\dfrac{m - 1}{5mv^2} + \dfrac{5}{3mv} + \dfrac{2}{15v^2}$

**29.** $\dfrac{2x^2}{x^2 + 4x + 4} - \dfrac{5x}{3x + 6}$                    **30.** $1 + \dfrac{3}{w + 1} - \dfrac{w + 1}{w - 1}$

**31.** $\dfrac{p/q - q/p}{(p + q)/q}$                    **32.** $\left(\dfrac{f + 1}{f - 1} + \dfrac{f + 1}{f}\right)\left(\dfrac{f - 1}{f + 1}\right)$

In problems 33 to 50, solve each equation.

**33.** $\dfrac{m}{8} + \dfrac{3m}{4} = \dfrac{7}{2}$                    **34.** $\dfrac{2h + 1}{4} - \dfrac{h - 1}{4} = 1$

**35.** $\dfrac{x + 1}{x} = \dfrac{2}{3x} + \dfrac{5}{2x}$                    **36.** $\dfrac{4}{y + 1} - \dfrac{7}{2y + 1} = 0$

**37.** $\dfrac{m + 6}{6m + 6} + \dfrac{m - 1}{2m + 2} = \dfrac{1}{3}$                    **38.** $\dfrac{4}{t^2 - 4} = \dfrac{t}{t - 2} - \dfrac{t}{t + 2}$

**39.** $x^2 - 4x - 5 = 0$                    **40.** $3x^2 + x - 2 = 0$

**41.** $7p = 8p^2$                    **42.** $2r^2 - 0.32 = 0$

**43.** $(d + 1)^2 + 2d^2 = 2$                    **44.** $16y^2 + 9 = 24y$

**45.** $x^2 + 3x + 1 = 0$                    **46.** $4w^2 - 4w = 1$

**47.** $2x^2 + 2x + 1 = 10$                    © **48.** $R^2 - 0.3R = 0.5$

**49.** $5n + 2 = 3n^2 - 4$                    **50.** $2x^2 + 3x + 3 = 0$

Solve each literal equation in problems 51 and 52 for $x$.

**51.** $a^2x + 5 = t^2x$                    **52.** $x^2 - kx = k^2$

In problems 53 to 62, solve each applied problem.

**53.** The kinetic energy of one particle is given by $\frac{1}{2}mv_1^2$, and the kinetic energy of a second particle is $\frac{1}{8}mv_2^2$. Factor the expression that represents the difference in energy between the two particles.

**54.** The force $F$ exerted by a gas on the inside of a spherical balloon is:

$$F = \frac{8\pi r^4 - 16\pi r^3}{2r^2 - 10r}$$

where $r$ = radius of the balloon. Pressure is defined as $P = F/A$, where $A$ = area. Express the pressure on the inside of the balloon in terms of $r$, and simplify the expression. (*Note:* Surface area of a sphere = $4\pi r^2$.)

**55.** The surface area of a box is given by $S = 4h^2 + 28h + 48$. Factor this expression.

**56.** A variable with a subscript such as $T_1$ is written in BASIC as T1. You are given the following expression in BASIC:

$$K*T \wedge 2 - K*T1 \wedge 2$$

Write this expression in algebraic notation and factor this expression. (See Exercise 1-1, problem 23, and Exercise 1-4, problem 51.)

**57.** The stress factor in a certain spring is given by:

$$\frac{3.0}{K} + \frac{2.2K - 3.9}{1.3K + 2.1}$$

Combine fractions and simplify the expression.

**58.** A technician takes 3.0 h to record all the necessary data in a chemical plant. Her assistant takes 4.0 h to record all the data. Assuming they work independently, how long will it take for both of them to record all the data? (See Example 7-33.)

**59.** The total resistance $R$ of two resistances $R_1$ and $R_2$ in parallel is given by $1/R = 1/R_1 + 1/R_2$. Find $R_1$ and $R_2$ if $R = 6.0\ \Omega$ and $R_2 = R_1 + 5.0\ \Omega$.

**60.** A car leaves a town traveling west at an average speed of 50 km/h. Two hours later another car leaves the town traveling north at an average speed of 70 km/h. How many hours after the second car leaves will the cars be 200 km apart? (*Hint:* Use the pythagorean theorem.)

**61.** The deflection $y$ of a beam is given by $y = 3.0x^2 + 2.0x - 1.5$, where $x$ = distance from one end. At what distance $x$ is the deflection zero?

$\boxed{C}$ **62.** A piece of metal 25 cm wide is to be bent into an open rectangular trough of cross-sectional area 75 cm², as shown in Fig. 7-3. What are the two possible values for the height $h$ of the trough?

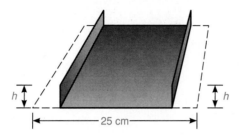

FIG. 7-3 *Trough, problem 62*

# 8

# Exponents and Radicals

## 8-1 | *Rules for Exponents*

The rules for exponents (Sec. 2-4) are as follows where $x$ or $y$ cannot equal zero:

**E**xponents (powers) and radicals (roots) are introduced in Chaps. 1 and 2. This chapter explores the subject further and shows the close relation between these two mathematical operations. It will provide you with the more complete understanding of exponents and radicals that is necessary for scientific and technical applications.

$$x^{-n} = \left(\frac{1}{x}\right)^n = \frac{1}{x^n} \qquad \left(\frac{x}{y}\right)^{-n} = \left(\frac{y}{x}\right)^n \quad (8\text{-}1)$$

$$x^0 = 1 \qquad\qquad\qquad (8\text{-}2)$$

$$(x^m)(x^n) = x^{m+n} \quad \text{addition rule} \quad (8\text{-}3)$$

$$\frac{x^n}{x^m} = x^{n-m} \quad \text{subtraction rule} \quad (8\text{-}4)$$

$$(x^m)^n = x^{mn} \quad \begin{array}{l}\text{multiplication} \\ \text{rule}\end{array} \quad (8\text{-}5)$$

$$(xy)^n = x^n y^n \qquad \left(\frac{x}{y}\right)^n = \frac{x^n}{y^n} \qquad (8\text{-}6)$$

While $x$ or $y$ cannot be equal to zero, there are no restrictions on $m$ and $n$. You will see that *these rules apply to exponents that are positive or negative integers or fractions,* that is, to any exponent that is a rational number. Study the following examples, which illustrate various applications of these rules. Expressions are considered simpler when all the exponents are positive.

### Example 8-1
Simplify and express with positive exponents:

**1.** $\dfrac{(x^2x^3)^2}{x^4} = \dfrac{(x^5)^2}{x^4} = \dfrac{x^{10}}{x^4} = x^6$

**2.** $\dfrac{-(2ab)^3}{2b^{-2}} = \dfrac{-8a^3b^3}{2b^{-2}} = -4a^3b^{3-(-2)} = -4a^3b^5$

**3.** $0.2w^4(0.1w)^{-2} = 0.2w^4\dfrac{1}{(0.1w)^2} = \dfrac{0.2w^4}{0.01w^2} = 20w^2$

**4.** $\left(\dfrac{3xy^2}{-5x^2y}\right)^{-2} = \left(\dfrac{-5x^2y}{3xy^2}\right)^2 = \left(\dfrac{-5x}{3y}\right)^2 = \dfrac{25x^2}{9y^2}$

**5.** $\dfrac{-5r^0}{(-r)^2(10r)^0} = \dfrac{-5(1)}{r^2(1)} = \dfrac{-5}{r^2}$

Observe in (5) that *an exponent applies to a coefficient or to a minus sign only if it is within parentheses.* For example, $(-2)^2 = 4$ but $-2^2 = -4$. Also observe that (2) can be done another way:

$$\dfrac{-(2ab)^3}{2b^{-2}} = \dfrac{-8a^3b^3}{2b^{-2}} = \dfrac{-8a^3b^3(b^2)}{2} = -4a^3b^5$$

Look at the arrows above. A *factor* can be switched from the denominator to the numerator or vice versa by changing the sign of the exponent:

$$\left(\dfrac{a^{-2}}{b^{-1}}\right) = \dfrac{b}{a^2}$$

## Example 8-2
Simplify and express with positive exponents.

**1.** $(3c + c^{-1})^{-1} = \dfrac{1}{3c + c^{-1}} = \dfrac{1}{3c + 1/c} = \dfrac{1}{(3c^2 + 1)/c} = \dfrac{c}{3c^2 + 1}$

Observe in (1) that you cannot move $c^{-1}$ to the numerator since it is not a factor of the denominator.

**2.** $\dfrac{0.5x^2 - 1.5 + 0.8x^{-2}}{0.1x^{-2}} = \dfrac{0.5x^2}{0.1x^{-2}} - \dfrac{1.5}{0.1x^{-2}} + \dfrac{0.8x^{-2}}{0.1x^{-2}}$

$$= 5x^4 - 15x^2 + 8$$

**3.** $e^{-at}(e^{at} + e^{bt}) = e^0 + e^{-at+bt} = 1 + e^{(b-a)t}$

In (3) you apply the rules to the literal exponents and express the final exponent in factored form. Note that many of these problems can be done correctly in more than one way.

**Exercise 8-1**  In problems 1 to 34, simplify and express with positive exponents.

**1.** $(x^{-2})(x^2)^3$

**2.** $\dfrac{y^3y^2}{y^{-2}}$

**3.** $\dfrac{(aa^2)^2}{a^2}$

**4.** $(d^2)(d^{-2})^{-2}$

**5.** $(2t^0)(2t)^0$

**6.** $\dfrac{3y^0}{(3y)^0}$

**7.** $(3pq^{-2})(3pq)^{-2}$

**8.** $(abc^{-1})^2(abc)^{-2}$

**9.** $\dfrac{-(1.5r)^2}{-0.5r^{-1}}$

**10.** $-\left(\dfrac{1}{4}x^3\right)\left(-\dfrac{1}{2}x\right)^{-2}$

**11.** $(0.3w)^2(0.1w)^{-1}$

**12.** $\dfrac{5.2(EI)^2}{1.3EI^{-1}}$

**13.** $\dfrac{a^{-1}b^2}{ab^{-2}}$

**14.** $\dfrac{2x^{-2}}{(3xy)^{-1}}$

**15.** $\left(\dfrac{-3Lx^0}{5L^2}\right)^{-2}$

**16.** $\left(\dfrac{2mn^{-2}}{7m^2}\right)^{-1}$

**17.** $\dfrac{-\frac{1}{5}sv^{-2}}{(\frac{1}{5}sav)^2}$

**18.** $\dfrac{5.0(mt)^{-1}}{(-0.1mt)^{-2}}$

**19.** $(10x^{-3}y)\left(\dfrac{5y}{3x^2}\right)^{-3}$

**20.** $\left(\dfrac{am^2}{bm}\right)^{-2}(abm^2)^2$

**21.** $\left(\dfrac{1}{m}\right)^{-1} + (mn)^{-2}$

**22.** $(2ab)^{-1} - 3ab^{-2}$

**23.** $(p + q^{-1})^{-1}x$

**24.** $(R_1^{-1} + 3R_2)^{-1}$

**25.** $(x^{-1} + y^{-1})^{-1}$

**26.** $(a^{-1} - b^{-1})(b - a)^{-1}$

**27.** $\dfrac{4\pi rh^2 + 2\pi rh^{-2}}{2\pi rh^{-2}}$

**28.** $\dfrac{6.3xy + 4.2xy^0}{2.1xy^{-1}}$

**29.** $e^{at}(e^{-at} - e^{bt})$

**30.** $r^{n\theta}(r^{2n\theta} + r^{n\theta})$

**31.** $(p^x + p^{-x})(p^x - p^{-x})$

**32.** $(x^n + x^{-n})^2$

**33.** $\dfrac{k^y + k^{-y}}{k^y}$

**34.** $\dfrac{v_0^{2x} - v_0^x}{v_0^x}$

**35.** The total resistance $R$ of the circuit in Fig. 8-1 is given by the equation $R = [R_1^{-1} + (R_2 + R_3)^{-1}]^{-1}$. Simplify this expression for $R$.

**36.** A circular coil of $n = 2 \times 10^2$ turns and radius $r = 10^{-1}$ m has a magnetic flux density of $B = 30 \times 10^{-5}$ T (tesla). The current $I$ required to produce this flux

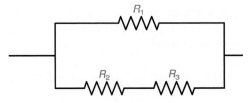

FIG. 8-1 *Parallel resistances, problem 35*

density in a vacuum is given by:

$$I = \frac{2rB}{(4\pi \times 10^{-7})(n)}$$

Substitute the values supplied and calculate the current in amperes.

**37.** In studying the vibration of a machine on spring mounts, the following expression arises:

$$\frac{e^{-\gamma(n+1)t}}{e^{-\gamma nt}}$$

Simplify the expression.

**38.** The amount of radiation from the sun that reaches the earth's surface is:

$$1.35 \text{ kW} \cdot \text{m}^{-2} = 3.94 \times 10^2 \text{ Btu} \cdot \text{h}^{-1} \cdot \text{ft}^{-2}$$

Express both sets of units without negative exponents.

**39.** The efficiency of a turbojet engine is given by the expression:

$$1 - \left(\frac{p_1}{p_2}\right)^{\frac{\alpha}{1-\alpha}}$$

where $p_1/p_2$ is the compression ratio. Simplify this expression for $\alpha = 1.5$.

**40.** To determine how much money $P$ (principal) you must save today so that it will amount to $A$ dollars in $n$ years at an interest rate of $r$ percent per year, apply the compound interest formula: $P = A(1 + r)^{-n}$. Find $P$ if $A = \$10{,}000$, $r = 10$ percent, and $n = 10$ years.

## 8-2 | *Fractional Exponents*

In Sec. 1-4, rule (1-10), the fractional exponent $1/n$ is defined:

$$x^{1/n} = \sqrt[n]{x} \qquad (8\text{-}7)$$

The exponent $1/n$ is defined in this way so that the rules of exponents apply. Observe how the rules are satisfied for $x^{1/2}$:

$$(x^{1/2})^2 = (x^{1/2})(x^{1/2}) = x^{1/2+1/2} = x^1 = x$$
$$\downarrow \qquad \downarrow \quad \downarrow \qquad\qquad\qquad \text{results agree}$$
$$(\sqrt{x})^2 = (\sqrt{x})(\sqrt{x}) = x$$

When the multiplication rule is used (8-5), it follows that:

$$(x^{1/n})^n = x^{n/n} = x \qquad \text{or} \qquad (\sqrt[n]{x})^n = x \qquad (8\text{-}8)$$

This is the same as rule (1-9) in Sec. 1-4.

Further extension of this idea gives the definition of a *fractional exponent:*

$$x^{m/n} = (x^{1/n})^m = (\sqrt[n]{x})^m \qquad \text{or} \qquad \sqrt[n]{x^m} \qquad (8\text{-}9)$$

All the rules for exponents, (8-1) to (8-6), apply to fractional exponents.

### Example 8-3

Evaluate or simplify and express with positive exponents.

**1.** $27^{2/3} = (\sqrt[3]{27})^2 = (3)^2 = 9$

Notice in (1) that $27^{2/3}$ can also be evaluated by first raising to the power and then taking the root: $\sqrt[3]{27^2} = \sqrt[3]{729} = 9$. However, it is easier to take the root first.

**2.** $16^{-3/4} = \dfrac{1}{16^{3/4}} = \dfrac{1}{(\sqrt[4]{16})^3} = \dfrac{1}{2^3} = \dfrac{1}{8}$

In (2) a negative fractional exponent is treated in the same way as a negative integral exponent using (8-1).

**3.** $(4 \times 10^{-4})^{3/2} = 4^{3/2} \times 10^{(-4)(3/2)} = 8 \times 10^{-6}$ or $\dfrac{8}{10^6}$

**4.** $(0.25x^2y^4)^{3/2} = (0.25)^{3/2}(x^{(2)(3/2)})(y^{(4)(3/2)}) = 0.125x^3y^6$

**5.** $\dfrac{(4v_0t)^{3/2}}{(4v_0t)^{1/2}} = (4v_0t)^{3/2-1/2} = (4v_0t)^1 = 4v_0t$

**6.** $(a^{1/2} + a^{-1/2})(a^{1/2} - a^{-1/2}) = (a^{1/2})(a^{1/2}) - (a^{-1/2})(a^{-1/2}) = a - a^{-1}$

$= a - \dfrac{1}{a} = \dfrac{a^2 - 1}{a}$

Observe in (6) that the binomial product has no middle term like the difference of two squares.

Rational exponents are readily done on a calculator as shown in the next example.

### ⊏ Example 8-4

The average velocity $\bar{v}$ of fluid flow in an open channel is given by the Chézy formula

$$\bar{v} = \frac{1.49r_H^{2/3}S_0^{1/2}}{n}$$

where $r_H$ = hydraulic radius, $S_0$ = slope of the channel, and $n$ = roughness factor. Calculate $\bar{v}$ (in meters per second) when $r_H = 5.4$ m, $S_0 = 0.0016$, and $n = 0.028$.

**Solution**

$$\bar{v} = \frac{1.49(5.4)^{2/3}(0.0016)^{1/2}}{0.028} = 6.6 \text{ m/s} \quad \text{(two figures)}$$

This is done on the calculator:

1.49 $\boxed{\times}$ $\boxed{(}$ 5.4 $\boxed{y^x}$ $\boxed{(}$ 2 $\boxed{\div}$ 3 $\boxed{)}$ $\boxed{)}$ $\boxed{\times}$ 0.0016 $\boxed{\sqrt{x}}$ $\boxed{\div}$ 0.028 $\boxed{=}$ → 6.6

**Exercise 8-2**

In problems 1 to 40, evaluate or simplify and express with positive exponents.

**1.** $25^{3/2}$

**2.** $16^{3/4}$

**3.** $(-27)^{4/3}$

**4.** $-81^{5/4}$

**5.** $100^{-1/2}$

**6.** $(1/32)^{1/5}$

**7.** $9^{-3/2}$

**8.** $(-64)^{-2/3}$

**9.** $0.216^{1/3}$

**10.** $0.512^{2/3}$

**11.** $(4 \times 10^2)^{5/2}$

**12.** $(8 \times 10^{-6})^{2/3}$

**13.** $0.001^{-2/3}$

**14.** $0.0016^{3/4}$

**15.** $\left(\dfrac{49}{16}\right)^{3/2}$

**16.** $\left(-\dfrac{27}{8}\right)^{-2/3}$

**17.** $8^{1/6}8^{1/2}$

**18.** $9^{-1/3}9^{5/6}$

**19.** $x^{-1/4}x^{-3/4}$

**20.** $y^{0.3}y^{0.2}$

**21.** $(-27a^3b^6)^{1/3}$

**22.** $(256h^4p^2)^{0.5}$

**23.** $(100w^2x^2)^{-3/2}$

**24.** $(0.001b^3y^6)^{2/3}$

**25.** $\left(\dfrac{10^{-6}}{I^2}\right)^{1/2}$

**26.** $\left(\dfrac{R^{-2}}{16}\right)^{-1/4}$

**27.** $\left(\dfrac{p^{1.5}}{p^{0.5}}\right)^{-2}$

**28.** $\left(\dfrac{V^{0.5}}{V^{-1.5}}\right)^{1/2}$

**29.** $(2x^{1/2})^2(3x^{4/3})^3$

**30.** $(-5y^{1/3})^{-3}(10y^{3/2})^2$

**31.** $\left(\dfrac{9a^{-2}}{4b^4}\right)^{-1/2}$

**32.** $\left(\dfrac{64k^{-3}}{k^6}\right)^{1/3}$

**33.** $\left(\dfrac{e^{4t}}{e^t}\right)^{1/3}$

**34.** $(k^m k^{-3m})^{-1/2}$

**35.** $\dfrac{a^{1/2}(a^{3/2} - a^{1/2})}{a}$

**36.** $\dfrac{(8t)^{4/3} + t^{1/3}}{(8t)^{1/3}}$

**37.** $(2^{1/2} + 2^{-1/2})(2^{1/2} - 2^{-1/2})$

**38.** $(4^{1/4} + 4^{-1/4})^2$

**39.** $(x^{3/2} + x^{1/2})^2$

**40.** $(y^{1/2} + 1)(y^{1/2} - 1)$

41. The behavior of iron in a magnetic field gives rise to the equation $W = nB^{1.6}$. Calculate $W$ if $n = 1.3$ and $B = 0.072$.

42. The maximum deflection $\delta$ (delta) of a beam with a concentrated off-center load is given by $\delta = (3.1 \times 10^{-6})(b/l)(l^2 - b^2)^{3/2}$. Find $\delta$ when $l = 25$ m and $b = 20$ m.

43. In calculating the heat transfer during a chemical reaction, the following expression arises: $R^{0.8}(C/R)^{0.4}$. Simplify this expression.

44. In calculating the stress in a soil sample under a concentrated load, the following expression arises:

$$\left[ (Z^{-4/5}) \left( \frac{Z^2 + r^2}{Z^2} \right) \right]^{5/2}$$

Simplify this expression.

45. At a certain reservoir, the time $t$ (in minutes) that it takes for the runoff of rainfall to reach a maximum is given by:

$$t = 7.0 \left( \frac{d}{Sr^2} \right)^{1/3}$$

where $d$ = distance (in meters) to the most remote area supplying the reservoir, $S$ = slope, and $r$ = rain intensity (in $cm/h$). Find $t$ when $d = 200$ m, $S = 0.20$, and $r = 8.0$ cm/h.

46. The average discharge velocity $\bar{v}$ of a large rectangular orifice in a water tank is given by:

$$\bar{v} = \frac{2}{3} C_Q b (h_2{}^{2/3} - h_1{}^{2/3})(2g)^{1/2}$$

Find $\bar{v}$ (in meters per second) when $C_Q = 0.600$, $b = 0.100$ m, $h_2 = 0.343$ m, $h_1 = 0.216$ m, and $g = 9.81$ m/s$^2$.

# 8-3 | *Basic Operations with Radicals*

From definition (8-7) you can see that fractional exponents and radicals are two ways of expressing the same idea. At times, radical notation is more convenient, especially when you work with square roots and cube roots, the most common radicals. The following rules for radicals follow from rule (8-6) for exponents:

$$\sqrt[n]{xy} = (\sqrt[n]{x})(\sqrt[n]{y}) \tag{8-10}$$

$$\sqrt[n]{\frac{x}{y}} = \frac{\sqrt[n]{x}}{\sqrt[n]{y}} \tag{8-11}$$

Rules (8-10) and (8-11) *do not apply to even roots of negative numbers,* which are imaginary.

### Example 8-5

Simplify by applying the rules for radicals.

**1.** $\sqrt{(5x)^2} = 5x$      by (8-8)

**2.** $\pi\sqrt[3]{r^6} = = \pi r^{6/3} = \pi r^2$      by (8-9)

Notice in (2) that you divide the index of the radical into the exponent under the radical.

**3.** $5\sqrt{18} = 5\sqrt{9}\sqrt{2} = 5(3)\sqrt{2} = 15\sqrt{2}$      by (8-10)

**4.** $\sqrt{a^3b^5} = \sqrt{a^2b^4}\sqrt{ab} = ab^2\sqrt{ab}$      by (8-10)

**5.** $\sqrt[3]{16} = \sqrt[3]{8}\sqrt[3]{2} = 2\sqrt[3]{2}$      by (8-10)

In (3), (4), and (5), you factor out perfect roots to simplify the radical. The only integral perfect square roots that should be factored out are $\sqrt{4}$, $\sqrt{9}$, $\sqrt{16}$, $\sqrt{25}$, $\sqrt{36}$, etc. In would not help to factor $\sqrt{18}$ as $\sqrt{6}\sqrt{3}$, since $\sqrt{6}$ and $\sqrt{3}$ are not perfect roots.

**6.** $\dfrac{\sqrt{6}}{\sqrt{24}} = \sqrt{\dfrac{6}{24}} = \sqrt{\dfrac{1}{4}} = \dfrac{1}{2}$      by (8-11)

Fractions with radicals are in a simpler form to work with when you *rationalize the denominator.* This means eliminating radicals (irrational quantities) from the denominator as the next example shows.

### Example 8-6

Rationalize the denominator for each radical:

**1.** $\sqrt{\dfrac{3}{5}}$      **2.** $\sqrt{\dfrac{25}{18}}$

**Solution**

**1.** $\sqrt{\dfrac{3}{5}}$

You can do the example two ways. Work inside the radical or work outside: Inside method:

$$\sqrt{\frac{3}{5}} = \sqrt{\frac{3(5)}{5(5)}} = \sqrt{\frac{15}{25}} = \frac{\sqrt{15}}{5} \quad \text{or} \quad \frac{1}{5}\sqrt{15}$$

Outside method:

$$\sqrt{\frac{3}{5}} = \frac{\sqrt{3}}{\sqrt{5}} = \frac{\sqrt{3}(\sqrt{5})}{\sqrt{5}(\sqrt{5})} = \frac{\sqrt{15}}{5} \quad \text{or} \quad \frac{1}{5}\sqrt{15}$$

In both methods multiply the top and bottom by the quantity that makes the *radicand* (the number inside the radical) in the denominator a perfect square. Radicals with fractions can usually be simplified correctly in more than one way.

**2.** $\sqrt{\dfrac{25}{18}}$

You can multiply top and bottom by 18 to obtain the perfect square of 324 in the denominator; however, it is easier to multiply by 2 and obtain 36:

$$\sqrt{\frac{25}{18}} = \sqrt{\frac{25(2)}{18(2)}} = \sqrt{\frac{50}{36}} = \frac{5\sqrt{2}}{6}$$

Another way to do the example is to first simplify the numerator and denominator:

$$\sqrt{\frac{25}{18}} = \frac{\sqrt{25}}{\sqrt{18}} = \frac{5}{3\sqrt{2}} = \frac{5\ (\sqrt{2})}{3\sqrt{2}(\sqrt{2})} = \frac{5\sqrt{2}}{3(2)} = \frac{5\sqrt{2}}{6}$$

*Radicals cannot be combined unless they are similar, meaning the radicands (and indices) are identical.* Remember that a radical symbol is a convenient way of representing an irrational number (or infinite decimal): $\sqrt{2} = 1.414213562 \cdots$, $\sqrt{7} = 2.645751311 \cdots$, etc. Similar radicals are treated like similar algebraic terms. Study the following example, which shows the comparison.

### Example 8-7
Combine similar radicals.      *Algebraic comparison*

**1.** $\sqrt{2} + \sqrt{3} + \sqrt{2} =$      $x + y + x =$      $(x = \sqrt{2},\ y = \sqrt{3})$
    $2\sqrt{2} + \sqrt{3}$            $2x + y$

**2.** $5\sqrt{7} + 2\sqrt{5} - 3\sqrt{5} =$     $5a + 2b - 3b =$     $(a = \sqrt{7},\ b = \sqrt{5})$
    $5\sqrt{7} - \sqrt{5}$             $5a - b$

**3.** $\dfrac{3\sqrt{2}}{2} - \dfrac{\sqrt{2}}{2} = \dfrac{2\sqrt{2}}{2} = \sqrt{2}$     $\dfrac{3x}{2} - \dfrac{x}{2} = \dfrac{2x}{2} = x$     $(x = \sqrt{2})$

Observe that only similar radicals are combined; radicals that are not similar remain as separate terms. Radicals that do not appear similar may sometimes be simplified and then combined as follows.

## Example 8-8

Simplify and combine similar radicals.

**1.** $2\sqrt{12} + 4\sqrt{3} - \sqrt{27} = 2\sqrt{4}\sqrt{3} + 4\sqrt{3} - \sqrt{9}\sqrt{3}$
$$= 2(2)\sqrt{3} + 4\sqrt{3} - 3\sqrt{3}$$
$$= 4\sqrt{3} + 4\sqrt{3} - 3\sqrt{3} = 5\sqrt{3}$$

**2.** $\sqrt{50x^3} + 3\sqrt{72x^3} - 5\sqrt{8x^2}$
$$= \sqrt{25x^2}\sqrt{2x} + 3\sqrt{36x^2}\sqrt{2x} - 5\sqrt{4x^2}\sqrt{2}$$
$$= 5x\sqrt{2x} + 3(6x)\sqrt{2x} - 5(2x)\sqrt{2}$$
$$= 5x\sqrt{2x} + 18x\sqrt{2x} - 10x\sqrt{2}$$
$$= 23x\sqrt{2x} - 10x\sqrt{2}$$

Notice in (2) that $\sqrt{2x}$ and $\sqrt{2}$ cannot be combined.

**3.** $\sqrt{\dfrac{2}{3}} + \dfrac{\sqrt{3}}{\sqrt{8}} = \sqrt{\dfrac{2(3)}{3(3)}} + \dfrac{\sqrt{3}(\sqrt{2})}{\sqrt{8}(\sqrt{2})} = \dfrac{\sqrt{6}}{\sqrt{9}} + \dfrac{\sqrt{6}}{\sqrt{16}}$
$$= \dfrac{\sqrt{6}}{3} + \dfrac{\sqrt{6}}{4} = \dfrac{7\sqrt{6}}{12}$$

In (3) you could multiply the top and bottom of the second fraction by $\sqrt{8}$ to rationalize the denominator; however, it is simpler to multiply by the smaller quantity, $\sqrt{2}$.

**4.** $2\sqrt[3]{16} + 4\sqrt[3]{2} = 2\sqrt[3]{8}\sqrt[3]{2} + 4\sqrt[3]{2}$
$$= 2(2)\sqrt[3]{2} + 4\sqrt[3]{2} = 8\sqrt[3]{2}$$

**Exercise 8-3**   In problems 1 to 36, simplify by applying the rules for radicals.

**1.** $\sqrt{(6x)^2}$                                **2.** $(\sqrt[3]{3y})^3$

**3.** $\sqrt{0.04v^2}$                              **4.** $\sqrt{\dfrac{R^4}{49}}$

**5.** $\sqrt{4x^4y^2}$                              **6.** $\sqrt[3]{8a^3b^6}$

**7.** $\sqrt{12}$                                     **8.** $\sqrt{50}$

**9.** $2\sqrt{27}$                                   **10.** $5\sqrt{200}$

**11.** $\dfrac{2\sqrt{18}}{3}$                       **12.** $\dfrac{5\sqrt{32}}{4}$

**13.** $\sqrt{x^5y^3}$                              **14.** $a\sqrt{ab^3}$

**15.** $\sqrt{32r^2s^5}$                            **16.** $\sqrt{64p^3q^3}$

**17.** $\sqrt[3]{24}$                                **18.** $\sqrt[4]{32}$

**19.** $\sqrt[3]{0.027v_0^4}$                        **20.** $\sqrt[4]{81d^5}$

**21.** $\sqrt{5}\sqrt{45}$                           **22.** $\sqrt{6}\sqrt{18}$

**23.** $\dfrac{\sqrt{2}\sqrt{27}}{\sqrt{6}}$

**24.** $\dfrac{\sqrt{50}}{\sqrt{5}\sqrt{40}}$

**25.** $\sqrt[4]{0.08}\sqrt[4]{0.02}$

**26.** $\dfrac{\sqrt[3]{16x^4}}{\sqrt[3]{2x}}$

**27.** $\sqrt{0.50h}\sqrt{2.0h^3}$

**28.** $\sqrt{0.2I^2R}\sqrt{0.05I^2R}$

**29.** $\sqrt{12 \times 10^4}$

**30.** $\sqrt{16 \times 10^{-5}}$

**31.** $\sqrt{1.44 \times 10^{-6}}$

**32.** $\sqrt{1.69 \times 10^3}$

**33.** $\sqrt[4]{x^{4n}}$

**34.** $\sqrt{b^{2n+4}}$

**35.** $\sqrt[n]{a^{n+1}}$

**36.** $\sqrt[2n]{r^{4n}t^{3n}}$

In problems 37 to 56, rationalize each denominator and simplify.

**37.** $\sqrt{\dfrac{1}{3}}$

**38.** $\dfrac{1}{\sqrt{2}}$

**39.** $\dfrac{\sqrt{5}}{\sqrt{6}}$

**40.** $\sqrt{\dfrac{2}{5}}$

**41.** $\dfrac{2}{\sqrt{3}}$

**42.** $\dfrac{\sqrt{7}}{\sqrt{12}}$

**43.** $\sqrt{\dfrac{9}{8}}$

**44.** $\sqrt{\dfrac{8}{27}}$

**45.** $\sqrt[3]{\dfrac{1}{4}}$

**46.** $\sqrt[3]{\dfrac{2}{3}}$

**47.** $\sqrt{\dfrac{1}{x}}$

**48.** $\sqrt{\dfrac{1}{a^3}}$

**49.** $\sqrt{\dfrac{2h}{g}}$

**50.** $\dfrac{\sqrt{0.02F_x}}{\sqrt{0.5F_y}}$

**51.** $\sqrt[3]{\dfrac{3x}{4y^2}}$

**52.** $\dfrac{8}{\sqrt[4]{8}}$

**53.** $\sqrt{\dfrac{1}{R} + \dfrac{1}{4}}$

**54.** $\sqrt{\dfrac{8m}{m_0 + m}}$

**55.** $\sqrt[3]{\dfrac{1000}{qH}}$

**56.** $\sqrt[4]{\dfrac{0.0001p}{t}}$

In problems 57 to 90, simplify and combine similar radicals.

**57.** $3\sqrt{5} + 2\sqrt{7} - \sqrt{5}$

**58.** $5\sqrt{3} - 7\sqrt{2} + 6\sqrt{2}$

**59.** $\sqrt{45} + 2\sqrt{20}$

**60.** $4\sqrt{24} - 5\sqrt{6}$

**61.** $3\sqrt{12} - \sqrt{48} + 2\sqrt{27}$

**62.** $8\sqrt{8} + \sqrt{50} - 3\sqrt{2}$

**63.** $\sqrt{125} + 2\sqrt{5} - \sqrt{500}$

**64.** $3\sqrt{63} - 8\sqrt{28} + 7\sqrt{7}$

**65.** $\dfrac{1}{2}\sqrt{32} + 2\sqrt{\dfrac{1}{2}}$

**66.** $6\sqrt{\dfrac{1}{3}} - 2\sqrt{12}$

**67.** $\sqrt[3]{16} + 2\sqrt[3]{2}$

**68.** $\sqrt[4]{32} - \sqrt[4]{2}$

**69.** $\sqrt{18x^3} + 2\sqrt{2x^3}$

**70.** $\sqrt{12ab^2} - \sqrt{48ab^2}$

**71.** $\sqrt{0.64\pi r^2 h} - \sqrt{0.16\pi r^2 h}$

**72.** $0.10\sqrt{n^2 C^3} - 0.05nC\sqrt{4C}$

**73.** $\sqrt{\dfrac{1}{2}} + \sqrt{\dfrac{1}{8}}$

**74.** $\dfrac{2}{\sqrt{8}} - \dfrac{1}{2\sqrt{2}}$

**75.** $\dfrac{3\sqrt{3}}{2} + \dfrac{2\sqrt{12}}{3}$

**76.** $\dfrac{3\sqrt{200}}{10} - \dfrac{2\sqrt{50}}{5}$

**77.** $6\sqrt{\dfrac{2}{5}} - \sqrt{10} + \dfrac{2\sqrt{40}}{5}$

**78.** $5\sqrt{20} - 20\sqrt{\dfrac{1}{5}} - \sqrt{5}$

**79.** $1.5t\sqrt{v} + \dfrac{6.6vt}{\sqrt{v}}$

**80.** $3.3a\sqrt{\dfrac{w^3}{a}} - 2.5w\sqrt{aw}$

**81.** $5x\sqrt[3]{xy^4} - 3y\sqrt[3]{x^4 y}$

**82.** $\sqrt[3]{81T^4} - T\sqrt[3]{24T}$

**83.** $\sqrt{1.44 \times 10^3} + \sqrt{1.21 \times 10^3}$

**84.** $\sqrt{2.25 \times 10^7} - \sqrt{1.00 \times 10^7}$

**85.** $\sqrt{4x - 4} + \sqrt{9x - 9}$

**86.** $\sqrt{16 + 16y} + 2\sqrt{1 + y}$

**87.** $\sqrt{\dfrac{9p}{2q}} + \sqrt{\dfrac{p}{8q}}$

**88.** $n\sqrt{\dfrac{m}{2n}} + m\sqrt{\dfrac{n}{2m}}$

**89.** $\dfrac{\sqrt{x + 1}}{\sqrt{x - 1}} + \dfrac{\sqrt{x - 1}}{\sqrt{x + 1}}$

**90.** $\sqrt{\dfrac{1}{r^2} + 1} - \dfrac{\sqrt{r^2 + 1}}{2r}$

**91.** The resultant $F$ of three forces is expressed by the equation $F = \sqrt{(3x)^2 + (4x)^2 + (5x)^2}$. Simplify the radical.

**92.** Figure 8-2 shows the vector diagram for the resultant velocity $v$ of a missile with a horizontal component $v_x$ and a vertical component $v_y$. If $v_x = 3t + 4$ and $v_y = 4t - 3$, express $v$ in terms of $t$ in simplest radical form. [*Hint:* Use the pythagorean theorem (4-6).]

**93.** In a steady-state $RC$ (resistance-capacitance) circuit the impedance $Z$ is given by:

$$Z = \sqrt{R^2 + \left(\dfrac{1}{2\pi fC}\right)^2}$$

Calculate $Z$ in ohms when $R = 40\ \Omega$, $C = 2 \times 10^{-5}$ F, and $f = 120$ Hz.

**94.** The diameter $d$ required for a circular shaft that transmits $H$ hp at $n$ revolutions per minute (r/min) with torsional stress $s$ is given by:

$$d = \sqrt[3]{\dfrac{(3.65 \times 10^5)H}{ns}}$$

Find $d$ (in centimeters) when $H = 650$ hp, $n = 200$ r/min, and $s = 800$ kgf/cm$^2$.

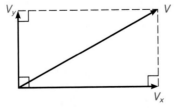

FIG. 8-2 *Velocity diagram, problem 92*

**95.** A circular disk of radius $r$ and weight $w$ rolling down an inclined plane reaches a maximum velocity given by:

$$v = \sqrt{\frac{2ghr^2w}{gI + r^2w}}$$

where $h$ = initial height, $g$ = gravitational acceleration, and $I$ = moment of inertia. Given $I = wr^2/(2g)$, simplify this radical expression for $v$.

**96.** The velocity $V$ (in miles per hour) of a boat is given by:

$$V = 1.1\sqrt{L}\sqrt[3]{\frac{1000H}{D}}$$

where $L$ = waterline length, $H$ = brake or shaft output horsepower, and $D$ = displacement (weight). Find $V$ when $L = 32$ ft, $H = 64$ hp, and $D = 27,000$ lb.

**97.** The impedance $Z$ of a certain circuit is given by:

$$Z = \frac{1}{\sqrt{(1/X)^2 + (1/R)^2}}$$

Simplify this expression by rationalizing the denominator.

**98.** The velocity $V$ of a small water wave is given by:

$$V = \sqrt{\frac{\pi}{4\lambda d}} + \sqrt{\frac{4\pi}{\lambda d}} \quad (\lambda = \text{lambda})$$

Simplify and combine, expressing $V$ with a rational denominator.

# 8-4 | *Binomial Radical Expressions*

The previous section shows how to use rules (8-10) and (8-11) to multiply and divide monomial radicals. When radical expressions contain more than one term, the same rules apply; but you must remember to treat the radicals as algebraic terms. Study the following example, which shows the procedure.

### Example 8-9
Multiply and simplify.

**1.** $\sqrt{2}(2\sqrt{3} + \sqrt{2}) = (\sqrt{2})(2\sqrt{3}) + (\sqrt{2})(\sqrt{2})$
$= 2\sqrt{2}\sqrt{3} + (\sqrt{2})^2 = 2\sqrt{6} + 2$

**2.** $\sqrt[3]{4}(2\sqrt[3]{2} + \sqrt[3]{16}) = 2\sqrt[3]{4}\sqrt[3]{2} + \sqrt[3]{4}\sqrt[3]{16}$
$= 2\sqrt[3]{8} + \sqrt[3]{64} = (2)(2) + 4 = 8$

**3.** $(\sqrt{3} + \sqrt{2})(2\sqrt{3} + \sqrt{2}) = (\sqrt{3})(2\sqrt{3}) + (\sqrt{3})(\sqrt{2})$
   $+ (\sqrt{2})(2\sqrt{3}) + (\sqrt{2})(\sqrt{2})$
   $= 2(\sqrt{3})^2 + \sqrt{6} + 2\sqrt{6} + (\sqrt{2})^2$
   $= 2(3) + 3\sqrt{6} + 2 = 8 + 3\sqrt{6}$

**4.** $(\sqrt{5} + \sqrt{3})(\sqrt{5} - \sqrt{3}) = (\sqrt{5})^2 - (\sqrt{3})^2 = 5 - 3 = 2$

Observe in (3) and (4) that the FOIL method of multiplying algebraic binomials is used (see Sec. 2-6). *Note particularly in (4) that there are no radicals in the result.* Binomial radical expressions of the form $a\sqrt{x} + b\sqrt{y}$ and $a\sqrt{x} - b\sqrt{y}$ are called *conjugates*. You can rationalize the denominator of a fraction containing a binomial radical expression by multiplying by the conjugate of the denominator as follows.

**Example 8-10**
Rationalize the denominators and simplify.

**1.** $\dfrac{2}{\sqrt{3} - 1} = \dfrac{2}{(\sqrt{3} - 1)} \dfrac{(\sqrt{3} + 1)}{(\sqrt{3} + 1)} = \dfrac{2\sqrt{3} + 2}{(\sqrt{3})^2 - (1)^2}$

   $= \dfrac{2\sqrt{3} + 2}{3 - 1} = \dfrac{2(\sqrt{3} + 1)}{2} = \sqrt{3} + 1$

**2.** $\dfrac{\sqrt{2} - 2\sqrt{3}}{\sqrt{2} + \sqrt{3}} = \dfrac{(\sqrt{2} - 2\sqrt{3})}{(\sqrt{2} + \sqrt{3})} \dfrac{(\sqrt{2} - \sqrt{3})}{(\sqrt{2} - \sqrt{3})}$

   $= \dfrac{(\sqrt{2})^2 - \sqrt{2}\sqrt{3} - 2\sqrt{2}\sqrt{3} + 2(\sqrt{3})^2}{(\sqrt{2})^2 - (\sqrt{3})^2}$

   $= \dfrac{2 - 3\sqrt{6} + 6}{2 - 3} = \dfrac{8 - 3\sqrt{6}}{-1} = -8 + 3\sqrt{6} \text{ or } 3\sqrt{6} - 8$

Notice in (1) and (2) that rationalizing the denominator greatly simplifies the expression.

**3.** $\dfrac{\sqrt{x} + \sqrt{y}}{\sqrt{x} - \sqrt{y}} = \dfrac{(\sqrt{x} + \sqrt{y})}{(\sqrt{x} - \sqrt{y})} \dfrac{(\sqrt{x} + \sqrt{y})}{(\sqrt{x} + \sqrt{y})}$

   $= \dfrac{(\sqrt{x})^2 + \sqrt{x}\sqrt{y} + \sqrt{x}\sqrt{y} + (\sqrt{y})^2}{(\sqrt{x})^2 - (\sqrt{y})^2}$

   $= \dfrac{x + 2\sqrt{xy} + y}{x - y}$

Study the multiplication in the numerator of (3) closely. Note that $(\sqrt{x} + \sqrt{y})(\sqrt{x} + \sqrt{y}) = (\sqrt{x} + \sqrt{y})^2 = x + 2\sqrt{xy} + y$. This is *not* the same as $(\sqrt{x})^2 + (\sqrt{y})^2 = x + y$. It is a common mistake to square a binomial by squaring the first and last terms and not multiplying the outer and inner products. Note also that $\sqrt{x^2 + y^2}$ is *not* the same as $\sqrt{x^2} + \sqrt{y^2}$. You can see this by substituting numbers into the expressions: $\sqrt{3^2 + 4^2} = \sqrt{9 + 16} = \sqrt{25} = 5$, which is not the same as $\sqrt{3^2} + \sqrt{4^2} = 3 + 4 = 7$. $\sqrt{x^2 + y^2}$ can not be simplified.

**Exercise 8-4**

In problems 1 to 30, perform the indicated operations and simplify.

1. $\sqrt{3}(\sqrt{6} - 1)$

2. $3\sqrt{2}(\sqrt{10} + \sqrt{2})$

3. $2\sqrt{5}(\sqrt{15} + 3\sqrt{2})$

4. $\sqrt{8}(4\sqrt{2} - 4)$

5. $\sqrt{\frac{1}{3}}\left(\sqrt{\frac{2}{3}} + \sqrt{6}\right)$

6. $\sqrt{\frac{1}{2}}\left(\sqrt{8} - \sqrt{\frac{3}{4}}\right)$

7. $\sqrt[3]{3}(\sqrt[3]{18} - 2\sqrt[3]{9})$

8. $\sqrt[4]{4}(\sqrt[4]{4} + 2\sqrt[4]{8})$

9. $\sqrt{0.1x}(\sqrt{0.9x} + \sqrt{0.1x})$

10. $\sqrt{2.0r}(\sqrt{0.72r} - \sqrt{0.32})$

11. $\sqrt{ab}(\sqrt{2a} + \sqrt{3b})$

12. $\sqrt{2v}(\sqrt{8v^3} - \sqrt{2v})$

13. $\dfrac{\sqrt{15} - \sqrt{20}}{\sqrt{5}}$

14. $\dfrac{\sqrt{10x^3} + \sqrt{6x}}{\sqrt{2x}}$

15. $(\sqrt{3} - 1)(\sqrt{3} + 3)$

16. $(\sqrt{5} - 4)(\sqrt{5} + 1)$

17. $(3\sqrt{6} + \sqrt{2})(\sqrt{6} + 3\sqrt{2})$

18. $(\sqrt{7} - 2\sqrt{3})(\sqrt{7} + \sqrt{3})$

19. $(\sqrt{2} - 3)^2$

20. $(\sqrt{3} + 2)^2$

21. $(\sqrt{2} - 3)(\sqrt{2} + 3)$

22. $(\sqrt{3} - \sqrt{2})(\sqrt{3} + \sqrt{2})$

23. $(2\sqrt{5} - \sqrt{7})(2\sqrt{5} + \sqrt{7})$

24. $(\sqrt{8} - 3\sqrt{2})(\sqrt{8} + 3\sqrt{2})$

25. $(\sqrt{3} + 7)(\sqrt{3} + 1)(\sqrt{3} - 2)$

26. $(3 - \sqrt{2})(1 + \sqrt{2})^2$

27. $(\sqrt{xy} - 1)(\sqrt{xy} + 1)$

28. $(\sqrt{m} - \sqrt{n})^2$

29. $\left(\sqrt{\frac{1}{2}} + \sqrt{\frac{1}{8}}\right)^2$

30. $\left(\sqrt{\frac{x}{3}} - \sqrt{\frac{y}{5}}\right)\left(\sqrt{\frac{x}{3}} + \sqrt{\frac{y}{5}}\right)$

In problems 31 to 42, rationalize denominators and simplify.

31. $\dfrac{1}{\sqrt{2} - 1}$

32. $\dfrac{1}{2 + \sqrt{5}}$

33. $\dfrac{5}{\sqrt{7} - \sqrt{2}}$

34. $\dfrac{6}{2\sqrt{3} + 3\sqrt{2}}$

35. $\dfrac{\sqrt{3} + 2}{\sqrt{3} - 2}$

36. $\dfrac{2 + \sqrt{7}}{5 + \sqrt{7}}$

37. $\dfrac{\sqrt{3} + \sqrt{5}}{3\sqrt{3} - \sqrt{5}}$

38. $\dfrac{\sqrt{6} - 2\sqrt{7}}{\sqrt{6} + \sqrt{7}}$

39. $\dfrac{x}{\sqrt{2x} + \sqrt{x}}$

40. $\dfrac{a}{\sqrt{ax} - \sqrt{a}}$

41. $\dfrac{\sqrt{p} + \sqrt{q}}{2\sqrt{p} - 3\sqrt{q}}$

42. $\dfrac{\sqrt{ac} + \sqrt{bc}}{\sqrt{a} + \sqrt{b}}$

43. Multiply and simplify the following expression from a problem involving the centroid (the center of mass) of a body:

$$\pi\sqrt{ab}(b\sqrt{a} - a\sqrt{b})$$

44. Simplify the following expression from a problem in fluid dynamics:

$$\frac{157}{1 + m/\sqrt{r}}$$

**45.** The following expression arises from a problem involving the expansion of a gas:

$$\frac{\sqrt{V_1}}{\sqrt{V_1} - \sqrt{V_2}}$$

Simplify the expression.

**46.** The total resistance $R$ of two resistances in parallel, $R_1$ and $R_2$, is given by:

$$R = \frac{R_1 R_2}{R_1 + R_2}$$

In a given circuit $R_1 = x$ and $R_2 = \sqrt{x}$. *(a)* Substitute for $R_1$ and $R_2$ and simplify the result. *(b)* Find $R$ when $x = 15\ \Omega$.

## 8-5 | *Equations with Radicals*

Applied problems often lead to equations with radicals. Radicals are eliminated by squaring (or raising to a higher power) both sides of the equation. Study the following examples, which show how to solve equations with radicals. The first example illustrates the important idea of an extraneous root.

**Example 8-11**
Solve and check:     $\sqrt{3x - 2} = 4$

**Solution**
Square both sides:

$$(\sqrt{3x - 2})^2 = (4)^2$$
$$3x - 2 = 16$$
$$x = 6$$

*Check:*

$$\sqrt{3(6) - 2} = 4?$$
$$\sqrt{16} = 4\ \checkmark$$

Squaring sometimes introduces roots that do *not* satisfy the original equation, called *extraneous roots*. It is therefore necessary to *check all answers*. Consider Example 8-11, with $-4$ instead of $+4$ on the right side of the equation. When you square both sides, $(\sqrt{3x-2})^2 = (-4)^2$, you get the same equation as with $+4$: $3x - 2 = 16$, and the same answer, $x = 6$. When you check this answer, it does not satisfy the equation:

$$\sqrt{3(6) - 2} = \sqrt{16} = 4 \neq -4$$

The problem arises because the radical $\sqrt{16}$ represents only the positive root and squaring introduces the negative root. The equation $\sqrt{3x - 2} = -4$ has no solution; $x = 6$ is an extraneous root.

### Example 8-12
Solve and check all roots:      $\sqrt{4x + 2} + 2x = 3$

### Solution
You must first *isolate* the radical; otherwise, squaring both sides will not eliminate it:

$$\sqrt{4x + 2} = 3 - 2x$$
$$(\sqrt{4x + 2})^2 = (3 - 2x)^2$$
$$4x + 2 = 9 - 12x + 4x^2$$

Note that you must square the entire left side and right side of the equation. This means squaring the binomial on the right, producing three terms and a quadratic equation:

$$4x^2 - 16x + 7 = 0$$
$$(2x - 1)(2x - 7) = 0$$
$$2x - 1 = 0 \qquad 2x - 7 = 0$$
$$x = \tfrac{1}{2} \quad \text{and} \quad x = \tfrac{7}{2}$$

*Check:*

$$x = \tfrac{1}{2} \qquad\qquad x = \tfrac{7}{2}$$
$$\sqrt{4(\tfrac{1}{2}) + 2} = 3 - 2(\tfrac{1}{2})? \qquad \sqrt{4(\tfrac{7}{2}) + 2} = 3 - 2(\tfrac{7}{2})?$$
$$\sqrt{4} = 3 - 1 \checkmark \qquad\qquad \sqrt{16} \neq 3 - 7 \checkmark$$

The check shows that $x = \tfrac{7}{2}$ is an extraneous root and the only solution is $x = \tfrac{1}{2}$.

### Example 8-13
The average velocity $\bar{v}$ of water draining slowly from the vessel in Fig. 8-3 is given by:

$$\bar{v} = \frac{C\sqrt{2g}}{2}(\sqrt{H} + \sqrt{h})$$

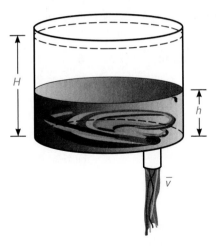

FIG. 8-3 *Velocity of water drain, Example 8-13*

where $C$ = constant, $g$ = gravitational acceleration, $H$ = initial height, and $h$ = final height. The velocity is to average 26 ft/s as the water level drops from full ($H$) to 25 percent full ($h$). If $C = 1.0$, to what height should the tank be filled?

**Solution**
Substitute $h = H/4$, $g = 32$ ft/s², $C = 1.0$, and $\bar{v} = 26$ into the formula:

$$26 = \frac{1.0\sqrt{64}}{2}\left(\sqrt{H} + \sqrt{\frac{H}{4}}\right)$$

Simplify and solve for $H$:

$$26 = 4\left(\sqrt{H} + \frac{\sqrt{H}}{2}\right) = 4\left(\frac{3\sqrt{H}}{2}\right) = 6\sqrt{H}$$

$$\sqrt{H} = \frac{26}{6} = 4.5$$

$$H = (4.5)^2 = 20 \text{ ft} \quad \text{(two figures)}$$

**Example 8-14**
Solve and check all roots: $\qquad \sqrt{5r - 1} - \sqrt{r} = 3$

**Solution**
You must square *twice* to eliminate all the radicals as follows:

Isolate the first radical: $\qquad\qquad\qquad \sqrt{5r - 1} = \sqrt{r} + 3$

Square carefully:                    $(\sqrt{5r-1})^2 = (\sqrt{r}+3)^2$

$$5r - 1 = r + 6\sqrt{r} + 9$$

Isolate the second radical:          $6\sqrt{r} = 4r - 10$

Divide out 2:                        $3\sqrt{r} = 2r - 5$

Square again:                        $(3\sqrt{r})^2 = (2r-5)^2$

$$9r = 4r^2 - 20r + 25$$

Solve the quadratic:            $4r^2 - 29r + 25 = 0$

$$(r-1)(4r-25) = 0$$

$$r = 1 \quad \text{and} \quad r = \frac{25}{4}$$

Both roots check in the original equation. Neither is an extraneous root.

**Exercise 8-5**     In problems 1 to 18, solve and check all roots.

1. $\sqrt{x+2} = 3$

2. $2\sqrt{y-5} = 4$

3. $\sqrt{5a-2} = 1/2$

4. $\sqrt{0.1 - 0.6b} = 0.2$

5. $\sqrt{3p+5} + 1 = 3p$

6. $\sqrt{T-7} + 1 = T/4$

7. $\sqrt{2w} + \sqrt{8w} = 4w + 1$

8. $\sqrt{5v} - 5 = \sqrt{20v} - 2v$

9. $\sqrt[3]{x} + 0.030 = 0.40$

10. $\sqrt[4]{y} - 0.010 = 0.30$

11. $\sqrt{m} + \sqrt{2m+1} = 5$

12. $\sqrt{3q+1} = \sqrt{q} - 1$

13. $\sqrt{c+2} + 1 = \sqrt{c+6}$

14. $\dfrac{\sqrt{d}}{2} + \dfrac{5}{2} = \sqrt{d+7}$

15. $\sqrt{L} + \sqrt{L+2} = \sqrt{3}$

16. $\sqrt{3h+5} = \sqrt{h} + \sqrt{5}$

17. $\sqrt{x} + \sqrt{2x} = 1$

18. $\sqrt{y} + \sqrt{\dfrac{y}{3}} = 2$

In problems 19 to 22, solve each literal equation for the given letter.

19. $R = N\sqrt{1 + u^2}$; $u$

20. $r = \sqrt[3]{\dfrac{3A}{4\pi}}$; $A$

21. $t = \dfrac{v_0}{g} + \sqrt{\dfrac{v_0^2}{g^2} - \dfrac{2s_0}{g}}$; $s_0$

22. $Q = \dfrac{2bH}{3}\sqrt{2gH}$; $H$

23. In Example 8-13, to what height should the tank be filled to average the same velocity while the water level drops from full to half-full? (*Hint:* Factor out $\sqrt{H}$.)

24. In Example 8-13, the time $t$ it takes to lower the water level from $H$ to $h$ is given approximately by:

$$t = 100\left(\sqrt{\dfrac{H}{2g}} - \sqrt{\dfrac{h}{2g}}\right)$$

If $H = 16$ ft and the water flows for 10 s, how high is the final water level $h$? Use $g = 32$ ft/s.

25. If at a point between two light sources the illuminance from each source is the same, the following equation holds:

$$\frac{d_1}{d_2} = \frac{\sqrt{I_1}}{\sqrt{I_2}}$$

Here $d_1$ = distance from the source whose luminous intensity is $I_1$ and $d_2$ = distance from the source whose luminous intensity is $I_2$. Find $I_1$ and $I_2$ in candelas (cd) when $d_1 = 13$ m, $d_2 = 6.0$ m, and $I_1 = 4I_2 + 20$.

26. In an ac circuit containing a resistance and inductance, the maximum current is given by:

$$I_{max} = \frac{V_{max}}{\sqrt{R^2 + (2\pi f L)^2}}$$

where $V_{max}$ = maximum voltage and $f$ = frequency. Solve this formula for the inductance $L$.

27. The diameter $d$ of a cylindrical column necessary for it to resist a crushing force $F$ is given by $d = (1.12 \times 10^{-2})\sqrt[3]{Fl}$, where $l$ is the length of the column. Solve for $F$.

28. In studying the molecular refraction of a substance, this formula arises:

$$n = \sqrt{\frac{M + 2Nd}{M - Nd}}$$

Solve for $N$.

29. The side $s$ of a cube is increased to $s'$, doubling the volume. (a) Express the ratio $s/s'$ in simplest radical form. (b) Find $s'$ when $s = 2.3$ cm.

30. For a regular hexagon the area $A = 3sr$, where $r$ = radius of the inscribed circle and $s$ = side. Given that $r = s\sqrt{3}/2$, solve the formula for $s$ and show that $A = 2r^2\sqrt{3}$. (This is the formula in Table 4-4.)

# 8-6 | *Review Questions*

In problems 1 to 12, evaluate or simplify and express with positive exponents.

1. $(x^2 x^{-3})^{-1}$

2. $\dfrac{ab^{-2}}{a^2 b^2}$

3. $\left(\dfrac{3n^0}{2p^{-2}}\right)^3$

4. $\left(\dfrac{\pi r^2}{r^3}\right)(2\pi r)^2$

5. $\left(\dfrac{x}{2}\right)^{-2} + \left(\dfrac{3x}{4}\right)^{-1}$

6. $\dfrac{k^y - k^{-y}}{k^{-y}}$

**7.** $27^{2/3}$

**8.** $(1.44 \times 10^4)^{3/2}$

**9.** $(8a^3b^6)^{1/3}$

**10.** $\left(\dfrac{x^2y^4}{16}\right)^{-1/2}$

**11.** $\left(\dfrac{25K^5}{4K^{-3}}\right)^{1/2}$

**12.** $p^{1/2}(p^{1/2} - p^{3/2})$

In problems 13 to 32, perform the indicated operations or rationalize denominators and simplify.

**13.** $\sqrt{18}$

**14.** $\dfrac{5\sqrt{12}}{2}$

**15.** $2x\sqrt{75x^3}$

**16.** $\sqrt[3]{2x}\sqrt[3]{32x^2}$

**17.** $\sqrt{1.21 \times 10^{-4}}$

**18.** $\dfrac{\sqrt{96}}{\sqrt{3}}$

**19.** $\dfrac{1}{\sqrt{8}}$

**20.** $\sqrt{\dfrac{12}{7}}$

**21.** $\sqrt{\dfrac{x}{y^3}}$

**22.** $\sqrt{\dfrac{1}{0.08F}}$

**23.** $\sqrt{6} - \sqrt{3} + 3\sqrt{6}$

**24.** $2\sqrt{50} - \sqrt{8} - \sqrt{18}$

**25.** $\sqrt[3]{2} + 2\sqrt[3]{16}$

**26.** $\dfrac{2}{\sqrt{3}} + \dfrac{1}{\sqrt{12}}$

**27.** $\sqrt{0.16x} + \sqrt{\dfrac{x}{0.25}}$

**28.** $3a\sqrt{at^3} - 2t\sqrt{a^3t}$

**29.** $\sqrt{12}(3\sqrt{2} - 2\sqrt{3})$

**30.** $(\sqrt{3} + 2\sqrt{5})(\sqrt{3} - \sqrt{5})$

**31.** $\dfrac{4}{\sqrt{6} - 2}$

**32.** $\dfrac{\sqrt{x} + 1}{\sqrt{x} - 1}$

In problems 33 to 36, solve and check all roots.

**33.** $\sqrt{2x + 1} = 4x$

**34.** $a - \sqrt{a - 1} = 7$

**35.** $\dfrac{\sqrt{h}}{2} + \dfrac{5\sqrt{h}}{6} = \dfrac{4}{5}$

**36.** $2\sqrt{w} = 2 + \sqrt{w + 7}$

In problems 37 to 44, solve each applied problem.

**37.** A power cable, telephone wire, or any line supported only at the ends will lie in a curve called a *catenary,* which is described by the function:

$$y = \dfrac{e^x + e^{-x}}{2}$$

where $e \approx 2.718$. Express this function without negative exponents by multiplying the top and bottom by $e^x$.

**38.** The approximate discharge $Q$ of a triangular dam used to divert water is given by $Q = mH^{5/2}$, where $H$ = height of the water and $m$ = constant. Find $Q$ (in meters cubed per second) if $m = 2.50$ and $H = 16.0$ m.

C **39.** When a gas expands adiabatically (with no heat added or removed from the system), temperatures and volumes are related:

$$\frac{T_1}{T_2} = \left(\frac{V_1}{V_2}\right)^{1-\alpha}$$

where $\alpha$ depends on the specific heat of the gas. *(a)* Show that this relation can be written:

$$\frac{T_1}{T_2} = \left(\frac{V_1}{V_2}\right)^{\alpha-1}$$

*(b)* Find $T_1/T_2$ when $V_1 = 2.30$ m³, $V_2 = 5.75$ m³, and $\alpha = 1.40$.

**40.** In an ac circuit the effective value of the current $I = \sqrt{I_{max}^2/2}$, where $I_{max} =$ maximum current. Simplify this expression for $I$.

**41.** Simplify the following expression for the average velocity of a fluid draining from a vessel:

$$\bar{v} = C\left(\sqrt{\frac{gH}{2}} + \sqrt{\frac{gh}{2}}\right)$$

C **42.** The maximum range of a ship's radar unit is given by $R_{max} = 2.2(\sqrt{h_1} + \sqrt{h_2})$, where $h_1 =$ height of the antenna and $h_2 =$ height of the target. A tower at a harbor entrance begins to show as an image on the radar screen at the maximum range of 16 mi. If the antenna height is 20 m, what is the height of the tower?

**43.** A submerged object appears to be closer to the surface than it actually is because of refraction. When the angle of incidence $\theta$ in Fig. 8-4 is 30°, the ratio of the true

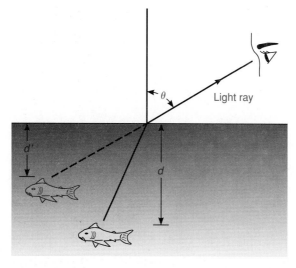

FIG. 8-4 *Refraction of light, problem 43*

depth $d$ to the apparent depth $d'$ is given by:

$$\frac{d}{d'} = \sqrt{\frac{4\mu^2 - 1}{3}}$$

where $\mu$ (mu) is the index of refraction of the liquid. *(a)* Solve this formula for $\mu$. *(b)* If an object 7.0 m below the surface appears to be only 5.0 m below it, find $\mu$ for the liquid.

44. Einstein's theory of relativity states that the mass $m$ of a body at a velocity $v$ close to the speed of light $c$ is given by:

$$m = \frac{m_0}{\sqrt{1 - v^2/c^2}}$$

where $m_0$ = rest mass. Find $v$ when $m = 2m_0$ (use $c = 3.0 \times 10^8$ m/s).

# CHAPTER

# 9

# Exponential and Logarithmic Functions

## 9-1 | *Exponential Functions*

**F**igure 9-1 shows world population growth from the beginning of modern science and technology to the present. If the population continues to increase at this rate, there will be over 6 billion people in the world before the year 2000. Population tends to increase *exponentially* when uncontrolled. Exponential functions and their inverse, logarithmic functions, are very important in technology. They are used in the study of electric circuits, electronics, energy exchange, earthquake measurement, chemical concentration, radioactive decay, and monetary interest. This chapter discusses these functions and many of their applications.

An *exponential function* is defined as:

$$y = b^x \qquad b > 0 \qquad (9\text{-}1)$$

where the *base b* is a positive constant and the exponent is a variable.

### Example 9-1

Graph $y = 2^x$ and compare the curve with that in Fig. 9-1.

### Solution

Choose integral values of $x$ for ease in graphing:

| $x$ | ← ... | $-4$ | $-3$ | $-2$ | $-1$ | $0$ | $1$ | $2$ | $3$ | $4$ | → ... |
|---|---|---|---|---|---|---|---|---|---|---|---|
| $y = 2^x$ | $0\leftarrow$ | $\dfrac{1}{16}$ | $\dfrac{1}{8}$ | $\dfrac{1}{4}$ | $\dfrac{1}{2}$ | $1$ | $2$ | $4$ | $8$ | $16$ | $\to \infty$ |

Notice that $2^x$ approaches positive infinity ($\infty$) as $x$ increases and approaches zero as $x$ decreases, but $2^x$ never equals zero. Remember: *A negative exponent means take the reciprocal and raise to the positive power*. For example, $2^{-2} =$

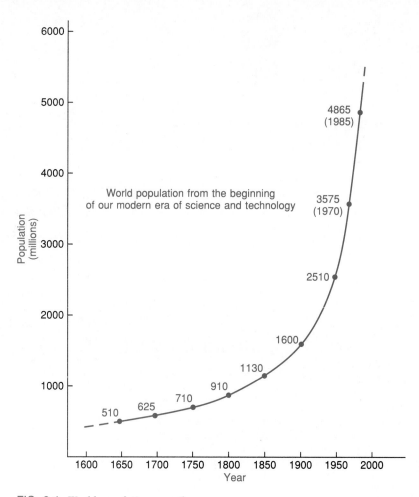

FIG. 9-1 *World population growth*

$1/2^2 = 1/4$. The graph of $y = 2^x$ (Fig. 9-2) is very similar in shape to the population graph in Fig. 9-1. These graphs represent *exponential growth functions,* which are functions that increase at faster and faster rates as they grow larger. For this reason world population growth cannot continue on its present path.

By using the curve in Fig. 9-2, a value of $y$ corresponding to *any real value* of the exponent $x$ can be approximated. For example, $2^{1.8} \approx 3.5$, as shown on the graph. The calculator provides a more precise answer:

$$2 \;\boxed{y^x}\; 1.8 \;\boxed{=}\; \longrightarrow 3.4822 \cdots$$

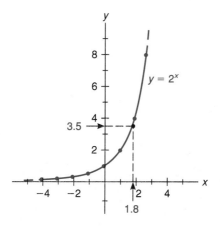

FIG. 9-2 *Exponential growth function, Example 9-1*

You can also attach meaning to an irrational exponent such as $\pi$: $2^{\pi} = 8.82497 \cdots$. Try $2^{\pi}$ on your calculator.

## Example 9-2
Graph $y = 2^{-x}$.

## Solution
Before you look at the table below, note that $y = 2^{-x} = 1/2^x = (\frac{1}{2})^x$. An exponential function with a negative exponent is equivalent to the reciprocal base with a positive exponent:

| $x$ | $\leftarrow$ ... | $-4$ | $-3$ | $-2$ | $-1$ | $0$ | $1$ | $2$ | $3$ | $4$ | $\rightarrow$ ... |
|---|---|---|---|---|---|---|---|---|---|---|---|
| $y = 2^{-x} = \left(\dfrac{1}{2}\right)^x$ | $\infty \leftarrow$ | $16$ | $8$ | $4$ | $2$ | $1$ | $\dfrac{1}{2}$ | $\dfrac{1}{4}$ | $\dfrac{1}{8}$ | $\dfrac{1}{16}$ | $\rightarrow 0$ |

The graph (Fig. 9-3) represents an *exponential decay function*. This graph is the mirror image of that in Fig. 9-2; the $y$ axis is the "mirror."

The base 2, used in Examples 9-1 and 9-2, is good for illustration but is not as useful as base 10 and base $e$ (discussed below). The graph of $10^x$ is similar to that of $2^x$; however, only a small part of the curve can be effectively shown on ordinary graph paper. The graph of $10^x$ is shown in Fig. 9-9 on semilogarithmic graph paper.

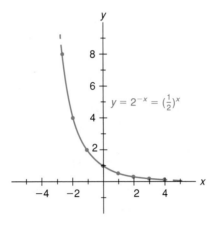

FIG. 9-3 *Exponential decay function, Example 9-2*

## *The Exponential Function* $e^x$

The number $e$, which equals $2.718281828459 \cdots$, is irrational like $\pi$ and is basic to scientific and technical work. The importance of $e$ is shown in examples and exercises in this chapter and in Chap. 13. The reason for its choice as a base is explained in calculus. The letter $e$ is chosen in honor of the Swiss mathematician Leonhard Euler (1707–1783), a great contributor to mathematics.

The expression $(1 + 1/n)^n$ approaches $e$ as $n$ becomes infinitely large. You can approximate $e$ on the calculator by letting $n$ equal a large number, say $10^6$. Then $1 + 1/n = 1 + 1/10^6 = 1.000001$, and $(1 + 1/n)^n$ is:

$$1.000001 \boxed{y^x} \; 1000000 \boxed{=} \rightarrow 2.71828$$

This is equal to $e$ to six figures. Values of $e^x$ are obtained directly on the calculator by using $\boxed{\text{INV}}$ $\boxed{\ln x}$, $\boxed{\text{2nd F}}$ $\boxed{\ln x}$ or $\boxed{e^x}$ (see Example 9-3 below and Sec. 9.6), or with the table in App. B-3.

The following example shows an application of the exponential function $e$ to radioactive decay.

### Example 9-3

Radioactive materials decay exponentially according to the law $m = m_0 e^{-kt}$, where $m$ = mass, $t$ = time, $m_0$ = initial mass (when $t = 0$), and $k$ = constant rate of decay.

1. Graph $m$ vs. $t$ for 1.0 g of tritium, a radioactive isotope of hydrogen, if $k = 0.06$ (6 percent) per year.

2. Determine the *half-life* of tritium from the graph. That is, after how many years is the mass reduced to one-half of its initial value ($m = 0.50$ g)?

## Solution

1. Graph $m$ vs. $t$ for 1.0 g of tritium, a radioactive isotope of hydrogen, if $k = 0.06$ (6 percent) per year.

   The equation is $m = 1.0e^{-0.06t}$. Testing values of $t$, one finds that when $t = 50$ yr, $m = 1.0e^{-0.06(50)} = e^{-3.0} = 0.050$ g, and only 5 percent of the original amount is left. This calculation is done by using the table or calculator:

$$0.06 \;\boxed{+/-}\; \boxed{\times}\; 50 \;\boxed{=}\; \overbrace{\boxed{\text{INV}}\; \boxed{\ln x}}^{*} \longrightarrow 0.050 \qquad (* \text{ or } \boxed{e^x})$$

   Therefore you can choose values of $t$ such as 0, 10, 20, 30, 40, and 50 to produce six points and show the shape of the curve (Fig. 9-4):

| $t$ | 0 | 10 | 20 | 30 | 40 | 50 |
|---|---|---|---|---|---|---|
| $m = e^{-0.06t}$ | 1.0 | 0.55 | 0.30 | 0.17 | 0.091 | 0.050 |

2. Determine the *half-life* of tritium from the graph. That is, after how many years is the mass reduced to one-half of its initial value ($m = 0.50$ g)?

   In Fig. 9-4 the value of $t$ corresponding to $m = 0.50$ g is shown to be approximately 12 yr.

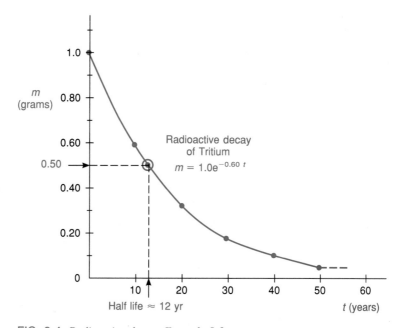

FIG. 9-4 *Radioactive decay, Example 9-3*

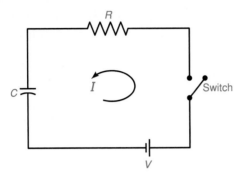

FIG. 9-5  *RC circuit, Example 9-4*

In electric circuits current and voltage undergo exponential growth and decay. The next example illustrates one of these situations.

### Example 9-4

In the circuit of Fig. 9-5, which contains a resistance $R$, a capacitance $C$, and a constant voltage $V$, the current at $t$ seconds after the switch is closed is given by $I = I_0 e^{-t/(RC)}$, where the initial current $I_0 = V/R$. Graph $I$ versus $t$ when $V = 100$ V, $R = 500$ $\Omega$, and $C = 200$ $\mu$F (0.000200 F).

### Solution

The equation for $I$ is:

$$I = \frac{V}{R}e^{-t/(RC)} = \frac{100}{500}e^{-t/(500)(0.000200)} = 0.20e^{-t/0.10} = 0.20e^{-10t}$$

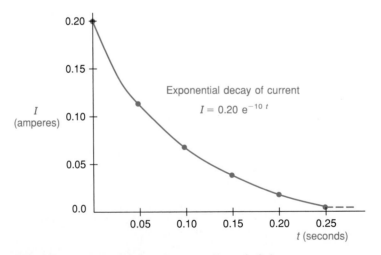

FIG. 9-6  *Exponential decay of current, Example 9-4*

When $t = 0.25$, $I = 0.20e^{-10(0.25)} = 0.20e^{-2.5} = 0.20(0.082) = 0.016$ A, which is 8 percent of the initial current. Therefore choose values of $t$ every 0.05 s to show the shape of the curve:

| $t$ | 0 | 0.05 | 0.10 | 0.15 | 0.20 | 0.25 |
|---|---|---|---|---|---|---|
| $I = 0.20e^{-10t}$ | 0.20 | 0.12 | 0.074 | 0.045 | 0.027 | 0.016 |

The graph in Fig. 9-6 shows the exponential decay of current. Theoretically, the current never reaches zero. Practically, however, it reaches zero in a few seconds.

**Exercise 9-1**

In Problems 1 to 24, graph $y$ vs. $x$, plotting at least five points. Adjust scales if necessary.

**1.** $y = 3^x$

**2.** $y = 2^{x+1}$

**3.** $y = 3(2)^{-x} = 3\left(\dfrac{1}{2}\right)^x$

**4.** $y = 2(3)^{-x} = 2\left(\dfrac{1}{3}\right)^x$

**5.** $y = \left(\dfrac{2}{3}\right)^x$

**6.** $y = 1.5^x$

**7.** $y = 2^{2x}$

**8.** $y = 3^{2x}$

**9.** $y = 2^{t/2}$

**10.** $y = 2^{2-t} = \left(\dfrac{1}{2}\right)^{t-2}$

**11.** $y = 2 - 2^{-x}$

**12.** $y = 3 - 3^x$

**13.** $y = 0.70^x$

**14.** $y = 1.2^x$

☐ **15.** $y = \pi^x$

☐ **16.** $y = \left(\dfrac{\pi}{2}\right)^{-x} = \left(\dfrac{2}{\pi}\right)^x$

Use a calculator or the table in App. B-3 for the following.

**17.** $y = e^x$

**18.** $y = e^{-x}$

**19.** $y = 1 - e^{-x}$

**20.** $y = 1 - e^x$

**21.** $y = e^{-0.05t}$

**22.** $y = e^{0.30t}$

**23.** $y = 10e^{0.10t}$

**24.** $y = 5e^{-0.01t}$

**25.** (a) Graph mass vs. time from $t = 0$ for the radioactive decay of 1.0 g of strontium 90 (Sr 90) if $k = 0.03$ per year. (Sr 90 is found in nuclear power plants.) Use $m = m_0e^{-kt}$ from Example 9-3, and plot values of $t$ for every 20 yr. (b) Find the half-life of Sr 90 from the graph.

**26.** In a certain chemical reaction, the quantity of salt remaining after $t$ hours is given by $Q = Q_0e^{-kt}$, where $Q_0 = $ initial amount. If $Q_0 = 25$ kg and $k = 0.40$ per hour, graph $Q$ vs. $t$ from $t = 0$ to $t = 5$.

☐ **27.** The world population was growing in 1985 at the rate of approximately 2 percent per year. This can be expressed mathematically: $A = A_0(1.02)^t$, where $A_0 = 4.87 \times 10^9$ people (1985 estimate) and $A = $ population $t$ years from 1985. (a) Graph the predicted mathematical curve for $A$ from 1985 to 2000 ($t = 0$ to $t = 15$). (b) Ap-

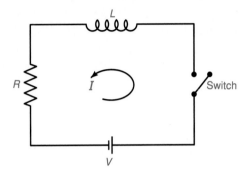

FIG. 9-7 *RL circuit, problem 30*

proximately when does the graph predict the population will reach 6 billion ($6 \times 10^9$)? (See Example 9-21 for the mathematical solution.)

C **28.** Inflation causes prices to increase according to the formula $P = P_0(1 + r)^n$, where $P$ = price $n$ years from now, $P_0$ = present price, and $r$ = rate of inflation (assumed to be constant). (*a*) If $P_0$ = \$5.00 is the present price of a movie and $r$ = 10 percent per year, graph $P$ vs. $n$ for $n = 0$ to $n = 8$. (*b*) After how many years will the price double?

**29.** Referring to Example 9-4, graph $I$ vs. $t$ from $t = 0$ when $V = 100$ V, $R = 500$ Ω, and $C = 400$ μF (0.000400 F). Use values of $t$ for every 0.1 s.

**30.** In the circuit in Fig. 9-7, which contains a resistance $R$, an inductance $L$, and a constant voltage $V$, the current $I$ at $t$ seconds after the switch is closed is given by $I = I_0(1 - e^{-Rt/L})$, where $I_0 = V/R$. Graph $I$ vs. $t$ from $t = 0$, when $V = 100$ V, $R = 40$ Ω, and $L = 2.0$ H. Use values of $t$ for every 0.05 s.

**31.** Graph $y = 2^x$ and $x = 2^y$ on the same graph, and compare the curves. (See Fig. 9-8 for the result.)

C **32.** In Example 16-15 the infinite series formula is shown:

$$e = 2 + \frac{1}{2} + \frac{1}{2 \cdot 3} + \frac{1}{2 \cdot 3 \cdot 4} + \frac{1}{2 \cdot 3 \cdot 4 \cdot 5} + \cdots$$

Calculate $e$ to four figures by adding the first six terms of this series.

# 9-2 | *Logarithmic Functions: Base 10, Base e*

*Logarithmic functions* are the inverse of exponential functions in the same way that division is the inverse of multiplication, and taking a root is the inverse of raising to a power. The logarithmic function defined below is equivalent to the exponential function shown in parentheses:

$$y = \log_b x \qquad b > 0, \ b \neq 1 \qquad \text{(equivalent to } b^y = x) \qquad (9\text{-}2)$$

Equation (9-2) says, "The *logarithm* of $x$ to the base $b$ is the *exponent y*," or, more briefly, "log $x$ to the base $b$ is $y$." *Logarithms are exponents*. Expressions in logarithmic form can be written in exponential form and vice versa. Study the next example.

### Example 9-5
Express each in exponential form.

| Logarithmic form | Exponential form |
|---|---|
| **1.** $\log_5 25 = 2$ | $5^2 = 25$ |
| **2.** $\log_4 64 = 3$ | $4^3 = 64$ |
| **3.** $\log_3 \dfrac{1}{3} = -1$ | $3^{-1} = \dfrac{1}{3}$ |
| **4.** $\log_2 \dfrac{1}{16} = -4$ | $2^{-4} = \dfrac{1}{16}$ |
| **5.** $\log_8 4 = \dfrac{2}{3}$ | $8^{2/3} = 4$ |
| **6.** $\log_9 \dfrac{1}{27} = -\dfrac{3}{2}$ | $9^{-3/2} = \dfrac{1}{27}$ |
| **7.** $\log_b 1 = 0$ | $b^0 = 1$ |

Notice in (7) that the log of 1 to any base $b$ is zero because any base $b$ to the zero power is 1. The base is always a positive number; therefore *the logarithm of a negative number or zero is not defined*. The number 1 is not used for a base because 1 to any power is 1. The bases that are used most often are 10 and $e$. Logarithms having base 10 are called *common logs,* and those having base $e$ are called *natural logs*. The word *log* written without the base *means base 10,* and *ln means base e*.

### Example 9-6
Express each in logarithmic form.

| Exponential form | Logarithmic form |
|---|---|
| **1.** $6^3 = 216$ | $\log_6 216 = 3$ |
| **2.** $10^2 = 100$ | $\log 100 = 2$ |
| **3.** $4^{5/2} = 32$ | $\log_4 32 = 5/2$ |
| **4.** $e^2 = 7.39$ | $\ln 7.39 = 2$ |
| **5.** $10^{-3} = 0.001$ | $\log 0.001 = -3$ |
| **6.** $10^{0.3010} = 2.00$ | $\log 2.00 = 0.3010$ |
| **7.** $e^{-0.6931} = 0.50$ | $\ln 0.50 = -0.6931$ |

It is necessary to understand the logarithmic form and be able to work with it comfortably. At first you may need to write expressions in exponential form. However, you will soon be able to visualize the exponential form and do the work in logarithmic form. The next example illustrates this process.

### Example 9-7
Find $x$ or $y$ in each of the following. More and more of the process is done mentally in each successive example.

1. $\log_2 x = 3$       Write in exponential form: $2^3 = x$. Then $x = 8$.

2. $\log x = 1$       Exponential form: $10^1 = x$. Then $x = 10$.

3. $\log_3 \dfrac{1}{9} = y$       $3^y = \dfrac{1}{9}$. Since $3^{-2} = \dfrac{1}{9}$, $y = -2$.

4. $\log 1000 = y$       $10^y = 1000$. Then $y = 3$.

5. $\log_4 x = \dfrac{1}{2}$       $4^{1/2} = x$ and $x = 2$.

6. $\log x = -2$       $10^{-2} = x$ and $x = 0.01$.

7. $\ln x = -1$       $e^{-1} = x$ and $x = \dfrac{1}{e}$.

8. $\log 100 = y$       $y = 2$.

Do you understand how to get all the answers? The more work you do mentally, the more you will understand the concept of logarithms. More difficult problems require a calculator or table and are discussed in the next section.

The graph of the logarithmic function is the inverse of the graph of the exponential function. This is shown in the next example.

### Example 9-8
Graph $y = \log_2 x$ and compare with $y = 2^x$.

### Solution
The table of values is equivalent to that for $x = 2^y$. Compare it with the table for $y = 2^x$ in Example 9-1:

| $x$ | $\frac{1}{16}$ | $\frac{1}{8}$ | $\frac{1}{4}$ | $\frac{1}{2}$ | 1 | 2 | 4 | 8 | 16 |
|---|---|---|---|---|---|---|---|---|---|
| $y = \log_2 x$ | $-4$ | $-3$ | $-2$ | $-1$ | 0 | 1 | 2 | 3 | 4 |

The graph of $y = \log_2 x$ is shown in Fig. 9-8 and is compared with the graph of $y = 2^x$, shown lightly. The curves are mirror images of each other. The line $y = x$ (dashed line) is the "mirror."

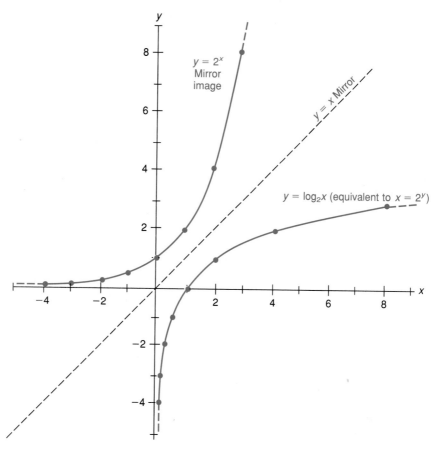

FIG. 9-8 *Logarithmic function, Example 9-8*

**Exercise 9-2**

Express problems 1 to 14 in exponential form (log means base 10, ln means base $e$).

**1.** $\log_2 16 = 4$         **2.** $\log_3 27 = 3$

**3.** $\log_5 125 = 3$         **4.** $\log_7 49 = 2$

**5.** $\log_4 \dfrac{1}{16} = -2$         **6.** $\log_8 2 = \dfrac{1}{3}$

**7.** $\log 1000 = 3$         **8.** $\log 0.01 = -2$

**9.** $\log 3.00 = 0.4771$         **10.** $\log 30.0 = 1.4771$

**11.** $\ln e = 1$         **12.** $\ln 1 = 0$

**13.** $\ln 5.1 = 1.629$         **14.** $\ln 0.20 = -1.609$

Express problems 15 to 24 in logarithmic form.

**15.** $2^5 = 32$         **16.** $3^4 = 81$

**17.** $10^4 = 10,000$         **18.** $2^{-2} = 0.25$

**19.** $4^{3/2} = 8$           **20.** $8^{-4/3} = 0.0625$

**21.** $10^{-2} = 0.01$        **22.** $e^3 = 20.1$

**23.** $e^{3.219} = 25.0$      **24.** $10^{-1.523} = 0.0300$

In problems 25 to 44, find $x$ or $y$ without a table or calculator. Express problems 35 and 36 in terms of $e$.

**25.** $\log_3 x = -2$         **26.** $\log_4 x = \dfrac{1}{2}$

**27.** $\log x = 5$            **28.** $\log x = -4$

**29.** $\log_5 125 = y$        **30.** $\log_2 \dfrac{1}{4} = y$

**31.** $\log_4 2 = y$          **32.** $\log_8 4 = y$

**33.** $\log 0.001 = y$        **34.** $\log 1 = y$

**35.** $\ln x = 2$             **36.** $\ln x = \dfrac{1}{2}$

**37.** $\ln e = y$             **38.** $\ln (1/e) = y$

**39.** $\log_2 0.5 = y$        **40.** $\log_5 0.04 = y$

**41.** $\log 10^{1.2} = y$     **42.** $\log \sqrt[3]{10} = y$

**43.** $\ln e^2 = y$           **44.** $\ln \sqrt{e} = y$

In problems 45 to 48, graph $y$ vs. $x$ for the given values.

**45.** $y = 2(\log_2 x);\ x = \dfrac{1}{8}, \dfrac{1}{4}, \dfrac{1}{2}, 1, 2, 4, 8$

**46.** $y = \log_3 x;\ x = \dfrac{1}{9}, \dfrac{1}{3}, 1, 3, 9$

**47.** $y = 2(\log_4 x);\ x = \dfrac{1}{8}, \dfrac{1}{4}, \dfrac{1}{2}, 1, 2, 4, 8$

**48.** $y = \ln x$; find $x$ in $e^y = x$ for $y = -3, -2, -1, 0, 1, 2, 3$

**49.** The Richter scale number $R$, used for measuring the magnitude of earthquakes, is defined as $R = \log I$, where $I$ = relative intensity of the shock. *(a)* Express this in exponential form. *(b)* How many times more intense was the Italian earthquake of November 1980, where $R$ was 7, than the San Francisco quake of August 1979, where $R$ was 6? (*Hint:* Find the ratio of $I$ when $R = 7$ to $I$ when $R = 6$.)

**50.** The hydrogen ion concentration [H$^+$] in moles per liter (mol/L) of a solution is defined in terms of pH: pH $= -\log$ [H$^+$]. *(a)* Express this in exponential form. (*Hint:* First multiply both sides by $-1$.) *(b)* Find [H$^+$] when pH $= 7$ (neutral solution).

**51.** Express the equation in Example 9-4, $I = I_0 e^{-t/(RC)}$, in logarithmic form by first solving for $I/I_0$.

**52.** The voltage $V$ in an $RL$ circuit is given by $V = V_0 e^{-Rt/L}$. Express this equation in logarithmic form by first solving for $V/V_0$.

**53.** The human ear detects the loudness of sounds in proportion to the logarithm of their actual intensity, or transmitted power. Therefore the difference in loudness $\Delta L$ between two sounds of intensities $I_1$ and $I_2$ is given by:

$$\Delta L = L_2 - L_1 = 10 \log \frac{I_2}{I_1}$$

where $L$ is measured in decibels (dB), named after Alexander Graham Bell, the inventor of the telephone. *(a)* Express this equation in exponential form by first solving for $\log (I_2/I_1)$. *(b)* If $L_1 = 60$ dB (the loudness of ordinary conversation) and $L_2 = 100$ dB (the loudness of a subway train), how many times more intense is the sound of a subway train than ordinary conversation?

**54.** The brightness of a light source seen by the human eye is proportional to the logarithm of its actual intensity. As a result, the magnitude $M$, or brightness, of a heavenly body is given by $M = 2.5 \log (I_0/I)$, where $I_0 =$ intensity of a body of zero magnitude. *(a)* Express this equation in exponential form by first solving for $\log (I_0/I)$. *(b)* The full moon has a brightness $M = -12.5$. How many times more intense is the full moon than a zero-magnitude body? *(c)* How many times more intense is Sirius, the brightest star seen from earth ($M = -1.4$), than Polaris, the North Star ($M = +2.0$)?

**55.** The equation for world population growth from problem 27, Exercise 9-1, can be expressed by using a base of 2: $A = A_0 2^{0.028t}$ . Express this equation in logarithmic form (base 2) by first solving for $A/A_0$.

**56.** The temperature loss of a certain warm object is expressed by $T = 83(2)^{-0.073t}$, where $t =$ time and $T =$ temperature difference between the object and its surroundings. Express this equation in logarithmic form (base 2) by first dividing by 83.

## 9-3 | *Rules for Logarithms: Common and Natural Logarithms*

It is difficult to imagine multiplying and dividing 10-digit numbers by hand, especially in today's age of electronic calculation. However, over 400 years ago, people did not have much choice when they encountered such problems in astronomy. This led to the discovery of logarithms in 1614 by the Scotchman John Napier. The theory of logarithms was used to build slide rules, calculating machines, computers, and most recently pocket calculators. Logarithms are used more today in working with exponential functions than in performing calculations. However, you can understand logarithms better by seeing how they are used in calculations. The following example illustrates how logarithms are used to simplify multiplication. Further calculations using logarithms are shown in Example 9-14, after the rules for logarithms are discussed.

## Example 9-9

Multiply $(2.86)$ $(3.19)$, using common logarithms (base 10).

## Solution

Use a calculator or table (App. B-4) to find the logarithm of each number as follows:

*Calculator.*   Enter the number and press $\boxed{\log}$:

$$2.86\boxed{\log} \rightarrow 0.4564 \qquad \text{and} \qquad 3.19\boxed{\log} \rightarrow 0.5038$$

*Table.*   The table gives the common logarithms of numbers from 1.00 to 9.99. The logarithms in the table are therefore numbers between 0 and 1. They are called *mantissas,* and all *decimal points are understood.* The correct row is found by locating the first two digits of the number on the left. The column is found by locating the third digit of the number at the top. The logarithms are log 2.86 = 0.4564 and log 3.19 = 0.5038. If you change the last two equations from logarithmic to exponential form:

$$10^{0.4564} = 2.86 \qquad \text{and} \qquad 10^{0.5038} = 3.19$$

you can express the multiplication as follows:

$$(2.86)(3.19) = (10^{0.4564})(10^{0.5038})$$

By using the addition rule for exponents, the *multiplication problem is reduced to a simpler addition problem*:

$$10^{0.4564+0.5038} = 10^{0.9602}$$

This is expressed in terms of logs as:

$$\log [(2.86)(3.59)] = \log 2.86 + \log 3.59$$
$$0.9602 = 0.4564 + 0.5038$$

The final answer is then $10^{0.9602} = 9.12$. It is obtained by using $\boxed{2^{nd}\ F}$ $\boxed{\log}$, $\boxed{INV}$ $\boxed{\log}$ or $\boxed{10^x}$ on the calculator:

$$0.9602\ \boxed{2^{nd}\ F}\ \boxed{\log} \rightarrow 9.12$$

or by reading the table backward and finding the closest mantissa (9600). This is equivalent to raising 10 to the power 0.9600. See Sec. 9-6 for more calculator information.

## *Rules for Logarithms*

Example 9-9 above illustrates the addition rule for logarithms. This is one of the rules for logarithms that follow directly from the rules for exponents (Sec. 8-1):

$$\log_b XY = \log_b X + \log_b Y \qquad \text{addition rule} \qquad (9\text{-}3)$$

$$\log_b \frac{X}{Y} = \log_b X - \log_b Y \qquad \text{subtraction rule} \qquad (9\text{-}4)$$

$$\log_b X^n = n \log_b X \qquad \text{multiplication rule} \qquad (9\text{-}5)$$

Rule (9-3) reduces multiplication to addition, and rule (9-4) reduces division to subtraction. Rule (9-5) is true for $n =$ any real number and reduces raising to a power to multiplication or taking a root to division. A special case of rule (9-5) is when $X = b$:

$$\log_b b^n = n \log_b b = n(1) = n \qquad (9\text{-}6)$$

The proof of (9-3) is as follows:

Let $\qquad\qquad\qquad\qquad X = b^x \qquad\qquad Y = b^y \qquad\qquad XY = b^{x+y}$

$$\downarrow \qquad\qquad\qquad \downarrow \qquad\qquad\qquad \downarrow$$

Then $\qquad\qquad\qquad \log_b X = x \qquad \log_b Y = y \qquad \log_b XY = x + y$

Rule (9-3) follows from the last three equations:

$$\log_b X + \log_b Y = \log_b XY \qquad (x + y = x + y)$$

The proofs of (9-4) and (9-5) are left as problems 69 and 70.

The next two examples illustrate how the rules are used to change expressions containing logarithms.

## Example 9-10

Express as a combination of logarithms and simplify.

**1.** $\log_b \dfrac{25b^3}{a} = \log_b (25b^3) - \log_b a$ $\qquad\qquad$ subtraction rule

$\qquad\qquad = \log_b 25 + \log_b b^3 - \log_b a \qquad$ addition rule

$\qquad\qquad = \log_b 25 + 3 - \log_b a \qquad\qquad$ rule (9-6)

**2.** $\log (100x^2y) = \log 100 + \log x^2 + \log y \qquad$ addition rule

$\qquad\qquad\quad = 2 + 2 \log x + \log y \qquad\qquad$ multiplication rule

Notice in (2) that the $\log 100 = 2$.

### Example 9-11

Express as a single logarithm and simplify.

**1.** $\log_2 24 + \log_2 12 - \log_2 18$

$\qquad = \log_2 [(24)(12)] - \log_2 18 \qquad$ addition rule

$\qquad = \log_2 \dfrac{(24)(12)}{18} = \log_2 16 = 4 \qquad$ subtraction rule

**2.** $2 \ln \sqrt{e} + \ln t - \ln e = \ln (\sqrt{e})^2 + \ln t + 1 \qquad$ multiplication rule

$\qquad = \ln e + \ln t + 1 = 1 + \ln t + 1 = \ln t + 2$

## Common Logarithms

Base 10 is used for logarithms because our number system is based on 10, and so it is easy to work with. Learn the following table of common logs which will help you when working with logarithms to base 10:

$$\log 0.001 = \log 10^{-3} = -3 \qquad \log 10 = \log 10^1 = 1$$
$$\log 0.010 = \log 10^{-2} = -2 \qquad \log 100 = \log 10^2 = 2$$
$$\log 0.100 = \log 10^{-1} = -1 \qquad \log 1000 = \log 10^3 = 3$$
$$\log 1.0 = \log 10^0 = 0 \qquad \log 10{,}000 = \log 10^4 = 4$$

Observe that the common logarithm of a number between two numbers in the table lies between the two powers. For example, log 50 is between 1 and 2, and log 0.5 is between −1 and 0. Knowing this table provides a quick check on results. The next two examples, 9-12 and 9-13, show how to find common logarithms and their inverse using a table. Use of the calculator is shown in Examples 9-27 and 9-28, Sec. 9-6.

### Example 9-12

Find each common logarithm using the table of mantissas in App. B-4.

**1.** log 28.6      **2.** log 286      **3.** log 0.286      **4.** log 0.0286

### Solution

**1.** log 28.6

For a number *not* between 1 and 10, write the number in scientific notation and apply the rules for logarithms:

$$\log (2.86 \times 10^1) = \log 2.86 + \log 10^1 = 0.4564 + 1 = 1.4564$$

The number 1 added to the mantissa 0.4564 is called the *characteristic*. The characteristic is the power of 10 when the number is written in scientific notation (Sec. 2-5).

**2.** log 286:

$$\log 286 = \log (2.86 \times 10^2) = \log 2.86 + \log 10^2 = 2.4564$$

The characteristic can be determined mentally and added to the value in the table, as shown in the next example.

**3.** log 0.286:

$$\log 0.286 = 0.4564 - 1 \quad \text{(since } 0.286 = 2.86 \times 10^{-1})$$

This is a negative logarithm equal to $-1.0000 + 0.4564 = -0.5436$, which is the value obtained with a calculator. When you use a table for computation, the logarithm is left in the form $0.4564 - 1$. This allows you to read the table backward since the table contains only positive mantissas. (See Example 9-14.)

**4.** log 0.0286:

$$\log 0.0286 = 0.4564 - 2 \quad \text{or} \quad -1.5436$$

To find the logarithm of a number with more than three figures, it is necessary to interpolate the values in the table or to use a calculator.

See Example 9-27 (Sec. 9-6) for the calculator solution of Example 9-12. The next example illustrates how to find the inverse of a logarithm, or antilog.

### Example 9-13
Use the table to find $X$ in each of the following.

**1.** log $X$ = 0.7649        **2.** log $X$ = 3.4250        **3.** log $X$ = 0.5038 − 2

### Solution
**1.** log $X$ = 0.7649

For log $X$ = 0.7649, read the table backward. The digits corresponding to the mantissa 7649 are 582. The characteristic (the power of 10) is zero, therefore $X = 5.82 \times 10^0 = 5.82$.

**2.** log $X$ = 3.4250

For log $X$ = 3.4250, the closest mantissa is 4249, which corresponds to 266. Therefore $X = 2.66 \times 10^3$ or 2660, to three figures.

**3.** log $X$ = 0.5038 − 2

From the table $X = 3.19 \times 10^{-2}$, or 0.0319.
See Example 9-28 for the calculator solution of Example 9-13.

The next example shows how many of the ideas in this section are used to perform a calculation using logarithms. See Example 9-29 for the calculator solution of Example 9-14.

### Example 9-14

Compute using logarithms: $\dfrac{0.350}{\sqrt[4]{50.0}}$

### Solution

Set the expression equal to $X$, and take the logarithm of both sides, applying the rules for logarithms:

$$X = \frac{0.350}{\sqrt[4]{50.0}}$$

$$\log X = \log 0.350 - \log (50.0)^{1/4} = \log 0.350 - \frac{\log 50.0}{4}$$

When the table is used:

$$\log X = 0.5441 - 1 - \frac{1.6990}{4}$$

$$= 0.5441 - 1 - 0.4248 = 0.1193 - 1$$

This answer corresponds to $X = 1.32 \times 10^{-1}$ or 0.132 (the nearest mantissa is 1206).

## *Natural Logarithms*

Base $e$ is used for scientific and technical work with exponential functions. Natural logarithms are obtained on a calculator by pressing $\boxed{\ln x}$ or $\boxed{\ln}$. They can also be obtained by using the table in App. B-5 or by using a table of common logarithms and the following conversion formula:

$$\ln X = 2.3026 \log X \tag{9-7}$$

Formula (9-7) derives from the fact that $e^{2.3026} = 10$. If you let $X = 10^x$, then $X = (e^{2.3026})^x = e^{2.3026x}$. In logarithmic form these two equations are $\log X = x$ and $\ln X = 2.3026x$. Substituting $\log X$ for $x$ in the last equation yields (9-7).

### Example 9-15

Find the natural logarithm, using the table in App. B-5.

**1.** $\ln 5.3$      **2.** $\ln 760$      **3.** $\ln 0.041$

### Solution

**1.** ln 5.3

Read the result directly from the table: ln $5.3 = 1.6677$. Using formula (9-7) and the table of common logarithms produces the same result:

$$\ln 5.3 = 2.3026(\log 5.3) = 2.3026(0.7243) = 1.6677$$

**2.** ln 760

Write the logarithm in scientific notation:

$$\ln 760 = \ln (7.6 \times 10^2) = 1n\ 7.6 + 2(\ln 10)$$
$$= 2.0281 + 2(2.3026) = 6.6333$$

Multiples of 1n 10 are given in the table.

**3.** ln 0.041:

$$1n\ (4.1 \times 10^{-2}) = \ln 4.1 - 2(\ln 10) = 1.4110 - 4.6052 = -3.1942$$

See Example 9-30 for the calculator solution of Example 9-15 and Example 9-31 for how to find the inverse of a natural logarithm. Natural logarithms can also be used for calculations instead of the common logarithms shown in Example 9-14.

When a logarithm to a base other than 10 or $e$ is desired, it can be calculated by the following formula:

$$\log_b a = \frac{\log a}{\log b} \tag{9-8}$$

Formula (9-8) comes from the following. If $u = \log_b a$, then by changing to exponential form:

$$b^u = a$$

Taking the common logarithm of both sides of this equation yields:

$$\log b^u = \log a$$

Then by rule (9-5):

$$u(\log b) = \log a$$

and:

$$u = \frac{\log a}{\log b}$$

or:

$$\log_b a = \frac{\log a}{\log b}$$

Note that log $a$/log $b$ is *not* the same as log $(a/b)$, which equals log $a$ − log $b$.

## Example 9-16

Find the following logarithms.

**1.** $\log_2 18$      **2.** $\log_5 3$

## Solution

**1.** $\log_2 18$

    Apply formula (9-8) where $a = 18$ and $b = 2$:

$$\log_2 18 = \frac{\log 18}{\log 2}$$

$$= \frac{1.255}{0.3010} = 4.170$$

**2.** $\log_5 3$

    Apply formula (9-8) where $a = 3$ and $b = 5$:

$$\log_5 3 = \frac{\log 3}{\log 5}$$

$$= \frac{0.4771}{0.6990} = 0.6826$$

The next example illustrates an application of natural logarithms to radiation intensity.

## Example 9-17

When radiation or light penetrates a substance, the light diminishes in intensity as the depth increases. This is expressed by ln $(I/I_0) = -kx$, where $I_0$ = initial intensity, $I$ = intensity at depth $x$, and $k$ is a coefficient of absorption, the value of which depends on the frequency of the radiation and the type of substance. Under certain conditions, for a lead shield, $k = 2.6$ per millimeter for x-rays.

**1.** At what depth will the intensity be reduced to 0.5 percent of its original intensity?

**2.** To what percentage will the intensity be reduced at a depth of 1.0 mm?

### Solution

1. At what depth will the intensity be reduced to 0.5 percent of its original intensity?

Let $I/I_0 = 0.5$ percent $= 0.005$. Then:

$$\ln 0.005 = -2.6x$$
$$-5.2983 = -2.6x$$
$$x = \frac{-5.2983}{-2.6} = 2.0 \text{ mm}$$

2. To what percentage will the intensity be reduced at a depth of 1.0 mm?

$$\ln \frac{I}{I_0} = -2.6(1.0) = -2.6$$

Change to exponential form:

$$e^{-2.6} = \frac{I}{I_0}$$

Use a table or calculator:

$$\frac{I}{I_0} = 0.074 = 7.4 \text{ percent}$$

**Exercise 9-3**

In problems 1 to 10, express as a combination of logarithms and simplify. (See Example 9-10.)

1. $\log_2 (4x)$

2. $\log_2 \dfrac{y}{16}$

3. $\log_3 \dfrac{27p^2}{q}$

4. $\log_5 (25\pi r^2)$

5. $\log \dfrac{v}{\sqrt{10}}$

6. $\log (0.01w^3)$

7. $\ln (5.5e^2)$

8. $\ln \dfrac{0.30t}{e}$

9. $\log_b (ab^x)$

10. $\log_b (bx^n)$

In problems 11 to 20, express as a single logarithm and simplify. (See Example 9-11.)

11. $\log_2 6 - \log_2 3$

12. $\log_2 10 + \log_2 0.10$

13. $\log_4 8 + 3 \log_4 2$

14. $\log_8 (2x) - 2 \log_8 \sqrt{x}$

15. $\log (1.0w) - \log (0.01w)$

16. $2 \log (5y) + \log (2y)$

**17.** $\ln (3e) + \ln e^2$          **18.** $\ln \dfrac{e}{2} - \ln (\sqrt{e})$

**19.** $\log_b x + \log_b y - \log_b (tx)$          **20.** $\log_b (3v) - 2 \log_b v$

In problems 21 to 40, find $X$, using a calculator or table.

**21.** $X = \log 86.0$          **22.** $X = \log 5.30$

**23.** $X = \log 0.105$          **24.** $X = \log 0.0484$

**25.** $X = \log 5600$          **26.** $X = \log 747$

**27.** $X = \log (5.2 \times 10^{-3})$          **28.** $X = \log (2.1 \times 10^4)$

**29.** $X = \log 0.00600$          **30.** $X = \log 4000$

**31.** $\log X = 2.5211$          **32.** $\log X = 1.6191$

**33.** $\log X = 0.7126 - 1$          **34.** $\log X = 0.9571 - 2$

**35.** $\log X = 0.8698$          **36.** $\log X = 0.7782$

**37.** $\log X = 0.8008 - 3$          **38.** $\log X = 0.2716 - 1$

**39.** $\log X = 3.3715$          **40.** $\log X = 2.9000$

Compute problems 41 to 46, using logarithms (see Example 9-14).

**41.** $(52.3)(0.812)$          **42.** $85.1/0.710$

**43.** $0.0795/1.29^3$          **44.** $(\sqrt{320})(0.0393)$

**45.** $(\sqrt[3]{2200})(0.612)$          **46.** $(0.0550)(1.91)^4$

In problems 47 to 60, find each logarithm, using a calculator or table.

**47.** $\ln 3.8$          **48.** $\ln 7.0$

**49.** $\ln 60$          **50.** $\ln 330$

**51.** $\ln 0.92$          **52.** $\ln 0.055$

**53.** $\ln (2.5 \times 10^4)$          **54.** $\ln (4.0 \times 10^{-3})$

**55.** $\log_2 20$          **56.** $\log_3 38$

**57.** $\log_5 4$          **58.** $\log_6 2$

**59.** $\log_3 0.4$          **60.** $\log_2 0.6$

**61.** During the power stroke in an internal-combustion engine, when the gas expands adiabatically (no loss or gain of heat), the following expression arises:

$$\log \frac{P_2 V_2{}^n}{V_1{}^n}$$

Express this as a combination of logarithms.

**62.** The following expression comes from a problem in fluid flow: $\log 1.486 - \log n + \log (r/6)$. Express this as a single logarithm.

**63.** When interest is compounded daily in a bank account at a rate of $r$ percent a year, the principal (the initial amount) will double in approximately $n$ years where $n = (\ln 2)/r$. Find $n$ when $r$ is *(a)* 6 percent and *(b)* 8 percent.

**64.** At an inflation rate of $r$ percent a year, prices will double in approximately $n$ years where $n = \log 2/\log (1 + r)$. Find $n$ when $r$ is (a) 8 percent and (b) 10 percent.

**65.** Under certain conditions the decibel gain of an amplifier in terms of voltage is given by $n = 20 \log (V_2/V_1)$, where $V_1 = $ input voltage and $V_2 = $ output voltage. Find $n$ when $V_1 = 0.10$ V and $V_2 = 25$ V.

**66.** The capacitance of a cylindrical capacitor in farads (F) is given by $C = cl/[(18 \times 10^9) \ln (R_2/R_1)]$, where $c = $ dielectric constant, $l = $ length, $R_1 = $ inside radius, and $R_2 = $ outside radius. Find $C$ in nanofarads (1 nF $= 10^{-9}$ F) if $c = 5.8$, $l = 0.15$ m, and $R_2/R_1 = 1.1$.

**67.** The German lightning calculator Zacharias Dase (1824–1861) was able to calculate *mentally* the product of two $n$-digit numbers in an incredibly small time. The time $t$ in seconds he took to calculate is given approximately by the formula $\log t = 2.72 \log n + \log 0.102$. For example, he could mentally multiply two 20-digit numbers in 360 s, or 6 min. How long did it take him to mentally multiply two (a) five-digit numbers and (b) 100-digit numbers? (See problem 28, Sec. 9-5.)

**68.** Because of pollution the population of fish in a certain river decreases according to the formula $\ln (P/P_0) = -0.0345t$, where $P = $ population after $t$ years and $P_0 = $ original population. (a) After how many years will there be only 50 percent of the original population left? (b) What percentage will die in the first year of pollution? (See Example 9-17.)

**69.** By letting $X = b^x$, $Y = b^y$, and $X/Y = b^{x-y}$, show that the subtraction rule for logarithms (9-4) is true.

**70.** By letting $X = b^x$ and $X^n = b^{nx}$, show that the multiplication rule (9-5) is true.

**71.** By letting $a^x = b$ and $b^{1/x} = a$, show that $\log_a b = 1/\log_b a$.

**72.** The following is the infinite series formula shown in calculus for $\ln 2$:

$$\ln 2 = 2 \left[ \frac{1}{3} + \frac{(\frac{1}{3})^3}{3} + \frac{(\frac{1}{3})^5}{5} + \cdots \right]$$

Calculate $\ln 2$ to five figures by adding the first five terms of this series. Check your result directly.

# 9-4 | *Exponential and Logarithmic Equations*

## *Exponential Equations*

Many applications of exponential functions have been shown in this chapter. When you work with exponential functions, you encounter exponential equations. An *exponential equation* is an equation in which the unknown is an exponent. Some exponential equations can be solved by expressing both sides in the same base and equating exponents, as the next example shows.

### Example 9-18

Solve the exponential equation:    $3^{x+2} = 9^x$

**Solution**

Express both sides in base 3:

$$3^{x+2} = (3^2)^x = 3^{2x}$$

When the bases are equal, you can equate the exponents:

$$x + 2 = 2x$$
$$x = 2$$

The general method for solving any exponential equation is to *take the logarithm of each side* as follows.

### Example 9-19

Solve the exponential equation:    $3^{2n} = 12$

**Solution**

Take the logarithm of each side:

$$\log(3^{2n}) = \log 12$$
$$2n(\log 3) = \log 12$$

Solve for $n$:

$$n = \frac{\log 12}{2 \log 3} = \frac{1.0792}{2(0.4771)} = 1.131$$

Example 9-18 can also be solved like Example 9-19, by taking the logarithm of each side.

The next example shows how to solve a literal exponential equation by using natural logarithms.

### Example 9-20

Solve for $t$:    $I = I_0 e^{(-Rt/L)}$

**Solution**

Take the natural logarithm of each side since the base is $e$:

$$\ln I = \ln I_0 + \ln e^{-Rt/L} = \ln I_0 - \frac{Rt}{L}$$

$$\frac{Rt}{L} = \ln I_0 - \ln I = \ln \frac{I_0}{I}$$

$$t = \frac{L}{R} \ln \frac{I_0}{I}$$

The following example shows an application of exponential equations to population growth discussed in the beginning of the chapter.

C **Example 9-21**

The exponential function describing world population increase in 1985 is $A = A_0 (1.02)^t$, where $A_0 = 4.87 \times 10^9$ (1985 population) and $A =$ population $t$ years from 1985. Based on this function,

**1.** In how many years from 1985 will the world population reach 6 billion $(6 \times 10^9)$?

**2.** In how many years from 1985 will the world population double?

**Solution**

**1.** In how many years from 1985 will the world population reach 6 billion $(6 \times 10^9)$?

Let $A = 6 \times 10^9$. Then:

$$6 \times 10^9 = 4.87 \times 10^9 (1.02)^t$$

$$(1.02)^t = \frac{6 \times 10^9}{4.87 \times 10^9} = 1.23$$

Take the logarithm of each side:

$$t \log 1.02 = \log 1.23$$

$$t = \frac{\log 1.23}{\log 1.02} = \frac{0.8990}{0.0086} = 10.5 \text{ yr}$$

The population will reach 6 billion during 1996.

**2.** In how many years from 1985 will the world population double?

Let $A = 2A_0$:

$$2A_0 = A_0 (1.02)^t$$

$$2 = (1.02)^t$$

Take logarithms and solve for $t$:

$$t = \frac{\log 2}{\log 1.02} = \frac{0.3010}{0.0086} = 35 \text{ yr}$$

The population will double in about the year 2020.

## Logarithmic Equations

*Logarithmic equations* are equations in which the unknown is part of a logarithmic expression. They are not all solved by the same method. The following two examples show the basic type, which is solved by combining logarithms and taking the inverse (antilog) of both sides.

### Example 9-22
Solve for $x$:     $\log (3x + 1) - 2 \log x = 1$

**Solution**

$$\log (3x + 1) - \log x^2 = 1 \qquad \text{multiplication rule}$$

$$\log \frac{3x + 1}{x^2} = 1 \qquad \text{subtraction rule}$$

$$\frac{3x + 1}{x^2} = 10 \qquad \text{take inverse of both sides}$$

This last step is equivalent to changing the equation to exponential form. The equation is quadratic, with two possible solutions:

$$10x^2 - 3x - 1 = 0$$
$$(5x + 1)(2x - 1) = 0$$
$$x = -\frac{1}{5} \qquad x = \frac{1}{2}$$

The only acceptable answer is $x = \frac{1}{2}$; the logarithm of a negative number is not defined.

### Example 9-23
Solve for $A$:     $\log A - \log P = n \log (1 + r)$

**Solution**

$$\log A = \log P + n \log (1 + r)$$
$$\log A = \log P + \log (1 + r)^n = \log [P(1 + r)^n]$$
$$A = P(1 + r)^n \qquad \text{take inverse of both sides}$$

**Exercise 9-4**

In problems 1 to 26, solve each equation for the unknown. Use a table or calculator if necessary.

1. $3^{x+3} = 81$

2. $2^{1-x} = \frac{1}{8}$

3. $5^{4x} = 25^{x-1}$

4. $4^{2x} = 16^{2x-2}$

5. $8^y = 32^{y+1}$

6. $9^{-y} + 3 = 30$

7. $4^t = 23$

8. $5^n = 2$

9. $3^x = 6.5$

10. $2^y = 8.1$

11. $(5.0)^{-3k} = 3.8$

12. $(8.3)^{2m} = 9.6$

13. $10^{x+1} = 20.3$

14. $10^{2n} = 0.15$

15. $1 - e^{-2t} = 0.30$

16. $e^{0.02x} = 6.0$

17. $(0.90)^n = 0.53$

18. $(0.030)^k = 1.31$

19. $10^y = 5.5 \times 10^3$

20. $e^t = 3.2 \times 10^{-5}$

21. $\log (x + 1) = 2$

22. $\log x^2 + 1 = \log x$

23. $\log x + \log (x + 1)$
$= \log (3x + 3)$

24. $\log (x + 1) = 1 - \log (x - 1)$

25. $\ln (x - 1) = 2$

26. $\ln x = 1 + 2 \ln x$

In problems 27 to 34, solve for the given letter.

27. $Q = 100(5^x)$; $x$

28. $P = P_0(10)^{n/2}$; $n$

29. $V = V_0 e^{-t/(RC)}$; $t$

30. $I = e^{-10^3 t}$; $t$

31. $\log (I/I_0) = L$; $I$

32. $\log T = \log T_0 - kt$; $T$

33. $\ln m_0 - \ln m = kt$; $m$

34. $\ln R_1 - \ln R_2 = 2\pi T$; $R_1$

35. In a chrome electroplating process, the mass $m$ in grams of the chrome plating increases according to the formula $m = 200 - 2^{t/2}$, where $t =$ time in minutes. How many minutes does it take to form 100 g of the plating?

36. Using the formula for pH:

$$pH = -\log [H^+]$$

find the hydrogen ion concentration $[H^+]$ in mol/L of *(a)* eggs, pH = 8.0, and *(b)* pickles, pH = 3.5. (When pH > 7, the solution is basic. When pH < 7, the solution is acidic.)

37. The formula for compound interest is $A = P(1 + i)^n$, where $A =$ amount after $n$ interest periods, $P =$ principal (the initial amount), and $i =$ interest rate *per period*. How many years will it take for money to double when it is compounded yearly at 6 percent per year? (*Hint*: Let $A = 2P$.)

38. Interest on savings accounts is compounded daily in most banks. An approximate formula for the amount $A$ after $n$ yr compounded daily is $A = Pe^{rn}$, where $P =$ principal and $r =$ interest rate *per year*. Find the answer to problem 37 if the interest is compounded daily. (This formula is for continuous compounding, that is, compounding every instant, which is very close to daily compounding. In Virginia banks compound continuously.)

**39.** In Example 9-3 the half-life of tritium was determined graphically to be approximately 12 yr. Using the formula for radioactive decay $m = m_0 e^{-kt}$, where $k = 0.055$ per year for tritium, find the half-life of tritium mathematically. Let $m/m_0 = 0.5$ and solve for $t$.

**40.** The bones of an old mathematics professor contain 99.50 percent of the original amount of carbon 14. How long ago did she die if $k = 1.238 \times 10^{-4}$ per year for C 14? (Use the formula in problem 39 and find $t$ when $m/m_0 = 0.9950$.)

**41.** In Example 9-4 the formula for the exponential decay of current is given as $I = I_0 e^{-t/(RC)}$, where $I_0 = 0.20$ A, $R = 500\ \Omega$, and $C = 200\ \mu F$. Find $t$ when $I = 0.050$ A.

**42.** In Example 9-17 the formula for the decrease in radiation intensity is given: $\ln (I/I_0) = -kx$. A part of the Sargasso Sea, which is a very clear portion of the Atlantic Ocean around Bermuda, has $k = 0.080$ per meter for a light ray whose intensity $I_0 = 100$ cd. *(a)* What is the intensity $I$ at a depth $x = 5.0$ m? *(b)* At what depth will the intensity be reduced by 90 percent ($I/I_0 = 0.10$)?

**43.** In the isothermal (constant temperature) expansion of a gas in a steam engine, the work done is expressed as $W = P_1 V_1 \ln (P_1/P_2)$. What is the final pressure $P_2$ if the initial pressure $P_1 = 1200$ lb/ft$^2$, the initial volume $V_1 = 0.50$ ft$^3$, and $W = 1800$ ft · lb?

**44.** The temperature $T$ of a hot object placed into cooling water is given by $T = T_w + (T_0 - T_w)10^{-0.08t}$, where $T_0 =$ initial temperature, $T_w =$ water temperature, and $t =$ time in minutes. A steel bar whose temperature $T_0 = 1200°C$ is placed in water whose temperature $T_w = 20°C$. How long will it take for the steel to cool to $T = 40°C$ (cool enough to touch)?

**45.** In problem 53, Sec. 9-2, the formula for the difference in loudness $\Delta L$ between two sounds of intensities $I_1$ and $I_2$ is given as $\Delta L = L_2 - L_1 = 10 \log (I_2/I_1)$. The lowest intensity of sound that the human ear can detect (the hearing threshold) is $I_1 = 10^{-12}$ W/m$^2$. This corresponds to $L_1 = 0$ dB. The threshold of pain for the human ear corresponds to $L_2 = 120$ dB. What is the intensity $I_2$ at the pain threshold?

**46.** The difference in sound intensity $\Delta I$ through a sound-absorbent material is given by $\Delta I = I_2 - I_1 = I_2(1 - 10^{-k})$, where $I_2 =$ approach intensity, $I_1 =$ exit intensity, and $k =$ constant, the value of which depends on the material and the thickness. Find $k$ for a glass window if $I_2 = 6.5 \times 10^{-5}$ W/m$^2$ (the intensity of loud street traffic) and $I_1 = 5.0 \times 10^{-6}$ W/m$^2$ (the intensity of ordinary conversation).

**47.** The formula for power gain or loss in decibels is given by $n = 10 \log (P_2/P_1)$. The input power to a telephone line is $P = 10$ mW, and the decibel loss is $n = -20$ dB. What is the output power $P_2$?

**48.** In problem 47 what input power $P_1$ is needed to produce an output power $P_2 = 10$ mW if the decibel loss is $n = -20$ dB?

**49.** In Example 9-21 how many years from 1985 will the world population reach 5.5 billion?

**50.** In Example 9-21 how many years from 1985 will the population increase by 50 percent?

## 9-5 | *Semilogarithmic and Logarithmic Graphs*

When you are graphing exponential functions, the variables can have a large range of values, making it difficult to plot many points. Semilogarithmic and logarithmic graph paper scales are proportional to the logarithms of numbers (base 10), which allows plotting a wide range of values.

### Semilogarithmic Graphs

*Semilogarithmic (semilog) graph paper* has a vertical logarithmic scale and a horizontal linear scale (Fig. 9-9). *The exponential function $y = b^x$ graphs as a straight line on semilog paper.* Study the following example.

**Example 9-24**
Graph $x = 10^x$ on semilog paper.

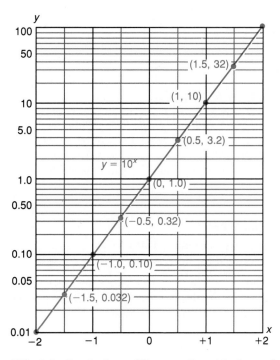

FIG. 9-9 *Graph of $y = 10^x$ on semilogarithmic graph paper (4 cycles), Example 9-24*

## Solution

Construct a table of values, using a calculator or table:

| $x$ | −2.0 | −1.5 | −1.0 | −0.5 | 0 | 0.5 | 1.0 | 1.5 | 2.0 |
|---|---|---|---|---|---|---|---|---|---|
| $y = 10^x$ | 0.010 | 0.032 | 0.10 | 0.32 | 1.0 | 3.2 | 10 | 32 | 100 |

Notice the wide range of $y$ values which could not be effectively plotted on ordinary rectangular graph paper. Semilog paper brings out the proportional relation between these values. The vertical scale is marked off in cycles. Each cycle corresponds to the range of a power of 10. The cycles are labeled in a manner depending on the values to be plotted. This example uses four cycles labeled from 0.01, or $10^{-2}$, to 100, or $10^2$. The graph of the straight line is shown in Fig. 9-9.

When you take logarithms of both sides of the exponential function $y = b^x$, you obtain $\log y = x \log b$. This shows that *log y is a linear function of x* (Sec. 5-1). The graph is therefore a straight line because the vertical scale is logarithmic and *log y* is plotted against $x$ on the horizontal scale. Notice the points on the $y$ axis corresponding to 0.05, 0.50, 5.0, and 50 are not four-ninths the distance between the powers of 10, as they would be on rectangular graph paper, but are $\log 5 \approx \frac{7}{10}$ of the distance between the powers.

## Logarithmic Graphs

On *logarithmic (log-log) graph paper,* both scales are logarithmic (Fig. 9-10). Log-log paper is useful when both variables have a large range of values. *A function of the type $y^b = x^a$, which can be written $y = x^{a/b}$, graphs as a straight line on log-log paper.* Taking logarithms yields $\log y = (a/b) \log x$, and *log y is a linear function of log x.*

### Example 9-25

Graph $y^2 = x^3$ on log-log paper.

## Solution

The equation can be written $y = x^{3/2}$. Choose positive values of $x$ that are perfect squares to readily compute $y$:

| $x$ | 0 | 1 | 4 | 9 | 16 | 25 | 64 | 100 |
|---|---|---|---|---|---|---|---|---|
| $y = x^{3/2}$ | 0 | 1 | 8 | 27 | 64 | 125 | 512 | 1000 |

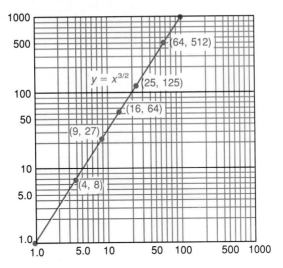

FIG. 9-10 *Graph of* $y = x^{3/2}$ *on logarithmic graph paper (3 × 3 cycles), Example 9-25*

Note that negative values of $x$ yield imaginary values for $y$. These values and the point (0, 0) cannot be plotted since a logarithmic scale contains only positive values and does not contain zero. The straight-line graph is shown in Fig. 9-10. It is plotted on 3 × 3 cycle log-log paper. The cycles on the $x$ and $y$ axes are labeled to include the range of values to be plotted.

Semilog and log-log paper are used for other functions that do not necessarily graph as straight lines but may have a large range of values, such as that of certain experimental data.

### Example 9-26

A concrete column is tested for compressive strength. The values of the stress $S$ at failure vs. length $L$ are tabulated below. Plot the graph of $S$ vs. $L$:

| $L$ (ft) | 2 | 4 | 6 | 8 | 10 | 12 | 14 | 16 |
|---|---|---|---|---|---|---|---|---|
| $S$ (lb/in²) | 5000 | 1800 | 900 | 600 | 380 | 280 | 200 | 150 |

### Solution

Since the values of $L$ are linear and the values of $S$ vary greatly, the points are plotted on semilog paper (Fig. 9-11). The curve is not a straight line but is shown more effectively than on rectangular graph paper.

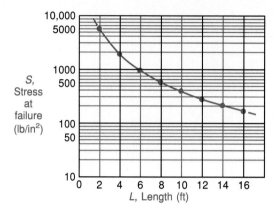

FIG. 9-11 *Compressive strength of concrete column on semilogarithmic paper (3 cycles), Example 9-26*

**Exercise 9-5**

In problems 1 to 10, graph on semilog paper that has at least 3 cycles.

**1.** $y = 4^x$

**2.** $y = 10^{-x} = \left(\dfrac{1}{10}\right)^x$

**3.** $y = 3(10)^x$

**4.** $y = 2(5)^x$

**5.** $y = x^4$ $(x > 0)$

**6.** $y = x^{-2} = (1/x)^2$ $(x > 0)$

**7.** $y = (5.5)^x$

**8.** $y = (0.20)^x$

**9.** $y = 10e^x$

**10.** $y = e^{2t}$

In problems 11 to 20, graph on log-log paper that has at least $3 \times 2$ cycles.

**11.** $y^3 = x^2$

**12.** $y^2 = 4x^3$

**13.** $3y^2 = x^3$

**14.** $y^2 = x$

**15.** $y = 0.10x^2$

**16.** $y = \pi x^3$

**17.** $y = x^{2.5}$

**18.** $y = x^{0.20}$

**19.** $x^2 y = 4$

**20.** $xy = 0.08$

**21.** The atmospheric pressure in kilopascals (kPa) is given approximately by $P = 100e^{-0.3h}$, where $h$ = altitude in kilometers. Graph $P$ vs. $h$ on semilog paper.

**22.** The amount of acid (grams) remaining after a neutralization reaction is given by $A = 400(0.5)^t$, where $t$ = time in minutes. Graph $A$ vs. $t$ on semilog paper.

**23.** A computer depreciates at a rate $r = 10$ percent per year. Using the formula for depreciation $P = P_0(1 - r)^n$, graph $P$ vs. $n$ on semilog paper if the initial value of the computer $P_0 = \$50,000$.

**24.** The resistance $R$ of a copper wire varies inversely as its cross-sectional area $A$: $R = K(1/A)$. If $K = 2.0 \times 10^{-3}\ \Omega \cdot \text{cm}^2$, graph $R$ vs. $A$ on log-log paper for $A = 0.001$ to $A = 1.0\ \text{cm}^2$.

25. *(a)* Using the formula for radioactive decay $m = m_0e^{-kt}$, graph $m$ vs. $t$ for germanium 68 (Ge 68) where $k = 2.8 \times 10^{-3}$ per day and $m_0 = 1000$ g. Use log-log paper and graph for $t = 1$ to $t = 1000$ days. *(b)* Determine the half-life of Ge 68 from the graph.

26. The viscosity of a fluid is a measure of its resistance when one plate slides over another with the fluid in between. It must be determined experimentally. Values of viscosity vs. temperature for a crude oil are given in the table below. Graph $v$ vs. $t$ on log-log paper:

| $t(°F)$ | 20 | 40 | 60 | 80 | 100 | 200 | 400 |
|---|---|---|---|---|---|---|---|
| $v$ (m$^2$/s) | 900 | 220 | 100 | 50 | 33 | 8.0 | 2.1 |

27. In calibrating a particular thermocouple, the table below of temperature vs. electromotive force (voltage) is used as a reference. Graph $t$ vs. emf on semilog paper:

| emf (mV) | 0 | 5 | 10 | 20 | 30 | 40 | 50 |
|---|---|---|---|---|---|---|---|
| $t$ (°C) | 10 | 250 | 500 | 900 | 1400 | 1800 | 2200 |

28. Human ability at times tends to be exponential. Easy tasks can be done quite rapidly, but the time required for more difficult ones increases exponentially. The time $t$ in seconds required for Zacharias Dase, the German speed calculator, to mentally calculate the product of two $n$-digit numbers was:

| $n$ | 5 | 8 | 20 | 40 | 100 |
|---|---|---|---|---|---|
| $t$ (s) | 8 | 54 | 360 | 2400 | 31,500 |

Graph $t$ vs. $r$ on log-log paper, and show it is approximately a straight line. The approximate equation of the line is $\log t = 2.72 \log n + \log 0.102$.

# 9-6 | *Hand Calculator Operations*

Exponents and logarithms are readily done on the calculator. However, it is important that you understand the theory behind such calculations and be able to interpret and judge results.

## *The Exponential Function $e^x$*

To obtain values of $e^x$ use $\boxed{\text{INV}}$ $\boxed{\ln x}$, $\boxed{\text{2nd F}}$ $\boxed{\ln x}$ or $\boxed{e^x}$. For example, $e^{2.5}$ is done:

$$2.5 \ \boxed{\text{2nd F}} \ \boxed{\ln x} \rightarrow 12.18$$

See Example 9-3 for another illustration.

## Common Logarithms

Common logarithms (base 10) are obtained by pressing $\boxed{\log}$.

### Example 9-27

Find each common logarithm. (This is Example 9-12.)

**1.** log 28.6:

$$28.6 \; \boxed{\log} \rightarrow 1.4564$$

**2.** log 286:

$$286 \; \boxed{\log} \rightarrow 2.4564$$

Notice that for a number larger than 1 the mantissa of the logarithm (the decimal part) is the same when the digits of the number are the same. See Example 9-12 for the explanation.

**3.** log 0.286:

$$0.286 \; \boxed{\log} \rightarrow -0.5436$$

**4.** log 0.0286:

$$0.0286 \; \boxed{\log} \rightarrow -1.5436$$

Notice that for a number smaller than 1 the logarithm is negative and that the negative mantissa is the same when the digits are the same.

If you enter zero or a negative number and press $\boxed{\log}$, the display will show error, since these logarithms are not defined.

To find the inverse of a common logarithm (the antilog), use $\boxed{INV}\;\boxed{\log}$ or $\boxed{10^x}$.

### Example 9-28

Find $X$ in each of the following. (This is Example 9-13.)

**1.** log $X = 0.7649$:

$$0.7649 \; \boxed{INV} \; \boxed{\log} \rightarrow 5.820$$

**2.** log $X = 3.4250$:

$$3.4250 \; \boxed{INV} \; \boxed{\log} \rightarrow 2660.7$$

**3.** $\log X = 0.5038 - 2$

Change to a negative logarithm first:

$$0.5038 \boxed{-} 2 \boxed{=} \boxed{\text{INV}} \boxed{\log} \rightarrow 0.03190$$

**Example 9-29**

Compute using logarithms: $\dfrac{0.350}{\sqrt[4]{50.0}}$

**Solution**

This is Example 9-14. Set $X$ equal to the expression, and take the logarithm of each side:

$$\log X = \log 0.350 - \frac{\log 50.0}{4}$$

Compute $\log X$:

$$0.350 \boxed{\log} - 50.0 \boxed{\log} \boxed{\div} 4 \boxed{=} \rightarrow -0.8807$$

Use the inverse to find $X$:

$$-0.8807 \boxed{\text{INV}} \boxed{\log} \rightarrow 0.1316$$

You can check this example directly on the calculator:

$$0.350 \boxed{\div} 50.0 \boxed{\text{INV}} \boxed{y^x} 4 \boxed{=} \rightarrow 0.1316$$

## *Natural Logarithms*

Natural logarithms (base $e$) are obtained by using $\boxed{\ln x}$ or $\boxed{\ln}$.

**Example 9-30**

Find each natural logarithm. (This is Example 9-15.)

**1.** $\ln 5.3$:

$$5.3 \boxed{\ln x} \rightarrow 1.6677$$

**2.** $\ln 760$:

$$760 \boxed{\ln x} \rightarrow 6.6333$$

**3.** ln 0.041:

$$0.041 \; \boxed{\ln x} \rightarrow 3.1942$$

To find the inverse of a natural logarithm, use $\boxed{\text{INV}} \; \boxed{\ln}$, $\boxed{\text{2nd F}} \; \boxed{\ln x}$ or $\boxed{e^x}$.

### Example 9-31
Find $X$ in each of the following.

**1.** ln $X = 2.800$:

$$2.800 \; \boxed{\text{INV}} \; \boxed{\ln} \rightarrow 16.44$$

**2.** ln $X = 1.321$:

$$1.321 \; \boxed{\text{INV}} \; \boxed{\ln} \rightarrow 3.747$$

**3.** ln $X = -0.2531$:

$$0.2531 \; \boxed{+/-} \; \boxed{\text{INV}} \; \boxed{\ln} \rightarrow 0.7764$$

**Exercise 9-6**   In problems 1 to 16, find $X$.

**1.** $X = \log 53.27$          **2.** $X = \log 1025$

**3.** $X = \log 0.05168$        **4.** $X = \log 0.1234$

**5.** $\log X = 0.6166$         **6.** $\log X = 2.1565$

**7.** $\log X = -2.3738$        **8.** $\log X = -0.9714$

**9.** $X = \ln 17.55$           **10.** $X = \ln 3.0131$

**11.** $X = 0.8308$             **12.** $X = \ln 0.00153$

**13.** ln $X = 1.520$           **14.** ln $X = 2.301$

**15.** ln $X = -2.8713$         **16.** ln $X = -0.1302$

In problems 17 to 20, compute, using logarithms. Check each answer directly.

**17.** $(5.812)(0.3151)^2$          **18.** $(\sqrt[3]{82.31})(0.1562)$

**19.** $\dfrac{\sqrt[4]{8.012}}{30.21}$          **20.** $\dfrac{\sqrt{0.08761}}{\sqrt[3]{3.142}}$

## 9-7 | *Review Questions*

In problems 1 to 4, graph $y$ vs. $x$, plotting at least five points.

**1.** $y = 3(2^x)$              **2.** $y = (0.80)^x$

**3.** $y = (0.5)^{x-1}$         **4.** $y = e^x + 1$

In problems 5 and 6, express in exponential form.

**5.** $\log_3 9 = 2$          **6.** $\log 0.10 = -1$

In problems 7 and 8, express in logarithmic form.

**7.** $5^{-2} = 0.04$          **8.** $10^2 = 100$

In problems 9 to 16, find $x$ or $y$ without a table or calculator.

**9.** $\log_4 x = -2$          **10.** $\log_8 x = \dfrac{1}{3}$

**11.** $\log 1000 = y$          **12.** $\log_5 0.2 = y$

**13.** $\ln x = 1$          **14.** $\ln x = 3$

**15.** $\log \sqrt{10} = y$          **16.** $\ln 1 = y$

In problems 17 and 18, express as a combination of logarithms.

**17.** $\log_2 \dfrac{x\sqrt{2}}{8}$          **18.** $\log (100a^3)$

In problems 19 and 20, express as a single logarithm.

**19.** $\log_4 2 + \log_4 32$      **20.** $3 \log 2x - 2 \log x$

In problems 21 to 26, find $X$, using a calculator or table.

**21.** $X = \log 13.2$          **22.** $X = \log 0.505$

**23.** $\log X = 3.4099$          **24.** $\log X = 0.8938 - 2$

**25.** $X = \ln 7.50$          **26.** $\ln X = 0.5306$

In problems 27 to 30, solve for $x$.

**27.** $2^{x-2} = 64$          **28.** $5^x = 30$

**29.** $2 \log (3x - 2) = 4$      **30.** $e^{2x} = 1.5$

**31.** The atmospheric pressure in kilopascals is given approximately by $P = 100(0.80)^h$, where $h$ is the height above sea level in miles. Graph $P$ vs. $h$ from $h = 0$ to $h = 6$ mi.

**32.** Depreciation of an item can be expressed by the relation $P = P_0(1 - r)^n$, where $P$ = value $n$ years from now, $P_0$ present value, and $r$ = rate of depreciation. *(a)* If a \$10,000 car depreciates 30 percent per year, graph $P$ vs. $n$ for $n = 0$ to $n = 5$ yr, the average life expectancy of a car. *(b)* After how many years will the value of the car be one-half the original value?

**33.** The half-life $t$ of a radioactive element is given by $t = (\ln 2)/k$, where $k$ is the rate of decay. Radioactive carbon 14 (C 14) is present at a constant level in all living things but decays slowly after death at a rate of $k = 1.238 \times 10^{-4}$ per year. The amount of C 14 present in organic material therefore determines its age. What is the half-life of C 14?

**34.** The formula for the decibel gain of an amplifier in terms of power is $n = 10 \log (P_2/P_1)$, where $P_1$ = power input and $P_2$ = power output. Find the power gain $P_2/P_1$ when $n = 25$ dB.

**35.** In problem 30, Sec. 9-1, the formula for the current in an *RL* circuit is given as $I = I_0 (1 - e^{-Rt/L})$.  *(a)* Show that this formula can be solved for $t = (L/R) \ln [I_0/(I_0 - I)]$. *(b)* Given $I_0 = 2.5$ A, $R = 40$ $\Omega$, and $L = 2.0$ H, find $t$ when $I = 2.0$ A.

**36.** A virus population increases according to $N = N_0 (2)^{0.20t}$, where $N_0 =$ number of viruses at $t = 0$ h and $N =$ number after $t$ hours. How many hours does it take for the population to *(a)* double and *(b)* triple?

**37.** A discharged battery is being charged by a solar cell. After $t$ hours the voltage $V = 13.2 (1 - e^{-0.1t})$. Graph $V$ vs. $t$ on semilog paper.

**38.** In a turbojet engine the pressure ratio $P_r$ and temperature ratio $T_r$ are related during the expansion phase by $P_r^{k-1} = T_r^k$, or $P_r = T_r^{k/k-1}$. If $k = 1.8$, graph $P_r$ vs. $T_r$ on log-log paper for $T_r = 1.0$ to $T_r = 20$.

# 10

# Trigonometry

## 10-1 | *Circular Trigonometric Functions: Definitions*

Consider the circle shown in Fig. 10-1, whose center is at the origin of a rectangular-coordinate system. The radius $r$ is considered a *vector,* a line segment having both length and direction. The direction of the radius vector is determined by an angle $\theta$ in the circle. The angle $\theta$ is measured from the positive $x$ axis, or *initial side,* either counterclockwise (positive direction) or clockwise (negative direction) to the *terminal side.* The circle in Fig. 10-1 contains four quadrants and has a positive angle $\theta_1$ in the first quadrant and a positive angle $\theta_2$ in the second quadrant.

**C**hapter 4, Sec. 4-8, defines the trigonometric functions for an angle in a right triangle. However, an angle measures rotation and is therefore more basic to a circle than to a triangle. Therefore this chapter defines the trigonometric functions for an angle in a circle and includes the right-triangle definitions as a special case. These new definitions extend the ideas of trigonometry to include the important concept of vectors, the law of sines, and the law of cosines. These ideas lead to many applications in science and technology.

The *six trigonometric functions* for an angle $\theta$ whose terminal side passes through the point $(x, y)$ on a circle of radius $r$ are defined as follows:

$$\sin \theta = \frac{y}{r} \qquad \csc \theta = \frac{1}{\sin \theta} = \frac{r}{y}$$

$$\cos \theta = \frac{x}{r} \qquad \sec \theta = \frac{1}{\cos \theta} = \frac{r}{x} \qquad (10\text{-}1)$$

$$\tan \theta = \frac{y}{x} \qquad \cot \theta = \frac{1}{\tan \theta} = \frac{x}{y}$$

Note that these definitions are the same as (4-8) and (4-9) in Sec. 4-8 when they are applied to the special case of an angle in the first quadrant, such as $\theta_1$ in

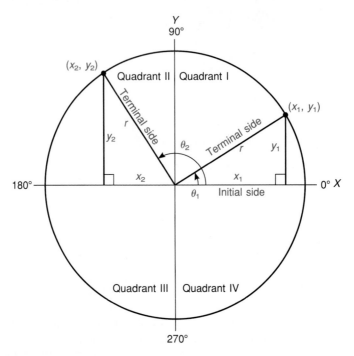

FIG. 10-1 *Trigonometry of the circle*

Fig. 10-1. It is important to realize, however, that the above definitions apply to an angle *in any quadrant*. Study the following example which considers four angles, one in each quadrant.

### Example 10-1

Draw each angle whose terminal side passes through the given point, and find the six trigonometric functions of the angle.

1. $\theta_1$: (4, 3)
2. $\theta_2$: (−4, 3)
3. $\theta_3$: (−4, −3)
4. $\theta_4$: (4, −3)

### Solution

The four cases are drawn in Fig. 10-2. For each case a perpendicular is drawn from the point to the *x axis*, not the *y axis*. The cases represent one angle in each quadrant, each having the same length radius. To find the six trigonometric functions for these angles, first note that for an angle in any quadrant, $x^2 + y^2 = r^2$ by the pythagorean theorem, *where r is always considered positive*. Therefore, for all four cases $r = 5$, since $x^2 + y^2 = 9 + 16 = 25 = (5)^2$ regardless of the sign of $x$ or $y$. According to definition (10-1) the trigonometric functions for each case are as follows:

**1.** Quadrant I: $x = 4$, $y = 3$, $r = 5$

$$\sin \theta_1 = \frac{3}{5} \qquad \csc \theta_1 = \frac{5}{3}$$

$$\cos \theta_1 = \frac{4}{5} \qquad \sec \theta_1 = \frac{5}{4}$$

$$\tan \theta_1 = \frac{3}{4} \qquad \cot \theta = \frac{4}{3}$$

**2.** Quadrant II: $x = -4$, $y = 3$, $r = 5$

$$+\blacklozenge \sin \theta_2 = \frac{3}{5} \qquad \csc \theta_2 = \frac{5}{3}$$

$$\cos \theta_2 = -\frac{4}{5} \qquad \sec \theta_2 = -\frac{5}{4}$$

$$\tan \theta_2 = -\frac{3}{4} \qquad \cot \theta_2 = -\frac{4}{3}$$

**3.** Quadrant III: $x = -4$, $y = -3$, $r = 5$

$$\sin \theta_3 = -\frac{3}{5} \qquad \csc \theta_3 = -\frac{5}{3}$$

$$\cos \theta_3 = -\frac{4}{5} \qquad \sec \theta_3 = -\frac{5}{4}$$

$$+\blacklozenge \tan \theta_3 = \frac{3}{4} \qquad \cot \theta_3 = \frac{4}{3}$$

**4.** Quadrant IV: $x = 4$, $y = -3$, $r = 5$

$$\sin \theta_4 = -\frac{3}{5} \qquad \csc \theta_4 = -\frac{5}{3}$$

$$+\blacklozenge \cos \theta_4 = \frac{4}{5} \qquad \sec \theta_4 = \frac{5}{4}$$

$$\tan \theta_4 = -\frac{3}{4} \qquad \cot \theta_4 = -\frac{4}{3}$$

Observe that the values are the same in each quadrant or differ only in sign. The signs are determined by the point $(x, y)$ on the terminal side. All the functions are positive in the first quadrant, and only two functions are positive

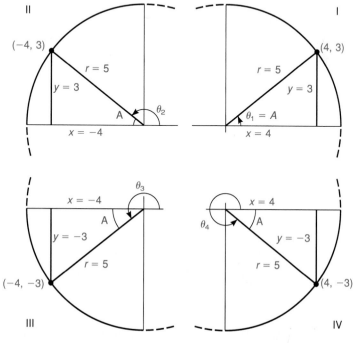

FIG. 10-2 *Angles in each quadrant, Example 10-1*

in any other quadrant (see arrows above). The *reference angle A* in Fig. 10-2 is equal to $\theta_1$ in each quadrant. The importance of the reference angle is discussed in the next section.

The positive functions for angles in each quadrant are shown in Fig. 10-3. One way to remember these signs is to memorize the word *CAST*. The letters in CAST are shown in Fig. 10-3, starting from the fourth quadrant and reading counterclockwise.

## Example 10-2

Given that $\tan \theta = -1$ and that $\cos \theta$ is positive:

1. Draw $\theta$, showing the values for $x$, $y$, and $r$.
2. Find the six trigonometric functions of $\theta$.

## Solution

1. Draw $\theta$, showing the values for $x$, $y$, and $r$.

   Since $\cos \theta$ is positive and $\tan \theta$ is negative, Fig. 10-3 shows that $\theta$ must be in the fourth quadrant. Any positive value for $x$ and the corresponding negative value for $y$ can be chosen such that $\tan \theta = y/x = -1$. Choose simple values; for example, $x = 1$, $y = -1$. Then $r = \sqrt{1^2 + (-1)^2} = \sqrt{2}$. Angle $\theta$ is shown in Fig. 10-4. The triangle shown is called the *reference triangle*. It is always drawn to the $x$ axis.

2. Find the six trigonometric functions of $\theta$.

   The values of the functions are determined from Fig. 10-4:

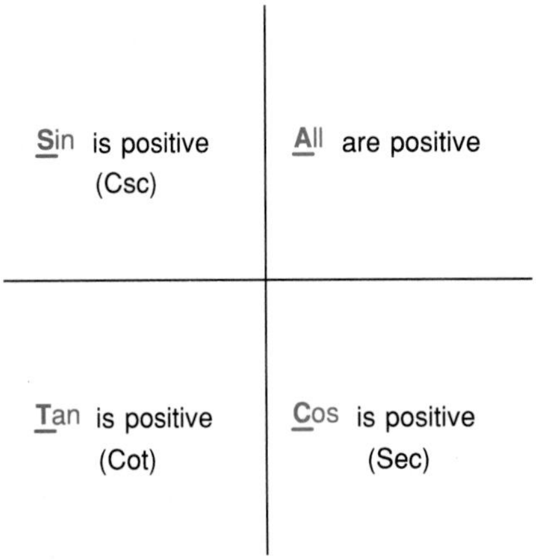

|  |  |
|---|---|
| Sin is positive | All are positive |
| (Csc) |  |
| Tan is positive | Cos is positive |
| (Cot) | (Sec) |

FIG. 10-3 *Positive trigonometric functions*

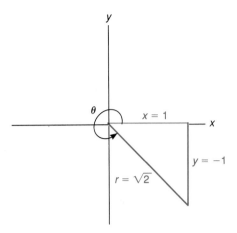

FIG. 10-4  *Fourth-quadrant angle, Example 10-2*

$$x = 1 \qquad y = -1 \qquad r = \sqrt{2}$$

$$\sin \theta = \frac{-1}{\sqrt{2}} = \frac{-\sqrt{2}}{2} \quad \text{or} \quad -0.7071 \qquad \csc \theta = -\sqrt{2} \quad \text{or} \quad -1.414$$

$$\cos \theta = \frac{1}{\sqrt{2}} = \frac{\sqrt{2}}{2} \quad \text{or} \quad 0.7071 \qquad \sec \theta = \sqrt{2} \quad \text{or} \quad 1.414$$

$$\tan \theta = \frac{-1}{1} = -1 \qquad\qquad\qquad \cot \theta = \frac{1}{-1} = -1$$

Notice that the denominator for sin $\theta$ and cos $\theta$ is rationalized by multiplying top and bottom by $\sqrt{2}$.

Given the values of two trigonometric functions of an angle $\theta$ that are not reciprocal functions, the values of the remaining functions can be found directly by using the reciprocal identities in (10-1) for csc $\theta$, sec $\theta$, and cot $\theta$ and by using the *ratio identities:*

$$\tan \theta = \frac{\sin \theta}{\cos \theta} \qquad \cot \theta = \frac{\cos \theta}{\sin \theta} \tag{10-2}$$

The proof of identities (10-2) follows from the definitions of tan $\theta$ and cot $\theta$:

$$\tan \theta = \frac{y}{x} = \frac{y/r}{x/r} = \frac{\sin \theta}{\cos \theta} \qquad \cot \theta = \frac{x}{y} = \frac{x/r}{y/r} = \frac{\cos \theta}{\sin \theta}$$

The following example shows how to apply identities (10-2).

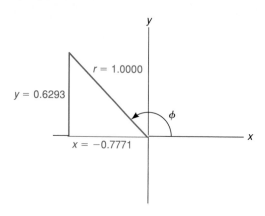

FIG. 10-5 *Second-quadrant angle, Example 10-3*

### Example 10-3

Given $\sin \phi = 0.6293$ and $\cos \phi = -0.7771$, find the other trigonometric functions of $\phi$ (phi).

### Solution

The functions can be found (1) by obtaining values for $x$, $y$, and $r$ or (2) by using the reciprocal identities in (10-1) and the ratio identities (10-2).

1. Since $\sin \phi$ is positive and $\cos \phi$ is negative, $\phi$ is in the second quadrant. Let $x = -0.7771$ and $y = 0.6293$. Therefore we have $r = \sqrt{(-0.7771)^2 + (0.6293)^2} = 1.000$. Angle $\phi$ is shown in Fig. 10-5. The other functions are then found by applying the definitions from (10-1) as in Examples 10-1 and 10-2.

2. Using the identities in (10-1) and (10-2) yields the following:

$$\csc \phi = \frac{1}{\sin \phi} = \frac{1}{0.6293} = 1.589$$

$$\sec \phi = \frac{1}{\cos \phi} = \frac{1}{-0.7771} = -1.287$$

$$\tan \phi = \frac{\sin \phi}{\cos \phi} = \frac{0.6293}{-0.7771} = -0.8098$$

$$\cot \phi = \frac{\cos \phi}{\sin \phi} = \frac{-0.7771}{0.6293} = -1.235$$

**Exercise 10-1**

In problems 1 to 12, draw the angle whose terminal side passes through the given point, and find the six trigonometric functions of the angle.

**1.** (3, 4)       **2.** (−6, 8)       **3.** (5, −12)

**4.** (−15, −8)     **5.** (−1, −1)      **6.** (1, $\sqrt{3}$)

**7.** $(-3.9, 5.2)$     **8.** $(16.8, -7.00)$     **9.** $\left(-\dfrac{1}{2}, -\dfrac{2}{3}\right)$

**10.** $\left(\dfrac{1}{2}, \dfrac{1}{2}\right)$     **11.** $(4x, 5x)$     **12.** $(-6k, -8k)$

In problems 13 to 24, draw the angle and find the remaining trigonometric functions.

**13.** $\sin \theta = -\frac{3}{5}$, $\tan \theta$ positive
**14.** $\cos \theta = \frac{12}{13}$, $\sin \theta$ negative
**15.** $\tan \theta = -\frac{4}{3}$, $\cos \theta$ positive
**16.** $\cot \theta = \frac{15}{8}$, $\sin \theta$ positive
**17.** $\cos \phi = -0.8$, $\sin \phi$ positive
**18.** $\sin \phi = 0.5$, $\cos \phi$ negative
**19.** $\sin A = \frac{1}{3}$, $A$ in quadrant I
**20.** $\cos B = \sqrt{2}/2$, $B$ in quadrant IV
**21.** $\cot x = -1$, $x$ in quadrant II
**22.** $\tan y = 0.750$, $y$ in quadrant III
**23.** $\sec \phi = 2.00$, $\phi$ in quadrant IV
**24.** $\csc \phi = 1.25$, $\phi$ in quadrant II

© In problems 25 to 30, find the remaining trigonometric functions.

**25.** $\sin \theta = 0.6000$, $\cos \theta = 0.8000$
**26.** $\sec \theta = 1.414$, $\csc \theta = 1.414$
**27.** $\sin A = -0.5000$, $\cos A = -0.8660$
**28.** $\sin B = 0.7071$, $\tan B = -1.000$
**29.** $\sin \phi = 0.9511$, $\cos \phi = -0.3090$
**30.** $\cos x = -0.6428$, $\tan x = -1.192$
**31.** If $\tan x = 1/\sqrt{3}$, what are the two possible values for $\csc x$?
**32.** If $\sec x = 2$, what are the two possible values for $\sin x$?
**33.** A force vector $F$ has components $F_x = -4.5$ lb and $F_y = 8.5$ lb, as shown in Fig. 10-6. Find $\sin \theta$ and $\cos \theta$.

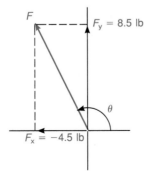

FIG. 10-6 *Force vectors, problem 33*

**34.** In an ac circuit $\tan \phi_z = X/R$, where $\phi_z$ is the phase angle of the impedance, and $\tan \phi_y = -X/R$, where $\phi_y$ is the phase angle of the admittance. If $X^2 + R^2 = Z^2$, find $\tan \phi_z$ and $\tan \phi_y$ when $R = 40.0 \ \Omega$ and $Z = 52.0 \ \Omega$.

# 10-2 | Reference Angles and Special Angles

## Reference Angles

Section 10-1 does not discuss how to find the trigonometric functions when you are given the angle and it is not in the first quadrant. For example, what is sin 110°? What is cos 230°? What is tan 315°? Such problems can arise when you work with vectors, triangles that have obtuse angles, and trigonometric curves.

Study Example 10-1 and Fig. 10-2 carefully. The values of the trigonometric functions of the four angles $\theta_1$, $\theta_2$, $\theta_3$, and $\theta_4$ are the same or differ only in sign. The absolute (positive) values of $x$, $y$, and $r$ are equal for each angle. The four triangles in Fig. 10-2 are all congruent, and the acute angles $A$ are all equal to $\theta_1$. These triangles are called *reference triangles* and the acute angles *reference angles*. For an angle in any quadrant the following rule applies:

> The absolute, or positive, value of the trigonometric function of an angle in any quadrant is equal to the trigonometric function of its reference angle.

That is, the value of the function is the same as that of its reference angle or differs only in sign. *The reference triangle must always be drawn to the x axis;* otherwise, *y* is not opposite to the reference angle, and the above rule does not apply. Study the next example carefully.

### Example 10-4
Find the value of each of the following:

**1.** sin 110°     **2.** cos 230°     **3.** tan 315°

For each angle *(a)* determine the reference angle, *(b)* determine the correct sign of the function, and *(c)* find the value of the function of the reference angle and attach the correct sign to it.

### Solution
Each of the angles and their reference triangles are shown in Fig. 10-7.

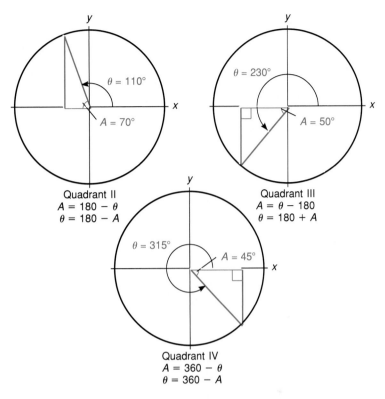

FIG. 10-7 *Reference angles, Example 10-4*

1. sin 110°. For an angle $\theta$ in quadrant II, the reference angle $A = 180° - \theta$. Therefore $A = 180° - 110° = 70°$. The sine is positive in quadrant II, so sin 110° = +sin 70° = 0.9397. The value of the trigonometric function is obtained with a table or calculator (Sec. 4-8). The calculator will also give the trigonometric function of the original angle without changing it to the reference angle. (See Sec. 10-7, Example 10-26, for how to use a hand calculator in these problems.) However, the reference angle must be understood for work with inverse functions, equations, and identities. Therefore examples are done by using the reference angle. The original angle can be used on the calculator to check the result.

2. cos 230°. For an angle $\theta$ in quadrant III, the reference angle $A = \theta - 180°$. Therefore $A = 230° - 180° = 50°$. The cosine is negative in quadrant III, so cos 230° = −cos 50° = −0.6428.

3. tan 315°. For an angle $\theta$ in quadrant IV, the reference angle $A = 360° - \theta$. Therefore $A = 360° - 315° = 45°$. The tangent is negative in quadrant IV, so tan 315° = −tan 45° = −1.000.

## Special Angles

The angles 30°, 45°, and 60° occur often in problems not only by themselves, but also as reference angles. Their trigonometric functions are easy to learn, so a calculator or table is not necessary to find them. It is useful to know the values of these functions for a better understanding of how trigonometric functions behave and to be able to check results. The following two facts should be memorized:

$$\sin 30° = \frac{1}{2} = 0.5000 \qquad \tan 45° = 1.000$$

With the aid of the above facts and the pythagorean theorem, the triangles in Fig. 10-8 can readily be drawn. (You can also just memorize the sides of each triangle.) The trigonometric functions of 30°, 45°, or 60° are then found from the triangles. For example, $\tan 30° = 1/\sqrt{3} = \sqrt{3}/3$ or 0.5774, $\sin 45° = 1/\sqrt{2} = \sqrt{2}/2$ or 0.7071, etc. The values for these functions are summarized in Table 10-1.

Study the next example, which illustrates the use of Table 10-1.

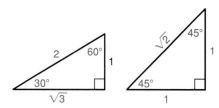

FIG. 10-8 *Special angles and triangles*

### Table 10-1  SPECIAL ANGLES

|  | 30° | 45° | 60° |
|---|---|---|---|
| sin | $\frac{1}{2}$ or 0.5000 | $\frac{1}{\sqrt{2}} = \frac{\sqrt{2}}{2}$ or 0.7071 | $\frac{\sqrt{3}}{2}$ or 0.8660 |
| cos | $\frac{\sqrt{3}}{2}$ or 0.8660 | $\frac{1}{\sqrt{2}} = \frac{\sqrt{2}}{2}$ or 0.7071 | $\frac{1}{2}$ or 0.5000 |
| tan | $\frac{1}{\sqrt{3}} = \frac{\sqrt{3}}{3}$ or 0.5774 | 1.000 | $\sqrt{3}$ or 1.732 |

### Example 10-5
Find the six trigonometric functions of 300° *without a table or calculator*.

### Solution
Figure 10-9 shows the reference triangle for 300° containing a reference angle of 60°. The functions can be obtained directly from the sides of the triangle or from the reference angle and Table 10-1 by attaching the correct signs:

$$\sin 300° = -\sin 60° = -\frac{\sqrt{3}}{2} \quad \text{or} \quad -0.8660$$

$$\cos 300° = \cos 60° = \frac{1}{2} \quad \text{or} \quad 0.5000$$

$$\tan 300° = -\tan 60° = -\sqrt{3} \quad \text{or} \quad -1.732$$

$$\csc 300° = \frac{-2}{\sqrt{3}} = -\frac{2\sqrt{3}}{3} \quad \text{or} \quad -1.155$$

$$\sec 300° = 2.000$$

$$\cot 300° = -1/\sqrt{3} = -\frac{\sqrt{3}}{3} \quad \text{or} \quad -0.5774$$

Note that by memorizing $\sqrt{2} = 1.414$ and $\sqrt{3} = 1.732$ you have a complete independent check on the calculator or table for the values of the functions of 30°, 45°, and 60°.

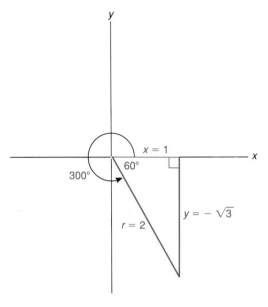

FIG. 10-9 *Special angle, Example 10-5*

## Quadrant Angles

Reference triangles cannot be drawn for the *quadrant angles* 0°, 90°, 180°, and 270°. The trigonometric functions of these angles are determined from the unit circle (Fig. 10-10), where $r = 1$. In the unit circle $\sin \theta = y/r = y/1 = y$ and $\cos \theta = x/r = x/1 = x$. Therefore the coordinates of a point $(x, y)$ on the terminal side of $\theta$ correspond to $(\cos \theta, \sin \theta)$. For example, $\sin 0° = 0$, $\cos 0° = 1$, $\sin 90° = 1$, $\cos 90° = 0$, etc. Applying $\tan \theta = y/x = \sin \theta/\cos \theta$, you readily obtain $\tan 0° = 0/1 = 0$, $\tan 90° = 1/0$ (undefined), etc. The values for the quadrant angles are summarized in Table 10-2.

As $\theta$ approaches 90° or 270°, $\tan \theta$ becomes infinitely large, which is indicated by the symbol $\infty$. The tangent is undefined at 90° and 270°: It does not have any value at these angles. When values are inverted in the table, the reciprocal functions are also undefined if the denominator is zero. For example,

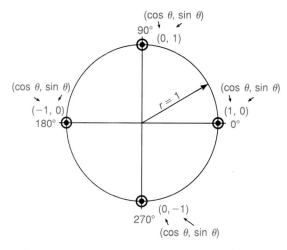

FIG. 10-10  *Quadrant angles in the unit circle*

**Table 10-2    QUADRANT ANGLES**

|  | 0° | 90° | 180° | 270° |
|---|---|---|---|---|
| sin | 0 | 1 | 0 | $-1$ |
| cos | 1 | 0 | $-1$ | 0 |
| $\tan\left(\dfrac{\sin}{\cos}\right)$ | $\dfrac{0}{1} = 0$ | $\dfrac{1}{0}$ $(\infty)$ <br> undefined | $-\dfrac{0}{1} = 0$ | $-\dfrac{1}{0}$ $(\infty)$ <br> undefined |

csc $180° = 1/\sin 180° = 1/0$ is undefined. The calculator display will indicate error when a function is undefined.

## Coterminal Angles

For angles of 360° or larger, or angles less than 0°, the values of the trigonometric functions are the same as those for the *coterminal angle* between 0° and 360°. The coterminal angle is the angle having the same terminal side as the given angle. For example, $\sin 390° = \sin (390° - 360°) = \sin 30°$, $\cos 900° = \cos (900° - 720°) = \cos 180°$, $\tan (-60°) = \tan (360° - 60°) = \tan 300°$, etc.

The next example brings together many of the ideas in this section. Study it carefully.

### Example 10-6
Find the value of each of the following:

**1.** $\sin 260.5°$     **2.** $\tan 150°10'$

**3.** $\cos 5.41$     **4.** $\csc 450°$

**5.** $\sec \dfrac{7\pi}{4}$     **6.** $\cot (-240°)$

### Solution
Change each function to a function of the reference angle, or quadrant angle. *If necessary, draw the reference triangle to determine the reference angle.* Find the value of the function with a calculator or by using the closest value in the table.

**1.** $\sin 260.5° = -\sin (260.5° - 180°) = -\sin 80.5° = -0.9863$

**2.** $\tan 150°10' = -\tan (180° - 150°10') = -\tan 29°50' = -0.5735$

In (3) there is no degree mark, hence the angle is in radian measure. Multiply by 57.30° to change it to degrees (see Sec. 4-3) or work in radians as follows. The angle is in quadrant IV, and $360° = 2\pi = 6.283$ rad, so

**3.** $\cos 5.41 = +\cos (6.283 - 5.41) = \cos 0.873 = 0.642$

**4.** $\csc 450° = \csc (450° - 360°) = \csc 90° = 1/(\sin 90°) = 1/1 = 1$

In (5) the angle is in radians in terms of $\pi$. It can readily be changed to degrees by multiplying by $180/\pi$ (see Sec. 4-3):

**5.** $\sec 7\pi/4 = \sec [(7\pi/4)(180/\pi)] = \sec 315° = +\sec (360° - 315°)$
    $= \sec 45° = \sqrt{2}$ or 1.414. You can also let $\pi = 3.142$ and work with the angle in radians.

**6.** cot (−240°) = cot (360° − 240°) = cot 120° = −cot (180° − 120°)

$$= -\cot 60° = -\frac{\sqrt{3}}{3} \text{ or } -0.5774.$$

A diagram is definitely helpful for Example 10-6. Draw a diagram for each case and see if you can obtain the reference angle in fewer steps.
    See Example 10-26 for the calculator solution of Example 10-6.

**Exercise 10-2**    In problems 1 to 16, express each as a function of the reference angle and find the value. Do special angles without a table or calculator.

1. sin 130°          2. cos 220°

3. tan 330°          4. cot 150°

5. cos 265°          6. sin 305°

7. sec 120°          8. csc 225°

9. sin $\dfrac{3\pi}{4}$    10. cos $\dfrac{4\pi}{3}$

11. cot 256°          12. tan 306°

13. cos (−30°)      14. sin (−45°)

15. tan 720°          16. cot (−405°)

Do problems 17 to 22 without a table or calculator.

17. Complete the table:

|       | 30°        | 45°        | 60°        |
|-------|------------|------------|------------|
| csc   |            | $\sqrt{2}$ |            |
| sec   | $2\sqrt{3}/3$ |         |            |
| cot   |            |            | $\sqrt{3}/3$ |

18. Find the six trigonometric functions of *(a)* 210° and *(b)* 135°.

19. Complete the table:

|      | 0° | 90°                  | 180° | 270° |
|------|----|----------------------|------|------|
| csc  |    |                      |      | −1   |
| sec  |    | undefined ($\infty$) |      |      |
| cot  |    |                      |      | 0    |

20. Show that *(a)* sin 90° ≠ 2(sin 45°) and *(b)* sin 90° = 2(sin 45°)(cos 45°).

21. Show that *(a)* $\sin^2(5\pi/6) + \cos^2(5\pi/6) = 1$  [*Note:*  $\sin^2\theta = (\sin\theta)^2$]  and *(b)* $\sec^2 210° - \tan^2 210° = 1$.

22. Show that *(a)* sin 60° ≠ sin 120°/2 and *(b)* $\sin 60° = \sqrt{(1 - \cos 120°)/2}$.

In problems 23 to 40, find the value of each function by any method.

**23.** sin 130.6°        **24.** cos 230.8°

**25.** cos 220°20′       **26.** sin 303°40′

**27.** tan 2.27          **28.** cot 6.02

**29.** sec 175.3°        **30.** csc 278.2°

**31.** $\sin \dfrac{7\pi}{2}$        **32.** $\cos(-5\pi)$

**33.** cot (−135.5°)     **34.** tan (−370.5°)

**35.** $\csc \dfrac{5\pi}{4}$        **36.** $\sec \dfrac{4\pi}{3}$

**37.** cos 420°30′       **38.** sin 700°50′

**39.** tan 3.00          **40.** cot 1.50

**41.** Show without a table or calculator that when $\theta$ = 30°, 45°, 60°, and 90°:

$$\sin(-\theta) = -\sin\theta \quad \text{and} \quad \cos(-\theta) = \cos\theta \qquad (10\text{-}3)$$

**42.** Prove (10-3) for any angle $\theta$ by using the definitions in (10-1) and Fig. 10-11.

**43.** The range $x$ of a projectile fired with velocity $v_0$ at an angle $\phi$ with the horizontal is given by $x = (v_0^2 \sin 2\phi)/g$, where $g$ = gravitational acceleration. Find $x$ in terms of $v_0$ and $g$ when (a) $\phi$ = 30°, (b) $\phi$ = 45°, (c) $\phi$ = 75°. (d) Show that $x$ is greatest at $\phi$ = 45°, which is the angle of maximum range.

**44.** The resultant velocity $v$ of a ship (in knots) is given by $v = \sqrt{67 - 28\cos 120°}$. Calculate $v$.

**45.** In an ac circuit containing a resistance $R$, the current $I = I_{max} \sin \omega t$, where $I_{max}$ = maximum current and $\omega t$ = angle of rotation. Find $I$ when $I_{max}$ = 10 A and (a) $\omega t = \pi/2$, (b) $\omega t = \pi/3$.

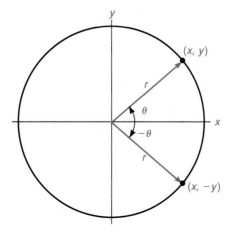

FIG. 10-11 *Problem 42*

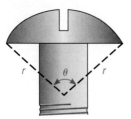

FIG. 10-12 *Bolt head (segment of circle), problem 49*

**46.** In a certain circuit containing a capacitance, the voltage $V = 45 \sin (\omega t - \pi/2)$. Find $V$ when (*a*) $\omega t = 3\pi/4$ and (*b*) $\omega t = 0$.

**47.** The index of refraction of a certain prism is $n = \sin 100°/\sin 47°$. Calculate the value of $n$.

**48.** A formula for the area $A$ of a parallelogram with sides $a$ and $b$ and included angle $\theta$ is $A = ab \sin \theta$. Find the area of a bridge section in the shape of a parallelogram where $a = 9.2$ m, $b = 7.6$ m, and $\theta = 130°$.

🖸 **49.** The cross section of the bolt head shown in Fig. 10-12 is in the shape of a segment of a circle. The area of a circular segment is given by $A = (r^2/2)(\theta - \sin \theta)$, where $\theta$ is in *radians*. Find $A$ for the bolt head if $r = 15$ mm and $\theta = 110°$.

🖸 **50.** One of the formulas used to calculate the value of $\sin x$ comes from calculus:

$$\sin x = x - \frac{x^3}{2 \cdot 3} + \frac{x^5}{2 \cdot 3 \cdot 4 \cdot 5} - \frac{x^7}{2 \cdot 3 \cdot 4 \cdot 5 \cdot 6 \cdot 7} + \cdots \qquad \text{(to infinity)}$$

The formula goes on forever, increasing in accuracy. The angle $x$ must be in radians. Using your calculator, check the formula with the four terms given when (*a*) $x = \pi/6$ and (*b*) $x = \pi/2$.

## 10-3 | *Inverse Trigonometric Functions*

Section 4-8 introduces the inverse trigonometric functions which are used to find an angle when two sides of a right triangle are known. This section focuses on the inverse functions more closely and shows you how to find what quadrant or quadrants an angle is in when trigonometric functions of the angle are known. Study the next example.

### Example 10-7

Given $\tan \theta = -1.600$, find $\theta$ to the nearest $0.1°$ for $0° \le \theta < 360°$ ($\theta$ greater than or equal to $0°$ and less than $360°$).

### Solution

Tangent is negative in the second and the fourth quadrant, so there are two possible answers (see Fig. 10-13). First find the reference angle $A$, using the

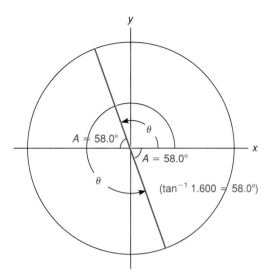

FIG. 10-13 *Inverse function, Example 10-7*

*positive* value of the function, and either reading the table backward or using [INV] [tan], [2nd F] [tan], or [tan⁻¹] on the calculator:

$$1.600 \;\boxed{INV}\; \boxed{tan} \to 58.0°$$

This is expressed mathematically as $\tan^{-1}(1.600) = 58.0°$ or Arctan $1.600 = 5.80°$ (read "angle whose tangent is $1.600 = 58.0°$"). Then from Fig. 10-13:

$$\theta = 180° - 58.0° = 122° \qquad \text{quadrant II}$$

or

$$\theta = 360° - 58.0° = 302° \qquad \text{quadrant IV}$$

It is important to realize that the calculator cannot necessarily be used directly to solve a problem like Example 10-7. If you enter the negative value $-1.600$ in the calculator:

$$1.600 \;\boxed{+/-}\; \boxed{INV}\; \boxed{tan} \to -58.0°$$

the result is a negative value in the fourth quadrant, that is, $\tan^{-1}(-1.600) = -58.0°$. The reason is that the *inverse trigonometric functions have only one principal value of the angle defined for each value of x:*

$$-90° \le \sin^{-1} \le 90° \qquad 0 \le \cos^{-1} \le 180° \qquad (10\text{-}4)$$
$$-90° < \tan^{-1} < 90°$$

The functions $\sin^{-1} x$ and $\tan^{-1} x$ assume values that are those of angles in the first quadrant from 0° to 90° or those of negative angles in the fourth quadrant from $0°$ to $-90°$. The function $\cos^{-1} x$ assumes values that are those of angles in the first and second quadrant from $0°$ to $180°$. The reason for these definitions is to provide a continuous range of values for each inverse function that *includes all possible values of x*. Note that *when x is positive, the values of the inverse functions are always positive acute angles or 90°*. Therefore, for ease in calculation, positive values of $x$ are always used with inverse functions in the examples in the text to provide values of the reference angle. See Sec. 10-7 for more calculator information on the subject.

### Example 10-8
Find angle $A = \cos^{-1}(\sqrt{2}/2)$ (read, "find the angle $A$ whose cosine is $\sqrt{2}/2$").

### Solution
This is one of the special angles in Table 10-1. Since $\cos 45° = \sqrt{2}/2$, angle $A = \cos^{-1}\sqrt{2}/2 = 45°$. Note that there is only one solution for angle $A$. The next example shows when there is more than one solution for the angle.

### Example 10-9
Given each of the following, find $\theta$ to the nearest $0.1°$ for $0° \le \theta < 360°$.

**1.** $\sin \theta = 0.3843$      **2.** $\cos \theta = -0.0157$      **3.** $\sec \theta = 1.553, \sin \theta < 0$

**4.** The terminal side of $\theta$ passes through point $(2, -1)$.

### Solution
**1.** $\sin \theta = 0.3843$

Angle $\theta$ is in quadrant I or II since the sine is positive. The reference angle $A = \sin^{-1} 0.3843 = 22.6°$, using a calculator or table. Then $\theta = 22.6°$ or $\theta = 180° - 22.6° = 157.4°$.

**2.** $\cos \theta = -0.0157$

Angle $\theta$ is in quadrant II or III. Reference angle $A = \cos^{-1} 0.0157 = 89.1°$. Then $\theta = 180° - 89.1° = 90.9°$ or $\theta = 180° + 89.1° = 269.1°$.

**3.** $\sec \theta = 1.553, \sin \theta < 0$

Find the reciprocal function: $\cos \theta = 1/1.553 = 0.6439$. Angle $\theta$ can only be in quadrant IV. Reference angle $A = \cos^{-1} 0.6439 = 49.9°$. Then $\theta = 360° - 49.9° = 310.1°$.

**4.** The terminal side of $\theta$ passes through point $(2, -1)$. Angle $\theta$ is in quadrant IV. The reference angle $A = \tan^{-1}(1/2) = 26.6°$. Then $\theta = 360° - 26.6° = 333.4°$.

See Example 10-27 for the use of inverse functions on the calculator.

The next example illustrates the solution of a basic trigonometric equation.

## Example 10-10
Given $2 \sin \theta + 1 = 0$, find $\theta$ in degrees and radians when $0 \le \theta < 2\pi$.

## Solution
This is a basic trigonometric equation. *First solve for sin θ and then for θ:*

$$2 \sin \theta + 1 = 0$$

$$2 \sin \theta = -1$$

$$\sin \theta = -\frac{1}{2} = -0.5000$$

Then reference angle $A = \sin^{-1} 0.5000 = 30°$ or $\pi/6$ and:

$$\theta = 210° \quad \text{or} \quad \frac{7\pi}{6} \qquad \text{quadrant III}$$

$$\theta = 330° \quad \text{or} \quad \frac{11\pi}{6} \qquad \text{quadrant IV}$$

For a review of radians, study the beginning of Sec. 4-3.

**Exercise 10-3** In problems 1 to 10, find the value of each angle. Do special angles without table or calculator.

**1.** $\sin^{-1} (\sqrt{3}/2)$  **2.** $\tan^{-1} \sqrt{3}$

**3.** $\tan^{-1} 1.192$  **4.** $\cos^{-1} 0.8192$

**5.** $\cos^{-1} 0.9890$  **6.** $\sin^{-1} 0.0610$

**7.** $\tan^{-1} (\sqrt{3}/3)$  **8.** $\cos^{-1} (1/\sqrt{2})$

**9.** $\tan (\sin^{-1} \frac{1}{2})$  **10.** $\sin (\tan^{-1} 1)$

In problems 11 to 26, find $\theta$ to the nearest 0.1 degree (or to the nearest 10 minutes if directed by the instructor) for $0° \le \theta < 360°$. For problems 21 to 26, the given point is on the terminal side of $\theta$.

**11.** $\cos \theta = 0.9063$  **12.** $\sin \theta = 0.6293$

**13.** $\tan \theta = -0.8451$  **14.** $\cot \theta = 6.174$

**15.** $\sin \theta = -0.5556$  **16.** $\cos \theta = -0.1334$

**17.** $\cot \theta = 1.104$  **18.** $\tan \theta = -0.4320$

**19.** $\sec \theta = 1.053$  **20.** $\csc \theta = 2.924$

**21.** $(-3, 3)$  **22.** $(\sqrt{3}, -1)$

**23.** $(-1, 2)$  **24.** $(-3, -1)$

**25.** $(-1.5, -0.81)$  **26.** $(3.2, -2.8)$

In problems 27 to 38, find $\theta$ in radians when $0 \le \theta < 2\pi$.

**27.** $\tan \theta = 1.000$  **28.** $\cot \theta = \sqrt{3}/3$

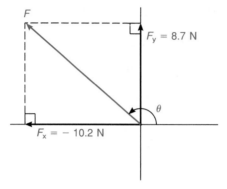

FIG. 10-14 *Force components, problem 39*

**29.** $\sin \theta = 0.8660$          **30.** $\cos \theta = -0.7071$

**31.** $\cos \theta = 0.5324$          **32.** $\sin \theta = -0.6316$

**33.** $\cos \theta + 1 = 0$          **34.** $\sin \theta - 1 = 0$

**35.** $2 \cot \theta = 3$          **36.** $3 \tan \theta = -1$

**37.** $\sin^2 \theta = \dfrac{1}{2}$          **38.** $\cot^2 \theta = 3$

**39.** The components $F_x$ and $F_y$ of a resultant force $F$ are shown in Fig. 10-14. Find $\theta$ if $F_x = -10.2$ N and $F_y = 8.7$ N.

**40.** The true heading of a certain air flight is given by $H = 40° - \sin^{-1} (0.40 \sin 65°)$. Determine $H$.

**41.** The voltage $V$ of ordinary house current is expressed as $V = 170 \sin 2\pi ft$, where $f = $ frequency $= 60$ Hz and $t = $ time in seconds. *(a)* Find the angle $2\pi ft$ in radians when $V = 120$ V and $0 \le 2\pi ft < 2\pi$. *(b)* Find $t$ when $V = 120$ V.

**42.** In a problem involving the free vibration of a beam, the following expression arises: $\alpha = \tan^{-1} x + \tan^{-1} (1/x)$. *(a)* Find $\alpha$ when $x = 0.50$. *(b)* Show that $\alpha = \pi/2$ for any value of $x$. (*Hint:* Draw a diagram.)

**43.** Find $\theta_1$, $\theta_2$, $\theta_3$, and $\theta_4$ in Example 10-1.

**44.** Find $\theta$ in Exercise 10-1, problem 33 (Fig. 10-6).

# 10-4 | *Vectors and Applications*

*Vectors* are quantities that have both *magnitude* and *direction,* where direction is usually specified by an angle. Two numbers must therefore be used to describe a vector. You are already familiar with some vector quantities, as the next example illustrates.

## Example 10-11

Describe some examples of vectors.

1. *Force:* A force exerted by a load on a beam is 100 N at an angle of 20° with the vertical.
2. *Velocity:* The velocity of a wind is 40 mi/h, blowing from the northwest.
3. *Radius vector:* A radius vector is 5 units long at an angle of 120°.
4. *Voltage vector:* A voltage vector in an ac circuit is 110 V, and its phase angle is 64°.
5. *Magnetic flux density:* The flux density of a magnetic field is $4 \times 10^{-5}$ T (tesla) in a direction perpendicular to the direction of the current flowing in the wire.

Most quantities can be described with only one number. They are called *scalars*. The magnitude of a vector is therefore a scalar. In Example 10-11(2), the magnitude or *speed* of the wind is 40 mi/h. The word *speed* is used to describe a scalar, and the word *velocity* is used to describe a vector.

## Adding Vectors

Graphically vectors are indicated by arrows and boldface letters. The length of the arrow represents the magnitude, and the arrowhead represents the direction. In formulas and equations, vectors can be represented in several different ways, such as boldface letter **A** or with arrows above $\vec{A}$. In this text boldface letters are used.

*Two vectors* **A** *and* **B** *are added to produce a resultant vector* **R** = **A** + **B** *by placing the tail of one vector on the head of the other without changing its direction* (Fig. 10-15). The resultant **R** goes from the tail of the first vector to the head of the second vector. To subtract two vectors **R** = **A** − **B**, reverse the direction of **B** and add.

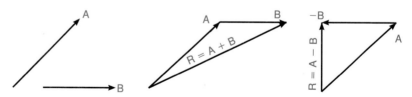

FIG. 10-15 *Tail-to-head method of adding vectors*

## Example 10-12

A body is acted on by a horizontal force $F_x = 15$ N acting to the right and a vertical force $F_y = 8$ N acting down.

1. Draw the vector diagram showing the resultant force **F**.
2. Find the magnitude and the direction of the resultant vector **F**.

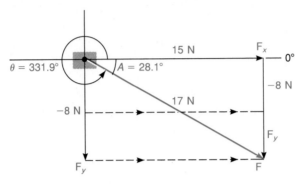

FIG. 10-16 *Force diagram, Example 10-12*

### Solution

**1.** Draw the vector diagram showing the resultant force **F**.

The diagram is shown in Fig. 10-16 with the body placed at the origin. Vector $\mathbf{F}_y$ is added to $\mathbf{F}_x$ by moving $\mathbf{F}_y$ parallel to its original direction and placing its tail on the head of $\mathbf{F}_x$.

**2.** Find the magnitude and the direction of the resultant vector **F**.

Find the magnitude by using the pythagorean theorem: $F = \sqrt{F_x^2 + F_y^2} = \sqrt{15^2 + (-8)^2} = \sqrt{289} = 17$ N. Find the direction $\theta$ by first finding the reference angle $A$: $A = \tan^{-1}(F_y/F_x) = \tan^{-1}(8/15) = \tan^{-1} 0.5333 = 28.1°$. Then $\theta = 360° - 28.1° = 331.9°$.

### Example 10-13

In a parallel *RC* (resistance-capacitance) circuit, the current $\mathbf{I}_1$ through the capacitance leads the current $\mathbf{I}_2$ through the resistance by 90°, as shown in Fig. 10-17. If $I_1 = 1.3$ A and $I_2 = 1.8$ A, find the vector sum **I** (the magnitude and the phase angle $\phi$).

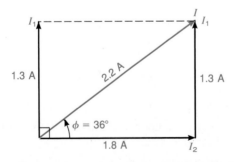

FIG. 10-17 *Electric current vectors, Example 10-13*

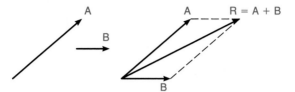

FIG. 10-18 *Parallelogram method of adding vectors*

**Solution**

$$I = \sqrt{(I_1)^2 + (I_2)^2} = \sqrt{(1.3)^2 + (1.8)^2} = \sqrt{4.9} = 2.2 \text{ A}$$

$$\phi = \tan^{-1} \frac{1.3}{1.8} = \tan^{-1} 0.7222 = 36°$$

Two vectors may also be added by the *parallelogram method* which is equivalent to the tail-to-head method. The vectors are placed tail to tail, and the resultant is the diagonal of the parallelogram, as shown in Fig. 10-18.

## *Vector Components*

When a vector is placed on a rectangular coordinate system with its tail at the origin, it can be resolved into *components* along the $x$ and $y$ axes. Figure 10-16 shows the vector **F** and its components $\mathbf{F}_x$ and $\mathbf{F}_y$. Study the next example, which shows how to resolve a force into horizontal and vertical components.

### Example 10-14

A Lear jet traveling with a speed of 700 mi/h is ascending at an angle of $\theta = 40°$ with the horizontal. Find the horizontal component $\mathbf{V}_x$ and the vertical component $\mathbf{V}_y$ of the velocity **V** (Fig. 10-19).

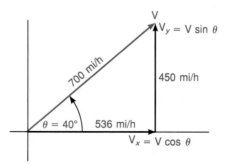

FIG. 10-19 *Velocity components, Example 10-14*

## Solution

The magnitudes of the components $V_x$ and $V_y$ equal the sides of the reference triangle:

$$\cos \theta = \frac{V_x}{V} \qquad \text{or} \qquad V_x = V \cos \theta \qquad x \text{ component}$$

and $\qquad \sin \theta = \dfrac{V_y}{V} \qquad \text{or} \qquad V_y = V \sin \theta \qquad y \text{ component}$

Then $\qquad V_x = 700 \cos 40° = 700(0.7660) = 536$ mi/h

$V_y = 700 \sin 40° = 700(0.6428) = 450$ mi/h

In general:

> To find the *x* component of a vector, multiply the magnitude by cos $\theta$, where $\theta$ is the angle the vector makes with the positive *x* axis. To find the *y* component, multiply the magnitude by sin $\theta$.

The two methods described for adding vectors (tail to head and parallelogram) do not lend themselves easily to computing the resultant of vectors that are not perpendicular. To add such vectors, you must resolve each vector into $x$ and $y$ components, as in Example 10-14. Add the $x$ components to get the $x$ component of the resultant, and add the $y$ components to get the $y$ component of the resultant. Then combine the components to get the resultant, as in Examples 10-12 and 10-13. Study the next two examples, which illustrate the procedure.

### Example 10-15

Carol is swimming across the Hudson River, which she knows has a 2.0-mi/h current. To compensate for the current she heads in a direction upstream at an angle of 110° with the current (Fig. 10-20$a$). If she swims through the water at a velocity $V = 4.0$ mi/h, what is her resultant velocity **R** (the magnitude of the velocity and the angle with the current)? (See Fig. 10-20$b$.)

### Solution

The vector representing the current **C** is placed on the $x$ axis and has no $y$ component. It is important to interpret the angle of 110° correctly. It is the angle between the vectors placed *tail to tail*. In Fig. 10-20$a$ notice that the supplement of 110°, which is 70°, is the reference angle that **V** makes with the $x$ axis. The components of **V** are then:

$$V_x = V \cos 70° = 4.0(0.3420) = 1.4$$
$$V_y = V \sin 70° = 4.0(0.9397) = 3.8$$

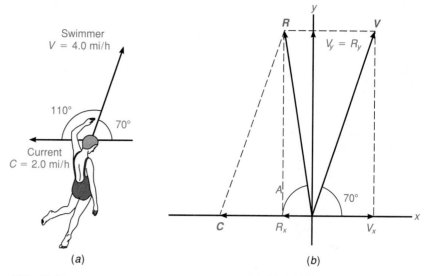

FIG. 10-20 *Velocity diagram of swimmer, Example 10-15*

The components of the resultant velocity are:

$$R_x = V_x - C = 1.4 - 2.0 = -0.6$$
$$R_y = V_y = 3.8$$

and the magnitude of $R$ is:

$$R = \sqrt{(R_x)^2 + (R_y)^2}$$
$$= \sqrt{(-0.6)^2 + (3.8)^2}$$
$$= \sqrt{14.8}$$
$$= 3.8 \text{ mi/h}$$

The angle $A$ that **R** makes with **C** is given by:

$$A = \tan^{-1} \frac{R_y}{R_x} = \tan^{-1} \frac{3.8}{0.6} = \tan^{-1} 6.33 = 81°$$

Notice that $A$ is a positive acute angle like a reference angle.

### Example 10-16
Find the resultant of the three forces in Fig. 10-21*a*.

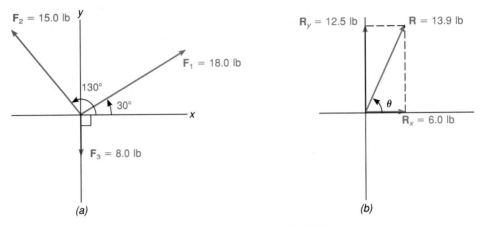

FIG. 10-21 *Resultant of three forces, Example 10-16*

### Solution

The $x$ and $y$ components of the forces and the resultants are conveniently calculated as follows:

|  | $x$ Component | $y$ Component |
|---|---|---|
| $\mathbf{F_1}$ | $18.0 \cos 30° = 15.6$ | $18.0 \sin 30° = \phantom{0}9.0$ |
| $\mathbf{F_2}$ | $15.0 \cos 130° = -9.6$ | $15.0 \sin 130° = 11.5$ |
| $\mathbf{F_3}$ | $0$ | $-8.0$ |
| $\mathbf{R}$ | $\mathbf{R}_x = \phantom{0}6.0$ | $\mathbf{R}_y = 12.5$ |

Then $\mathbf{R} = \sqrt{(6.0)^2 + (12.5)^2} = \sqrt{192} = 13.9$ lb. (See Fig. 10-21*b*.) Since both components are positive, the angle $\theta$ that $\mathbf{R}$ makes with the $x$ axis is in the first quadrant:

$$\theta = \tan^{-1} \frac{12.5}{6.0} = \tan^{-1} 2.08 = 64°$$

## Angular Velocity

Many applied problems involve rotation such as those describing drive shafts, generators, motors, pulleys, wheels, etc. When a body moves in a circle, the arc length $s$ traversed is related to the angle $\theta$ in radians by formula (4-4), Sec. 4-5: $s = r\theta$. Dividing both sides of this formula by the time $t$ results in the relation $s/t = r\theta/t$. This is written:

$$v = r\omega \tag{10-5}$$

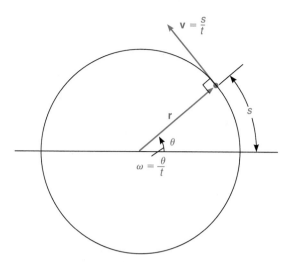

FIG. 10-22 *Linear and angular velocity*

where $v = s/t$ is the *linear velocity* and $\omega = \theta/t$ is the *angular velocity* in radians per unit time (Fig. 10-22). The linear velocity vector **v** is always perpendicular to the rotating radius vector **r**.

### Example 10-17
An engine operating at 1500 revolutions per minute (r/min) drives a belt on a pulley whose diameter is 30.0 cm.

**1.** What is the angular velocity of the pulley?

**2.** What is the linear velocity of the belt in centimeters per second?

### Solution
**1.** What is the angular velocity of the pulley?
Since $2\pi$ rad $= 1$ r, $\omega = 1500(2\pi) = 3000\pi$ rad/min.

**2.** What is the linear velocity of the belt in centimeters per second?
The linear velocity of the belt is equal to the velocity of a point on the rim of the pulley. From (10-5):

$$v = r\omega = \left(\frac{30.0}{2}\right)(3000\pi) = 141{,}400 \text{ cm/min}$$

$$= 2360 \text{ cm/s} \quad \text{(three figures)}$$

**Exercise 10-4** In problems 1 to 8, find the $x$ and $y$ components for each vector. Angle $\theta$ is measured counterclockwise from the positive $x$ axis.

**1.** $A = 17$, $\theta = 60°$      **2.** $B = 28$, $\theta = 45°$

**3.** $R = 10$, $\theta = 110°$      **4.** $H = 12$, $\theta = 290°$

**5.** $F = 1.13$, $\theta = 195°$     **6.** $E = 3.66$, $\theta = 220°$

**7.** $I = 0.021$, $\theta = 6.0$ rad     **8.** $V = 0.38$, $\theta = 2.5$ rad

In problems 9 to 16, find the vector sum (the magnitude and angle) for each set of components.

**9.** $R_x = 16$, $R_y = 30$     **10.** $V_x = 18$, $V_y = 24$

**11.** $F_x = -7.5$, $F_y = -18$     **12.** $E_x = -12$, $E_y = -22.5$

**13.** $A_x = 530$, $A_y = -250$     **14.** $B_x = 76.2$, $B_y = -89.1$

**C 15.** $H_x = -1.58$, $H_y = 1.03$   **C 16.** $I_x = -0.400$, $I_y = 0.599$

In problems 17 to 22, find the vector sum for each set of vectors. The angles $\theta$ and $\phi$ are measured counterclockwise from the positive $x$ axis.

**17.** $A = 13$, $\theta = 30°$     **18.** $F_1 = 5.3$, $\theta = 20°$
  $B = 15$, $\phi = 60°$       $F_2 = 4.8$, $\phi = 100°$

**C 19.** $R = 1.25$, $\theta = 130°$   **C 20.** $C = 89.5$, $\theta = 31°$
  $X = 2.37$, $\phi = 200°$       $D = 138$, $\phi = 140°$

**C 21.** $I_1 = 0.355$, $\theta = -64°$   **C 22.** $F_1 = 5.80$, $\theta = 385°$
  $I_2 = 0.533$, $\phi = 36°$       $F_2 = 9.10$, $\phi = 190°$

**23.** A horizontal force of 23 N acts to the left on a body, and a vertical force of 35 N acts upward on the body. Find the magnitude and direction of the resultant vector.

**24.** A child is pulling a toy along the ground with a rope that makes a 60° angle with the horizontal. If the tension in the rope (the force exerted by the child) is 5.0 lb, what are the horizontal and vertical forces acting on the toy?

**C 25.** Find the vector sum **I** (magnitude and phase angle) in Example 10-13 if $I_1 = 0.677$ A and $I_2 = 1.28$ A.

**26.** In a parallel $RL$ (resistance-inductance) circuit, the current $I_1$ through the inductance lags the current $I_2$ through the resistance by 90°, as shown in Fig. 10-23. Find the magnitude of **I** and the *negative* phase angle $\theta$ when $I_1 = 5.8$ A and $I_2 = 7.9$ A.

**27.** A rocket traveling at 1000 km/h ascends at an angle of 80° with the horizontal. What are the horizontal and vertical components of the velocity?

**28.** A New York City taxi driver traveling at 10 mi/h throws a cigarette out the window perpendicular to his direction at a speed of 4 mi/h. What is the velocity of the

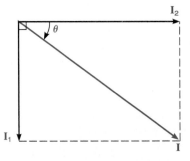

FIG. 10-23 *Current diagram, problem 26*

cigarette as it leaves his hand (the speed and direction with respect to the taxi's direction)?

29. A ship travels at a speed of 10 kn (nmi/h) on a southeast heading (45° east of south) for 2 h and then travels southwest for 3 h. How far south is the ship from its starting point? (*Hint:* Add southerly components.)

30. A light airplane flying due north at 180 km/h experiences a 40-km/h head wind from the northwest (45° west of north). How far north does the plane travel in 2 h? (*Hint:* Subtract northerly components.)

☐ 31. On a rectangular coordinate grid, a target has coordinates (−38.2 m, 46.9 m). What is the displacement vector (radius vector) of the target?

32. If the target of problem 31 moves to coordinates (−18.2 m, 60.9 m), how far and in what direction has the target moved? Find the angle measured from the positive $x$ axis. (*Hint:* Subtract displacement vectors.)

33. If Carol in Example 10-15 becomes tired and can swim only 3.0 mi/h, what is her resultant velocity?

34. The heading of a boat traveling 7.0 mi/h upriver makes an angle of 120° with the current, which is 3.0 mi/h. What is the boat's resultant velocity (the speed and angle with respect to its heading)?

35. A trailer truck weighing 5000 kg (11,000 lb) must be pulled up a ramp that is inclined 8° with the horizontal (Fig. 10-24). If friction is neglected, what force in kilograms and in pounds must be exerted to just move the truck? (*Hint:* Find the component of the weight **W** parallel to the ramp.)

36. An aerial cable car weighing 1 t (1000 kg) is being pulled slowly up a Swiss mountainside (Fig. 10-25). If the cable makes an angle of 20° with the horizontal, what is the component $T$ of the weight parallel to the cable?

☐ 37. Two forces, 500 and 200 lb, act on the end of a cantilever beam as shown in Fig. 10-26. What is the resultant force? Find the angle measured, the same as shown for each force.

38. Two ships leave the same port, one ship sailing to point $A$ and the other to point $B$ (Fig. 10-27). What is the displacement (the magnitude and direction $\theta$) of the ship at $B$ relative to the ship at $A$? (*Hint:* Vector **AB** = vector **OB** − vector **OA**.)

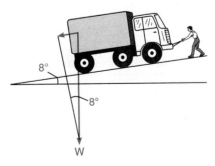

FIG. 10-24 *Trailer truck on incline, problem 35*

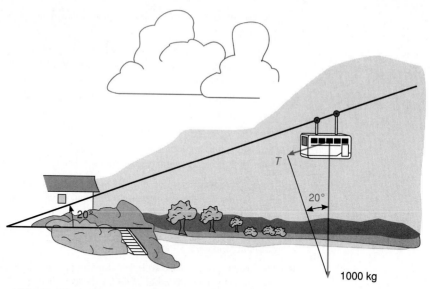

FIG. 10-25 *Cable car, problem 36*

**39.** A strong wind force is accelerating a sky diver upward 2.2 m/s² at an angle of 10° with the horizontal, while gravity is accelerating her down 9.8 m/s². What is her resultant acceleration? Find the angle measured below the horizontal.

**40.** A weight of 100 kg is supported by two steel bars, as shown in Fig. 10-28. Find the magnitudes of the tensile force $\mathbf{F}_1$ and the compressive force $\mathbf{F}_2$. Note that the system is in static equilibrium and $\mathbf{W} = \mathbf{F}_1 + \mathbf{F}_2$. Solve simultaneous equations for $F_1$ and $F_2$.

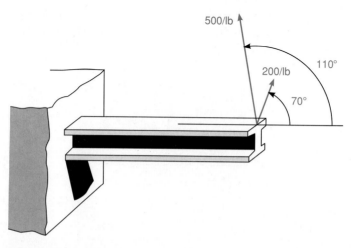

FIG. 10-26 *Cantilever beam, problem 37*

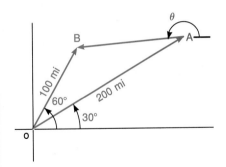

FIG. 10-27 *Ship displacement diagram, problem 38*

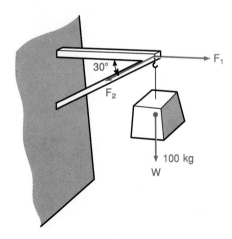

FIG. 10-28 *Weight and supports, problem 40*

**41.** At a certain point in a magnetic field the magnetic flux density $\mathbf{B}_1$ due to the current in one wire is $4 \times 10^{-5}$ T, and the flux density $\mathbf{B}_2$ due to the current in another wire is $6 \times 10^{-5}$ T. If $\mathbf{B}_1$ and $\mathbf{B}_2$ act together at an angle of $25°$, what is the resultant flux density? Find the angle measured from $\mathbf{B}_1$.

**42.** At a point in an electric field, two electric charges produce the intensity vectors shown in Fig. 10-29. If $E_1 = 1.9 \times 10^2$ newton/coulomb (N/C), $E_2 = 2.3 \times 10^2$ N/C, $\alpha = 130°$, and $\beta = 200°$, what is the resultant intensity?

**43.** Find the resultant of the three forces shown in Fig. 10-30 if $F_1 = 80$ N, $F_2 = 60$ N, $F_3 = 40$ N, $\theta_1 = 100°$, $\theta_2 = 200°$, and $\theta_3 = 320°$. (See Example 10-16.)

□ **44.** Find the resultant of the three forces shown in Fig. 10-30 if $F_1 = 5.5$ lb, $F_2 = 9.8$ lb, $F_3 = 6.2$ lb, $\theta_1 = 110°$, $\theta_2 = 225°$, and $\theta_3 = 300°$. (See Example 10-16.)

**45.** The total impedance $Z$ of a series ac circuit is the resultant of the resistance $R$, the inductive reactance $X_L$, and the capacitive reactance $X_C$ (Fig. 10-31). Find $Z$ when $R = 18$ Ω, $X_L = 60$ Ω, and $X_C = 40$ Ω.

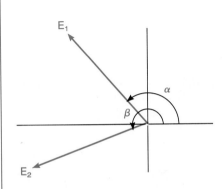

FIG. 10-29 *Electric intensity vectors, problem 42*

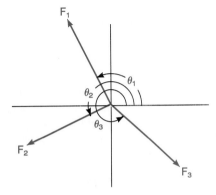

FIG. 10-30 *Three forces, problems 43, 44*

FIG. 10-31  *AC circuit vectors, problem 45*

**46.** Find $Z$ in problem 45 when $R = 5.0 \times 10^2 \, \Omega$, $X_L = 2.2 \times 10^3 \, \Omega$, and $X_C = 2.9 \times 10^3 \, \Omega$.

**47.** The engine in Example 10-17 increases its revolutions per minute to 2000. *(a)* Find the angular velocity of the pulley. *(b)* Find the linear velocity of the belt in centimeters per second.

**48.** The belt in Example 10-17 drives a generator pulley whose diameter is 8.0 cm. If the linear velocity of the belt is 1700 cm/s, find *(a)* the angular velocity of the generator pulley in radians per minute and *(b)* the revolutions per minute of the pulley.

**49.** A centrifugal pump operates at 100 r/min. Find the linear velocity in centimeters per second of a point on the tip of the impeller if its diameter is 1.5 m.

**50.** The linear velocity of a sunspot on the sun's equator is measured to be $2.0 \times 10^3$ m/s. *(a)* If the radius of the sun is $7.0 \times 10^8$ m, find the angular velocity of the sunspot. *(b)* What is the period of solar rotation (the number of earth days per rotation) at the equator?

# 10-5 | *Oblique Triangles: Law of Sines*

An *oblique* triangle is a triangle that is not a right triangle. Two important laws, the *law of sines* and the *law of cosines,* can be used to find the sides or angles of an oblique triangle when you know certain other sides and angles. Given triangle *ABC* (Fig. 10-32), the *law of sines* states:

$$\frac{a}{\sin A} = \frac{b}{\sin B} = \frac{c}{\sin C} \qquad (10\text{-}6)$$

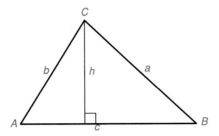

FIG. 10-32 *Law of sines, Example 10-18*

The law of sines says that the side of a triangle is proportional to the sine of the opposite angle. For example, sin 74° = 0.94 and is approximately twice sin 28° = 0.47. If angle $A$ in Fig. 10-32 increases from 28° to 74° and side $b$ and angle $B$ do not change, then side $a$ will approximately double in length.

The law of sines can be used to find the missing parts of a triangle when either of the following is true:

**1.** *Two angles and a side opposite one of the angles are known.*

**2.** *Two sides and an angle opposite one of the sides are known.*

### Example 10-18
In Fig. 10-32, $A = 40.5°$, $B = 60.5°$, and $c = 5.0$. Find the three missing parts of triangle $ABC$.

### Solution
The only side known is $c$. The angle opposite, angle $C = 180° - A - B = 180° - 40.5° - 60.5° = 79°$. Use part of the law of sines (10-6) to find $b$:

$$\frac{b}{\sin B} = \frac{c}{\sin C}$$

Solve for $b$ and substitute:

$$b = \frac{c(\sin B)}{\sin C} = \frac{5.0(\sin 60.5°)}{\sin 79°} = \frac{5.0(0.8704)}{0.9816} = 4.4$$

Use formula (10-6) again to find $a$:  $\dfrac{a}{\sin A} = \dfrac{c}{\sin C}$

$$a = \frac{c(\sin A)}{\sin C} = \frac{5.0(\sin 40.5°)}{\sin 79°} = \frac{5.0(0.6494)}{0.9816} = 3.3$$

## *Proof of the Law of Sines*

The proof of the law of sines applies to the acute triangle $ABC$ in Fig. 10-32. The proof for an obtuse triangle is similar. In triangle $ABC$:

$$\sin A = \frac{h}{b} \qquad \text{and} \qquad h = b(\sin A)$$

$$\sin B = \frac{h}{a} \qquad \text{and} \qquad h = a(\sin B)$$

Equate the two expressions for $h$:

$$a(\sin B) = b(\sin A)$$

Divide by $(\sin A)(\sin B)$ to obtain:

$$\frac{a}{\sin A} = \frac{b}{\sin B}$$

Similarly it can be shown that:

$$\frac{b}{\sin B} = \frac{c}{\sin C}$$

Therefore any two of the three ratios in (10-6) are equal to each other.

### Example 10-19

An oil tanker heads northeast (45° east of north) from port for 2.0 mi to avoid a shoal and then turns due east until it reaches a loading platform 2.5 mi from port (Fig. 10-33). A small tug leaves port to assist the tanker. If the tug does not need to avoid the shoal, how many degrees north of east should the tug head to travel directly to the loading platform?

### Solution

The angle $\theta$ in Fig. 10-33 represents the tug's heading. Angle $\theta$ equals angle $C$ ($\angle ACB$) because $BC$ is parallel to $AE$ and they are alternate interior angles. Angle $B = 135°$, and angle $C$ is found by using the law of sines (10-6):

$$\frac{2.0}{\sin C} = \frac{2.5}{\sin 135°}$$

$$\sin C = \frac{2.0(\sin 135°)}{2.5} = \frac{2.0(0.7071)}{2.5} = 0.5657$$

$$C = \sin^{-1} 0.5657 = 34.4°$$

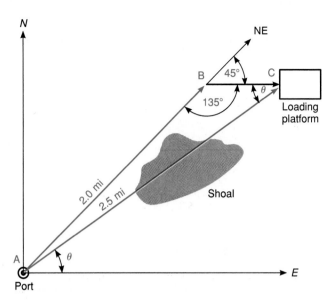

FIG. 10-33 *Oil tanker route, Example 10-19*

Consider the equation above: sin $C$ = 0.5657. There are actually two solutions to this equation, $C$ = 34.4° and $C$ = 180° − 34.4° = 145.6°. However, 145.6° cannot be a solution to the problem since triangle $ABC$ can have only one obtuse angle. The next example illustrates a situation in which the law of sines is used and two solutions are possible.

### Example 10-20
In triangle $ABC$, $A$ = 30°, $a$ = 2.0 cm, and $b$ = 3.0 cm. Find angle $B$.

### Solution
When you carefully draw the triangle to scale (Fig. 10-34), there are two possible solutions: acute triangle $ABC$ or obtuse triangle $AB'C$. Such a situation is called the *ambiguous case*. It can occur when two sides and an angle opposite

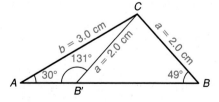

FIG. 10-34 *Law of sines (ambiguous case), Example 10-20*

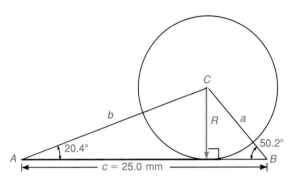

FIG. 10-35 *Design problem, Example 10-21*

are given. More information is necessary to determine a unique solution. The possible values for angle $B$ are:

$$\sin B = \frac{b(\sin A)}{a} = \frac{3.0(0.50)}{2.0} = 0.75$$

$$B = 49° \quad \text{or} \quad 131°$$

In Fig. 10-34 angle $AB'C$ corresponds to $B = 131°$.

### Example 10-21

Find the radius of the circle in the design problem shown in Fig. 10-35.

### Solution

To find $R$, you must first find $a$ or $b$. Angle $C = 180° - 20.4° - 50.2° = 109.4°$ and:

$$a = \frac{c(\sin A)}{\sin C} = \frac{25.0(\sin 20.4°)}{\sin 109.4°} = 9.24 \text{ mm}$$

Then:

$$\sin 50.2° = \frac{R}{a} \quad \text{and} \quad R = a(\sin 50.2°) = 7.10 \text{ mm}$$

**Exercise 10-5**    In problems 1 to 14, find the three missing parts of each triangle. Express angles in the measure given.

1. $A = 65°$, $B = 75°$, $b = 8.0$
2. $A = 100°$, $C = 35°$, $a = 5.5$
3. $B = 98.5°$, $C = 39.6°$, $a = 14.3$
4. $A = 57.2°$, $B = 48.1°$, $c = 1.07$
5. $a = 5.0$, $b = 4.0$, $A = 57°$

**6.** $b = 865$, $c = 621$, $B = 87.2°$

**7.** $a = 0.456$, $c = 0.567$, $C = 105.2°$

**8.** $a = 11.1$, $b = 15.8$, $B = 150.7°$

**9.** $A = 1.5$ rad, $B = 1.2$ rad, $a = 60$

**10.** $B = 1.0$ rad, $C = 0.50$ rad, $b = 0.045$

**11.** $A = 110°30'$, $C = 30°40'$, $b = 528$

**12.** $B = 99°40'$, $C = 20°50'$, $a = 0.0132$

**13.** $b = 5.0 \times 10^{-3}$, $c = 6.0 \times 10^{-3}$, $C = 37.9°$

**14.** $a = 1.5 \times 10^4$, $c = 1.2 \times 10^4$, $A = 79.6°$

In problems 15 and 16, find the two solutions for angle $B$ and carefully draw each solution. (See Example 10-20.)

**15.** $A = 60°$, $a = 8.0$ cm, $b = 9.0$ cm

**16.** $A = 45°$, $a = 2.5$ in., $b = 3.0$ in.

Use the law of sines (10-6) to solve problems 17 to 30.

**17.** Two high-tension cables are to be strung across a river. One each is to go from towers $B$ and $C$, which are 840 ft apart, to tower $A$ on the other bank. A surveyor measures angle $ABC$ to be $50°30'$ and angle $BCA$ to be $95°30'$. What are the distances $AB$ and $AC$?

**18.** Two fire lookout towers $A$ and $B$ are 17.0 km apart. Tower $A$ spots a fire at $C$ and measures angle $CAB$ to be $45.2°$, whereas tower $B$ measures angle $ABC$ to be $38.3°$. Which tower is closer to the fire and by how much?

**19.** Two forces $\mathbf{F}_1$ and $\mathbf{F}_2$ act on a column to produce a resultant $\mathbf{R}$, as shown in Fig. 10-36. If $\mathbf{R} = 1350$ N, $\theta = 70.2°$, and $\mathbf{F}_1 = 885$ N, find $\mathbf{F}_2$ and the angle it makes with $\mathbf{R}$.

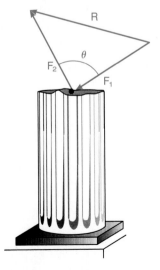

FIG. 10-36  *Forces on column, problems 19 and 20*

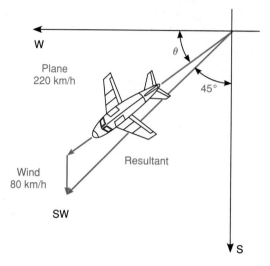

FIG. 10-37   *Flight direction, problem 22*

**20.** In Fig. 10-36 find $\mathbf{F}_1$ and the angle it makes with $\mathbf{R}$ if $\mathbf{F}_2 = 19.5$ lb, $\theta = 102°$, and $\mathbf{R} = 25.0$ lb.

**21.** In Example 10-19, find the distance the oil tanker travels to the loading platform after turning east.

**22.** Luis, a pilot, wishes to fly southwest. If his airspeed is 220 km/h and there is an 80-km/h wind from due north, how many degrees south of west should he fly so that the resultant direction will be southwest? Find $\theta$ in Fig. 10-37.

**23.** The currents in a three-phase alternator give rise to the vector diagram shown in Fig. 10-38. Find the line current $I$ when the phase current $I_p = 25$ A.

**24.** In Fig. 10-38 find the formula for $I$ in terms of $I_p$.

**25.** Find the radius $R$ of the circle in the design problem shown in Fig. 10-39 if $AB = 15.0$ mm. (*Hint:* First find $AC$ or $BC$.)

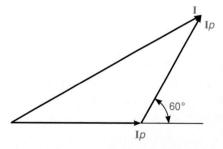

FIG. 10-38   *Three-phase alternator currents, problem 23*

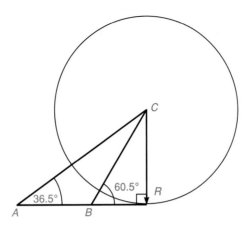

FIG. 10-39 *Design problem, problem 25*

☐ **26.** Find $R$ in Fig. 10-39 if $AB = 0.1013$ in.

**27.** From the top of a lighthouse the angles of depression of two ships sighted in the same direction are $3°10'$ and $5°20'$. If the ships are 2.3 mi apart, how far is the closer ship from the lighthouse?

**28.** An antenna whose length is 55 m is on the top edge of a building. From a point on the ground, the angles of elevation of the top and bottom of the antenna are $66.0°$ and $59.0°$. How high is the building?

**29.** In a triangle $ABC$ explain why there is no solution for angle $B$ when $A = 30°$, $a = 4.0$ cm, and $b = 10.0$ cm. Use the law of sines (10-6) and draw a careful diagram.

**30.** What happens in problem 29 when $a = 5.0$ cm?

# 10-6 | *Law of Cosines*

For any triangle $ABC$ (Fig. 10-40) the *law of cosines* states:

$$c^2 = a^2 + b^2 - 2ab \cos C \qquad (10\text{-}7)$$

The law of cosines is a generalization of the pythagorean theorem. When $C = 90°$, $\cos 90° = 0$ and (10-7) becomes $c^2 = a^2 + b^2$.

The law of cosines is used to find the parts of an oblique triangle when the law of sines does not apply. The law of cosines applies when either of the following is true:

**1.** *Two sides and the included angle are known.*

**2.** *Three sides are known.*

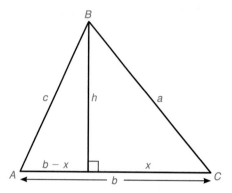

FIG. 10-40 *Law of cosines (10-7)*

### Example 10-22

In Fig. 10-40 given $C = 55.0°$, $a = 12.0$, and $b = 15.0$, find the three missing parts of triangle $ABC$.

### Solution

Because two sides and the included angle are known, you use the law of cosines (10-7) to find $c$:

$$c^2 = a^2 + b^2 - 2ab \cos C$$
$$c^2 = (12.0)^2 + (15.0)^2 - 2(12.0)(15.0) \cos 55.0°$$
$$c^2 = 144 + 225 - 360(0.5736) = 162.5$$
$$c = \sqrt{162.5} = 12.75$$

Use the law of sines (10-6) to find $A$ (or $B$):

$$\sin A = \frac{a(\sin C)}{c} = \frac{12.0(\sin 55.0°)}{12.75} = 0.7710$$

$$A = \sin^{-1} 0.7710 = 50.4° \quad \text{and} \quad B = 180 - 55.0° - 50.4° = 74.6°$$

## *Proof of the Law of Cosines*

The proof of the law of cosines (10-7) applies to the acute triangle $ABC$ in Fig. 10-40. The proof for an obtuse triangle is similar. Apply the pythagorean theorem to each of the right triangles in triangle $ABC$:

$$c^2 = (b - x)^2 + h^2 \quad \text{and} \quad a^2 = x^2 + h^2$$

Subtract the two equations:

$$c^2 - a^2 = (b - x)^2 + \cancel{h^2} - x^2 - \cancel{h^2}$$
$$c^2 - a^2 = b^2 - 2bx + \cancel{x^2} - \cancel{x^2}$$
$$c^2 = a^2 + b^2 - 2bx$$

But $\cos C = x/a$ and $x = a \cos C$. Substituting $a \cos C$ for $x$ produces (10-7):

$$c^2 = a^2 + b^2 - 2ab \cos C$$

By switching the letters in Fig. 10-40 the law of cosines (10-7) can be written:

or
$$b^2 = a^2 + c^2 - 2ac \cos B$$
$$a^2 = b^2 + c^2 - 2bc \cos A \tag{10-8}$$

However, it is only necessary to memorize formula (10-7). By switching the letters or labeling the triangle accordingly, the angle in the formula can be made to correspond to the given angle, or to the angle to be found, as follows.

### Example 10-23
Given $B = 20.0°$, $a = 0.453$, and $c = 0.681$, find the three missing parts of triangle $ABC$.

### Solution
In (10-7) switch $b$ with $c$ (and $B$ with $C$) to get the first of formulas (10-8):

$$b^2 = a^2 + c^2 - 2ac \cos B$$
Then
$$b^2 = (0.453)^2 + (0.681)^2 - 2(0.453)(0.681)(\cos 20.0°)$$
$$b^2 = 0.2052 + 0.4638 - 0.6170(0.9397) = 0.0892$$
$$b = \sqrt{0.0892} = 0.299 \quad \text{(three significant figures)}$$

Use the law of sines (10-6) to find $C$:

$$\sin C = \frac{c(\sin B)}{b} = \frac{(0.681)(0.3420)}{0.299} = 0.7790$$

At this point a word of caution is necessary: *C can be 51.2° or 128.8°. Which answer is correct?* If you choose $C = 51.2°$, then $A = 180° - 51.2° - 20.0° = 108.8°$. That is not possible. Angle *C must be the largest angle since it is opposite the largest side.* Hence the only solution is $C = 128.8°$ and $A = 180° - 128.8° - 20.0° = 31.2°$.

*Note:* To avoid this situation you should find the smaller angle first if you have two unknown angles. If you have three unknown angles such as those in Example 10-25, find the largest angle first.

### Example 10-24

Two magnets exert forces on a body at an angle of 72°, as shown in Fig. 10-41. Find the resultant force **R** (the magnitude and angle *A*).

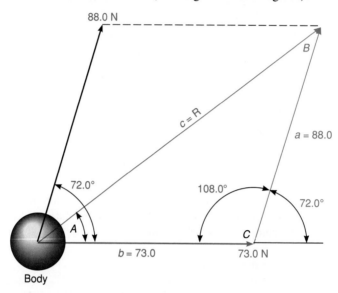

88.0 N

*B*

*c = R*

*a* = 88.0

72.0°          108.0°

72.0°

*A*

*b* = 73.0          *C*          73.0 N

Body

FIG. 10-41  *Magnetic force vectors, Example 10-24*

### Solution

The law of cosines (10-7) can be used to find the resultant of two vectors as an alternative to solving by components. Add the two vectors by the parallelogram method, and find the diagonal of the parallelogram. In Fig. 10-41 the known angle is labeled *C*, which is supplementary to 72°, and side *c* is the resultant. Apply formula (10-7):

$$c^2 = a^2 + b^2 - 2ab \cos C$$
$$c^2 = (88.0)^2 + (73.0)^2 - 2(88.0)(73.0)(\cos 108°)$$
$$c^2 = 7744 + 5329 - (12,848)(-0.3090) = 17,043$$
$$c = \sqrt{17,043} = 131 \text{ N}$$

Notice that cos 108° is negative, which results in a positive sign in front of $2ab \cos C$. Use the law of sines (10-6) to find *A*:

$$\sin A = \frac{a(\sin C)}{c} = \frac{88.0(\sin 108°)}{131} = 0.6389 \qquad A = 39.7°$$

Study the next example, which shows how to find an angle in a triangle when three sides are known.

**Example 10-25**
A triangular brace has sides 33.0, 42.0, and 68.0 ft. Find the largest angle of the brace.

**Solution**
The largest angle is opposite the largest side. Therefore let $c = 68.0$, $a = 33.0$, and $b = 42.0$. Solve for cos $C$ in (10-7):

$$\cos C = \frac{a^2 + b^2 - c^2}{2ab}$$

Then

$$\cos C = \frac{(33.0)^2 + (42.0)^2 - (68.0)^2}{2(33.0)(42.0)}$$

$$= \frac{-1771}{2772} = -0.6389$$

Since cos $C$ is negative, $C$ is obtuse. The reference angle $= \cos^{-1}(0.6389) = 50.3°$ and:

$$C = 180° - 50.3° = 129.7°$$

Note that because of the definition of $\cos^{-1}$ it is not necessary to find the reference angle in this type of problem when you are using the calculator:

$$0.6389 \boxed{+/-} \boxed{\text{INV}} \boxed{\cos} \rightarrow 129.7°$$

**Exercise 10-6**    In problems 1 to 16, find the three missing parts of each triangle. Express angles in the measure given.

1. $C = 70°$, $a = 5.0$, $b = 7.0$
2. $C = 55°$, $a = 12$, $b = 18$
3. $B = 145°$, $a = 0.13$, $c = 0.31$
4. $A = 110°$, $b = 180$, $c = 230$
5. $a = 2.0$, $b = 3.0$, $c = 4.0$
6. $a = 39$, $b = 29$, $c = 40$
7. $a = 10.3$, $b = 11.2$, $c = 19.9$
8. $a = 421$, $b = 222$, $c = 268$
9. $A = 16°$, $b = 1.3$, $c = 1.5$
10. $B = 27°$, $a = 0.011$, $c = 0.022$
11. $C = 10°10'$, $a = 2300$, $b = 5500$
12. $C = 100°20'$, $a = 3.12$, $b = 5.31$
13. $B = 2.3$ rad, $a = 0.35$, $c = 0.54$
14. $A = 0.60$ rad, $b = 11$, $c = 15$

**15.** $a = 4.5 \times 10^{-6}$, $b = 5.4 \times 10^{-6}$, $c = 8.2 \times 10^{-6}$

**16.** $a = 1.4 \times 10^3$, $b = 1.6 \times 10^3$, $c = 1.8 \times 10^3$

Use the law of cosines (10-7) to solve problems 17 to 30.

**17.** Two forces $F_1 = 750$ lb and $F_2 = 1100$ lb act on a point of a bridge girder at an angle of 100°. Find the resultant force $R$ (the magnitude and direction measured from $F_1$).

**18.** Two magnetic forces of 55 and 72 N act on an object to produce a resultant force of 100 N. Find the angle that the resultant makes with each force.

**19.** A sailboat moving due east at 4.5 kn encounters a favorable 1.5-kn current whose direction is 20° north of east. What is the sailboat's resultant velocity? Find the angle measured from the heading of due east.

**20.** A Concorde jet traveling due west at 1300 km/h (808 mi/h) encounters a head wind coming from 35° north of west due to a 160-km/h (100-mi/h) jetstream. Find the jet's resultant speed in kilometers per hour and in miles per hour and its direction measured from the heading of due west.

**21.** Find $x$ in the design problem shown in Fig. 10-42. (*Hint:* First find angle $A$.)

**22.** Using the answer to $x$ from problem 21, find $\theta$ in the design problem of Fig. 10-42.

**23.** Solve Example 10-15, using the law of cosines.

**24.** Solve problem 42 of Sec. 10-4, using the law of cosines.

**25.** Each of the sides of the Pentagon is 281 m (921 ft) long. How far is it in meters and feet from one corner of the building to a nonadjacent corner? *Note:* The interior angle of a regular pentagon is 108°.

**26.** A triangular circuit board has dimensions 3.20, 5.30, and 4.40 mm. (*a*) Find the smallest angle of the board. (*b*) Find the area $A$ of the board, using the formula $A = \frac{1}{2}ab \sin C$.

**27.** A tugboat is towing a barge, as shown in Fig. 10-43. The tension in one cable is 1200 lb, and the tension in the other cable is 1000 lb. What is the resultant force exerted by the tug?

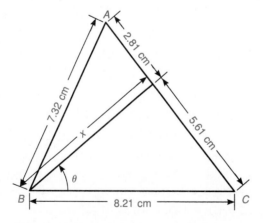

FIG. 10-42 *Design problem, problem 21*

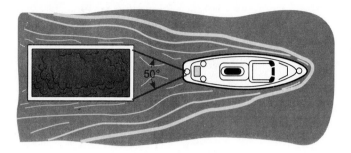

FIG. 10-43 *Tug pulling barge, problem 27*

**28.** Don hikes into the woods in a northeast direction (45° E of N) for 5 mi then turns due east and hikes 6 mi. How far is he from his starting point?

**29.** Given triangle $ABC$ with $a = 6.78$, $b = 9.04$, and $c = 11.3$, find the largest angle. What kind of a triangle is $ABC$?

**30.** Given an isosceles triangle $ABC$ with angle $C = 120°$ and $a = b$, show that $c = a\sqrt{3}$ by using the law of cosines.

# 10-7 | *Hand Calculator Operations*

## *Trigonometric Functions*

The calculator gives the value of the trigonometric functions of any positive or negative angle.

### Example 10-26
Find the value of each of the following. (This is Example 10-6 in Sec. 10-2.)

**1.** sin 260.5°

$$260.5 \ \boxed{\text{sin}} \ \rightarrow \ -0.9863$$

**2.** tan 150°10′

$$150 \ \boxed{+} \ 10 \ \boxed{\div} \ 60 \ \boxed{=} \ \boxed{\text{tan}} \ \rightarrow \ -0.5735$$

**3.** cos 5.41

$$5.41 \ \boxed{\text{DRG}} \ \boxed{\text{cos}} \ \rightarrow \ 0.6424$$

**4.** csc 450°

$$450 \ \boxed{\text{sin}} \ \boxed{1/x} \ \rightarrow \ 1$$

**5.** $\sec \dfrac{7\pi}{4}$

$$7 \boxed{\times} \boxed{\pi} \boxed{\div} 4 \boxed{=} \boxed{\text{DRG}} \boxed{\cos} \boxed{1/x} \to 1.414$$

**6.** $\cot(-240°)$

$$240 \boxed{+/-} \boxed{\tan} \boxed{1/x} \to -0.5774$$

## Inverse Trigonometric Functions

The inverse trigonometric functions $\sin^{-1}$ etc. correspond to the keys $\boxed{\text{INV}}$ $\boxed{\sin}$ or $\boxed{\sin^{-1}}$ etc. The calculator is programmed to display certain values of the angle for these functions, as explained in Sec. 10-3 and definition (10-4). For positive values entered, the calculator displays positive angles in the first quadrant. For negative values entered, the calculator displays negative angles in the fourth quadrant for $\boxed{\text{INV}}$ $\boxed{\sin}$ and $\boxed{\text{INV}}$ $\boxed{\tan}$ and positive angles in the second quadrant for $\boxed{\text{INV}}$ $\boxed{\cos}$. You can always get the reference angle by entering the positive value in the calculator and obtain solutions in other quadrants as shown in the text. The following example illustrates a method for using negative values with the inverse trigonometric functions on the calculator.

### Example 10-27
Find $\theta$ for each of the following for $0° \leq \theta < 360°$.

**1.** $\cos \theta = -0.0157$.      **2.** $\sin \theta = -0.3843$.      **3.** $\tan \theta = -1.087$.

### Solution
**1.** $\cos \theta = -0.0157$. [This is Example 10-9(2) in Sec. 10-3.]

A negative value for $\boxed{\text{INV}}$ $\boxed{\cos}$ produces a second-quadrant angle on the display:

$$0.0157 \boxed{+/-} \boxed{\text{INV}} \boxed{\cos} \to 90.9$$

To get the other solution, which is in the third quadrant, subtract the angle from $360°$: $360° - 90.9° = 269.1°$.

**2.** $\sin \theta = -0.3843$.

A negative value for $\boxed{\text{INV}}$ $\boxed{\sin}$ produces a negative angle in the fourth quadrant:

$$0.3843 \boxed{+/-} \boxed{\text{INV}} \boxed{\sin} \to -22.6$$

To get the positive solution in the third quadrant, subtract the angle from $180°$: $180° - (-22.6°) = 202.6°$. For the fourth-quadrant solution, add $360°$: $360° - 22.6° = 337.4°$.

**3.** tan $\theta = -1.087$.

A negative value for INV tan produces a negative angle in the fourth quadrant:

$$1.087 \boxed{+/-} \boxed{\text{INV}} \boxed{\text{tan}} \rightarrow -47.4$$

To get the positive solution in the second quadrant, add 180°: $180° - 47.4° = 132.6°$. For the fourth quadrant solution, add 360°: $360° - 47.4° = 312.6°$.

Note that sine and cosine cannot be greater than 1 or less than $-1$. If a value greater than 1 or less than $-1$ is entered and INV sin or INV cos is pressed, the display will show error.

**Exercise 10-7**  In problems 1 to 14, find the value of each of the following.

**1.** sin 330.6°                    **2.** cos 285.5°

**3.** cos 100°20′                  **4.** sin 300°30′

**5.** tan $(-36°)$                  **6.** cot $(-95°)$

**7.** cos 426°                      **8.** sin 500°

**9.** cot 2.52                      **10.** tan $(-1.82)$

**11.** sin $(-\pi/8)$               **12.** cos $(3\pi/5)$

**13.** csc 172°                     **14.** sec 295°

In problems 15 to 22, find $\theta$ for each of the following for $0° \leq \theta < 360°$. (*Note:* Some angles are undefined.)

**15.** sin $\theta = -0.2021$       **16.** cos $\theta = -0.5113$

**17.** tan $\theta = -3.102$        **18.** cot $\theta = -0.0115$

**19.** sec $\theta = -2.413$        **20.** csc $\theta = -3.128$

**21.** sin $\theta = -1.213$        **22.** sec $\theta = -0.5050$

# 10-8 | *Review Questions*

In problems 1 to 4, find all the trigonometric functions of angle $\theta$.

**1.** The terminal side passes through $(-5, 12)$.

**2.** The terminal side passes through $(-0.8, -0.6)$.

**3.** sin $\theta = -15/17$, cos $\theta$ positive

**4.** tan $\theta = 2$, $\theta$ in quadrant III

In problems 5 to 12, find the value of the trigonometric function.

**5.** tan 300°                      **6.** sin 270°

**7.** cos 190.5°                    **8.** cot 95°50′

**9.** $\sin(-50°)$            **10.** $\cos(-135°)$

**11.** $\sec \pi$            **12.** $\csc(5\pi/2)$

In problems 13 to 18, find the value of $\theta$ to the nearest 0.1 degree (and in radians if directed by the instructor) for $0° \le \theta < 360°$.

**13.** $\sin \theta = \dfrac{1}{2}$            **14.** $\cos \theta = -1$

**15.** $\tan \theta = -0.9657$        **16.** $\cot \theta = \sqrt{3}$

**17.** $\sec \theta + 2 = 0$         **18.** $\csc \theta = -1.743$

In problems 19 to 22, find the vector sum for each set of vectors.

**19.** $X = 16,\ \theta = 0°$         **20.** $X = 3.2,\ \theta = 180°$
      $Y = 12,\ \phi = 90°$             $Y = 1.1,\ \phi = 270°$

**21.** $A = 9.80,\ \theta = 45°$      **22.** $A = 0.75,\ \theta = 200°$
      $B = 7.10,\ \phi = 120°$          $B = 0.90,\ \phi = 320°$

In problems 23 to 28, find the missing parts of each triangle.

**23.** $A = 85°,\ B = 50°,\ a = 3.0$

**24.** $C = 110°,\ b = 12,\ c = 18$

**25.** $C = 60°,\ a = 10,\ b = 15$

**26.** $C = 95°,\ a = 5.5,\ b = 6.5$

**27.** $a = 30.0,\ b = 40.0,\ c = 60.0$

**28.** $a = 7.1,\ b = 6.2,\ c = 8.8$

**29.** The velocity $v$ in centimeters per second of a piston is given by:

$$v = (1000 \sin \theta)\left[1 + \cos \frac{\theta}{3}\right]$$

where $\theta$ is the angle of rotation of the drive shaft. Find $v$ when $\theta$ is (*a*) $3\pi/4$ and (*b*) $4\pi/3$.

**30.** The maximum velocity a block sliding down an inclined plane attains is given by $v = \sqrt{2gh(1 - \mu \cot \theta)}$, where $g$ = gravitational acceleration, $h$ = initial height, $\mu$ = coefficient of kinetic friction, and $\theta$ = angle of the plane. Find $v$ when $g = 9.8$ m/s², $h = 10$ m, $\mu = 0.10$, and $\theta = 20°$.

C **31.** The angle $\alpha$ of a laser beam is expressed as:

$$\alpha = 2 \tan^{-1} \frac{w}{2d}$$

where $w$ = width of the beam (the diameter) and $d$ = distance from the source. Find $\alpha$ if $w = 1.00$ m when $d = 1000$ m.

**32.** A woman is pushing a lawn mower with a force of 15 N. If the handle makes a 50° angle with the ground, what are the horizontal and vertical forces exerted on the lawn mower?

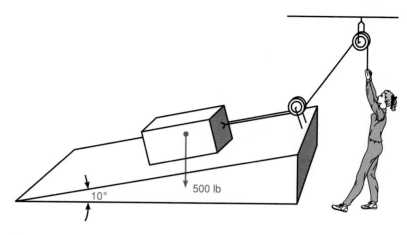

FIG. 10-44 *Woman pulling weight, problem 34*

**33.** A body is acted on by a horizontal force of 8.5 lb acting to the left and a vertical force of 5.4 lb acting down. Find the magnitude and direction of the resultant force.

**34.** In Fig. 10-44, neglecting friction, what force does the person have to exert to move the weight up the plane? (*Hint:* Find the component parallel to the plane.)

**35.** A small plane heading north with an airspeed of 200 mi/h experiences a tail wind from the southwest (blowing toward the northeast—halfway between north and east). Ground radar reports that the ground speed is 220 mi/h. What are the wind speed and the direction of flight relative to the heading? [*Hint:* Use the law of sines (10-6).]

**36.** In Fig. 10-35, Example 10-21, find $R$ if $A = 33°$, $B = 48°$, and $c = 13$ cm.

**37.** To find the length to dig a railroad tunnel, a surveyor measures two distances from the ends of the tunnel and the included angle as shown in Fig. 10-45. What should be the length of the tunnel?

**38.** Jean measures the three sides of her triangular garden to be 10.5, 9.5, and 13 m. She believes it is a right triangle. Is she correct? If not, what are the three angles of the triangle?

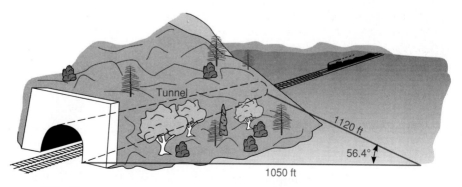

FIG. 10-45 *Tunnel survey, problem 37*

# C H A P T E R

# 11

# Trigonometric Graphs

## 11-1 | *Sine and Cosine Curves*

Sine and cosine curves occur in applications of mathematics involving circular motion, such as those describing alternating-current (ac) electricity, which is generated when a coil of wire rotates in a magnetic field.

Further applications are listed below.

**T**he trigonometric functions are periodic functions. They repeat their values after a certain number of degrees called their *period*. Many physical and technical phenomena are periodic and can be described by the graphs of the trigonometric functions. The sine and cosine curves have the most applications, and they are studied in depth throughout the chapter. The other trigonometric curves are discussed in Sec. 11-3. Sec. 11-4 studies Lissajous figures, which are used in electronics, and Sec. 11-5 presents polar coordinates, a coordinate system used in electricity, mechanics, physics, and calculus.

### Example 11-1
List some important applications of sine and cosine curves.

1. Alternating-current electricity
2. Motion of a reciprocating piston
3. Light waves, radio waves, and other electromagnetic waves
4. Mechanical vibration of a spring, wire, beam, etc.
5. Fluid motion and tidal phenomena
6. Motion of a pendulum
7. Sound waves and acoustical phenomena
8. Orbital motion of the earth, moon, sun, etc.

The relationship of the sine curve to circular motion is shown in the next example.

### Example 11-2
Graph $y = \sin x$ from $x = 0$ to $x = 2\pi$.

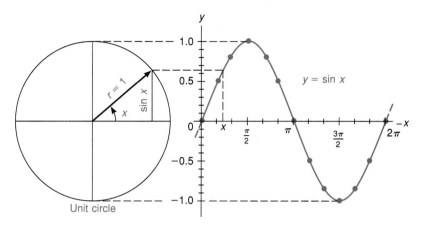

FIG. 11-1 *Sine curve showing projection of the unit circle, Example 11-2*

## Solution

Construct a table of values at a convenient interval, using a table or calculator. The angle $\pi/6$ (30°) is chosen as the interval because it is one of the special angles and provides sufficient points for the graph:

| $x$, degrees | 0 | 30 | 60 | 90 | 120 | 150 | 180 | 210 | 240 | 270 | 300 | 330 | 360 |
|---|---|---|---|---|---|---|---|---|---|---|---|---|---|
| $x$, rad | 0 | $\dfrac{\pi}{6}$ | $\dfrac{\pi}{3}$ | $\dfrac{\pi}{2}$ | $\dfrac{2\pi}{3}$ | $\dfrac{5\pi}{6}$ | $\pi$ | $\dfrac{7\pi}{6}$ | $\dfrac{4\pi}{3}$ | $\dfrac{3\pi}{2}$ | $\dfrac{5\pi}{3}$ | $\dfrac{11\pi}{6}$ | $2\pi$ |
| $y = \sin x$ | 0 | 0.50 | 0.87 | 1 | 0.87 | 0.50 | 0 | $-0.50$ | $-0.87$ | $-1$ | $-0.87$ | $-0.50$ | 0 |

The graph of the equation is shown in Fig. 11-1, which also shows how the points on the graph can be obtained as projections of points on the unit circle. Notice in the unit circle in Fig. 11-1 that the height of the point is equal to $\sin x$. As the radius vector rotates, the height of the vector corresponds to the $y$ coordinate on the graph for each angle $x$.

The next example shows the cosine curve, which is very similar to the sine curve.

### Example 11-3
Graph $y = \cos x$ from $x = 0$ to $x = 2\pi$.

### Solution
The table of values and the graph (Fig. 11-2) are given below. In this and further examples, the angle is expressed only in radians since this lends more readily to scientific and technical applications.

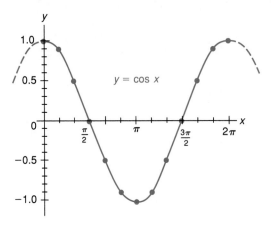

FIG. 11-2 *Cosine curve, Example 11-3*

| $x$ | 0 | $\dfrac{\pi}{6}$ | $\dfrac{\pi}{3}$ | $\dfrac{\pi}{2}$ | $\dfrac{2\pi}{3}$ | $\dfrac{5\pi}{6}$ | $\pi$ | $\dfrac{7\pi}{6}$ | $\dfrac{4\pi}{3}$ | $\dfrac{3\pi}{2}$ | $\dfrac{5\pi}{3}$ | $\dfrac{11\pi}{6}$ | $2\pi$ |
|---|---|---|---|---|---|---|---|---|---|---|---|---|---|
| $y = \cos x$ | 1 | 0.87 | 0.50 | 0 | −0.50 | −0.87 | −1 | −0.87 | −0.50 | 0 | 0.50 | 0.87 | 1 |

Compare the table and the graph with those for sin $x$ in Example 11-2. Cos $x$ is 90° out of phase with sin $x$. Study Fig. 11-3, which illustrates this relationship more clearly by showing both curves on the same graph. If you move sin $x$ to the left 90° or cos $x$ to the right 90°, both curves will match exactly.

The *period,* which is the number of radians in one cycle, is $2\pi$, for both sin $x$ and cos $x$. The *amplitude,* which is the maximum height, is 1. Both curves repeat continuously in either direction.

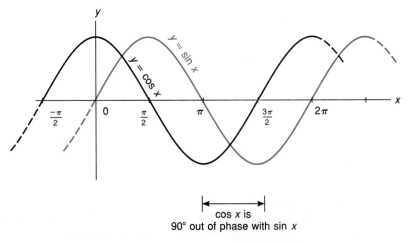

cos $x$ is
90° out of phase with sin $x$

FIG. 11-3 *Sine and cosine curves compared, Example 11-3*

## $y = a \sin bx$ and $y = a \cos bx$

The amplitude of the sine or cosine curve can be changed by a coefficient in front of the function, whereas the period can be changed by a coefficient in front of the angle.

Given $y = a \sin bx$ or $y = a \cos bx$, then:

$$\text{amplitude} = |a| \qquad \text{period} = \frac{2\pi}{b} \qquad \text{frequency} = \frac{b}{2\pi} = \frac{1}{\text{period}} \qquad (11\text{-}1)$$

The expression $|a|$ means the absolute (positive) value of $a$.

Study the following examples, which illustrate definitions (11-1).

### Example 11-4

Graph $y = 3 \sin x$ from $x = 0$ to $x = 2\pi$. Find the amplitude, period, and frequency.

### Solution

The values for $y$ are 3 times the values shown in the table for $\sin x$ in Example 11-2. Compare the graph in Fig. 11-4 with that in Fig. 11-1. For $y = 3 \sin x$, $a = 3$ and $b = 1$ therefore applying definitions (11-1), the amplitude $= |3| = 3$, and the period $= 2\pi/1 = 2\pi$. The frequency, which is the reciprocal of the period, equals $1/(2\pi)$. This means there is one cycle every $2\pi$ rad.

### Example 11-5

Graph $y = \sin 2x$ from $x = 0$ to $x = 2\pi$. Find the amplitude, period, and frequency.

### Solution

To calculate the table of values, you *first multiply the angle by 2* and then find the sine. For example, when $x = \pi/6$, $2x = 2(\pi/6) = \pi/3$ and $y = \sin 2x =$

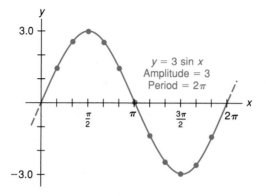

FIG. 11-4 *Graph of $y = 3 \sin x$, Example 11-4*

sin $\pi/3 = 0.87$. The table below is set up in this way. *To obtain enough points, multiples of $\pi/4$ are added to the values used in previous examples:*

| $x$ | 0 | $\dfrac{\pi}{6}$ | $\dfrac{\pi}{4}$ | $\dfrac{\pi}{3}$ | $\dfrac{\pi}{2}$ | $\dfrac{2\pi}{3}$ | $\dfrac{3\pi}{4}$ | $\dfrac{5\pi}{6}$ | $\pi$ | $\dfrac{7\pi}{6}$ | $\cdots$ | $2\pi$ |
|---|---|---|---|---|---|---|---|---|---|---|---|---|
| $2x$ | 0 | $\dfrac{\pi}{3}$ | $\dfrac{\pi}{2}$ | $\dfrac{2\pi}{3}$ | $\pi$ | $\dfrac{4\pi}{3}$ | $\dfrac{3\pi}{2}$ | $\dfrac{5\pi}{3}$ | $2\pi$ | $\dfrac{7\pi}{3}$ | table repeats | $4\pi$ |
| $y = \sin 2x$ | 0 | 0.87 | 1 | 0.87 | 0 | $-0.87$ | $-1$ | $-0.87$ | 0 | 0.87 | $\cdots$ | 0 |

The amplitude $= |1| = 1$, and the period $= 2\pi/2 = \pi$. The frequency $= 2/(2\pi)$, which means 2 cycles every $2\pi$ rad. Notice that the table repeats values for $y$ after $x = \pi$. See the graph in Fig. 11-5.

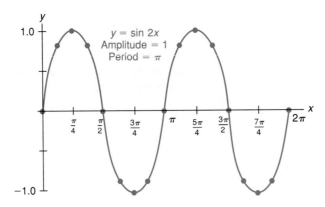

FIG. 11-5  *Graph of y = sin 2x, Example 11-5*

When you know the general shape of the sine or cosine curve, you can sketch the graph from the amplitude and period by plotting the *x intercepts* and *maximum* and *minimum points*. The next example shows how to sketch one of these curves without constructing a table.

### Example 11-6
Find the amplitude, period, and frequency and sketch the curve of $y = 2 \cos 3x$ from $x = 0$ to $x = 2\pi$.

### Solution
The amplitude $= |2| = 2$, the period $= 2\pi/3$, and the frequency $= 3/(2\pi)$. From this information you know the following: The maximum point is 2, and the minimum point is $-2$. The curve passes through one cycle in $2\pi/3$ rad, and there are 3 cycles in $2\pi$ rad. You should *plot points every quarter of a period =*

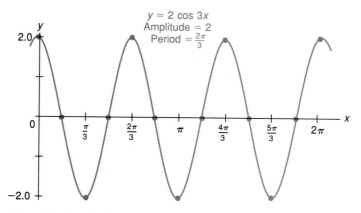

FIG. 11-6 *Sketch of y = 2 cos 3x, Example 11-6*

¼(2π/3) = π/6, which is where the curve intercepts the x axis or has a maximum or a minimum point (Fig. 11-6). Since it is a cosine curve, it starts at its maximum point and drops to zero in one-quarter of a period. It passes through its minimum point after another quarter period, then returns to zero and increases to its maximum point. The cycle then begins repeating.

## Harmonic Motion

Consider a "perfect" sine wave generated in a body of water in which you are floating on a raft (Fig. 11-7). Your up-and-down vertical motion is called simple harmonic motion. *Simple harmonic motion* is the motion of the *projection* of the tip of the radius vector on the y axis (or any diameter) as the radius rotates with constant angular velocity. See Fig. 11-8.

Since angular velocity $\omega = \theta t$, where $\theta$ is the angle of the radius vector and $t$ is time, $\theta = \omega t$. Compare the circle and the sine curve in Fig. 11-1 to Fig. 11-8.

Simple harmonic motion

FIG. 11-7 *Simple harmonic motion on water wave*

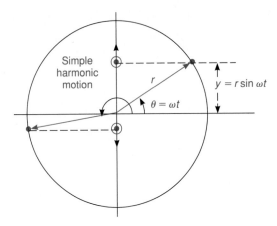

FIG. 11-8 *Simple harmonic motion in a circle*

The angle $x$ is replaced by $\omega t$, and the amplitude of the motion is the radius $r$. It follows that *the equation for the simple harmonic motion of the projected point on the y axis* is:

$$y = r \sin \omega t \qquad (11\text{-}2)$$

When $t = 0$, $\omega t = 0$ and the point is at the beginning of the cycle ($y = 0$). When $t = 2\pi/\omega$, $\omega t = 2\pi$ and the point has completed 1 cycle of the motion. The period is therefore $2\pi/\omega$. Examples of harmonic motion are given in Example 11-1 such as the motion of a pendulum or of an object vibrating on the end of a spring.

## Application to AC Circuits

Alternating current is generated when a coil of wire rotates in a magnetic field. In a simple ac circuit, the instantaneous sinusoidal current $I$ is given by an equation describing simple harmonic motion:

$$I = I_{max} \sin \omega t \qquad (11\text{-}3a)$$

where $I_{max}$ = maximum current in amperes (A), $\omega$ = angular velocity of the coil in radians per second (rad/s), and $t$ = time in seconds (s). The *frequency f* of the current is given by the relation $f = \omega/(2\pi)$ and is measured in hertz or cycles per second. Since $\omega = 2\pi f$, the current can also be expressed in terms of the frequency:

$$I = I_{max} \sin 2\pi f t \qquad (11\text{-}3b)$$

The *period* $= 2\pi/\omega = 1/f$ and is measured in seconds per cycle. Similarly, voltage in an ac circuit is given by $V = V_{max} \sin \omega t$ or $V = V_{max} \sin 2\pi ft$. Study the next example, which illustrates these ideas.

### Example 11-7
Given household current where $f = 60$ Hz and $I_{max} = 2.5$ A.

**1.** Find the amplitude, period, and angular velocity of the current.

**2.** Sketch current vs. time for 1 cycle from $t = 0$.

**3.** Find $t$ when $I = 2.0$ A.

### Solution
**1.** Find the amplitude, period, and angular velocity of the current.

The amplitude $= |I_{max}| = |2.5| = 2.5$ A, and the period $= 1/f = \frac{1}{60}$ seconds/cycle. The angular velocity $\omega = 2\pi f = 2\pi(60) = 120\pi$ rad/s.

**2.** Sketch current vs. time for 1 cycle from $t = 0$.

By applying (11-3b), the equation of the current is $I = I_{max} \sin 2\pi ft = 2.5 \sin 120\pi t$. Plot $t$ every quarter of a period $= (\frac{1}{4})(\frac{1}{60}) = \frac{1}{240}$ s. See the curve in Fig. 11-9, which shows the intercepts and maximum and minimum points.

**3.** Find $t$ when $I = 2.0$ A.

When $I = 2.0$ A:

$$2.5 \sin 120\pi t = 2.0$$

$$\sin 120\pi t = \frac{2.0}{2.5} = 0.80$$

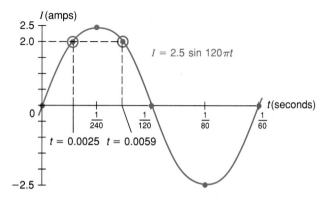

FIG. 11-9 *Household alternating current, Example 11-7*

The reference angle is then:

$$120\pi t = \sin^{-1} 0.80$$

$$120\pi t = 0.93 \text{ rad}$$

One solution for $t$ in the first quadrant is

$$t = 0.93/120\pi = 0.0025 \text{ s}$$

In the second quadrant, the angle is

$$120\pi t = 3.142 - 0.93 = 2.21 \text{ rad}$$

and          $$t = \frac{2.21}{120\pi} = 0.0059 \text{ s}$$

The solutions are shown on the graph in Fig. 11-9.

**Exercise 11-1**

In problems 1 to 12, graph $y$ vs. $x$, plotting at least 12 points from $x = 0$ to $x = 2\pi$. Find the amplitude, period and frequency of each curve. See Examples 11-4 and 11-5.

**1.** $y = 2 \sin x$        **2.** $y = 3 \cos x$

**3.** $y = 0.5 \cos x$      **4.** $y = 1.5 \sin x$

**5.** $y = -3 \sin x$      **6.** $y = -2 \cos x$

**7.** $y = \cos 2x$        **8.** $y = \sin \dfrac{x}{2}$

**9.** $y = 3 \sin 2x$       **10.** $y = -2 \sin 2x$

**11.** $y = 2 \sin \dfrac{3x}{2}$     **12.** $y = 3 \cos \dfrac{2x}{3}$

In problems 13 to 22, find the amplitude, period and frequency and sketch each curve, showing intercepts and maximum and minimum points from $x = 0$ to $x = 2\pi$. See Example 11-6.

**13.** $y = 2 \sin 3x$      **14.** $y = 3 \cos 3x$

**15.** $y = 4 \cos 4x$      **16.** $y = 2 \sin 4x$

**17.** $y = -10 \cos 3x$    **18.** $y = -5 \sin 3x$

**19.** $y = 5 \sin \dfrac{5x}{2}$     **20.** $y = 4 \cos \dfrac{7x}{2}$

**21.** $y = 2.5 \sin 5x$     **22.** $y = 0.5 \sin 6x$

In problems 23 to 30, find the amplitude, period and frequency of each curve. Sketch one cycle from $t = 0$, showing intercepts and maximum and minimum points. See Example 11-7.

**23.** $y = \sin \pi t$        **24.** $y = \cos 2\pi t$

**25.** $y = 3 \cos 2\pi t$       **26.** $y = 2 \sin 4\pi t$

**27.** $I = 10 \sin 120\pi t$       **28.** $I = 12 \sin 100\pi t$

**29.** $E = 120 \sin 100\pi t$       **30.** $E = 110 \sin 120\pi t$

**31.** A weight vibrates on a spring in simple harmonic motion. The amplitude = 12 cm, and $\omega = 2.0$ rad/s. Write the equation for the weight's position $y$ if $y = 0$ when $t = 0$. [See formula (11-2).]

**32.** A float bobs on a water wave in simple harmonic motion. Write the equation for the float's position $y$ if the amplitude = 0.5 m, $\omega = \pi/6$ rad/s, and $y = 0$ when $t = 0$. [See formula (11-2).]

□ **33.** The acceleration of a pendulum in its direction of motion is given by $a = -g \sin \theta$, where $\theta$ is the angular displacement from the vertical position and $g$ is gravitational acceleration. Graph $a$ vs. $\theta$ from $x = 0$ to $x = 2\pi$, plotting at least 12 points. Use $g = 9.8$ m/s$^2$.

□ **34.** A weight hanging from a spring vibrates in simple harmonic motion, expressed by $y = 5.5 \sin 1.5t$, where $y$ = position of the weight in centimeters and $t$ = time in seconds. *(a)* Find the amplitude and period. *(b)* Graph $y$ vs. $t$ from $t = 0$ to $t = 2\pi$, plotting at least 12 points.

**35.** In an ac circuit the current $I$ is given by the relation $I = 8 \sin 60\pi t$. Find the amplitude, period, frequency, and angular velocity. (See Example 11-7.)

**36.** The voltage in an ac circuit containing a constant resistance is given by $V = V_{max} \sin \omega t$. If $V_{max} = 240$ V and $f = 30$ Hz, *(a)* find the amplitude, period, and angular velocity. *(b)* Sketch $V$ vs. $t$ for one cycle from $t = 0$. (See Example 11-7.)

**37.** The height of a water wave is expressed as $y = (3/2) \sin (t/3)$, where $y$ is in meters and $t$ is in seconds. *(a)* Find the amplitude and period, and *(b)* sketch the curve for 1 cycle from $t = 0$.

**38.** Under certain conditions the height of the tide above its mean level is given by $y = 7.2 \cos 0.50t$, where $y$ is in feet and $t$ in hours. *(a)* Find the amplitude and period. *(b)* Sketch $y$ vs. $t$ for 1 cycle from $t = 0$.

□ **39.** In Example 11-7, *(a)* sketch $I$ vs. $t$ for one complete cycle starting from $t = 1/80$ s, and *(b)* find $t$ during this interval when $I = 1.0$ A.

□ **40.** If in Example 11-7 the circuit contains only a constant capacitance, then the voltage $V = -V_{max} \cos \omega t$. *(a)* If $V_{max} = 120$ V, sketch $V$ vs. $t$ for one cycle from $t = 0$. *(b)* Find $t$ when $V = -100$ V.

**41.** The horizontal velocity of a point on the rim of a rolling wheel is given by $v_x = r\omega + r\omega \cos \theta$, where $r$ = radius, $\omega$ = angular velocity, and $\theta$ = angle of rotation. If $r = 2.0$ ft and $\omega = 12$ rad/s, sketch $v_x$ vs. $\theta$ for 1 cycle from $\theta = 0$.

□ **42.** For small angles, the angular displacement of a swinging pendulum from its vertical position can be expressed as $\theta = \theta_a \cos (t\sqrt{g/l})$, where $\theta_a$ = initial displacement, $g$ = gravitational acceleration, $l$ = length, and $t$ = time. Given $\theta_a = 0.30$ rad, $l = 1.0$ m, and $g = 9.8$ m/s$^2$, *(a)* sketch $\theta$ vs. $t$ for one cycle from $t = 0$, and *(b)* find $t$ when $\theta = 0.10$ rad.

## 11-2 | *Phase Angle:* $y = a \sin (bx + \phi)$, $y = a \cos (bx + \phi)$

Study Fig. 11-3 which shows that the function $\cos x$ is 90° out of phase with $\sin x$ and that if you shift the curve for $\sin x$ to the left 90°, it will match $\cos x$ exactly. This shift is done by adding a *phase angle* $\phi = \pi/2$ to $\sin x$:

$$\sin\left(x + \frac{\pi}{2}\right) = \cos x$$

Cos $x$ is therefore said to *lead* $\sin x$ by 90°, as shown in the next example.

### Example 11-8

Sketch $y = \sin (x + \pi/2)$ for 1 cycle from $x = 0$.

### Solution

When $x = 0$, $\sin (0 + \pi/2) = \sin (\pi/2) = 1 = \cos 0$ and the curve is identical to that of $\cos x$ (Fig. 11-10). The table of values for the intercepts and maximum and minimum points is:

| $x$ | $0$ | $\dfrac{\pi}{2}$ | $\pi$ | $\dfrac{3\pi}{2}$ | $2\pi$ |
|---|---|---|---|---|---|
| $x + \dfrac{\pi}{2}$ | $\dfrac{\pi}{2}$ | $\pi$ | $\dfrac{3\pi}{2}$ | $2\pi$ | $\dfrac{5\pi}{2}$ |
| $y = \sin\left(x + \dfrac{\pi}{2}\right)$ | $1$ | $0$ | $-1$ | $0$ | $1$ |

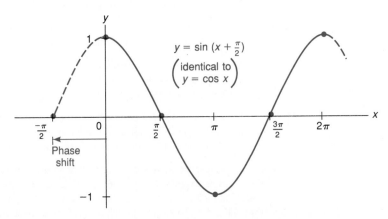

FIG. 11-10 *Phase angle, Example 11-8*

The general expression for a sinusoidal curve is:

$$y = a \sin (bx + \phi) \quad \text{or} \quad y = a \cos (bx + \phi) \tag{11-4}$$

where
$$\text{amplitude} = |a|$$

$$\text{period} = \frac{2\pi}{b}$$

$$\text{phase angle} = \phi$$

$$\text{frequency (1/period)} = \frac{b}{2\pi}$$

The period is measured in radians per cycle and the frequency in cycles per radian when $x$ is in radians. In applications to harmonic motion and ac circuits, when the variable is time and the angle is $\omega t$, the period is measured in seconds per cycle and the frequency in hertz, or cycles per second. The *phase shift* is related to the phase angle and is the value of $x$ when $bx + \phi = 0$, which is:

$$x = \frac{-\phi}{b} \tag{11-5}$$

In Example 11-8, $b = 1$ and the phase shift is $-\pi/2$, or 90° to the left.
Study the next two examples, which apply formulas (11-4) and (11-5).

### Example 11-9
Given the curve $y = \cos (2x + \pi/4)$:

**1.** Find the amplitude, period, frequency, phase angle, and phase shift.

**2.** Sketch the curve for at least 1 cycle from $x = 0$.

### Solution
**1.** Find the amplitude, period, frequency, phase angle, and phase shift.

Apply formulas (11-4) and (11-5): Amplitude $= |1| = 1$, period $= 2\pi/2 = \pi$, frequency $= 1/\pi$, phase angle $= \pi/4$, and phase shift $= -(\pi/4)/2 = -\pi/8$ or 22.5° to the left.

**2.** Sketch the curve for at least 1 cycle from $x = 0$.

This cosine curve begins its cycle at $-\pi/8$, where it has its maximum value. To plot the intercepts and maximum and minimum points, it is necessary to choose points starting with $x = -\pi/8$ and then every quarter of a period ($\pi/4$):

| $x$ | $-\dfrac{\pi}{8}$ | $\dfrac{\pi}{8}$ | $\dfrac{3\pi}{8}$ | $\dfrac{5\pi}{8}$ | $\dfrac{7\pi}{8}$ | $\dfrac{9\pi}{8}$ |
|---|---|---|---|---|---|---|
| $2x + \pi/4$ | $0$ | $\dfrac{\pi}{2}$ | $\pi$ | $\dfrac{3\pi}{2}$ | $2\pi$ | $\dfrac{5\pi}{2}$ |
| $\cos(2x + \pi/4)$ | $1$ | $0$ | $-1$ | $0$ | $1$ | $0$ |

Figure 11-11 shows the curve $y = \cos(2x + \pi/4)$ from $x = -\pi/8$ to $x = 9\pi/8$ with the phase shift of $\pi/8$ to the left.

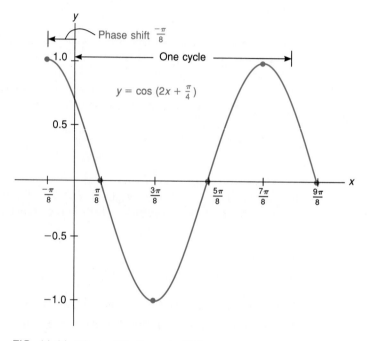

FIG. 11-11  *Phase shift, Example 11-9*

## Example 11-10

Given the curve $y = 3\sin(4x - \pi)$:

1. Find the amplitude, period, frequency, phase angle, and phase shift.
2. Sketch the curve for at least 1 cycle from $x = 0$ to $x = \pi$.

### Solution

1. Find the amplitude, period, frequency, phase angle, and phase shift.

   Amplitude $= |3| = 3$, period $= 2\pi/4 = \pi/2$, frequency $= 2/\pi$, phase angle $= -\pi$, and phase shift $= -(-\pi)/4 = \pi/4$, or $45°$ to the right.

**2.** Sketch the curve for at least 1 cycle from $x = 0$ to $x = \pi$.

Plot points every quarter of a period, which equals $\pi/8$. The table shows 2 cycles of the curve from $x = 0$ to $x = \pi$:

| $x$ | $0$ | $\dfrac{\pi}{8}$ | $\dfrac{\pi}{4}$ | $\dfrac{3\pi}{8}$ | $\dfrac{\pi}{2}$ | $\dfrac{5\pi}{8}$ | $\dfrac{3\pi}{4}$ | $\dfrac{7\pi}{8}$ | $\pi$ |
|---|---|---|---|---|---|---|---|---|---|
| $4x - \pi$ | $-\pi$ | $-\dfrac{\pi}{2}$ | $0$ | $\dfrac{\pi}{2}$ | $\pi$ | $\dfrac{3\pi}{2}$ | $2\pi$ | $\dfrac{5\pi}{2}$ | $3\pi$ |
| $y = 3 \sin (4x - \pi)$ | $0$ | $-3$ | $0$ | $3$ | $0$ | $-3$ | $0$ | $3$ | $0$ |

Figure 11-12 shows $y = 3 \sin (4x - \pi)$ with the curve shift of $\pi/4$ to the right compared to $y = 3 \sin 4x$.

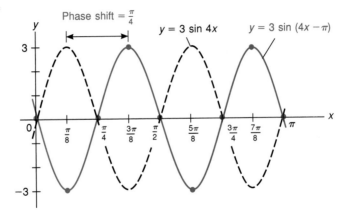

FIG. 11-12 *Phase shift, Example 11-10*

The following example illustrates an application of phase angle to ac circuits.

## Example 11-11

In an ac circuit containing only a constant inductance, the voltage *leads* the current by 90°. That is, if $I = I_{max} \sin \omega t$, then $V = V_{max} \sin (\omega t + \pi/2)$. If $f = 60$ Hz, sketch $I$ vs. $t$ and $V$ vs. $t$ on the same graph for one cycle from $t = 0$.

## Solution

When the relation $\omega = 2\pi f = 120\pi$ is used, the equations of the curves are $I = I_{max} \sin 120\pi$ and $V = V_{max} \sin (120\pi t + \pi/2)$. The period for both is $2\pi/120\pi t = 1/60$ s. The phase angle for $V$ is $\pi/2$, and the phase shift for $V$ is $(-\pi/2)/(120\pi) = -1/240$ s. The sketch is shown in Fig. 11-13 with separate scales for $I$ and $V$.

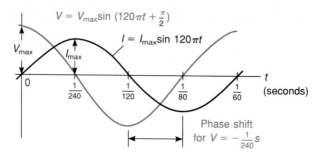

FIG. 11-13 *Voltage leading current 90°, Example 11-11*

**Exercise 11-2**

In problems 1 to 20, find the amplitude, period, frequency, phase angle, and phase shift of each curve. Sketch each curve for at least 1 cycle from $x$ or $t = 0$.

1. $y = \sin \left( x - \dfrac{\pi}{2} \right)$

2. $y = \cos (x - \pi)$

3. $y = 2 \cos (x + \pi)$

4. $y = 3 \sin \left( 2x + \dfrac{\pi}{2} \right)$

5. $y = 4 \sin \left( 3x + \dfrac{\pi}{2} \right)$

6. $y = 5 \sin \left( 4x - \dfrac{\pi}{2} \right)$

7. $y = 3 \cos \left( 3x - \dfrac{\pi}{3} \right)$

8. $y = 3 \sin \left( 2x + \dfrac{\pi}{4} \right)$

9. $y = 10 \sin (2\pi t + \pi)$

10. $y = 5 \cos \left( \pi t - \dfrac{\pi}{6} \right)$

11. $y = 3.2 \cos (5\pi t + 1.6)$

12. $y = 2.2 \sin (3\pi t - 0.40)$

13. $I = 8 \sin \left( 120\pi t + \dfrac{\pi}{2} \right)$

14. $I = 14 \sin \left( 120\pi t + \dfrac{\pi}{3} \right)$

15. $V = 240 \sin \left( 100\pi t - \dfrac{\pi}{3} \right)$

16. $V = 120 \sin \left( 120\pi t - \dfrac{\pi}{4} \right)$

17. $I = 10^{-3} \sin \left( 10^3 t - \dfrac{\pi}{4} \right)$

18. $V = 10^{-2} \left( 10^4 t + \dfrac{\pi}{6} \right)$

19. $V = 29 \cos (314t + 1.0)$

20. $I = 38 \cos (377t + 1.5)$

21. Write the equation of the sine curve for harmonic motion due to free linear vibration when  (a) $a = 0.5$ m,  period $= 0.25$ s,  $\phi = -\pi/4$;  (b) $a = 10^{-4}$ m,  period $= 10^{-3}$ s, $\phi = 0.80$ rad. [See formula (11-4).]

22. In an ac circuit, write the equation of the sine curve for (a) the current $I$ when $I_{max} = 5.3$ A, $f = 80$ Hz, $\phi = \pi/6$ and (b) the voltage $V$ when $V_{max} = 110$ V, $f = 30$ Hz, $\phi = -\pi/3$. [See formula (11-4).]

23. The angular acceleration of a simple pendulum is given by $\alpha = 2 \sin (2t + \pi/2)$, where $t =$ time in seconds. (a) Find the period of the pendulum and sketch $\alpha$ vs. $t$ for 1 cycle. (b) Express $\alpha$ in terms of cosine.

24. The height of a point on the tip of a rotating gear is expressed as $y = 10 \cos(\theta - \pi/2)$, where $\theta$ = angle of rotation. *(a)* Sketch $y$ vs. $\theta$ for 1 cycle. *(b)* Express $y$ in terms of sine.

25. An x-ray is described by the equation:

$$y = 10^{-10} \sin(2\pi \times 10^{18}t + \pi)$$

   Find the frequency and sketch 1 cycle.

26. The position in meters of a point on a vibrating cable is given by $y = 10^{-2} \cos(50\,t - \pi/4)$. Find the frequency and sketch 1 cycle.

27. In the ac circuit containing only a constant capacitance, the current leads the voltage by 90°. *(a)* If $V = V_{max} \sin 2\pi ft$, write the equation for $I$. (See Example 11-11.) *(b)* If $f = 60$ Hz, sketch 1 cycle of $I$ vs. $t$ and $V$ vs. $t$ on the same graph.

28. If $I = I_{max} \sin 120\pi t$ in problem 27, *(a)* write the equation for $V$ and *(b)* sketch 1 cycle of $I$ vs. $t$ and $V$ vs. $t$ on the same graph.

C 29. The height of a point on the rim of a rolling wheel is given by $y = r + r \sin(\theta - \pi/4)$, where $r$ = radius and $\theta$ = angle of rotation. If $r = 25$ in., sketch $y$ vs. $\theta$ for 1 cycle.

C 30. In an *RC* (resistance-capacitance) circuit $I = I_{max} \sin(2\pi ft + \phi)$. If $I_{max} = 3.2$ A, $f = 60$ Hz, and $\phi = 0.56$ rad, sketch $I$ vs. $t$ for 1 cycle.

31. A rotating coil of wire in a magnetic field develops an electromotive force (emf), which is expressed by emf $= nBA \cos(\omega t - \pi/2)$, where $n$ = no. of turns of the coil, $B$ = magnetic density, and $A$ = area of the coil. If $n = 100$, $B = 3 \times 10^{-3}$ T, $A = 2$ m$^2$, and $\omega = 60$ rad/s, *(a)* sketch emf vs. $t$ for 1 cycle, and *(b)* express emf in terms of sine.

32. Show that $\cos(x + \pi/2) = -\sin x$ by sketching 1 cycle of each curve.

# 11-3 | *Composite Curves and Other Trigonometric Curves*

## *Composite Curves*

Consider a spring fixed at one end with a weight vibrating on the other end in a sinusoidal motion: $y_1 = 2 \sin x$. If the fixed end is now moved with a sinusoidal motion $y_2 = \cos x$, the movement of the weight will be the sum of the two sinusoidal motions: $y = y_1 + y_2 = 2 \sin x + \cos x$ (Fig. 11-14). The next example shows how to sketch the curve that results from combining these two sinusoidal motions.

### Example 11-12

Sketch the composite curve $y = 2 \sin x + \cos x$ from $x = 0$ to $x = 2\pi$.

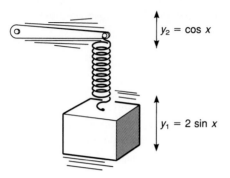

FIG. 11-14  *Composite sinusoidal motion*

## Solution

One way to sketch the curve is to plot points every $\pi/4$ rad to determine its shape. More points can be chosen if necessary:

| $x$ | 0 | $\dfrac{\pi}{4}$ | $\dfrac{\pi}{2}$ | $\dfrac{3\pi}{4}$ | $\pi$ | $\dfrac{5\pi}{4}$ | $\dfrac{3\pi}{2}$ | $\dfrac{7\pi}{4}$ | $2\pi$ |
|---|---|---|---|---|---|---|---|---|---|
| $2 \sin x$ | 0 | 1.4 | 2 | 1.4 | 0 | $-1.4$ | $-2$ | $-1.4$ | 0 |
| $\cos x$ | 1 | 0.71 | 0 | $-0.71$ | $-1$ | $-0.71$ | 0 | 0.71 | 1 |
| $y = 2 \sin x + \cos x$ | 1 | 2.1 | 2 | 0.7 | $-1$ | $-2.1$ | $-2$ | $-0.7$ | 1 |

The composite curve is shown in Fig. 11-15 on top of the curves for $2 \sin x$ and $\cos x$. The curve $y = 2 \sin x + \cos x$ can also be sketched by graphically adding the $y$ values (ordinates) of $2 \sin x$ and $\cos x$ as shown for $x = \pi/4$ in Fig. 11-15. Note that when one curve is zero, the other curve has a point in common

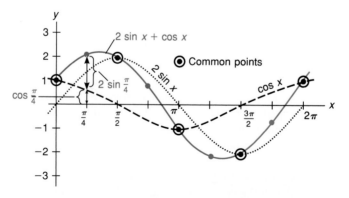

FIG. 11-15  *Composite sinusoidal curve, Example 11-12*

with the composite curve. You can use this fact to help sketch the composite curve.

Observe that $y = 2 \sin x + \cos x$ is also a sine curve.

*When two sinusoidal functions that have the same period are combined, the resulting function is a sinusoidal function that has the same period.*

The amplitude and phase angle of the resulting function are determined by adding radius *vectors* (see problems 35 and 36 at the end of this section for the formulas.) When you combine two sinusoidal functions that have different periods, the resulting function is not necessarily sinusoidal but is periodic, as the next example shows. Periodic functions are encountered in the study of electricity, electronics, mechanical vibration, sound, fluid motion, optics, and servomechanisms.

### Example 11-13

Sketch 1 cycle of $y = \sin 4x + 2 \sin x$ by graphically adding ordinates ($y$ values).

### Solution

The composite curve is shown in Fig. 11-16. The points where $\sin 4x$ or $2 \sin x$ equals zero are marked on the curve. At these points one of the curves has a

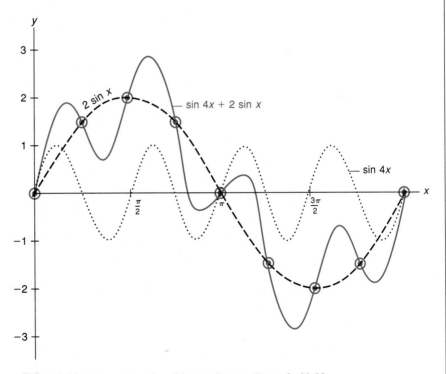

FIG. 11-16 *Composition by adding ordinates, Example 11-13*

point in common with the composite curve. Other points are determined approximately by graphically adding $y$ values. When necessary, points can be obtained by calculation. The composite curve will repeat after $2\pi$ and is therefore a periodic function whose period is $2\pi$.

## Other Trigonometric Curves

The other trigonometric functions are periodic, but they approach infinite values at certain points. The curves have breaks at these points and lack continuity. The graphs of tan $x$ and sec $x$ are shown below. Cot $x$ and csc $x$ are left as problems 19 and 21.

### Example 11-14

Graph $y = \tan x$ from $x = 0$ to $x = 2\pi$.

### Solution

Plot points every 30° to determine the shape of the curve:

| $x$ | 0 | $\dfrac{\pi}{6}$ | $\dfrac{\pi}{3}$ | $\dfrac{\pi}{2}$ | $\dfrac{2\pi}{3}$ | $\dfrac{5\pi}{6}$ | $\pi$ | $\dfrac{7\pi}{6}$ | |
|---|---|---|---|---|---|---|---|---|---|
| $y = \tan x$ | 0 | 0.58 | 1.7 | $\infty$* | $-1.7$ | 0.58 | 0 | 0.58 repeats | |

*Note:* As $x$ approaches $\pi/2$ or $3\pi/2$, tan $x$ becomes infinitely large ($\infty$). This is shown in Fig. 11-17 by the curve approaching a vertical line at $\pi/2$ and $3\pi/2$.

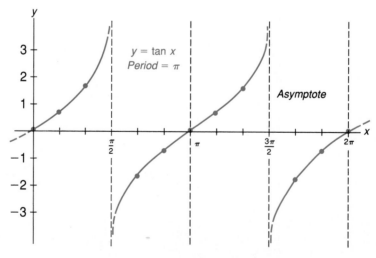

FIG. 11-17 *Tangent curve, Example 11-14*

The curve never touches this line, which is called an *asymptote*, but gets closer and closer to it.

Observe that the period of the tangent is $\pi$, and not $2\pi$ like that of the sine and cosine.

### Example 11-15

Graph $y = \sec x$ from $x = 0$ to $x = 2\pi$.

### Solution

Use $\sec x = 1/\cos x$ to obtain values with a table or calculator every $\dfrac{\pi}{6}$ rad:

| $x$ | 0 | $\dfrac{\pi}{6}$ | $\dfrac{\pi}{3}$ | $\dfrac{\pi}{2}$ | $\dfrac{2\pi}{3}$ | $\dfrac{5\pi}{6}$ | $\pi$ | $\dfrac{7\pi}{6}$ | $\dfrac{4\pi}{3}$ | $\dfrac{3\pi}{2}$ | $\dfrac{5\pi}{3}$ | $\dfrac{11\pi}{6}$ | $2\pi$ |
|---|---|---|---|---|---|---|---|---|---|---|---|---|---|
| $\cos x$ | 1 | 0.87 | 0.50 | 0 | −0.50 | −0.87 | −1 | −0.87 | −0.50 | 0 | 0.50 | 0.87 | 1 |
| $y = \sec x$ | 1 | 1.2 | 2 | $\infty$ | −2 | −1.2 | −1 | −1.2 | −2 | $\infty$ | 2 | 1.2 | 1 |

The graph of $\sec x$ is shown in Fig. 11-18. It can be sketched graphically by inverting the values of $\cos x$ (dashed curve). As $\cos x$ decreases to zero, $\sec x$

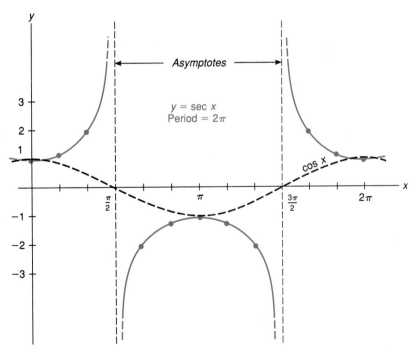

FIG. 11-18 *Secant curve, Example 11-15*

increases to infinity. As cos $x$ approaches 1 or $-1$, so does sec $x$. The vertical lines at $\pi/2$ and $3\pi/2$ are the asymptotes. The period of sec $x$ is $2\pi$, the same as that of the reciprocal function, cos $x$.

**Exercise 11-3**

In problems 1 to 18, sketch the composite curve for 1 cycle and state the period.

**1.** $y = \sin x + \cos x$

**2.** $y = \sin x - \cos x$

**3.** $y = \sin x + 2 \cos x$

**4.** $y = 3 \sin x + \cos x$

**5.** $y = \sin \left(x + \dfrac{\pi}{2}\right) - \sin x$

**6.** $y = \cos \left(x + \dfrac{\pi}{2}\right) - \cos x$

**7.** $y = 2 \cos x - \cos \left(x + \dfrac{\pi}{4}\right)$

**8.** $y = \sin x + 2 \sin \left(x + \dfrac{\pi}{4}\right)$

**9.** $y = \sin x + \sin 2x$

**10.** $y = 2 \cos x + \cos 2x$

**11.** $y = 2 \cos x + \cos 3x$

**12.** $y = \sin 3x + 3 \sin x$

**13.** $y = \sin 4x + \sin x$

**14.** $y = \cos x + 2 \sin 4x$

**15.** $y = \sin 2\pi t + \sin 4\pi t$

**16.** $y = \cos \pi t - \cos 2\pi t$

**17.** $I = 5 \sin 120\pi t + 3 \cos 120\pi t$

**18.** $I = 2 \sin 120\pi t - \sin \left(120\pi t + \dfrac{\pi}{2}\right)$

In problems 19 to 26, graph $y$ vs. $x$, plotting at least 12 points from $x = 0$ to $x = 2\pi$, and state the period.

**19.** $y = \cot x$

**20.** $y = -\tan x$

**21.** $y = \csc x$

**22.** $y = \sec 2x$

**23.** $y = \tan 2x$

**24.** $y = \tan \left(x + \dfrac{\pi}{2}\right)$

**25.** $y = \sec 2\pi x$

**26.** $y = \csc \pi x$

**27.** Show that csc $x = $ sec $(x - \pi/2)$ by sketching both curves.

**28.** Show that cot $x = -$tan $(x + \pi/2)$ by sketching both curves.

**29.** When a weight hanging on a spring is given an initial displacement $y_0$ and an initial velocity $v_0$, its position is given by $y = y_0 \cos \omega t + (v_0/\omega) \sin \omega t$, where the angular frequency $\omega$ depends on the weight and elasticity of the spring. Sketch $y$ vs. $t$ for 1 cycle when $y_0 = 0.50$ m, $v_0 = 1.0$ m/s, and $\omega = 2.0$ rad/s.

**30.** The instantaneous power in an ac circuit containing an inductance is given by $P = VI$, where $V = 120 \cos \omega t$ and $I = 5 \sin \omega t$. (a) Sketch $P$ vs. the angle $\omega t$ for 1 cycle by finding values for $V$ and $I$ and computing the product $VI$. (b) Show that the curve in (a) is equivalent to the one described by $P = 300 \sin 2\omega t$ by sketching this curve and comparing it with the curve in (a).

**31.** Two waves in a medium will reinforce or weaken each other depending on their phase difference. Show that when one water wave described by $y_1 = 2 \sin x$ meets another water wave $y_2 = \cos (\pi/2 - x)$, the result is a reinforced wave (constructive interference), by graphing $y = y_1 + y_2$.

32. Show that when one light wave described by $y_1 = \cos 2x - 1$ meets another light wave $y_2 = 2 \sin^2 x$, they cancel (the phenomenon of destructive interference). Graph $y = y_1 + y_2$. *Note:* $\sin^2 x = (\sin x)^2$.

33. The horizontal position of a point on a rolling wheel is given by $x = r\theta + r \sin \theta$, where $r$ = radius and $\theta$ = angle of rotation in radians. Sketch $x$ vs. $\theta$ for one rotation when $r = 0.10$ m.

34. The acceleration of a piston is expressed as $a = 10 \, \omega^2 (\cos \omega t + \frac{1}{2} \cos 2 \, \omega t)$, where $\omega = 10^2$ rad/s. Sketch $a$ vs. $t$ for 1 cycle.

35. The composite function $y = a \sin x + b \cos x$ is equivalent to the sine function $y = A \sin (x + \phi)$, $a > 0$, where $A = \sqrt{a^2 + b^2}$ and $\phi = \tan^{-1} (b/a)$. *(a)* Find the equivalent sine function for the curve in Example 11-12. *(b)* Show that the table of values is the same and that the curve for the equivalent sine function is the same as the curve in Fig. 11-15, by sketching this function.

36. Given $y = 4 \sin x + 3 \cos x$: *(a)* Use the formula in problem 35 to find the equivalent sine function. *(b)* Sketch the composite curve and the equivalent sine function, and show that they are the same.

37. The safe angle of banking $\theta$ for a highway curve is related to the radius $R$ of the curve by the equation $R = (v^2/g) \cot \theta$, where $v$ = velocity of a vehicle and $g$ = gravitational acceleration. If $v = 25$ m/s (55 mi/h) and $g = 9.8$ m/s$^2$, graph $R$ vs. $\theta$ for $\theta = 0°$ to $\theta = 90°$. Plot points every 10°.

38. If a container of liquid moves with a constant acceleration $a$ along a horizontal plane, the angle $\theta$ the surface of the liquid makes with the horizontal is related to the acceleration by the equation $a = g \tan \theta$, where $g$ = gravitational acceleration. For $g = 32$ ft/s$^2$, graph $a$ vs. $\theta$ for $\theta = 0°$ to $\theta = 90°$. Plot points every 10°.

# 11-4 | *Parametric Equations: Lissajous Figures*

The coordinates $(x, y)$ of a point on a curve can be represented by a pair of equations such as the line $x = t + 1$, $y = 2t$. These equations express each of the variables $x$ and $y$ as a function of a third variable, or *parameter*. They are therefore called *parametric equations*.

### Example 11-16
Graph the straight line represented by the parametric equations $x = t + 1$ and $y = 2t$.

### Solution
Assign simple values to $t$ to make a table of values of $x$ and $y$:

| $t$ | $-3$ | $-2$ | $-1$ | 0 | 1 | 2 | 3 |
|---|---|---|---|---|---|---|---|
| $x = t + 1$ | $-2$ | $-1$ | 0 | 1 | 2 | 3 | 4 |
| $y = 2t$ | $-6$ | $-4$ | $-2$ | 0 | 2 | 4 | 6 |

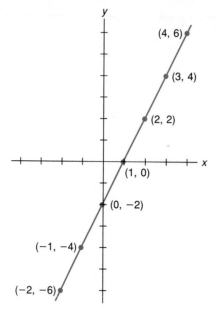

FIG. 11-19 *Parametric equations of a line, Example 11-16*

The graph of the line is shown in Fig. 11-19.

In this example you can eliminate the parameter $t$ by substituting $t = x - 1$ into the equation for $y$:

$$y = 2t$$
$$y = 2(x - 1)$$
$$y = 2x - 2$$

This shows that $y$ is a linear, or first-degree, function of $x$ and so the graph is a straight line. The parameter $t$ cannot always be easily eliminated, as in the case of the parametric equations $x = \cos t$, $y = \cos 2t$ which is the next example. Parametric equations involving trigonometric functions occur often in technical and scientific applications.

### Example 11-17

Sketch the portion of the *parabola* given by the parametric equations $x = \cos t$ and $y = \cos 2t$.

### Solution

Choose values of $t$ at least every 30°, or $\pi/6$ rad:

| $t$ | $0$ | $\dfrac{\pi}{6}$ | $\dfrac{\pi}{3}$ | $\dfrac{\pi}{2}$ | $\dfrac{2\pi}{3}$ | $\dfrac{5\pi}{6}$ | $\pi$ | $\dfrac{7\pi}{6}$ | $\dfrac{4\pi}{3}$ |
|---|---|---|---|---|---|---|---|---|---|
| $x = \cos t$ | $1$ | $0.87$ | $0.50$ | $0$ | $-0.50$ | $-0.87$ | $-1$ | $-0.87$ | $-0.50$ |
| $y = \cos 2t$ | $1$ | $0.50$ | $-0.50$ | $-1$ | $-0.50$ | $0.50$ | $1$ | $0.50$ | $-0.50$ |

repeats in reverse →

| $t$ | $\dfrac{3\pi}{2}$ | $\dfrac{5\pi}{3}$ | $\dfrac{11\pi}{6}$ | $2\pi$ |
|---|---|---|---|---|
| $x = \cos t$ | $0$ | $0.50$ | $0.87$ | $1$ |
| $y = \cos 2t$ | $-1$ | $-0.50$ | $0.50$ | $1$ |

. . .
. . .

Observe that the points repeat in reverse order after $t = \pi$. The curve has period $\pi$ and is shown in Fig. 11-20. The parabola and the other conic sections (the circle, ellipse, and hyperbola) are studied in more detail in Chap. 14. Problems 21 and 22 show an application of the parabola.

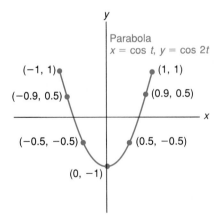

FIG. 11-20 *Parametric equations of a parabola, Example 11-17*

## Lissajous Figures

When the parametric equations of a point describe simple harmonic motions, the resulting curve is called a *Lissajous figure*. Figure 11-20 is therefore a Lissajous figure since the parametric equations are sinusoidal functions which describe simple harmonic motions (Sec. 11-1). When voltages of different frequencies are applied to the vertical and horizontal plates of an oscilloscope, Lissajous figures result, allowing an unknown frequency to be determined.

### Example 11-18
Sketch the Lissajous figure $x = \cos \omega t$ and $y = \sin 3\omega t$.

### Solution
Choose the angle $\omega t$ as the parameter and plot points every $\pi/6$ radians:

| $\omega t$ | 0 | $\dfrac{\pi}{6}$ | $\dfrac{\pi}{3}$ | $\dfrac{\pi}{2}$ | $\dfrac{2\pi}{3}$ | $\dfrac{5\pi}{6}$ | $\pi$ | $\dfrac{7\pi}{6}$ | $\dfrac{4\pi}{3}$ |
|---|---|---|---|---|---|---|---|---|---|
| $x = \cos \omega t$ | 1 | 0.87 | 0.50 | 0 | −0.50 | −0.87 | −1 | −0.87 | −0.50 |
| $y = \sin 3\omega t$ | 0 | 1 | 0 | −1 | 0 | 1 | 0 | −1 | 0 |

| $\omega t$ | $\dfrac{3\pi}{2}$ | $\dfrac{5\pi}{3}$ | $\dfrac{11\pi}{6}$ | $2\pi$ |
|---|---|---|---|---|
| $x = \cos \omega t$ | 0 | 0.50 | 0.87 | 1 |
| $y = \sin 3\omega t$ | 1 | 0 | −1 | 0 |

repeats

The graph (Fig. 11-21) is a smooth curve of period $2\pi$ that starts at (1, 0) when $\omega t = 0$ and traces the path indicated by the arrows.

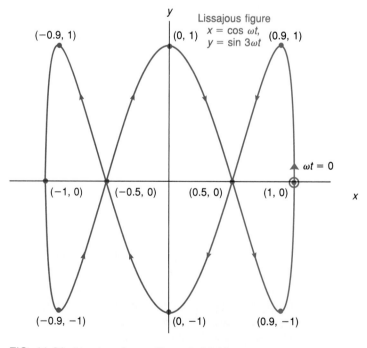

FIG. 11-21 *Lissajous figure, Example 11-18*

**Exercise 11-4**   In problems 1 to 12, graph each curve given by the parametric equations. Plot at least six points for each.

**1.** $x = t$, $y = 2t$ (line)

**2.** $x = 2t$, $y = t + 1$ (line)

**3.** $x = t$, $y = t^2$ (parabola)

**4.** $x = t$, $y = 1 - t^2$ (parabola)

**5.** $x = 3 \cos t$, $y = 3 \sin t$ (circle)

**6.** $x = 3 \cos t$, $y = \sin t$ (ellipse)

**7.** $x = \cos 2t$, $y = \cos t$ (parabola)

**8.** $x = \sin t$, $y = \cos 2t$ (parabola)

**9.** $x = t - \sin t$, $y = 1 - \cos t$ (cycloid)

**10.** $x = 2 \sin t$, $y = 4 \cos t$ (ellipse)

**11.** $x = \tan \theta$, $y = 12 \cot \theta$ (hyperbola)

**12.** $x = \sec \theta$, $y = \tan \theta$ (hyperbola)

In problems 13 to 20, sketch the Lissajous figures.

**13.** $x = \sin \omega t$, $y = \sin 2\omega t$        **14.** $x = \sin 2\omega t$, $y = 2 \cos \omega t$

**15.** $x = \sin \left( \omega t + \dfrac{\pi}{4} \right)$, $y = \cos \omega t$        **16.** $x = \cos \left( 2\omega t + \dfrac{\pi}{4} \right)$, $y = \sin \omega t$

**17.** $x = \sin \omega t$, $y = \cos 3\omega t$        **18.** $x = \sin 3\omega t$, $y = \cos \omega t$

**19.** $x = \cos 3\omega t$, $y = \cos 2\omega t$        **20.** $x = \sin 2\omega t$, $y = \cos 3\omega t$

**21.** The path of a projectile shot at an angle $\theta$ above the horizontal is a parabola described by the parametric equations $x = v_0 t \cos \theta$ and $y = v_0 t \sin \theta - \frac{1}{2}gt^2$, where $v_0 =$ initial velocity, $t =$ time, and $g =$ gravitational acceleration. Graph $y$ vs. $x$ when $v_0 = 100$ m/s and $\theta = 30°$. Use $g = 9.8$ m/s$^2$.

**22.** Applying the equations in problem 21, plot the path of a missile fired at an angle $\theta = 45°$ with a velocity $v_0 = 1500$ ft/s. Use $g = 32$ ft/s$^2$. What is the horizontal range of the missile?

**23.** Two voltages $V_1 = 110 \cos 120\pi t$ and $V_2 = 110 \cos (120\pi t + \pi/2)$ are applied, respectively, to the horizontal and vertical plates of an oscilloscope. Sketch the resulting Lissajous figure.

**24.** If the frequency of $V_2$ in problem 23 is doubled, what will be the resulting Lissajous figure? Sketch the figure.

# 11-5 | *Polar Coordinates*

Suppose you were piloting a plane from your town or city to a nearby town or city. What kind of coordinate system would you use to determine your flight path? You would assign two coordinates—*distance* and *direction*—to the nearby town or city. This is equivalent to assigning a *radius vector* from your location to your destination. Points located by radius vectors use a system of *polar coordinates*. Polar coordinates are used not only in air, marine, and space navigation but also in applications to alternating current (Sec. 13-5), rotating machinery, electric and magnetic fields, atomic physics, and calculus.

The origin of the polar coordinate system is called the *pole,* and the positive *x* axis is the *polar axis* (Fig. 11-22). Each point has coordinates $(r, \theta)$, where $r$ is the length of the radius vector and $\theta$ is the angle, measured counterclockwise from the polar axis (this is the same as for trigonometric angles). However note that *both coordinates assume positive and negative values,* which allows a point to have *more than one* set of coordinates. Study the following example carefully.

### Example 11-19
Plot the following points on a polar coordinate graph.

1. $\left(3, \dfrac{\pi}{4}\right)$     2. $\left(3, \dfrac{9\pi}{4}\right)$

3. $\left(2, \dfrac{3\pi}{2}\right)$     4. $\left(2, -\dfrac{\pi}{2}\right)$

5. $(4, 0)$     6. $(-4, \pi)$

7. $\left(3, \dfrac{3\pi}{4}\right)$     8. $\left(-3, -\dfrac{\pi}{4}\right)$

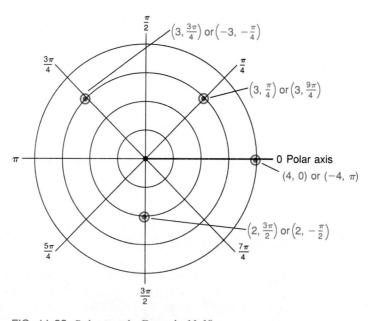

FIG. 11-22 *Polar graph, Example 11-19*

## Solution

The points are plotted on the polar graph in Fig. 11-22, where the circles represent radii of length 1, 2, 3, and 4. The eight coordinate pairs actually represent only four points. **1** and **2** are the same point as are **3** and **4**, because the angles have the same terminal side. **5** and **6** are the same point as are **7** and **8**, because *a negative value of r is measured in the opposite direction, or π rad from θ.* For example, the point $(-3, -\pi/4)$ is located by measuring a radius vector of 3 opposite in direction to $-\pi/4$, which is equivalent to the direction $\theta = 3\pi/4$. Note that for trigonometric angles $r$ is always positive; however, in polar coordinates $r$ can be positive or negative. Note also that the pole has coordinates $(0, \theta)$ for *any* angle $\theta$.

The following relations between polar and rectangular coordinates come from the definitions of the trigonometric functions and from the pythagorean theorem (see Fig. 11-23):

$$x = r \cos \theta \qquad\qquad r = \pm\sqrt{x^2 + y^2}$$

$$y = r \sin \theta \qquad\qquad \tan \theta = \frac{y}{x} \tag{11-6}$$

The following examples show how to apply formulas (11-6).

## Example 11-20
Change each to polar coordinates.

**1.** $(3, -4)$      **2.** $(0, 5)$

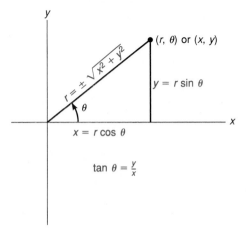

FIG. 11-23 *Polar and rectangular coordinates*

**Solution**

**1.** $(3, -4)$

Apply (11-6): $r = \pm\sqrt{(3)^2 + (-4)^2} = \pm\sqrt{25} = \pm 5$. Angle $\theta$ is in the fourth quadrant. The reference angle $= \tan^{-1}(4/3) = 53°$. Therefore $\theta = 360° - 53° = 307°$, and one set of polar coordinates is $(5, 307°)$. It is possible to find many sets of polar coordinates. Another set is $(-5, 127°)$.

**2.** $(0, 5)$

The polar coordinates can be determined graphically. Since $(0, 5)$ lies on the $y$ axis 5 units above the origin, one set of polar coordinates is $(5, \pi/2)$. Another set is $(5, 5\pi/2)$.

### Example 11-21
Change each to rectangular coordinates.

**1.** $(5, 108°)$      **2.** $(3, 2\pi)$

**Solution**

**1.** $(5, 108°)$

Apply (11-6): $x = 5\cos 108° = -1.5$ and $y = 5\sin 108° = 4.8$. The rectangular coordinates are therefore $(-1.5, 4.8)$.

**2.** $(3, 2\pi)$

The rectangular coordinates can be determined graphically. Since $(3, 2\pi)$ lies on the polar axis 3 units to the right of the origin, the rectangular coordinates are $(3, 0)$.

Many calculators have keys that convert rectangular coordinates to polar coordinates and vice versa.

The next two examples show how to construct polar graphs of certain lines and circles.

### Example 11-22
Change the equation of the line $y = x$ to polar coordinates.

**Solution**

Substitute, using (11-6), and simplify:

$$y = x$$
$$r\sin\theta = r\cos\theta$$
$$\sin\theta = \cos\theta$$
$$\frac{\sin\theta}{\cos\theta} = \tan\theta = 1$$
$$\theta = \frac{\pi}{4}$$

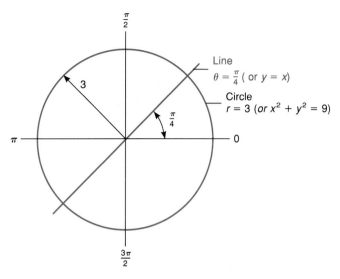

FIG. 11-24 *Line and circle on polar graph, Examples 11-22 and 11-23*

The equation $\theta = \pi/4$ is one equation of the line in polar coordinates that uses an angle in the first quadrant. Another equation is $\theta = 5\pi/4$. The line passes through the pole and makes a 45° angle with the polar axis (Fig. 11-24).

### Example 11-23
Change the equation of the circle $x^2 + y^2 = 9$ to polar coordinates.

### Solution
Substitute $r^2$ for $x^2 + y^2$: $r^2 = 9$, and then $r = 3$ (or $r = -3$) is the equation of the circle in polar coordinates. The circle has its center at the pole and a radius of 3 (Fig. 11-24).

From Example 11-22 we can see that *the equation of a line through the pole has the form:*

$$\theta = \text{constant (in degrees or radians)} \qquad (11\text{-}7)$$

where the constant is the angle the line makes with the polar axis.

From Example 11-23 we see that *the equation of a circle whose center is the pole has the form:*

$$r = \text{constant} \qquad (11\text{-}8)$$

where the absolute value of the constant is the radius of the circle. Equations of lines and circles are studied in more detail in Chap. 14.

Certain curves, like those described by formulas (11-7) and (11-8), can be represented and graphed more easily in polar coordinates than in rectangular coordinates. The following examples illustrate some of these curves.

### Example 11-24
Draw the graph of the *circle* $r = 2 \cos \theta$.

### Solution
Choose values of $\theta$ at least every $\pi/6$ rad, and compute $r$ for each value of $\theta$:

| $\theta$ | 0 | $\dfrac{\pi}{6}$ | $\dfrac{\pi}{3}$ | $\dfrac{\pi}{2}$ | $\dfrac{2\pi}{3}$ | $\dfrac{5\pi}{6}$ | $\pi$ | $\dfrac{7\pi}{6}$ | $\dfrac{4\pi}{3}$ | $\dfrac{3\pi}{2}$ | $\dfrac{5\pi}{3}$ | $\dfrac{11\pi}{6}$ | $2\pi$ |
|---|---|---|---|---|---|---|---|---|---|---|---|---|---|
| $r = 2 \cos \theta$ | 2 | 1.7 | 1 | 0 | −1 | −1.7 | −2 | −1.7 | −1 | 0 | 1 | 1.7 | 2 |

$\overrightarrow{\text{points repeat}}$

Notice in the table and Fig. 11-25 that the points repeat after $\theta = 5\pi/6$. *Remember negative values of r are plotted in the opposite direction of θ.*

### Example 11-25
Draw the graph of the *cardioid* (heart-shaped curve) $r = 2(1 - \sin \theta)$.

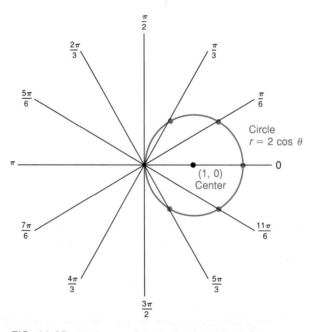

FIG. 11-25 *Circle on polar graph, Example 11-24*

### Solution
Values of $\theta$ every $\pi/6$ rad are used, as in Example 11-24:

| $\theta$ | 0 | $\dfrac{\pi}{6}$ | $\dfrac{\pi}{3}$ | $\dfrac{\pi}{2}$ | $\dfrac{2\pi}{3}$ | $\dfrac{5\pi}{6}$ | $\pi$ | $\dfrac{7\pi}{6}$ | $\dfrac{4\pi}{3}$ | $\dfrac{3\pi}{2}$ | $\dfrac{5\pi}{3}$ | $\dfrac{11\pi}{6}$ | $2\pi$ |
|---|---|---|---|---|---|---|---|---|---|---|---|---|---|
| $\sin\theta$ | 0 | 0.50 | 0.87 | 1 | 0.87 | 0.50 | 0 | $-0.50$ | $-0.87$ | $-1$ | $-0.87$ | $-0.50$ | 0 |
| $r = 2(1-\sin\theta)$ | 2 | 1 | 0.27 | 0 | 0.27 | 1 | 2 | 3 | 3.7 | 4 | 3.7 | 3 | 2 |

repeats $\longrightarrow$

The cardioid is shown in Fig. 11-26.

### Example 11-26
Draw the graph of the *parabola*:

$$r = \frac{2}{1 - \sin\theta}$$

### Solution

| $\theta$ | 0 | $\dfrac{\pi}{6}$ | $\dfrac{\pi}{3}$ | $\dfrac{\pi}{2}$ | $\dfrac{2\pi}{3}$ | $\dfrac{5\pi}{6}$ | $\pi$ | $\dfrac{7\pi}{6}$ | $\dfrac{4\pi}{3}$ | $\dfrac{3\pi}{2}$ | $\dfrac{5\pi}{3}$ | $\dfrac{11\pi}{6}$ | $2\pi$ |
|---|---|---|---|---|---|---|---|---|---|---|---|---|---|
| $r = \dfrac{2}{1-\sin\theta}$ | 2 | 4 | 15 | $\infty$ | 15 | 4 | 2 | 1.3 | 1.1 | 1 | 1.1 | 1.3 | 2 |

repeats $\longrightarrow$

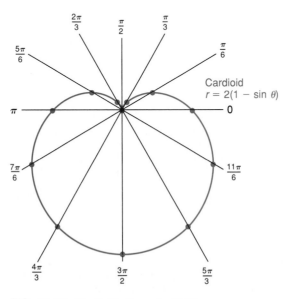

FIG. 11-26 *Cardioid, Example 11-25*

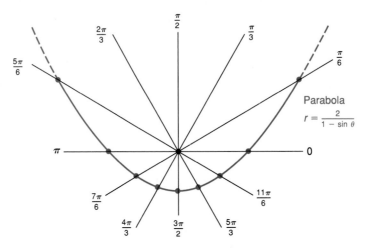

FIG. 11-27 *Parabola, Example 11-26*

Notice that $r$ approaches infinity ($\infty$) as $\theta$ approaches $\pi/2$. The parabola is shown in Fig. 11-27. The points where $r$ equals 15 are omitted.

### Example 11-27
Draw the graph of the *three-leaved rose* $r = 4 \sin 3\theta$.

### Solution
The period of $\sin 3\theta$ is 120°, and so it is necessary to plot values of $\theta$ closer together than $\pi/6$ rad to determine the shape of the curve. Choose values every $\pi/12$ rad:

| $\theta$ | 0 | $\dfrac{\pi}{12}$ | $\dfrac{\pi}{6}$ | $\dfrac{\pi}{4}$ | $\dfrac{\pi}{3}$ | $\dfrac{5\pi}{12}$ | $\dfrac{\pi}{2}$ | $\dfrac{7\pi}{12}$ | $\dfrac{2\pi}{3}$ | $\dfrac{3\pi}{4}$ | $\dfrac{5\pi}{6}$ | $\dfrac{11\pi}{12}$ | $\pi$ | $\cdots$ |
|---|---|---|---|---|---|---|---|---|---|---|---|---|---|---|
| $3\theta$ | 0 | $\dfrac{\pi}{4}$ | $\dfrac{\pi}{2}$ | $\dfrac{3\pi}{4}$ | $\pi$ | $\dfrac{5\pi}{4}$ | $\dfrac{3\pi}{2}$ | $\dfrac{7\pi}{4}$ | $2\pi$ | $\dfrac{9\pi}{4}$ | $\dfrac{5\pi}{2}$ | $\dfrac{11\pi}{4}$ | $3\pi$ | $\cdots$ |
| $r = 4 \sin 3\theta$ | 0 | 2.8 | 4 | 2.8 | 0 | $-2.8$ | $-4$ | $-2.8$ | 0 | 2.8 | 4 | 2.8 | 0 | |

repeats $\longrightarrow$

The rose is shown in Fig. 11-28. The next point after $\theta = \pi$ is $(13\pi/12, -2.8)$. This is the same as what point in the table?

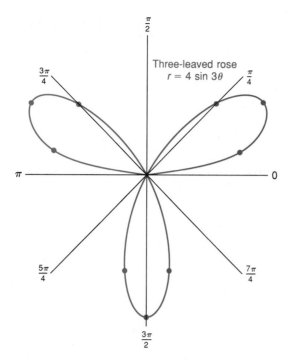

FIG. 11-28 *Three-leaved rose, problem 11-27*

**Exercise 11-5** In problems 1 to 10, plot the points on polar coordinate paper.

**1.** $\left(2, \dfrac{\pi}{3}\right)$      **2.** $\left(1, \dfrac{5\pi}{6}\right)$

**3.** $(0, 3\pi)$      **4.** $(0, -\pi)$

**5.** $\left(-2, \dfrac{4\pi}{3}\right)$      **6.** $\left(-1, \dfrac{17\pi}{6}\right)$

**7.** $\left(-3, -\dfrac{\pi}{4}\right)$      **8.** $(-2.0, -2.0)$

**9.** $(1.5, 10°)$      **10.** $(-3.8, 95°)$

In problems 11 to 14, change to polar coordinates.

**11.** $(-5, 12)$      **12.** $(2, -2)$

**13.** $(-0.4, 0)$      **14.** $(-4.5, -6.0)$

In problems 15 to 18, change to rectangular coordinates.

**15.** $(3, 72°)$      **16.** $\left(2.5, \dfrac{7\pi}{4}\right)$

**17.** $(1.0, 1.0)$      **18.** $(-1.7, -3\pi)$

In problems 19 to 26, change the equation to polar coordinates and draw the graph of the curve on polar coordinate paper. (See Examples 11-22 and 11-23.)

**19.** $y = 2x$

**20.** $y = -x$

**21.** $x = 3$

**22.** $y = 5$

**23.** $x^2 + y^2 = 4$

**24.** $x^2 + y^2 = 25$

**25.** $x^2 + y^2 = 5x$ (circle)

**26.** $x^2 + y^2 = 4y$ (circle)

In problems 27 to 42, draw each graph on polar coordinate paper.

**27.** $r = 5 \sin \theta$ (circle)

**28.** $r = 6 \cos \theta$ (circle)

**29.** $r = 3(1 + \cos \theta)$ (cardioid)

**30.** $r = 2(1 + \sin \theta)$ (cardioid)

**31.** $r = \dfrac{1}{1 - \sin \theta}$ (parabola)

**32.** $r = \dfrac{3}{2 + 2 \cos \theta}$ (parabola)

**33.** $r = 3 \cos 3\theta$
(three-leaved rose)

**34.** $r = 4 \sin 2\theta$
(four-leaved rose)

**35.** $r = \dfrac{4}{2 + \cos \theta}$ (ellipse)

**36.** $r = \dfrac{2}{1 + 2 \cos \theta}$ (hyperbola)

**37.** $r = 3 \sin \theta + 2$ (limaçon)

**38.** $r = 3 - 2 \sin \theta$ (limaçon)

**39.** $r = \dfrac{\theta}{\pi}$ (archimedean spiral—plot $\theta$ in radians)

**40.** $r = 2^{\theta/(2\pi)}$ (logarithmic spiral: natural growth pattern found in seashells, in flowers, and in other living things; plot $\theta$ in radians)

**41.** $r = 5 \sin 5\theta$ (five-leaved rose—plot $\theta$ every $\dfrac{\pi}{20}$ rad)

**42.** $r^2 = 9 \cos 2\theta$ (lemniscate—use positive values of $\cos 2\theta$)

**43.** In an ac circuit, polar coordinates are used to represent the impedance vector $(Z, \phi)$ (Fig. 11-29). Find $(Z, \phi)$ when the resistance $R = 5.3\Omega$ and the reactance $X = 8.0\ \Omega$.

**44.** In Fig. 11-29 find $R$ and $X$ when $Z = 3.0\ \Omega$ and $\phi = 0.33$ rad.

**45.** The location of points on earth involves a polar coordinate system of latitude and longitude measured from the north pole (or south pole) (Fig. 11-30). Latitude corre-

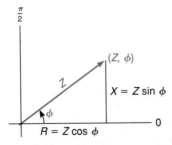

FIG. 11-29 *Impedance vector, problems 43 and 44*

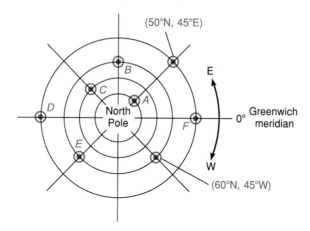

FIG. 11-30 *Latitude and longitude, problems 45 and 46*

sponds to the radius and decreases from 90° as one moves away from the pole. Longitude corresponds to the angle $\theta$, with east of the Greenwich meridian positive and west negative. If the circles of latitude in Fig. 11-30 differ by 10°, find the latitude and longitude of the six lettered points.

**46.** Find the polar and rectangular coordinates of the two points in Fig. 11-30 whose latitude and longitude are given. Assume that the radii of the circles differ by 1 unit and that the angle is measured as in a polar graph.

**47.** An engine flywheel is tested for dynamic balance by rotating it and tracing a polar graph of a point on its rim. The equation of the graph is $r = 0.1 \cos \theta + 3.0$. Draw the graph and observe the state of balance of the flywheel.

**48.** An engine gear is tested for dynamic balance in the same way as the flywheel in problem 47. Draw the graph given by $r = 4.0/(1 - 0.05 \sin \theta)$ and observe the state of balance of the gear.

**49.** Artillery located 5 km behind the front (the pole) is fired into enemy lines on a trajectory given by $r = 5/(1 + \sin \theta)$, where $\theta$ is measured from $-\pi$ to $-2\pi$. Draw the polar graph of the trajectory (a parabola) for these values of $\theta$.

**50.** An atomic particle traces the following path in a particle accelerator, hitting the target at $\theta = \pi$: $r = 3.8 - 2.2 \cos \theta$. Graph $r$ vs. $\theta$ from $\theta = 0$ to $\theta = \pi$.

**51.** A comet enters the solar system, traveling in an elliptical orbit given by $r = 10^7/(1.1 - \cos \theta)$, where $r$ is in miles and the sun is at the pole. Draw the graph of the comet's path. Use a suitable scale for $r$.

**52.** Two electrons travel in an electric field in near-circular paths approximated in rectangular coordinates by $x^2 + y^2 = 3.5x$ and $x^2 + y^2 = 4.8y$. Change these equations to polar coordinates, and draw the two paths on the same graph.

**53.** An antenna radiation pattern in a horizontal plane is described by the polar equation $r = 5 \sin 2\theta$ (a four-leaved rose). Antenna radiation patterns can take the form of polar roses and cardioids. Graph this radiation pattern.

C **54.** A signal tower transmits signals from two antennas whose radiation patterns are described by $r_1 = 2.7(1 + \cos \theta_1)$ and $r_2 = 2.7 (1 - \cos \theta_2)$. Plot these two curves on the same polar graph and observe the antenna patterns.

## 11-6 | *Review Questions*

In problems 1 to 6, find the amplitude, period, frequency, phase angle, and phase shift. Sketch for 1 cycle from $x$ or $t = 0$.

**1.** $y = 3 \sin x$            **2.** $y = 2 \cos 3x$

**3.** $y = 2 \cos (x - \pi)$      **4.** $y = 1.5 \sin \left(2x + \dfrac{\pi}{2}\right)$

**5.** $I = 5 \sin 120\pi t$       **6.** $V = 120 \cos \left(60\pi t - \dfrac{\pi}{3}\right)$

In problems 7 to 10, sketch for 1 cycle and state the period.

**7.** $y = \cos x - \sin x$

**8.** $y = \sin x + 3 \cos x$

**9.** $y = 2 \sin x + \sin 2x$

**10.** $y = \tan (x + \pi)$

In problems 11 and 12, graph each curve given by the parametric equations.

**11.** $x = t - 1$, $y = 2t$

**12.** $x = \cos 2t$, $y = \sin t$

In problems 13 and 14, graph each curve on polar coordinate paper.

**13.** $r = 6 \sin \theta$

**14.** $y = 3(1 - \cos \theta)$

**15.** A weight vibrates on a spring in simple harmonic motion. The amplitude = 3.0 in. and $\omega = 2.0$ rad/s. *(a)* Write the equation for the weight's position $y$ if $y = 0$ when $t = 0$. *(b)* Find the period and graph the equation for 1 cycle from $t = 0$.

**16.** The height of a wave (in meters) traveling in a cable is given by $y = 1.5 \cos 0.5t$. Graph this equation for 1 cycle from $t = 0$.

**17.** In an *RL* (resistance-inductance) circuit $I = I_{max} \sin (2\pi ft - \phi)$. If $I_{max} = 2.6$ A, $f = 60$ Hz and $\phi = \pi/2$, sketch $I$ vs. $t$ for 1 cycle.

C **18.** The acceleration of a piston is given by $a = r \, \omega^2 \cos (\omega t - \pi/4)$, where $r = 8.5$ cm and $\omega = 20$ rad/s. Find the period and sketch $a$ vs. $t$ for 1 cycle.

**19.** The velocity of a pendulum is expressed as $v = v_0 \cos \omega t - \omega y_0 \sin \omega t$, where $v_0 =$ initial velocity and $y_0 =$ initial displacement. *(a)* Sketch $v$ vs. $t$ for 1 cycle when $v_0 = 1.0$ m/s, $y_0 = 0.5$ m, and $\omega = 2.0$ rad/s. *(b)* Find the period of the curve.

C **20.** The voltage in an ac circuit is given by $V = 120 \cos \omega t + 120 \cos (\omega t + \pi/3)$. Sketch $V$ vs. $\omega t$ for 1 cycle.

21. Two voltages represented by $V_1 = \sin 2\omega t$ and $V_2 = \sin \omega t$ are applied, respectively, to the horizontal and vertical plates of an oscilloscope. Sketch the resulting Lissajous figure.

22. Given the equation $x^2 + y^2 = 9$, (a) change the equation to polar coordinates and (b) identify and sketch the curve on polar coordinate paper.

23. A communications center (the pole) receives signals from two antennas. The first antenna is 2 mi north of the center, and the second is 3 mi east. The shape of the radio waves received from the first antenna in a horizontal plane is given by $r_1 = 4 \sin \theta_1$ and the shape of those received from the second antenna by $r_2 = 6 \cos \theta_2$. Plot these two curves on the same polar graph, and observe how they meet at the communications center.

24. An airplane propeller is tested for dynamic balance by rotating it and tracing a polar graph of a point on its rim. The equation of the graph is $r = 6.0 - 0.2 \cos \theta$. Draw the graph on polar coordinate paper and observe the state of balance of the propeller.

C H A P T E R

# 12

# Trigonometric Identities and Equations

## 12-1 | *Basic Identities*

An *identity* is an equation that is true for *all values* of the variable. The definitions of the trigonometric functions give rise to eight basic identities:

**E**quations and problems involving trigonometric functions are found in almost all fields. Trigonometric identities are useful in solving such problems and provide the means for evaluating trigonometric functions. In calculus and other subjects, mathematical expressions arise that are not necessarily trigonometric but that can be simplified by substituting trigonometric expressions and using identities. In Chap. 10 the trigonometric functions and some basic identities are presented. In this chapter more trigonometric identities are explored, and their application to solving trigonometric equations is studied.

Reciprocal

$$\csc \theta = \frac{1}{\sin \theta}$$

$$\sec \theta = \frac{1}{\cos \theta} \qquad (12\text{-}1)$$

$$\cot \theta = \frac{1}{\tan \theta}$$

Ratio

$$\tan \theta = \frac{\sin \theta}{\cos \theta}$$

$$\cot \theta = \frac{\cos \theta}{\sin \theta} \qquad (12\text{-}2)$$

Pythagorean

$$\sin^2 \theta + \cos^2 \theta = 1$$

$$\tan^2 \theta + 1 = \sec^2 \theta \quad (12\text{-}3)$$

$$\cot^2 \theta + 1 = \csc^2 \theta$$

Note that variations of the above are also true. For example, $\sin \theta = 1/\csc \theta$, $\tan \theta \cos \theta = \sin \theta$, $\sin^2 \theta = 1 - \cos^2 \theta$, and so on.

The reciprocal and ratio identities are shown in Sec. 10-1. The pythagorean identities result from the pythagorean theorem as follows. Consider the unit circle (Fig. 12-1), where $\sin \theta = y/1 = y$ and $\cos \theta = x/1 = x$. By the pythagorean theorem, $x^2 + y^2 = 1^2 = 1$. Therefore by substitution, $\sin^2 \theta + \cos^2 \theta = 1$. The proofs of the other two pythagorean identities are left as problems 1 and 2.

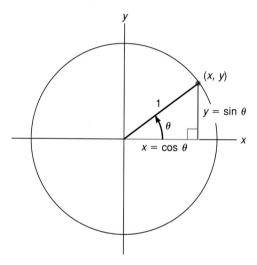

FIG. 12-1  *Unit circle*

The basic identities can be used to develop and prove other identities. To prove an identity is true, you transform one side, or both sides, to show that the expressions are identical. The following examples are transformed on one side only to make them easier to follow.

**Example 12-1**
Prove the identity:     $\tan \theta \csc \theta = \sec \theta$

**Solution**
Transform the left side so that it is identical to the right side:

$$\tan \theta \csc \theta = \left(\frac{\sin \theta}{\cos \theta}\right)\left(\frac{1}{\sin \theta}\right) \qquad \text{by (12-1) and (12-2)}$$

$$= \left(\frac{\cancel{\sin \theta}}{\cos \theta}\right)\left(\frac{1}{\cancel{\sin \theta}}\right) = \frac{1}{\cos \theta}$$

$$= \sec \theta \qquad \text{by (12-1)}$$

Each identity is different, and you have to try various substitutions and algebraic techniques to prove some of them. Study each of the following examples.

### Example 12-2

Prove the identity: $\quad \sec^2 \theta \sin^2 \theta + 1 = \sec^2 \theta$

### Solution

Transform the left side so that it is identical to the right side:

$$\sec^2 \theta \sin^2 \theta + 1 = \left(\frac{1}{\cos^2 \theta}\right)(\sin^2 \theta) + 1 \qquad \text{by (12-1)}$$

$$= \frac{\sin^2 \theta}{\cos^2 \theta} + 1 = \tan^2 \theta + 1 \qquad \text{by (12-2)}$$

$$= \sec^2 \theta \qquad \text{by (12-3)}$$

There is usually more than one way to prove an identity. Example 12-2 can also be done by substituting $1 - \cos^2 \theta$ for $\sin^2 \theta$ and $1/\cos^2 \theta$ for $\sec^2 \theta$ on the left side:

$$\left(\frac{1}{\cos^2 \theta}\right)(1 - \cos^2 \theta) + 1 = \frac{1}{\cos^2 \theta} - 1 + 1 \qquad \text{by (12-1) and (12-3)}$$

$$= \frac{1}{\cos^2 \theta} = \sec^2 \theta \qquad \text{by (12-1)}$$

*A trigonometric expression can always be changed to sines and cosines,* as in Examples 12-1 and 12-2, by using the basic identities. Changing to sines and cosines is a useful procedure in proving many identities.

### Example 12-3

Prove the identity: $\quad \csc x \sec x - \cot x = \tan x$

### Solution

Change the left side to sines and cosines:

$$\frac{1}{\sin x}\frac{1}{\cos x} - \frac{\cos x}{\sin x} = \frac{1}{\sin x \cos x} - \frac{\cos x}{\sin x} \qquad \text{by (12-1) and (12-2)}$$

$$= \frac{1 - (\cos x)(\cos x)}{\sin x \cos x}$$

$$= \frac{1 - \cos^2 x}{\sin x \cos x} \qquad \text{combine fractions}$$

$$= \frac{\sin^2 x}{\cancel{\sin x} \cos x} \qquad \text{by (12-3)}$$

$$= \frac{\sin x}{\cos x} = \tan x \qquad \text{by (12-2)}$$

## Example 12-4

Prove the identity: $\quad \dfrac{\sin \phi}{1 - \cos \phi} = \dfrac{1 + \cos \phi}{\sin \phi}$

## Solution

You can change either side; however, both are in terms of sine and cosine. It is necessary to try some algebraic techniques. You can multiply the top and bottom of any fraction by the same quantity without changing the value of the fraction. Multiply the left side by $\sin \phi/\sin \phi$ so that the denominator has the same factor as the right side:

$$\left(\frac{\sin \phi}{\sin \phi}\right)\left(\frac{\sin \phi}{1 - \cos \phi}\right) = \frac{\sin^2 \phi}{(\sin \phi)(1 - \cos \phi)}$$

$$= \frac{1 - \cos^2 \phi}{(\sin \phi)(1 - \cos \phi)} \qquad \text{by (12-3)}$$

Factor the numerator and simplify:

$$= \frac{(1 + \cos \phi)(\cancel{1 - \cos \phi})}{(\sin \phi)(\cancel{1 - \cos \phi})} = \frac{1 + \cos \phi}{\sin \phi}$$

Notice that the denominator is not multiplied out because the $\sin \phi$ is what you are trying to obtain in the result. This identity can be proved in several ways. Another way is to multiply the left side by $(1 + \cos \phi)/(1 + \cos \phi)$.

## Example 12-5

Prove the identity: $\quad \dfrac{\tan A + \cot A}{\tan A} = \sec^2 A \cot^2 A$

## Solution

This identity can be done by changing it to sines and cosines. However, you can also work with the given functions and change the left side:

$$\frac{\tan A + \cos A}{\tan A} = \frac{\tan A + 1/\tan A}{\tan A} \qquad \text{by (12-1)}$$

$$= \frac{(\tan A)}{(\tan A)}\left(\frac{\tan A + 1/\tan A}{\tan A}\right)$$

$$= \frac{\tan^2 A + 1}{\tan^2 A}$$

$$= \frac{\sec^2 A}{\tan^2 A} = \sec^2 A \left( \frac{1}{\tan A} \right)^2 \qquad \text{by (12-3)}$$

$$= \sec^2 A \cot^2 A \qquad \text{by (12-1)}$$

Proving identities requires a good working knowledge of the basic identities and of algebraic techniques. You will develop this knowledge as you do more and more exercises.

**Exercise 12-1**

1. Prove the pythagorean identity $\tan^2 \theta + 1 = \sec^2 \theta$ by starting with $\sin^2 \theta + \cos^2 \theta = 1$.

2. Prove the pythagorean identity $\cot^2 \theta + 1 = \csc^2 \theta$ by starting with $\sin^2 \theta + \cos^2 \theta = 1$.

In problems 3 to 28, prove each identity.

3. $\cos \theta \csc \theta = \cot \theta$

4. $\sin \theta \sec \theta = \tan \theta$

5. $\dfrac{\sec \theta}{\csc \theta} = \tan \theta$

6. $\dfrac{\sec \theta}{\tan \theta} = \csc \theta$

7. $\tan x \cos x \csc x = 1$

8. $\cot y \sec y = 1/\sin y$

9. $\cos^2 x - \sin^2 x = 2 \cos^2 x - 1$

10. $\sin^2 t - \cos^2 t = 2 \sin^2 t - 1$

11. $\sin x \cos x (\cot x + \tan x) = 1$

12. $\sec^2 x + \csc^2 x = \sec^2 x \csc^2 x$

13. $\csc^2 A \cos^2 A = \csc^2 A - 1$

14. $\cot A - \cot A \cos^2 A = \sin A \cos A$

15. $\sin^2 x \cot^2 x = 1 - \sin^2 x$

16. $\sec \phi - \sin \phi \tan \phi = \cos \phi$

17. $\tan^2 y + \csc^2 y = \cot^2 y + \sec^2 y$

18. $\tan \phi + \cot \phi = \tan \phi \csc^2 \phi$

19. $\sin \theta \sec \theta + \cot \theta = \sec \theta \csc \theta$

20. $\csc \theta (\sec \theta - \cos \theta) = \tan \theta$

21. $\dfrac{\cos^2 A}{1 - \sin A} = 1 + \sin A$

22. $\dfrac{\sin^2 A}{1 - \cos A} = \cot A (\sin A + \tan A)$

23. $\dfrac{\tan x}{\sec x + 1} = \dfrac{\sec x - 1}{\tan x}$

24. $\dfrac{\tan^2 x - 1}{\tan^2 x + 1} = \sin^2 x - \cos^2 x$

25. $1 - \sin^4 \theta = \cos^2 \theta + \sin^2 \theta \cos^2 \theta$

26. $\cos^4 \theta - \sin^4 \theta = 1 - 2 \sin^2 \theta$

27. $\dfrac{\tan x - \cot x}{\sin x \cos x} = \sec^2 x - \csc^2 x$

28. $\dfrac{\sin y}{1 - \sin y} + \dfrac{\sin y}{1 + \sin y} = 2 \tan y \sec y$

29. Prove Example 12-4 by changing the right side to the left side.

30. Prove Example 12-5 by changing the left side to sines and cosines.

31. Prove that each of the following are *not* identities by letting $A = 30°$ and $B = 45°$ and evaluating each side: (a) $\sin (A + B) \neq \sin A + \sin B$, (b) $2 \sin A \neq \sin 2A$, (c) $(\tan B)/2 \neq \tan (B/2)$.

**32.** Do the same as in problem 31 for $A = 0.6$ rad and $B = 1.0$ rad.

**33.** In calculating the moment about a point, the following expression arises: $\sqrt{2.0(1.0 + \cos \alpha)} - (1.0 + \cos \alpha)^2$. Show that the expression is equal to $\sin \alpha$.

**34.** The angle $\phi$ of the acceleration vector for a certain object in circular motion is given by $\tan \phi = (\sin \theta + \cos \theta)^2 - 2.0(\sin \theta \cos \theta + 0.5)$. Simplify the expression and demonstrate that $\phi = 0$ or $\pi$ by showing that $\tan \phi = 0$.

**35.** Refraction causes a submerged object at an actual depth $d$ to appear to be at a depth $d'$, which can be expressed as:

$$\frac{d}{d'} = \sqrt{\frac{\mu^2 - \sin^2 \theta_i}{\cos^2 \theta_i}}$$

where $\mu$ = index of refraction of the liquid and $\theta_i$ = angle of incidence. When $d = 5.0$ m and $d' = 3.0$ m, show that $\cos \theta_i = 0.75\sqrt{\mu^2 - 1}$. (See Fig. 8-4.)

**36.** The displacement $x$ of a piston is given by:

$$x = \sqrt{0.25 - 0.25 \sin^2 \omega t + 2.0 \cos^2 \omega t}$$

Show that $x = 1.5 \cos \omega t$.

**37.** The following expression results from a problem in electric circuit theory:

$$\frac{(1.2 \sin \omega t - 1.6 \cos \omega t)^2 + (1.6 \sin \omega t + 1.2 \cos \omega t)^2}{2L}$$

Show that this expression can be simplified to $2.0/L$.

**38.** Show that the following expression from a problem in fluid flow:

$$\sqrt{\frac{\csc \phi_x - 1}{\csc \phi_x + 1}}$$

is equal to $(\csc \phi_x - 1)/\cot \phi_x$. (*Hint:* Multiply the top and bottom of the fraction.)

**39.** Show that the parametric equations $x = a \cos t$ and $y = b \sin t$ are the equations of an ellipse (Sec. 14-4):

$$\frac{x^2}{a^2} + \frac{y^2}{b^2} = 1$$

(*Hint:* Substitute for $x$ and $y$.)

**40.** Using formula (16-11) for the sum of an infinite geometric series (see Sec. 16-2), show that $\tan^2 x = \sin^2 x + \sin^4 x + \sin^6 x + \cdots$.

**41.** In calculus the formula for the derivative of $\tan u$ results from the expression:

$$\frac{(\cos u)(\cos u) - (\sin u)(-\sin u)}{(\cos u)^2}$$

Show that this expression equals $\sec^2 u$.

**42.** In evaluating a certain integral in calculus, the following expression arises:

$$\frac{16 \sec^2 \theta}{[(2 \tan \theta)^2 + 4]^2}$$

Show that this expression equals $\cos^2 \theta$. (*Hint:* Factor the denominator.)

# 12-2 | *Sum and Difference Identities*

Identities involving the sum and the difference of two angles are very useful in geometric applications, electric circuit theory, wave mechanics, and calculus. In addition, they lead to further important identities involving double angles and half angles.

The six *sum and difference identities* are:

$$\sin (A \pm B) = \sin A \cos B \pm \cos A \sin B \qquad (12\text{-}4)$$

$$\cos (A \pm B) = \cos A \cos B \mp \sin A \sin B \qquad (12\text{-}5)$$

$$\tan (A \pm B) = \frac{\tan A \pm \tan B}{1 \mp \tan A \tan B} \qquad (12\text{-}6)$$

The top signs on the left of each identity belong with the top signs on the right, and likewise for the bottom signs. These formulas can be used to find the trigonometric functions of $A + B$ or $A - B$, without a table or calculator, when the functions of $A$ and $B$ are known as follows.

### Example 12-6
Given $\sin A = \frac{3}{5}$, $\cos B = \frac{5}{13}$, and angles $A$ and $B$ in the first quadrant.

**1.** Find $\sin (A + B)$, using identity (12-4).
**2.** Find $\cos (A + B)$, using identity (12-5).
**3.** Find $\tan (A + B)$, using identity (12-6).
**4.** In what quadrant is angle $A + B$?

### Solution
**1.** Find $\sin (A + B)$, using identity (12-4).

First find $\cos A$ and $\sin B$. Using reference triangles in the first quadrant and the pythagorean theorem (Fig. 12-2), $\cos A = 4/5$ and $\sin B = 12/13$. Then substitute the values in (12-4):

$$\sin (A + B) = \sin A \cos B + \cos A \sin B$$

$$= \left(\frac{3}{5}\right)\left(\frac{5}{13}\right) + \left(\frac{4}{5}\right)\left(\frac{12}{13}\right) = \frac{15}{65} + \frac{48}{65} = \frac{63}{65}$$

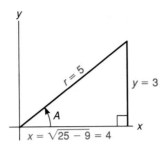 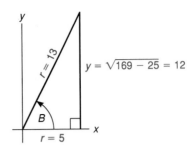

FIG. 12-2 *Example 12-6*

**2.** Find cos $(A + B)$, using identity (12-5):

$$\cos (A + B) = \cos A \cos B - \sin A \sin B$$

$$= \left(\frac{4}{5}\right)\left(\frac{5}{13}\right) - \left(\frac{3}{5}\right)\left(\frac{12}{13}\right) = -\frac{16}{65}$$

**3.** Find tan $(A + B)$, using identity (12-6).

First find tan $A = 3/4$ and tan $B = 12/5$. Then:

$$\tan (A + B) = \frac{\tan A + \tan B}{1 - \tan A \tan B}$$

$$= \frac{(3/4) + (12/5)}{1 - (3/4)(12/5)} = \frac{(3/4) + (12/5)}{1 - (36/20)}$$

Multiply each term in the top and bottom of the fraction by the lcd = 20:

$$= \frac{15 + 48}{20 - 36} = -\frac{63}{16}$$

You can also find tan $(A + B)$ by using the identity:

$$\tan (A + B) = \frac{\sin (A + B)}{\cos (A + B)} = \frac{63/65}{-16/65} = -\frac{63}{16}$$

You can check the results in (1), (2), and (3) with a calculator as follows. First find angles $A$, $B$, and $A + B$. Then find sin $(A + B)$, cos $(A + B)$, and tan $(A + B)$ and compare with above.

**4.** In what quadrant is angle $A + B$? Since sin $(A + B)$ is positive and cos $(A + B)$ is negative, angle $A + B$ is in the second quadrant.

## *Proof of Formula for sin (A + B)*

Consider Fig. 12-3, in which $A$ and $B$ are acute angles shown at the origin. The hypotenuse of the right triangle containing angle $A + B$ is drawn equal to 1. The angle in the small right triangle (upper right) is equal to $A$, since it is complementary to angle $\theta$, to which $A$ is complementary. Then:

$$\sin A = \frac{p}{OM} \qquad \cos B = \frac{OM}{1} \qquad \sin B = \frac{r}{1}$$

and in the small right triangle $\cos A = q/r$. It follows that:

$$\sin A \cos B + \cos A \sin B = \left(\frac{p}{OM}\right)\left(\frac{OM}{1}\right) + \left(\frac{q}{r}\right)\left(\frac{r}{1}\right) = p + q$$

But $\sin (A + B) = (p + q)/1 = p + q$. Therefore:

$$\sin (A + B) = \sin A \cos B + \cos A \sin B$$

The formula for $\cos (A + B)$ can be derived in a similar way. Both formulas can be extended for angles in any quadrant. By using identities (10-3), $\sin (-\theta) = -\sin \theta$ and $\cos (-\theta) = \cos \theta$ (problem 42, Exercise 10-2), the formula for $\sin (A - B)$ follows from that for $\sin (A + B)$ by substituting $-B$ for $B$:

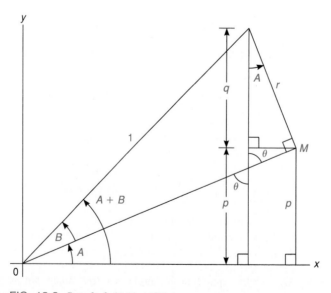

FIG. 12-3 *Proof of sin (A + B)*

$$\sin [A + (-B)] = \sin A \cos(-B) + \cos A \sin(-B)$$
$$= \sin A(\cos B) + (\cos A)(-\sin B)$$
$$= \sin A \cos B - \cos A \sin B$$

The formula for $\cos (A - B)$ can be shown in a similar way. The formula for $\tan (A + B)$ is left as problem 21.

Using the values for the special angles $30°$, $45°$, and $60°$ from Table 10-1 and the sum and the difference formulas, you can find the trigonometric functions of $15°$ and $75°$ without a table or calculator as follows.

### Example 12-7

Using the formula for $\cos (A + B)$, find $\cos 75°$ without a table or calculator in radical form and as a decimal.

### Solution

$$\cos 75° = \cos (30° + 45°) = \cos 30° \cos 45° - \sin 30° \sin 45°$$
$$= \left(\frac{\sqrt{3}}{2}\right)\left(\frac{\sqrt{2}}{2}\right) - \left(\frac{1}{2}\right)\left(\frac{\sqrt{2}}{2}\right)$$
$$= \frac{\sqrt{6} - \sqrt{2}}{4}$$

or
$$\frac{2.45 - 1.41}{4} = 0.259$$

### Example 12-8

Prove the identity:     $\tan (x - \pi) = \tan x$

### Solution
Use (12-6):

$$\tan (x - \pi) = \frac{\tan x - \tan \pi}{1 + \tan x \tan \pi} = \frac{\tan x - (0)}{1 + \tan x(0)} = \tan x$$

The sum and difference formulas can be used to prove certain sum and product identities that are useful in calculus and other advanced applications. One of these is shown in the next example, two more in problems 19 and 20, and the fourth in problem 18, Section 12-5.

### Example 12-9

**1.** Prove $2 \cos A \sin B = \sin (A + B) - \sin (A - B)$.

**2.** Using the result from (1), show that:

$$\sin X - \sin Y = 2 \cos \tfrac{1}{2}(X + Y) \sin \tfrac{1}{2}(X - Y)$$

### Solution

**1.** Prove $2 \cos A \sin B = \sin (A + B) - \sin (A - B)$.
Change the right side:

$$\sin (A + B) - \sin (A - B) = \sin A \cos B + \cos A \sin B$$
$$- (\sin A \cos B - \cos A \sin B) = 2 \cos A \sin B$$

**2.** Using the result from (1), show that:

$$\sin X - \sin Y = 2 \cos \tfrac{1}{2}(X + Y) \sin \tfrac{1}{2}(X - Y)$$

Let $A + B = X$ and $A - B = Y$. Then $\tfrac{1}{2}(X + Y) = \tfrac{1}{2}(A + B + A - B)$
$= 2A/2 = A$ and $\tfrac{1}{2}(X - Y) = B$. Substitute into (1):

$$\sin X - \sin Y = 2 \cos \frac{1}{2}(X + Y) \sin \frac{1}{2}(X - Y)$$

**Exercise 12-2**    Do problems 1 to 16, using identities (12-4), (12-5), and (12-6) without a table or calculator except to check results.

1. Given $\sin A = \frac{8}{17}$, $\cos B = \frac{3}{5}$, and $A$ and $B$ in the first quadrant. Find *(a)* $\sin (A - B)$, *(b)* $\cos (A - B)$, *(c)* $\tan (A - B)$; *(d)* tell which quadrant angle $(A - B)$ is in.

2. Given $\tan A = -\frac{3}{4}$, $A$ in the second quadrant, and $\sin B = \frac{5}{13}$, $B$ in the first quadrant. Find *(a)* $\sin (A + B)$, *(b)* $\cos (A + B)$, *(c)* $\tan (A + B)$; *(d)* tell which quadrant angle $(A + B)$ is in.

3. Given $\tan A = 2.5$, $\tan B = 0.40$, and $A$ and $B$ in the first quadrant. Find $\tan (A - B)$. *(b)* Using a calculator or table, find angle $A$, angle $B$, and $\tan (A - B)$. Check that the last result agrees with the result in *(a)*.

4. Given $\sin A = \frac{4}{5}$, $\sin (A + B) = \frac{63}{65}$, $\sin (A - B) = \frac{33}{65}$, and $A$ and $B$ in the first quadrant. *(a)* Find $\sin B$ and $\cos B$. *(Hint:* Solve the two equations simultaneously.) *(b)* Using a calculator or table, find angle $A$, angle $(A + B)$, angle $B$, and then $\sin B$ and $\cos B$. Check that the last two results agree with those in *(a)*.

In problems 5 to 10 use Table 10-1 to find each value in radical form and as a decimal. Check with a table or calculator. (See Example 12-7.)

5. $\sin 75°$                    6. $\cos 15°$

7. $\tan 105°$                8. $\cot 195°$

9. $\cos \dfrac{7\pi}{12}$            10. $\sec \dfrac{5\pi}{12}$

In problems 11 to 16, prove each identity.

11. $\sin \left( \theta + \dfrac{\pi}{2} \right) = \cos \theta$        12. $\sin (\theta - \pi) = -\sin \theta$

13. $\cos (\theta - \pi) = -\cos \theta$        14. $\cos (\theta + 2\pi) = \cos \theta$

15. $\tan (\theta + \pi) = \tan \theta$          16. $\tan (\theta - \pi) = \tan \theta$

17. Show that sin $(A + B)$ is *not* the same as sin $A$ + sin $B$ by using a calculator or table and letting $A = 20°$ and $B = 40°$.

18. Do the same as in problem 17 for sin $AB$ and (sin $A$)(sin $B$) by letting $A = 0.50$ rad and $B = 0.60$ rad.

19. *(a)* Prove 2 sin $A$ cos $B$ = sin $(A + B)$ + sin $(A - B)$. *(b)* Using the formula in *(a)*, show that sin $X$ + sin $Y$ = 2 sin ½$(X + Y)$ cos ½ $(X - Y)$. (See Example 12-9.)

20. *(a)* Prove 2 cos $A$ cos $B$ = cos $(A + B)$ + cos $(A - B)$. *(b)* Using the formula in *(a)*, show that cos $X$ + cos $Y$ = 2 cos ½ $(X + Y)$ cos ½ $(X - Y)$. (See Example 12-9.)

21. *(a)* By expressing tan $(A + B)$ as sin $(A + B)$/cos $(A + B)$, prove identity (12-6) for tan $(A + B)$. (*Hint:* Divide the top and bottom by cos $A$ cos $B$.)

22. Derive the identity for:

$$\cot (A \pm B) = \frac{\cot A \cot B \mp 1}{\cot B \pm \cot A}$$

from the identity for tan $(A \pm B)$.

In problems 23 to 28, prove each identity.

23. cos $(A - B)$ = cos $(B - A)$

24. tan $(A - B)$ = $-$tan $(B - A)$

25. sin $(A + A)$ = sin $2A$ = 2 sin $A$ cos $A$

26. cos $(A + A)$ = cos $2A$ = $\cos^2 A - \sin^2 A$

27. sin $(x + y)$ sin $x$ + cos $(x + y)$ cos $x$ = cos $y$

28. cos $(x + y)$ cos $(x - y)$ = $\cos^2 x - \sin^2 y$

29. A spring vibrating in harmonic motion described by the equation $y_1 = A_1 \cos (\omega t + \pi)$ is subjected to another harmonic motion described by $y_2 = A_2 \cos (\omega t - \pi)$. Show that the resultant motion $y_1 + y_2 = -(A_1 + A_2) \cos \omega t$ by using identity (12-5).

30. Show that if a light wave described by $y_1 = \sin (x - 3\pi/2)$ meets another light wave $y_2 = \sin (x + 3\pi/2)$, they will cancel each other. Find $y_1 + y_2$ by using identity (12-4).

31. The current in an ac circuit is given by $I = I_{max} \sin (120\pi t + \pi/3) - I_{max} \cos (120\pi t + \pi/6)$, where $I_{max}$ = maximum current. Express $I$ in terms of functions of the angle $120\pi t$.

32. The instantaneous power in an ac circuit is given by $P = [V_{max} \sin (\omega t + \phi)] \times [I_{max} \sin (\omega t - \phi)]$. Show that $P = V_{max}I_{max} (\sin^2 \omega t - \sin^2 \phi)$.

33. When a light ray passes through a pane of glass, the displacement is given by $d = t \sin (\theta_i - \theta_r)/\cos \theta_r$, where $t$ = thickness, $\theta_i$ = angle of incidence, and $\theta_r$ = angle of refraction. Show that $d = t (\sin \theta_i - \cos \theta_i \tan \theta_r)$.

34. In Sec. 11-3 a composite trigonometric function $y = a \sin x + b \cos x$ is shown to be a sinusoidal function $y = A \sin (x + \phi)$, $a > 0$. By applying (12-4) show that the two functions are equivalent where tan $\phi = b/a$ and $A = \sqrt{a^2 + b^2}$.

**35.** In analytic geometry the angle between two lines whose slopes are $m_1 = \tan \theta_1$ and $m_2 = \tan \theta_2$ is found using identity (12-6):

$$\tan (\theta_2 - \theta_1) = \frac{\tan \theta_2 - \tan \theta_1}{1 + \tan \theta_2 \tan \theta_1} = \frac{m_2 - m_1}{1 + m_2 m_1}$$

Find $\theta_2 - \theta_1$ for lines $l_1$: $y = 2x + 1$ ($m_1 = 2$) and $l_2$: $y = -x + 3$ ($m_2 = -1$).

**36.** Show that the formula in problem 35 does not apply to lines $l_1$: $y = -2x + 1$ ($m_1 = -2$) and $l_2$: $y = \frac{1}{2}x + 2$ ($m_2 = \frac{1}{2}$). Can you explain why?

# 12-3 | *Double- and Half-Angle Identities*

The following double- and half-angle identities are very useful for working with trigonometric functions.

## *Double-Angle Identities*

$$\sin 2A = 2 \sin A \cos A \tag{12-7}$$
$$\cos 2A = \cos^2 A - \sin^2 A$$

or
$$= 2 \cos^2 A - 1 \tag{12-8}$$

or
$$= 1 - 2 \sin^2 A$$

$$\tan 2A = \frac{2 \tan A}{1 - \tan^2 A} \tag{12-9}$$

The double-angle identities are special cases of the sum identities, when $A = B$. For example, for identity (12-8):

$$\cos 2A = \cos (A + A) = (\cos A)(\cos A) - (\sin A)(\sin A)$$
$$= \cos^2 A - \sin^2 A$$

By using the identity $\sin^2 A + \cos^2 A = 1$, the above identity can be changed to either of the other two forms shown for $\cos 2A$ (left as problems 17 and 18).

### Example 12-10

Given $\sin A = -0.60$ and $A$ in quadrant III, find each of the following by using the double-angle identities.

**1.** $\sin 2A$     **2.** $\cos 2A$     **3.** $\tan 2A$

**4.** Tell which quadrant angle $2A$ is in and check the results in (1), (2), and (3) with a calculator or table by finding angle $A$.

## Solution

**1.** $\sin 2A$

Using a reference triangle in the third quadrant, find $\cos A = -0.80$. Then:

$$\sin 2A = 2 \sin A \cos A = 2(-0.60)(-0.80) = 0.96$$

**2.** $\cos 2A$

Any of the three forms of identity (12-8) can be used:

$$\cos 2A = \cos^2 A - \sin^2 A = (-0.80)^2 - (0.60)^2 = 0.28$$

**3.** $\tan 2A$

Find $\tan A = 0.75$. Then:

$$\tan 2A = \frac{2(0.75)}{1 - (0.75)^2} = 3.43$$

or

$$\tan 2A = \frac{\sin 2A}{\cos 2A} = \frac{0.96}{0.28} = 3.43$$

**4.** Tell which quadrant angle $2A$ is in and check the results in (1), (2), and (3) with a calculator or table by finding angle $A$.

Since $\sin 2A$ and $\cos 2A$ are both positive, angle $2A$ is in quadrant I. To check the results, find the reference angle $= \sin^{-1} 0.60 = 37°$ and $A = 180° + 37° = 217°$. Then $2A = 434°$, and the results can be verified. For example, $434°$ is in the first quadrant, and $\sin 434° = \sin 74° = 0.96$.

## Example 12-11
Prove the identity:    $\sin 4x = 4 \sin x \cos^3 x - 4 \sin^3 x \cos x$

## Solution
Change the left side, using identities (12-7) and (12-8):

$$\sin 4x = \sin 2(2x) = 2(\sin 2x)(\cos 2x)$$
$$= 2(2 \sin x \cos x)(\cos^2 x - \sin^2 x)$$
$$= 4 \sin x \cos^3 x - 4 \sin^3 x \cos x$$

## Example 12-12
Prove the identity:    $\tan\left(\theta + \dfrac{\pi}{4}\right) = \dfrac{1 + \sin 2\theta}{\cos 2\theta}$

## Solution

Change the left side, using identity (12-6):

$$\tan\left(\theta + \frac{\pi}{4}\right) = \frac{\tan\theta + \tan(\pi/4)}{1 - \tan\theta\tan(\pi/4)} = \frac{\tan\theta + 1}{1 - \tan\theta}$$

Then change to sines and cosines and simplify:

$$= \frac{[(\sin\theta/\cos\theta) + 1]\ (\cos\theta)}{[1 - (\sin\theta/\cos\theta)]\ (\cos\theta)} = \frac{\sin\theta + \cos\theta}{\cos\theta - \sin\theta}$$

Multiply the top and bottom by $\cos\theta + \sin\theta$ to produce the answer:

$$\frac{(\cos\theta + \sin\theta)\ (\cos\theta + \sin\theta)}{(\cos\theta - \sin\theta)\ (\cos\theta + \sin\theta)} = \frac{\sin^2\theta + \cos^2\theta + 2\sin\theta\cos\theta}{\cos^2\theta - \sin^2\theta}$$

$$= \frac{1 + \sin 2\theta}{\cos 2\theta} \qquad \text{by (12-7) and (12-8)}$$

## Half-Angle Identities

$$\sin\frac{x}{2} = \pm\sqrt{\frac{1 - \cos x}{2}} \qquad\qquad (12\text{-}10)$$

$$\cos\frac{x}{2} = \pm\sqrt{\frac{1 + \cos x}{2}} \qquad\qquad (12\text{-}11)$$

$$\tan\left(\frac{x}{2}\right) = \frac{1 - \cos x}{\sin x} \quad\text{or}\quad \frac{\sin x}{1 + \cos x} \qquad (12\text{-}12)$$

Identities (12-10) and (12-11) are obtained from the double-angle identity (12-8) by writing it in a different form. If you let $2A = x$ and $A = x/2$ in the identity $\cos 2A = 1 - 2\sin^2 A$, then:

$$\cos x = 1 - 2\sin^2\frac{x}{2}$$

and solving for $\sin x/2$ yields:

$$2\sin^2\frac{x}{2} = 1 - \cos x$$

$$\sin\frac{x}{2} = \pm\sqrt{\frac{1 - \cos x}{2}} \qquad \text{(take square roots)}$$

Formulas (12-11) and (12-12) are left as problems 31 and 32.

## Example 12-13

Given $\cos x = -\frac{7}{9}$ and angle $x$ in quadrant II, find each of the following by using the half-angle identities (12-10) to (12-12).

**1.** $\sin \dfrac{x}{2}$     **2.** $\cos \dfrac{x}{2}$     **3.** $\tan \dfrac{x}{2}$

**4.** Check the results directly with a calculator or table.

## Solution

**1.** $\sin \dfrac{x}{2}$

You must choose the correct sign when using identity (12-10) or (12-11). Since half of a second-quadrant angle is in the first quadrant, the sign is positive:

$$\sin \frac{x}{2} = \sqrt{\frac{1 - \cos x}{2}} = \sqrt{\frac{1 + \frac{7}{9}}{2}} = \sqrt{\frac{8}{9}}$$

$$= \frac{2\sqrt{2}}{3} \quad \text{or} \quad 0.943$$

**2.** $\cos \dfrac{x}{2}$

$$\cos \frac{x}{2} = \sqrt{\frac{1 + \cos x}{2}} = \sqrt{\frac{1 - \frac{7}{9}}{2}} = \frac{1}{3} \quad \text{or} \quad 0.333$$

**3.** $\tan \dfrac{x}{2}$

First find $\sin x = \sqrt{32}/9 = 4\sqrt{2}/9$. Then:

$$\tan \frac{x}{2} = \frac{1 - \cos x}{\sin x} = \frac{1 + 7/9}{4\sqrt{2}/9} = \frac{16}{4\sqrt{2}} = \frac{4}{\sqrt{2}} = 2\sqrt{2} \quad \text{or} \quad 2.83$$

$$\text{or} \quad \tan \frac{x}{2} = \frac{\sin (x/2)}{\cos (x/2)} = \frac{2\sqrt{2}/3}{1/3} = 2\sqrt{2} \quad \text{or} \quad 2.83$$

**4.** Check the results directly with a calculator or table.

Angle $x = \cos^{-1}(-7/9) = 141°$. Then $x/2 = 70.5°$, $\sin 70.5° = 0.943$, and so on.

The trigonometric functions of many angles can be calculated by using the special angles 30°, 45°, and 60° and the half-angle identities repeatedly, as shown in the next example.

### Example 12-14
Find cos 7.5°, using identity (12-11) and the special angle 30°.

### Solution
Use the formula twice:

$$\cos 15° = \cos \frac{30°}{2} = \sqrt{\frac{1 + \cos 30°}{2}} = \sqrt{\frac{1 + 0.866}{2}} = 0.966$$

$$\cos 7.5° = \cos \frac{15°}{2} = \sqrt{\frac{1 + \cos 15°}{2}} = \sqrt{\frac{1 + 0.966}{2}} = 0.991$$

Check this result directly, using your calculator or a table.

**Exercise 12-3**

Do problems 1 to 16 without a table or calculator unless indicated. In problems 1 to 4, find each of the following by using the double-angle identities: *(a)* sin 2A; *(b)* cos 2A; *(c)* tan 2A; and *(d)* tell which quadrant angle 2A is in. Check your results directly with a calculator or table by finding angle A. (See Example 12-10.)

**1.** cos A = 5/13, angle A in quadrant I

**2.** sin A = −4/5, angle A in quadrant III

**3.** tan A = −0.75, angle A in quadrant IV

**4.** csc A = 2.60, angle A in quadrant II

In problems 5 to 8, find each of the following by using the half-angle identities: *(a)* sin (x/2), *(b)* cos (x/2), *(c)* tan (x/2). *(d)* Check your results directly with a calculator or table by finding angle x.

**5.** sin x = 3/5, angle x in quadrant I

**6.** cos x = 1/9, angle x in quadrant IV (*Note: x/2 is in quadrant II.*)

**7.** sec x = −1.25, angle x in quadrant II

**8.** cot x = 0.75, angle x in quadrant III (*Note: x/2 is in quadrant II.*)

**9.** Given cos x = 4/5, x in quadrant I, find sin 4x by using identity (12-7) twice. Check directly with a calculator or table.

**10.** Given tan x = 24/7, x in quadrant I, find tan (x/4) by using identity (12-12) twice. Check directly with a calculator or table.

In problems 11 to 16, find each of the following, using the half-angle identities and functions of the special angles 30°, 45°, and 60°. Check directly with a calculator or table. (See Example 12-14.)

| | |
|---|---|
| **11.** cos 22.5° | **12.** sin 15° |
| **13.** cot 75° | **14.** tan 22.5° |
| **15.** csc 15° | **16.** sec 7.5° |

In problems 17 to 30, prove each identity.

| | |
|---|---|
| **17.** $\cos 2A = 2 \cos^2 A - 1$ | **18.** $\cos 2A = 1 - 2 \sin^2 A$ |
| **19.** $(\sin x - \cos x)^2 = 1 - \sin 2x$ | **20.** $\cos x(1 - \cos 2x) = \sin x \sin 2x$ |

**21.** $\tan 2x = \dfrac{2 \sin x}{2 \cos x - \sec x}$

**22.** $\cot 2x = \dfrac{\cot^2 x - 1}{2 \cot x}$

**23.** $\tan x(1 + \cos 2x) = \sin 2x$

**24.** $2 \csc 2x = \tan x + \cot x$

**25.** $\cos 3x = 4 \cos^3 x - 3 \cos x$

**26.** $\sin 3x = 3 \sin x - 4 \sin^3 x$

**27.** $\tan \dfrac{x}{2} + \cot \dfrac{x}{2} = 2 \csc x$

**28.** $\cos 4x = 1 - 8 \sin^2 x \cos^2 x$

**29.** $\tan \left( \theta - \dfrac{\pi}{4} \right) = \dfrac{\sin 2\theta - 1}{\cos 2\theta}$

**30.** $\tan \left( \theta + \dfrac{\theta}{2} \right) = \dfrac{\sin 2\theta + \sin \theta}{\cos 2\theta + \cos \theta}$

**31.** Using the derivation of $\sin (x/2)$ in the text as a guide, derive the identity for $\cos (x/2)$ from $\cos 2A = 2 \cos^2 A - 1$.

**32.** Starting with $\tan (x/2) = \sin (x/2)/\cos (x/2)$, derive identity (12-12) for $\tan (x/2)$. (*Hint:* Multiply the top and bottom of the radicand by $1 - \cos x$ or $1 + \cos x$.)

**33.** Show that the path of a projectile described by the parametric equations $x = \sin t$ and $y = \cos 2t$ is the parabola $y = 1 - 2x^2$.

**34.** A cable vibrates with decreasing amplitude, given by the equation $y = \sqrt{e^{-2x}(1 + \sin 2x)}$. Show that $y = e^{-x}(\sin x + \cos x)$.

**35.** In an ac circuit containing a reactance, the instantaneous power is given by $P = V_{max} I_{max} \cos \omega t \sin \omega t$. Show that $P = (V_{max} I_{max}/2) \sin 2\omega t$.

**36.** The instantaneous power in an ac circuit containing a resistive load is given by $P = 2I^2R \cos^2 (\omega t + \alpha)$. Show that $P = I^2R \; [(1 + \cos (2\omega t + 2\alpha)]$. [*Hint:* $\cos (2\omega t + 2\alpha) = \cos 2(\omega t + \alpha)$.]

**37.** If a light ray passes through a triangular prism and the angle of incidence equals the angle of emergence, then $\sin \theta_i = \mu \sin (\alpha/2)$, where $\theta_i = $ angle of incidence, $\mu = $ index of refraction, and $\alpha = $ apex angle of the prism. Show that $\cos \alpha = 1 - 0.89 \sin^2 \theta_i$ when $\mu = 1.5$.

**38.** In problem 37, the angle that the incident ray makes with the emerging ray is called the *minimum deviation angle D*. It is related to $\alpha$ by:

$$\mu = \frac{\sin (\alpha + D/2)}{\sin (\alpha/2)}$$

When $\alpha = D$, show that $\mu = \sqrt{2 + 2 \cos \alpha}$.

**39.** The approximate discharge $Q$ in a triangular weir (a flow control for liquid in an open channel) is related to the pressure head $H$ by:

$$2.22C (\sin \theta)H^{5/2} - Q \cos \theta - Q = 0$$

where $\theta$ is the angle of the channel and $C$ is a correction factor. Show that $Q = 2.22C \tan (\theta/2) H^{5/2}$.

**40.** In computing the probability of measurement errors in celestial mechanics, the following formula is encountered:

$$\sigma_x^2 = \frac{\sigma^2}{1 - \cos \alpha}$$

Using identity (12-10), show that this formula can be expressed as

$$\sigma_x = \frac{\sigma\sqrt{2}}{2}\,\csc\frac{\alpha}{2}$$

**41.** Find tan 15° without a table or calculator (*a*) using identity (12-6) for tan (60° − 45°) and (*b*) using identity (12-12) for tan (30°/2). (*c*) Show that the results agree *exactly* by expressing each in simplest radical form.

**42.** Find cos 3*A*, given cos *A* = ⅗, without a table or calculator.

**43.** Find sin (2*A* + 3π/4), given sin *A* = ⅘ and angle *A* in quadrant I, without a table or calculator.

**44.** Find cos (2*A* − π/3), given cos *A* = −⅗ and angle *A* in quadrant II, without a table or calculator.

# 12-4 | *Trigonometric Equations*

Chapter 10 contains elementary trigonometric equations where the value of a trigonometric function is given and the angle is found by using an inverse trigonometric function. The methods for reducing more complex trigonometric equations to this elementary form include those for solving algebraic equations (linear, quadratic, and higher degree) and those for proving identities. The following examples illustrate basic types of these equations. All *solutions are for nonnegative angles less than 2π or 360°*.

**Example 12-15**
Solve for θ:     $\tan^2\theta - 3 = 0$ when $0 \le \theta < 2\pi$.

**Solution**
This equation is a simple quadratic in tan θ. First solve for tan θ and then find θ:

$$\tan^2\theta = 3$$
$$\tan\theta = \pm\sqrt{3}$$

Since tan θ is positive or negative, θ can be in any quadrant. The reference angle = π/3, or 60°. The four solutions are then:

$$\theta = \frac{\pi}{3}, \frac{2\pi}{3}, \frac{4\pi}{3}, \frac{5\pi}{3}$$

or                    $\theta = 60°, 120°, 240°, 300°$

### Example 12-16
Solve for $x$: $\quad 2\sin^2 x = \sin x + 1, \ 0 \leq x < 2\pi$.

### Solution
This equation is a quadratic in $\sin x$. Solve it by putting the terms on one side and factoring:

$$2\sin^2 x - \sin x - 1 = 0$$

$$(\sin x - 1)(2\sin x + 1) = 0$$

$$\sin x = 1 \quad \text{and} \quad \sin x = -\tfrac{1}{2}$$

$\sin x = 1$ yields one solution between 0 and $2\pi$:

$$x = \pi/2 \quad \text{or} \quad 90°$$

$\sin x = -\tfrac{1}{2}$ yields two more solutions in quadrants III and IV:

$$x = 7\pi/6 \quad \text{or} \quad 210° \qquad x = 11\pi/6 \quad \text{or} \quad 330°$$

Some equations can be solved by factoring when the functions are not all the same. Study the next example.

### Example 12-17
Solve for $t$: $\quad 2\sin 2t = \cos t, \ 0 \leq t < 360°$.

### Solution
You need to make the angles the same (both equal to $t$) to solve this equation. Use identity (12-7) on $\sin 2t$, transpose, and factor:

$$2(2\sin t \cos t) = \cos t$$

$$4\sin t \cos t - \cos t = 0$$

$$(\cos t)(4\sin t - 1) = 0$$

$$\cos t = 0 \quad \text{and} \quad \sin t = 1/4 = 0.250$$

Then $\qquad\qquad t = 90°, 270° \qquad\qquad t = 14.5°, 165.5°$

### Example 12-18
Solve for $\phi$: $\quad \cos 2\phi + 2\cos\phi = 0, \ 0 \leq \phi < 360°$.

### Solution
Make the angles the same by substituting $2\cos^2\phi - 1$ for $\cos 2\phi$ and arrange terms:

$$2\cos^2\phi + 2\cos\phi - 1 = 0$$

This equation cannot be factored. You need to use the quadratic formula to find $\cos \phi$. Substitute $a = 2$, $b = 2$, and $c = -1$:

$$\cos \phi = \frac{-2 \pm \sqrt{(2)^2 - 4(2)(-1)}}{2(2)} = \frac{-2 \pm \sqrt{12}}{4} = \frac{-1 \pm \sqrt{3}}{2}$$

$$\cos \phi = 0.366 \qquad \text{and} \qquad \cos \phi = -1.37$$

Since $-1 \leq \cos \phi \leq +1$, the only acceptable answer is:

$$\cos \phi = 0.366 \qquad \text{and} \qquad \phi = 68.5°, \ 291.5°$$

### Example 12-19

Solve for $y$: $\tan 2y + \cot 2y = 2$, $0 \leq y < 2\pi$.

### Solution

The angles are the same, but you need to make both functions the same:

$$\tan 2y + \frac{1}{\tan 2y} = 2$$

Multiply by $\tan 2y$ and solve by factoring:

$$\tan^2 2y - 2\tan 2y + 1 = 0$$
$$(\tan 2y - 1)^2 = 0$$
$$\tan 2y = 1$$

Since $y$ is the unknown and can be between 0 and $2\pi$, we know $2y$ can be between 0 and $4\pi$:

$$2y = \frac{\pi}{4}, \ \frac{5\pi}{4}, \ \frac{9\pi}{4}, \ \frac{13\pi}{4}$$

Then:

$$y = \frac{\pi}{8}, \ \frac{5\pi}{8}, \ \frac{9\pi}{8}, \ \frac{13\pi}{8}$$

The last example can be solved in at least two ways and requires some manipulation.

### Example 12-20

Solve: $\qquad \dfrac{4 \sin x}{1 + \cos x} = 3 \qquad 0 \leq x < 360°$

**Solution**

*Method 1.*    Change to one function by clearing the fraction and squaring both sides:

$$(4 \sin x)^2 = (3)^2(1 + \cos x)^2$$
$$16 \sin^2 x = 9 + 18 \cos x + 9 \cos^2 x$$
$$16(1 - \cos^2 x) = 9 + 18 \cos x + 9 \cos^2 x$$
$$25 \cos^2 x + 18 \cos x - 7 = 0$$
$$(25 \cos x - 7)(\cos x + 1) = 0$$

$$\cos x = \tfrac{7}{25} = 0.280 \qquad \text{and} \qquad \cos x = -1$$
$$x = 73.7°, \ 286.3° \qquad\qquad\qquad x = 180°$$

Multiplying by the unknown and squaring may introduce extraneous roots, and *you need to check these answers:*

When $x = 73.7°$, $\cos x = \tfrac{7}{25}$ and $\sin x = \tfrac{24}{25}$. Substitute into the original equation:

$$\frac{4(\tfrac{24}{25})}{1 + \tfrac{7}{25}} = 3 \qquad \text{(answer checks)}$$

When $x = 286.3°$, $\cos x = \tfrac{7}{25}$ and $\sin x = -\tfrac{24}{25}$:

$$\frac{4(-\tfrac{24}{25})}{1 + \tfrac{7}{25}} = -3 \qquad \text{(reject)}$$

When $x = 180°$, $\cos x = -1$ and $\sin x = 0$:

$$\frac{4(0)}{1 + (-1)} = \frac{0}{0} \quad \text{or} \quad \text{undefined} \qquad \text{(reject)}$$

The only solution therefore is $x = 73.7°$.

*Method 2.*    You can solve the equation more directly by dividing by 4 and using the identity for $\tan (x/2)$:

$$\frac{\sin x}{1 + \cos x} = \frac{3}{4}$$
$$\tan \frac{x}{2} = \frac{3}{4} = 0.750$$
$$\frac{x}{2} = 36.9°$$
$$x = 73.7°$$

**Exercise 12-4**    In problems 1 to 26, solve each equation for nonnegative angles less than $2\pi$, or $360°$.

1. $2 \cos \theta + 1 = 0$          2. $2 \sin \theta + 2 = 0$

3. $3 \tan \theta = 5$          4. $0.01 \cot \theta = 0.10$

5. $4 \sin^2 x = 3$          6. $25 \cos^2 y - 9 = 0$

7. $5.5 \cot^2 \theta - 22 = 0$          8. $1.5 \tan^2 \theta = 3.0$

9. $3 \cos^2 x - \cos x = 0$          10. $\sin x \tan x = \tan x$

11. $\tan^2 y + 4 \tan y + 3 = 0$          12. $4 \sin^2 y + 5 \sin y = 6$

13. $\sin \phi - \cos \phi = 0$          14. $\tan \phi = \csc \phi$

15. $\cos 2t + 3 \sin t = 2$          16. $\sin^2 t + \cos 2t + \cos t = 0$

17. $2 \sin 2x + \cos x = 0$          18. $2 \sin^2 x = \sin 2x$

19. $4 \sin^2 x + 2 \sin x - 1 = 0$          20. $2 \sec^2 x - 3 = 2 \tan x$

21. $4 \sin^2 x + 8 \cos x = 3$          22. $\tan^2 x + 2 = 2 \tan x$

23. $\tan 2y - 3 \cot 2y = 0$          24. $3 \csc 2y = 4 \sin 2y$

25. $\dfrac{1 - \cos x}{\sin x} = 2$          26. $\tan \dfrac{x}{2} + \cot \dfrac{x}{2} = 4$

In problems 27 to 32, find the angle in radians to the nearest hundredth.

27. The range $R$ of a projectile that is shot at an angle $\alpha$ with the horizontal is given approximately by:

$$v_0^2 \tan \alpha = 0.50R \sec^2 \alpha$$

where $v_0 =$ initial velocity (m/s). At what angle has a projectile been fired if $v_0 = 100$ m/s and $R = 10{,}000$ m?

28. The orbit of a satellite is given by the parametric equations $x = 4.3 \cos u$ and $y = 4.1 \sin u$. At what two positive values of $u$ less than $2\pi$ does $y = x$?

29. In the circuit in Fig. 12-4, which contains a resistance and a reactance, the reactive power is given by $P_R = VI \cos \phi$, and the active power is given by $P_X = VI \sin \phi$, where $\phi$ is the phase angle by which the current lags the voltage. If $P_R = 476$ W and $P_X = 420$ W, what is the value of the phase angle? (*Hint:* Divide equations.)

30. The instantaneous current in one ac circuit is given by $I_1 = 2 \cos \omega t$ and in a second circuit by $I_2 = \cos 2\omega t$. At what positive value of $\omega t$ less than $\pi$ is the sum of these currents equal to 2.00 A?

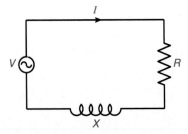

FIG. 12-4 *AC circuit, problem 29*

**31.** The displacement $x$ of a piston is given by $x = \sin \omega t + \frac{1}{2} \sin 2\omega t$. At what positive values of $\omega t$ less than $2\pi$ is the displacement zero?

**32.** The movement of a wheel on an inclined plane gives rise to the equations:

$$\sin A + \sin B = 1$$
$$2 \sin A + \cos B = 1$$

where $A$ and $B$ are angles associated with the plane. Solve these equations simultaneously for $A$ and $B$ less than $\pi/2$.

# 12-5 | *Review Questions*

In problems 1 to 12, prove each identity.

**1.** $\cos \theta \tan \theta = \sin \theta$

**2.** $\dfrac{\cos \theta}{\cot \theta} = \sin \theta$

**3.** $\sec^2 \theta - \tan^2 \theta = \cos^2 \theta + \sin^2 \theta$

**4.** $\sin \theta + \cot \theta \cos \theta = \csc \theta$

**5.** $\dfrac{\sin x}{1 + \cos x} = \dfrac{1 - \cos x}{\sin x}$

**6.** $\cot x (\sec x - 1) = \dfrac{\tan x}{\sec x + 1}$

**7.** $\cos (\pi - \theta) = -\cos \theta$

**8.** $\tan \left(\theta - \dfrac{\pi}{4}\right) = \dfrac{\tan \theta - 1}{\tan \theta + 1}$

**9.** $(\sin x + \cos x)^2 = 1 + \sin 2x$

**10.** $\tan 2x = \dfrac{2 \cot x}{\cot^2 x - 1}$

**11.** $\cot \dfrac{x}{2} = \dfrac{1 + \cos x}{\sin x}$

**12.** $\sin 3x = (\sin x)(4 \cos^2 x - 1)$

Do problems 13 to 18 without a calculator or table.

**13.** Given $\sin A = \frac{4}{5}$, $\cos B = \frac{5}{13}$, and angles $A$ and $B$ in the first quadrant, find *(a)* $\sin (A + B)$, *(b)* $\cos (A - B)$, and *(c)* $\tan (A + B)$.

**14.** Given $\tan A = -0.5$ and angle $A$ in the second quadrant, find *(a)* $\sin 2A$, *(b)* $\cos 2A$, and *(c)* $\tan 2A$.

**15.** Find $\tan 75°$.

**16.** Find $\sin 22.5°$.

**17.** Given $\cos x = -\frac{3}{5}$ and $x$ in quadrant II, find *(a)* $\sin x/2$, *(b)* $\cos x/2$, and *(c)* $\tan x/2$.

**18.** *(a)* Prove $2 \sin A \sin B = \cos (A - B) - \cos (A + B)$. *(b)* Using *(a)*, show that $\cos X - \cos Y = -2 \sin \frac{1}{2}(X + Y) \sin \frac{1}{2}(X - Y)$.

In problems 19 to 24, solve each equation for nonnegative angles less than $2\pi$ or $360°$.

**19.** $1 - 2 \sin \theta = 0$

**20.** $3 \tan^2 x = 1$

**21.** $3 \cos^2 x - \sin^2 x = 0$

**22.** $\sin^2 y + 3 \sin y = 4$

**23.** $\cos 2x = \cos^2 x$

**24.** $4 \cos^2 x + 2 \cos x = 1$

**25.** Simplify the following expression from a problem in light reflection:

$$(1 - \cos\theta)^2 + \sin^2\theta$$

**26.** Show that the following expression from a problem in mechanics can be simplified to $2\sin\omega t$:

$$\sqrt{(3 - 2\cos\omega t)(3 + 2\cos\omega t) - 5}$$

**27.** The voltage in an ac circuit is given by:

$$V = V_{max}\cos\left(60\pi t - \frac{\pi}{3}\right) + V_{max}\cos\left(60\pi t + \frac{\pi}{3}\right)$$

where $V_{max}$ = maximum voltage. Express $V$ in terms of functions of the angle $60\pi t$.

**28.** A vibrating column gives rise to the expression $\sqrt{16(\cos 2x + 1)}$. Show that it is equal to $4\sqrt{2}\cos x$.

**29.** The acceleration of a piston is given by $a = 5.0(\sin\omega t + \cos 2\omega t)$. At what positive values of $\omega t$ less than $2\pi$ does $a = 0$?

**30.** The instantaneous power in an ac circuit containing an inductance is given by $P = VI$, where $V = 120\sin\omega t$ and $I = 5\cos\omega t$. At what positive values of $\omega t$ less than $\pi$ does $P = 150$ W? [*Hint:* Use identity (12-7).]

# 13

# Complex Numbers

## 13-1 | *Imaginary and Complex Numbers*

### *Imaginary Numbers*

The *imaginary unit j* is defined as:

$$j = \sqrt{-1}$$
$$\text{or} \qquad\qquad (13\text{-}1)$$
$$j^2 = (\sqrt{-1})^2 = -1$$

In Chap. 7 square roots of negative numbers are shown to be roots of a quadratic equation (Example 7-39). When square roots of negative numbers first appeared about 500 years ago, they were found to have little application. The word *imaginary* unfortunately was used to describe them. Today imaginary numbers are not just found in dreams but are considered as "real" as real numbers. Imaginary numbers are used in many scientific and technical applications, especially in the theory of alternating current. This chapter explores the system of real and imaginary numbers, called *complex numbers,* and some of their applications.

Note that in mathematics *i* is used to represent the imaginary unit. However, since *i* represents current also, in science and technology *j* is generally used for the imaginary unit. It is also called the *j operator*. Definition (13-1) allows you to simplify square roots of negative numbers by separating *j* as shown in the following example.

### Example 13-1

Simplify and express each radical in terms of *j*.

1. $\sqrt{-4} = \sqrt{4}\sqrt{-1} = 2j$
2. $\sqrt{-0.25} = \sqrt{0.25}\sqrt{-1} = 0.5j$
3. $-\sqrt{-7} = -\sqrt{7}\sqrt{-1} = -\sqrt{7}j$ or $-j\sqrt{7}$
4. $2\sqrt{-8} = 2\sqrt{8}\sqrt{-1} = 2\sqrt{4}\sqrt{2}\sqrt{-1} = 2(2)\sqrt{2}j = 4j\sqrt{2}$

*To simplify square roots of negative numbers, you first separate the imaginary unit:*

$$\sqrt{-x} = j\sqrt{x} \qquad x > 0 \qquad\qquad (13\text{-}2)$$

The rules for radicals (8-9), (8-10) and (8-11), *do not apply* to imaginary numbers except for the basic rule $(\sqrt{-x})^2 = -x \ (x > 0)$, which is the definition of the square root symbol.

### Example 13-2
Simplify each of the following:

**1.** $(\sqrt{-9})^2 = -9$

**2.** $\sqrt{-9}\sqrt{-4} = (3j)(2j) = 6j^2 = 6(-1) = -6$

Notice in (2) that if the numbers under the radicals are multiplied before the separation of $j$, the result would be $\sqrt{(-9)(-4)} = \sqrt{36} = 6$, which is incorrect.

**3.** $\sqrt{-0.2}\sqrt{0.8} = j\sqrt{0.2}\sqrt{0.8} = j\sqrt{(0.2)(0.8)} = j\sqrt{0.16} = 0.4j$

**4.** $\sqrt{-3}\sqrt{-5} = (j\sqrt{3})(j\sqrt{5}) = j^2\sqrt{15} = (-1)\sqrt{15} = -\sqrt{15}$

**5.** $(-5\sqrt{-2})(3\sqrt{-6}) = (-5j\sqrt{2})(3j\sqrt{6}) = -15j^2\sqrt{12}$
$$= -15(-1)(2)\sqrt{3} = 30\sqrt{3}$$

Multiplying more than two imaginary numbers produces higher powers of $j$. However, $j^n$ has the cyclic property shown in Fig. 13-1. To obtain the values of $j^n$, *treat $j$ as any algebraic variable, but replace $j^2$ by $-1$.* For example, $j^3 = j^2 \cdot j = (-1)j = -j$, $j^4 = j^2 \cdot j^2 = (-1)(-1) = 1$, $j^5 = j^4 \cdot j = (1)j = j$, etc. Notice that $j^n = 1$ whenever $n$ is a multiple of 4. The values therefore repeat after $j$ is multiplied by itself 4 times. Any product of imaginary numbers can be simplified to a real number or an imaginary number in terms of $j$.

### Example 13-3
Simplify each of the following.

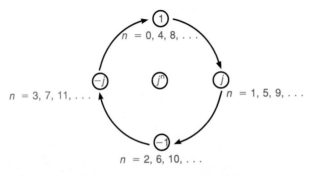

FIG. 13-1 *Cyclic property of $j^n$, Example 13-3*

**1.** $(\sqrt{-4})(\sqrt{-9})(\sqrt{-16}) = (2j)(3j)(4j) = 24j^3 = 24(-j) = -24j$

**2.** $(2j)^5(-j)^4 = (32j^5)(j^4) = (32j)(1) = 32j$

**3.** $(\sqrt{-2})^5(\sqrt{-3})^3 = (j\sqrt{2})^5(j\sqrt{3})^3 = (j^5)(4\sqrt{2})(j^3)(3\sqrt{3})$
$$= (j^8)(12\sqrt{6}) = (1)(12\sqrt{6})$$
$$= 12\sqrt{6}$$

**4.** $j^{18} = (j^{16})(j^2) = (1)(-1) = -1$

## Complex Numbers

An imaginary number combined with a real number is called a *complex number*. It has the form $a + bj$, where $a$ and $b$ are real numbers. Complex numbers include all the numbers you need for mathematical applications, as shown in Fig. 13-2. When $b = 0$, $a + 0j = a$, which is a real number. When $a = 0$, $0 + bj = bj$, which is a *pure imaginary* number. Two complex numbers $a + bj$ and $x + yj$ are equal only if $a = x$ and $b = y$. When multiplying or dividing a complex number by a real or a pure imaginary number, you multiply or divide the real and imaginary parts separately as follows.

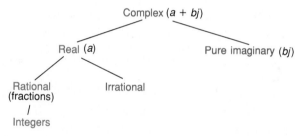

FIG. 13-2  *Number systems*

## Example 13-4
Simplify and express in the complex form $a + bj$.

**1.** $5(2 - 3j) = 10 - 15j$

**2.** $-j(1.5 + j) = -1.5j - j^2 = -1.5j - (-1) = 1 - 1.5j$

**3.** $\dfrac{2 + \sqrt{-8}}{2} = \dfrac{2 + 2j\sqrt{2}}{2} = \dfrac{2}{2} + \dfrac{2j\sqrt{2}}{2} = 1 + j\sqrt{2}$

**4.** $\dfrac{3 + 3j}{j} = \dfrac{3}{j} + \dfrac{3j}{j} = \dfrac{3}{j}\left(\dfrac{j}{j}\right) + 3 = \dfrac{3j}{-1} + 3 = 3 - 3j$

Notice in (4) that $j$ is eliminated from the denominator by multiplying the top and bottom of the fraction by $j$. This is the same method used to eliminate a real radical from a denominator (Chap. 8, Sec. 8-3).

Complex numbers that occur as roots of quadratic and higher-degree equations always occur in conjugate pairs. The *conjugate* of the complex number $a + bj$ is $a - bj$. Study the next example.

### Example 13-5
Solve $5x^2 - 2x + 1 = 0$. Simplify the roots and express as complex conjugates.

### Solution
Use the quadratic formula:

$$x = \frac{-b \pm \sqrt{b^2 - 4ac}}{2a}$$

Substitute $a = 5$, $b = -2$, and $c = 1$:

$$x = \frac{-(-2) \pm \sqrt{(-2)^2 - 4(5)(1)}}{2(5)} = \frac{2 \pm \sqrt{4 - 20}}{10}$$

$$= \frac{2 \pm \sqrt{-16}}{10} = \frac{2 \pm 4j}{10} = \frac{2}{10} \pm \frac{4j}{10} = \frac{1}{5} \pm \frac{2}{5}j \quad \text{or} \quad \frac{1 \pm 2j}{5}$$

The roots $\frac{1}{5} + \frac{2}{5}j$ and $\frac{1}{5} - \frac{2}{5}j$ are complex conjugates.

### Exercise 13-1

Simplify and express problems 1 to 10 in terms of the imaginary unit.

1. $\sqrt{-9}$                     2. $\sqrt{-49}$
3. $\sqrt{-0.04}$                  4. $\sqrt{-1.21}$
5. $-\sqrt{-18}$                   6. $\sqrt{-50}$
7. $2\sqrt{-10}$                   8. $-3\sqrt{-20}$
9. $\sqrt{-1/4}$                   10. $\sqrt{-1/3}$

Simplify each of problems 11 to 28.

11. $(\sqrt{-7})^2$                12. $(2\sqrt{-3})^2$
13. $(\sqrt{-25})(\sqrt{-4})$      14. $(\sqrt{-2})(\sqrt{-8})$
15. $(\sqrt{-0.1})(\sqrt{0.1})$    16. $(\sqrt{-\frac{1}{2}})(\sqrt{\frac{1}{2}})$
17. $(-\sqrt{-5})(\sqrt{-15})$     18. $(-\sqrt{10})(\sqrt{-2})$
19. $(2\sqrt{-4})(-\sqrt{11})$     20. $(1.5\sqrt{3})(1.8\sqrt{-6})$
21. $(-\sqrt{-3})^2(\sqrt{-16})$   22. $(\sqrt{-4})(\sqrt{-25})(\sqrt{-1})$
23. $(\sqrt{-5})^3(-\sqrt{-3})^3$  24. $(2\sqrt{2})(\sqrt{-2})^5$
25. $(-3j)^4$                      26. $(-j^2)(2j)^5$
27. $j^{20}$                       28. $j^{41}$

Simplify and express problems 29 to 36 in the form $a + bj$.

**29.** $3(5 + 2j)$

**30.** $-4(\sqrt{3} - j)$

**31.** $-2j(1 - j)$

**32.** $0.3j(-0.5 + 0.5j)$

**33.** $\dfrac{4 + \sqrt{-12}}{2}$

**34.** $\dfrac{3 - \sqrt{-18}}{15}$

**35.** $\dfrac{1 + j}{j}$

**36.** $\dfrac{2 - \sqrt{-2}}{\sqrt{-2}}$

Solve problems 37 to 44, using the quadratic formula. Simplify the roots and express as complex conjugates.

**37.** $x^2 - 2x + 2 = 0$

**38.** $x^2 + 4x + 8 = 0$

**39.** $x^2 + 9 = 0$

**40.** $x^2 = -25$

**41.** $9x^2 - 12x + 8 = 0$

**42.** $8x^2 + 4x + 1 = 0$

**43.** $x^2 + 3 = 2x$

**44.** $x^2 + 16 = 4x$

**45.** The impedance $Z$ in an ac circuit is given by $Z = V/I$, where $V$ = voltage and $I$ = current. Find $Z$ when $V = 1.5 - 0.5j$ and $I = -1.0j$.

**46.** Find $V$ in problem 45 if $Z = -0.8 + 1.6j$ and $I = 2.0j$.

**47.** What type of a number is *(a)* the sum of a complex number and its conjugate and *(b)* the difference of a complex number and its conjugate?

**48.** What is the result when you divide a complex number $x + yj$ by *(a)* $j$ and *(b)* $-j$?

# 13-2 | *Operations with Complex Numbers*

*Combine complex numbers* by combining the real and imaginary parts separately:

$$(a + bj) + (c + dj) = (a + c) + (b + d)j \qquad (13\text{-}3)$$

### Example 13-6
Combine and express in the form $a + bj$.

**1.** $(2 + 3j) + (1 - 4j) = (2 + 1) + (3 - 4)j = 3 - j$

**2.** $(1.5 + 0.5j) - (2.0 - 2.0j) = (1.5 - 2.0) + (0.5 + 2.0)j$
$$= -0.5 + 2.5j$$

**3.** $(1 - 2j) - (3 - 5j) + (-2 + 3j) = (1 - 3 - 2) + (-2 + 5 + 3)j$
$$= -4 + 6j$$

*Multiply complex numbers* as you would two algebraic binomials, using the FOIL method (Sec. 2-6), and *replace $j^2$ by $-1$:*

$$(a + bj)(c + dj) = ac + adj + bcj + bdj^2 = ac + bd(-1) + adj + bcj$$
$$= (ac - bd) + (ad + bc)j \tag{13-4}$$

### Example 13-7
Multiply and express in the form $a + bj$.

**1.** $(3 + 4j)(2 - j) = 6 - 3j + 8j - 4j^2 = 6 - 4(-1) - 3j + 8j$
$$= 10 + 5j$$

**2.** $(1 + j)^2 = (1 + j)(1 + j) = 1 + 2j + j^2 = 1 + 2j + (-1) = 2j$

**3.** $(x + yj)(x - yj) = x^2 - y^2j^2 = x^2 - y^2(-1) = x^2 + y^2$

Notice in (3) that *the product of two conjugate complex numbers is a real number*.

    *Divide complex numbers* by applying this last idea and multiplying the top and bottom of the fraction by the conjugate of the denominator:

$$\frac{a + bj}{c + dj} \frac{(c - dj)}{(c - dj)} = \frac{(ac + bd) + (bc - ad)j}{c^2 + d^2} \tag{13-5}$$

The above (13-5) is the same procedure as that for rationalizing a denominator containing a binomial radical expression (Sec. 8-4).

### Example 13-8
Divide and express in the form $a + bj$.

**1.** $\dfrac{4 - 2j(1 - j)}{1 + j\ (1 - j)} = \dfrac{4 - 6j + 2j^2}{1 + 1} = \dfrac{2 - 6j}{2} = 1 - 3j$

**2.** $\dfrac{2 + 3j(5 + 4j)}{5 - 4j(5 + 4j)} = \dfrac{10 + 23j + 12j^2}{25 + 16} = \dfrac{-2 + 23j}{41} = \dfrac{-2}{41} + \dfrac{23}{41}j$

**3.** $\dfrac{0.20}{0.70 + 0.60j} \dfrac{(0.70 - 0.60j)}{(0.70 - 0.60j)} = \dfrac{0.14 - 0.12j}{0.49 + 0.36} = \dfrac{0.14 - 0.12j}{0.85}$
$$= 0.16 - 0.14j$$

**Exercise 13-2**

In problems 1 to 38, perform the operations and express answers in the form $a + bj$.

**1.** $(1 - 2j) + (3 + 4j)$                 **2.** $(5 + j) + (3 - 2j)$

**3.** $(3 + 2j) - (3 - 3j)$                **4.** $(4 - 2j) - (6 - 2j)$

**5.** $(5.3 + 9.1j) + (3.1 - 6.4j)$        **6.** $(1.0 - 1.3j) - (0.9 + 0.7j)$

**7.** $(1 - j) - (2 + j) + (1 - 2j)$       **8.** $(5 + 3j) - 2(5 - 3j)$

**9.** $(3 - 2j)(5 + j)$                    **10.** $(2 - 5j)(3 - 4j)$

**11.** $(2 + 2j)^2$

**12.** $(1 - j)^2$

**13.** $(1.4 - 7.1j)(3.2 + 5.6j)$

**14.** $(-0.8 + 0.6j)(0.3 - 0.5j)$

**15.** $(5 + 6j)(5 - 6j)$

**16.** $(-7 + 2j)(-7 - 2j)$

**17.** $(2 + \sqrt{-3})(2 - \sqrt{-3})$

**18.** $(\sqrt{3} + \sqrt{-2})(\sqrt{3} - \sqrt{-2})$

**19.** $(3)(5 - 2j)(1 + 4j)$

**20.** $(6j)(1 - j)(2 + j)$

**21.** $(3 - 3j)^3$

**22.** $(1 + j)^4$

**23.** $\dfrac{5 + 5j}{1 - 2j}$

**24.** $\dfrac{6 - 4j}{1 - j}$

**25.** $\dfrac{1}{2 + 3j}$

**26.** $\dfrac{1}{5 - 4j}$

**27.** $\dfrac{2j}{-1 + j}$

**28.** $\dfrac{10}{3 + j}$

**29.** $\dfrac{0.05}{0.01 + 0.02j}$

**30.** $\dfrac{1.1j}{1.3 - 1.9j}$

**31.** $\dfrac{3 + 2j}{2 + 5j}$

**32.** $\dfrac{4 - 3j}{2 - 7j}$

**33.** $\dfrac{(1 + j)(2 - j)}{2 + 3j}$

**34.** $\dfrac{3 - 2j}{(-2j)(5 + 3j)}$

**35.** $\dfrac{2 - \sqrt{-2}}{2 + \sqrt{-2}}$

**36.** $\dfrac{\sqrt{3} + j}{\sqrt{3} - j}$

**37.** $\dfrac{4}{\sqrt{2} + \sqrt{-2}}$

**38.** $\dfrac{\sqrt{3} - \sqrt{-6}}{\sqrt{-3}}$

**39.** The total impedance $Z$ of an ac circuit containing two impedances $Z_1$ and $Z_2$ in series is given by $Z = Z_1 + Z_2$. Find $Z$ when $Z_1 = 0.15 + 0.20j$ and $Z_2 = 0.05 - 0.15j$.

**40.** The total impedance $Z$ of an ac circuit containing two impedances $Z_1$ and $Z_2$ in parallel is given by:

$$Z = \frac{Z_1 Z_2}{Z_1 + Z_2}$$

Find $Z$ when $Z_1 = 4$ and $Z_2 = 2j$.

**41.** The voltage $V$ in an ac circuit is given by $V = IZ$, where $I =$ current and $Z =$ impedance. Find $V$ when $I = 15.3 + 8.8j$ and $Z = 18.0 - 9.2j$.

**42.** In problem 41 find $I$ when $V = 0.2 - 0.5j$ and $Z = 0.3 + 0.1j$.

**43.** Show that $\sqrt{j} = (1 + j)/\sqrt{2}$ by computing $[(1 + j)/\sqrt{2}]^2$ and obtaining $j$.

**44.** Any quadratic equation $ax^2 + bx + c = 0$ can be factored if the factors are allowed to be complex numbers. Show that the factors of $x^2 - 2x + 2 = 0$ are $[x - (1 + j)][x - (1 - j)] = 0$ by multiplying out the factors. (*Note:* The roots are the negative of the factors: $x = 1 + j$ and $x = 1 - j$.)

# 13-3 | *Complex Vectors and Polar Form*

## *Complex Vectors*

Complex numbers can be represented as vectors, a form in which they are very useful in applications. The point $(x, y)$ and its radius vector are associated with the complex number $x + yj$ as shown in Fig. 13-3. The $x$ axis represents real numbers and the $y$ axis pure imaginary numbers. Figure 13-3 represents what is called the *complex plane*, or *gaussian plane*, after Karl Friedrich Gauss (1777–1855), the foremost German mathematician of the nineteenth century.

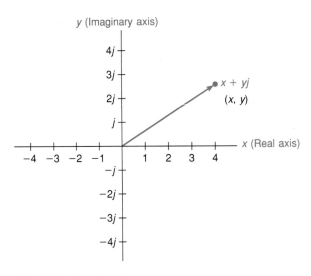

FIG. 13-3  *Complex plane*

### Example 13-9
Graph each complex number as a vector in the complex plane.

**1.** $3 + j$     **2.** $2 - 3j$     **3.** $-1 - j$     **4.** $-4$     **5.** $2j$

### Solution
The vectors are shown in Fig. 13-4. Notice that a real number is a horizontal vector and a pure imaginary number is a vertical vector.

When two complex numbers are added algebraically, the sum is the same as the resultant obtained by adding their two complex vectors.

### Example 13-10
Add the complex numbers $2 + j$ and $1 + 3j$ algebraically and graphically as vectors.

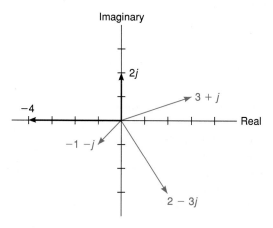

FIG. 13-4  *Complex vectors, Example 13-9*

### Solution

The sum $(2 + j) + (1 + 3j) = 3 + 4j$ is shown graphically in Fig. 13-5. The components of the resultant vector $3 + 4j$ consist of the sum of the $x$ components (real numbers) and the sum of the $y$ components (imaginary numbers). Therefore algebraic addition of the two complex numbers is the same as vector addition.

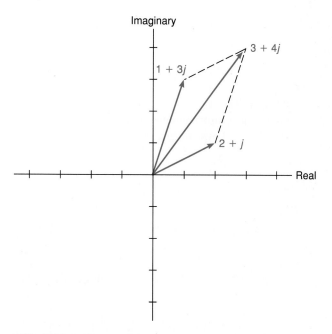

FIG. 13-5  *Adding complex vectors, Example 13-10*

## Polar Form

Complex numbers or vectors can be written in polar form (also called *trigonometric form*) by using the formulas for polar coordinates (11-6). From Fig. 13-6 it follows that:

$$x + yj = r \cos \theta + (r \sin \theta)j = r(\cos \theta + j \sin \theta) \quad \text{or} \quad r \angle \theta \quad (13\text{-}6)$$

The notation $x + yj$ is called *rectangular form,* and $r(\cos \theta + j \sin \theta)$, abbreviated $r \angle \theta$, is called *polar form.* The radius $r$, also called the *modulus* or *absolute value,* is always positive. The angle $\theta$, also called the *argument,* is usually expressed between 0° and 360°. Another notation used for polar form is r cis $\theta$. The $i$ in r cis $\theta$ represents $\sqrt{-1}$ as in the mathematics usage.

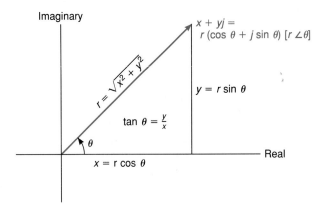

FIG. 13-6  *Polar form of complex vector*

### Example 13-11

Graph each complex number as a vector and change to polar form. The vectors for each are shown in Fig. 13-7.

**1.** $1 + j$     **2.** $-4 + 3j$     **3.** $-2.6j$

**Solution**

**1.** $1 + j$

To change to polar form, use the trigonometric formulas from Fig. 13-6: $r = \sqrt{x^2 + y^2} = \sqrt{1^2 + 1^2} = \sqrt{2}$. Since $\theta$ is in the first quadrant, $\theta = \tan^{-1}(1/1) = 45°$. Then $1 + j = \sqrt{2}(\cos 45° + j \sin 45°)$ or $\sqrt{2} \angle 45°$. This is the same as finding polar coordinates, which is shown in Sec. 11-5.

**2.** $-4 + 3j$

Compute $r = \sqrt{(-4)^2 + (3)^2} = 5$. Angle $\theta$ is in the second quadrant. The

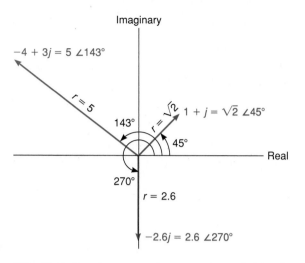

FIG. 13-7 *Complex vectors in rectangular and polar form, Example 13-11*

reference angle = $\tan^{-1} \frac{3}{4} = 37°$ and $\theta = 180° - 37° = 143°$. Then $-4 + 3j = 5(\cos 143° + j \sin 143°)$ or $5 \angle 143°$.

**3.** $-2.6j$

The values of $r$ and $\theta$ can be obtained graphically from Fig. 13-7: $r = 2.6$ and $\theta = 270°$. Then $-2.6j = 2.6 (\cos 270° + j \sin 270°)$ or $2.6 \angle 270°$. This can also be expressed as $2.6 \angle(-90°)$. Negative angles in the fourth quadrant are common in applications of complex numbers.

### Example 13-12

Graph each complex number as a vector and change to rectangular form. The vectors for each are shown in Fig. 13-8.

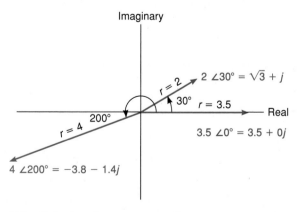

FIG. 13-8 *Complex vectors in polar and rectangular form, Example 13-12*

**1.** $2 \angle 30°$     **2.** $4 \angle 200°$     **3.** $3.5 \angle 0°$

### Solution

**1.** $2 \angle 30°$

To change to rectangular form, write out in polar form and compute $x$ and $y$:
$2(\cos 30° + j \sin 30°) = 2 \cos 30° + 2j \sin 30° = 2(\sqrt{3}/2) + 2j(\frac{1}{2}) = \sqrt{3} + j$.

**2.** $4 \angle 200°$

$$4(\cos 200° + j \sin 200°) = 4 \cos 200° + 4j \sin 200°$$
$$= 4(-0.9397) + 4j(-0.3420)$$
$$= -3.8 - 1.4j \quad \text{(two figures)}$$

**3.** $3.5 \angle 0°$

$3.5 \cos 0° + 3.5j \sin 0° = 3.5(1) + 3.5j(0) = 3.5 + 0j$    or    $3.5$

Polar form is very useful for multiplying and dividing complex numbers. The following formulas apply:

$$(r_1 \angle \theta_1)(r_2 \angle \theta_2) = r_1 r_2 \angle (\theta_1 + \theta_2) \tag{13-7}$$

$$\frac{r_1 \angle \theta_1}{r_2 \angle \theta_2} = \frac{r_1}{r_2} \angle (\theta_1 - \theta_2) \tag{13-8}$$

To *multiply two complex numbers in polar form*, multiply the radii and *add* the angles. To *divide two complex numbers in polar form*, divide the radii and *subtract* the angles.

The proof of the product formula for complex numbers in polar form [Eq. (13-7)] is shown below. This proof uses the identities for $\sin (A + B)$ and $\cos (A + B)$ from Sec. 12-2.

### Example 13-13

Prove the product formula for complex numbers in polar form:

$$[r_1(\cos \theta_1 + j \sin \theta_1)][r_2(\cos \theta_2 + j \sin \theta_2)]$$
$$= r_1 r_2[\cos (\theta_1 + \theta_2) + j \sin (\theta_1 + \theta_2)]$$

### Solution

The product on the left is:

$$r_1 r_2(\cos \theta_1 \cos \theta_2 + j^2 \sin \theta_1 \sin \theta_2 + j \sin \theta_1 \cos \theta_2 + j \cos \theta_1 \sin \theta_2)$$
$$= r_1 r_2[(\cos \theta_1 \cos \theta_2 - \sin \theta_1 \sin \theta_2) + j(\sin \theta_1 \cos \theta_2 + \cos \theta_1 \sin \theta_2)]$$

Apply identities (12-4) and (12-5):

$$= r_1r_2[\cos (\theta_1 + \theta_2) + j \sin (\theta_1 + \theta_2)]$$

The quotient formula for complex numbers in polar form is left as problem 65.

### Example 13-14
1. Multiply $(2 + 5j)(1 - j)$ in rectangular and polar form, and show that the answers agree.
2. Divide the following fraction in rectangular and polar form, and show that the answers agree:

$$\frac{2 + 5j}{1 - j}$$

### Solution
1. Multiply $(2 + 5j)(1 - j)$ in rectangular and polar form, and show that the answers agree.

*Rectangular form:*

$$(2 + 5j)(1 - j) = 2 + 3j - 5j^2 = 7 + 3j$$

*Polar form:*

$$2 + 5j = 5.4 \angle 68° \quad \text{and} \quad 1 - j = 1.4 \angle 315°$$

Apply (13-7):

$$(5.4 \angle 68°)(1.4 \angle 315°) = (5.4)(1.4) \angle (68° + 315°)$$
$$= 7.6 \angle 383°$$
$$= 7.6 \angle 23° \quad \text{(change to angle } <360°)$$

The answers agree to two figures as follows:

$$7.6 \angle 23° = 7.6 \cos 23° + 7.6j \sin 23°$$
$$= 7.6(0.9205) + 7.6j(0.3907)$$
$$= 7.0 + 3.0j$$

2. Divide the following fraction in rectangular and polar form, and show that the answers agree:

$$\frac{2 + 5j}{1 - j}$$

*Rectangular form:*

$$\frac{2 + 5j\,(1 + j)}{1 - j\,(1 + j)} = \frac{-3 + 7j}{2} = -1.5 + 3.5j$$

*Polar form:*　Apply (13-8):

$$\frac{2 + 5j}{1 - j} = \frac{5.4\ \angle 68°}{1.4\ \angle 315°} = \frac{5.4}{1.4}\ \angle(68° - 315°) = 3.9\ \angle(-247°)$$

$$= 3.9\ \angle 113°$$

The answers agree as follows:

$$3.9\ \angle 113° = 3.9\cos 113° + 3.9j\sin 113°$$
$$= 3.9(-0.3907) + 3.9(0.9205) = -1.5 + 3.6j$$

There is not exact agreement because values are rounded. This example illustrates that formulas (13-7) and (13-8) yield correct results. A very useful application of these formulas is shown in the next section.

**Exercise 13-3**

In problems 1 to 10, graph each complex number as a vector in the complex plane.

**1.** $1 + 2j$　　　　　　　　　　**2.** $-4 + j$

**3.** $2 - 3j$　　　　　　　　　　**4.** $-5 - 5j$

**5.** $3j$　　　　　　　　　　　　**6.** $-6j$

**7.** $-4$　　　　　　　　　　　　**8.** $5$

**9.** $-1.4 + 3.4j$　　　　　　　　**10.** $4.6 - 2.3j$

In problems 11 to 16, add algebraically and graphically as vectors.

**11.** $(1 + 2j) + (3 + j)$　　　　　**12.** $(3 + 4j) + (2 + j)$

**13.** $(2 + 3j) + (2 - 3j)$　　　　**14.** $(4 + 2j) + (-4 + 2j)$

**15.** $(2j) + (-3 - j) + (3 + j)$　　**16.** $(-4) + (2 - j) + (2 + j)$

In problems 17 to 28, graph each as a vector and change to polar form.

**17.** $2 + 2j$　　　　　　　　　　**18.** $1 - j$

**19.** $-2 + 4j$　　　　　　　　　**20.** $-3 - 4j$

**21.** $\sqrt{3} - j$　　　　　　　　　**22.** $3 + 3j\sqrt{3}$

**23.** $-5.5j$　　　　　　　　　　**24.** $2.2$

**25.** $6.0 + 2.5j$　　　　　　　　**26.** $0.8 - 1.5j$

**27.** $-0.05 + 0.04j$　　　　　　**28.** $-0.25 - 0.75j$

In problems 29 to 42, graph each as a vector and change to rectangular form.

**29.** $3\ \angle 45°$　　　　　　　　　　**30.** $2\ \angle 150°$

**31.** $3 \angle 300°$                               **32.** $4 \angle 225°$

**33.** $5.5 \angle 250°$                            **34.** $7.2 \angle 295°$

**35.** $2.8 \angle 180°$                            **36.** $1.9 \angle 90°$

**37.** $4.70 \angle 120.5°$                        **38.** $3.87 \angle 37.8°$

**39.** $\sqrt{2} \angle (3\pi/4)$                   **40.** $2 \angle (\pi/6)$

**41.** $6 \angle 460°$                              **42.** $15 \angle (-90°)$

In problems 43 to 54, perform the indicated operations, using formulas (13-7) and (13-8). Express answers in polar form with $0° \le \theta < 360°$.

**43.** $(2 \angle 40°)(3 \angle 60°)$              **44.** $(4 \angle 150°)(\angle 240°)$

**45.** $(1.5 \angle 162°)(3.0 \angle 310°)$        **46.** $(8.2 \angle 36.2°)(7.6 \angle 83.1°)$

**47.** $\dfrac{6 \angle 200°}{4 \angle 120°}$      **48.** $\dfrac{17 \angle 350°}{8 \angle 50°}$

**49.** $\dfrac{7.3 \angle 73.0°}{8.7 \angle 97.9°}$  **50.** $\dfrac{0.50 \angle 235°}{0.10 \angle 115°}$

**51.** $\left(12 \angle \dfrac{\pi}{2}\right)\left(8.0 \angle \dfrac{\pi}{4}\right)$   **52.** $\dfrac{15 \angle (3\pi/2)}{7.5 \angle (\pi/3)}$

**53.** $(5 \angle 300°)(\angle 27°)(9 \angle 95°)$   **54.** $\dfrac{(3 \angle 59°)(6 \angle 138°)}{\angle 22°}$

In problems 55 to 60, perform the indicated operations for the complex numbers in rectangular form and polar form. Show that the answers agree (see Example 13-14).

**55.** $(1 + j)(2 - 2j)$                          **56.** $(4 - 3j)(3 + 4j)$

**57.** $\dfrac{6 + 8j}{12 - 9j}$                   **58.** $\dfrac{-2 - j}{-1 + j}$

**59.** $(2j)(1 - j)^2$                            **60.** $\dfrac{\sqrt{3} - j}{(\sqrt{3} + j)}$

**61.** A force vector in the complex plane is represented by $5.80 - 3.50j$. What are the magnitude (absolute value) and direction (angle) of this vector?

**62.** A force vector in the complex plane is represented by $15.8 (\cos 160° + j \sin 160°)$. What are the real and imaginary components of this vector?

**63.** The *active power* of an ac circuit whose current $I = a + bj$ and voltage $V = c + dj$ is given by $P = ac + bd$. Find $P$ when $I = 5(\cos 45° + j \sin 45°)$ and $V = 110(\cos 30° + j \sin 30°)$ by first changing the numbers to rectangular form.

**64.** Find the impedance $Z$ in polar form of the circuit in problem 63. *Note:* $Z = V/I$.

**65.** Prove the quotient formula for complex numbers in polar form (see Example 13-13):

$$\frac{r_1(\cos \theta_1 + j \sin \theta_1)}{r_2(\cos \theta_2 + j \sin \theta_2)} = \frac{r_1}{r_2} \cos (\theta_1 - \theta_2) + j \sin (\theta_1 - \theta_2)$$

(*Hint:* Multiply the top and bottom by $\cos \theta_2 - j \sin \theta_2$.)

# 13-4 | *Powers and Roots: De Moivre's Theorem*

When formula (13-7) is applied to the square of a complex number, you obtain

$$(r \angle \theta)^2 = (r \angle \theta)(r \angle \theta) = r^2 \angle 2\theta$$

If you extend this formula to any power $n$, the result is *De Moivre's theorem:*

$$(x + yj)^n = (r \angle \theta)^n = r^n \angle n\theta \qquad (13\text{-}9)$$

Formula (13-9) applies when *n is any rational number.* It simplifies not only raising to powers but also taking roots when the exponent is a fraction. Study the following examples closely to learn the procedure.

## *Powers*

### Example 13-15
Evaluate $(1 + 2j)^3$, using De Moivre's theorem (13-9), and express the answer in rectangular form.

### Solution
First change to polar form and then apply formula (13-9):

$$(1 + 2j)^3 = (\sqrt{5} \angle 63.4°)^3 = (\sqrt{5})^3 \angle [(3)(63.4°)]$$
$$= 11.2 \angle 190.2° = 11.2 \cos 190.2° + 11.2j \sin 190.2°$$
$$= 11.2(-0.9842) + 11.2j(-0.1771) = -11.0 - 2.0j$$

Check this example by multiplying out $(1 + 2j)^3$ in rectangular form.

### Example 13-16
Evaluate $(-1 + j)^6$ and express the answer in rectangular form.

### Solution

$$(-1 + j)^6 = (\sqrt{2} \angle 135°)^6 = (\sqrt{2})^6 \angle [(6)(135°)]$$
$$= 8 \angle 810° = 8 \angle 90°$$
$$= 8 \cos 90° + 8j \sin 90° = 8(0) + 8j(1)$$
$$= 8j$$

Notice that 810° is changed to 90°, the coterminal angle less than 360°. This example is done more readily in polar form (as shown above) than in rectangular form.

## *Roots*

De Moivre's theorem is most useful for taking roots. Taking roots of complex numbers is not done as directly as raising to a power. Study the following elementary example.

### Example 13-17

Find the two roots of $\sqrt{-9}$, using De Moivre's theorem.

### Solution

Use De Moivre's theorem to find the first root $w_1$:

$$w_1 = \sqrt{-9} = (-9)^{1/2} = (-9 + 0j)^{1/2} = (9 \angle 180°)^{1/2}$$

$$= 9^{1/2} \angle \left[\frac{180°}{2}\right] = 3 \angle 90° = 3 \cos 90° + 3j \sin 90°$$

$$= 3(0) + 3j(1) = 3j$$

The number $w_1$ is called the *principal root*. To obtain the other root (or roots), you can add 360° (or multiples of 360°) to the original angle:

$$w_2 = [9 \angle (180° + 360°)]^{1/2} = 9^{1/2} \angle \left(\frac{180°}{2} + \frac{360°}{2}\right)$$

$$= 3 \angle (90° + 180°) = 3 \angle 270° = 3 \cos 270° + 3j \sin 270°$$

$$= 3(0) + 3j(-1) = -3j$$

Note that adding 360° to the angle is equivalent to adding 360°/2, or 180°, to the principal root to obtain the other root. *The other root (or roots) can be obtained directly from the principal root by adding the necessary fractional part of 360°.* The *n*th root of $x + yj$ has *n* distinct answers. The formula for obtaining the *n*th roots is:

$$\text{*n*th roots of } x + yj$$

$$(x + yj)^{1/n} = r^{1/n} \angle \left(\frac{\theta}{n} + k\frac{360°}{n}\right) \qquad k = 0, 1, 2, \ldots, n - 1 \quad (13\text{-}10)$$

Study the next example to learn how to use formula (13-10).

### Example 13-18

Find all the roots of $\sqrt[4]{16}$. Express answers in rectangular form.

### Solution

Use (13-10) to find the principal root ($k = 0$).

$k = 0$:          $w_1 = 16^{1/4} = (16 + 0j)^{1/4} = (16 \angle 0°)^{1/4}$

$$= 16^{1/4} \angle \frac{0°}{4} = 2 \angle 0° = 2 \cos 0° + 2j \sin 0° \qquad \text{by (13-10)}$$

$$= 2(1) + 2j(0) = 2$$

As expected 2 is one of the roots. You obtain the other three roots by using formula (13-10) and adding multiples of $360°/n = 360°/4 = 90°$ to the principal root:

$k = 1$:     $w_2 = 2 \angle [0° + (1)(90°)] = 2 \angle 90°$

$$= 2 \cos 90° + 2j \sin 90°$$

$$= 2(0) + 2j(1) = 2j$$

$k = 2$:          $w_3 = 2 \angle [0° + (2)(90°)] = 2 \angle 180° = -2$

$k = 3$:          $w_4 = 2 \angle [0° + (3)(90°)] = 2 \angle 270° = -2j$

The four roots are $w_1 = 2$, $w_2 = 2j$, $w_3 = -2$, and $w_4 = -2j$. Adding more multiples of $90°$ causes the roots to repeat.

C **Example 13-19**

Find all the roots of $\sqrt[3]{3 + j}$. Express answers in rectangular form.

**Solution**

$k = 0$:          $w_1 = (3 + j)^{1/3} = (3.16 \angle 18.4°)^{1/3}$

$$= (3.16)^{1/3} \angle \frac{18.4°}{3} = 1.47 \angle 6.1°$$

$$= 1.47 \cos 6.1° + 1.47j \sin 6.1°$$

$$= 1.46 + 0.157j$$

Add multiples of $360°/3 = 120°$ to obtain the other two roots:

$k = 1$:     $w_2 = 1.47 \angle [6.1° + (1)(120°)] = 1.47 \angle 126.1°$

$$= -0.867 + 1.19j$$

$k = 2$:     $w_3 = 1.47 \angle [6.1° + (2)(120°)] = 1.47 \angle 246.1°$

$$= -0.595 - 1.34j$$

**Exercise 13-4**

In problems 1 to 14, evaluate using De Moivre's theorem and express each answer in rectangular form.

**1.** $(5 \angle 45°)^2$          **2.** $(3 \angle 120°)^3$

**3.** $[2 \angle (-30°)]^3$          **4.** $(\sqrt{2} \angle 315°)^4$

**5.** $(1 + j)^3$          **6.** $(2 - j)^4$

**7.** $(-3 - j)^3$         **8.** $(-1 + 4j)^3$

**9.** $(1.5 + 2.0j)^4$     **10.** $(0.8 + 0.6)^5$

**11.** $(\sqrt{3} - j)^6$       **12.** $(-1 - j\sqrt{3})^4$

**13.** $[(2 - 2j)/(1 + j)]^3$    **14.** $(1 - j)^3(3 + 3j)^3$

In problems 15 to 28, find all the roots and express in rectangular form.

**15.** $\sqrt{25} \angle 0°$         **16.** $\sqrt{0.36} \angle 270°$

**17.** $\sqrt[3]{\angle 180°}$       **18.** $\sqrt[3]{27} \angle 90°$

**19.** $\sqrt[3]{1}$           **20.** $\sqrt[3]{-8}$

**21.** $\sqrt[4]{-16}$         **22.** $\sqrt{j}$

**23.** $\sqrt[3]{27}$         **24.** $\sqrt[3]{-8j}$

⊏ **25.** $(1 - j)^{1/3}$     ⊏ **26.** $(1 + 3j)^{1/4}$

**27.** $(2 + 1.5j)^{1/2}$    **28.** $(1 - j\sqrt{3})^{1/2}$

**29.** Find the four roots of the equation $x^4 - 1 = 0$ and check each root in the equation. (*Note*: $x = \sqrt[4]{1}$.)

**30.** Find the three roots of the equation $x^3 + 64 = 0$ and check each root in the equation.

**31.** The impedance in a series *RLC* (resistance-inductance-capacitance) circuit is given by:

$$Z = \frac{(1 + j)^2(1 - j)^2}{(3 + 4j)^2}$$

Evaluate $Z$ by changing the expression to polar form.

**32.** In an ac circuit the admittance $Y$ is given by $Y = Z^{-1}$, where $Z$ is the impedance. If $Z = 3 + 4j$, find $Y$ *(a)* by using polar form and De Moivre's theorem and *(b)* by using rectangular form. Check that the answers in *(a)* and *(b)* agree.

# 13-5 | *Exponential Form and AC Circuits*

In applications of the theory of complex numbers, the *exponential form* is commonly used:

$$re^{j\theta} = r(\cos \theta + j \sin \theta) = r \angle \theta \qquad (13\text{-}11)$$

The number $e = 2.7182\cdots$ is the base of natural logarithms, described in Chap. 9. The angle $\theta$ is expressed in radians, and all the rules of exponents apply to $e^{j\theta}$. The justification for the exponential form is shown in calculus. You will see in what follows that the exponential form satisfies the rules for complex numbers.

## Example 13-20
Express each complex number in exponential form.

**1.** $2.2(\cos 45° + j \sin 45°)$      **2.** $2.0 + 1.0j$      **3.** $3.0 - 4.0j$

### Solution
**1.** $2.2(\cos 45° + j \sin 45°)$

The angle $45° = 45(\pi/180) = \pi/4$ or $0.785$ rad. Then:

$$2.2(\cos 45° + j \sin 45°) = 2.2e^{j(\pi/4)} \text{ or } 2.2e^{0.785j}$$

**2.** $2.0 + 1.0j$

First find $r$ and $\theta$. Calculate $r = \sqrt{(2.0)^2 + (1.0)^2} = \sqrt{5}$ or $2.24$. Angle $\theta$ is in the first quadrant, and $\theta = \tan^{-1} \frac{1}{2} = 0.464$ rad. Then $2.0 + 1.0j = \sqrt{5}e^{0.464j}$ or $2.24e^{0.464j}$

**3.** $3.0 - 4.0j$

Now $r = \sqrt{(3.0)^2 + (-4.0)^2} = 5.0$. Angle $\theta$ is in the fourth quadrant. The reference angle $\tan^{-1}(4/3) = 0.927$ rad, and $\theta = 2\pi - 0.927 = 5.36$ rad. Then $3.0 - 4.0j = 5.0e^{5.36j}$. This can also be expressed with a negative angle in the fourth quadrant as $5.0e^{-0.927j}$, which is more meaningful in applications. *To reverse the above procedure,* change from exponential form to polar form first and then to rectangular form.

The rules for multiplying and dividing complex numbers in exponential form follow directly from the addition and subtraction rules for exponents:

$$(r_1e^{j\theta_1})(r_2e^{j\theta_2}) = r_1r_2e^{j\theta_1 + j\theta_2} = r_1r_2e^{j(\theta_1 + \theta_2)} \qquad (13\text{-}12)$$

$$\frac{r_1e^{j\theta_1}}{r_2e^{j\theta_2}} = \frac{r_1}{r_2}e^{j\theta_1 - j\theta_2} = \frac{r_1}{r_2}e^{j(\theta_1 - \theta_2)} \qquad (13\text{-}13)$$

Notice that formulas (13-12) and (13-13) agree with (13-7) and (13-8), respectively. That is, to multiply two complex numbers, multiply the radii and add the angles. To divide two complex numbers, divide the radii and subtract the angles. De Moivre's theorem follows from the multiplication rule for exponents:

$$(re^{j\theta})^n = r^ne^{jn\theta} \qquad (13\text{-}14)$$

## Example 13-21
Calculate each using exponential form. Express answers in rectangular form.

**1.** $(2 + 2j)(1 - j)$      **2.** $\dfrac{2 + 2j}{1 - j}$

**Solution**

**1.** $(2 + 2j)(1 - j) = [2\sqrt{2}e^{j(\pi/4)}][\sqrt{2}e^{j(-\pi/4)}] = (2\sqrt{2})(\sqrt{2})e^{\pi/4 - \pi/4}$
$$= 4e^{j(0)} = 4e^0 = 4(1) = 4$$

Note that a negative angle $(-\pi/4)$ is used instead of $7\pi/4$, which is common in applications of complex numbers.

**2.** $\dfrac{2 + 2j}{1 - j} = \dfrac{2\sqrt{2}e^{j(\pi/4)}}{\sqrt{2}e^{j(-\pi/4)}} = \dfrac{2\sqrt{2}}{\sqrt{2}}e^{j(\pi/4 + \pi/4)} = 2e^{j(\pi/2)}$

$$= 2\cos\frac{\pi}{2} + 2j\sin\frac{\pi}{2}$$

$$= 2(0) + 2j(1)$$

$$= 2j$$

## Application to AC Circuits

In dc circuit theory, the basic relation between voltage and current is stated by Ohm's law: $E = IR$. In ac circuit theory the counterpart of Ohm's law is:

$$V = IZ \tag{13-15}$$

where $V$ = voltage, $I$ = current, and $Z$ = impedance in ohms. *Impedance* is the opposition to current produced by a resistance $R$, an inductance $L$, a capacitance $C$, or any combination of these. Consider the series *RLC* circuit with a sinusoidal ac voltage shown in Fig. 13-9. The opposition to the current produced by the inductance is called the *inductive reactance* $X_L$, and the opposition produced by the capacitance is called the *capacitive reactance* $X_C$. The voltage across the *resistance is in phase* with the current. The voltage across the *inductance leads* the current by 90°, and the voltage across the *capacitance lags* the current by 90°. The relation of the vectors $R$, $X_L$, and $X_C$ is linked to the voltage across each and is shown in Fig. 13-10a. Because of this vector relation it is useful to represent the impedance $Z$ as a complex number or vector in which $R$

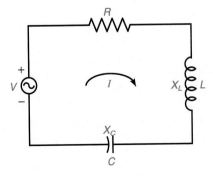

FIG. 13-9 *Series RLC circuit*

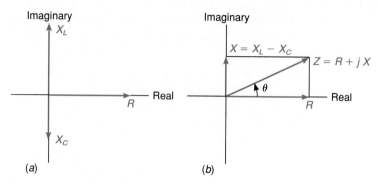

FIG. 13-10  *AC circuit vectors*

is the real part and the reactance $X = X_L - X_C$ is the imaginary part (Fig. 13-10*b*):

$$Z = R + jX = R + j(X_L - X_C) \qquad (13\text{-}16)$$

### Example 13-22
The impedance of an ac circuit is given by $Z = 5.0e^{0.60j}\ \Omega$. Find the resistance $R$ and the reactance $X$.

### Solution
Change $Z$ to rectangular form:

$$Z = 5.0e^{0.60j} = 5.0\cos 0.60 + 5.0j\sin 0.60$$
$$= 5.0(0.8253) + 5.0(0.5646) = 4.1 + 2.8j\ \ \Omega$$

Then the resistance $R = 4.1\ \Omega$ and the reactance $X = 2.8\ \Omega$.

The *magnitude* (absolute value) of $Z$ (written $|Z|$) and the *phase angle* $\theta$ derive from the formulas for complex vectors:

$$|Z| = \sqrt{R^2 + X^2} \qquad \tan\theta = \frac{X}{R} \quad \text{or} \quad \theta = \tan^{-1}\frac{X}{R} \qquad (13\text{-}17)$$

The phase angle $\theta$ is generally expressed as a positive first-quadrant angle or a negative fourth-quadrant angle, which illustrates the idea of the voltage leading (positive angle) or lagging (negative angle) the current. Because of this you can write $\theta = \tan^{-1}(X/R)$ since $\tan^{-1}$ is defined specifically for these angles. The calculator is programmed for these angles when you press $\boxed{\text{INV}}\ \boxed{\text{tan}}$ or $\boxed{\text{tan}^{-1}}$.

### Example 13-23
In the *RLC* circuit of Fig. 13-9, $R = 40\ \Omega$, $X_L = 55\ \Omega$, and $X_C = 25\ \Omega$.

**1.** Find the magnitude and phase angle of $Z$.

**2.** If the current $I = 3.2$ A, find the voltage across the circuit.

### Solution

**1.** Find the magnitude and phase angle of $Z$.

By (13-16) the reactance $X = X_L - X_C = 55 - 25 = 30$ $\Omega$, and $Z = R + jX = 40 + 30j$ $\Omega$. Then, using (13-17):

$$|Z| = \sqrt{40^2 + 30^2} = 50 \ \Omega$$

and
$$\theta = \tan^{-1} \frac{30}{40} = 36.9° \quad \text{or} \quad 0.644 \text{ rad}$$

**2.** If the current $I = 3.2$ A, find the voltage across the circuit.

Apply (13-15):

$$V = IZ = (3.2)(50) = 160 \text{ V}$$

### Example 13-24

Using the series $RC$ circuit in Fig. 13-11, in which $R = 4.5$ $\Omega$ and $X_C = 6.0$ $\Omega$:

**1.** Draw the vector diagram for the impedance $Z$.

**2.** Find the magnitude and phase angle of $Z$.

**3.** If the voltage $V = 110$ V, find the magnitude of the current $I$.

### Solution

**1.** Draw the vector diagram for the impedance $Z$.

The vector diagram is shown in Fig. 13-12. In this circuit $X_L = 0$ and $X = X_L - X_C = 0 - 6.0 = -6.0$ $\Omega$.

**2.** Find the magnitude and phase angle of $Z$.

The magnitude of $Z = |Z| = \sqrt{(4.5)^2 + (-6.0)^2} = \sqrt{56.25} = 7.5$ $\Omega$. The

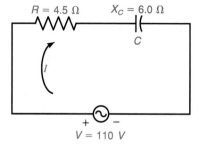

FIG. 13-11 *Series RC circuit, Example 13-24*

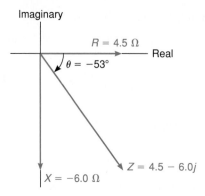

FIG. 13-12 *Voltage lagging current, Example 13-24*

phase angle $\theta = \tan^{-1}(-6.0/4.5) = -53°$ or $-0.925$ rad. This means the voltage lags the current by $53°$.

3. If the voltage $V = 110$ V, find the magnitude of the current $I$:

$$I = \frac{V}{Z} = \frac{110}{7.5} = 14.7 \text{ A}$$

Equation (13-15) holds when $V$, $I$, and $Z$ are scalar *or vector* quantities, as shown in the next example.

**Example 13-25**

In an ac circuit the voltage $V = 120e^{j(\pi/3)}$, and the current $I = 10e^{j(\pi/4)}$. Find the impedance $Z$.

**Solution**

Use Eq. (13-15) for the vector quantities:

$$Z = \frac{V}{I} = \frac{120e^{j(\pi/3)}}{10e^{j(\pi/4)}} = 12e^{j(\pi/3 - \pi/4)} = 12e^{j(\pi/12)}$$

## *Series and Parallel Impedances*

When two impedances are connected in *series*, as in Fig. 13-11, *the total impedance is their sum.* This is the same relation as that of the total resistance in a dc circuit to two resistances connected in series. When two impedances $Z_1$ and $Z_2$ are connected in parallel, the total impedance $Z$ is given by:

$$\frac{1}{Z} = \frac{1}{Z_1} + \frac{1}{Z_2} \qquad \text{or} \qquad Z = \frac{Z_1 Z_2}{Z_1 + Z_2} \tag{13-18}$$

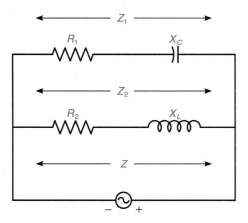

FIG. 13-13 *Impedances in parallel, Example 13-26*

Equation (13-18) is the same relation as the total resistance in a dc circuit to two resistances connected in parallel.

### Example 13-26
Given the parallel circuit in Fig. 13-13, find the total impedance $Z$ when $R_1 = 2.0\ \Omega$, $R_2 = 6.0\ \Omega$, $X_C = 4.0\ \Omega$, and $X_L = 2.0\ \Omega$.

### Solution
First determine $Z_1$ and $Z_2$:

$$Z_1 = R_1 - jX_C = 2.0 - 4.0j$$
$$Z_2 = R_2 + jX_L = 6.0 + 2.0j$$

Then apply formula (13-18):

$$Z = \frac{Z_1 Z_2}{Z_1 + Z_2} = \frac{(2.0 - 4.0j)(6.0 + 2.0j)}{(2.0 - 4.0j) + (6.0 + 2.0j)} = \frac{20 - 20j}{8.0 - 2.0j}$$

$$= \frac{10 - 10j}{4.0 - 1.0j}\frac{(4.0 + 1.0j)}{(4.0 + 1.0j)} = \frac{50 - 30j}{17} = 2.9 - 1.8j$$

Notice that the common factor of 2 is divided out before you do the division.

**Exercise 13-5**

In problems 1 to 8, express each number in exponential form.

| | |
|---|---|
| **1.** $1.1 \angle 30°$ | **2.** $6.3 \angle 60°$ |
| **3.** $0.32 \angle 40°$ | **4.** $15 \angle (-55°)$ |
| **5.** $2.4 - 2.4j$ | **6.** $3.6 + 4.8j$ |
| **7.** $0.60 + 1.44j$ | **8.** $17.6 - 23.0j$ |

Perform the indicated operations in problems 9 to 16 by first changing to exponential form. Express answers in rectangular form.

**9.** $(1 + j)(3 + 3j)$        **10.** $(2 - j)(1 + 2j)$

**11.** $(1.5 - 2.0j)(4.0 - 3.0j)$        **12.** $(0.50 + 0.30j)(0.20 + 0.40j)$

**13.** $\dfrac{2 - 3j}{3 - 2j}$        **14.** $\dfrac{1 + j}{1 - j}$

**15.** $\dfrac{6.0 + 6.5j}{3.5 + 3.0j}$        **16.** $\dfrac{4.8 - 9.0j}{7.2 + 3.0j}$

In problems 17 to 22, for each impedance Z, find the resistance R and the reactance X.

**17.** $Z = 3.0e^{1.5j}$        **18.** $Z = 1.2e^{0.10j}$

**19.** $Z = 0.24e^{-0.53j}$        **20.** $Z = 4.3e^{-1.21j}$

**21.** $Z = 5.6e^{j(-\pi/2)}$        **22.** $Z = 0.12e^{j(\pi/2)}$

Given the values for the series $RLC$ circuit of Fig. 13-9, in problems 23 to 28 find the magnitude and phase angle of Z.

**23.** $R = 3.0\ \Omega,\ X_L = 6.0\ \Omega,\ X_C = 5.0\ \Omega$

**24.** $R = 0.40\ \Omega,\ X_L = 0.90\ \Omega,\ X_C = 0.50\ \Omega$

**25.** $R = 28.0\ \Omega,\ X_L = 14.0\ \Omega,\ X_C = 35.0\ \Omega$

**26.** $R = 3.0\ \Omega,\ X_L = 12.5\ \Omega,\ X_C = 25.0\ \Omega$

**27.** $R = 0.36\ \Omega,\ X_L = 0.25\ \Omega,\ X_C = 0.43\ \Omega$

**28.** $R = 1.38\ \Omega,\ X_L = 2.42\ \Omega,\ X_C = 1.10\ \Omega$

In problems 29 to 32, for each series ac circuit, *(a)* draw the vector diagram for the impedance Z and *(b)* find the magnitude and phase angle of Z.

**29.** RC circuit: $R = 3.5\ \Omega$ and $X_C = 2.5\ \Omega$

**30.** RC circuit: $R = 33\ \Omega$ and $X_C = 44\ \Omega$

**31.** RL circuit: $R = 0.025\ \Omega$ and $X_L = 0.035\ \Omega$

**32.** LC circuit: $X_C = 2.13\ \Omega$ and $X_L = 1.23\ \Omega$

In problems 33 to 38, for each ac circuit, find the indicated quantity and express in the form of the given quantities.

**33.** $V = 10 + 20j,\ Z = 20 + 40j;\ I$

**34.** $V = 220 + 120j,\ I = 20.0 - 10.0j;\ Z$

**35.** $V = 2.1 + 1.5j,\ Z = 1.8 + 1.2j;\ I$

**36.** $I = 0.80 + 0.0j,\ Z = 0.40 - 0.30j;\ V$

**37.** $I = 2.0e^j,\ Z = 3.0e^{-j};\ V$

**38.** $V = 115e^{0.32j},\ I = 2.5e^{0.28j};\ Z$

In problems 39 to 42, given the values for the parallel circuit in Fig. 13-13, find the total impedance Z.

**39.** $R_1 = 0.0\ \Omega,\ R_2 = 3.0\ \Omega,\ X_C = 2.0\ \Omega,\ X_L = 0.0\ \Omega$

**40.** $R_1 = 1.0\ \Omega,\ R_2 = 0.0\ \Omega,\ X_C = 0.0\ \Omega,\ X_L = 0.80\ \Omega$

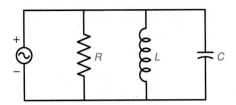

FIG. 13-14  *Parallel RLC circuit, problems 43 and 44*

**41.** $R_1 = 0.0\ \Omega$, $R_2 = 0.0\ \Omega$, $X_C = 6.0\ \Omega$, $X_L = 8.0\ \Omega$

**42.** $R_1 = 2.0\ \Omega$, $R_2 = 3.0\ \Omega$, $X_C = 1.0\ \Omega$, $X_L = 2.0\ \Omega$

**43.** For the *RLC* parallel circuit shown in Fig. 13-14, find the total impedance $Z$ when $R = 6.0\ \Omega$, $X_L = 3.0\ \Omega$, and $X_C = 2.0\ \Omega$. (*Note:* For three parallel impedances $1/Z = 1/Z_1 + 1/Z_2 + 1/Z_3$.)

**44.** Find the total impedance in Fig. 13-14 when $R = 3.0\ \Omega$, $X_L = 4.5\ \Omega$, and $X_C = 9.0\ \Omega$. See problem 43.

**45.** In a series ac circuit whose frequency is $f$, $X_L$ and $X_C$ are given by $X_L = 2\pi f L$ and $X_C = 1/(2\pi f C)$, where $L$ is the inductance in henrys (H) and $C$ is the capacitance in farads (F). Assuming $R = 0\ \Omega$, find $X_L$, $X_C$, and the total series impedance $Z$ when $f = 60$ Hz, $L = 0.05$ H, and $C = 3.0 \times 10^{-5}$ F.

**46.** Do the same as in problem 45 when $f = 60$ Hz, $L = 0.10$ H, and $C = 15\ \mu$F.

**47.** The total impedance of three impedances in parallel is $Z = 2.0\ \Omega$. If $Z_2 = -2Z_1$ and $Z_3 = 1 - Z_1$, find $Z_1$. See problem 43.

# 13-6 │ *Review Questions*

In problems 1 to 4, simplify and express in terms of $j$.

**1.** $\sqrt{-25}$    **2.** $(\sqrt{-3})(\sqrt{-12})$

**3.** $(2\sqrt{-5})(\sqrt{10})$    **4.** $(2j^3)^2$

In problems 5 to 12, perform the indicated operations. Express answers in the form $a + bj$.

**5.** $3j(2 - j)$    **6.** $\dfrac{6 + \sqrt{-8}}{2}$

**7.** $(4 - 2j) + (5 + j)$    **8.** $(1 - 3j)(2 + 2j)$

**9.** $(2 + j)^2$    **10.** $\dfrac{6 - 2j}{1 + j}$

**11.** $\dfrac{4 - 3j}{4 + 3j}$    **12.** $\dfrac{1}{\sqrt{2} - j}$

In problems 13 to 20, graph each complex number as a vector and change from rectangular to polar form, or vice versa.

**13.** $3 + 3j$

**14.** $1.5 - 2.0j$

**15.** $3.2j$

**16.** $1 + j\sqrt{3}$

**17.** $5 \angle 120°$

**18.** $6 \angle 315°$

**19.** $2.8 \angle 25°$

**20.** $4 \angle 270°$

In problems 21 to 24, perform the indicated operations and express the answer in rectangular form.

**21.** $(\angle 60°)(5 \angle 120°)$

**22.** $\dfrac{3.0 \angle 200°}{0.5 \angle 140°}$

**23.** $(4 \angle 10°)^3$

**24.** $(1 + j)^5$

In problems 25 to 28, find all the roots and express in rectangular form.

**25.** $\sqrt{9 \angle 60°}$

**26.** $\sqrt[3]{-1}$

**27.** $\sqrt[3]{8}$

**28.** $\sqrt{4j}$

In problems 29 and 30, change to exponential form and perform the indicated operations.

**29.** $(3 + j)(2 - j)$

**30.** $\dfrac{2 - 2j}{1 + j}$

**31.** A force vector in the complex plane is given by $6.2 + 1.3j$. What are the magnitude and direction (the angle) of this vector?

**32.** Using De Moivre's theorem, solve the equation $x^4 = 81$.

**33.** Given the series $RLC$ circuit in Fig. 13-9 with $R = 2.0 \ \Omega$, $X_L = 4.0 \ \Omega$, and $X_C = 3.0 \ \Omega$, find the magnitude and phase angle of $Z$.

**34.** Given the series $RC$ circuit of Fig. 13-11 with $R = 4.5 \ \Omega$ and $X_C = 5.0 \ \Omega$, (a) draw the vector diagram for the impedance $Z$ and (b) find the magnitude and phase angle of $Z$.

**35.** In an ac circuit the voltage $V = 120e^{j(\pi/4)}$ and the current $I = 12e^{j(\pi/2)}$. Find the impedance $Z$.

**36.** In the parallel circuit of Fig. 13-13, $R_1 = 1.0 \ \Omega$, $R_2 = 0.0 \ \Omega$, $X_C = 0.0 \ \Omega$, and $X_L = 2.0 \ \Omega$. Find the total impedance $Z$.

# 14

# Analytic Geometry and Quadratic Systems

## 14-1 | *Straight Line: Linear Function*

### *Distance Formula*

The formula for the distance, or length of the straight-line segment, between two points $(x_1, y_1)$ and $(x_2, y_2)$ is fundamental to analyzing all curves. When the pythagorean theorem is applied to the right triangle in Fig. 14-2, the distance $d$ is given by $d^2 = (x_2 - x_1)^2 + (y_2 - y_1)^2$ or:

$$d = \sqrt{(x_2 - x_1)^2 + (y_2 - y_1)^2} \quad (14\text{-}1)$$

**Example 14-1**

Find the distance between each set of points.

**1.** $(-3, 4)$ and $(-1, -2)$

**2.** $(0, 4)$ and $(4, 0)$

**Solution**

**1.** $(-3, 4)$ and $(-1, -2)$

*A**nalytic geometry** links algebra and geometry, providing a better understanding of both. It is essential background for many ideas, especially calculus. The linear function is covered first in detail because it is basic to the graphs of almost all functions. The parabola or quadratic function is the next analytic step, followed by the other second degree curves—the circle, ellipse, and hyperbola. These four curves are called the *conic sections* because they are obtained by cutting a cone in different ways (Fig. 14-1). Conic sections are important in the laws of motion and many other scientific and technical applications.

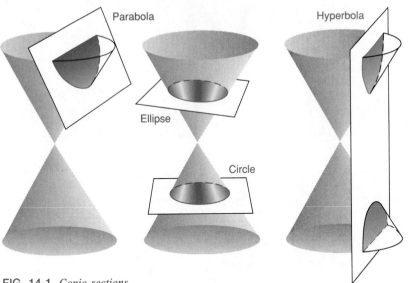

FIG. 14-1 *Conic sections*

You can let $(x_1, y_1)$ in formula (14-1) be either point. Let $(x_1, y_1)$ be $(-3, 4)$ and $(x_2, y_2)$ be $(-1, -2)$:

$$d = \sqrt{[-1 - (-3)]^2 + (-2 - 4)^2} = \sqrt{2^2 + (-6)^2} = \sqrt{4 + 36}$$
$$= \sqrt{40} = 2\sqrt{10} \quad \text{or} \quad 6.32 \text{ (three figures)}$$

**2.** $(0, 4)$ and $(4, 0)$.

Let $(x_1, y_1)$ be $(4, 0)$:

$$d = \sqrt{(0 - 4)^2 + (4 - 0)^2} = \sqrt{16 + 16} = \sqrt{32} = 4\sqrt{2} \quad \text{or} \quad 5.66$$

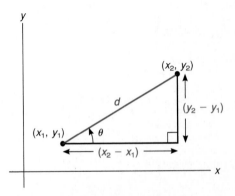

FIG. 14-2 *Distance and slope formulas*

## Slope Formula

The slope of a straight line is a measure of its steepness with respect to the $x$ axis. In Fig. 14-2, as you travel on the line from the lower point to the upper point, the *slope* is the ratio of the vertical change $y_2 - y_1$ to the horizontal change $x_2 - x_1$. The *slope* is also equal to tan $\theta$, where $\theta$ is the counterclockwise angle the line makes with the positive $x$ direction, called the *inclination:*

$$\text{Slope} = m = \frac{y_2 - y_1}{x_2 - x_1} \qquad \text{or} \qquad m = \tan \theta \qquad 0° \le \theta < 180° \quad (14\text{-}2)$$

### Example 14-2
Find the slope and inclination of each line (Fig. 14-3).

**1.** Line joining $(-3, 2)$ and $(5, 3)$

**2.** Line joining $(-3, 2)$ and $(3, -3)$

**3.** Line joining $(-3, 2)$ and $(4, 2)$

### Solution
**1.** Line joining $(-3, 2)$ and $(5, 3)$

Either point can be $(x_1, y_1)$ in formula (14-2). Let $(5, 3)$ be $(x_1, y_1)$:

$$m = \frac{2 - 3}{-3 - 5} = \frac{-1}{-8} = \frac{1}{8} \quad \text{or} \quad 0.125$$

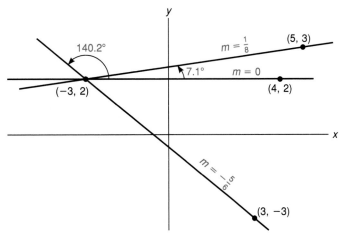

FIG. 14-3 *Slope and inclination, Example 14-2*

Then because $m = \tan \theta$:

$$\theta = \tan^{-1} 0.125 = 7.1°$$

2. Line joining $(-3, 2)$ and $(3, -3)$

$$m = \frac{2 - (-3)}{-3 - 3} = \frac{5}{-6} \quad \text{or} \quad -0.833$$

$$\text{Reference angle} = \tan^{-1} 0.833 = 39.8°$$

and

$$\theta = 180° - 39.8° = 140.2°$$

3. Line joining $(-3, 2)$ and $(4, 2)$

$$m = \frac{2 - 2}{-3 - 4} = \frac{0}{-7} = 0$$
$$\theta = \tan^{-1} 0 = 0°$$

In Fig. 14-3 you can see that a line rising to the right has a positive slope, a line falling to the right has a negative slope, and a horizontal line has a slope of zero. Knowing that $\tan 0° = 0$, $\tan 45° = 1$, $\tan 90° = \infty$ (infinity), and $\tan 135° = -1$, you can visualize how the slope changes with $\theta$ by studying Fig. 14-4. As $\theta$ increases from $0°$ to $45°$, $m$ increases from 0 to 1. As $\theta$ increases from $45°$ to $90°$, $m$ increases from 1 to positive infinity, that is, $m$ becomes infinitely large. *The slope of a vertical line is undefined.* As $\theta$ decreases from $180°$ to $135°$, $m$ decreases from 0 to $-1$, and as $\theta$ decreases from $135°$ to $90°$, $m$ decreases from $-1$ to negative infinity.

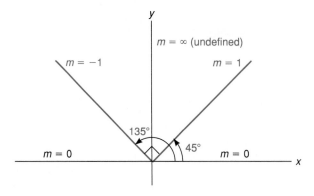

FIG. 14-4 *Slope and inclination*

## Parallel and Perpendicular Lines

*Parallel lines have the same slope.* This follows from the fact that their inclinations are equal. *Two perpendicular lines have negative reciprocal slopes* as follows. Consider the congruent right triangles in Fig. 14-5. Line $L_1$ is perpendicular to line $L_2$ because the two angles, $\theta$ and $90° - \theta$, are complementary. When the slope formula for the two points on each line is applied:

$$\text{Slope of } L_1 = \frac{m - 0}{1 - 0} = m$$

$$\text{Slope of } L_2 = \frac{1 - 0}{-m - 0} = -\frac{1}{m}$$

$L_1, L_2$ perpendicular lines    (14-3)
$m \neq 0$

Now any two perpendicular lines that are parallel to $L_1$ and $L_2$, respectively, and do not pass through the origin, will have the same slopes $m$ and $-1/m$. Therefore formula (14-3) is true for *any* two perpendicular lines, except a horizontal and a vertical line.

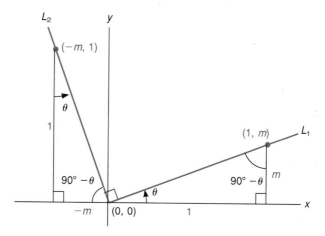

FIG. 14-5 *Perpendicular lines*

## Example 14-3

Given the quadrilateral with vertices $A(2, 4)$, $B(7, 0)$, $C(3, -5)$, and $D(-2, -1)$, show that the figure is a square.

## Solution

The quadrilateral is shown in Fig. 14-6. The slopes of the sides are:

$$m_{AB} = \frac{4 - 0}{2 - 7} = -\frac{4}{5} \qquad m_{BC} = \frac{0 - (-5)}{7 - 3} = \frac{5}{4}$$

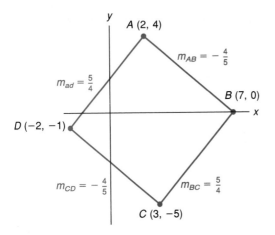

FIG. 14-6 *Square, Example 14-3*

$$m_{CD} = \frac{-5 - (-1)}{3 - (-2)} = -\frac{4}{5} \qquad m_{AD} = \frac{4 - (-1)}{2 - (-2)} = \frac{5}{4}$$

Since $m_{AB} = m_{CD}$ and $m_{BC} = m_{AD}$, the opposite sides are parallel. Since $m_{AB} = -1/m_{BC}$ and $m_{CD} = -1/m_{AD}$, the adjacent sides are perpendicular.

The quadrilateral is therefore a rectangle. It is now necessary to show that all sides are equal by applying the distance formula (14-1). It is readily shown that each side is equal to:

$$d = \sqrt{16 + 25} = \sqrt{41}$$

and the figure is therefore a square.

## Linear Function and Slope-Intercept Form

Consider any straight line whose slope is $m$ and intercepts the $y$ axis at the point $(0, b)$ (Fig. 14-7). Between any point $(x, y)$ on this line and $(0, b)$ the slope is:

$$m = \frac{y - b}{x - 0} = \frac{y - b}{x}$$

Then:

$$mx = y - b$$

and solving for $y$, you obtain the *slope intercept form* of the equation of a straight line:

$$y = mx + b \tag{14-4}$$

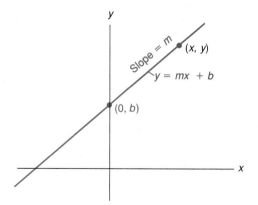

FIG. 14-7 *Slope-intercept equation of a straight line*

*This equation expresses y as a linear function of x and is a very useful form of* the straight line equation. Linear functions are introduced in Sec. 5-1.

### Example 14-4
Find the slope and $y$ intercept of each line.

**1.** $3x - y = 4$     **2.** $4x + 2y - 5 = 0$

### Solution
**1.** $3x - y = 4$

Solve for $y$: $y = 3x - 4$. Then $m = 3$ and $b = -4$.

**2.** $4x + 2y - 5 = 0$

Solve for $y$: $2y = -4x + 5$ and $y = -2x + 5/2$. Then $m = -2$ and $b = 5/2$ or 2.5.

### Example 14-5
Write the equation of each line.

**1.** Inclination $\theta = 45°$ and $y$ intercept $= -2$
**2.** Slope is 0 and passes through $(0, 4)$

### Solution
**1.** Inclination $\theta = 45°$ and $y$ intercept $= -2$

The slope $m = \tan 45° = 1$ and $b = -2$. The equation is $y = x - 2$.

**2.** Slope is 0 and passes through $(0, 4)$

The slope $m = 0$ and $b = 4$. The equation is $y = 4$. An equation of this form $y = b$ *is always a horizontal line* parallel to the $x$ axis and is a special case of

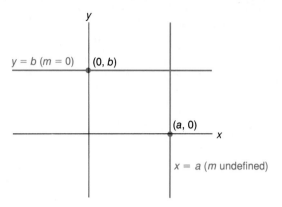

FIG. 14-8  *Equations of horizontal and vertical lines, Example 14-5(2)*

Eq. (14-4). Similarly $x = a$ *is always a vertical line* parallel to the $y$ axis, and the point $(a, 0)$ is the $x$ intercept (Fig. 14-8).

## Point-Slope Form and General Form

If one is given the coordinates of a point on a line, say $(x_1, y_1)$, then the slope $m$ between the given point and any point $(x, y)$ is:

$$m = \frac{y - y_1}{x - x_1}$$

Multiplying both sides by $x - x_1$, you obtain the *point-slope form* of the equation of a straight line:

$$y - y_1 = m(x - x_1) \qquad\qquad (14\text{-}5)$$

This equation is useful if you know a point on a line and its slope and want to write the equation. Study the next example.

### Example 14-6
Write the equation of each line in the slope-intercept form.

**1.** $m = -3$ and passes through $(3, -4)$

**2.** Line passes through $(-1, -2)$ and $(6, 4)$

**3.** Line perpendicular to $2x - 3y = 5$ and passing through $(2, 1)$

### Solution
**1.** $m = -3$ and passes through $(3, -4)$

Let $x_1 = 3$, $y_1 = -4$, and $m = -3$ in Eq. (14-5):

$$y - (-4) = -3(x - 3)$$
$$y + 4 = -3x + 9$$

The equation in slope intercept form is:

$$y = -3x + 5$$

Notice that the coefficient of $x$ is the slope $m = -3$. This line intercepts the $y$ axis at $(0, 5)$.

**2.** Line passes through $(-1, -2)$ and $(6, 4)$

First find the slope:

$$m = \frac{4 - (-2)}{6 - (-1)} = \frac{6}{7}$$

You can now let either point be $(x_1, y_1)$ in Equation 14-5. Let $(6, 4)$ be $(x_1, y_1)$:

$$y - 4 = \frac{6}{7}(x - 6)$$

Then:

$$y = \frac{6}{7}x - \frac{36}{7} + 4$$

and the equation in slope-intercept form is:

$$y = \frac{6}{7}x - \frac{8}{7}$$

**3.** Line perpendicular to $2x - 3y = 5$ and passing through $(2, 1)$

Put the given line in the slope-intercept form:

$$3y = 2x - 5$$
$$y = \frac{2}{3}x - \frac{5}{3}$$

The slope of the given line is 2/3. The slope of the perpendicular line is the negative reciprocal, $-3/2$. Use Eq. (14-5) with $(x_1, y_1) = (2, 1)$ and $m = -3/2$ to obtain the equation of the perpendicular line:

$$y - 1 = -\frac{3}{2}(x - 2)$$

$$y = -\frac{3}{2}x + 4$$

This equation can also be written without fractions in what is called the general form:

$$3x + 2y - 8 = 0$$

The *general form* of the equation of a straight line is:

$$Ax + By + C = 0 \qquad (14\text{-}6)$$

where $A$, $B$, and $C$ are integers. The general form is useful when solving linear systems (Chap. 5).

### Example 14-7

The length $l$ of a metal rod is a linear function of the temperature $t$.

1. If $l_0$ = length at 0°C, express $l$ as a linear function of $t$.
2. An aluminum rod 10.000 cm long at 0°C increases to 10.024 cm at 100°C. Find the coefficient $a$ and express $l$ in terms of $t$.

### Solution

1. If $l_0$ = length at 0°C, express $l$ as a linear function of $t$.

   The linear function takes the form of the slope-intercept equation (14-4): $l = at + l_0$, where $l_0$ is the equivalent of the $y$ intercept ($t = 0$) and $a$ is the slope (commonly used instead of $m$ in this application).

2. An aluminum rod 10.000 cm long at 0°C increases to 10.024 cm at 100°C. Find the coefficient $a$ and express $l$ in terms of $t$.

   Substitute $l_0 = 10.000$, $l = 10.024$, and $t = 100$ and solve for $a$:

   $$10.024 = a(100) + 10.000$$

   $$a = \frac{0.024}{100} = 0.00024 \text{ cm/°C} \quad \text{or} \quad 2.4 \times 10^{-4} \text{ cm/°C}$$

   Then the linear function is:

   $$l = (0.00024)t + 10.000 \qquad \text{or} \qquad l = (2.4 \times 10^{-4})t + 10.000$$

### Exercise 14-1

In problems 1 to 8, for each pair of points find the distance between them, the slope, and the inclination of the line connecting the points.

1. (2, 3), (5, 7)
2. (−4, 5), (−1, 8)
3. (−2, −2), (4, −4)
4. (3, 4), (5, −1)
5. (1.5, 3.5), (5.0, −4.5)
6. (1/2, 1/3), (7/2, 4/3)
7. (−2, 6), (−1, 6)
8. (−5, −2), (−5, −4)

9. Prove that the three points $A(-1, -2)$, $B(3, 0)$, and $C(9, 3)$ are collinear (lie on the same line) by showing the distance relation: $d_{AB} + d_{BC} = d_{AC}$.

10. Prove problem 9 by showing the slope relation: $m_{AB} = m_{BC} = m_{AC}$.

11. Prove that the three points $A(2, 5)$, $B(3, -4)$, and $C(-2, 0)$ form a right triangle by showing that the slopes of the legs are negative reciprocals.

12. Prove problem 11 by showing the distance relation: $(d_{AC})^2 + (d_{BC})^2 = (d_{AB})^2$. (This is the converse of the pythagorean theorem.)

13. Prove that the points $A(3, 7)$, $B(5, 4)$, $C(2, 2)$, and $D(0, 5)$ form a rectangle.

14. Prove that the points $A(4.0, 2.0)$, $B(1.5, -1.0)$, $C(-0.5, -1.5)$, and $D(2.0, 1.5)$ form a parallelogram.

15. What kind of a triangle is formed by the three points $A(2, 5)$, $B(1, 6)$, and $C(-2, 2)$?

16. In problem 15 show that the line from $C$ to the midpoint of $AB$ is perpendicular to $AB$. *Note:* The coordinates of the *midpoint* are:

$$x_m = \frac{x_1 + x_2}{2} \qquad y_m = \frac{y_1 + y_2}{2}$$

where $(x_1, y_1)$ and $(x_2, y_2)$ are the endpoints.

In problems 17 to 20, find the slope and $y$ intercept of each line.

17. $y = 3x - 2$    18. $y - 4x = 5$
19. $2x - 3y = 6$    20. $3x + 2y = 1$

In problems 21 to 38, write the equation of each line in slope-intercept form.

21. $m = 0$, line passes through $(0, 2)$

22. $m = 1/2$, $y$ intercept $= 0$

23. Inclination $= 135°$, $y$ intercept $= 3$

24. Inclination $= 45°$, line passes through $(0, -1)$

25. $m = 3$, line passes through $(-4, 3)$

26. $m = -1/3$, line passes through $(1, 1)$

27. Line passes through $(-2, 0)$ and $(4, 4)$

28. Line passes through $(-3, -2)$ and $(-1, 2)$

29. Line parallel to $3x - 4y = 12$ and passes through $(1, -1)$

30. Line parallel to line joining $(1, 3)$ and $(-1, 7)$ and passes through $(2, -3)$

31. Line perpendicular to $3x - 4y = 12$ and passes through $(1, -1)$

32. Line perpendicular to line joining $(1, 3)$ and $(-1, 7)$ and passes through $(-2, -3)$

33. Horizontal line passing through $(6, 5)$

34. Vertical line passing through $(6, 5)$

35. Perpendicular bisector of line joining $(-2, -2)$ and $(2, 2)$ (See problem 16.)

36. Perpendicular bisector of line joining $(-1, 3)$ and $(3, 5)$ (See problem 16.)

37. Line passes through point $(2, 1)$ and intersection of lines $x + y = 5$ and $x + 2y = 7$

38. Slope $= -2$ and passes through intersection of $x - y = 2$ and $x$ axis

39. The resistance $R$ of a material is a linear function of the temperature $t$ in degrees Celsius. *(a)* If $R_0$ is the resistance at 0°C, express $R$ as a linear function of $t$, using $a$ as the slope. *(b)* If for an aluminum wire $R_0 = 0.300$ Ω and $R = 0.369$ Ω at 60°C, find the coefficient $a$ and express $R$ in terms of $t$. (See Example 14-7.)

40. In a dc circuit, when the internal resistance of the voltage source is taken into account, the voltage $E$ is a linear function of the current $I$ given by $E = IR + Ir$, where $R$ = circuit resistance and $r$ = internal resistance. If the resistance of the circuit $R = 4.0$ Ω and $I = 2.5$ A when $E = 12.0$ V, *(a)* find $r$ and *(b)* express $E$ as a function of $I$.

41. The velocity of sound $v$ in air is approximately a linear function of the temperature $t$. If $v_1 = 338$ m/s at $t_1 = 10$°C and $v_2 = 350$ m/s at $t_2 = 30$°C, express $v$ as a linear function of $t$. *(Hint:* First find the slope.)

42. The cost $C$ for use of a certain computer is a linear function of the time $t$. If 5 min costs $1000 and 15 min costs $2000, *(a)* express $C$ as a function of $t$ and *(b)* find the cost for 1 hour of computer time.

43. Hooke's law states that stress is proportional to strain. When applied to torque on a circular shaft, the law means that the angle of twist $\theta$ is a linear function of the stress $s$. If $\theta = 0$° when $s = 0$ and $\theta = 1.23$° when $s = 500$ lb/ft$^2$, *(a)* express $\theta$ as a linear function of $s$ and *(b)* find $\theta$ when $s = 700$ lb/ft$^2$.

44. The volume $V$ of a liquid is approximately a linear function of the temperature $t$. *(a)* Express $V$ as a function of $t$, using $a$ as the slope, if $V = V_0$ when $t = 0$°C. *(b)* If 4.00 L of lubricating oil at 0°C expands to 4.43 L at 120°C, find the slope and express $V$ in terms of $t$. (See Example 14-7.)

# 14-2 | *Parabola: Quadratic Function*

The path of a projectile in a gravitational field is a parabola, whose equation is a quadratic function. Satellite antennas, radio telescopes, and light reflectors are parabolic in shape. Cables and beams, under certain loads, assume the shape of parabolas, and the motion of a fluid flowing in a pipe forms a parabolic pattern. These are some examples of the importance of the parabola, which is the next analytic step after the straight line.

## *Quadratic Function*

The equation of a *parabola with a vertical axis* is given by the *quadratic function* (Fig. 14-9):

$$y = ax^2 + bx + c \tag{14-7}$$

The *axis of symmetry* divides it into two halves, which are mirror images of each other. *The parabola opens up when a > 0 and down when a < 0.* The

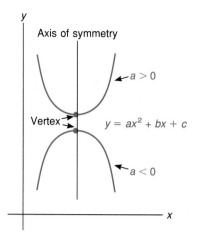

FIG. 14-9 *Parabola—graph of the quadratic function*

turning point is called the *vertex*. The curve opens wider and wider with distance from the vertex, and the sides approach the shape of vertical lines. The $x$ *coordinate of the vertex* is given by:

$$x = \frac{-b}{2a} \qquad\qquad (14\text{-}8)$$

### Example 14-8
Draw the graph of the parabola $y = x^2 - 4x + 3$, showing the vertex, the axis of symmetry, and at least four other points.

### Solution
The parabola opens up since $a = 1$ is positive. Find the $x$ coordinate of the vertex, using Eq. (14-8):

$$x = \frac{-b}{2a} = \frac{-(-4)}{2(1)} = 2$$

Then find the $y$ coordinate of the vertex by substituting the $x$ value into the given function: $y = (2)^2 - 4(2) + 3 = 4 - 8 + 3 = -1$. Since the parabola is symmetric on both sides of the vertex, choose values for $x$ on either side of the vertex, find the corresponding $y$ values, and draw the curve (Fig. 14-10):

vertex
↓

| $x$ | $-1$ | 0 | 1 | 2 | 3 | 4 | 5 |
|---|---|---|---|---|---|---|---|
| $y$ | 8 | 3 | 0 | $-1$ | 0 | 3 | 8 |

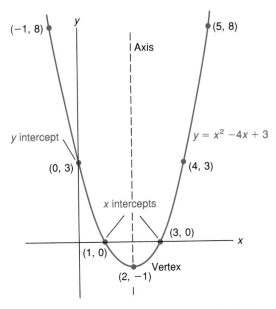

FIG. 14-10 *Parabola opening up, Example 14-8*

Notice that the $y$ intercept ($y = 3$) is where $x = 0$ and is the value of the constant term $c$ in the quadratic function (14-7). *The x intercepts represent points where y = 0 and are the solutions of the quadratic equation ax$^2$ + bx + c = 0.* In this example the equation is $x^2 - 4x + 3 = 0$, which is solved by factoring: $(x - 1)(x - 3) = 0$. The solutions $x = 1$ and $x = 3$ are the $x$ intercepts shown on the graph. The parabola can therefore be used to solve a quadratic equation by graphically determining its $x$ intercepts.

### Example 14-9
Draw the graph of the parabola $y = -2x^2 - 4x + 1$. Plot at least five points, including the vertex, $y$ intercept, and $x$ intercepts.

### Solution
The parabola opens down since $a = -2$ is negative. The vertex is $x = -(-4)/2(-2) = -1$, and $y = -2(-1)^2 - 4(-1) + 1 = 3$. The $y$ intercept is $c = 1$. The $x$ intercepts are the solutions of the equation $-2x^2 - 4x + 1 = 0$, which is solved by the quadratic formula:

$$x = \frac{-(-4) \pm \sqrt{(-4)^2 - 4(-2)(1)}}{2(-2)} = \frac{4 \pm \sqrt{24}}{-4} = \frac{4 \pm 2\sqrt{6}}{-4} = \frac{2 \pm \sqrt{6}}{-2}$$

$$x = -2.2 \quad \text{and} \quad x = 0.22$$

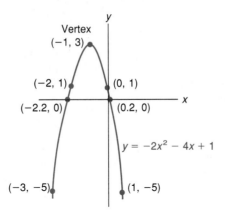

FIG. 14-11 *Parabola opening down, Example 14-9*

Choose values for $x$ on either side of the vertex, and draw the curve (Fig. 14-11):

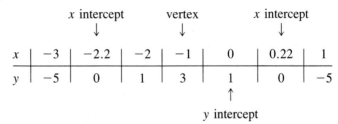

|  | *x* intercept |  | vertex |  | *x* intercept |  |
|---|---|---|---|---|---|---|
| $x$ | $-3$ | $-2.2$ | $-2$ | $-1$ | $0$ | $0.22$ | $1$ |
| $y$ | $-5$ | $0$ | $1$ | $3$ | $1$ | $0$ | $-5$ |

*y* intercept

The next example shows an important application of the quadratic function to motion in a gravitational field.

## Example 14-10

The height $s$ ft above the ground of a submarine-launched missile is given by the quadratic function $s = -16t^2 + 96t - 80$, where $t$ = time in seconds.

**1.** Find the times at which the missile leaves and returns to the water.

**2.** Draw the graph of $s$ vs. $t$, showing the maximum height and the points where $s = 0$.

## Solution

**1.** Find the times at which the missile leaves and returns to the water.

These are the values of $t$ when $s = 0$ (equivalent to the $x$ intercepts):

$$-16t^2 + 96t - 80 = 0$$
$$t^2 - 6t + 5 = 0 \quad \text{(divide by } -16\text{)}$$
$$(t - 1)(t - 5) = 0$$
$$t = 1 \qquad t = 5$$

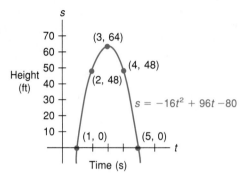

FIG. 14-12  *Height of missile, Example 14-10*

The missile leaves the water at $t = 1$ s and returns at $t = 5$ s.

**2.** Draw the graph of $s$ vs. $t$, showing the maximum height and the points where $s = 0$.

The parabola opens down ($a = -16$), and the maximum height is at the vertex:

$$t = \frac{-(96)}{2(-16)} = 3 \qquad \text{and} \qquad s = -16(3)^2 + 96(3) - 80 = 64 \text{ ft}$$

The table of values is given below and the graph is shown in Fig. 14-12:

| $t$ | 0 | 1 | 2 | 3 | 4 | 5 | 6 |
|---|---|---|---|---|---|---|---|
| $s$ | $-80$ | 0 | 48 | 64 | 48 | 0 | $-80$ |

## Definition of the Parabola and Standard Equation

The parabola is defined as the set of points equidistant from a fixed point called the *focus* and a fixed line called the *directrix*. Consider the parabola in Fig. 14-13, with its vertex at the origin and opening upward. The vertex is a point on the parabola and is therefore halfway between the focus and the directrix. The focus is $p$ units above the vertex at $(0, p)$. The directrix is perpendicular to the axis and $p$ units below the vertex. Its equation is $y = -p$. For every point $(x, y)$ on the parabola, the distance to the focus $d_1$ equals the distance to the directrix $d_2$. Applying the distance formula (14-1) leads to the equation of the parabola:

$$d_1 = d_2$$
$$\sqrt{(x - 0)^2 + (y - p)^2} = y + p$$

Square both sides: $\quad (x - 0)^2 + (y - p)^2 = (y + p)^2$

$$x^2 + \cancel{y^2} - 2py + \cancel{p^2} = \cancel{y^2} + 2py + \cancel{p^2}$$

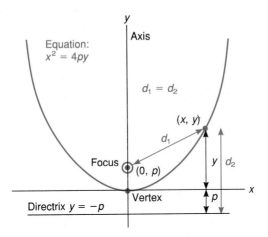

FIG. 14-13 *Definition of the parabola*

Therefore:

$$x^2 = 4py \qquad (14\text{-}9)$$

This is the *standard equation of the parabola with vertex (0, 0), axis the y axis, focus (0, p), and directrix the line $y = -p$*. This equation is useful when the focus or directrix is known or when either needs to be determined. Further treatment of the standard equation when the vertex is not at (0, 0) is presented in Sec. 14-7. Note that (14-9) can be written as $y = [1/(4p)]x^2$. This is the form of a quadratic function (14-7), where $a = 1/(4p)$, $b = 0$, and $c = 0$. Therefore when $p > 0$, the parabola opens up, and when $p < 0$, the parabola opens down.

### Example 14-11
Write the equation of a parabola with a vertex of (0, 0) and a focus of (0, 3).

### Solution
Since $p = 3$, $4p = 12$ and the equation is $x^2 = 12y$.

### Example 14-12
Find the focus and equation of the directrix, and sketch the parabola $x^2 = -4y$.

### Solution
Since $4p = -4$, $p = -1$. The focus is at $(0, -1)$, and the equation of the directrix is $y = 1$. The sketch is shown in Fig. 14-14.

Similarly it can be shown that the *standard equation of a parabola with vertex (0, 0), axis the x axis, focus (p, 0), and directrix the line $x = -p$* is:

$$y^2 = 4px \qquad (14\text{-}10)$$

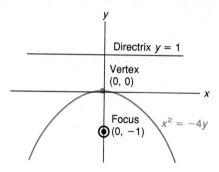

FIG. 14-14 *Parabola on the y axis, Example 14-12*

In this equation, when $p > 0$, the parabola opens to the right; and when $p < 0$, the parabola opens to the left.

### Example 14-13

Write the equation and sketch the parabola that has vertex $(0, 0)$ and an equation of the directrix $x = 3$.

### Solution

The directrix is perpendicular to the positive $x$ axis, and this parabola opens to the left with $p = -3$ and $4p = -12$. The equation is $y^2 = -12x$ (Fig. 14-15).

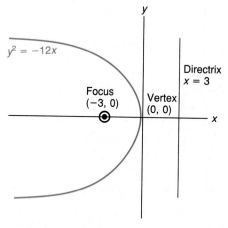

FIG. 14-15 *Parabola on the x axis, Example 14-13*

### Example 14-14

Given two parabolas whose vertices are $(0, 0)$ and pass through $(4, -2)$:

1. Sketch the two parabolas.
2. Find the equation of each parabola.

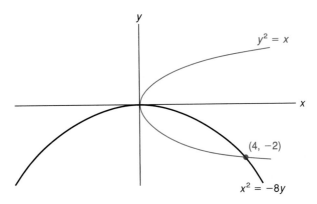

FIG. 14-16 *Parabolas with common point, Example 14-14*

## Solution

**1.** Sketch the two parabolas.

The sketch is shown in Fig. 14-16.

**2.** Find the equation of each parabola.

The parabola whose axis is the $y$ axis has an equation of the form $x^2 = 4py$, and the other parabola, $y^2 = 4px$. Substitute the coordinates of the point in each equation and solve for $p$:

$$x^2 = 4py \qquad\qquad y^2 = 4px$$
$$(4)^2 = 4p(-2) \qquad (-2)^2 = 4p(4)$$
$$p = \frac{16}{-8} = -2 \qquad p = \frac{4}{16} = \frac{1}{4}$$

The equations are then $x^2 = -8y$ and $y^2 = x$.

**Exercise 14-2**  In problems 1 to 16, draw the graph of each parabola. Plot at least five points, including the vertex, $y$ intercept, and $x$ intercepts.

**1.** $y = x^2 - 6x + 8$            **2.** $y = x^2 - 2x - 3$

**3.** $y = x^2 - 4$                  **4.** $y = x^2 - 4x$

**5.** $y = -x^2 + 6x - 5$        **6.** $y = -x^2 + 2x + 3$

**7.** $y = 3x - x^2$                 **8.** $y = 9 - x^2$

**9.** $y = 4x^2 - 4x + 1$        **10.** $y = x^2 + 2x + 3$

**11.** $y = 4x^2 - 8x + 3$       **12.** $y = 3x^2 + 2x - 1$

**13.** $y = 2x^2 - 4x + 1$       **14.** $y = x^2 + 2x - 1$

**15.** $y = -x^2 + 3x + 3$      **16.** $y = -3x^2 + 6x - 1$

In problems 17 to 26, for each parabola with vertex at the origin, find the focus and equation of the directrix, and sketch the curve.

**17.** $x^2 = 4y$          **18.** $x^2 = 12y$

**19.** $x^2 = -10y$         **20.** $x^2 = -6y$

**21.** $y^2 = 16x$          **22.** $y^2 = 8x$

**23.** $y^2 = -12x$         **24.** $y^2 = -2x$

**25.** $y = -x^2$           **26.** $y = 0.25x^2$

In problems 27 to 34, find the equation of each parabola with vertex at the origin.

**27.** Focus $(0, 3)$        **28.** Directrix $y = 4$

**29.** Directrix $x = -1$     **30.** Focus $(-2, 0)$

**31.** Focus $(0, -0.5)$      **32.** Focus $(1.5, 0)$

**33.** Axis on the $x$ axis, passes through $(-9, 3)$

**34.** Axis on the $y$ axis, passes through $(2, 1)$

**35.** A rocket travels in a trajectory given by the quadratic function $y = 1.2x - 0.010x^2$, where $x$ = horizontal distance in meters and $y$ = vertical distance in meters. Draw the graph of the trajectory showing the points where $y = 0$ and the maximum height of the rocket.

**36.** A baseball is thrown horizontally. Its height $h$ in feet above the ground is given by the quadratic function $h = 4.9 - 0.001x^2$, where $x$ is the horizontal distance, $x >$ 0. Draw the graph of $h$ vs. $x$ from $x = 0$ ft to when the ball hits the ground ($h = 0$).

**37.** An automobile alternator generates power at 14 V and has an internal resistance of 0.20 $\Omega$. The power output is given by the quadratic function $P = 14I - 0.20I^2$, where $I$ = current. *(a)* At what current does the alternator generate maximum power, and what is the maximum power? *(b)* Find the points where $P = 0$, and draw the graph of $P$ vs. $I$ between these points.

**38.** The heat in joules (J) produced by a voltage $V$ across a heating coil is given by $H = V^2 t/R$, where $t$ = time and $R$ = resistance. Given $t = 1$ h and $R = 10 \ \Omega$, draw the graph of $H$ vs. $V$ from $V = 0$ to 120 V.

**39.** A communications satellite reflector is parabolic, so that all radio waves from space entering parallel to its axis are reflected to the antenna at its focus (Fig. 14-17). If the reflector's vertical cross section is given by the equation $x^2 = 100y$, where $x$ and $y$ are in meters, *(a)* find the height of the antenna at the focus and *(b)* find the coordinates of the point $(x, y)$, shown in Fig. 14-17, at which radio waves are reflected at right angles to the axis of the parabola. (*Hint:* Find the distance to the directrix.)

**40.** The velocity distribution of natural gas flowing smoothly in a pipeline is given by $V = 6.6x - x^2$, where $V$ = velocity in meters per second and $x$ = distance in meters from the inside wall of the pipe. *(a)* Draw the graph of $V$ vs. $x$ for the positive range of $V$. *(b)* What is the maximum velocity of the gas?

**41.** *(a)* Sketch the parabola $y = x^2$ and the line $y = x$ on the same graph, and find the points of intersection. *(b)* Determine the points algebraically by substituting $x$ for $y$ in $y = x^2$ and solving the system simultaneously.

**42.** *(a)* Sketch the two parabolas $y = x^2$ and $x = y^2$ on the same graph, and find the

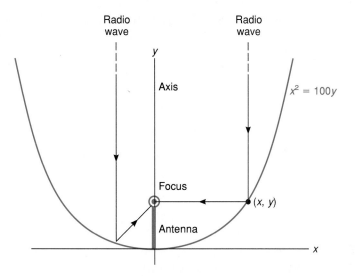

FIG. 14-17 *Communications satellite antenna, problem 39*

points of intersection. *(b)* Determine the points algebraically by substituting $y^2$ for $x$ in $y = x^2$ and solving the system simultaneously.

**43.** The parabola $y = x^2 - 6x + 9$ has only one $x$ intercept. Draw the graph and explain what happens algebraically when you set $y = 0$ to find the $x$ intercepts.

**44.** The parabola $y = x^2 - 6x + 10$ has no $x$ intercepts. Draw the graph and explain what happens algebraically when you set $y = 0$ to find the $x$ intercepts.

**45.** *(a)* Sketch the two parabolas whose vertices are $(0, 0)$ and pass through $(-2, -2)$. *(b)* Find the equation of each curve.

**46.** Using the derivation of Eq. (14-9) in the text as a guide, derive Eq. (14-10) for the parabola with vertex $(0, 0)$ and focus $(p, 0)$.

# 14-3 | *Circle*

The circle is the most familiar curve, and its importance is seen everywhere. It is defined as the set of points a fixed distance from a fixed point. The fixed point is the *center* of the circle, and the fixed distance is the *radius*.

## *Standard Equation*

Consider the circle in Fig. 14-18 with center at $(h, k)$ and radius $r$. Applying the distance formula (14-1) to the center and any point $(x, y)$ yields $r = \sqrt{(x - h)^2 + (y - k)^2}$. This leads to the *standard equation of the circle:*

$$(x - h)^2 + (y - k)^2 = r^2 \qquad (14\text{-}11)$$

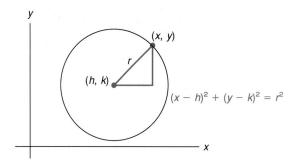

FIG. 14-18 *Standard equation of the circle*

Study the following examples, which show how to apply Eq. (14-11).

### Example 14-15
Write the equation of each circle.

**1.** Center $(1, 4)$; $r = 2$
**2.** Center $(-2, 5)$; $r = 3$
**3.** Center $(0, 0)$; $r = 1.5$
**4.** Center $(1, 3)$ and passes through $(-2, -1)$

### Solution
**1.** Center $(1, 4)$; $r = 2$

Apply Eq. (14-11) with $h = 1$, $k = 4$, and $r^2 = 2^2 = 4$:

$$(x - 1)^2 + (y - 4)^2 = 4$$

**2.** Center $(-2, 5)$; $r = 3$

Apply Eq. (14-11) with $h = -2$, $k = 5$, and $r = 3$:

$$[x - (-2)]^2 + (y - 5)^2 = 3^2$$
$$(x + 2)^2 + (y - 5)^2 = 9$$

**3.** Center $(0, 0)$; $r = 1.5$

Apply Eq. (14-11) with $h = 0$, $k = 0$, and $r = 1.5$:

$$(x - 0)^2 + (y - 0)^2 = (1.5)^2$$
$$x^2 + y^2 = 2.25$$

Notice in (3) that a *circle with center (0, 0)* has an equation of the form

$$x^2 + y^2 = r^2 \qquad (14\text{-}12)$$

**4.** Center $(1, 3)$ and passes through $(-2, -1)$

The equation has the form $(x - 1)^2 + (y - 3)^2 = r^2$. Substitute $-2$ for $x$ and $-1$ for $y$, since the values must satisfy the equation, and solve for $r^2$:

$$(-2 - 1)^2 + (-1 - 3)^2 = r^2$$
$$r^2 = 9 + 16 = 25$$

The equation is then:

$$(x - 1)^2 + (y - 3)^2 = 25$$

### Example 14-16
Sketch the circle $(x + 1)^2 + (y + 3)^2 = 16$, showing the center and four points.

### Solution
The center $(h, k) = (-1, -3)$, and $r = \sqrt{16} = 4$. Plot the center and the following four points: above the center, below the center, and to the right and left of it (Fig. 14-19). The four points are plotted by measuring distances equal to the radius in the $x$ and $y$ directions.

When you multiply out the standard equation of the circle (14-11), you obtain:

$$x^2 - 2hx + h^2 + y^2 - 2ky + k^2 = r^2$$

If you order the terms on one side:

$$x^2 + y^2 + (-2h)x + (-2k)y + (h^2 + k^2 - r^2) = 0$$

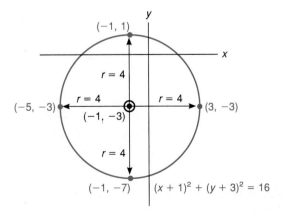

FIG. 14-19 *Circle, Example 14-16*

you have the *general equation of the circle*:

$$x^2 + y^2 + Dx + Ey + F = 0 \qquad (14\text{-}13)$$

where $D = -2h$, $E = -2k$, and $F = h^2 + k^2 - r^2$. When you are given the general equation of a circle and want to find the center and radius, you have to reverse the above process and change the general equation to the standard equation. This is done by *completing the square,* which is the process used to derive the quadratic formula (Sec. 7-7). Study the next example closely to learn the procedure.

## Example 14-17
Find the center and radius of the circle $x^2 + y^2 - 4x - 2y - 4 = 0$.

## Solution
Order the $x$ and $y$ terms separately, move the constant to the right side, and express the equation:

$$(x^2 - 4x + ?) + (y^2 - 2y + ?) = 4$$

The question marks represent values to be added to the left side of the equation to make each expression in parentheses a perfect square. To obtain the necessary value in each set of parentheses, *halve the middle coefficient and square it.* For the $x$ terms, the value to be added is $(-4/2)^2 = 4$, and for the $y$ terms $(-2/2)^2 = 1$. The values must be added to both sides to balance the equation:

$$(x^2 - 4x + 4) + (y^2 - 2y + 1) = 4 + (4) + (1)$$

The left side can now be factored:

$$(x - 2)^2 + (y - 1)^2 = 9$$

The reason this works is as follows. If the first coefficient is 1, the second number in each factor is half the middle coefficient and when squared equals the value that was added. The last equation is the standard equation of the circle. The center is therefore $(2, 1)$, and the radius is $\sqrt{9} = 3$.

## Example 14-18
Sketch the circle $4x^2 + 4y^2 - 12x + 24y + 20 = 0$.

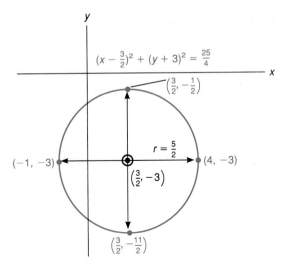

FIG. 14-20  *Circle, Example 14-18*

**Solution**

Change the equation to the standard form by first dividing out the common coefficient of $x^2$ and $y^2$, 4, and then completing the square:

$$x^2 + y^2 - 3x + 6y + 5 = 0$$
$$(x^2 - 3x + ?) + (y^2 + 6y + ?) = -5$$
$$\left(x^2 - 3x + \frac{9}{4}\right) + (y^2 + 6y + 9) = -5 + \left(\frac{9}{4}\right) + (9) = \frac{25}{4}$$
$$\left(x - \frac{3}{2}\right)^2 + (y + 3)^2 = \left(\frac{5}{2}\right)^2$$

The center is therefore (3/2, −3) and $r = 5/2$, or 2.5. The sketch (Fig. 14-20) shows four points plotted by measuring the radius in the $x$ and $y$ directions from the center.

**Exercise 14-3**   In problems 1 to 12, write the equation of each circle.

**1.** Center (2, 3); $r = 3$          **2.** Center (4, 4); $r = 5$

**3.** Center (−1, 2); $r = 1$          **4.** Center (−3, −5); $r = 4$

**5.** Center (1/2, 0); $r = 7/2$          **6.** Center (0, 1.5); $r = 2.2$

**7.** Center (0, 0); passes through (1, 3)

**8.** Center (8, −6); passes through (5, −2)

**9.** Center (0, −3); tangent to $x$ axis at origin

10. Tangent to $y$ axis at $(0, 7)$ and $x$ axis at $(7, 0)$

11. Center on $x$ axis; tangent to line $y = 3$ at $(3, 3)$

12. Concentric with $x^2 + y^2 = 8$ (same center); $r = \sqrt{10}$

In problems 13 to 18, sketch each circle, showing the center and four points. (See Example 14-16.)

13. $(x - 4)^2 + (y - 1)^2 = 16$     14. $(x - 2)^2 + (y + 5)^2 = 9$

15. $x^2 + y^2 = 25$                  16. $9x^2 + 9y^2 = 36$

17. $x^2 + (y - 2)^2 = 12$            18. $(x - 3.5)^2 + y^2 = 4$

In problems 19 to 28, find the center and radius of each circle by changing the general equation to the standard equation. Sketch each circle. (See Examples 14-17 and 14-18.)

19. $x^2 + y^2 - 2x - 4y + 4 = 0$

20. $x^2 + y^2 - 4x - 8y + 16 = 0$

21. $x^2 + y^2 + 6x + 8y = 0$

22. $x^2 + y^2 + 6x + 6y + 9 = 0$

23. $x^2 + y^2 - x - 2 = 0$

24. $x^2 + y^2 - 5y = 0$

25. $3x^2 + 3y^2 + 12x - 18y + 12 = 0$

26. $5x^2 + 5y^2 - 10x + 10y = 0$

27. $4x^2 + 4y^2 - 12x - 12y - 7 = 0$

28. $3x^2 + 3y^2 + 2y - 1 = 0$

29. Figure 14-21 shows a pulley belt drive drawn on a coordinate system. If the radius of the large pulley is 4.2 cm and the radius of the small pulley is 2.1 cm, find the equation of each circular pulley. (*Hint:* Use the pythagorean theorem.)

30. A cylindrical capacitor consists of two concentric cylinders separated by a distance of 2.5 mm. The cross section of the capacitor is drawn on a coordinate system with the outer circle in the first quadrant tangent to the $x$ and $y$ axes. If the radius of the inner circle is 4.5 mm, find the equation of each circle.

31. An archway is in the shape of a rectangle with a semicircle on top. The archway is drawn on a coordinate system such that the four vertices of the rectangle are $(2, 0)$, $(6, 0)$, $(6, 5)$, and $(2, 5)$. What is the equation of the circle that the semicircle is part of?

32. The earth's orbit around the sun and the moon's orbit around the earth are approximate circles of radii $1.5 \times 10^8$ and $3.8 \times 10^5$ km, respectively, where the sun is at the center of the earth's orbit and the earth is at the center of the moon's orbit. If the sun is placed at the origin of a coordinate system, what is the equation of the moon's orbit when the center of the earth is *(a)* on the positive $x$ axis and *(b)* in the first quadrant on the line $y = x$?

33. *(a)* Sketch the circle $x^2 + y^2 = 16$ and the line $y = x$ on the same graph, and estimate the points of intersection. *(b)* Determine the points of intersection exactly by substituting $x$ for $y$ in the equation $x^2 + y^2 = 16$ and solving the system algebraically.

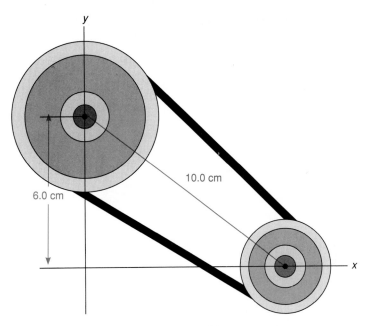

FIG. 14-21 *Pulley belt drive, problem 29*

**34.** *(a)* Sketch the circles $x^2 + (y - 3)^2 = 25$ and $x^2 + (y + 3)^2 = 25$ on the same graph, and estimate the points of intersection. *(b)* Determine the points of intersection by subtracting the two equations and solving the system algebraically.

**35.** Find the equation of the circle passing through the points $(2, 3)$, $(-1, 4)$, and $(3, -4)$. Use the general equation to set up three equations in three unknowns $(D, E, F)$ by substituting the coordinates of each point, and solve the system.

**36.** Equation (14-13) does not always represent a circle. Complete the square for the following, and explain why each does not represent a circle: *(a)* $x^2 + y^2 + 4x + 6y + 13 = 0$; *(b)* $x^2 + y^2 - 2x + 8y + 18 = 0$.

# 14-4 | *Ellipse*

The circle is actually a special case of the ellipse. Satellites, planets, and most heavenly bodies travel in elliptical orbits. Cams and gears are made in elliptical shapes, and ellipses are used in the design of arches and other structures in buildings and bridges.

The *ellipse* is defined as a set of points such that the sum of the distances from each point to two fixed points is constant. Each fixed point is called a *focus*. The *major axis* of the ellipse passes through the foci, and its length is $2a$. Figure 14-22 shows an ellipse with its center at the origin and foci at $(c, 0)$ and $(-c, 0)$. The ends of the major axis are called *vertices*. The *minor axis* is

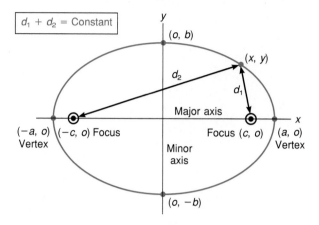

FIG. 14-22 *Definition of the ellipse*

perpendicular to the major axis, and its length is $2b$. Consider the vertex $(a, 0)$ of the ellipse in Fig. 14-23. The sum of the distances from the vertex to the foci is $d_1 + d_2 = (a - c) + (a + c) = 2a$, or the major axis. Therefore for any point, the constant sum $d_1 + d_2 = 2a$. Consider the end of the minor axis $(0, b)$ in Fig. 14-23. The distances to the foci are equal, and the sum is $2a$, so each distance is equal to $a$ as shown. The pythagorean relation follows where $a$, not $c$, is the largest quantity:

$$b^2 + c^2 = a^2 \qquad (14\text{-}14)$$

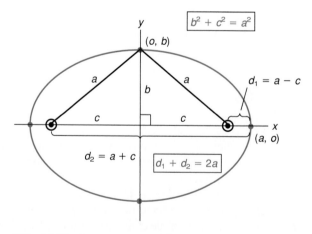

FIG. 14-23 *Geometry of the ellipse*

## Standard Equation

From the definition and Fig. 14-22, the following is true for any point $(x, y)$ on the ellipse:

$$d_1 + d_2 = \text{constant} = 2a$$
$$\sqrt{(x - c)^2 + (y - 0)^2} + \sqrt{(x + c)^2 + (y - 0)^2} = 2a$$

The equation of the ellipse follows from the above by isolating the radicals two times, squaring, and simplifying:

$$\sqrt{(x - c)^2 + y^2} = 2a - \sqrt{(x + c)^2 + y^2}$$

$$\left[\sqrt{(x - c)^2 + y^2}\right]^2 = \left[2a - \sqrt{(x + c)^2 + y^2}\right]^2$$

$$(x - c)^2 + y^2 = 4a^2 - 4a\sqrt{(x + c)^2 + y^2} + (x + c)^2 + y^2$$

$$4a\sqrt{(x + c)^2 + y^2} = 4cx + 4a^2$$

$$\left[a\sqrt{(x + c)^2 + y^2}\right]^2 = (cx + a^2)^2$$

$$a^2[(x + c)^2 + y^2] = c^2x^2 + 2a^2cx + a^4$$

$$a^2x^2 + 2a^2cx + a^2c^2 + a^2y^2 = c^2x^2 + 2a^2cx + a^4$$

$$(a^2 - c^2)x^2 + a^2y^2 = a^2(a^2 - c^2)$$

From Eq. (14-14), $b^2 = a^2 - c^2$:

$$b^2x^2 + a^2y^2 = a^2b^2$$

Dividing by $a^2b^2$ leads to the *standard equation of the ellipse with center (0, 0) and major axis the x axis:*

$$\frac{x^2}{a^2} + \frac{y^2}{b^2} = 1 \qquad (14\text{-}15)$$

### Example 14-19
Sketch the ellipse $x^2/25 + y^2/16 = 1$, showing the vertices, foci, and ends of the minor axis.

### Solution
From the equation, $a^2 = 25$, $b^2 = 16$, and $c^2 = a^2 - b^2 = 25 - 16 = 9$. Then $a = 5$, $b = 4$, and $c = 3$. Plot the vertices and ends of the minor axis. Sketch the ellipse through the four points (Fig. 14-24).

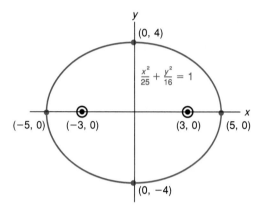

FIG. 14-24 *Ellipse on the x axis, Example 14-19*

When the *major axis is the y axis, the standard equation of the ellipse with center (0, 0) is*:

$$\frac{x^2}{b^2} + \frac{y^2}{a^2} = 1 \qquad (14\text{-}16)$$

Equation (14-16) is equivalent to Eq. (14-15) with $x$ and $y$ interchanged.

### Example 14-20

Sketch the ellipse $25x^2 + 16y^2 = 400$, showing the vertices, foci, and ends of the minor axis.

### Solution

Divide by 400 to change the equation to the standard equation:

$$\frac{\cancel{25}x^2}{\cancel{400}_{16}} + \frac{\cancel{16}y^2}{\cancel{400}_{25}} = \frac{\cancel{400}}{\cancel{400}}$$

$$\frac{x^2}{16} + \frac{y^2}{25} = 1$$

*The larger denominator represents the major axis,* which is the $y$ axis. Then $a^2 = 25$, $b^2 = 16$, $c^2 = 25 - 16 = 9$, and $a = 5$, $b = 4$, $c = 3$. The ellipse is shown in Fig. 14-25. Compare it with the ellipse in Fig. 14-24. Notice that the ellipses are the same curve but the major axis has been rotated 90°.

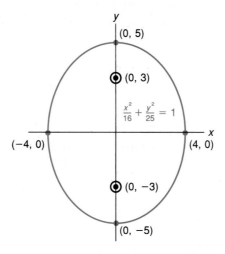

FIG. 14-25 *Ellipse on the y axis, Example 14-20*

The four intercepts of the ellipse can also be found by setting $x = 0$ and $y = 0$. For example, for the above ellipse $25x^2 + 16y^2 = 400$, when $x = 0$:

$$16y^2 = 400$$
$$y^2 = 25$$
$$y = \pm 5$$

and when $y = 0$:

$$25x^2 = 400$$
$$x^2 = 16$$
$$x = \pm 4$$

Observe that when $a = b$, the ellipse is a circle: $x^2/a^2 + y^2/a^2 = 1$, or $x^2 + y^2 = a^2$. The radius of the circle is then $a$.

### Example 14-21
Find the equation of the following ellipses with center $(0, 0)$.

**1.** Vertices $(\pm 3, 0)$, foci $(\pm 2, 0)$
**2.** Foci $(0, \pm\sqrt{3})$, ends of the minor axis $(\pm 1, 0)$
**3.** Vertex $(2, 0)$, passes through $(1, 1)$

### Solution
**1.** Vertices $(\pm 3, 0)$, foci $(\pm 2, 0)$

The vertices are on the major axis, which is the $x$ axis, and $a = 3$, $c = 2$. Then $b^2 = a^2 - c^2 = 9 - 4 = 5$. The equation is:

$$\frac{x^2}{9} + \frac{y^2}{5} = 1$$

**2.** Foci $(0, \pm\sqrt{3})$, ends of the minor axis $(\pm 1, 0)$

The foci are on the major axis, which is the $y$ axis, and $b = 1$, $c = \sqrt{3}$. Then $a^2 = b^2 + c^2 = 1 + 3 = 4$. The equation is:

$$\frac{x^2}{1} + \frac{y^2}{4} = 1$$

**3.** Vertex $(2, 0)$, passes through $(1, 1)$

The major axis is the $x$ axis with $a^2 = 4$. The equation has the form $x^2/4 + y^2/b^2 = 1$. Substitute the coordinates of the point, and solve for $b^2$:

$$\frac{(1)^2}{4} + \frac{(1)^2}{b^2} = 1$$

$$\frac{1}{b^2} = 1 - \frac{1}{4} = \frac{3}{4}$$

$$b^2 = \frac{4}{3}$$

The equation is then $x^2/4 + y^2/(4/3) = 1$ or $x^2 + 3y^2 = 4$.

The next example shows an application of the ellipse to the orbit of the earth.

© **Example 14-22**
The earth's orbit is an ellipse with a semimajor axis of $a = 1.49596 \times 10^8$ km ($9.2955 \times 10^7$ mi) and a semiminor axis $b = 1.49575 \times 10^8$ km ($9.2942 \times 10^7$ mi). The orbit is almost a circle since $a$ is close in value to $b$. If the sun is at one focus, find the closest distance of the earth to the sun (perihelion) and the farthest distance (aphelion) in kilometers and in miles.

**Solution**
The distance from the center of the orbit to the sun (the focus) is:

$$c = \sqrt{a^2 - b^2} = \sqrt{(1.49596 \times 10^8)^2 - (1.49575 \times 10^8)^2}$$
$$= \sqrt{6.28259 \times 10^{12}} = 2.5065 \times 10^6 \text{ km}$$

The perihelion is then:

$$a - c = 1.49596 \times 10^8 - 2.5065 \times 10^6$$
$$= 1.47090 \times 10^8 \text{ km}$$

or $\qquad (1.47090 \times 10^8 \text{ km})(0.62137 \text{ mi/km}) = 9.1397 \times 10^7 \text{ mi}$

The aphelion is:

$$a + c = 1.49596 \times 10^8 + 2.5065 \times 10^6$$
$$= 1.52103 \times 10^8 \text{ km} \quad \text{or} \quad 9.4512 \times 10^7 \text{ mi}$$

The equation of an ellipse whose center is not at the origin is discussed in Sec. 14-7.

**Exercise 14-4**

In problems 1 to 12, sketch each ellipse, showing the vertices, foci, and ends of the minor axis.

**1.** $\dfrac{x^2}{25} + \dfrac{y^2}{9} = 1$  $\qquad$ **2.** $\dfrac{x^2}{36} + \dfrac{y^2}{100} = 1$

**3.** $\dfrac{x^2}{25} + \dfrac{y^2}{36} = 1$  $\qquad$ **4.** $\dfrac{x^2}{16} + \dfrac{y^2}{4} = 1$

**5.** $x^2 + 4y^2 = 4$  $\qquad$ **6.** $4x^2 + 9y^2 = 36$

**7.** $4x^2 + y^2 = 36$  $\qquad$ **8.** $25x^2 + 9y^2 = 225$

**9.** $3x^2 + 4y^2 = 12$  $\qquad$ **10.** $5x^2 + y^2 = 20$

**11.** $9x^2 + 4y^2 = 16$  $\qquad$ **12.** $16x^2 + 25y^2 = 25$

In problems 13 to 22, find the equation of each ellipse with center $(0, 0)$.

**13.** Vertices $(\pm 5, 0)$, foci $(\pm 3, 0)$

**14.** Foci $(\pm 8, 0)$, ends of minor axis $(0, \pm 6)$

**15.** Vertex $(0, 10)$, distance from focus to vertex $= 2$

**16.** Focus $(0, 1)$, major axis $= 6$

**17.** Vertex $(0, 5)$, passes through $(9/5, 4)$

**18.** End of minor axis $(0, 1)$, passes through $(-1, \frac{1}{2})$

**19.** Focus $(0, \sqrt{10})$, vertex $(0, 4)$

**20.** Vertex $(1.5, 0)$, minor axis $= 2$

**21.** Same foci as $9x^2 + 25y^2 = 225$, minor axis $= 8$

**22.** Same vertices as $2x^2 + y^2 = 16$, same foci as $25x^2 + 16y^2 = 400$

**23.** The end of a pipe whose inside diameter is 10 cm is cut at a 60° angle as shown in Fig. 14-26. What are the lengths of the major and minor axes of the elliptical opening?

**24.** An elliptical cam rotates about its focus and moves a lever up and down, which rests on the cam as shown in Fig. 14-27. If the major axis of the cam is 7.0 in. and the minor axis is 5.6 in., how far does point $A$ on the lever move from one extreme position to the other?

FIG. 14-26 *Cut pipe, problem 23*

C **25.** The moon's orbit is an ellipse with a semimajor axis of $3.8200 \times 10^5$ km ($2.3736 \times 10^5$ mi) and a semiminor axis of $3.8118 \times 10^5$ km ($2.3686 \times 10^5$ mi). If the earth is at one focus, find the closest distance of the moon to the earth (perigee) and the farthest distance (apogee) in kilometers and in miles. Disregard the diameters of the bodies. (See Example 14-22.)

**26.** The design of an automobile cam is programmed by using the equation of an ellipse and of a circle: $0.16x^2 + 0.25y^2 = 0.04$ and $(x - 0.30)^2 + y^2 = 0.01$. Sketch these two curves on the same graph to show the design of the cam. Use suitable scales for $x$ and $y$.

**27.** The whispering gallery in the Capitol in Washington, D.C., has a domed ceiling in the shape of an ellipsoid, the surface formed by rotating an ellipse about one of its axes. A whisper at one focus of the gallery can be clearly heard at the other focus (an ellipse reflects waves from one focus to the other focus). If the distance between the foci is 48 ft and the length of the room (the major axis) is 52 ft, what is the equation of the ellipse that forms the gallery? Assume that the major axis is the $x$ axis and the center is $(0, 0)$.

**28.** An ellipse can be drawn by attaching a string at two ends so that it forms a loop and, holding the string taut with a pencil, moving the pencil along the string. If the string is 1.00 m long and the distance between its ends is 0.80 m, what is the equation of the ellipse? Let the center be the origin and the $x$ axis the major axis. (*Hint:* The length of the string is equal to the sum of the distances from a point to the foci.)

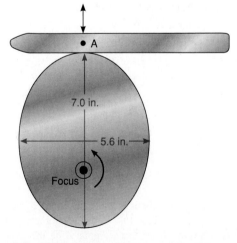

FIG. 14-27 *Cam and lever, problem 24*

29. *(a)* Sketch the ellipse $2x^2 + y^2 = 12$ and the line $y = -x$ on the same graph, and estimate the points of intersection. *(b)* Determine the points exactly by substituting $-x$ for $y$ in $2x^2 + y^2 = 12$ and solving the system algebraically.

30. *(a)* Sketch the ellipses $x^2 + 2y^2 = 12$ and $2x^2 + y^2 = 12$ on the same graph, and estimate the four points of intersection. *(b)* Determine the points exactly by solving the system algebraically. Eliminate $x^2$ or $y^2$ by combining equations.

31. The *eccentricity e* of an ellipse is a measure of its elongation and is defined as $e = c/a$. For a circle $e = 0$, whereas $0 < e < 1$ for an ellipse that is not a circle. What is the equation of an ellipse whose eccentricity $e = 0.60$, center is $(0, 0)$, and vertex is $(0, 10)$?

32. Derive Eq. (14-16) by following a procedure similar to that shown in the text for Eq. (14-15).

# 14-5 | *Hyperbola*

Certain heavenly bodies, such as comets, can travel in orbits that are hyperbolas. A shock wave or sonic boom is hyperbolic, and inverse proportional relations graph as hyperbolas.

The *hyperbola* is closely related to the ellipse in its definition and equation. It is defined as a set of points such that the difference of the distances from each point to two fixed points, or *foci*, is constant. The *transverse axis* of the hyperbola passes through the foci, and its length is $2a$. Figure 14-28 shows a hyper-

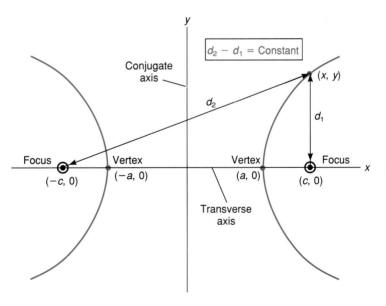

FIG. 14-28 *Definition of the hyperbola*

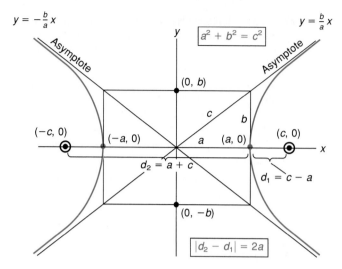

FIG. 14-29 *Geometry of the hyperbola*

bola with its center at the origin and foci at $(c, 0)$ and $(-c, 0)$. The ends of the transverse axis are the *vertices*. The *conjugate axis* is perpendicular to the transverse axis, and its length is $2b$; its meaning is shown in Fig. 14-29. The diagonals of the rectangle whose dimensions are $2a$ and $2b$ are called *asymptotes*. The hyperbola approaches the asymptotes with distance from the vertex but never touches them. The asymptotes are the sides of the cone shown in Fig. 14-1 when it is cut to form the hyperbola.

Drawing the asymptotes enables you to readily sketch the curve. Consider the vertex $(a, 0)$ in Fig. 14-29. The difference of the distances from this vertex to the foci is $d_2 - d_1 = (a + c) - (c - a) = 2a$, which is the same constant as in the ellipse. For a point on the left branch of the hyperbola, $d_1 > d_2$ and the absolute (positive) value of this difference is used. It can be shown that the diagonal of the rectangle is $2c$ and therefore the pythagorean relation between $a$, $b$, and $c$ is

$$a^2 + b^2 = c^2 \qquad (14\text{-}17)$$

Note that *c is greater than a or b* and that *b can be greater than, equal to, or less than a*. This is not the same as in the ellipse, where $a > b$ and $a > c$.

## Standard Equation

The equation of the hyperbola is derived by a procedure similar to that shown for the ellipse, (14-15). By starting with:

$$|d_2 - d_1| = 2a$$
$$\sqrt{(x + c)^2 + y^2} - \sqrt{(x - c)^2 - y^2} = 2a$$

the equation leads to *the standard equation of the hyperbola with center (0, 0) and transverse axis the x axis:*

$$\frac{x^2}{a^2} - \frac{y^2}{b^2} = 1 \tag{14-18}$$

Notice that the equation is similar to that of the ellipse but has a minus sign. *The positive term indicates the transverse axis.*

### Example 14-23
Sketch the hyperbola $x^2/9 - y^2/16 = 1$, showing the vertices, foci, and asymptotes.

### Solution
Notice that the denominator of $x^2$ is smaller than $y^2$, but the $x$ term is positive, so the hyperbola opens in the $x$ direction. From the equation, $a^2 = 9$, $b^2 = 16$, and $c^2 = a^2 + b^2 = 9 + 16 = 25$. Then $a = 3$, $b = 4$, and $c = 5$. Plot the vertices and ends of the conjugate axis. Draw the rectangle and asymptotes lightly, and sketch the hyperbola within the asymptotes (Fig. 14-30).

*When the transverse axis is the y axis, the standard equation of the hyperbola with center (0, 0) is:*

$$\frac{y^2}{a^2} - \frac{x^2}{b^2} = 1 \tag{14-19}$$

Notice that the positive term is the $y$ term.

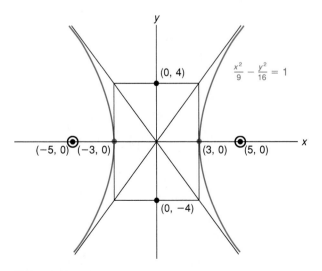

FIG. 14-30 *Hyperbola on the x axis, Example 14-23*

### Example 14-24

Sketch the hyperbola $9y^2 - 16x^2 = 144$, showing the vertices, foci, and asymptotes.

### Solution

Divide by 144 to change the equation to the standard equation:

$$\frac{\cancel{9}y^2}{\cancel{144}} - \frac{\cancel{16}x^2}{\cancel{144}} = \frac{\cancel{144}}{\cancel{144}}$$
$$\quad\ \ 16 \qquad 9$$

$$\frac{y^2}{16} - \frac{x^2}{9} = 1$$

The $y$ term is positive, so the hyperbola opens in the $y$ direction. Then $a^2 = 16$, $b^2 = 9$, $c^2 = a^2 + b^2 = 25$, and $a = 4$, $b = 3$, $c = 5$. Compare this hyperbola (Fig. 14-31) with the one from Fig. 14-30, shown dashed. They are *not* the same curve but *share* the same asymptotes. Two hyperbolas that share the same asymptotes are called *conjugate hyperbolas*. The equation of the conjugate hyperbola can be obtained by changing the signs of the $x$ and $y$ terms.

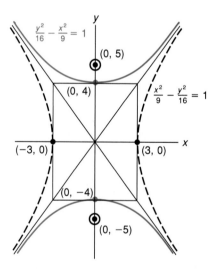

FIG. 14-31 *Conjugate hyperbolas, Example 14-24*

### Example 14-25

Find the equation of the following hyperbolas with center $(0, 0)$.

**1.** Vertices $(\pm 3, 0)$, foci $(\pm 4, 0)$

**2.** Foci $(0, \pm 3)$, ends of conjugate axis $(\pm 2, 0)$

**3.** Conjugate hyperbola to $9x^2 - 16y^2 = 144$

**4.** Vertex $(3, 0)$, passes through $(5, 4)$

**Solution**

**1.** Vertices $(\pm 3, 0)$, foci $(\pm 4, 0)$

The transverse axis is the $x$ axis, and $a = 3$, $c = 4$. Then $b^2 = c^2 - a^2 = 16 - 9 = 7$. Applying (14-19) the equation is:

$$\frac{x^2}{9} - \frac{y^2}{7} = 1$$

**2.** Foci $(0, \pm 3)$, ends of conjugate axis $(\pm 2, 0)$

The transverse axis is the $y$ axis, and $c = 3$, $b = 2$. Then $a^2 = c^2 - b^2 = 9 - 4 = 5$. Applying (14-19) the equation is:

$$\frac{y^2}{5} - \frac{x^2}{4} = 1$$

**3.** Conjugate hyperbola to $9x^2 - 16y^2 = 144$

The conjugate hyperbola can be obtained by changing the signs of the $x$ and $y$ terms:

$$16y^2 - 9x^2 = 144 \qquad \text{or} \qquad \frac{y^2}{9} - \frac{x^2}{16} = 1$$

**4.** Vertex $(3, 0)$, passes through $(5, 4)$

The transverse axis is the $x$ axis, and $a^2 = 9$. The equation has the form:

$$\frac{x^2}{9} - \frac{y^2}{b^2} = 1$$

Substitute the coordinates of the point and solve for $b^2$:

$$\frac{(5)^2}{9} - \frac{(4)^2}{b^2} = 1$$
$$25b^2 - 144 = 9b^2$$
$$b^2 = 9$$

The equation is:

$$\frac{x^2}{9} - \frac{y^2}{9} = 1 \qquad \text{or} \qquad x^2 - y^2 = 9$$

## Equilateral Hyperbola

Notice in the above equation that $a = b = 3$. When $a = b$, the hyperbola is a special case called an *equilateral hyperbola*. The asymptotes are perpendicular since the rectangle is a square. An equilateral hyperbola can be rotated 45° so that the *coordinate axes are the asymptotes*. Its equation then becomes:

$$xy = k \qquad k = \text{constant} \qquad (14\text{-}20)$$

When $k > 0$, the hyperbola is in the first and third quadrants, and when $k < 0$, it is in the second and fourth quadrants (Fig. 14-32).

### Example 14-26

Draw the equilateral hyperbolas on the same graph: $xy = 12$ and $xy = -12$.

### Solution

Plot points by constructing tables of values and draw each curve (Fig. 14-32):

$xy = 12$

| $x$ | $-6$ | $-4$ | $-3$ | $-2$ | $2$ | $3$ | $4$ | $6$ |
|---|---|---|---|---|---|---|---|---|
| $y$ | $-2$ | $-3$ | $-4$ | $-6$ | $6$ | $4$ | $3$ | $2$ |

$xy = -12$

| $x$ | $-6$ | $-4$ | $-3$ | $-2$ | $2$ | $3$ | $4$ | $6$ |
|---|---|---|---|---|---|---|---|---|
| $y$ | $2$ | $3$ | $4$ | $6$ | $-6$ | $-4$ | $-3$ | $-2$ |

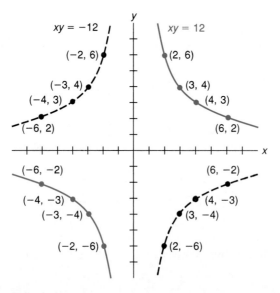

FIG. 14-32 *Conjugate equilateral hyperbolas, Example 14-26*

Notice in an equilateral hyperbola as $x$ approaches infinity $(x \to \infty)$, $y$ approaches zero $(y \to 0)$ and vice versa.

Equation (14-20) states that $y$ is inversely proportional to $x$ (Sec. 3-3). An equilateral hyperbola is therefore the graph of inverse variation and has many applications as shown below and in the problems.

### Example 14-27

For a gas at a constant temperature, Boyle's law states that $PV = k$, where $P$ = pressure and $V$ = volume. If 5.00 m³ of nitrogen is under a pressure of 800 kPa, draw the graph of $V$ vs. $P$ at constant temperature.

### Solution

Find $k$ using Boyle's law: $k = (800)(5.00) = 4000$. The equation is then $PV = 4000$, and the table of values is:

| $P$ | 32 | 40 | 50 | 80 | 100 | 125 |
|---|---|---|---|---|---|---|
| $V$ | 125 | 100 | 80 | 50 | 40 | 32 |

Negative values do not apply. The graph is one branch of an equilateral hyperbola (Fig. 14-33).

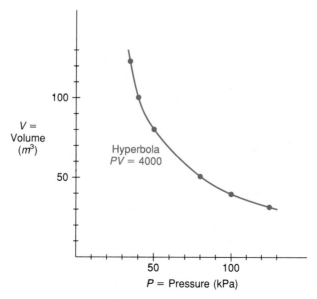

FIG. 14-33 *Boyle's law, Example 14-27*

**Exercise 14-5**    In problems 1 to 12, sketch each hyperbola, showing the vertices, foci, and asymptotes.

1. $\dfrac{x^2}{16} - \dfrac{y^2}{9} = 1$      2. $\dfrac{x^2}{36} - \dfrac{y^2}{64} = 1$

3. $\dfrac{y^2}{9} - \dfrac{x^2}{4} = 1$      4. $\dfrac{y^2}{25} - \dfrac{x^2}{16} = 1$

5. $25x^2 - 9y^2 = 225$      6. $4x^2 - 4y^2 = 16$

7. $y^2 - x^2 = 49$      8. $4y^2 - 9x^2 = 36$

9. $9x^2 - 2y^2 = 18$      10. $3x^2 - 4y^2 = 12$

11. $4y^2 - 25x^2 = 25$      12. $5y^2 - 4x^2 = 20$

In problems 13 to 20, find the equation of each hyperbola with center $(0, 0)$.

13. Vertices $(\pm 4, 0)$, foci $(\pm 6, 0)$

14. Foci $(0, \pm 5)$, ends of conjugate axis $(\pm 2, 0)$

15. Focus $(0, 6)$, transverse axis $= 10$

16. Vertex $(3, 0)$, distance from focus to vertex $= 2$

17. Conjugate hyperbola to $2x^2 - 3y^2 = 6$

18. Vertex $(2, 0)$, same foci as $16x^2 - 9y^2 = 144$

19. End of conjugate axis $(0, 2\sqrt{2})$, passes through $(6, 8)$

20. Vertex $(0, 6)$, passes through $(8, 10)$

In problems 21 to 26, draw each equilateral hyperbola.

21. $xy = 24$      22. $xy = 15$

23. $xy = -8$      24. $xy = -10$

25. $Fd = 80$      26. $IR = 4.8$

27. When a plane breaks the sound barrier, the shock wave is the shape of a cone, which intersects the ground in one branch of a hyperbola. The shock is felt at the same time at every point on the hyperbola. A woman at the *vertex* of a hyperbolic shock wave 1 mi north of the center experiences it at the same time as a man 1 mi west and 2 mi north of the center. *(a)* Sketch the branch of the hyperbola with center $(0, 0)$ that represents the shock wave, using the $y$ axis as the transverse axis. *(b)* Find the equation of the entire hyperbola.

28. A long-range hyperbolic navigation system transmits radio signals from two towers, which are the *foci* of a series of hyperbolas. Two ships receiving the signals determine their position on the same hyperbola if the time difference in arrival of the signals is the same. The transmitting towers are 1000 mi apart on an east-west line. A ship on the line between the two towers determines its position to be 200 mi from the east tower. *(a)* Sketch the signal hyperbola with center $(0, 0)$ that the ship is located on, using the $x$ axis as the transverse axis. *(b)* Find the equation of the hyperbola.

29. Ohm's law for a dc circuit states that when the voltage $E$ in an electric circuit is constant, the current $I$ is inversely proportional to the resistance $R$: $E = IR$. Sketch the curve of $I$ vs. $R$ if $E = 120$ V.

30. When the power in an electric circuit is constant, the voltage $E$ is inversely propor-

tional to the current $I$ expressed by $P = EI$. Sketch the curve of $E$ vs. $I$ for a power output of 100 W.

**31.** If $10.0 \text{ m}^3$ of propane is under a pressure of 240 kPa, find the equation and sketch the curve of volume vs. pressure for a constant temperature. (See Example 14-27.)

**32.** A comet travels in a hyperbolic path with the sun at its focus. At the comet's closest point to the sun, it is calculated to be $2 \times 10^8$ km from the center of the sun and $8 \times 10^8$ km from the center of its path. Find the equation of its path if the sun is on the $x$ axis and the center is $(0, 0)$.

**33.** Find the equations of the two hyperbolas with centers at the origin and semimajor axes $= 1$ that pass through $(5, 4)$.

**34.** *(a)* Sketch the hyperbola $25x^2 - 9y^2 = 225$ and the line $y = -x$ on the same graph, and estimate the points of intersection. *(b)* Determine the points exactly by solving the system algebraically. (*Hint:* Substitute $y = -x$ in the first equation.)

**35.** *(a)* Sketch the equilateral hyperbola $x^2 - y^2 = 18$ and the circle $x^2 + y^2 = 32$ on the same graph, and estimate the points of intersection. *(b)* Determine the points exactly by solving the system algebraically. (*Hint:* Combine equations.)

**36.** Sketch the ellipse $x^2 + 4y^2 = 16$ and its *associated conjugate hyperbolas* $x^2 - 4y^2 = 16$ and $4y^2 - x^2 = 16$ on the same graph.

# 14-6 | *Solution of Quadratic Systems*

Chapter 5 shows how to solve systems containing two linear equations graphically and algebraically. Now that you are familiar with the graphs of quadratic equations, this section shows how to solve systems of two equations, one or both of which are quadratic.

## *Linear Equation and Quadratic Equation*

A straight line can intersect a conic section in two, one, or zero points. This produces real unequal, real equal, or imaginary roots, respectively, when the linear and the quadratic equations are solved simultaneously. The following two examples illustrate the cases of real unequal and real equal roots. Imaginary roots are not considered.

### Example 14-28
Solve the system graphically and algebraically:

$$xy = 12 \qquad 2x - 3y = 1$$

### Solution
This system consists of an equilateral hyperbola and a straight line. Both curves are sketched by plotting values for $x$ and $y$ (Fig. 14-34). The graphical solution is the points of intersection: $(-4, -3)$ and $(9/2, 8/3)$. Most often the graphical

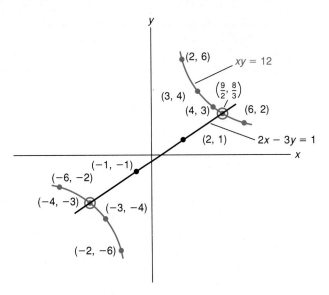

FIG. 14-34 *Hyperbola and line, Example 14-28*

solution is an estimation that serves to predict precise algebraic results and helps you to interpret them. However, at times graphical solutions are more efficient and preferred to algebraic ones (Sec. 15-6). The algebraic solution of a linear and quadratic system is obtained by *substitution*. Solve the linear equation for $y$ (or $x$) and substitute this expression into the other equation:

$$y = \frac{2x - 1}{3}$$

$$xy = 12$$

$$x\left(\frac{2x - 1}{3}\right) = 12$$

$$\frac{2x^2 - x}{3} = 12$$

$$2x^2 - x - 36 = 0$$

$$(x + 4)(2x - 9) = 0$$

$$x = -4 \qquad \text{and} \qquad x = \frac{9}{2}$$

The corresponding $y$ values are obtained from the expression for $y$:

$$y = \frac{2(-4) - 1}{3} = -3 \qquad \text{and} \qquad y = \frac{2(9/2) - 1}{3} = \frac{8}{3}$$

The algebraic solution is therefore $(-4, -3)$ and $(9/2, 8/3)$, which agrees with the graphical solution.

### Example 14-29

Solve the system graphically and algebraically:

$$y = x^2 - 4x + 2 \qquad 2x + y = 1$$

### Solution

This system is a parabola and a straight line. The parabola is sketched by plotting the vertex and four points (Fig. 14-35). Notice that the line is tangent to the parabola. The solution is only one point, $(1, -1)$. *When a line is tangent to a conic section, there are equal roots in the algebraic solution:*

$$y = (1 - 2x)$$
$$y = x^2 - 4x + 2$$
$$(1 - 2x) = x^2 - 4x + 2$$
$$x^2 - 2x + 1 = 0$$
$$(x - 1)(x - 1) = 0$$
$$x = 1 \qquad \text{and} \qquad x = 1$$

The corresponding solution for $y$ is $y = -1$, which agrees with the graphical solution.

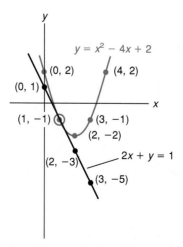

FIG. 14-35 *Parabola and line, Example 14-29*

## *Two Quadratic Equations*

Two conic sections can intersect in four or fewer points. The following examples illustrate some elementary cases.

### Example 14-30

Solve graphically and algebraically:

$$x^2 + 5y^2 = 9 \qquad x^2 - 2y^2 = 2$$

### Solution

The ellipse and hyperbola are shown in Fig. 14-36. Irrational points, such as the ends of the minor axis of the ellipse $(0, \pm\sqrt{9/5})$, are plotted to the nearest tenth. The graphical solution shows four points of intersection. These are shown exactly, but in general they would be estimated. The algebraic solution of two quadratic equations, when there are no first-degree terms, is obtained by combining equations to eliminate $x$ or $y$ (as with two linear equations):

$$
\begin{aligned}
x^2 + 5y^2 &= 9 \\
-1(x^2 - 2y^2 = 2) \rightarrow \quad -x^2 + 2y^2 &= -2 \\
\hline
7y^2 &= 7 \\
y &= \pm 1
\end{aligned}
$$

Each value of $y$ when substituted into one of the original equations produces two values for $x$: $\pm 2$. Hence there are four solutions: $(2, 1)$, $(2, -1)$, $(-2, -1)$, and $(-2, 1)$.

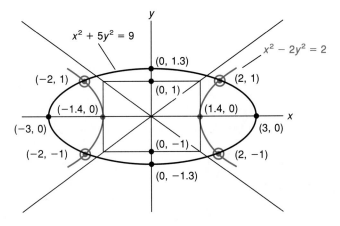

FIG. 14-36 *Ellipse and hyperbola, Example 14-30*

**Example 14-31**

Solve graphically and algebraically:

$$x^2 = 2y \qquad x^2 + y^2 = 3$$

**Solution**

The vertex of the parabola and the center of the circle are both $(0, 0)$. The graphical solution is shown in Fig. 14-37. The algebraic solution is found exactly by substitution:

$$x^2 = 2y$$
$$x^2 + y^2 = 3$$
$$2y + y^2 = 3$$
$$y^2 + 2y - 3 = 0$$
$$(y - 1)(y + 3) = 0$$
$$y = 1 \qquad \text{and} \qquad y = -3$$

When $y = 1$, $x = \pm\sqrt{2}$, which represents the two real solutions shown on the graph: $(\pm\sqrt{2}, 1)$ or $(\pm 1.4, 1)$. When $y = -3$, $x = \pm j\sqrt{6}$, which represents two imaginary solutions that cannot be shown graphically. Notice that the problem can also be solved by subtracting equations and eliminating $x^2$.

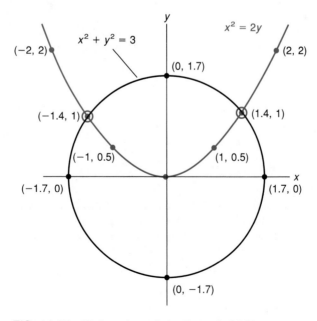

FIG. 14-37 *Circle and parabola, Example 14-31*

The next example shows an application to electric circuits.

### Example 14-32

The total resistance of two resistances in series is 5.30 $\Omega$, and the total resistance in parallel is 0.80 $\Omega$. What are the two resistances?

### Solution

The total resistance for two resistances $R_1$ and $R_2$ in series is $R = R_1 + R_2$. The total resistance in parallel is $R = R_1R_2/(R_1 + R_2)$. Substitute the given information into the two equations:

$$R_1 + R_2 = 5.30 \qquad \frac{R_1R_2}{R_1 + R_2} = 0.80$$

When 5.30 is substituted for $R_1 + R_2$ in the equation on the right:

$$\frac{R_1R_2}{5.30} = 0.80 \qquad \text{or} \qquad R_1R_2 = 4.24$$

The last equation is an equilateral hyperbola, and the first equation is a straight line. This system is similar to Example 14-28. The algebraic solution is found by substituting for $R_2$ and solving for $R_1$:

$$R_2 = 5.30 - R_1$$
$$R_1R_2 = 4.24$$
$$R_1(5.30 - R_1) = 4.24$$
$$R_1{}^2 - 5.30\,R_1 + 4.24 = 0$$

Use the quadratic formula:

$$R_1 = \frac{5.30 \pm \sqrt{(5.30)^2 - 4(1)(4.24)}}{2(1)} = \frac{5.30 \pm \sqrt{11.1}}{2}$$
$$R_1 = 4.32 \qquad \text{and} \qquad 0.98\ \Omega$$

This yields $R_2 = 0.98$ and 4.32 $\Omega$, respectively. Thus, there are two mathematical solutions but only one physical solution. Draw the graphical solution, and see how the two mathematical solutions represent one physical solution.

### Exercise 14-6

In problems 1 to 10, solve each system of a linear and a quadratic equation graphically and algebraically.

1. $x^2 + y^2 = 25$
   $x - y = 1$

2. $x^2 + y^2 = 5$
   $2x - y = -5$

**3.** $y = x^2 - 1$
    $x + y = 1$

**4.** $xy = 8$
    $x - 3y = 2$

**5.** $xy = -6$
    $3x - 8y = 24$

**6.** $y = x^2 - 4x + 3$
    $2x - y = 6$

**7.** $4x^2 + 9y^2 = 45$
    $2x - y = 5$

**8.** $x^2 - 2y^2 = 1$
    $2x - y = 4$

**9.** $3y^2 - 4x^2 = 12$
    $2x + y = -1$

**10.** $3x^2 + 2y^2 = 6$
    $x - y = 1$

In problems 11 to 20, solve each system of two quadratic equations graphically and algebraically.

**11.** $x^2 + y^2 = 25$
    $x^2 - y^2 = 7$

**12.** $2x^2 + y^2 = 6$
    $y^2 - x^2 = 3$

**13.** $x^2 + 4y^2 = 4$
    $x^2 - 2y^2 = 1$

**14.** $x^2 + y^2 = 12$
    $3x^2 - 5y^2 = 12$

**15.** $x^2 + y^2 = 8$
    $y^2 = 2x$

**16.** $x^2 + 3y^2 = 12$
    $y = -x^2 + 2$

**17.** $2x^2 + 3y^2 = 6$
    $3x^2 - 2y^2 = 9$

**18.** $4x^2 + 3y^2 = 19$
    $3x^2 - 4y^2 = 8$

**19.** $y^2 - 2x^2 = 4$
    $y = x^2 - 2$

**20.** $y = x^2 - 4x + 4$
    $y = -x^2 + 4x - 2$

In problems 21 to 28, solve each applied problem algebraically unless otherwise indicated.

**21.** A computer circuit chip consists of two rectangles with the proportions shown in Fig. 14-38. The total area is to be 9.0 mm², and the perimeter is to be 16 mm. What are the possible sets of dimensions for the chip?

**22.** Two communications satellites are both in circumpolar orbits in the same plane. The orbits are described by the equations $20u^2 + 30v^2 = 6000$ and $40u^2 + 10v^2 =$

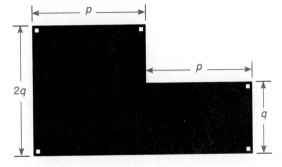

FIG. 14-38 *Computer circuit chip, problem 21*

4000. Sketch their orbits on the same graph, and find their common points graphically and algebraically.

**23.** The total resistance of two resistances in series is 6.50 Ω and in parallel, the total resistance is 0.60 Ω. What are the two resistances? (See Example 14-32.)

**24.** Two resistances connected in parallel have a total resistance of 7.5 Ω. When they are connected in series with a third resistance of 10 Ω, the total resistance is 50 Ω. What are the two resistances? (See Example 14-32.)

[C] **25.** The time $t$ in minutes it takes for one step in a chemical process is a function of the temperature of the chemicals $T$ (in degrees Celsius), given by $t = 6000/T$. A simultaneous step in the process is governed by the equation $t = -3.10T + 505$. If both steps are to take the same time, at what temperature should they be conducted to take the shortest time?

**26.** The boundary curve of a stationary cold front is approximated by the parabola $h^2 = 4.8s$, where $h$ = height in miles and $s$ = horizontal distance from the front's contact with the ground (the origin). A weather balloon is launched within the cold front and travels in a path approximated by the line $h = -0.5s + 3.0$. Sketch the two curves and find the point where the balloon makes contact with the cold-air boundary graphically and algebraically.

**27.** A 300-kg weight 0.50 m from the fulcrum of a lever can just be lifted by a force at the other end. When the same force is applied 1.5 m closer to the fulcrum, the force must be increased by 50 kgf to just lift the weight. What are the original force and length of the lever? (*Note:* Force is inversely proportional to distance from the fulcrum.)

[C] **28.** A ferry travels 5.5 mi every day between an island and the mainland. One day Liz, the captain, increases the ferry's speed by 6.0 mi/h and finds it takes 20 min less time to make the trip. Find the normal time to make the trip and the normal speed of the ferry.

# 14-7 | *General Quadratic Equation*

This section discusses the standard equations of the conic sections whose center, or vertex, is at any point $(h, k)$ and whose axes are parallel to the coordinate axes. The standard equation of the circle whose center is at $(h, k)$, Eq. (14-11), is examined first, for it leads to the form for the equations of the other conic sections.

## *Standard Equations, Center or Vertex at (h, k)*

Figure 14-39 contrasts the standard equation of a circle whose center is (0, 0) with the same circle whose center is $(h, k)$. When the circle moves from (0, 0) to $(h, k)$, every point on the circle goes through a *translation* of $h$ units in the $x$ direction and of $k$ units in the $y$ direction. The effect of this translation on the equation is that $x$ *is replaced by* $(x - h)$ and $y$ *is replaced by* $(y - k)$. This

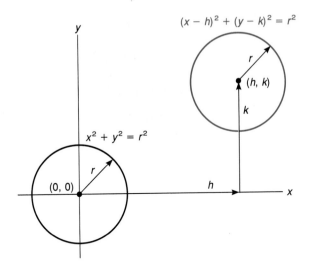

FIG. 14-39 *Translation of the circle*

"balances" the effect of the translation and refers all the translated points to the new center, $(h, k)$.

The standard equations of the parabola, ellipse, and hyperbola are changed in the same way by such a translation and take the following forms:

*Parabola with vertex at (h, k):*

Axis parallel to $y$ axis and focus $(h, k + p)$:

$$(x - h)^2 = 4p(y - k) \qquad \text{[see (14-9)]} \qquad (14\text{-}21)$$

Axis parallel to $x$ axis and focus $(h + p, k)$:

$$(y - k)^2 = 4p(x - h) \qquad \text{[see (14-10)]} \qquad (14\text{-}22)$$

*Ellipse with center (h, k):*

Major axis parallel to $x$ axis and vertices $(h \pm a, k)$:

$$\frac{(x - h)^2}{a^2} + \frac{(y - k)^2}{b^2} = 1 \qquad \text{[see (14-15)]} \qquad (14\text{-}23)$$

Major axis parallel to $y$ axis and vertices $(h, k \pm a)$:

$$\frac{(x - h)^2}{b^2} + \frac{(y - k)^2}{a^2} = 1 \qquad \text{[see (14-16)]} \qquad (14\text{-}24)$$

*Hyperbola with center (h, k):*

Transverse axis parallel to *x* axis and vertices $(h \pm a, k)$:

$$\frac{(x - h)^2}{a^2} - \frac{(y - k)^2}{b^2} = 1 \qquad \text{[see (14-18)]} \qquad \text{(14-25)}$$

Transverse axis parallel to *y* axis and vertices $(h, k \pm a)$:

$$\frac{(y - k)^2}{a^2} - \frac{(x - h)^2}{b^2} = 1 \qquad \text{[see (14-19)]} \qquad \text{(14-26)}$$

The following examples show how to apply the above equations.

### Example 14-33

Sketch the parabola $(x - 5)^2 = 8(y - 3)$, showing the vertex and focus.

### Solution

Compare the equation with Eq. (14-21). This parabola has its axis parallel to the *y* axis. The vertex is (5, 3); $4p = 8$ and $p = 2$. The focus is therefore (5, 5). The sketch is shown in Fig. 14-40.

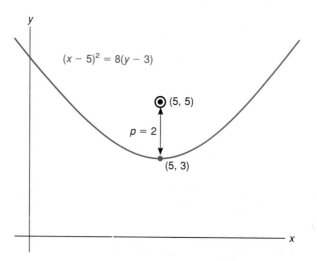

FIG. 14-40 *Parabola, Example 14-33*

## Example 14-34

Sketch the hyperbola:

$$\frac{(x + 2)^2}{9} - \frac{(y + 3)^2}{16} = 1$$

showing the center, vertices, foci, ends of conjugate axis, and asymptotes.

## Solution

Compare the equation to Eq. (14-25). This hyperbola has its transverse axis parallel to the $x$ axis. The center is $(-2, -3)$; $a^2 = 9$, $b^2 = 16$, $c^2 = a^2 + b^2 = 25$, and $a = 3$, $b = 4$, $c = 5$. Measure $a$ and $b$ from the center and plot the vertices $(1, -3)$ and $(-5, -3)$ and the ends of the conjugate axis $(-2, 1)$ and $(-2, -7)$. Draw the rectangle and the asymptotes, and sketch the hyperbola within the asymptotes (Fig. 14-41).

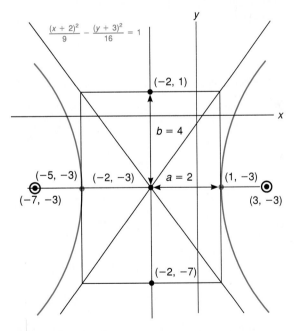

FIG. 14-41 *Hyperbola, Example 14-34*

## Example 14-35

Find the equation of each conic section.

**1.** Ellipse: center $(-2, 4)$, vertex $(-2, 10)$, focus $(-2, 8)$

**2.** Parabola: vertex $(-1, -1)$, focus $(2, -1)$

**3.** Hyperbola: center $(4, -3)$, focus $(4, 1)$, end of conjugate axis $(2, -3)$

### Solution

**1.** Ellipse: center $(-2, 4)$, vertex $(-2, 10)$, focus $(-2, 8)$

This ellipse has its major axis parallel to the $y$ axis since the $x$ coordinates of the center, vertex, and focus are the same. Then $a = 10 - 4 = 6$, $c = 8 - 4 = 4$, and $b^2 = a^2 - c^2 = 20$. The equation is:

$$\frac{(x + 2)^2}{20} + \frac{(y - 4)^2}{36} = 1$$

Sketch this ellipse or either of the next two curves to help you see how each equation is determined.

**2.** Parabola: vertex $(-1, -1)$, focus $(2, -1)$

This parabola has its axis parallel to the $x$ axis, and $p = 2 - (-1) = 3$. The equation is:

$$(y + 1)^2 = 12(x + 1)$$

**3.** Hyperbola: center $(4, -3)$, focus $(4, 1)$, end of conjugate axis $(2, -3)$

This hyperbola has its transverse axis parallel to the $y$ axis: $c = 1 - (-3) = 4$, $b = 4 - 2 = 2$, and $a^2 = c^2 - b^2 = 12$. The equation is:

$$\frac{(y - 4)^2}{12} - \frac{(x + 3)^2}{4} = 1$$

## *General Quadratic Equation*

When the standard equation of a conic section is expanded, it takes the form of the *general quadratic equation in two variables:*

$$Ax^2 + Bxy + Cy^2 + Dx + Ey + F = 0 \qquad (14\text{-}27)$$

For the standard equations (14-21) to (14-26), $B = 0$, and $A$ and $C$ have the following relation for each conic section:

| | |
|---|---|
| $A$ or $C = 0$ | Parabola |
| $A = C$ | Circle |
| $A$ and $C > 0$ (or $A$ and $C < 0$) | Ellipse |
| $A$ or $C < 0$ (but not both) | Hyperbola |

$(14\text{-}28)$

For the special case of the equilateral hyperbola, given by Eq. (14-20), $A = 0$, $C = 0$, and $B \neq 0$.

To sketch a conic section given in the form of the general equation (14-27), it is first necessary to complete the square and change it to the standard equation, as is done for the circle in Sec. 14-3 (Examples 14-17 and 14-18).

### Example 14-36

Identify and sketch the conic section, showing the basic features of the curve:

$$9x^2 + 25y^2 - 54x + 100y - 44 = 0$$

### Solution

This equation is an ellipse, since $A = 9$ and $B = 25$ are both positive. To complete the square, group the $x$ and $y$ terms and put the constant on the other side:

$$9x^2 - 54x + 25y^2 + 100y = 44$$

Factor the coefficients of $x^2$ and $y^2$ separately, and complete the square for each:

$$9(x^2 - 6x + 9) + 25(y^2 + 4y + 4) = 44 + 9(9) + 25(4)$$

Notice that the term added in each set of parentheses is affected by the factor in front and must be multiplied by that factor when added to the other side. Factor each square and divide by the constant term:

$$9(x - 3)^2 + 25(y + 2)^2 = 44 + 81 + 100 = 225$$

$$\frac{(x - 3)^2}{25} + \frac{(y + 2)^2}{9} = 1$$

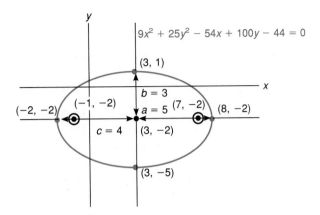

FIG. 14-42 *Ellipse, Example 14-36*

This ellipse has its major axis parallel to the $x$ axis with center at $(3, -2)$; $a^2 = 25$, $b^2 = 9$, $c^2 = a^2 - b^2 = 16$, and $a = 5$, $b = 3$, $c = 4$. The sketch in Fig. 14-42 shows the vertices, foci, and ends of the minor axis.

**Exercise 14-7**

In problems 1 to 8, sketch each conic section, showing the basic features of each curve.

**1.** $(x - 3)^2 = 4(y - 4)$        **2.** $(y + 2)^2 = 8(x - 2)$

**3.** $\dfrac{(x + 4)^2}{16} + \dfrac{(y - 2)^2}{25} = 1$      **4.** $\dfrac{(x - 1)^2}{9} + \dfrac{(y + 1)^2}{4} = 1$

**5.** $\dfrac{(x - 2)^2}{9} - \dfrac{(y - 5)^2}{16} = 1$      **6.** $\dfrac{(y + 3)^2}{4} - \dfrac{(x + 4)^2}{1} = 1$

**7.** $(y - 2)^2 = -16(x + 1)$      **8.** $4(x + 5)^2 + 9(y + 6)^2 = 36$

In problems 9 to 16, find the equation of each conic section.

**9.** Parabola: vertex $(2, 2)$, focus $(3, 2)$

**10.** Parabola: vertex $(5, 3)$, focus $(5, 1)$

**11.** Ellipse: center $(5, -5)$, vertex $(0, -5)$, focus $(2, -5)$

**12.** Ellipse: vertices $(2, \pm 6)$, foci $(2, \pm 4)$

**13.** Hyperbola: vertices $(3, -2)$ and $(7, -2)$, foci $(1, -2)$ and $(9, -2)$

**14.** Hyperbola: center $(-3, 4)$, vertex $(-3, 1)$, focus $(-3, -1)$

**15.** Parabola: focus $(-2, 3)$, directrix $y = -3$

**16.** Ellipse: center $(-1, -1)$, focus $(-3, -1)$, minor axis $= 6$

In problems 17 to 22, identify and sketch each conic section, showing the basic features of each curve. See Example 14-36.

**17.** $x^2 + 4x - 4y + 4 = 0$

**18.** $y^2 - 2x - 2y + 5 = 0$

**19.** $9x^2 + 4y^2 - 18x - 24y + 9 = 0$

**20.** $x^2 + 4y^2 + 4x = 0$

**21.** $x^2 - y^2 + 4y - 5 = 0$

**22.** $16x^2 - 9y^2 + 32x + 18y + 151 = 0$

In problems 23 to 26, solve each quadratic system graphically and algebraically.

**23.** $4x^2 + y^2 - 2y - 3 = 0$        **24.** $x^2 - y^2 - 4x + 3 = 0$
     $x - y = 0$                           $x + y = 0$

**25.** $x^2 - 4x + y + 2 = 0$        **26.** $x^2 - 9y^2 + 2x - 8 = 0$
     $x^2 + y^2 - 4x + 2 = 0$               $x^2 + 9y^2 + 2x - 8 = 0$

**27.** The cable of a suspension bridge approximates a parabola. The horizontal distance between the highest points is 1000 m, and the lowest point (the vertex) is 125 m below the highest points. If the vertex is placed on the negative $y$ axis and the highest points are placed on the $x$ axis, what is the equation of the parabola?

**28.** A tunnel entrance is in the shape of a semiellipse with a maximum clearance of 50 ft and a maximum width of 200 ft at the ground. If the left end (the vertex) of the ellipse is placed at the origin, what is the equation of the ellipse?

**29.** A cantilever beam (a beam fixed at one end) under a uniform load assumes a parabolic curve, with the fixed end being the vertex (Fig. 14-43). For a cantilever beam 1.00 m long, the equation of the load parabola is approximately $x^2 + 100y - 1.00 = 0$, where $x$ is the distance in meters from the fixed end and $y$ is the displacement. *(a)* Change this equation to the standard equation, and sketch the load parabola. *(b)* How far is the free end of the beam displaced from its no-load position?

**30.** Two impedances in parallel are given by $X_L - X_C$ and $X_L + X_C$, where $X_L =$ inductive reactance and $X_C =$ capacitive reactance. The total impedance $Z$ of the circuit is constant. Using Eq. (13-19), *(a)* write the relation between $X_L$ and $X_C$ and change it to the form of Eq. (14-27); *(b)* determine which conic section is represented by this relation.

**31.** A patrol boat travels 16 mi downstream and 12 mi upstream in 5 h. On another patrol the boat travels 32 mi downstream and 4 mi upstream in the same time and at the same speed relative to the water. What are the speed of the boat and the speed of the current? *(Hint:* Use time = distance/rate and set up two equations for the total time.)

**32.** A circular lawn is surrounded by a circular path. The area of the circular path is 100 ft². Increasing the width of the path by 2 ft doubles its area. What are the inner and outer radii of the path?

**33.** The paths of two satellites in the same plane are described by:

$$p^2 + \frac{(q - 2)^2}{5} = 1 \qquad \text{and} \qquad p^2 + \frac{(q + 2)^2}{5} = 1$$

where the center of the earth is at their common focus. *(a)* Sketch the two curves on the same graph, and show that the common focus is the origin. *(b)* Find their common points algebraically.

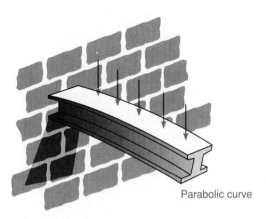

Parabolic curve

FIG. 14-43 *Cantilever beam, problem 29*

**34.** Given the parabola described by the quadratic function $y = ax^2 + bx + c$, change it to the standard equation by completing the square and show that the coordinates of the vertex are:

$$\left(\frac{-b}{2a},\ c - \frac{b^2}{4a}\right)$$

## 14-8 | *Review Questions*

In problems 1 and 2, find the distance, slope, and inclination for each pair of points.

**1.** (3, 2), (5, 5)

**2.** (−0.5, 4.0), (1.5, 3.0)

In problems 3 to 6, write the equation of each line.

**3.** $m = \frac{1}{3}$, $y$ intercept $= -2$

**4.** $m = -1$, passes through (2, 1)

**5.** Parallel to $4x - 3y = 12$, and passes through (4, 3)

**6.** Perpendicular to the line joining (1, 1) and (−3, 2) and passes through (−2, −5)

In problems 7 to 20, sketch each conic section, showing the basic features of each curve.

**7.** $y = x^2 - 8x + 12$　　　　　　**8.** $y = 8 - 2x^2$

**9.** $x^2 = 8y$　　　　　　　　　　　**10.** $y^2 = -4x$

**11.** $(x - 2)^2 + (y + 1)^2 = 9$　　**12.** $3x^2 + 3y^2 = 12$

**13.** $x^2/9 + y^2/25 = 1$　　　　　**14.** $4x^2 + 16y^2 = 64$

**15.** $4x^2 - 25y^2 = 100$　　　　　**16.** $xy = 18$

**17.** $(x - 2)^2 = 8(y + 1)$　　　　**18.** $x^2 - (y - 1)^2 = 4$

**19.** $x^2 + y^2 + 4x - 2y + 4 = 0$　　**20.** $4x^2 + 9y^2 + 24x - 18y + 9 = 0$

In problems 21 and 22 solve each system graphically and algebraically.

**21.** $y^2 - 2x^2 = 1$　　　　　　　**22.** $x^2 + y^2 = 4$
　　$x - 2y = -4$　　　　　　　　　　$3x^2 + 4y^2 = 16$

In problems 23 to 30, find the equation of each conic section.

**23.** Parabola: vertex (0, 0), focus (3, 0)

**24.** Parabola: vertex (1, 2), focus (1, 4)

**25.** Circle: center (−2, 1), radius $= 3$

**26.** Circle: center (4, 0), tangent to $y$ axis at origin

**27.** Ellipse: vertices (±5, 0), foci (±4, 0)

**28.** Ellipse: center (0, 0), focus (0, −1), minor axis $= 4$

**29.** Hyperbola: vertices (±3, 0), passes through (5, 4)

**30.** Hyperbola: center (5, 2), vertex (0, 2), conjugate axis $= 4$

31. When Hooke's law is applied to a spring, it means that the force $F$ is a linear function of the distance $x$ that the spring stretches. A certain spring has an initial force of 12 lb applied to it, which corresponds to $x = 0$. (a) Express $F$ as a linear function of $x$, using $k$ as the slope (the constant of proportionality). (b) If $x = 2.0$ in. when $F = 25$ lb, what is the slope?

32. A cable supported at its ends approximates the shape of a parabola under a uniformly distributed load. When the lowest point of a suspension bridge cable is placed at the origin, the coordinates of one end are (500 m, 50 m). Assuming the shape of the cable is a parabola, find its equation.

33. Michael wants to enclose a rectangular yard with 200 ft of fencing. If $x =$ length, then $100 - x =$ width and the area $A = x(100 - x)$. (a) Draw the graph of $A$ vs. $x$ and identify the curve. (b) What is the maximum area Michael can enclose?

34. A telecommunications satellite has a maximum altitude of 1300 km (808 mi) and a minimum altitude of 700 km (435 mi). If the center of the earth is at one focus, find the semimajor and semiminor axes of the elliptical orbit in kilometers and in miles. Use 6400 km for the earth's radius.

35. When a source of light is placed at the focus of an elliptical mirror, all the reflected rays converge at the other focus. The distance between the foci is 40 cm for such a mirror, and a light ray travels 50 cm from one focus to the other. What is the equation of the ellipse that contains the light ray? Let the center be the origin and the $x$ axis be the major axis. (*Note:* The path of the light ray is equivalent to the sum of the distances from a point on the ellipse to the foci.)

36. Under certain economic conditions, the demand for an item is inversely proportional to the price. A particular home microcomputer when offered for sale at $1200 produces 5000 orders. (a) If price and demand are inversely proportional, what is the equation relating them? (b) Draw the graph of this equation, using suitable scales.

37. (a) Sketch the hyperbola having the same vertices as the ellipse $x^2 + 9y^2 = 9$ and the same foci as the ellipse $x^2 + 26y^2 = 26$. (b) What is the equation of this hyperbola?

38. A communications system component is to be made in the form of two right triangles sharing a common side (Fig. 14-44). What should be the dimensions $b$ and $c$ of the component?

39. A missile is launched on a parabolic path calculated to be $y = x - 0.10x^2$. A few seconds later an antimissile is launched on a path calculated to be $y = 2.50 - 0.10x^2$ and intended to intercept the missile. Sketch the two missiles' paths, and determine the coordinates of their common point graphically and algebraically.

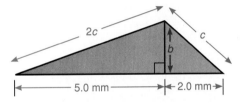

FIG. 14-44 *Communications system component, problem 38*

**40.** A concrete bridge for a highway overpass is constructed in the shape of an elliptic arch as shown in Fig. 14-45. If the beginning of the arch is placed at the origin with the center on the positive $x$ axis, what is the equation of the ellipse of which the arch is half?

FIG. 14-45 *Highway bridge with elliptic arch, problem 40*

# 15

# Polynomial Functions and Higher-Degree Equations

## 15-1 *Function Notation and Polynomial Functions*

**F**irst-degree (linear) and second-degree (quadratic) equations belong to the general class of *polynomial equations,* which includes third-degree, fourth-degree, etc. Higher-degree equations occur in many scientific and technical fields. This chapter builds on the methods used for solving first- and second-degree equations to develop procedures for solving higher-degree equations. The *function notation f(x)* is discussed first and is very useful in applying the ideas of this chapter and in the study of all functions, particularly those in calculus. Polynomial functions are studied next, and this leads to the solution of higher-degree equations.

### Function Notation f(x)

The *symbol* $f(x)$ is read "function of $x$" or "$f$ of $x$" and is used in place of $y$ to represent a function of $x$. A function is defined in Sec. 5-1. Definition (5-1) states that: Given a function $f(x)$ and a set of values for $x$, there corresponds one and only one value of $y$ for each value of $x$. Examples of functions are $f(x) = mx + b$ (linear function), $f(x) = ax^2 + bx + c$ (quadratic function), $f(x) = \sin x$ (trigonometric function), and $f(x) = \ln x$ (natural logarithmic function). *When the x in parentheses is replaced by a constant, such as $f(2)$, $f(0)$, $f(-1)$, or a variable, such as $f(a)$, the symbol represents the value of the function when the constant or variable is substituted for x.* Study the next example, which illustrates this idea.

**Example 15-1**

For each function find $f(2)$, $f(0)$, and $f(a)$.

**1.** $f(x) = 2x + 3$

**2.** $f(x) = x^2 - 3x - 4$

**3.** $f(x) = \sin x$

**4.** $f(x) = \ln x$

**Solution**

**1.** $f(x) = 2x + 3$

$$f(2) = 2(2) + 3 = 7 \qquad f(0) = 2(0) + 3 = 3 \qquad f(a) = 2a + 3$$

**2.** $f(x) = x^2 - 3x - 4$

$$f(2) = (2)^2 - 3(2) - 4 = -6 \qquad f(0) = (0)^2 - 3(0) - 4 = -4$$
$$f(a) = a^2 - 3a - 4 \quad \text{or} \quad (a + 1)(a - 4)$$

**3.** $f(x) = \sin x$

Assume $x$ is in radians. Then:

$$f(2) = \sin 2 = 0.9093 \qquad f(0) = \sin 0 = 0 \qquad f(a) = \sin a$$

**4.** $f(x) = \ln x$

$$f(2) = \ln 2 = 0.6931 \qquad f(0) = \ln 0 = \text{undefined} \qquad f(a) = \ln a$$

Other symbols, such as $g(x)$, $F(x)$, $\phi(x)$, etc., are also used to represent functions.

## Polynomial Functions

A *polynomial function of degree n* is defined as follows:

$$f(x) = a_n x^n + a_{n-1} x^{n-1} + \cdots + a_1 x + a_0 \qquad a_n \neq 0 \qquad (15\text{-}1)$$

where the $a$'s are constants. It is true that

**1.** A first-degree polynomial function is a linear function and graphs as a straight line.

**2.** A second-degree polynomial function is a quadratic function and graphs as a parabola (one turning point).

**3.** Higher-degree polynomial functions tend to have more turning points as the degree increases (Figs. 15-1 and 15-2).

### Example 15-2

Graph the third-degree polynomial function $f(x) = x^3 - 2x^2 - x + 2$ from $x = -2$ to $x = 3$.

### Solution

Substitute the values of $x$ and construct a table of values of $f(x)$:

| $x$ | $-2$ | $-1$ | 0 | 1 | 2 | 3 |
|---|---|---|---|---|---|---|
| $f(x)$ | $-12$ | 0 | 2 | 0 | 0 | 8 |

For example, $f(-2) = (-2)^3 - 2(-2)^2 - (-2) + 2 = -8 - 8 + 2 + 2 = -12$. Plot the points and connect them with a smooth curve (Fig. 15-1). The curve has two turning points. *A third-degree polynomial has two or no turning*

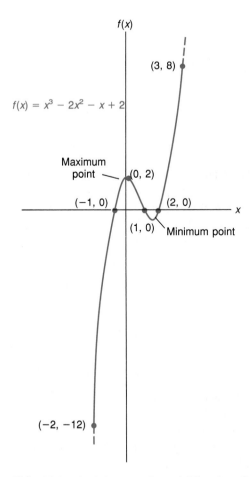

FIG. 15-1 *Third-degree polynomial function, Example 15-2*

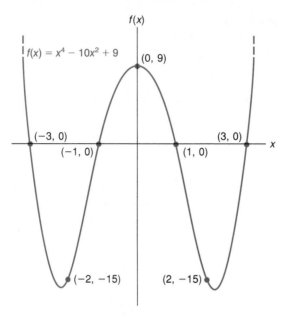

FIG. 15-2 *Fourth-degree polynomial function, Example 15-3*

*points*. The higher turning point is a relative *maximum point,* and the lower turning point is a relative *minimum point.* In calculus it is shown how to find these points exactly [$(-0.2, 2.1)$ is the maximum point]. Notice the $x$ intercepts of the graph: $(-1, 0)$, $(1, 0)$, and $(2, 0)$. *The x intercepts, or zeros, of the function are the roots of the polynomial equation $f(x) = 0$.* This particular polynomial can be factored: $f(x) = (x + 1)(x - 1)(x - 2)$. When you set $f(x) = 0$ in factored form as $(x + 1)(x - 1)(x - 2) = 0$, the roots are obtained in the same way as in a quadratic equation, by setting each factor equal to zero:

$$x + 1 = 0 \qquad x - 1 = 0 \qquad x - 2 = 0$$
$$x = -1 \qquad x = 1 \qquad x = 2$$

The factors of $f(x)$ are therefore important in finding the $x$ intercepts, or the roots of $f(x) = 0$. The procedures for factoring polynomials are discussed more in the following sections.

### Example 15-3
Graph the fourth-degree polynomial function $f(x) = x^4 - 10x^2 + 9$ from $x = -4$ to $x = +4$.

### Solution
The table of values is:

| $x$ | $-4$ | $-3$ | $-2$ | $-1$ | 0 | 1 | 2 | 3 | 4 |
|---|---|---|---|---|---|---|---|---|---|
| $f(x)$ | 105 | 0 | $-15$ | 0 | 9 | 0 | $-15$ | 0 | 105 |

The graph is shown in Fig. 15-2, with different scales for the $x$ and $f(x)$ axes. The points $(-4, 105)$ and $(4, 105)$ are not included but help in sketching the curve. This curve has three turning points. *A fourth-degree polynomial function has one or three turning points.* The $x$ intercepts $x = -3$, $x = -1$, $x = 1$, and $x = 3$ are the roots of $f(x) = 0$. Note that by changing the signs of the roots you can write the factors of the polynomial: $f(x) = (x + 3)(x + 1)(x - 1)(x - 3)$.

**Example 15-4**
Graph $f(x) = (x)(x - 2)(x + 2)$, showing the intercepts.

**Solution**
The intercepts are obtained by setting the factors equal to zero or by changing the signs: $x = 0$, $x = 2$, $x = -2$. The table of values is set up around those intercepts:

| $x$ | $-3$ | $-2$ | $-1$ | 0 | 1 | 2 | 3 |
|---|---|---|---|---|---|---|---|
| $f(x)$ | $-15$ | 0 | 3 | 0 | $-3$ | 0 | 15 |

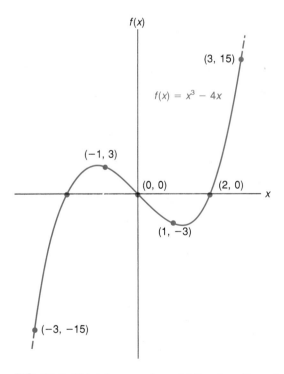

FIG. 15-3 *Third-degree polynomial function, Example 15-4*

The values of $f(x)$ can be obtained by using $f(x)$ in factored form or by multiplying out $f(x) = x^3 - 4x$. The graph is shown in Fig. 15-3.

Polynomial graphs become increasingly complex as the degree increases; they are studied further in calculus. The above elementary examples introduce some of the basic ideas used throughout this chapter.

## Exercise 15-1

In problems 1 to 14, for each function find the given values.

1. $f(x) = 3x - 1$; $f(-4)$, $f(\frac{1}{3})$
2. $f(x) = 2x^2 + 3x + 1$; $f(-1)$, $f(2)$
3. $f(x) = x^3 - x^2 - x + 1$; $f(-2)$, $f(3)$
4. $f(x) = 2x^3 - 3x^2 + x - 5$; $f(0)$, $f(2)$
5. $f(x) = x^5 + x - 1$; $f(-1)$, $f(-x)$
6. $f(x) = 4x^4 - 4x^2 - 4$; $f(-1)$, $f(-x)$
7. $F(x) = \log x$; $F(1)$, $F(\frac{1}{10})$
8. $F(x) = \tan x$; $F(\pi/4)$, $F(\pi)$
9. $g(y) = \sqrt{y^2 + 2}$; $g(-1)$, $g(2)$
10. $h(v) = v^{2/3}$; $h(1)$, $h(8)$
11. $\phi(t) = 1.6t^2 - 3.2t + 0.8$; $\phi(1.0)$, $\phi(1.5)$
12. $\phi(R) = 1.5R/(2.0R + 3)$; $\phi(3.0)$, $\phi(1.8)$
13. $f(x) = (x + 1)^2$; $f(2x)$, $f(x + h)$
14. $f(x) = x^3$; $f(x + 1)$, $f(x + \Delta x)$

In problems 15 to 28, graph each polynomial function for the given range of values.

15. $f(x) = x^3 - 6x^2 + 11x - 6$; $x = 0$ to $x = 4$
16. $f(x) = x^3 - x^2 - 4x + 4$; $x = -3$ to $x = 3$
17. $f(x) = 2x^3 - 3x^2 - 3x + 2$; $x = -2$ to $x = 3$
18. $f(x) = x^3 - 3x - 2$; $x = -2$ to $x = 3$
19. $f(x) = x^3 + 3x^2 + 3x + 1$; $x = -3$ to $x = 1$
20. $f(x) = x^3 - 4x + 4$; $x = -3$ to $x = 2$
21. $f(x) = 4x^2 - x^4$; $x = -3$ to $x = 3$
22. $f(x) = x^3 + 2x^2 - 5x - 3 = 0$; $x = -4$ to $x = 2$
23. $f(x) = 1 - x^4$; $x = -2$ to $x = 2$
24. $f(x) = x^4 - 8x^3 + 14x^2 + 8x - 15$; $x = -1$ to $x = 5$
25. $F(x) = x^4 - 3x^3$; $x = -1$ to $x = 3$
26. $F(x) = x^5 - 16x$; $x = -2$ to $x = 2$
27. $h(v) = v/2 - 2v^3$; $v = -2$ to $v = 2$
28. $\phi(t) = t^4 - 0.25t^2$; $t = -2$ to $t = 2$

In problems 29 to 34, graph each polynomial function, showing the intercepts.

**29.** $f(x) = (x + 1)(x + 2)(x + 3)$

**30.** $f(x) = (x)(x - 1)(x + 2)$

**31.** $f(x) = (x)^2(x^2 - 4)$

**32.** $f(x) = (x - 1)^2(x - 3)$

**33.** $f(x) = (x)^2(x + 2)^2$

**34.** $f(x) = (x^2 - 1)(x^2 - 4)$

**35.** The bending moment (twisting force) on a beam at a distance $d$ cm from the center is given by $M(d) = 12d - d^3/3$. A negative value of $d$ means the moment is left of center, and a positive value of $d$ means it is right of center. Graph $M(d)$ for $d = -7$ to $d = 7$. Use an appropriate scale for $M(d)$.

☐ **36.** The time rate of radiation in W/(m² · h) of a solar heater is given by $R(T) = 5.7 \times 10^{-8} (T + 273)^4$, where $T$ is the temperature in degrees Celsius. Graph $R(T)$ for $T = -50$ to $T = 50$. Use appropriate scales for $T$ and $R(T)$.

**37.** The volume of a box $s$ meters high with a square base $s - 2$ meters on a side is given by $V(s) = (s)(s - 2)^2$. Graph $V(s)$, showing the points where $V(s) = 0$.

**38.** The temperature $T$ in degrees Celsius of an electrochemical reaction is given by $T(t) = t^3 - 7t^2 + 10t + 8$, where $t$ = time in minutes. Graph $T(t)$ from $t = 0$ to $t = 5$.

## 15-2 | *Polynomial Division and Synthetic Division*

In the last section it is shown that the factors of a polynomial function $f(x)$ lead to the roots of the polynomial equation $f(x) = 0$. When one factor is known, the other factor is obtained by dividing polynomials. Polynomial division and a shortcut method, synthetic division, is therefore studied first before the solution of higher degree equations.

### Example 15-5
Divide $2x^3 - x^2 - 8x + 5$ by $x - 2$.

### Solution
Polynomial division is similar to long division in arithmetic. It is set up with the terms in descending order:

$$
\begin{array}{r}
2x^2 + 3x - 2 \qquad \text{quotient} \\
x - 2 \overline{\smash{\big)}\ 2x^3 - x^2 - 8x + 5} \qquad \text{dividend} \\
\underline{-(2x^3 - 4x^2)} \phantom{xxxxxxxx} \\
3x^2 - 8x \phantom{xxxx} \\
\underline{-(3x^2 - 6x)} \phantom{xx} \\
-2x + 5 \\
\underline{-(-2x + 4)} \\
1 \qquad \text{remainder}
\end{array}
$$

The basic steps in polynomial division are these:

1. Divide the first term of the divisor $(x - 2)$ into the first term of the dividend $(2x^3 - x^2 - 8x + 5)$. The answer $(2x^2)$ is the first term of the quotient.
2. Multiply this term of the quotient by the divisor: $(2x^2)(x - 2) = 2x^3 - 4x^2$.
3. Subtract the product from the dividend.
4. Bring down (add) the next term of the dividend to this difference.
5. Repeat steps 1 to 4, using the last result instead of the dividend, until all the terms of the dividend have been used. The last subtraction produces the remainder.

Notice that the first term of each subtraction in polynomial division always cancels and that the coefficients (shown in boldface) are the same as the coefficients in the quotient. The important steps that lead to the quotient are the multiplications (step 2) and the subtractions (step 3). The process is done in an abbreviated form called *synthetic division,* by using just these two steps and the coefficients. The sign of the constant term in the divisor is changed (from $-2$ to $+2$), making the subtractions additions as follows:

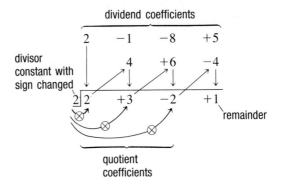

The basic steps in the *synthetic division* above are these:

1. Write down the dividend coefficients in descending order (use 0 for a missing term) and the constant term of the divisor with the sign changed, as shown above.
2. Bring down the first coefficient (this is the first coefficient of the quotient).
3. Multiply this quotient coefficient by the divisor constant ($\otimes$ above).
4. Add the product to the next coefficient of the dividend. The result is the next coefficient of the quotient.
5. Repeat steps 3 and 4 until all the dividend coefficients have been used. The last coefficient is the remainder.

Study the following examples to learn the procedure.

### Example 15-6

Divide $3x^4 + 5x^3 - 2x^2 - 4x + 3$ by $x + 1$

### Solution

Change the sign of the divisor constant from $+1$ to $-1$, and write down the dividend coefficients. The synthetic division is then a repeated process of multiplying and adding:

$$
\begin{array}{r}
3 \; +5 \; -2 \; -4 \; +3 \\
-3 \; -2 \; +4 \quad 0 \\
\hline
-1 \; \big| \; 3 \; +2 \; -4 \; +0 \; +3
\end{array}
$$

*The degree of the quotient is 1 less than that of the dividend:* $3x^3 + 2x^2 - 4x$. There is no constant term since the coefficient is zero. The remainder is 3, or more precisely $3/(x + 1)$, as in arithmetic.

### Example 15-7

Divide $x^4 - 4x^3 + 7x + 6$ by $x - 3$, using synthetic division.

### Solution

*A zero must be used to represent the missing term* $(0x^2)$:

$$
\begin{array}{r}
1 \; -4 \; +0 \; +7 \; +6 \\
3 \; -3 \; -9 \; -6 \\
\hline
+3 \; \big| \; 1 \; -1 \; -3 \; -2 \; +0
\end{array}
$$

The quotient is $x^3 - x^2 - 3x - 2$. The remainder is zero, which means that $x - 3$ is a factor and $f(x) = (x - 3)(x^3 - x^2 - 3x - 2)$.

### Example 15-8

Determine if $x - \frac{1}{2}$ is a factor of $f(x) = 2x^3 - x^2 + 2x - 1$. If so, express $f(x)$ in factored form.

### Solution

Use synthetic division:

$$
\begin{array}{r}
2 \; -1 \; +2 \; -1 \\
1 \; +0 \; +1 \\
\hline
\frac{1}{2} \; \big| \; 2 \; +0 \; +2 \; +0
\end{array}
$$

Since the remainder is zero, $x - \frac{1}{2}$ is a factor and $f(x) = (2x^2 + 2)(x - \frac{1}{2})$ or $f(x) = (x^2 + 1)(2x - 1)$.

Note that synthetic division is used for linear divisors while polynomial division can be used for any divisor. (See Example 15-16.)

**Exercise 15-2**

In problems 1 to 8, find the quotient and remainder by polynomial long division, and check by synthetic division. (See Example 15-5.)

1. $(x^2 - 3x + 3) \div (x + 3)$
2. $(2x^2 + 5x - 7) \div (x + 4)$
3. $(x^3 + 2x^2 - 4x + 1) \div (x - 1)$
4. $(3x^3 + 6x^2 - x - 4) \div (x + 2)$
5. $(2x^3 + 5x^2 - 4x - 3) \div (x + 3)$
6. $(x^4 + 3x^3 - 2x^2 - 1) \div (x - 1)$
7. $(3h^4 + 5h^3 - h^2 + 4h + 5) \div (h + 2)$
8. $(t^5 - 3t^4 + t^2 - 3t) \div (t - 3)$

In problems 9 to 24, find the quotient and remainder by synthetic division.

9. $(x^3 + 3x^2 - 2x - 4) \div (x + 2)$
10. $(2x^3 - 3x^2 + 4x - 1) \div (x - 1)$
11. $(3x^3 - 7x^2 - 5x - 3) \div (x - 3)$
12. $(4x^3 + 6x^2 - x - 3) \div (x + 1)$
13. $(2y^4 - 4y^3 + y^2 - 3) \div (y - 2)$
14. $(3y^4 + 8y^3 - 5y^2 + y - 2) \div (y + 3)$
15. $(x^5 + x^3 - x) \div (x + 1)$
16. $(x^5 - x^4 + x^3 - x^2 + x - 1) \div (x - 1)$
17. $(2x^3 + 3x^2 - 2x + 4) \div (x - 1/2)$
18. $(3x^3 - 5x^2 + 7x + 3) \div (x + 1/3)$
19. $(6w^4 - 7w^3 - w^2 + 5w - 2) \div (w - 2/3)$
20. $(2p^4 + p^3 - 3p - 5) \div (p - 3/2)$
21. $(4t^3 + 2t^2 - 7t + 1) \div (t - 1.5)$
22. $(2r^4 + 3r^3 + 4r^2 - 5r + 1) \div (r - 0.5)$
23. $(x^5 - a^5) \div (x - a)$
24. $(x^5 + a^5) \div (x + a)$

In problems 25 to 30, for each function determine whether the given divisor is a factor. If so, express the function in factored form.

25. $f(x) = 2x^3 - 3x^2 + x - 6; \ (x - 2)$
26. $f(x) = 4x^3 + 2x^2 + 3x + 5; \ (x + 1)$
27. $f(x) = 4x^3 - 4x^2 + x + 3; \ (x + 1/2)$
28. $f(x) = 6x^3 - x^2 + 7x - 6; \ (x - 2/3)$
29. $g(y) = 5y^4 - 2y^3 - 5y + 2; \ (y - 0.4)$
30. $\phi(s) = 2s^3 - s^2 - 8s + 4; \ (s - 2.5)$

31. Using synthetic division, divide $f(x) = x^3 + 2x^2 - x + 1$ by $x - 3$ and show that the remainder equals $f(3)$.

32. Using synthetic division, divide $f(x) = x^3 + a^3$ by $x - a$ and show that the remainder equals $f(a)$.

33. To divide by $2x + 1$ using synthetic division, it is necessary to use a divisor of $(\frac{1}{2})(2x + 1) = x + \frac{1}{2}$ and then multiply the quotient *and* remainder by $\frac{1}{2}$ to obtain the correct answer. Using synthetic division, find the quotient and remainder of $(2x^3 + 3x^2 + 3x + 1) \div (2x + 1)$. Check with long division.

34. Using synthetic division, find the quotient and remainder of $(6x^4 - 2x^3 - 3x + 1) \div (3x - 1)$. Check with long division. (See problem 33.)

# 15-3 | *Remainder and Factor Theorems*

## *Remainder Theorem*

Consider the polynomial $f(x) = 2x^3 - x^2 - 8x + 5$ from Example 15-5. When $f(x)$ is divided by $x - 2$, the quotient is $2x^2 + 3x - 2$ and the remainder is $1/(x - 2)$. This can be expressed in terms of multiplication as:

$$f(x) = (2x^2 + 3x - 2)(x - 2) + 1$$

If you let $x = 2$, then $x - 2 = 0$ and the result is the remainder:

$$f(2) = 0 + 1 = 1$$

That is, the value of $f(x)$ when $x = 2$ is the same as the remainder when $f(x)$ is divided by $x - 2$. This is an example of the *remainder theorem:*

> When a polynomial $f(x)$ is divided by $x - r$, the
> remainder $R = f(r)$. (15-2)

The proof is similar to that in the above example. Let $q(x) =$ quotient polynomial. Then in terms of multiplication:

$$f(x) = q(x)(x - r) + R$$

Let $x = r$:  $\quad f(r) = q(r)(0) + R = 0 + R$

and  $\quad f(r) = R$

### Example 15-9
Given $f(x) = 3x^4 + 4x^3 - x^2 + 2x - 1$, find $f(-2)$ by using the remainder theorem and synthetic division. Check by finding $f(-2)$ directly.

**Solution**

To use the remainder theorem, let $r = -2$; then $x - r = x - (-2) = x + 2$. However, to use synthetic division, you change the sign of $+2$ in the divisor. Therefore, *the sign of the divisor constant is the same as the sign of r:*

$$
\begin{array}{r}
3\ +4\ -1\ +2\ -1 \\
-6\ +4\ -6\ +8 \\
\hline
-2\,\overline{)\,3\ -2\ +3\ -4\ +7}
\end{array}
\qquad \boxed{R = 7}
$$

*Check:*     $f(-2) = 3(-2)^4 + 4(-2)^3 - (-2)^2 + 2(-2) - 1$
$$= 48 - 32 - 4 - 4 - 1 = 7$$

The remainder theorem and synthetic division are used to determine values of $f(x)$ for graphing and solving equations (Sec. 15-6).

## Factor Theorem

An important special case of the remainder theorem is the *factor theorem:*

When a polynomial $f(x)$ is divided by $x - r$ and the remainder $R = 0$, then all the following are true and equivalent:

(a) $(x - r)$ is a factor of $f(x)$                                        (15-3)

(b) $R = f(r) = 0$ and $r$ is a zero of $f(x)$

(c) $r$ is a root of the equation $f(x) = 0$

The factor theorem shows clearly the relation between a factor of a polynomial, a zero of a polynomial function, and a root of a polynomial equation.

### Example 15-10

For each function $f(x)$ determine if the given divisor is a factor. If so, express $f(x)$ in factored form.

**1.** $f(x) = x^3 - 1; \ (x - 1)$

**2.** $f(x) = x^3 - 1; \ (x + 1)$

**Solution**

**1.** $f(x) = x^3 - 1; \ (x - 1)$

To use the factor theorem, find $f(r) = f(1) = 1^3 - 1 = 0$. Since $f(1) = 0$, $x - 1$ is a factor of $f(x)$. Find the other factor by synthetic division:

$$
\begin{array}{r}
1\ +0\ +0\ -1 \\
+1\ +1\ +1 \\
\hline
1\,\overline{)\,1\ +1\ +1\ +0}
\end{array}
\qquad \boxed{R = 0}
$$

Then:

$$x^3 - 1 = (x - 1)(x^2 + x + 1)$$

**2.** $f(x) = x^3 - 1; (x + 1)$

Find $f(-1) = (-1)^3 - 1 = -1 - 1 = -2$. Therefore $x + 1$ is not a factor.

### Example 15-11

For each function $f(x)$ and $r$, determine if $r$ is a zero of $f(x)$ [a root of $f(x) = 0$].

**1.** $f(x) = 2x^3 - 3x^2 - 3x + 2; r = -2$
**2.** $f(x) = 3x^3 - x^2 - 3x + 1; r = \frac{1}{3}$

### Solution

**1.** $f(x) = 2x^3 - 3x^2 - 3x + 2; r = -2$

You can find $f(-2)$ directly or use synthetic division, which is usually quicker:

$$
\begin{array}{r}
2 -3 - 3 + 2 \\
-4 +14 -22 \\
\hline
-2\,\lfloor 2\ -7\ +11\ -20
\end{array}
\qquad \boxed{R = -20}
$$

Since $R \neq 0$, $r = -2$ is not a zero of $f(x)$.

**2.** $f(x) = 3x^3 - x^2 - 3x + 1; r = \frac{1}{3}$:

$$
\begin{array}{r}
3 -1 -3 +1 \\
+1 +0 -1 \\
\hline
\tfrac{1}{3}\,\lfloor 3\ +0\ -3\ +0
\end{array}
\qquad \boxed{R = 0}
$$

Since $R = 0$, $r = \frac{1}{3}$ is a zero of $f(x)$, or a root of $f(x) = 0$. Also $x - \frac{1}{3}$ is a factor and $f(x) = (x - \frac{1}{3})(3x^2 - 3)$ or $f(x) = (3x - 1)(x^2 - 1)$.

### Example 15-12

Find the roots of $(x + 3)(x + 2)(x - 1) = 0$.

### Solution

Each factor gives rise to a root. Therefore the three roots are $r_1 = -3$, $r_2 = -2$, and $r_3 = 1$.

**Exercise 15-3**

In problems 1 to 12, for each $f(x)$ and value of $r$, find $f(r)$ by using the remainder theorem and synthetic division. Check by finding $f(r)$ directly.

**1.** $f(x) = 2x^2 - 5x - 10; r = 5$
**2.** $f(x) = 3x^2 + 7x - 3; r = -3$

**3.** $f(x) = x^3 + 2x^2 - 3x + 4;\ r = -2$

**4.** $f(x) = x^3 - 2x^2 - 5x + 7;\ r = 4$

**5.** $f(x) = 3x^3 - 2x + 3;\ r = 3$

**6.** $f(x) = 2x^3 + 5x^2 - 4;\ r = 1$

**7.** $f(x) = 2x^4 + 3x^3 + 2x^2 - x - 1;\ r = -1$

**8.** $f(x) = x^4 - 3x^3 - 4x^2 + x + 2;\ r = 2$

**9.** $f(x) = 3x^3 - 4x^2 + 7x - 3;\ r = 1/3$

**10.** $f(x) = 2x^4 + x^3 + x^2 - 8;\ r = -3/2$

C **11.** $f(x) = x^3 + x - 3;\ r = 1.3$

C **12.** $f(x) = 2x^4 - 6x^2 + 1;\ r = 0.6$

In problems 13 to 18, for each $f(x)$ determine if the given divisor is a factor by using the factor theorem. If so, express $f(x)$ in factored form.

**13.** $f(x) = 3x^3 - 5x^2 + x - 6;\ (x - 2)$

**14.** $f(x) = 2x^3 - 5x^2 - 9;\ (x - 3)$

**15.** $f(x) = x^4 - 4x^2 + 3;\ (x + 1)$

**16.** $f(x) = 2x^4 - 4x^3 + x^2 - 2x + 1;\ (x - 2)$

**17.** $f(x) = x^4 - 16;\ (x - 2)$

**18.** $f(x) = x^4 + 16;\ (x - 2)$

In problems 19 to 26, for each $f(x)$ and $r$, determine if $r$ is a zero of $f(x)$ [a root of $f(x) = 0$].

**19.** $f(x) = 3x^3 - 5x^2 + x - 6;\ r = 2$

**20.** $f(x) = 2x^3 - 3x^2 + x + 4;\ r = -3$

**21.** $f(x) = 3x^4 + 2x^2 + 3x + 1;\ r = -1$

**22.** $f(x) = x^4 - 4x^3 + 7x + 6;\ r = 3$

**23.** $f(x) = 2x^3 - x^2 - 2x - 1;\ r = 1/2$

**24.** $f(x) = 4x^4 - 3x^3 + 7x^2 - 2x - 1;\ r = -1/4$

**25.** $f(x) = 2x^3 - x^2 - 7x + 5;\ r = -2.5$

**26.** $f(x) = 4x^5 - 9x^3 - 2x + 3;\ r = 1.5$

In problems 27 to 32, find the roots of each equation.

**27.** $(x - 4)(x + 3)(x - 1) = 0$

**28.** $(x + 2)(x + 7)(x - 1) = 0$

**29.** $(w)(w - 1.1)(w + 1.1) = 0$

**30.** $3(t - 1/3)(t + 3)(t - 5) = 0$

**31.** $(y - 2)^2(y + 4) = 0$

**32.** $4(y)^2(y - 1/2)^2 = 0$

**33.** Using the factor theorem, show that $2x - 1$ is a factor of $f(x) = 2x^3 - 3x^2 - x + 1$. *Note*: Since $2x - 1 = 2(x - 1/2)$, show that $x - 1/2$ is a factor and then show that 2 is a factor of the quotient.

**34.** Using the factor theorem, show that $2x + 1$ is not a factor of $f(x) = 2x^3 - 3x^2 - x + 1$. (See problem 33.)

**35.** Using the factor theorem, show that $x - a$ is a factor of $x^n - a^n$ for any positive integer $n$.

**36.** Show that $x - 2$ is a double factor of $f(x) = x^4 - 4x^3 + 5x^2 - 4x + 4$ by applying the factor theorem to $f(x)$ and to the quotient after $f(x)$ is divided by $x - 2$.

# 15-4 | *Higher-Degree Equations*

Higher-degree equations are most effectively solved by indirect methods. There are direct formulas for solving third- and fourth-degree equations, but the formulas are involved and time-consuming to use. Direct formulas for solving fifth- and higher-degree equations do not exist. This startling discovery was demonstrated by a young Frenchman Evariste Galois (1811–1832). Fearing his fate during a time of political turmoil, he recorded his results on the eve of his twenty-first birthday. The next morning he was killed in a duel. His contribution has had a significant effect on twentieth-century mathematics. This section describes some of the basic theory used to solve higher-degree equations.

Consider a polynomial equation of degree $n$ with integral coefficients:

$$f(x) = a_n x^n + a_{n-1} x^{n-1} + \cdots + a_1 x + a_0 = 0 \qquad a_n \neq 0 \qquad (15\text{-}4)$$

The *fundamental theorem of algebra* states that every polynomial equation has at least one real or complex root. (This important result, which we accept without proof, was also proved by a young man, the great mathematician Gauss, at the age of 20.) Let this one root be $r_1$. Then by the factor theorem $x - r_1$ is a factor of $f(x)$ and $f(x) = (x - r_1)q_1(x) = 0$, where $q_1(x)$ is the quotient polynomial of degree $n - 1$ after division by $x - r_1$: As in the solution of a quadratic equation, either factor can be set equal to zero. This gives rise to the *depressed equation*, $q_1(x) = 0$, of degree $n - 1$. Applying the fundamental theorem again to the depressed equation yields $q_1(x) = (x - r_2)q_2(x) = 0$, where $q_2(x)$ is the quotient polynomial of degree $n - 2$. The original function can then be written:

$$f(x) = (x - r_1)(x - r_2)q_2(x)$$

Continuing in this way for $n$ divisions, one reaches the final quotient, the constant $a_n$. The function $f(x)$ *can then be expressed as a product of n linear factors* and the constant $a_n$:

$$f(x) = a_n(x - r_1)(x - r_2) \cdots (x - r_n) \qquad (15\text{-}5)$$

This result leads to the following important theorem:

A polynomial equation of degree $n$ has exactly
$n$ linear factors and $n$ roots.                    (15-6)

Study the following examples which illustrate the use of theorem (15-6).

### Example 15-13
Given that $r_1 = 1$ is a root of $f(x) = 2x^3 + 3x^2 - 3x - 2 = 0$, find the three roots and the three factors.

### Solution
Since $r_1 = 1$ is a root, $x - 1$ is a factor. Divide the polynomial by this factor:

$$\begin{array}{r} 2 +3 -3 -2 \\ 2 +5 +2 \\ \hline 1\,|\,2 +5 +2 +0 \quad \boxed{R = 0} \end{array}$$

Notice that *the root $r_1$ is the same as the divisor constant.* Then $f(x) = (x - 1)(2x^2 + 5x + 2) = 0$, and you obtain the depressed equation, $2x^2 + 5x + 2 = 0$. The depressed equation can be solved by factoring: $(x + 2)(2x + 1) = 0$. The three roots of $f(x)$ are then $r_1 = 1$, $r_2 = -2$, and $r_3 = -\frac{1}{2}$. The factors are $f(x) = (x - 1)(x + 2)(2x + 1)$ or $f(x) = (2)(x - 1)(x + 2)(x + \frac{1}{2})$, where the constant 2 is $a_n$, the first coefficient of $f(x)$.

### Example 15-14
Find the four roots of $9x^4 + 21x^3 + x^2 - 9x + 2 = 0$ given that $r_1 = -2$ and $r_2 = \frac{1}{3}$.

### Solution
Use synthetic division twice to divide out each root:

$$\begin{array}{r} 9 +21 +1 - 9 +2 \\ -18 -6 +10 -2 \\ \hline -2\,|\,9 + 3 -5 + 1 +0 \\ 3 +2 - 1 \\ \hline \tfrac{1}{3}\,|\,9 + 6 -3 + 0 \end{array}$$

Set the final quotient equal to zero, and solve the depressed equation:

$$9x^2 + 6x - 3 = 0$$

Divide by 3:                    $$3x^2 + 2x - 1 = 0$$

$$(3x - 1)(x + 1) = 0$$

$$x = \frac{1}{3} \quad \text{and} \quad x = -1$$

The four roots are then $r_1 = -2$, $r_2 = \frac{1}{3}$, $r_3 = \frac{1}{3}$, and $r_4 = -1$. Notice that *a root may occur more than once* (a multiple root).

### Example 15-15
Find the roots of the polynomial equation:

$$(2x^2 - 13x + 15)(x^2 - 2x + 3) = 0$$

### Solution
Set each factor equal to zero. The roots of a quadratic factor can then be found by factoring or by using the quadratic formula:

$$2x^2 - 13x + 15 = 0 \qquad\qquad x^2 - 2x + 3 = 0$$

$$(2x - 3)(x - 5) = 0 \qquad\qquad x = \frac{2 \pm \sqrt{-8}}{2}$$

$$x = \frac{3}{2} \quad \text{and} \quad x = 5 \qquad\qquad x = 1 \pm j\sqrt{2}$$

The four roots are then $r_1 = 3/2$, $r_2 = 5$, $r_3 = 1 + j\sqrt{2}$, and $r_4 = 1 - j\sqrt{2}$.

Notice that *the complex roots are a conjugate pair.* This leads to another basic theorem:

> Complex roots of polynomial equations with integral coefficients always occur in the conjugate pairs $a + bj$ and $a - bj$.          (15-7)

The reason for (15-7) lies in Eqs. (15-4) and (15-5). The factors of Eq. (15-5) produce Eq. (15-4) when multiplied out. The product of all the roots in (15-5) is the *real number* $a_0/a_n$ or $-a_0/a_n$. A complex root when multiplied by its conjugate produces a real number which helps to satisfy this condition. A consequence of (15-7) is that *every cubic (third-degree) equation has at least one real root.*

### Example 15-16
Find the remaining roots of $x^4 - x^2 - 2 = 0$ given one root $r_1 = j$.

### Solution
Applying (15-7) yields a second root $r_2 = -j$. The equation therefore has the linear factors $x - j$ and $x + j$ or their product, the quadratic factor $x^2 + 1$. You

can find the real roots by using synthetic division twice and dividing out each linear factor or by using long division and dividing out the quadratic factor. In this example the long division is no more difficult than the synthetic:

$$1 + 0 - 1 + 0 - 2$$

$$
\begin{array}{r}
x^2 - 2 \\
x^2 + 1\overline{)x^4 - x^2 - 2} \\
\underline{x^4 + x^2} \\
-2x^2 - 2 \\
\underline{-2x^2 - 2} \\
0
\end{array}
$$

$$
\begin{array}{r|l}
 & j - 1 - 2j + 2 \\
j & 1 + j - 2 - 2j + 0 \\
 & \underline{- j + 0 + 2j} \\
-j & 1 + 0 - 2 + 0
\end{array}
$$

In the long division, the missing terms do not need to be replaced by zeros because the divisor and dividend contain only even powers of $x$. The other two roots are then:

$$x^2 - 2 = 0$$

$$r_3 = \sqrt{2} \quad \text{and} \quad r_4 = -\sqrt{2}$$

Note that irrational roots do *not* always occur in conjugate pairs, as complex roots do. A cubic equation can have two imaginary roots and one irrational root or three irrational roots.

**Exercise 15-4**

In problems 1 to 6, find the roots of each equation.

**1.** $(x - 1)(x^2 + 2x - 3) = 0$

**2.** $(2x + 5)(x^2 - 4x + 1) = 0$

**3.** $(x + 2)^2(x^2 - 9) = 0$

**4.** $(x + 1)^3(x^2 - 8x + 16) = 0$

**5.** $(3k^2 - 5k - 2)(k^2 + 2k + 4) = 0$

**6.** $(p)(p^2 - 0.16)(p^2 - 0.25) = 0$

In problems 7 to 22, find the remaining roots of each equation.

**7.** $x^3 + x^2 - 14x - 24 = 0; \, r_1 = 4$

**8.** $x^3 - 7x + 6 = 0; \, r_1 = -3$

**9.** $x^3 - 2x - 1 = 0; \, r_1 = -1$

**10.** $x^3 - 2x^2 + 2x - 1; \, r_1 = 1$

**11.** $2x^3 + 7x^2 + 2x - 3 = 0; \, r_1 = 1/2$

**12.** $3x^3 + x^2 - 3x - 1 = 0; \, r_1 = -1/3$

**13.** $x^4 + 2x^3 - 7x^2 - 8x + 12 = 0; \, r_1 = 1, \, r_2 = 2$

**14.** $2x^4 + 3x^3 - 11x^2 - 3x + 9 = 0; \, r_1 = 1, \, r_2 = 3/2$

**15.** $x^4 - x^3 - 2x^2 + 3x - 1 = 0$; $r_1 = 1$, $r_2 = 1$

**16.** $5x^4 + 4x^3 - 11x^2 - 8x + 2 = 0$; $r_1 = 1/5$, $r_2 = -1$

**17.** $x^4 + 3x^2 - 4 = 0$; $r_1 = 2j$

**18.** $x^4 - x^2 + 2x + 2 = 0$; $r_1 = 1 + j$

**19.** $x^5 - 4x^4 + 3x^3 + 4x^2 - 4x = 0$; $r_1 = 2$, $r_2 = 2$

**20.** $x^5 - x^4 - 4x + 4 = 0$; $r_1 = 1$, $r_2 = j\sqrt{2}$

**21.** $2h^3 - h^2 - 8h - 5 = 0$; $r_1 = 2.5$

**22.** $4Q^4 - 5Q^3 + 11Q^2 + 15Q - 3$; $r_1 = 1.0$, $r_2 = 0.25$

**23.** The deflection of a simple beam (a beam supported at each end) is given by $d(x) = (x)(x - L)(9x^2 - 9xL + 2L^2)$, where $L$ is the length of the beam and $x$ is the distance from one end. Find the values of $x$, in terms of $L$, for which the deflection is zero.

**24.** The acceleration of a machine part in an assembly plant is given by $a(t) = (t - 3.5)(t^2 - 2.3t - 1.7)$, where $t$ is the time in seconds measured from the start of the motion. At what positive values of $t$ is the acceleration zero? (When the acceleration is zero, no force is exerted in the direction of motion.)

**25.** Find the fourth-degree polynomial equation with integral coefficients that has a double root of ½ and a double root of $-2$. (*Hint:* Determine the factors and multiply them out.)

**26.** Find the fourth-degree polynomial equation that has the roots $r_1 = j$ and $r_2 = 2j$. (See hint, problem 25.)

**27.** A spherical buoy of diameter $d$ and specific gravity 0.5 (one-half the density of water) will sink to a depth $x$ given by $4x^3 - 6 dx^2 + d^3 = 0$. Show that a solution to this equation is $x = d/2$.

**28.** Find the other two solutions for $x$, in terms of $d$, for the equation in problem 27.

## 15-5 | *Rational Roots*

The previous section shows how to solve a higher-degree equation when you know enough roots to obtain a quadratic depressed equation. However, it is unlikely you will know any roots when trying to solve such an equation. This section shows techniques for finding rational roots. Finding irrational roots is shown in the next section.

Consider the polynomial equation with integral coefficients:

$$a_n x^n + a_{n-1} x^{n-1} + \cdots + a_1 x + a_0 = 0$$

Divide each term by $a_n$:

$$x^n + \frac{a_{n-1}}{a_n} x^{n-1} + \cdots + \frac{a_1}{a_n} x + \frac{a_0}{a_n} = 0 \qquad (15\text{-}8)$$

Equation (15-8) can be expressed as a product of $n$ linear factors:

$$(x - r_1)(x - r_2) \cdots (x - r_{n-1})(x - r_n) = 0 \qquad (15\text{-}9)$$

where:

$$\left| r_1 \cdot r_2 \cdots r_{n-1} \right| = \left| \frac{a_0}{a_n} \right| \qquad (15\text{-}10)$$

That is, the absolute (positive) value of the product of the roots equals the absolute value of the constant term divided by the coefficient of the first term. This relation leads to the important theorem (which is presented without proof):

> If $p/q$ is a rational root of a polynomial
> equation in lowest terms, then $p$ is a fac- $\qquad$ (15-11)
> tor of $a_0$ and $q$ is a factor of $a_n$.

That is, for any fractional root the *numerator is a divisor of the last coefficient* and the *denominator is a divisor of the first coefficient*. Study the next example, which shows the application of (15-11).

### Example 15-17

Find the roots of $4x^3 - 5x^2 - 7x + 2 = 0$.

### Solution

Consider the term:

$$\left| \frac{a_0}{a_n} \right| = \frac{2}{4}$$

*Do not reduce* the fraction, but write all the factors of the top over all the factors of the bottom:

$$\frac{1,\ 2}{1,\ 2,\ 4}$$

The *possible rational roots* are the positive and negative combinations of top factors over bottom factors:

$$\pm 1 \qquad \pm 2 \qquad \pm \frac{1}{2} \qquad \pm \frac{1}{4}$$

Notice that the combination 2/4 is the same root as 1/2. Since a cubic equation has three roots, you need find only one rational root to obtain a quadratic depressed equation. Using synthetic division, test each possible root (beginning with the simpler ones) until an actual root is found:

$$\begin{array}{r} 4\ -5\ -7\ +2 \\ 4\ -1\ -8 \\ \hline 1\, |\, 4\ -1\ -8\ -6 \end{array} \qquad \begin{array}{r} 4\ -5\ -7\ +2 \\ -4\ +9\ -2 \\ \hline -1\, |\, 4\ -9\ +2\ +0 \end{array} \quad \boxed{Root}$$

Therefore $r_1 = -1$ is a root. The quadratic depressed equation $4x^2 - 9x + 2 = 0$ yields the other two roots, $r_2 = 2$ and $r_3 = \frac{1}{4}$. Note that the absolute value of the product of the roots $|(-1)(2)(\frac{1}{4})| = \frac{2}{4}$, which is $|a_0/a_n|$.

Another technique, *Descartes' rule of signs*, is helpful in finding roots:

> The number of positive roots of $f(x) = 0$ is not
>
> greater than the number of sign changes in $f(x)$.
>
> The number of negative roots is not greater than
>
> the number of sign changes in $f(-x)$.

(15-12)

A *sign change* is a change from $\oplus$ to $\ominus$ or $\ominus$ to $\oplus$, moving in order from one term to the next. A missing term does not affect the application of (15-12).

### Example 15-18

For each equation, apply (15-12) to find the maximum number of positive and negative roots.

**1.** $x^3 + x^2 - x - 1 = 0$

**2.** $x^3 + 2x^2 + 3x + 1 = 0$

**3.** $x^3 - 3x^2 + 4x - 2 = 0$

**4.** $x^4 + x^2 + 5 = 0$

### Solution

**1.** $x^3 + x^2 - x - 1 = 0$

In this equation $f(x) = x^3 + x^2 - x - 1$ has one sign change. Therefore the equation has a maximum of one positive root. The function $f(-x) = (-x)^3 + (-x)^2 - (-x) - 1 = -x^3 + x^2 + x - 1$ has two sign changes, and the equation has a maximum of two negative roots.

**2.** $x^3 + 2x^2 + 3x + 1 = 0$

The function $f(x)$ has no sign changes, and the equation has *no positive roots*. The function $f(-x) = -x^3 + 2x^2 - 3x + 1$ has three sign changes, and the equation has a maximum of three negative roots.

**3.** $x^3 - 3x^2 + 4x - 2 = 0$

This equation has a maximum of three positive roots. The function $f(-x) = -x^3 - 3x^2 - 4x - 2$, and the equation has *no negative roots*.

**4.** $x^4 + x^2 + 5 = 0$

This equation has no positive or negative roots: it has four complex roots. Find $f(-x)$ and show it has no sign changes.

Equations (2) and (3) are special cases of (15-12):

*If the signs in an equation are all the same,*

*there are no positive roots. If the signs alternate, with*

*no missing terms, there are no negative roots.*

Study the next example, which combines all the techniques for finding rational roots and introduces a shortcut way of testing roots.

### Example 15-19
Find the roots of $2x^4 + 7x^3 + 4x^2 + 2x - 3 = 0$

### Solution
This equation has four roots. Applying (15-12) shows that there is no more than one positive root and no more than three negative roots. The term $|a_0/a_n| = 3/2$, and the factors are:

$$\frac{1,\ 3}{1,\ 2}$$

The possible rational roots are $\pm 1$, $\pm 3$, $\pm \frac{1}{2}$, $\pm \frac{3}{2}$. The *synthetic division (shown below) is set up in the form of a table* to test the roots. The top line contains the coefficients of the equation. The *second line is omitted* and is done mentally. The third line is shown next to each possible root with the remainder in the last column. Test each possible root (starting with the simpler ones) until an actual root is found:

$$
\begin{array}{r|rrrrr}
 & 2 + & 7 + & 4 + & 2 - & 3 \\
\hline
1 & 2 + & 9 + 13 + & 15 + & 12 \\
-1 & 2 + & 5 - & 1 + & 3 - & 6 \\
3 & 2 + 13 + 43 + 131 + 390 \\
-3 & 2 + & 1 + & 1 - & 1 + & 0 \\
\end{array}
\quad \boxed{Root}
$$

At this point continue testing roots, but *use the last set of quotient coefficients,* since they are the coefficients of the depressed equation:

$$
\begin{array}{r|rrrr}
 & 2 & +1 & +1 & -1 \\
\hline
\tfrac{1}{2} & 2 & +2 & +2 & +0 \\
\end{array}
\quad \boxed{Root}
$$

The calculator can readily be used to perform the above calculations (see Example 15-21). The quadratic depressed equation is $2x^2 + 2x + 2 = 0$, or $x^2 + x + 1 = 0$. Its solution is $x = -\frac{1}{2} \pm j\sqrt{3}/2$, from the quadratic formula. The four roots are therefore $r_1 = -3$, $r_2 = \frac{1}{2}$, $r_3 = -\frac{1}{2} + j\sqrt{3}/2$, and $r_4 = -\frac{1}{2} - j\sqrt{3}/2$.

Notice that the quotient coefficients in Example 15-19 are all positive after dividing by the first possible root 1 and increase after dividing by a larger possible root, 3. The following two rules apply for synthetic division:

> When the quotient coefficients are all positive, there are
>
> no positive roots larger than the divisor.          (15-13)

> When the signs of the quotient coefficients alternate, there are
>
> no negative roots smaller than the divisor.          (15-14)

The following summarizes the techniques for finding the roots of a higher-degree equation:

1. An equation of degree $n$ has $n$ roots.
2. Complex roots occur in conjugate pairs.
3. Rational roots must satisfy theorem (15-11).
4. The maximum number of positive and negative roots is determined by Descartes' rule (15-12).
5. Rules (15-13) and (15-14) determine upper and lower bounds, respectively, for the roots.

The next example illustrates an interesting application of a higher-degree equation.

### Example 15-20

A rectangular piece of plastic 7 cm by 2 cm is to be molded into an open box with no waste of material. The box is to have a square base and the same thickness as the original piece. If the volume is to be 5 cm³, what are the possible dimensions for the box (Fig. 15-4)?

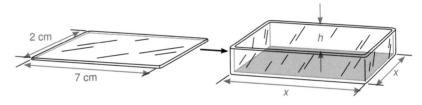

FIG. 15-4 *Molded plastic box, Example 15-20*

### Solution

Let:

$$x = \text{side of the square base}$$
$$h = \text{height of the box}$$

Then the volume is given by:

$$x^2h = 5$$

Since the box and the original piece contain the same amount of material, *the surface area of the box equals the area of the rectangular piece.* The surface area of the box consists of the base $x^2$ and the four sides, each of area $xh$. The total surface area is then given by $x^2 + 4xh$ and is equal to the area of the original rectangular piece $(2)(7) = 14$. The second equation is then:

$$x^2 + 4xh = (2)(7) = 14$$

To solve the two equations, solve the first equation for $h$:

$$h = \frac{5}{x^2}$$

Then substitute this expression for $h$ into the area equation:

$$x^2 + 4x\left(\frac{5}{x^2}\right) = 14$$

This simplifies to the cubic equation as follows:

$$x^2 + \frac{20}{x} = 14$$
$$x^3 + 20 = 14x$$
$$x^3 - 14x + 20 = 0$$

The possible rational roots are $\pm1$, $\pm2$, $\pm4$, $\pm5$, $\pm10$, and $\pm20$. After testing, 2 is found to be a root:

$$
\begin{array}{r}
1 \;+0 \;-14 \;+20 \\
2 \;+\; 4 \;-20 \\
\hline
2\,)\,\overline{1 \;+2 \;-10 \;+\; 0}
\end{array}
$$

Using the quadratic formula gives the other two roots as approximately 2.32 and $-4.32$. A negative answer does not apply, so there are *two* possible solutions: $x = 2$ cm, $h = 1.25$ cm, and $x = 2.32$ cm, $h = 0.93$ cm. The first solution produces a high box with a small base; the second solution, a lower box with a larger base. Both boxes, however, contain the same volume, and depending on the application, one shape may be more suitable than the other.

**Exercise 15-5**

In problems 1 to 18, find the roots of each equation. (Remember that a root may occur more than once.)

1. $x^3 + 6x^2 + 11x + 6 = 0$

2. $x^3 - x^2 - 14x + 24 = 0$

3. $6x^3 + x^2 - 4x + 1 = 0$

4. $8x^3 - 12x^2 + 6x - 1 = 0$

5. $2x^3 - 7x^2 - 6x + 8 = 0$

6. $6x^3 + 2x^2 - 7x + 2 = 0$

7. $4x^3 + 7x^2 + 11x + 6 = 0$

8. $3x^3 + 13x^2 - 9x + 5 = 0$

9. $x^4 - 2x^3 - 3x^2 + 8x - 4 = 0$

10. $x^4 - 13x^2 + 36 = 0$

11. $x^4 - 4x^3 - x^2 + 16x - 12 = 0$

12. $2x^4 - 4x^3 - 9x^2 + 20x - 4 = 0$

13. $4x^4 + 3x^3 + 3x^2 - 13x + 3 = 0$

14. $12x^4 - 25x^3 + 20x^2 + 10x - 12 = 0$

15. $x^5 - 5x^3 + 2x^2 + 6x - 4 = 0$

16. $2x^5 - x^4 + 2x^3 - 4x^2 - 24x = 0$

17. $4K^4 - 17K^2 + 4 = 0$

18. $d^5 - d^4 + d^2 - d = 0$

19. The width of the piece of plastic in Example 15-20 is increased to 3 cm, and the length remains the same. What are the possible dimensions for the box if it is to have the same volume?

20. The box in Example 15-20 is to be enclosed with a top and made from the piece of plastic 7 cm by 2 cm. If the volume must be reduced to 3 cm³, what are the possible dimensions for the box?

□ 21. The bending moment of a beam is given by $M(d) = 0.1d^4 - 1.8d^3 + 10.4d^2 - 19.2d$, where $d$ is the distance in meters from one end. Find the values of $d$ where the bending moment is zero. (*Hint:* Multiply by 10.)

22. The total impedance $Z$ of three impedances in series is given by $Z = Z_1 + Z_2 + Z_3$. If $Z = 3.0\ \Omega$, $Z_2 = Z_1^3$, and $Z_3 = 2 - Z_1^2$, what are the possible values for $Z_1$?

23. A square piece of steel $s$ in. on a side is to be made into a pan by cutting equal squares out of the corners, folding up the sides, and welding the seams (Fig. 15-5). If $s = 6$ in., what are the *two* possible solutions for $x$ that will produce a volume of 8 in³?

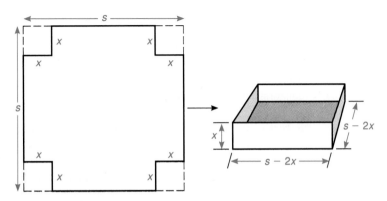

FIG. 15-5 *Steel pan, problem 23*

24. The maximum volume that can be obtained for the pan in problem 23 in 16 in$^3$. What should the side of the cut-out square be to produce this volume?

C 25. In an experiment to determine a point in a magnetic field having a certain magnetic intensity, the following equation arises: $0.1r^4 - 2.4r^2 + 5.5r - 3.0 = 0$, where $r$ is the distance in millimeters from the center of the field. What are the possible solutions for $r$?

26. Find the points of intersection of the equilateral hyperbola $xy = 3$ and the circle $x^2 + y^2 = 10$ by using substitution to solve the equations simultaneously.

27. To solve the cubic equation $x^3 + ax + b = 0$ by using the cubic formula, it is first necessary to find the real value of $k$:

$$k = \sqrt[3]{\frac{-b}{2} + \sqrt{\frac{b^2}{4} + \frac{a^3}{27}}}$$

The roots are then $r_1 = -k$, $r_2 = -k$, and $r_3 = 2k$. If $a = -3$ and $b = 2$, find the three roots by this method and show that they satisfy the equation.

28. The equation $a_n x^3 - 15x^2 - x + a_0 = 0$ has the roots $r_1 = \frac{1}{2}$, $r_2 = \frac{2}{3}$, and $r_2 = -\frac{1}{3}$. What are the values of $a_n$ and $a_0$?

29. (a) Show that $x^4 - 2x^2 - 3 = 0$ has no rational roots. (b) Solve the equation by factoring it into two quadratic factors.

30. Solve $x^4 - 1 = 0$ in three ways: (a) Divide out rational roots, (b) factor into two quadratic factors, and (c) find the four fourth roots of 1, using De Moivre's theorem (Sec. 13-4).

In problems 31 and 32, solve each equation for $\theta$ by first finding the rational values of the trigonometric function.

31. $2 \sin^3 \theta + \sin^2 \theta - 2 \sin \theta = 1$      32. $3 \tan^3 \theta - \tan^2 \theta - \tan \theta = 0$

## 15-6 | *Irrational Roots by Linear Approximation*

When a polynomial equation cannot be reduced to a quadratic factor by dividing out rational roots, you need a method for finding irrational roots. The method of linear approximation described below is one of several methods developed for this purpose. It is a straightforward procedure of testing roots by using synthetic division and can be readily done on any hand calculator. Since equations encountered in practice are very likely to have irrational roots, you should be familiar with a method for finding such roots. Linear approximation is best understood by studying the following examples.

### Example 15-21
Given $x^3 - x^2 + 2x - 4 = 0$, show that there is an irrational root between 1 and 2, and find this root to the nearest tenth.

## Solution

When the value of $f(x)$ changes sign between two values of $x$, the graph of $f(x)$ crosses the $x$ axis. This means there is a zero of $f(x)$, or a root of $f(x) = 0$, between the two values. Using the remainder theorem and synthetic division, compute $f(1)$ and $f(2)$ and show that there is a sign change:

$$
\begin{array}{r}
1 \ -1 \ +2 \ -4 \\
\end{array}
$$

Root between → $\underline{1\, |\, 1 \ +0 \ +2 \ -2}$ ← Sign change
→ $\underline{2\, |\, 1 \ +1 \ +4 \ +4}$ ←

By Descartes' rule of signs, this equation can only have positive roots since the signs alternate and there are no missing terms. The possible rational roots are $+1$, $+2$, and $+4$. The numbers 1 and 2 are not roots as shown above, and there can be no roots larger than 2 since the quotient coefficients are all positive. Hence there is an irrational root between 1 and 2 that cannot be obtained by first dividing out a rational root.

Obtain a *first linear approximation* by graphing the positive and negative values of $f(x)$ and approximating the curve as a straight line (Fig. 15-6). Each box on the $x$ axis is set equal to 0.1 so that the $x$ intercept, which is the approximate root, can be read to the nearest tenth. Since the line crosses the $x$ axis closer to 1.3 than to 1.4, test 1.3 by using synthetic division:

$$
\begin{array}{r}
1 \ - \ 1 \ + \ \ 2 \ -4 \\
1.3 \ +0.39 \ +3.107 \\
\hline
\underline{1.3}\, |\, 1 \ +0.3 \ +2.39 \ -0.893
\end{array}
$$

This division can be done on the calculator by storing the divisor:

$$
1.3 \ \boxed{M_{in}} \ \boxed{\times} \ 1 \ \boxed{-} \ 1 \ \boxed{=} \ \boxed{\times} \ \boxed{MR} \ \boxed{+} \ 2 \ \boxed{=} \ \boxed{\times} \ \boxed{MR} \ \boxed{-} \ 4 \ \boxed{=} \ \longrightarrow \ -0.893
$$

$\quad\ \downarrow \qquad\qquad\qquad\qquad\qquad \uparrow \qquad\qquad\qquad \uparrow$

$\quad 1.3 \qquad\qquad\qquad\qquad\quad 1.3 \qquad\qquad\quad 1.3$

The coefficients of the quotient appear after each use of the $\boxed{=}$ key. The remainder, $f(1.3) = -0.893$, represents a point on the curve of $f(x)$. Since this point is below the $x$ axis, the actual curve looks like the dotted line in Fig. 15-6. This tells you that 1.3 is less than the actual root because $f(x)$ is increasing as $x$ increases. Continue testing larger values until the remainder *changes sign:*

$$
\begin{array}{r}
1 \ - \ 1 \ + \ 2 \ - \ 4 \\
\end{array}
$$

Root between → $\underline{1.4\, |\, 1 \ +0.4 \ +2.56 \ -0.416}$ ← Sign change
→ $\underline{1.5\, |\, 1 \ +0.5 \ +2.75 \ +0.125}$ ←

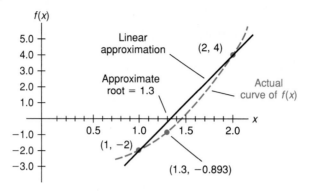

FIG. 15-6 *First linear approximation, Example 15-21*

You can now determine the irrational root to the nearest tenth. *In general, the closer the remainder is to zero, the closer the tested value is to the actual root.* Therefore $x = 1.5$ is the root to the nearest tenth.

The above process can be continued to obtain further accuracy, as shown in the next example.

### Example 15-22
Solve $x^4 - 2x^3 + 2x^2 - 8 = 0$ for all real roots to the nearest hundredth.

### Solution
First test for rational roots. There is one rational root, $r_1 = 2$, which is divided out:

$$
\begin{array}{r}
1 \ -2 \ +2 \ +0 \ -8 \\
2 \ +0 \ +4 \ +8 \\
\hline
2\overline{)\,1 \ +0 \ +2 \ +4 \ +0}
\end{array}
$$

The depressed equation $x^3 + 2x + 4 = 0$ has no positive roots by Descartes' rule (15-12). Calculate negative values of $f(x)$, beginning with zero, to determine where the root lies:

$$
\begin{array}{rl}
& 1 \ +0 \ + \ 2 \ + \ 4 \\
0 & 1 \ +0 \ + \ 2 \ + \ 4 \\
-1 & 1 \ -1 \ + \ 3 \ + \ 1 \\
-2 & 1 \ -2 \ + \ 6 \ - \ 8 \\
-3 & 1 \ -3 \ +11 \ -29
\end{array}
$$

Root between — Sign change / Signs alternate

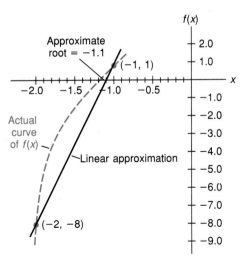

FIG. 15-7 *First linear approximation, Example 15-22*

There is a root between $-1$ and $-2$. The signs alternate after $-2$; therefore, by (15-14) there are no negative roots less than $-2$, and the other two roots are complex. The graph and first linear approximation $(-1.1)$ are shown in Fig. 15-7. Test roots until the remainder changes sign:

$$
\begin{array}{r}
1 + 0 + 2 + 4
\end{array}
$$

Root between

$$
\begin{array}{r|rrr}
-1.1 & 1 & -1.1 & +3.21 & +0.469 \\
-1.2 & 1 & -1.2 & +3.44 & -0.128
\end{array}
$$
Sign change

Graph $f(-1.1) = 0.469$ and $f(-1.2) = -0.128$ to obtain a *second linear approximation* of $-1.18$ (Fig. 15-8). Set each box on the $x$ axis equal to $0.01$ to

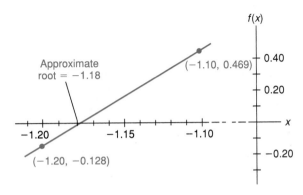

FIG. 15-8 *Second linear approximation, Example 15-22*

approximate the root to the nearest hundredth. Test roots until the remainder changes sign:

$$
\begin{array}{c}
\phantom{xxxxxxxx} 1 \;+0 \quad\; +2 \qquad +4 \\
\text{Root} \qquad\qquad \dfrac{-1.18}{} \; \big| \; 1 \; -1.18 \; +3.3924 \; -0.003032 \\
\text{between} \qquad\quad \dfrac{-1.17}{} \; \big| \; 1 \; -1.17 \; +3.3689 \; +0.058387
\end{array}
\qquad
\begin{array}{l}
\text{Sign} \\
\text{change}
\end{array}
$$

Notice that since the remainder is negative for $-1.18$, a larger value, $-1.17$, is tested to move in the direction of zero, as indicated by the graph. If the graph crossed the $x$ axis while decreasing, a smaller value would be tested, to move closer to zero. Refer to the graph to determine in which direction to test. The root is $-1.18$ to the nearest hundredth since the remainder is closer to zero. The equation $x^4 - 2x^3 + 2x^2 - 8 = 0$ therefore has one rational root, $r_1 = 2$, one irrational root, $r_2 = 1.18$, and two complex roots.

The method of linear approximation can actually be used to solve any equation because it is basically a process of carefully choosing and testing roots that are closer and closer to the solution.

## Exercise 15-6

In problems 1 to 6, find the irrational root between the given values to the nearest tenth.

1. $x^3 - x^2 + x - 2 = 0$; 1, 2
2. $x^3 + 2x^2 + x + 1 = 0$; $-1$, $-2$
3. $x^3 + 4x^2 - 2 = 0$; 0, $-1$
4. $2x^3 + x^2 - x - 1 = 0$; 0, 1
5. $x^4 - 6x^3 + 7x^2 + 5x - 1 = 0$; 2, 3
6. $x^4 + 4x^3 + 2x^2 - 4x - 2 = 0$; $-2$, $-3$

[C] In problems 7 to 10, find the irrational root between the given values to the nearest hundredth.

7. $2x^3 + 3x^2 - x + 1 = 0$; $-1$, $-2$
8. $3x^3 - 11x^2 - x + 5 = 0$; 3, 4
9. $2x^4 - 3x^3 + 5x^2 - x - 2 = 0$; 0, 1
10. $3x^4 + 13x^3 + 6x^2 - 2x - 2 = 0$; $-3$, $-4$

[C] In problems 11 to 14, solve each equation for all *real* roots to the nearest hundredth.

11. $x^3 - 3x^2 + 4x - 3 = 0$      12. $2h^3 + 8h^2 + 2h + 1 = 0$
13. $K^4 + 2K^3 + 2K^2 + 3K + 2 = 0$      14. $x^4 - 2x^3 + 2x^2 - 6x - 3 = 0$

[C] In problems 15 and 16, solve for all real roots to the nearest thousandth.

15. $w^4 - 3w^3 + 4w^2 - 9w + 3 = 0$
16. $n^5 + 3n^4 + 4n^3 + 5n^2 + 5n + 2 = 0$

[C] In applied problems 17 to 24, solve each to the nearest hundredth.

17. The deflection of a cantilever beam (a beam supported at one end) is given by $d(x) = x^4 - 6x^3 + 10x^2 - 6x$, where $x$ is the distance in meters from the fixed end. At what two values of $x$ is the deflection zero?

18. The total capacitance $C$ of three capacitances connected in series is given by:

$$\frac{1}{C} = \frac{1}{C_1} + \frac{1}{C_2} + \frac{1}{C_3}$$

If $C = 0.5 \ \mu F$, $C_2 = C_1 - 1.0 \ \mu F$, and $C_3 = C_1 + 1.0 \ \mu F$, what are the two possible values for $C_1$?

19. For the pan in problem 23, Sec. 15-5, what are the two solutions for $x$ if the volume is to be 12 in.$^3$ and $s = 6$ in.? (*Note:* $x$ must be less than 3.)

20. A cubic container measures 2.0 cm on a side. By how much should each side be reduced to produce a cube that has one-half the volume?

21. The temperature $T$ (in degrees Celsius) in the refrigeration car of a freight train during a 6½-h trip varies according to the function $T(t) = t^3 - 7t^2 + 4t + 10$, where $t = $ time in hours from the start of the trip. For how many hours is the temperature below freezing? (*Hint:* Find the difference of the two positive roots.)

22. One thousand dollars is deposited at the beginning of every year for 4 years, with interest compounded annually. At the end of the fourth year it amounts to $5000. The interest rate is given by the equation:

$$\frac{(r + 1)^4 - 1}{r} = 5$$

What is the interest rate to the nearest 1 percent?

23. Show that the equation $x^4 - 7x^2 + 1 = 0$ can be factored as $(x^2 + 3x + 1)(x^2 - 3x + 1) = 0$, and find the four irrational roots from the factors.

24. To solve the quartic equation $x^4 + 2x^3 + 2x + 1 = 0$ by using the quartic formula, it is first necessary to find the real root of the cubic equation $k^3 - 8 = 0$. The original equation can then be expressed as:

$$[x^2 + (1 + \sqrt{k + 1})x + 1][x^2 + (1 - \sqrt{k + 1})x + 1] = 0$$

Using this method, find the four roots of the original equation.

$\boxed{\text{C}}$ In problems 25 to 28, solve each equation for $\theta$ in radians between the given values to the nearest hundredth, by first finding the values of the trigonometric function.

25. $\tan^3 \phi - \tan^2 \phi + 2 \tan \phi - 4 = 0$; $0 < \phi < \pi/2$

26. $\sin^3 \phi + 4 \sin^2 \phi - 2 = 0$; $-\pi/2 < \phi < 0$

27. $\sin \phi + 1 = \phi$; $0 < \phi < \pi$

28. $\tan \phi - 2\phi = 0$; $0 < \phi < \pi/2$

$\boxed{\text{C}}$ 29. In studying the forces on a certain structure, the following equation arises: $\cos 2\theta - F\theta = 0$. If $F = 3.0$, using linear approximation, find $\theta$ in radians, $0 < \theta < \pi/2$, to the nearest hundredth.

□ **30.** A cycloid is a curve traced by a point on a circle that rolls along a straight line. It is used in the construction of gears and is described by the parametric equations $x = \theta - \sin \theta$ and $y = 1 - \cos \theta$. For what value of $\theta$ less than $\pi/2$ does $x + y = 1$?

## 15-7 | *Review Questions*

In problems 1 and 2, for each function find the given values.

**1.** $f(x) = 2x^3 - x^2 + 3x + 4$; $f(2)$, $f(-1/2)$

**2.** $f(x) = x^3 + a^3$; $f(0)$, $f(-a)$

In problems 3 to 6, graph each polynomial function for the given range of values.

**3.** $f(x) = x^3 - 3x^2 + 2x$;
$x = -1$ to $x = 3$

**4.** $f(x) = x^3 - 5x^2 + 3x + 9$;
$x = -2$ to $x = 4$

**5.** $f(x) = 2x^4 - 4x^2$;
$x = -2$ to $x = 2$

**6.** $f(x) = (x - 2)(x + 1)^2$;
$x = -2$ to $x = 3$

In problems 7 to 10, for each $f(x)$ and value of $r$ find $f(r)$, using the remainder theorem and synthetic division.

**7.** $f(x) = x^3 - 2x^2 + 3x + 4$; $r = 2$

**8.** $f(x) = 3x^4 - 2x^3 + x^2 - x + 1$; $r = -1$

**9.** $f(x) = x^4 + 3x^3 - 2x - 6$; $r = -3$

**10.** $f(x) = 4x^3 - 8x^2 + x + 1$; $r = 1/2$

In problems 11 to 14, for each $f(x)$ determine if the given divisor is a factor. If so, express $f(x)$ in factored form.

**11.** $f(x) = 2x^3 + 5x^2 - 4$; $(x + 2)$

**12.** $f(x) = x^4 - 6x^2 - 8x - 3$; $(x - 3)$

**13.** $f(x) = 2x^3 - 3x^2 + x + 4$; $(x - 1.5)$

**14.** $f(x) = x^4 + a^4$; $(x - a)$

In problems 15 to 20, find the roots of each equation.

**15.** $(x + 3)(2x^2 + 4x + 2) = 0$

**16.** $x^3 + x^2 - 10x + 8 = 0$

**17.** $2x^3 - 9x^2 + 10x - 3 = 0$

**18.** $9x^3 + 21x^2 - 17x + 3 = 0$

**19.** $x^4 - 4x^3 + 4x - 1 = 0$

**20.** $2x^4 - 5x^3 - x^2 + 11x + 5 = 0$

**21.** Find the irrational root between 2 and 3 to the nearest tenth for $x^3 - x^2 - 2x - 5 = 0$.

22. Find the negative irrational root to the nearest hundredth for $x^4 - 2x^3 - 3x^2 + 4x - 5 = 0$.

23. The deflection of a simple beam is given by $d(x) = (0.01x)(x - 10) \times (x^2 - 9x + 18)$. Use suitable scales and graph the polynomial function, showing the four intercepts.

24. The velocity of the tidal current in a certain river is given approximately by $v(t) = 2t^3 - 3t^2 - 59t + 30$, where $t$ = time in hours from high tide and varies from $t = -6$ to $t = 7$. At what times is the tidal current zero?

25. The total resistance of three resistances in parallel is given by $1/R = 1/R_1 + 1/R_2 + 1/R_3$. If $R = 0.5$ $\Omega$, $R_2 = R_1 + 2.0$ $\Omega$, and $R_3 = R_1 + 0.5$ $\Omega$, what is the value of $R_1$?

26. A rectangular box with a square base 1 ft on a side is 2 ft high. By how much should each side be increased to double the volume, if the increase is to be the same for each side? Find the answer to the nearest hundredth.

# C H A P T E R

# 16

# Series and the Binomial Formula

## 16-1 | *Arithmetic Series*

A *sequence* of $n$ terms is the arrangement:

$$a_1, a_2, a_3, \ldots, a_{n-1}, a_n \quad (16\text{-}1)$$

**M**any natural and physical patterns can be described by a sequence of numbers, such as the patterns found in the periodic table of the elements, in electric circuits, in population changes, or in the price of an item every year: $2.75, $3.00, $3.50, etc. Mathematical terms and functions can be defined in terms of series (sums) of numbers, for example, $\frac{1}{3} = 0.3 + 0.03 + 0.003 + \cdots$ or $e = 2 + \frac{1}{2} + \frac{1}{6} + \cdots$. Three basic types of sequences and series—arithmetic, geometric, and binomial—and their applications are studied in this chapter. Mathematical sequences and series are covered in greater detail in calculus.

An *arithmetic sequence*, or *arithmetic progression*, is a sequence where each term is obtained from the preceding term by *adding* a constant $d$ called the *common difference:*

$$a_1, a_1 + d, a_1 + 2d, \ldots , \\ a_1 + (n-2)d, a_1 + (n-1)d \quad (16\text{-}2)$$

Notice that the $n$th term $a_n$ is obtained from the first term $a_1$ by adding the common difference $n - 1$ times:

$$a_n = a_1 + (n-1)d \quad (16\text{-}3)$$

### Example 16-1
For the given value of $n$, find $d$ and $a_n$ for each arithmetic sequence.

**1.** $0, 3, 6, \ldots ; n = 9$

**2.** $5, \dfrac{9}{2}, 4, \ldots ; n = 12$

**3.** $0.31, 0.38, 0.45, \ldots ; n = 7$

## Solution
**1.** 0, 3, 6, . . . ; $n = 9$

Find the common difference by subtracting any term from the next term: $d = 6 - 3 = 3 - 0 = 3$. The ninth term is then obtained using Eq. (16-3), where $n = 9$, $a_1 = 0$, and $d = 3$:

$$a_9 = 0 + (9 - 1)(3) = 24$$

**2.** $5, \dfrac{9}{2}, 4, \ldots ; n = 12$:

$$d = 4 - \frac{9}{2} = -\frac{1}{2} \qquad \text{and} \qquad a_{12} = 5 + (12 - 1)\left(-\frac{1}{2}\right) = -\frac{1}{2}$$

**3.** 0.31, 0.38, 0.45, . . . ; $n = 7$:

$$d = 0.45 - 0.38 = 0.07 \qquad \text{and} \qquad a_7 = 0.31 + (7 - 1)(0.07) = 0.73$$

An *arithmetic series* is the sum of the terms of an arithmetic sequence:

$$S_n = a_1 + (a_1 + d) + (a_1 + 2d) + \cdots + [a_1 + (n - 1)d] \qquad (16\text{-}4)$$

A formula for the sum of the first $n$ terms is obtained by first writing the sum of the terms in reverse order, using $a_n$:

$$S_n = a_n + (a_n - d) + (a_n - 2d) + \cdots + [a_n - (n - 1)d]$$

and then adding the two equations for $S_n$; all the terms containing $d$ cancel:

$$2S_n = (a_1 + a_n) + (a_1 + a_n) + \cdots + (a_1 + a_n) = n(a_1 + a_n)$$

Therefore:

$$S_n = \frac{n}{2}(a_1 + a_n) \qquad (16\text{-}5)$$

## Example 16-2
For the given value of $n$, find $S_n$ for each arithmetic series:

**1.** $1 + 2 + 3 + \cdots ; n = 100$

**2.** $-\dfrac{4}{3} - \dfrac{2}{3} + 0 + \cdots ; n = 10$

**3.** $3.5 + 2.5 + 1.5 + \cdots ; n = 8$

## Solution

**1.** $1 + 2 + 3 + \cdots$; $n = 100$

Use $a_1 = 1$ and $d = 1$ to first find $a_{100} = 1 + (99)(1) = 100$. Then apply Eq. (16-5):

$$S_{100} = \frac{100}{2}(1 + 100) = 5050$$

**2.** $-\dfrac{4}{3} - \dfrac{2}{3} + 0 + \cdots$; $n = 10$:

$$a_{10} = -\frac{4}{3} + (9)\left(\frac{2}{3}\right) = \frac{14}{3} \qquad \text{and} \qquad S_{10} = \frac{10}{2}\left(-\frac{4}{3} + \frac{14}{3}\right) = \frac{50}{3}$$

**3.** $3.5 + 2.5 + 1.5 + \cdots$; $n = 8$:

$$a_8 = 3.5 + (7)(-1) = -3.5 \qquad \text{and} \qquad S_8 = \frac{8}{2}(3.5 - 3.5) = 0$$

### Example 16-3

Two companies offer you a job. Company X offers you $20,000 a year with a raise of $1000 per year, and company Y offers you $20,500 with a raise of $800 per year.

**1.** What would your salary be at each company in your 10th year?

**2.** At which company would your total earnings be greater after 10 years?

### Solution

**1.** What would your salary be at each company in your 10th year?

For company X, $a_{10} = 20,000 + 9(1000) = \$29,000$. For company Y, $a_{10} = 20,500 + 9(800) = \$27,700$.

**2.** At which company would your total earnings be greater after 10 years?

At company X, $S_{10} = (10/2)(20,000 + 29,000) = \$245,000$. At company Y, $S_{10} = (10/2)(20,500 + 27,700) = \$241,000$. Your total earnings at company X are greater, however, note that your earnings are greater at company Y for the first six years. This should be taken into account when you decide which offer to accept, because interest can be earned on the extra money in a savings account or you may decide to switch jobs after 5 years.

**Exercise 16-1**

In problems 1 to 10, for the given value of $n$, find $d$ and $a_n$ for each arithmetic sequence.

**1.** 5, 9, 13, . . . ; $n = 13$

**2.** $-1$, 2, 5, . . . ; $n = 12$

**3.** 5, 4, 3, . . . ; $n = 20$

**4.** 40, 32, 24, . . . ; $n = 16$

**5.** $-\dfrac{2}{5}, \dfrac{1}{5}, \dfrac{4}{5}, \ldots \,; n = 11$    **6.** $8, \dfrac{13}{2}, 5, \ldots \,; n = 8$

**7.** $-1.9, -1.6, -1.3, \ldots \,; n = 9$    **8.** $0.10, 0.15, 0.20, \ldots \,; n = 15$

**9.** $0, x + 1, 2x + 2, \ldots \,; n = 10$    **10.** $-e, k, 2k + e, \ldots \,; n = 7$

In problems 11 to 20, for the given value of $n$, find $S_n$ for each arithmetic series.

**11.** $0 + 7 + 14 + \cdots ; n = 10$

**12.** $1 + 3 + 5 + \cdots ; n = 30$

**13.** $-4 - 6 - 8 - \cdots ; n = 15$

**14.** $37 + 26 + 15 + \cdots ; n = 13$

**15.** $\dfrac{27}{4} + 6 + \dfrac{21}{4} + \cdots ; n = 9$

**16.** $\dfrac{1}{2} + \dfrac{2}{3} + \dfrac{5}{6} + \cdots ; n = 8$

**17.** $1.00 + 1.01 + 1.02 + \cdots ; n = 20$

**18.** $6.5 + 6.0 + 5.5 + \cdots ; n = 25$

**19.** $6bc + 4bc + 2bc + \cdots ; n = 12$

**20.** $\dfrac{\pi}{4} + \dfrac{\pi}{2} + \dfrac{3\pi}{4} + \cdots ; n = 10$

**21.** In an arithmetic series $a_1 = 18$ and $a_5 = 6$. Find $d$ and $S_5$.

**22.** In an arithmetic series $a_{10} = 8$ and $d = 0.4$. Find $a_1$ and $S_{10}$.

**23.** Find the sum $0.1 + 0.2 + 0.3 + \cdots + 10.0$. [*Hint:* Use Eq. (16-3) to first find $n$.]

**24.** Find the sum: $\frac{1}{8} + \frac{1}{4} + \frac{3}{8} + \cdots + 4$. (See hint in problem 23.)

**25.** Show that the sum of the first $n$ integers is $n(n + 1)/2$.

⊏ **26.** The reciprocals of an arithmetic series form a *harmonic series*. (a) Find the 10th term in the harmonic series $1 + \frac{1}{2} + \frac{1}{3} + \cdots$. (b) Find the sum of the first 10 terms in the series by using a calculator.

**27.** The spark-gap distance for a set of electrodes is 1.8 mm at 5 kV and increases 1.6 mm for each increase of 5 kV. What is the spark-gap distance at 35 kV?

**28.** A chemical process takes 50 min at 21°C but takes 20 s less for every 1° increase in temperature. How long will the process take at 60°C?

**29.** A bridge truss is in the shape of a right triangle with legs 30 ft long and 40 ft high. If it contains nine equally spaced supports, as shown in Fig. 16-1, what is the total length of the 10 vertical members? (*Hint:* All the triangles are similar.)

**30.** A ball thrown into the air from a cliff travels 16 ft up during the first second, 16 ft down during the second second, 48 ft down during the third second, etc. (a) How far does it fall during the eighth second? (b) What is the total distance it has fallen from its original height after 8 s?

**31.** Gravitational acceleration decreases by 0.003 m/s² for every kilometer of altitude. If $g = 9.793$ m/s² at 1 km above Los Angeles, what is the gravitational acceleration 9.0 km above Los Angeles?

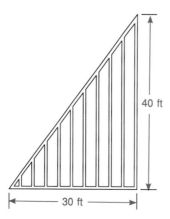

FIG. 16-1 *Bridge truss, problem 29*

**32.** Dockage fees for a boat at a marina are $3.00 for the first night, $3.25 for the second night, $3.50 for the third night, etc. How much does it cost to dock for 2 weeks?

**33.** The Chinese calendar has 12 animal designations, which repeat every 12 years. Beginning with 1868, a Year of the Dragon, how many ''dragon years'' will have occurred by the end of the Year of the Dragon, 2000? [*Hint:* Use Eq. (16-3) and find *n*.]

**34.** Two jobs are offered to you for 7 years in a foreign country: job A paying $22,000 per year with a $1500 per year raise and job B paying $23,000 per year with a $1000 per year raise. Which job pays more total salary and how much more?

**35.** As a result of inflation, the cost of a new-model automobile increased $100 after the first year of production, an *additional* $120 after the second year, an *additional* $140 after the third year, etc. If the car originally sold for $6,000, what was its price 6 years later?

**36.** In the Christmas carol ''The Twelve Days of Christmas,'' how many gifts are given on the twelfth day?

# 16-2 | *Geometric Series*

Geometric series are found in many applications, such as population growth, compound interest, radioactive decay, electric circuits, and in the concept of rational numbers. Geometric series are related to exponential functions (Sec. 9-1), and many of the applications are similar.

A *geometric sequence*, or *geometric progression*, is a sequence where each term is obtained from the preceding term by *multiplying* by a constant *r* called the *common ratio:*

$$a_1, a_1 r, a_1 r^2, \ldots, a_1 r^{n-2}, a_1 r^{n-1} \qquad (16\text{-}6)$$

The $n$th term is then:

$$a_n = a_1 r^{n-1} \tag{16-7}$$

### Example 16-4
For the given value of $n$, find $r$ and $a_n$ for each geometric sequence.

**1.** 3, 6, 12, . . . ; $n = 9$

**2.** 4, $\dfrac{8}{3}$, $\dfrac{16}{9}$, . . . ; $n = 7$

⌷ **3.** 10.0, 9.00, 8.10, . . . ; $n = 8$

### Solution
**1.** 3, 6, 12, . . . ; $n = 9$

Find the common ratio by dividing any term by the previous term: $r = 12/6 = 6/3 = 2$. Then apply Eq. (16-7):

$$a_9 = 3(2)^{9-1} = 3(2)^8 = 3(256) = 768$$

**2.** 4, $\dfrac{8}{3}$, $\dfrac{16}{9}$, . . . ; $n = 7$:

$$r = \frac{8/3}{4} = \frac{2}{3} \quad \text{and} \quad a_7 = 4\left(\frac{2}{3}\right)^6 = 4\left(\frac{64}{729}\right) = \frac{256}{729}$$

⌷ **3.** 10.0, 9.00, 8.10, . . . ; $n = 8$:

$$r = \frac{9.00}{10.0} = 0.90 \quad \text{and} \quad a_8 = 10.0(0.90)^7 = 4.78 \quad \text{(three figures)}$$

### Example 16-5
In a dc circuit a charged capacitor is short-circuited, causing its charge to decrease by 30 percent every 0.20 s. What percentage of charge is left after 1 s?

### Solution
Every 0.20 s the charge in the capacitor will be $100 - 30$ percent, or 70 percent, of the charge 0.20 s ago. Therefore the fraction of charge left after each 0.20 s interval forms a geometric sequence in which $r = 0.70$: 0.700, 0.490, 0.343, . . . . The fifth term represents the charge left after 1 s:

$$a_5 = 0.700(0.70)^4 = 0.168 \quad \text{or} \quad 16.8\% \quad \text{(three figures)}$$

A *geometric series* is the sum of the terms of a geometric sequence:

$$S_n = a_1 + a_1r + a_1r^2 + \cdots + a_1r^{n-1} \tag{16-8}$$

A formula for $S_n$ is obtained by multiplying (16-8) by $r$:

$$rS_n = a_1r + a_1r^2 + a_1r^3 + \cdots + a_1r^n$$

and subtracting equations:

$$S_n - rS_n = a_1 + (a_1r - a_1r) + (a_1r^2 - a_1r^2) + \cdots + (a_1r^{n-1} - a_1r^{n-1}) - a_1r^n$$

Then $S_n(1 - r) = a_1 - a_1r^n$ and:

$$S_n = \frac{a_1 - a_1r^n}{1 - r} \qquad \text{or} \qquad S_n = \frac{a_1r^n - a_1}{r - 1} \tag{16-9}$$

### Example 16-6
For the given value of $n$, find $S_n$ for each geometric sequence.

**1.** $1 + 3 + 9 + \cdots;\ n = 6$
**2.** $8 - 4 + 2 - \cdots;\ n = 9$
□ **3.** $50 + 60 + 72 + \cdots;\ n = 5$

### Solution
**1.** $1 + 3 + 9 + \cdots;\ n = 6$

First find $r = 9/3 = 3$. Then apply Eq. (16-9):

$$S_6 = \frac{1 - 1(3)^6}{1 - 3} = \frac{1 - 729}{-2} = 364$$

**2.** $8 - 4 + 2 - \cdots;\ n = 9$:

$$r = -\frac{4}{8} = -\frac{1}{2} \qquad \text{and} \qquad S_9 = \frac{8 - 8(-\frac{1}{2})^9}{1 - (-\frac{1}{2})} = \frac{8 + \frac{1}{64}}{\frac{3}{2}}$$

$$= \frac{171}{32} \quad \text{or} \quad 5.34$$

Notice in (2) that $r$ is negative, which causes the signs to alternate.

□ **3.** $50 + 60 + 72 + \cdots;\ n = 5$:

$$r = \frac{60}{50} = 1.2 \qquad \text{and} \qquad S_5 = \frac{50 - 50(1.2)^5}{1 - 1.2} = 372.08$$

### Example 16-7

Suppose you receive a chain letter and mail copies to four friends. Each of your friends then mails copies to four other friends. If this process continues unbroken for 6 cycles, how much money is spent in postage at 25 cents per letter?

### Solution

The total number of letters mailed is the geometric series $4 + 16 + 64 + \cdots$. After 6 cycles the sum is:

$$S_6 = \frac{4 - 4(4)^6}{1 - 4} = \frac{4 - 16,384}{-3} = 5460$$

The total postage is $5460(0.25) = \$1365.00$. Despite the postage the government might gain, chain letters that require giving money to a sender are illegal.

## Compound Interest

Suppose you deposit $100 at an interest rate of 10 percent per year compounded annually. The amount of money you will have after 1 year is:

$$100 + 100(0.10) = 100(1 + 0.10) = \$110$$

After 2 years the amount will be:

$$110 + 110(0.10) = 110(1 + 0.10) = \$121$$

The amounts after each year form a geometric sequence:

$$110, \ 121, \ 133.10, \ \ldots$$

where $r = 121/110 = 1.10$, or $1 + 0.10$. That is, $r = 1 + i$, where $i =$ interest rate. This leads to the formula for compound interest:

$$A = P(1 + i)^n \tag{16-10}$$

where $A =$ amount after $n$ interest periods, $P =$ principal (initial amount), and $i =$ interest rate *per interest period*. Equation (16-10) is an application of Eq. (16-7) in which $n$ is used instead of $n - 1$ since $P$ is equivalent to the "zero term," or the term before the first term. Interest is always stated on a yearly, or annual, basis. However, the interest rate $i$ is per interest period, so for interest compounded more often than yearly, $i$ is a fraction of the annual rate. Study the next example.

☐ **Example 16-8**

At the beginning of the year, $1000 is deposited in a savings bank at 8 percent annual interest.

1. If the interest is compounded quarterly, what will be the amount at the end of the year?
2. If the interest is compounded daily, what will be the amount at the end of the year?
3. If four additional deposits of $1000 are made at the end of each quarter, what will be the total amount at the end of the year?

**Solution**

1. If the interest is compounded quarterly, what will be the amount at the end of the year?

   The interest rate per interest period $i = 0.08/4 = 0.02$, $P = 1000$, and $n = 4$. Apply Eq. (16-10):

   $$A = 1000(1 + 0.02)^4 = 1000(1.02)^4 = \$1082.43$$

2. If the interest is compounded daily, what will be the amount at the end of the year?

   The interest rate $i = 0.08/365 = 0.0002191781$, $n = 365$, and:

   $$A = 1000(1 + 0.0002191781)^{365} = \$1083.28$$

   Observe the small difference between compounding quarterly ($1082.43) and daily ($1083.28). Also note that to obtain 6 figure accuracy you must calculate with 7 figures and then round off.

3. If four additional deposits of $1000 are made at the end of each quarter, what will be the total amount at the end of the year?

   There are five deposits in all. The fifth deposit of $1000 earns no interest. The fourth deposit earns interest for one quarter, and the amount is $1000(1.02)$; the third deposit earns interest for two quarters, and the amount is $1000(1.02)^2$; and so on. The total amount of the five deposits is the sum of a geometric series:

   $$S_5 = 1000 + 1000(1.02) + 1000(1.02)^2 + 1000(1.02)^3 + 1000(1.02)^4$$

   Apply Eq. (16-9):

   $$S_5 = \frac{1000 - 1000(1.02)^5}{1 - 1.02} = \$5204.04$$

## *Infinite Geometric Series*

Imagine yourself moving at 1 mi/h over a distance of 1 mi. It would take you 1 h to cover the distance. Reasoning another way, it would take you ½ h to cover half the distance, ¼ h to cover half the *remaining* distance, ⅛ h to cover half the distance *still remaining,* etc. The total time $T$ to travel the entire distance is an infinite geometric series:

$$T = \frac{1}{2} + \frac{1}{4} + \frac{1}{8} + \frac{1}{16} + \cdots \qquad r = \frac{1}{2}$$

To Zeno, the ancient Greek philosopher, it appeared that the sum would be infinite. He concluded that this reasoning leads to the paradox that motion is impossible, since the argument can be applied to any distance. However, if you apply the sum formula (16-9) to this series:

$$S_n = \frac{\frac{1}{2} - (\frac{1}{2})(\frac{1}{2})^n}{1 - \frac{1}{2}}$$

the term $(\frac{1}{2})^n$ becomes infinitely small as $n$ becomes infinitely large. For example, $(\frac{1}{2})^{20} = 0.000001$. This term can be ignored (assumed to be zero) without any loss of accuracy for large values of $n$. The ''infinite sum'' $S_\infty$, which is the value to which $S_n$ *becomes infinitesimally close as n becomes infinitely large,* is then:

$$S_\infty = \frac{\frac{1}{2}}{1 - \frac{1}{2}} = 1 \text{ h}$$

The answer agrees with what you know to be correct, and the reasoning is valid.

For an infinite geometric series where $-1 < r < 1$, or $|r| < 1$, you can apply the same reasoning to Eq. (16-9) and assume the term $a_1 r^n$ to be zero. The formula for *the sum of an infinite geometric series with $-1 < r < 1$, or $|r| <$ 1,* is then:

$$S_\infty = \frac{a_1}{1 - r} \tag{16-11}$$

Any repeating decimal can be shown to be a fraction or rational number by applying (16-11) as follows.

### Example 16-9
Express each repeating decimal as a fraction.

**1.** $0.2222\cdots$

**2.** $7.454545\cdots$

### Solution

**1.** $0.2222\cdots$

Let $0.2222\cdots = 0.2 + 0.02 + 0.002 + \cdots$. The repeating decimal is then an infinite geometric series with $a_1 = 0.2$ and $r = 0.1$. Use Eq. (16-11) to find the fraction:

$$S_\infty = \frac{0.2}{1 - 0.1} = \frac{0.2}{0.9} = \frac{2}{9}$$

**2.** $7.454545\cdots$

Let $7.454545\cdots = 7 + 0.45 + 0.0045 + 0.000045 + \cdots$, which is then:

$$7 + S_\infty = 7 + \frac{0.45}{1 - 0.01} = 7 + \frac{45}{99} = \frac{82}{11}$$

**Exercise 16-2**

In problems 1 to 12, for the given value of $n$, find $r$ and $a_n$ for each geometric sequence.

**1.** $1, 3, 9, \ldots$ ; $n = 7$      **2.** $5, 10, 20, \ldots$ ; $n = 8$

**3.** $27, 18, 12, \ldots$ ; $n = 5$      **4.** $16, 4, 1, \ldots$ ; $n = 6$

**5.** $250, -100, 40, \ldots$ ; $n = 6$      **6.** $\dfrac{1}{6}, -\dfrac{1}{2}, \dfrac{3}{2}, \ldots$ ; $n = 8$

© **7.** $1.6, 2.4, 3.6, \ldots$ ; $n = 7$      **8.** $70, 7, 0.7, \ldots$ ; $n = 7$

**9.** $e^3, e^2, e, \ldots$ ; $n = 6$      **10.** $\dfrac{\pi}{12}, \dfrac{\pi}{6}, \dfrac{\pi}{3}, \ldots$ ; $n = 9$

**11.** $7, -7, 7, \ldots$ ; $n = 18$      **12.** $\dfrac{1}{x+1}, 1, x+1, \ldots$ ; $n = 5$

In problems 13 to 20, for the given value of $n$, find $S_n$ for each geometric series.

**13.** $1 + 2 + 4 + \cdots$ ; $n = 8$

**14.** $2 + 6 + 18 + \cdots$ ; $n = 6$

© **15.** $243 + 324 + 432 + \cdots$ ; $n = 6$

**16.** $192 - 96 + 48 - \cdots$ ; $n = 7$

© **17.** $25.6 + 6.4 + 1.6 + \cdots$ ; $n = 5$

© **18.** $0.96 + 1.44 + 2.16 + \cdots$ ; $n = 5$

© **19.** $6.4\pi + 3.2\pi + 1.6\pi + \cdots$ ; $n = 7$

**20.** $x - 5x + 25x - \cdots$ ; $n = 6$

In problems 21 to 26, find the sum of each infinite geometric series.

**21.** $3 + 1 + \dfrac{1}{3} + \cdots$      **22.** $8 + 6 + \dfrac{9}{2} + \cdots$

**23.** $50 - 20 + 8 - \cdots$      **24.** $\dfrac{1}{2} - \dfrac{1}{4} + \dfrac{1}{8} - \cdots$

**25.** $0.100 + 0.090 + 0.081 + \cdots$      **26.** $10 + 7.0 + 4.9 + \cdots$

In problems 27 to 32, express each repeating decimal as a fraction. (See Example 16-9.)

**27.** $0.5555\cdots$                          **28.** $1.3333\cdots$

**29.** $0.09090909\cdots$                      **30.** $2.41414141\cdots$

**31.** $0.123123123\cdots$                     **32.** $0.357777\cdots$

**33.** Find the amount after 3 years of a principal of $100 earning 6 percent annual interest *(a)* compounded annually and *(b)* compounded semiannually.

**34.** Find the amount after 2 years of a principal of $500 earning 8 percent interest *(a)* compounded annually and *(b)* compounded quarterly.

**35.** Which provides a higher annual yield, 12.5 percent interest compounded annually or 12 percent interest compounded quarterly? (*Hint:* Find the amount for a principal of $1000 after 1 year.)

**36.** What is the difference in the interest after 1 year for a principal of $10,000 earning 10 percent interest compounded quarterly, compared with the interest for a principal of the same amount compounded daily? (See Example 16-8.)

**37.** Danielle decides to deposit $1000 at the beginning of each year to provide for her retirement. If her money earns 11 percent interest compounded annually, how much will she have at the beginning of the fifth year (immediately after her fifth deposit)? (See Example 16-8.)

**38.** Alan invests $500 at the beginning of each month to pay for his son's college education in 4 years. If the money earns 12 percent interest compounded monthly, how much will he have at the beginning of the fifth year (immediately after his 48th deposit)? (See Example 16-8.)

**39.** In a dc circuit an inductance is short-circuited, causing the current to decrease by one-half every 0.40 s. What fraction of the original current is left after 2 s? (See Example 16-5.)

**40.** A constant voltage is applied to a dc circuit, causing a capacitor to increase to 60 percent of its total charge in ¼ s. Every ¼ s thereafter, the charge increases by 40 percent of the increase during the last quarter of a second. The total charge $Q$ is then given by the infinite geometric series

$$Q = 0.600Q + 0.240Q + 0.096Q + \cdots$$

Each term represents the fractional increase in charge each ¼ s. *(a)* What fraction of the total charge is present after 2 s? *(b)* Show that the infinite sum equals $Q$, as indicated.

**41.** The atmospheric pressure at sea level is approximately 100 kPa and decreases by 20 percent for each 1-mi increase in altitude. What is the atmospheric pressure on the top of the highest unclimbed mountain, Zemu Gap Peak in the Himalayas, which is 5 mi high? (*Note:* $r = 0.80$.)

**42.** The area $A$ of a square stainless-steel plate increases by $0.000020A$ for every 1°C increase in temperature. What is the area in terms of the original area $A_0$ if the temperature increases by 101°C?

C **43.** If you have two children and each of your children has two children, etc., how many descendants will you have 14 generations from now (about 400 years)?

**44.** A commercial jet in a circular holding pattern over Miami International Airport flies at an altitude of 12,500 ft for 10 min and then descends 40 percent of this altitude into another holding pattern. Ten minutes later the jet descends 40 percent of the *distance of its first descent.* Ten minutes later it descends 40 percent of the distance of its second descent, etc. *(a)* How many feet does it descend after the fifth 10-min period? *(b)* What is the total decrease in altitude after 50 min?

**45.** After a major earthquake in an undeveloped country, an outbreak of typhoid spreads rapidly. The number of people who succumb each day forms a geometric sequence: 64, 96, 144, . . . . What is the total number of people afflicted after 1 week?

C **46.** In a chemical neutralization reaction, the number of grams of salt produced each minute forms a geometric sequence: 300, 270, 243, . . . . What is the total amount of salt produced in 0.5 h?

**47.** A boat at a calm anchorage experiences a series of waves, each wave having 20 percent less amplitude (height) than the previous one. If the amplitude of the first wave is 1 m, that is, the boat rises and falls 1 m, how much vertical distance does the boat travel before coming to rest?

C **48.** If the Indians who sold Manhattan Island to the Dutch in 1626 for \$24 had invested this money at 6 percent interest compounded annually, approximately how much would their investment have been worth at the end of the bicentennial year 1976?

**49.** The population of a third world country doubles every 20 years. If the present population is 30 million, approximately how many years will it take to reach 1 billion people?

C **50.** Advances in computer technology have been so rapid that the number of components on a microprocessor chip has approximately doubled every year since 1960. If this doubling rate is assumed, a chip with three components in 1960 would have how many components by 1990? (Answer to 3 figures)

**51.** Show that *(a)* $1/0.027027027 \cdots = 37$ and *(b)* $1/0.037037037 \cdots = 27$.

**52.** Show that the infinite sum $1/n + 1/n^2 + 1/n^3 + \cdots = 1/(n - 1)$, where $n$ is an integer greater than 1.

C **53.** Express $0.012345679012 \cdots$ as a fraction with a numerator of 1.

# 16-3 | *Binomial Formula*

The binomial formula is very useful in evaluating many mathematical expressions, including the base of natural logarithms $e$. Applications of the formula are found in calculus and in probability and statistics, in which it is of basic importance.

Examine the pattern of the expansions of the binomial $a + b$:

$$(a + b)^0 = \qquad\qquad\qquad 1$$
$$(a + b)^1 = \qquad\qquad\quad a \quad + \quad b$$
$$(a + b)^2 = \qquad\qquad a^2 \quad + \quad 2ab \quad + \quad b^2$$
$$(a + b)^3 = \qquad\quad a^3 \quad + \quad 3a^2b \quad + \quad 3ab^2 \quad + \quad b^3$$
$$(a + b)^4 = \qquad a^4 \quad + \quad 4a^3b \quad + \quad 6a^2b^2 \quad + \quad 4ab^3 \quad + \quad b^4$$
$$(a + b)^5 = \quad a^5 + 5a^4b \quad + \quad 10a^3b^2 \quad + \quad 10a^2b^3 \quad + \quad 5ab^4 + b^5$$

The following applies to the expansion of $(a + b)^n$:

**1.** The powers of $a$ decrease by 1 for each term, beginning with $a^n$ and ending with $a^0$.

**2.** The powers of $b$ increase by 1 for each term, starting with $b^0$ and ending with $b^n$.

**3.** The powers of $a$ and $b$ total to $n$ in each term.

**4.** There are $n + 1$ terms. The coefficients are the same starting from either end and working toward the middle, with the first coefficient equal to 1 and the second coefficient equal to $n$.

**5.** The binomial coefficients form a pattern known as the *Chinese triangle,* or *Pascal's triangle,* in which each coefficient is the sum of the two coefficients immediately above:

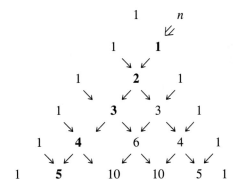

Compare Pascal's triangle with the binomial expansions of $a + b$ above. Notice that the second coefficient in each row (boldface) is $n$.

## Example 16-10
Expand $(a + b)^6$.

## Solution
The first and last coefficients are 1. The other coefficients are obtained by adding adjacent coefficients in row $n = 5$ of Pascal's triangle.

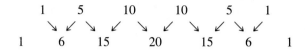

Apply rules 1, 2, and 3 above for the powers:

$$(a + b)^6 = a^6 + 6a^5b + 15a^4b^2 + 20a^3b^3 + 15a^2b^4 + 6ab^5 + b^6$$

### Example 16-11
Expand $(2x - 3y)^5$.

### Solution
Obtain the coefficients from Pascal's triangle and write the expansion, substituting $2x$ for $a$ and $-3y$ for $b$:

$$(2x - 3y)^5 = (2x)^5 + 5(2x)^4(-3y) + 10(2x)^3(-3y)^2 + 10(2x)^2(-3y)^3$$
$$+ 5(2x)(-3y)^4 + (-3y)^5$$

Multiply out each term to get the final result:

$$(2x - 3y)^5 = 32x^5 - 240x^4y + 720x^3y^2 - 1080x^2y^3 + 810xy^4 - 243y^5$$

Notice that the signs alternate when the second term $b$ in the binomial is negative.

The expansion $(a + b)^n$ is expressed mathematically by the *binomial formula:*

$$(a + b)^n = a^n + \binom{n}{1} a^{n-1}b + \binom{n}{2} a^{n-2}b^2 +$$

$$\binom{n}{3} a^{n-3}b^3 + \cdots + \binom{n}{r} a^{n-r}b^r + \cdots + b^n \quad (16\text{-}12)$$

where $\binom{n}{1}, \binom{n}{2}, \binom{n}{3}$, etc. represent the binomial coefficients. The coefficient $\binom{n}{1} = n$, and in general:

$$\binom{n}{r} = \frac{n!}{r!(n - r)!} \quad (16\text{-}13)$$

where the expansion $n!$, read *n factorial,* is the product:

$$n! = (n)(n - 1)(n - 2) \cdots (2)(1) \quad (16\text{-}14)$$

For example, $5! = 5 \cdot 4 \cdot 3 \cdot 2 \cdot 1 = 120$. Note that $0! = 1$. Some calculators have a key $\boxed{n!}$ or $\boxed{x!}$ for this function.

### Example 16-12
Find the first four terms of $(a + b)^{10}$, using the binomial formula (16-12).

### Solution
The first two coefficients are 1 and 10. The third and fourth coefficients are, respectively, applying formula (16-13):

$$\binom{10}{2} = \frac{10!}{2!8!} = \frac{10 \cdot 9 \cdot 8 \cdots 2 \cdot 1}{(2 \cdot 1) \cdot (8 \cdot 7 \cdots 2 \cdot 1)} = \frac{10 \cdot 9}{2} = 45$$

$$\binom{10}{3} = \frac{10!}{3!7!} = \frac{10 \cdot 9 \cdot 8 \cdots 2 \cdot 1}{(3 \cdot 2 \cdot 1)(7 \cdot 6 \cdots 2 \cdot 1)} = \frac{10 \cdot 9 \cdot 8}{3 \cdot 2} = 120$$

Notice that $8!$ divides into $10!$ for $\binom{10}{2}$, leaving $\dfrac{10 \cdot 9}{2 \cdot 1}$, and $7!$ divides into $10!$ for $\binom{10}{3}$, leaving $\dfrac{10 \cdot 9 \cdot 8}{3 \cdot 2 \cdot 1}$. This leads to another way of expressing formula (16-13):

$$\binom{n}{r} = \frac{n(n - 1)(n - 2) \cdots (n - r + 1)}{r!} \tag{16-15}$$

The first four terms of $(a + b)^{10}$ are then:

$$(a + b)^{10} = a^{10} + 10a^9b + 45a^8b^2 + 120a^7b^3 + \cdots$$

### Example 16-13
Find the seventh term of $(\pi + 1)^{11}$.

### Solution
You need to determine the value of $r$ before you can use formula (16-13) or (16-15) to find the coefficient. Notice in the binomial formula, (16-12), that $r$ *is the exponent of b and is 1 less than the number of the term.* Thus $r = 6$ for the seventh term. With $n = 11$, $a = \pi$, and $b = 1$, the seventh term is:

$$\binom{n}{r} a^{n-r}b^r = \binom{11}{6} (\pi)^{11-6}(1)^6 = \frac{11!}{6!5!} \pi^5$$

$$= \frac{11 \cdot 10 \cdot 9 \cdot 8 \cdot 7 \cdot 6}{6 \cdot 5 \cdot 4 \cdot 3 \cdot 2} \pi^5 = 462\pi^5$$

## Binomial Series

The binomial formula (16-12) can be shown to be true when the exponent $n$ is *any rational number* and $a = 1$, $|b| < 1$. Under these conditions, if the exponent is negative or a fraction, the formula becomes an infinite series that has a finite sum, as does an infinite geometric series. If (16-15) is used to express the coefficients, the *binomial series* is for $n = $ any rational number:

$$(1 + b)^n = 1 + nb + \frac{n(n-1)}{2!} b^2 + \frac{n(n-1)(n-2)}{3!} b^3 + \cdots \quad (16\text{-}16)$$

### Example 16-14
Find the value of $\sqrt[5]{0.830}$ to three significant figures, using the binomial series (16-16). Check the result directly with a calculator.

### Solution
Express $\sqrt[5]{0.830}$ as $(1 - 0.170)^{1/5}$. Use the binomial series, (16-16), with $b = -0.170$ and $n = \frac{1}{5}$:

$$\sqrt[5]{0.830} = (1 - 0.170)^{1/5} = 1 + \frac{1}{5}(-0.170) + \frac{(\frac{1}{5})(-\frac{4}{5})}{2}(-0.170)^2$$

$$+ \frac{(\frac{1}{5})(-\frac{4}{5})(-\frac{9}{5})}{3 \cdot 2}(-0.170)^3 + \cdots$$

The terms after the fourth term are not significant and can be omitted:

$$\sqrt[5]{0.830} = 1 - 0.034 - 0.002312 - 0.000235824$$
$$= 0.963 \quad \text{(three significant figures)}$$

This result checks directly on the calculator:

$$0.830 \;\boxed{\text{INV}}\; \boxed{y^x}\; 5 \;\boxed{=}\; \longrightarrow \; 0.9634\cdots$$

### Example 16-15
Show that $e = 1 + 1 + 1/2! + 1/3! + \cdots$ by using the binomial series and the fact that $(1 + 1/n)^n$ approaches $e$ as $n$ becomes infinitely large.

### Solution
Apply formula (16-16) with $b = 1/n$ (Note $n > 1$ so $|b| < 1$):

$$\left(1 + \frac{1}{n}\right)^n = 1 + n\left(\frac{1}{n}\right) + \frac{n(n-1)}{2!}\left(\frac{1}{n}\right)^2 + \frac{n(n-1)(n-2)}{3!}\left(\frac{1}{n}\right)^3 + \cdots$$

$$= 1 + 1 + \frac{1}{2!}\left(1 - \frac{1}{n}\right) + \frac{1}{3!}\left(1 - \frac{1}{n}\right)\left(1 - \frac{2}{n}\right) + \cdots$$

As $n$ becomes infinitely large, $1/n$, $2/n$, etc., approach zero and can be ignored. Therefore:

$$e = 1 + 1 + \frac{1}{2!} + \frac{1}{3!} + \cdots = 2.718\cdots$$

**Exercise 16-3**

In problems 1 to 14, expand each of the binomials, using Pascal's triangle or the binomial formula.

1. $(x + h)^3$                                      2. $(2a + 1)^4$

3. $(a + b)^7$                                      4. $(x - y)^6$

5. $(v - 2t)^4$                                     6. $(3p + q)^5$

7. $(T^2 + 4)^5$                                    8. $(5 - e^3)^3$

9. $\left(\pi r - \dfrac{1}{r}\right)^6$            10. $\left(x + \dfrac{2}{x}\right)^5$

11. $(\sqrt{x} + \sqrt{y})^4$                       12. $(x^{-2} + y^{-2})^3$

13. $(I + 2.5)^3$                                   14. $(1 - 0.9k)^4$

In problems 15 to 18, find the first four terms of each binomial expansion.

15. $(x + y)^{12}$                                  16. $(r - 2s)^{15}$

17. $(1 + y)^{-3}$ [use Eq. (16-16)]               18. $(1 + x)^{1/2}$ [use Eq. (16-16)]

□ In problems 19 to 22, find the value of each expression to three significant figures, using the binomial series (16-16). Check your result directly with a calculator.

19. $(1.01)^4 = (1 + 0.01)^4$                       20. $(0.920)^{-3} = (1 - 0.08)^{-3}$

21. $\sqrt[3]{0.880}$                               22. $\sqrt[5]{1.03}$

In problems 23 to 26, find the indicated term in the binomial expansion.

23. $(t - 1)^{10}$; seventh term                    24. $(u + v)^{14}$; fifth term

25. $\left(e + \dfrac{1}{e}\right)^{12}$; sixth term  26. $(\pi r^2 - 2)^9$; fourth term

27. In computing the torisonal stress on a shaft, the following expression arises:

$$\frac{\pi}{2}\left[(R + r)^4 - (R - r)^4\right]$$

Expand and simplify the expression.

28. At a point in a magnetic field the field strength is given by:

$$H = \frac{2ml}{(r^2 + l^2)^{3/2}}$$

Using the binomial series (16-16), find the first three terms of the expansion $(r^2 + l^2)^{3/2}$ by expressing it as:

$$(r^2)^{3/2}\left(1 + \frac{l^2}{r^2}\right)^{3/2}$$

29. The binomial coefficients $\binom{n}{r}$ represent the number of ways of choosing a sample of $r$ objects out of a collection of $n$ objects. In a transistor production line, a sample of 5 transistors is chosen to be tested out of a collection of 20. How many different samples is it possible to choose?

30. When a coin is tossed $n$ times, the binomial coefficients $\binom{n}{r}$ represent the number of ways of getting $r$ heads and $n - r$ tails. (a) How many ways are there of getting 5 heads and 5 tails when a coin is tossed 10 times? ($n = 10$, $r = 5$) (b) If you add all the binomial coefficients in row $n$ of Pascal's triangle, it equals the total possible outcomes, which is $2^n$. Divide the answer to (a) by $2^{10}$ to find the probability of getting 5 heads and 5 tails.

31. In the formula for a derivative in calculus, the following expression arises:

$$\frac{(x + h)^5 - x^5}{h}$$

Expand and simplify the expression.

32. Show that $(1 + j)^4 = -4$, where $j^2 = -1$, in two ways: (a) by using the binomial formula and (b) by using De Moivre's theorem (Sec. 13-4).

33. Show that $e^x = 1 + x + x^2/2! + x^3/3! + \cdots$ by showing that $(1 + 1/n)^{nx}$ approaches this expression as $n$ becomes infinitely large. (See Example 16-15.)

34. When $x = j\theta$ in problem 33, $e^{j\theta} = 1 + j\theta + (j\theta)^2/2! + (j\theta)^3/3! + \cdots$. Show that $\cos \theta = 1 - \theta^2/2! + \theta^4/4! - \cdots$ by using the formula $e^{j\theta} = \cos \theta + j \sin \theta$ and equating the real parts of both expressions.

## 16-4 | *Review Questions*

In problems 1 to 4, for the given value of $n$, find $d$, $a_n$, and $S_n$ for each arithmetic series.

1. $4 + 7 + 10 + \cdots$; $n = 12$
2. $15 + 11 + 7 + \cdots$; $n = 10$
3. $\dfrac{1}{5} + \dfrac{3}{5} + 1 + \cdots$; $n = 8$
4. $0.10 + 0.12 + 0.14 + \cdots$; $n = 9$

In problems 5 to 8, for the given value of $n$, find $r$, $a_n$, and $S_n$ for each geometric series.

5. $3 + 6 + 12 + \cdots$; $n = 7$
6. $192 + 96 + 48 + \cdots$; $n = 8$

**7.** $\dfrac{3}{8} + \dfrac{3}{2} + 6 + \cdots$; $n = 5$

C **8.** $3.2 + 8 + 20 + \cdots$; $n = 6$

**9.** Find the sum of the infinite geometric series:

$$10 + 2 + 0.4 + \cdots$$

**10.** Express the repeating decimal $5.7777\cdots$ as a fraction.

In problems 11 to 14, expand each binomial.

**11.** $(x + 2y)^4$ **12.** $(T - 3)^5$

**13.** $(x^2 + 1)^6$ **14.** $(a - b)^8$

**15.** Find the first four terms of $(1 + x)^{1/2}$.

**16.** Find the sixth term of $(1/x + 2x)^9$.

**17.** A labor union forfeits 1 week's wages for the first week it is on strike, $1\frac{1}{2}$ weeks' wages for the second strike week, 2 weeks' wages for the third strike week, etc. What are the total weeks' wages forfeited if the strike lasts 10 weeks?

**18.** Seawater pressure increases by $0.444$ lb/in$^2$ for every foot of depth. If atmospheric pressure is $14.7$ lb/in$^2$, at what seawater depth will the total pressure by $20.0$ lb/in$^2$?

**19.** Find the amount after 1 year of a principal of $1000.00 earning 10 percent interest *(a)* compounded semiannually and *(b)* compounded quarterly.

**20.** Helen paid $300 for her television set, which depreciates 30 percent in value every year. How much will it be worth 5 years after she purchased it?

C **21.** Radioactive nitrogen-13 decays 7 percent each minute. What fraction of the original amount will be left *(a)* after 10 min and *(b)* after 1 h?

**22.** The middle of a vibrating guitar string travels 98 percent of its previous distance during each cycle. If it travels 0.05 mm during the first cycle, what is the total distance it travels before coming to rest?

# 17

# Inequalities and Introduction to Linear Programming

## 17-1 *Properties of Inequalities: Linear Inequalities*

An *inequality* is an expression that contains an inequality sign ($>$ or $<$).

Inequalities are important for certain problems in technology, industry and business, and in the further study of mathematics, particularly probability and calculus. This chapter covers the properties of inequalities and how to solve them. The solution of inequalities in one variable is similar to the solution of equations in one variable. Inequalities with more than one variable are most effectively solved by using graphical techniques. An important application of inequalities in the field of linear programming is shown at the end of the chapter.

**Example 17-1**

List examples of inequalities, and state each example in words.

1. $5 > 4$     5 is greater than 4
2. $-\frac{1}{4} < -\frac{1}{5}$     $-\frac{1}{4}$ is less than $-\frac{1}{5}$
3. $a^2 + b^2 > 0$     $a^2 + b^2$ is greater than zero or $a^2 + b^2$ is positive
4. $r \geq \sqrt{10}$     $r$ is greater than or equal to $\sqrt{10}$

Notice in (4) that the sign $\geq$ means greater than or equal to. Similarly $\leq$ means less than or equal to.

5. $3.141 < \pi < 3.142$     $\pi$ is greater than 3.141 but less than 3.142

Notice in (5) that you can use two inequality signs in one statement. This is a common form of inequality. An equivalent of (5) would be: $\pi$ is between 3.141 and 3.142.

**6.** $x < -1$   or   $x > 1$     $x$ is either less than $-1$ or greater than 1

Problem (6) can be expressed by using the *absolute value* (positive-value) symbol: $|x| > 1$. In general for $a > 0$:

$$
\begin{aligned}
|x| < a &\quad \text{means} \quad -a < x < a \\
|x| > a &\quad \text{means} \quad x < -a \quad \text{or} \quad x > a
\end{aligned}
\qquad (17\text{-}1)
$$

## Properties of Inequalities

Three basic properties of inequalities are as follows:

Addition property:
$$\text{If } a > b, \text{ then } a + c > b + c. \qquad (17\text{-}2)$$

Multiplication and division properties:

$$\text{If } a > b \text{ and } c > 0, \text{ then } ac > bc \text{ and } \frac{a}{c} > \frac{b}{c}. \qquad (17\text{-}3)$$

$$\text{If } a > b \text{ and } c < 0, \text{ then } ac < bc \text{ and } \frac{a}{c} < \frac{b}{c}. \qquad (17\text{-}4)$$

Note that properties (17-2) and (17-3) are the same for equations as they are for inequalities but that property (17-4) is not the same. *When an inequality is multiplied or divided by a negative number, the inequality sign is reversed.*

### Example 17-2
Given that $a - b > 2b$, prove each of the following.

**1.** $a > 3b$

**2.** $\dfrac{a - b}{2} > b$

**3.** $b - a < -2b$

### Solution
**1.** $a > 3b$

Use property (17-2) and add $b$ to both sides:

$$a - b + (b) > 2b + (b)$$
$$a > 3b$$

2. $\dfrac{a - b}{2} > b$

Use property (17-3) and divide both sides by 2:

$$\frac{a - b}{2} > \frac{\cancel{2}b}{\cancel{2}}$$

3. $b - a < -2b$

Use property (17-4) and multiply both sides by $-1$:

$$(-1)(a - b) < (-1)2b$$
$$b - a < -2b$$

Notice in (3) that the inequality sign is reversed.

## Linear Inequalities

The properties of inequalities enable you to solve *conditional inequalities,* that is, inequalities that are true for certain values of the variable. Study the following examples of linear, or first-degree, inequalities.

### Example 17-3
Solve the linear inequality          $5x + 2 > 4 - x$

### Solution
Solve a linear inequality in the same way as a linear equation, by isolating the unknown:

$$5x + 2 + (x) > 4 - x + (x) \qquad \text{by (17-2)}$$
$$6x + 2 + (-2) > 4 + (-2) \qquad \text{by (17-2)}$$
$$6x > 2$$
$$x > \tfrac{1}{3} \qquad \text{by (17-3)}$$

### Example 17-4
Solve the linear inequality:

$$0.0 < 0.5(1.0 - y) \leq 2.0$$

### Solution
Apply the properties of inequalities to each of the three members of the inequality. The properties are valid for inequalities involving the signs $\geq$ and $\leq$:

$$\frac{0.0}{0.5} < 1.0 - y \leq \frac{2.0}{0.5} \qquad \text{by (17-3)}$$

$$0.0 < 1.0 - y \le 4.0$$
$$-1.0 < -y \le 3.0 \qquad \text{by (17-3)}$$
$$1.0 > y \ge -3.0 \qquad \text{by (17-4)}$$
or
$$-3.0 \le y < 1.0$$

Notice that the inequality signs reverse in the next-to-last step. The last form of the inequality is preferred since it increases from left to right.

### Example 17-5
Solve the linear inequality:

$$\left| t - \frac{1}{2} \right| > \frac{1}{4}$$

**Solution**
By property (17-1) this inequality means:

$$t - \frac{1}{2} < -\frac{1}{4} \qquad \text{or} \qquad t - \frac{1}{2} > \frac{1}{4}$$

Solve each inequality separately to get the solution:

$$t < \frac{1}{4} \qquad \text{or} \qquad t > \frac{3}{4}$$

The steps can also be reversed to express the two inequalities as one inequality.

### Example 17-6
A certain welding operation must be performed between 3500 and 4000°F. What is this temperature range in degrees Celsius?

**Solution**
Use the formula $F = \frac{9}{5}C + 32$ to solve the inequality:

$$3500 < F < 4000$$
$$3500 < \frac{9}{5}C + 32 < 4000$$
$$3468 < \frac{9}{5}C < 3968$$
$$1927 < C < 2204$$

The range is therefore between 1927 and 2204°C.

**Exercise 17-1**

In problems 1 to 8, express each statement by using inequality signs.

**1.** $-1$ is greater than $-2$

**2.** 9.9 is less than 10

**3.** $2ab$ is less than or equal to $a^2 + b^2$

**4.** $x - x^2$ is negative when $x$ is greater than 1

**5.** Any positive integer $n$ plus its reciprocal $1/n$ is greater than or equal to 2

**6.** $e$ is greater than 2.71 and less than 2.72

Express problems 7 and 8 in two ways: with and without the absolute value symbol.

**7.** $r$ is between $-1$ and 1

**8.** $h$ is either greater than ½ or less than $-½$

In problems 9 to 22, solve each linear inequality.

**9.** $2x + 1 < 5$          **10.** $5x > 3 - x$

**11.** $1 - x \geq 4$         **12.** $2 < 5 - 3x$

**13.** $y - \dfrac{1}{2} > 1 - \dfrac{y}{2}$      **14.** $\dfrac{y - 2}{5} \leq \dfrac{3}{10}$

**15.** $0.7(t - 0.5) < 2.1$      **16.** $\dfrac{1}{3}(0.01R - 0.06) > 0.18$

**17.** $-1 \leq 7 - 2q < 3$      **18.** $-\pi < 2\theta + \dfrac{\pi}{3} < 2\pi$

**19.** $|x - 1| > 2$         **20.** $|2 - x| > 1$

**21.** $|1 - 2e| < 5$        **22.** $|2w + 3| < 0.4$

**23.** Show that if $a > 1$, then $a^2 > a$.

**24.** Show that if $0 < a < 1$, then $a^2 < a$.

For each of problems 25 to 34, express the answer in terms of an inequality.

**25.** The tolerance for the radius $r$ of an automobile shaft is specified as $5.00 \pm 0.03$ mm. *(a)* What is the range for $r$? *(b)* What is the range for the diameter $d$?

**26.** A telecommunication satellite has a minimum altitude of 900 km and a maximum altitude of 1500 km. What is the range of altitude $h$?

**27.** The instrument error $\epsilon$ of a wattmeter is less than 0.5 W. *(a)* Express the error as an inequality, using absolute value. *(b)* When the pointer is exactly on the 7-W mark, a technician records the power $P$ as 7.0 W. What is the possible range of values for $P$?

**28.** The instantaneous current in an ac circuit is given by $I = 3.0 + 2.0 \sin \omega t$. What is the range of $I$?

**29.** An antifreeze solution has a boiling point of 230°F and a freezing point of $-10$°F. What is the temperature range in degrees Celsius through which the solution may be used? (See Example 17-6.)

**30.** The temperature $T$ of a battery acid must be kept such that $0°C < T \le 55°C$. What is this temperature range in degrees Fahrenheit? [*Note:* $C = \frac{5}{9}(F - 32)$.]

**31.** A sensitive microcomputer has a power fuse which will blow if the current exceeds 5.5 A. The computer requires more than 105 V to operate properly. If the total resistance of the circuit is constant at 23 $\Omega$, what should be the range of the voltage for the unit to operate properly?

**32.** The vertical component of velocity (in meters per second) of a projectile is given by $v_y = 360 - 9.8t$. For what values of $t$ is the projectile decreasing in altitude ($v_y < 0$)?

**33.** The minimum speed for vehicles on a highway is 40 mi/h; the speed limit is 55 mi/h. What is the range of speed in kilometers per hour? (See App. A.)

**34.** An open interval $(a, b)$ means all numbers $x$ that satisfy the inequality $a < x < b$. A closed interval $[a, b]$ means all numbers $x$ that satisfy the inequality $a \le x \le b$. *(a)* If $|x| < 2$, express the range of $2x$ as an open interval. *(b)* If $|x| \le 2$, express the range of $2x$ as a closed interval.

## 17-2 | *Quadratic Inequalities*

Quadratic inequalities are solved by using the methods for solving quadratic equations. The power and root property below is useful for higher-degree inequalities.

Power and root property:

If $a$, $b$, and $n > 0$ and $a > b$, then:

$$a^n > b^n \qquad \text{and} \qquad \sqrt[n]{a} > \sqrt[n]{b} \qquad\qquad (17\text{-}5)$$

Notice that property (17-5) applies only when all quantities are positive.

### Example 17-7
Solve the quadratic inequality $x^2 - x > 6$.

### Solution
Proceed as you would for a quadratic equation by putting all the terms on one side and factoring:

$$x^2 - x - 6 > 0 \qquad \text{by (17-2)}$$
$$(x + 2)(x - 3) > 0$$

The inequality is true when both factors have the same sign and their product is positive. The values that make the inequality zero are called *critical values*. The solution is found by examining the signs of the inequality on either side of the critical values. The critical values are $x = -2$ and $x = 3$. The signs of the factors and their product are tabulated for values of $x$ on either side of the critical values:

| Interval | $x + 2$ | $x - 3$ | $(x + 2)(x - 3)$ |
|---|---|---|---|
| $x < -2$ | $-$ | $-$ | $\oplus$ |
| $-2 < x < 3$ | $+$ | $-$ | $-$ |
| $x > 3$ | $+$ | $+$ | $\oplus$ |

The circled signs indicate the solution:

$$x < -2 \quad \text{or} \quad x > 3$$

The next two examples show inequalities that are similar to a quadratic and are solved by using the method of critical values.

### Example 17-8

Solve the inequality $\sqrt{d^2 + 9} \le 5$.

### Solution

Since the quantity $\sqrt{d^2 + 9}$ is always positive, you can apply property (17-5) and square both sides:

$$d^2 + 9 \le 25$$

Then
$$d^2 - 16 \le 0$$
$$(d + 4)(d - 4) \le 0$$

The critical values are $d = -4$ and $d = 4$. The signs of the factors and their product are tabulated for values of $d$ on either side of the critical values:

| Interval | $d + 4$ | $d - 4$ | $(d + 4)(d - 4)$ |
|---|---|---|---|
| $d < -4$ | $-$ | $-$ | $+$ |
| $-4 < d < 4$ | $+$ | $-$ | $\ominus$ |
| $d > 4$ | $+$ | $+$ | $+$ |

The circled sign indicates when the product is less than zero. Since the inequality includes equals, the critical values are included in the solution:

$$-4 \le d \le 4$$

The solution can also be expressed as $|d| \le 4$.

Example 17-8 could be solved by taking square roots after squaring:

$$d^2 + 9 \le 25$$
$$d^2 \le 16$$
$$d \le 4 \qquad d > 0 \qquad \text{by (17-5)}$$

However, this solution applies only when $d$ is positive. To obtain the solution for negative values, reverse the inequality: $-d \ge -4$, $d > 0$. The complete solution for $d$, positive and negative, is then $-4 \le d \le 4$.

### Example 17-9
Solve:

$$\frac{x + 3}{x - 1} \ge 0$$

### Solution
You cannot multiply both sides by the denominator since $x - 1$ could be positive or negative. The critical values lead to the solution. They are the values that make the numerator or the denominator zero: $x = -3$ and $x = 1$. The signs of the numerator, denominator, and quotient are tabulated for values of $x$ on either side of the critical values:

| Interval | $x + 3$ | $x - 1$ | $\dfrac{x + 3}{x - 1}$ |
|---|---|---|---|
| $x < -3$ | $-$ | $-$ | $\oplus$ |
| $-3 < x < 1$ | $+$ | $-$ | $-$ |
| $x > 1$ | $+$ | $+$ | $\oplus$ |

The solution is indicated by the circled signs: $x \le -3$ or $x > 1$. When $x = -3$, the numerator is zero and the fraction is zero, so it is included in the solution. When $x = 1$, the denominator is zero and the fraction is undefined, so it is not part of the solution.

### Exercise 17-2

In problems 1 to 20, solve each inequality.

1. $x^2 - 5x + 4 < 0$     2. $x^2 + 3x - 10 > 0$
3. $2x^2 + 3x > 2$      4. $3x^2 - 2 < x$
5. $x^2 + 10x \le 0$     6. $2x^2 \ge 4x$
7. $y^2 > 16$       8. $4y^2 - 1 \le 0$
9. $a^2 < 0.01$      10. $b^2 > 2.5$
11. $\sqrt{v^2 + 25} \ge 13$    12. $\sqrt{w^2 + 1} < \sqrt{2}$

13. $\dfrac{x-1}{x} > 0$     14. $\dfrac{x}{x+1} \le 0$

15. $\dfrac{3x+1}{x-4} \le 0$     16. $\dfrac{2x+3}{2x-3} > 0$

17. $h^3 - h > 0$     18. $\dfrac{p^2+p}{p+2} < 0$

19. $\left|\dfrac{1}{x}\right| > 1$     20. $|x^2 + 5x| < 6$

21. Show that the solution to $x^2 + 3 > 2 - 2x$ is $x =$ any real number.

22. Show that there is no solution to $x^2 + 1 < 2x$.

23. Show that $a^2 + b^2 \ge 2ab$. [*Hint:* Start with $(a - b)^2 \ge 0$.]

24. Using the result of problem 23, show for $a \ge b > 0$ that the *arithmetic mean* $(a + b)/2$ is greater than or equal to the *geometric mean* $\sqrt{ab}$.

Express the answer to problems 25 to 28 in terms of an inequality.

25. The deflection of a beam is given by $d = x^2 - 0.3x + 0.03$, where $x$ is the distance from one end. For what values of $x$ is $d > 0.01$?

26. The power output of a hydroelectric plant varies as a function of time during a 24-h period given by $P = -5t^2 + 120t + 1700$, where $t =$ time (h), $0 \le t \le 24$, and $P =$ power (MW). For what values of $t$ is $P > 2100$ MW?

27. The weight of a buoy (in kilograms) is given by the formula $W = 900 + 1600V$, where $V =$ volume (in cubic meters). The buoyant force $B$ acting on the buoy is given by $B = 100 + 1000V^2$. A body will float when $W < B$. For what values of the volume will the buoy float?

28. In calculus it is shown that $\tan x$ can be represented by an infinite series: $\tan x = x^3/3 + 2x^5/15 + 17x^7/315 + \cdots$ when $x^2 < \pi^2/4$. Solve this inequality for $x$, and express the answer with and without the absolute value symbol.

# 17-3 | *Graphs of Inequalities with Two Variables*

An equation in two variables represents a line or a curve on a graph, whereas an inequality in two variables represents an *area* on a graph. The first example shows some basic inequalities and the area each inequality represents.

### Example 17-10
Show the graph of the area that each inequality represents.

1. $y > 3$

2. $y \le -2$

3. $y > x$

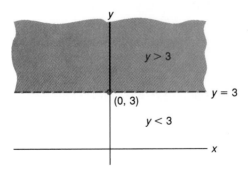

FIG. 17-1  *Graph of y > 3, Example 17-10(1)*

## Solution

**1.** $y > 3$

The inequality represents the area above the line $y = 3$ (Fig. 17-1). The line is shown dashed because it is not included in the inequality but is the border of the area. The area below the line is $y < 3$.

**2.** $y \leq -2$

The inequality includes the border line $y = -2$ and the area below (Fig. 17-2). The line is therefore shown solid. The area above the line is $y > -2$.

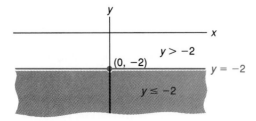

FIG. 17-2  *Graph of y ≤ -2, Example 17-10(2)*

**3.** $y > x$

The border of the area is the line $y = x$. The inequality represents the area above the line but does not include the line (Fig. 17-3), so it is shown dashed. The area below the line is $y < x$.

Example 17-10 illustrates the *basic procedure for graphing inequalities:*

**1.** Change the inequality to an equation and graph the equation. The equation is the border of the inequality area and is shown dashed for $>$ or $<$ and solid for $\geq$ or $\leq$.

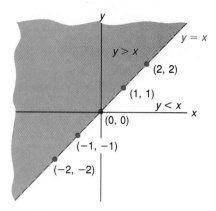

FIG. 17-3  *Graph of y > x, Example 17-10(3)*

2. Shade the area above the graph of the equation if the inequality is of the form $y > f(x)$ or $y \geq f(x)$. Shade the area below the graph of the equation if the inequality is of the form $y < f(x)$ or $y \leq f(x)$.

or

3. Pick any point $(x, y)$ not on the graph of the equation and substitute it into the inequality. If the point satisfies the inequality, it is in the shaded area, and vice versa.

### Example 17-11
Graph $3x + y > 3$.

### Solution
Graph the equation $3x + y = 3$ and show it as a dashed line (Fig. 17-4). Solve the inequality for $y$:

$$y > 3 - 3x$$

This is of the form $y > f(x)$. Therefore shade the area above the line. You can also pick a test point, say $(0, 0)$: $3(0) + (0) = 0 < 3$. Since this point does not satisfy the inequality and is below the line, the area above the line is the shaded area.

### Example 17-12
Graph $2x - 3y \leq 6$.

### Solution
Graph the equation $2x - 3y = 6$ and show it as a solid line (Fig. 17-5). Solve the inequality for $y$:

$$y \geq \frac{2}{3}x - 2$$

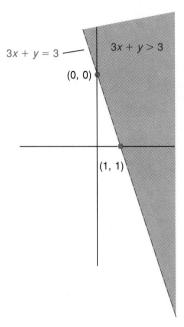

$3x + y = 3$ ——
$3x + y > 3$
(0, 0)

(1, 1)

FIG. 17-4  *Graph of 3x + y > 3, Example 17-11*

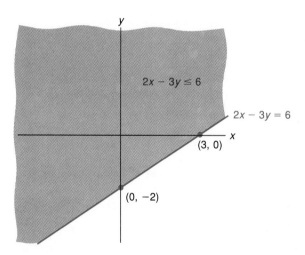

$y$

$2x - 3y \leq 6$

$2x - 3y = 6$

$x$
(3, 0)

(0, -2)

FIG. 17-5  *Graph of 2x − 3y ≤ 6, Example 17-12*

This is of the form $y \geq f(x)$. Therefore shade the area above the line. Pick the test point (0, 0): $2(0) - 3(0) = 0$. Since this satisfies the inequality, the point is in the shaded area.

## Example 17-13
Graph $y \geq x^2 - 4x$, where $y \leq 0$.

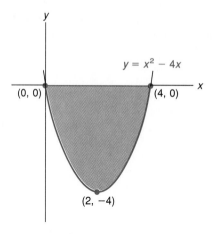

FIG. 17-6  *Graph of y ≥ x² − 4x where y ≤ 0, Example 17-13*

## Solution

The graph of $y = x^2 - 4x$ is the parabola shown as a solid line in Fig. 17-6. The shaded area above the parabola and below the $x$ axis including the borders satisfies both inequalities.

**Exercise 17-3**

In problems 1 to 20, graph each inequality.

1. $y > 1$        2. $y < -2$

3. $y \leq -3$        4. $y \geq 4$

5. $x + y > 0$        6. $y \leq x$

7. $x + 2y \leq 4$        8. $y - 2x > 2$

9. $4x - 3y \leq 12$        10. $3x + 4y > 12$

11. $x - 5y > 5$        12. $y < \dfrac{x}{2} + 1$

13. $0.5x + 0.2y \geq 1.0$        14. $1.5x - 1.0y \leq 3.0$

15. $y > x^2$        16. $y < 4 - x^2$

17. $y \leq 6x - x^2$        18. $y \leq x^2 + 3x$

19. $x^2 + y^2 < 9$        20. $4x^2 + 9y^2 \leq 9$

In problems 21 to 24, graph the area that satisfies each system of inequalities.

21. $y \leq x, \ x \geq 0$        22. $x + 2y \geq 0, \ y \leq 0$

23. $y \leq x^2, \ y \geq 0$        24. $y \geq x^2, \ y \leq x$

25. The temperature (°C) and pressure (kPa) of a controlled chemical reaction must be kept such that $1.5P + 2T < 300$. Graph this inequality for $T \geq 0$ and $P \geq 0$.

26. To keep the resistance of a sensitive electric circuit as constant as possible during temperature changes, a variable resistance must be adjusted such that $12 \leq T + R \leq 24$, where $T \geq 0$°C and $R \geq 0.0$ Ω. Graph this inequality.

**27.** To make enough income to maintain a small business, the owner calculates that the expenses $C$ should not be more than 70 percent of the sales $S$. *(a)* Express this relation as an inequality. *(b)* Graph the inequality.

**28.** A car manufacturing company has to recall some of its new-model cars because of defects that make them unsafe. The company will just break even if the total number of cars sold exceeds the number recalled by 3000. *(a)* Express the inequality that represents a profit for the company. *(b)* Graph the inequality.

# 17-4 | *Systems of Inequalities: Linear Programming*

Systems of inequalities arise when one is solving technological problems and problems in industry and business that are subject to various limitations or constraints. The first two examples illustrate the procedure for solving such a system, and the last two examples show important applications of such systems in the field of linear programming.

### Example 17-14
Solve the system of inequalities: $x + y \leq 10$, $x + 3y \leq 15$, $x \geq 0$, and $y \geq 0$.

### Solution
The solution is the area that satisfies all the inequalities. Since $x \geq 0$ and $y \geq 0$, the solution is in the first quadrant. Graph the two equations $x + y = 10$ and $x + 3y = 15$ as solid lines in the first quadrant (Fig. 17-7). Each line and

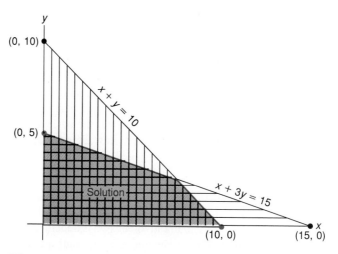

FIG. 17-7  *System of inequalities, Example 17-14*

the area below it satisfy the respective inequality. The area common to both is the solution.

## Example 17-15
Solve the system of inequalities: $y \geq 3x^2$, $y \leq 4 - x^2$, and $x \geq 0$.

### Solution
The graphs of $y = 3x^2$ and $y = 4 - x^2$ are the parabolas shown in Fig. 17-8. The area in the first quadrant above $y = 3x^2$ and below $y = 4 - x^2$ is the solution. The points of intersection of the parabolas, $(1, 3)$ and $(-1, 3)$, are found by solving the two equations simultaneously.

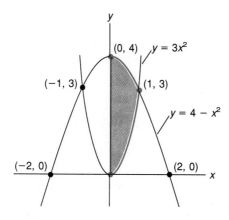

FIG. 17-8 *System of inequalities, Example 17-15*

## *Linear Programming*

Linear programming is the branch of mathematics concerned with finding the best, or optimum, solution to a problem whose variables are subject to certain constraints. These problems include maximizing of production or profit, minimizing of cost or waste, developing technical or corporate strategy, and many other types. Example 17-16 illustrates the basic approach to a maximizing type of problem. The second example illustrates the approach to a minimizing type of problem.

## Example 17-16
The ICTV company manufactures audio tape players and audio disk players. The creation of each unit involves two production processes: parts manufacture and assembly. The production data per unit are as follows:

**Work Hours Required**

|       | Parts | Assembly |
|-------|-------|----------|
| Tape  | 4     | 1        |
| Disk  | 3     | 2        |

The company's space and labor conditions allow a maximum of 120 h per week for parts manufacture and a maximum of 60 h per week for assembly.

1. What are the *feasible* numbers of units of each player that the company can produce in 1 week?
2. The profit is $60 for each tape player and $80 for each disk player. How many units of each type should the company produce each week to maximize profit?

**Solution**

1. What are the *feasible* numbers of units of each player that the company can produce in 1 week?

   Let:

$$x = \text{no. of tape players}$$
$$y = \text{no. of disk players}$$

The number of hours required to manufacture the parts for all the units produced in 1 week is then $4x + 3y$. This cannot be greater than the maximum number of hours available per week:

$$4x + 3y \leq 120$$

The number of hours required to assemble the units produced in 1 week is $x + 2y$. This cannot be greater than the maximum number of hours available:

$$x + 2y \leq 60$$

Furthermore, the number of units of each player produced must be greater than or equal to zero:

$$x \geq 0 \qquad y \geq 0$$

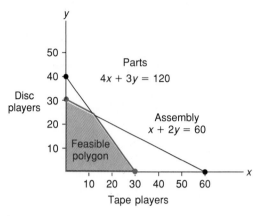

FIG. 17-9  *Feasible polygon, Example 17-16(1)*

The solution of the above four inequalities is called the *feasible area* or the *feasible polygon* (Fig. 17-9). This area represents the possible combinations of tape players and disk players that the company can produce in 1 week.

2. The profit is $60 for each tape player and $80 for each disk player. How many units of each type should the company produce each week to maximize profit?

The profit per week is:

$$P = 60x + 80y$$

This equation is called the *objective function*. It leads to the major purpose of the problem: *Find the maximum value of the objective function that lies within the feasible polygon.* That value can be found graphically by drawing lines for several values of $P$ and finding the maximum line that contains a point in the feasible polygon. Figure 17-10 shows profit lines for $P = 1200$, $1920$, $2640$, and $3120$. Notice that all these lines are parallel (slope $= -\frac{3}{4}$) and that $P$ increases as the distance from the origin increases. It can be seen that the line of maximum profit ($P = 2640$) passes through the intersection of the parts and assembly lines. This is the basis of the theory of linear programming: *The optimum solution lies on the boundary of the feasible polygon.* The solution is therefore found algebraically by solving $4x + 3y = 120$ and $x + 2y = 60$ simultaneously. That yields $x = 12$ and $y = 24$. The maximum profit is then:

$$P = 60(12) + 80(24) = \$2640$$

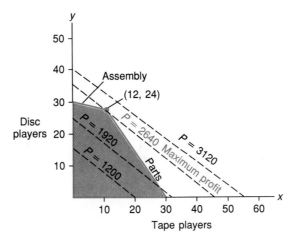

FIG. 17-10 *Profit lines, Example 17-16(2)*

## Example 17-17

A company that mines sand for use in concrete has two sand pits. Pit A contains 40 percent fine sand, and pit B contains 80 percent fine sand. The company has an order for at least 1 metric ton (t), or 1000 kg, of sand but not more than 1.25 t. The sand is to contain at least 500 kg of fine sand.

1. What feasible amounts from each pit can be mixed to fulfill the order?
2. The sand from pit A costs $20 per kilogram to mine, and the sand from pit B costs $30 per kilogram to mine. How much should be mixed from each pit to minimize the cost?

## Solution

1. What feasible amounts from each pit can be mixed to fulfill the order?

    Let:

$$x = \text{no. of kilograms from pit A}$$
$$y = \text{no. of kilograms from pit B}$$

Then $x \geq 0$, $y \geq 0$, and $1000 \leq x + y \leq 1250$. Since the mixture must contain 500 kg or more of fine sand,

$$0.40x + 0.80y \geq 500$$

or:

$$x + 2y \geq 1250 \qquad \text{(multiply by 2.5)}$$

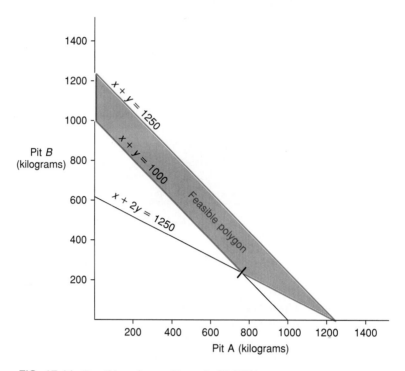

FIG. 17-11   *Feasible polygon, Example 17-17(1)*

The solution of the above five inequalities is the feasible polygon shown in Fig. 17-11. Any point in this area represents feasible amounts of $x$ and $y$ that will satisfy the constraints.

2. The sand from pit A costs \$20 per kilogram to mine, and the sand from pit B costs \$30 per kilogram to mine. How much should be mixed from each pit to minimize the cost?

The objective function is the cost:

$$C = 20x + 30y$$

Figure 17-12 shows the feasible polygon and the cost lines for $C = 15{,}000$, 22,500, and 30,000. The cost lines are all parallel (slope $= -\tfrac{2}{3}$), and $C$ decreases as the distance from the origin decreases. The minimum-cost line passes through the intersection of the lines $x + y = 1000$ and $x + 2y = 1250$. The optimum solution is therefore on the boundary of the feasible polygon, as in Example 17-16. The solution is found algebraically to be $x = 750$ kg and $y = 250$ kg, and the minimum cost is:

$$C = 20(750) + 30(250) = \$22{,}500$$

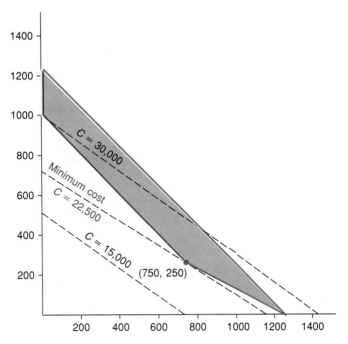

FIG. 17-12  *Cost lines, Example 17-17(2)*

Linear programming problems can involve a large number of variables and rely heavily on computers for their solution. However, the concepts and procedures used in the solution of the complex problems are similar to those shown above. The basic ideas must be understood to use the computer and interpret the results.

**Exercise 17-4**  In problems 1 to 12, solve each system of inequalities.

1. $2x + y \leq 10$, $x + 2y \leq 12$, $x \geq 0$, $y \geq 0$
2. $3x + 2y \leq 24$, $3x + 4y \leq 36$, $x \geq 0$, $y \geq 0$
3. $x + y \leq 20$, $2x + y \leq 30$, $x \geq 0$, $y \geq 0$
4. $x + y \leq 50$, $x + 3y \leq 90$, $x \geq 0$, $y \geq 0$
5. $25 \leq x + y \leq 35$, $3x + y \geq 30$, $x \geq 0$, $y \geq 0$
6. $30x + 20y \geq 600$, $10x + 20y \geq 400$, $x \geq 0$, $y \geq 0$
7. $x + y \leq 18$, $3x + 4y \geq 60$, $x \geq 0$, $y \geq 0$
8. $x + 2y \leq 15$, $2x + 3y \geq 27$, $x \geq 0$, $y \geq 0$
9. $3x + y \geq 9$, $y \leq 9 - x^2$
10. $x + y \leq 2$, $y \geq x^2$, $x \geq 0$
11. $y \geq 2x^2$, $y \leq 2 - x^2$, $x \geq 0$
12. $y \geq x^2$, $x \geq y^2$

For each linear programming problem in 13 to 20, sketch the feasible polygon and answer the given questions.

**13.** The Time-Out watch company produces an analog (dial) watch and a digital watch. The production data for parts manufacture, assembly, and the profit per unit are as follows:

|  | Hours Required | | |
|---|---|---|---|
|  | Parts | Assembly | Profit ($) |
| Analog | 20 | 2 | 30 |
| Digital | 30 | 2 | 40 |

The company can spend a maximum of 900 h per week for parts manufacture and 80 h per week for assembly. *(a)* How many watches of each type should be produced each week to maximize profit? *(b)* What is the maximum profit?

**14.** The Orange Computer company manufactures a home microcomputer and a business microcomputer. Each computer uses circuit chips $P$ and $Q$. The number of each chip used and the profit per unit are as follows:

|  | P | Q | Profit ($) |
|---|---|---|---|
| Home | 2 | 2 | 220 |
| Business | 1 | 4 | 330 |

During a certain month there are only $100P$ and $250Q$ circuit chips available. *(a)* How many computers of each type should be produced that month to maximize profit? *(b)* What is the maximum profit?

**15.** A manufacturer of automobile engines produces gasoline and diesel engines. The daily production data per unit are:

|  | Hours Required | | |
|---|---|---|---|
|  | Assembly | Finishing | Profit ($) |
| Diesel | 5 | 1 | 300 |
| Gas | 3 | 1.5 | 300 |

The manufacturer can allot 120 h per day for assembly and 33 h per day for finishing. *(a)* How many engines of each type should be produced daily to maximize profit? *(b)* What is the maximum profit?

**16.** In Example 17-16, the ICTV company finds they have to allow 3 h per week for assembly of the audio disk players. *(a)* If all other conditions remain the same, how many units of each type—tape players and disk players—should the company produce per week to maximize profit? *(b)* What is the maximum profit? (*Note:* Answers must be whole numbers.)

17. An oil refinery produces two basic types of fuel, low-octane and high-octane, and mixes them to create various grades of fuel. Low-octane fuel contains 2 percent additives by volume, and high-octane fuel contains 6 percent additives by volume. The refinery has an order for at least 100,000 L but not more than 120,000 L of fuel. The fuel is to contain at least 3000 L of additives. The low-octane fuel costs 20 cents a liter to produce, and the high octane costs 30 cents a liter to produce. *(a)* How much of each type should the refinery produce to minimize cost? *(b)* What is the minimum cost?

18. *(a)* If in problem 17 the fuel is to contain at least 4000 L of additives, how much of each type should the refinery produce to minimize cost? *(b)* What is the minimum cost?

19. A drug for headaches is to contain a mixture of two compounds. Compound X consists of 60 percent aspirin and 40 percent antacid. Compound Y consists of 80 percent aspirin and 20 percent antacid. Each pill of the drug must contain at least 240 mg of aspirin and at least 100 mg of antacid. Compound X costs $4.00 per kilogram to produce, and compound Y costs $3.00 per kilogram to produce. *(a)* How much of each compound should be used in each pill to minimize cost? *(b)* What is the minimum cost per kilogram to produce the drug?

20. *(a)* If in Example 17-17 the mixture of sand is to contain at least 400 kg of fine sand, how much sand should be mined from each pit to minimize cost? *(b)* What is the minimum cost? (*Hint:* Draw the feasible polygon carefully.)

## 17-5 | *Review Questions*

In problems 1 to 10, solve each inequality.

1. $3x + 1 \geq 2$
2. $\dfrac{x + 1}{2} < 7$
3. $3 < y - 1 < 5$
4. $|3x + 2| \leq 5$
5. $|1 - r| > 4$
6. $0.9t \leq 1.1(t + 2)$
7. $x^2 + 4x + 3 > 0$
8. $x^2 \leq 5x$
9. $y^2 \leq 25$
10. $\dfrac{x}{x - 1} > 0$

Express the answer to each of problems 11 to 14 in terms of an inequality.

11. A micrometer can be read to within one-thousandth of an inch. What is the possible range of values for a reading $x = 0.135$ in.?

12. The length of a computer chip is specified as 5.0 mm $< l <$ 5.5 mm and the width as 1.0 mm $< w <$ 1.2 mm. What is the range of the area $A$?

13. The instantaneous velocity of a piston is given by $v = 100(1 - \cos \theta)$. What is the range of $v$?

14. Two resistances $R_1$ and $R_2$ connected in parallel must have a total resistance $R$ greater than 1.5 $\Omega$. If $R_2 = 8 - R_1$, what is the permissible range for $R_1$? [*Note:* $R = R_1 R_2 / (R_1 + R_2)$.]

In problems 15 to 22, graph each inequality or system of inequalities.

**15.** $y \le 2$

**16.** $2x + y > 3$

**17.** $2y < x + 1$

**18.** $y \ge x^2 - 1$

**19.** $2x - y \ge 0$, $y \ge 0$, $x \le 2$

**20.** $x + y \le 5$, $2x + y \le 6$, $x \ge 0$, $y \ge 0$

**21.** $x + 2y \le 100$, $3x + y \le 90$, $x \ge 0$, $y \ge 0$

**22.** $2x + y \ge 4$, $y \le 4 - x^2$

For each linear programming problems 23 and 24, sketch the feasible polygon and answer the questions.

**23.** The Evinson Company manufactures 3- and 6-hp outboard motors. The production data for parts manufacture, assembly, and the profit per unit are as follows:

**Hours Required**

|       | Parts | Assembly | Profit ($) |
|-------|-------|----------|------------|
| 3 hp  | 10    | 3        | 100        |
| 6 hp  | 20    | 3        | 150        |

The company can provide a maximum of 200 h per week for parts manufacture and 45 h per week for assembly. *(a)* How many motors of each type should the company produce per week to maximize profit? *(b)* What is the maximum profit?

**24.** Burger Queen fast-food restaurants mix two types of meat in their hamburgers. Meat A contains 10 percent protein, and meat B contains 20 percent protein. Each "quarter pounder" must weigh at least 3 oz and contain at least 0.5 oz of protein to satisfy minimum requirements. Meat A costs 8 cents an ounce, and meat B costs 10 cents an ounce. *(a)* How many ounces of each meat should be used in each quarter pounder to minimize cost? *(b)* What is the minimum cost per quarter pounder?

CHAPTER

# 18

# Introduction to Computers and BASIC

## 18-1 | *How a Computer Works: Signing On*
### *Hardware*

A computer system consists of four basic physical units called the *hardware:* the input unit, the central processing or system unit, the memory unit, and the output unit (Fig. 18-1). The input unit is usually a keyboard or a tape or disk drive. The central processing unit performs the arithmetic calculations and controls the flow of data. The memory unit consists of primary storage, which is a magnetic core attached to the central processing unit, and secondary storage, in the form of magnetic tapes or disks. The output unit is sometimes the same as the input unit and can be a printer, a video display, or a tape or disk drive.

A large computer system (mainframe) has many inputs and outputs called *timesharing terminals*. They can all operate simultaneously, depending on the capacity of the central unit. The *microcomputer* contains the four basic units in one small package. It consists of a single keyboard terminal and/or a

**A** computer is a complex calculating machine that works with numbers and *alphanumeric* data (words, addresses, equations, etc.). The first electronic computer, ENIAC, was built in 1946 and cost $500,000. Today a $50 calculator has almost the computing power of ENIAC! Computers are everywhere, and some understanding of them is essential in science and technology. This chapter provides a brief study of computers and computer programming using the BASIC language. It will enable you to write a simple program on a computer and can be used to supplement other instructional material on computers and computer programming.

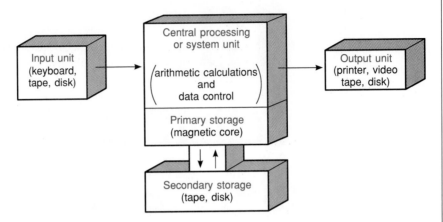

FIG. 18-1 *Computer system*

cursor control (mouse) attached to a system unit and the primary storage. Secondary storage is in the form of a tape cassette or a small disk system. The primary storage contains the random access memory (RAM), where programs and results are stored while the computer is operating. Memory capacity is measured in terms of characters, or bytes: 1 kilobyte (KB) = 1024 bytes ($2^{10}$). Microcomputers can contain from 64 KB to 1 MB (megabyte = 1000 KB) or more of RAM. Although the microcomputer can handle only one or a few inputs at a time, it is becoming more common than large terminal systems because of its low price and accessibility.

## Software

*Software* refers to the programs that allow the operator to communicate with the computer system. Communication takes place as follows. The operator uses a *programming language* (such as BASIC or FORTRAN) containing familiar verbal and mathematical statements. A translator program stored in the computer, called the *systems software,* translates the programming language to machine language. Permanently stored programs are also called the *read-only memory* (ROM). *Machine language* consists of numerically coded instructions that the computer understands. For example, 19 might represent multiplication, 27 a command to read data from the input, 23 to print data on the output, 65 to store data in the memory, etc. The data to be used are stored in memory locations that are numbered like mailboxes. The memory capacity is in the thousands for microcomputers and in the billions for large systems.

Software varies from computer to computer, with programming languages having various "dialects." Programs that work on one type of computer may not work on another type. However, the dialects of BASIC are similar and can be readily modified when you are switching from one type of computer to another.

## Keyboard Operation

The keyboard input on a timesharing terminal or microcomputer is similar to a conventional typewriter but contains additional special features and keys. These differ from unit to unit, and you must receive instructions or read the operating manual for the unit you are using. Here are some common features that are found on many keyboards:

| | |
|---|---|
| RETURN or ENTER | Returns to the beginning of the next line and signals the computer to record the typed information or respond to a command |
| LINE FEED or ↓ | Advances to the next line |
| ← or BACKSPACE | Backspace |
| BREAK or ESC | Stops any computer execution |
| CLEAR | Clears the display |

Some timesharing terminals print only capital letters, which are used in BASIC. The SHIFT key on these units is not for upper- and lowercase letters but for the symbols or commands shown above the letters and numbers on the keyboard. The SHIFT key can also have other functions when it is used in conjunction with certain keys. Note that some computers print $\emptyset$ for zero to distinguish it from the letter O. To *correct an error,* you simply backspace, ←, and type over the error. For example, if you type:

```
HELLO COMPUTET←R
```

the R replaces the T, and it is the same as if you had typed HELLO COM-PUTER. On a video display, the backspace automatically erases the last character typed.

## Signing On

Signing on to a microcomputer is done by turning on the unit and responding to a few questions, usually concerning the memory or software. The computer then types READY, OK, or its equivalent and you can begin use.

Signing on to a timesharing terminal is more involved and is known as *logging in.* Procedures differ for various terminals, and you must be instructed how to sign on to the terminal you are using. After you turn on the terminal, the computer asks you a series of questions, which usually involve what language you want to use, certain identification numbers, and a password. The password prevents improper use of the terminal and protects your file. If you enter all the proper information correctly, the computer types READY or its equivalent and you can begin. Some terminals use telephone couplers, and it is necessary to first dial the computer's number to begin the procedure. The following is an example of logging in on a large university computer system. The first question refers to the language you want to use. Callos is a code name for several languages, including BASIC. After you type the answer to each question (answers are underlined), you press the [RETURN] key:

```
TYPE: A FOR APL, W FOR WILBUR, O FOR CALLOS
O
PROJECT NUMBER, ID? 14939,076446350
PASSWORD ████████
ON AT 14:43 04/07/89 UCNY LINE 20
USER NUMBER? NYA031,DP7213
READY
```

The password is obscured by the computer so that no one can change any programs stored in your file. When you are finished using the computer, *log off* by typing BYE or OFF.

**Exercise 18-1**

1. List three uses of computers in your field of study.

2. List three uses of computers in fields other than your own.

3. What is the difference between computer hardware and computer software?

4. List the hardware components of a computer system.

5. What is the difference between a large computing system and a microcomputer?

6. What is the difference between programming language and machine language?

7. Assuming your brain corresponds to the central processing unit of a complex computer system, what would correspond to the input unit, the primary storage, the secondary storage, and the output unit?

8. List the differences and similarities between a calculator and computer.

9. The speed of a typist in words per minute is given by $s = n/6$, where $n$ = number of characters (including blanks) typed per minute. A modern high-speed printer prints 1200 lines per minute. If each line contains 130 characters, what is the speed of the printer, in words per minute?

10. Magnetic tape can store 2460 characters per centimeter. If a standard typewritten page contains 3000 characters, how many pages can be stored on a 500-m reel of tape?

11. In the memory unit, one magnetic core is called a *bit,* and a set of bits make up a *byte,* or character. *(a)* If 8 bits make up a byte and 4 bytes make up a computer "word", how many bits are in a memory unit containing 48K of RAM (1K = 1024 bytes)? *(b)* How many words can be stored in this memory unit?

12. Retrieval from memory can be extremely fast. If 1 bit can be retrieved in 80 ns (1 ns = $10^{-9}$ s), how many words can be retrieved in 1 s? (See problem 11.)

13. What will the video display show if you type the following?

```
MY NANR←←ME ISS← MS←AROS←←IA
```

**14.** When you type a series of numbered lines, the computer prints them *in order* if you type the command LIST and press ENTER or RETURN. What will the computer print if you give the command LIST after typing the following?

```
50 TJ0←←HIM←NK LIKE YOU
20 VIOLETS ARE BLUE
40 BUT I CANNOT
10 ROSES ARE RED
30 I MAY BE FS←ASR←T
```

**15.** Computer machine language uses the *binary system,* which contains only two digits, 0 and 1. A tiny magnetic core can be magnetized counterclockwise (binary 0) or clockwise (binary 1). Complete the table below, which shows the relation between binary and decimal numbers.

| Decimal | 0 | 1 | 2 | 3 | 4 | 5 | 6 | 7 | 8 |
|---|---|---|---|---|---|---|---|---|---|
| Binary | 000 | 001 | 010 | 011 | 100 | 101 | | | |

**16.** If you are using a timesharing terminal, list the steps for logging in on that terminal.

## 18-2 | *Introduction to BASIC*

The language of BASIC (beginners all-purpose symbolic instruction code) was developed in the 1960s at Dartmouth College, and it is one of the more popular computer languages. Dialects and higher levels of BASIC have been developed, but much of the original BASIC remains a part of them. The commands and statements discussed in this section are common to most BASIC dialects; some variations may apply to your particular system. This material provides an introduction to computer programming.

### *Program Format*

The BASIC program consists of a series of numbered lines, which you do not necessarily have to type in order. The computer automatically orders the program when it is entered. The lines are usually numbered with gaps of 5 or 10 to allow for insertion of additional lines if necessary. When you type two or more lines with the same number, the last typed line replaces any previous line. To erase a line, you type the number of the line and leave the space to the right blank. When you type the command LIST and press ENTER or RETURN, the computer prints the program with the lines in order and containing all the corrections.

### Example 18-1

What does the computer print if you type the following and then give the command LIST?

```
10  IN A RIGHT TRIANGLE
20  THE SUM OF THE SIDES
25  OF THE SQUARES EQUALS
20  THE SUM OF THE SQUARES
40  THE SQUARE OF
50  THE HYPOTENUSE
25
30  OF THE SIDES EQUALS
 5  PYTHAGOREAN THEOREM
```

### Solution

The computer prints:

```
 5  PYTHAGOREAN THEOREM
10  IN A RIGHT TRIANGLE
20  THE SUM OF THE SQUARES
30  OF THE SIDES EQUALS
40  THE SQUARE OF
50  THE HYPOTENUSE
```

The lines are printed in order, with line 25 not shown since it was typed blank the last time. Notice that the second line 20 replaces the first line 20.

## System Commands

The command LIST is a *system command*. It is not used in a program but is part of the communication between you and the computer. When you type in a system command and then press ENTER or RETURN, the computer will respond. Common system commands are:

DELETE: Erases specific lines in a program
LIST: Lists the program
NAME or RENAME: Names or renames a program
NEW: Erases any program in the memory to prepare for a new program
RUN: Executes program

The following system commands usually apply to storage when a *timesharing terminal* is used:

SAVE: Stores a program
PURGE or UNSAVE: Erases a stored program
OLD or LOAD: Retrieves a stored program
CATALOG: Lists stored programs

Storage commands for a *microcomputer* are the same or similar and apply to cassette or disk storage. For example, for some microcomputers:

CSAVE: Stores a program on cassette tape
CLOAD: Loads a program from cassette tape

### Example 18-2

What does the computer print after you enter the following? (READY is a response printed by the computer.)

```
NEW
READY
 1 CHARGE
 2 IN AN ELECTRIC CIRCUIT
 3 CURRENT VARIES DIRECTLY
DELETE 1-3
10 POTENTIAL
20 IN A DC CIRCUIT
30 CURRENT VARY DIRECTLY
40 AS VOLTAGE
30 CURRENT VARIES DIRECTLY
LIST
```

### Solution

The computer prints:

```
10 POTENTIAL
20 IN A DC CIRCUIT
30 CURRENT VARIES DIRECTLY
40 AS VOLTAGE
```

The computer deletes lines 1 to 3. The DELETE command must be followed by the numbers of the lines to be deleted. Notice that the second line 30 replaces the first line 30.

## LET and PRINT Statements

*LET* is a program statement that assigns values to a variable. These values are stored in the memory. For example:

```
10 LET X = 3.14
20 LET X = X + 1
```

Any letter can be used as a variable. When *LET* is used in more than one statement for the same variable as above, the last value is assigned and the previous value is erased. Note that the command *LET* can be omitted in most BASIC programs. That is, the above can be written:

```
10  X = 3.14
20  X = X + 1
```

The *PRINT* statement instructs the computer to print the value of a constant, a variable, or an arithmetic expression, for example:

```
30  PRINT X
40  PRINT X - 2.12
```

The BASIC symbols for addition and subtraction are the same as those in arithmetic. If you combine the above four statements and add the statement *END*, which tells the computer the program is finished, you have a *simple program:*

```
10  LET X = 3.14
20  LET X = X + 1
30  PRINT X
40  PRINT X - 2.12
50  END
```

### Example 18-3

When you enter the above program and give the command RUN, what does the computer print?

### Solution

The computer prints:

```
4.14
2.02
```

The first number is the last value assigned to $X$: $3.14 + 1 = 4.14$. The second number is the value: $X - 2.12 = 4.14 - 2.12 = 2.02$. The computer prints one value on each line. If you want both results printed on the same line, separate the variables with a comma or a semicolon:

```
40  PRINT X, X - 2.12
```

The computer then prints:

```
4.14    2.02
```

*If a comma is used, the results are printed separately in columns. If a semicolon is used, the results are printed next to each other.*

When PRINT is followed by a statement in quotes, the computer prints the statement exactly.

### Example 18-4

What does the computer print upon execution of the following program?

```
10 PRINT "EQUATION SOLVING"
20 LET X = 4 - 1
30 PRINT "IF X + 1 = 4,";
40 PRINT "THE SOLUTION IS" X
50 END
```

### Solution

The computer prints:

```
EQUATION SOLVING
IF X + 1 = 4, THE SOLUTION IS 3
```

The semicolon at the end of line 30 tells the computer to continue printing on the same line. The $X$ in line 40 is not in quotes because *the value of X,* not the letter X, is to be printed.

## Arithmetic Operations and Relations

The BASIC symbols for arithmetic operations are:

| | | |
|---|---|---|
| Addition | $+$ | |
| Subtraction | $-$ | |
| Multiplication | $*$ | (18-1) |
| Division | $/$ | |
| Power | $\wedge$ | |

For example, the expression $x^2 + 5x - \frac{1}{2}$ is written in BASIC as

$$X \wedge 2 + 5*X - 1/2$$

Variables in BASIC can be any letter or any letter followed by a single digit: A, N, Y, A3, B4, etc.

The order of operations in BASIC is the same as that in arithmetic: (1) raising to powers, (2) multiplication or division, and (3) addition or subtraction. Expressions in parentheses are evaluated first, as in your calculator. The BASIC symbols for the arithmetic relations, equals ($=$), greater than ($>$), and less than ($<$) are the same as those in algebra. The combined symbols in BASIC are:

$$\text{Greater than or equal to } (\geq) \qquad > =$$
$$\text{Less than or equal to } (\leq) \qquad < = \qquad\qquad (18\text{-}2)$$

### Example 18-5
Write each expression in BASIC.

1. $\dfrac{5 + xy}{2}$     $(5 + X * Y)/2$

2. $3(a - 2)^3$     $3 * (A - 2) \wedge 3$

3. $\dfrac{(0.13)^2}{i_2 - i_1}$     $0.13 \wedge 2/(I2 - I1)$

Note in (3) that a variable with a subscript is written in BASIC with the number after the letter.

4. $\sqrt{x} \geq 1$     $X \wedge 0.5 >= 1$

In (4) the 0.5 power is used for the square root. The ½ power can also be used or a special *square root function* SQR (X): SQR (X) $>= 1$. See Example 18-7 for another use of SQR(X).

The next two examples illustrate many of the ideas presented so far. Study them carefully.

### Example 18-6
What does the computer print upon execution of the following program?

```
10 PRINT "PROPORTION"
20 LET A = 1
30 LET B = 4
40 LET C = 5
50 LET D = B*C/A
60 PRINT "D =" D, "A/B =" A/B, "C/D =" C/D
70 END
```

### Solution
This program solves the proportion $A/B = C/D$ for $D$ when $A = 1$, $B = 4$, and $C = 5$. The computer prints:

```
PROPORTION
D = 20 A/B = 0.25 C/D = 0.25
```

The results show that the two fractions have the same ratio, that is, they are in proportion.

### Example 18-7

What does the computer print upon execution of the following program?

```
10 PRINT "PYTHAGOREAN THEOREM"
20 LET A = 3
30 LET B = 4
40 LET C = SQR (A ∧ 2 + B ∧ 2)
50 PRINT "A", "B", "C"
60 PRINT A, B, C
70 END
```

### Solution

This program solves for the hypotenuse of a right triangle $C = \sqrt{A^2 + B^2}$ when $A = 3$ and $B = 4$. The computer prints:

```
PYTHAGOREAN THEOREM
A           B           C
3           4           5
```

The commas in lines 50 and 60 cause the results to be printed neatly in columns. Notice in line 40 that the square root function SQR (X) is used. SQR (X) uses a special stored program in the computer and is faster than $X \wedge 0.5$ or $X \wedge (1/2)$.

### Exercise 18-2

What does the computer print when given the command LIST for problems 1 to 4?

```
1. 20 IS APPROXIMATELY
   10 THE SQUARE ROOT OF 2
   30 1.414213
   30 1.414214
2.  5 WORK
    5
   10 WORK EQUALS THE
   20 PORDUCT OF FOREC
   30 TIMES DISTANCE
   20 PRODUCT OF FORCE
3.  1 LINCOLN
   10 WITH CHARITY FOR ALL
   20 WITH MALICE
   DELETE 10-20
   10 WITH MALICE TOWARD NINE
```

```
20 WITH CHARITY FOR ALL
10 WITH MALICE TOWARD NONE
```
4.
```
 1 DC CIRCUIT
 2 THE RATIO OF VOLTAGE
20 TO CURRENT IS RESISTANCE
 1 OHM'S LAW
 2
10 THE RATIO OF VOLTAGE
```

**5.** What is the difference between a system command and a program statement?

**6.** Give three examples of system commands and three examples of program statements and tell the function of each.

In problems 7 to 10, what does the computer print upon execution of each program?

**7.**
```
10 LET E = 2.72
20 LET E = E - 2.00
30 PRINT E
```

**8.**
```
10 LET A = 0.125
20 LET B = 0.876
30 PRINT A, B, A + B
```

**9.**
```
 5 PRINT "METRIC"
10 LET C = 2.54
20 PRINT C "CM = 1 IN"
30 LET K = 0.453
40 PRINT K "KG = 1 LB"
```

**10.**
```
10 PRINT "AREA OF A SQUARE"
20 LET S = 5.2
30 PRINT "IF SIDE =" S;
40 PRINT "THEN AREA =" S ∧ 2
```

In problems 11 to 16, write each expression in BASIC.

**11.** $\dfrac{a + 3}{b - 2}$        **12.** $y = \dfrac{5}{2}x + \dfrac{1}{4}$

**13.** $(f_2 - f_1)^3$        **14.** $3.1(IR^2 + E)$

**15.** $\sqrt{2p} \le 6$        **16.** $\dfrac{t_3 + t_2}{2} > 5$

In problems 17 to 22, write each BASIC expression in algebraic notation.

**17.** $Y = A * X \wedge 2 + B * X + C$     **18.** $K * T = (P * V) \wedge N$

**19.** $(E1 + E2)/R - I$           **20.** $(2 + 3 * J)/(1 - J)$

**21.** $9.9 * (R + (Q * H) \wedge 3)$      **22.** $X > = 10 \wedge (1/2)$

In problems 23 to 28, what does the computer print upon execution of each program?

**23.**
```
10 PRINT "AVERAGE"
20 LET X = 89.5
```

```
          30 LET Y = 76.8
          40 LET A = (X + Y)/2
          50 PRINT "X", "Y", "A"
          60 PRINT X, Y, A
24.       10 PRINT "PI"
          20 LET A = 27.9
          30 LET R = 3.00
          40 LET P = A/R ∧ 2
          50 PRINT "A", "R", "P"
          60 PRINT A, R, P
25.       10 PRINT "PARALLEL CIRCUIT"
          20 LET R1 = 3.5
          30 LET R2 = 4.2
          40 LET R = R1*R2/(R1 + R2)
          50 PRINT "R1", "R2", "R TOTAL"
          60 PRINT R1, R2, R
26.       10 PRINT "DEFLECTION"
          20 LET P = 2.4*10 ∧ 3
          30 LET L = 2.0*10 ∧ 2
          40 LET D = (P*L ∧ 3)/(4.8*10 ∧ 10)
          50 PRINT "D =" D
27.       10 PRINT "SLOPE"
          20 LET X1 = 1
          30 LET Y1 = -2
          40 LET X2 = 3
          50 LET Y2 = -1
          60 LET M = (Y2 - Y1)/(X2 - X1)
          70 PRINT "(1, -2) AND (3, -1)"
          80 PRINT "SLOPE =" M
28.       10 PRINT "Y = 2X ∧ 2 - 3X + 1"
          20 LET A = 2
          30 LET B = -3
          40 LET C = 1
          50 LET X1 = (-B + SQR(B ∧ 2 - 4*A*C))/2*A
          60 LET X2 = (-B - SQR(B ∧ 2 - 4*A*C))/2*A
          70 PRINT "X =" X1 "AND X =" X2
```

**29.** Write a BASIC program that averages the numbers 2.78, 3.12, and 5.01 and prints the average.

**30.** Write a BASIC program that solves $3x - 2.5 = 8$ and prints the solution.

**31.** Write a BASIC program that uses Ohm's law and prints the voltage of a dc circuit with a resistance of 10.5 Ω and a current of 2.40 A.

**32.** Write a BASIC program that converts 20°F to degrees Celsius and prints the result. [*Note:* $C = \frac{5}{9}(F - 32)$.]

## 18-3 | *Flowcharts*

Before you can effectively write a computer program, you must clearly outline the step-by-step procedure that the computer will follow. A useful method for doing this is to construct a *flowchart,* which is a diagram of the logical sequence of steps. Study the following examples of flowcharts; they are used to write BASIC programs in Sec. 18-4.

### Example 18-8

1. Construct a flowchart for the following program: Print the three test grades 60, 75, and 83 and their average.

2. Add to the program a decision step that prints PASS if the average of the test grades is greater than or equal to 70 and FAIL if not.

### Solution

1. Construct a flowchart for the following program: Print the three test grades 60, 75, and 83 and their average.

   The flowchart is shown in Fig. 18-2. Observe the geometric shapes. An *oval* is used for the start and the stop operation. A *rectangle* represents an arithmetic procedure. A *parallelogram* is used for an output statement (such as PRINT) or an input statement.

2. Add to the program a decision step that prints PASS if the average of the test grades is greater than or equal to 70 and FAIL if not.

   The *diamond* is used for the decision step, as shown in the flowchart in Fig. 18-3.

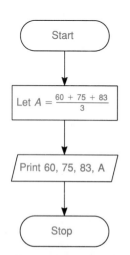

FIG. 18-2  *Flowchart, Example 18-8(1)*

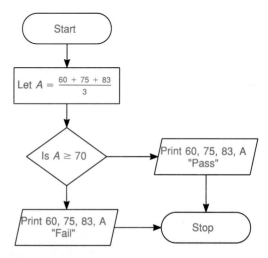

FIG. 18-3  *Flowchart, Example 18-8(2)*

### Example 18-9
Construct a flowchart for the following program: Print the first 10 whole numbers.

### Solution
The flowchart is shown in Fig. 18-4. It contains a *loop*. Study the flowchart carefully. To help you see how the program works, the first 10 steps the computer does are outlined below:

1. $N = 1$
2. PRINT 1
3. $N = 1 + 1 = 2$ } loop
4. YES
5. PRINT 2
6. $N = 2 + 1 = 3$ } loop
7. YES
8. PRINT 3
9. $N = 3 + 1 = 4$ } loop
10. YES

&#8942;

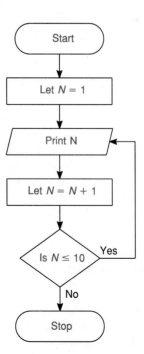

FIG. 18-4 *Flowchart, Example 18-9*

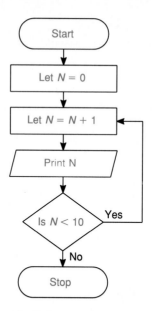

FIG. 18-5 *Flowchart, Example 18-9*

The groups of steps (2, 3, 4), (5, 6, 7), (8, 9, 10), etc. represent the loop. This procedure continues until:

**29.** PRINT 10
**30.** N = 10 + 1 = 11
**31.** NO
**32.** END

The BASIC program for the flowchart in Fig. 18-4 is shown in Example 18-13 (Sec. 18-4). You can construct more than one flowchart for the same program. Another one for Example 18-9 is shown in Fig. 18-5. Notice that this flowchart begins with $N = 0$. List the steps the computer takes in following this flowchart to help see how the program works.

The next example shows how you can build on the flowcharts in Example 18-9 to construct a flowchart for a similar but more involved program.

### Example 18-10
Construct a flowchart for the following program: Print the sum of the first 10 whole numbers.

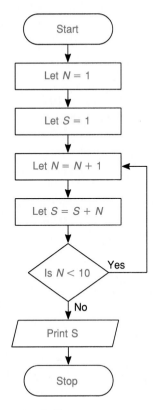

FIG. 18-6 *Flowchart, Example 18-10*

### Solution
This flowchart builds on the one in Fig. 18-4. You need to add another variable *S* to represent the sum of the numbers as shown in Fig. 18-6. The computer takes these steps:

1. N = 1
2. S = 1
3. N = 1 + 1 = 2 ⎫
4. S = 1 + 2 = 3 ⎬ loop
5. YES ⎭
6. N = 2 + 1 = 3 ⎫
7. S = 3 + 3 = 6 ⎬ loop
8. YES ⎭

9. $N = 3 + 1 = 4$
10. $S = 6 + 4 = 10$ } loop
11. YES

⋮

27. $N = 9 + 1 = 10$
28. $S = 45 + 10 = 55$
29. NO
30. PRINT 55
31. END

The flowchart would also work if you began with $N = 0$ as in Fig. 18-5, but it would have more steps. Note that you can modify the flowchart in Fig. 18-6 to compute the sum from 1 to 100, 1 to 1000, and so on.

### Example 18-11

Construct a flowchart for the following program: The distance an object falls from rest is given by $s = \frac{1}{2}gt^2$, where $g$ = gravitational acceleration (9.8 m/s²)

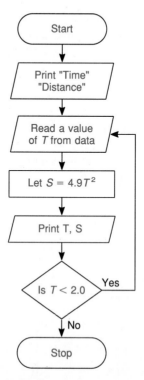

FIG. 18-7 *Flowchart, Example 18-11*

and $t$ = time in seconds. Print the values of time and distance for $t$ = 0.0, 0.5, 1.0, 1.3, 1.8, and 2.0 s in two columns with headings.

### Solution
This program is done by first entering the data for $t$ into the computer. The computer then reads the values for $t$ one at a time and does each calculation. The steps are accomplished by using READ and DATA statements, explained further in the next section. The flowchart is shown in Fig. 18-7. The BASIC program for the flowchart is shown in Example 18-14.

### Example 18-12
Construct a flowchart for the following program: For a given input of voltage and resistance in a dc circuit, print the words *current, voltage,* and *resistance* and their values in three columns.

### Solution
This program uses an INPUT statement which allows data to be entered for the variables $E$ (voltage) and $R$ (resistance). Ohm's law, $I = E/R$, is used to compute the current (Fig. 18-8). The BASIC program for this flowchart is shown in Example 18-17.

Flowcharts are not only useful for writing computer programs. They can be applied to technical problems and other problems whenever a diagram showing the logical sequence of steps will help in the solution.

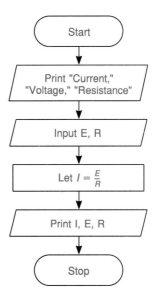

FIG. 18-8 *Flowchart, Example 18-12*

**Exercise 18-3**

1. Describe four geometric shapes used in a flowchart and the function of each.

2. Construct a flowchart that shows the steps you go through before a date or going out for the evening.

In problems 3 to 18, construct a flowchart for each program.

3. Print the result of (2.3)(5.8)/1.4.

4. Print the diameter and circumference of a circle whose radius is 4.5 in.

5. Given a weight $W$ in kilograms, print $W$ in pounds.

6. Given a distance traveled and the time elapsed, print the average rate of speed.

7. Given a square whose side is 5 and a circle whose radius is 3, print the name and the area of the figure that has the greater area.

8. Given $a = 4.2$ and $b = 4.8$, print the greater value: $2ab$ or $a^2 + b^2$.

9. Print the first 10 whole numbers, starting with $N = 10$. (See Example 18-9.)

10. Print the squares of the first 10 whole numbers.

11. Print the sum of the first 50 whole numbers. (See Example 18-10.)

12. Print the product of the first nine whole numbers.

13. Print the sum of the first 20 terms of the arithmetic progression 0.5, 1.0, 1.5, . . . .

14. Print the sum of the first 10 terms of the geometric progression 2, 4, 8, . . . .

15. Given the price of an item, print the final price, which includes a 25 percent discount and 7 percent sales tax.

16. Print the interest earned on a given principal after 1 year at 10 percent interest compounded semiannually. (See Sec. 16-2.)

17. Print the first 10 odd numbers, and next to each number print the sum of the odd numbers from 1 to that number.

18. Given three different numbers, print the smallest number.

19. Explain what the flowchart in Fig. 18-9 does.

20. Explain what the flowchart in Fig. 18-10 does.

21. Simplify the flowchart in Fig. 18-11 to produce the same result with one PRINT statement.

22. Explain what the flowchart in Fig. 18-12 does.

In problems 23 to 30, construct a flowchart for each program.

23. For a given input of two resistances in parallel in a dc circuit, print the value of each resistance and the total resistance in three columns with headings. (See Example 18-12.)

24. For a given input of current and resistance in a dc circuit, print the value of the current, resistance, and power in three columns with headings. (See Example 18-12.)

25. The velocity of a falling object is given by $v = v_0 + gt$, where $v_0$ = initial velocity, $g$ = gravitational acceleration, and $t$ = time. Given $v_0 = 10$ ft/s and $g = 32$ ft/s$^2$ for an object thrown from a cliff, print the values of time and velocity for $t = 0.2$, 0.5, 0.9, 1.5, and 2.1 s in two columns with headings. (See Example 18-11.)

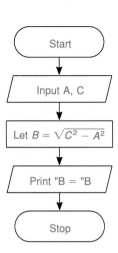

FIG. 18-9 *Flowchart, problem 19*

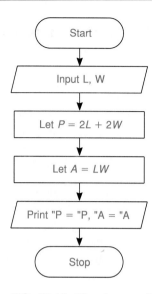

FIG. 18-10 *Flowchart, problem 20*

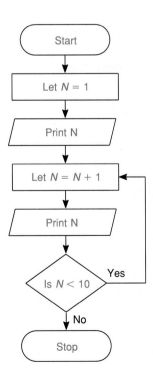

FIG. 18-11 *Flowchart, problem 21*

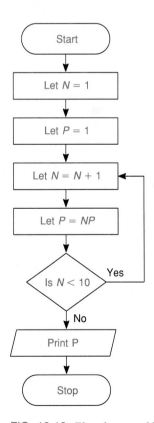

FIG. 18-12 *Flowchart, problem 22*

26. Print the temperatures in degrees Fahrenheit and degrees Celsius in two columns with headings, for the following values: $-18$, $-10$, 0, 12, 20, and 30°C. (See Example 18-11.) (*Note: F = 1.8C + 32.*)

27. A dc circuit has a resistance of 15.0 $\Omega$. The voltage is initially set at 10 V and is increased in steps of 10 V until it reaches 110 V. Print the voltage and current for each value of the voltage in two columns with headings.

28. A weight of 5.0 kg is placed 50 cm from a fulcrum. The weight is increased 5.0 kg at a time until the total weight is 100 kg. Print the weight and moment for each value of the weight in two columns with headings. (*Note:* Moment = force $\times$ distance.)

29. For a given input of two forces $F_1$ and $F_2$ acting at right angles, print the two forces and the resultant force $R$ in three columns with headings. (*Note:* $R = \sqrt{F_1^2 + F_2^2}$.)

30. For a given input of resistance $R$ and reactance $X$ in an ac circuit, print the resistance, reactance, and impedance $Z$ in three columns with headings. (*Note:* $Z = \sqrt{R^2 + X^2}$.)

# 18-4 | *Writing BASIC Programs*

Section 18-2 illustrates BASIC programs using the program statements LET and PRINT. This section introduces more program statements and shows how to use them to write programs that solve a variety of problems.

## GOTO and IF-THEN Statements

The *GOTO* statement tells the computer to go to a specific line. The *IF-THEN* statement is a decision statement that tells the computer to go to a specific line if a certain condition is true. If the condition is not true, the computer continues to the next line. For example, consider the following:

```
50 IF T >= 100 THEN 70
60 GOTO 30
```

When the computer reads line 50, it will go to line 70 if $T$ is greater than or equal to 100. If $T$ is less than 100, the computer will read line 60 and then go back to line 30.

### Example 18-13

Write a BASIC program for Example 18-9: Print the first 10 whole numbers.

### Solution

The program follows from the flowchart in Fig. 18-4:

```
10 LET N = 1
```

```
20 PRINT N
30 LET N = N + 1
40 IF N < = 10 THEN 20
50 END
```

The IF-THEN statement in line 40 corresponds to the diamond decision step in the flowchart. There are usually many ways to write a program, some more efficient than others. Another, but *less efficient,* way to write this program is:

```
10 LET N = 1
20 PRINT N
30 LET N = N + 1
40 IF N > 10 THEN 60
50 GOTO 20
60 END
```

Construct a flowchart to see how the above program works.

## READ, DATA, and INPUT Statements

READ, DATA, and INPUT statements are all used to assign values to variables. The *READ* and *DATA* statements assign values to one or more variables at a time. Consider the following:

```
30 READ A, B, C
40 DATA 3, 4, 5
```

When the computer reads lines 30 and 40, it assigns $A = 3$, $B = 4$, and $C = 5$. You can assign more than one set of values to the variables by using more data:

```
30 READ A, B, C
40 DATA 3, 4, 5, 6, 8, 10
50 GOTO 30
```

The first time the computer reads lines 30 and 40, it assigns $A = 3$, $B = 4$, and $C = 5$. The second time the computer reads lines 30 and 40, it assigns $A = 6$, $B = 8$, and $C = 10$. It will keep assigning in this way until there are no more data available.

### Example 18-14
Write a BASIC program for Example 18-11.

### Solution
The program follows from the flowchart in Fig. 18-7:

```
10 PRINT "TIME", "DISTANCE"
20 DATA 0.0, 0.5, 1.0, 1.3, 1.8, 2.0
30 READ T
40 LET S = 4.9*T ∧ 2
50 PRINT T, S
60 IF T < 2.0 THEN 30
70 END
```

Notice that the DATA statement appears before the READ statement. The DATA statement can appear anywhere that is convenient in a program. The computer scans the program for the DATA statement when it first comes to the READ statement and stores the data.

### Example 18-15
Write a BASIC program that prints the volume of a rectangular box having each set of dimensions: 1, 3, 3; 2, 4, 3; and 5, 5, 1.

### Solution

```
10 READ L, W, H
20 PRINT L * W * H
30 IF L = 5 THEN 60
40 GOTO 10
50 DATA 1,3,3,2,4,3,5,5,1
60 END
```

Notice that the DATA statement is at the end of the program. Line 30 ends the program when the data are finished.

The *INPUT* statement causes the computer to print a question mark and allows the operator to type in values for one or more variables.

### Example 18-16
Write a BASIC program that prints a person's age at the end of the present year.

### Solution
Use the INPUT statement in the program:

```
10 PRINT "WHAT YEAR WERE YOU BORN?"
20 INPUT B
30 PRINT "WHAT YEAR IS IT NOW?"
40 INPUT N
50 LET A = N - B
60 PRINT "CONGRATULATIONS,";
70 PRINT "YOU WILL BE" A "YEARS OLD";
```

```
80 PRINT "AT THE END OF THIS YEAR."
90 END
```

When the computer reads line 20, it prints a question mark under the question in line 10. The operator then types in his or her age and presses ENTER or RETURN. At line 40 the computer prints another question mark under the question in line 30. The operator enters the present year, and the computer prints lines 60, 70, and 80 on one line because of the semicolons.

The INPUT statement can ask for the value of more than one variable at a time, as shown in the next example.

**Example 18-17**
Write a BASIC program for Example 18-12.

**Solution**
If it is assumed that the values of the voltage and the resistance are supplied by the operator, the program is as follows, given the flowchart in Fig. 18-8:

```
10 PRINT "ENTER VOLTAGE, RESISTANCE"
20 INPUT E, R
30 LET I = E/R
40 PRINT "CURRENT", "VOLTAGE", "RESISTANCE"
50 PRINT I, E, R
60 END
```

The computer prints one question mark at line 20, and the operator must enter two values, separated by a comma, for the program to continue.

## Error Messages and REM Statement

In Example 18-17, if the operator does not enter two inputs, forgets to type in a comma, or makes some other error, the computer prints an error message such as:

```
        ? REDO
or      ? NOT ENOUGH DATA
```

Each system has its own error messages for various types of errors. One of the most common types is a format error in a program statement, which, upon execution of the program, causes a message like:

```
        SYNTAX ERROR IN LINE 30
or      ILLEGAL INSTRUCTION IN 50
```

Debugging a program can be very tedious. To help identify a program or parts of a program, the *REM* (remark) statement is used. The computer ignores any information printed in a REM statement. It is for the use of the programmer only.

## Mathematical Functions

BASIC contains statements for trigonometric and logarithmic functions. In Table 18-1 the statement produces the value of the function for the argument $X$ in parentheses. Values for the functions are usually given to a minimum of six significant figures. The computer can be programmed for more accuracy if desired. For SIN (X), COS (X), and TAN (X) the value of X must be in radians. ATN (X) gives the value of the angle between $-\pi/2$ and $\pi/2$. LOG (X) is the inverse of EXP (X), that is, LOG [EXP (X)] = X. See Chap. 9 for more information on $e^x$ and $\ln x$.

**Table 18-1**　MATHEMATICAL FUNCTIONS IN BASIC

| BASIC Statement | Function | Comment |
|---|---|---|
| SIN (X) | $\sin x$ | $x$ in radians |
| COS (X) | $\cos x$ | $x$ in radians |
| TAN (X) | $\tan x$ | $x$ in radians |
| ATN (X) | $\tan^{-1} x$ | Principal value in radians |
| EXP (X) | $e^x$ | $e = 2.71828$ |
| LOG (X) | $\ln x$ | Natural logarithm |

### Example 18-18
What does the computer print upon execution of the following program?

```
10 REM TRIG FUNCTIONS
20 PRINT "ANGLE", "RADIAN", "SIN", "COS", "TAN"
30 READ X
40 DATA 0, 30, 60, 90
50 LET R = 0.174529 * X
60 PRINT X, R, SIN (R), COS (R), TAN (R)
70 IF X = 90 THEN 90
80 GOTO 30
90 END
```

### Solution
The computer ignores line 10 and prints the table of values:

| ANGLE | RADIAN | SIN | COS | TAN |
|-------|--------|-----|-----|-----|
| 0 | 0 | 0 | 1 | 0 |
| 30 | 0.523599 | 0.5 | 0.866025 | 0.577350 |
| 60 | 1.04720 | 0.866025 | 0.5 | 1.73205 |
| 90 | 1.57080 | 1 | 0 | * |

*Some computers may print a large invalid value such as 890060 because of computer approximation. Others may print an error message such as DIVISION BY ZERO.

### Example 18-19

Write a BASIC program that converts rectangular coordinates to polar coordinates or, equivalently, converts a complex vector in rectangular form to polar form.

### Solution

Both problems involve the same formulas that are given in Secs. 11-5 and 13-3. The rectangular coordinates of a point $(X, Y)$ determine a complex vector in rectangular form $X + Yj$. The polar coordinates of a point $(R, G)$ determine a complex vector in polar form $R(\cos G + j \sin G)$ or $R \angle G$. Here $R$ is the radius or modulus, and $G$ is the angle or argument. The following program prints the pair $R$, $G$ for a given input $X$, $Y$:

```
10 REM POLAR CONVERSION
20 PRINT "ENTER X, Y, X NOT ZERO"
30 INPUT X, Y
40 PRINT "RADIUS", "ANGLE"
50 LET R = SQR(X ∧ 2 + Y ∧ 2)
60 LET G = ATN (Y/X)
70 IF X < 0 THEN 90
80 GOTO 100
90 LET G = 3.14159 + G
100 PRINT R, G
110 END
```

The program prints $G$ in the range $-\pi/2 < G < \pi/2$ and $\pi/2 < G < 3\pi/2$. Additional steps are necessary for the program to convert points on the $Y$ axis ($G = \pi/2$ or $3\pi/2$) since ATN (Y/X) is not defined when $X = 0$. Notice in line 70 that if $X < 0$, ATN(Y/X) is negative when $Y > 0$ and positive when $Y < 0$. However, line 90 produces the correct result in either case. Choose values and check for yourself, using a table or a calculator.

**Exercise 18-4**

1. Explain the difference between the GOTO and IF-THEN statements.
2. Explain the difference between the READ, DATA, and INPUT statements.

In problems 3 to 6, tell what the computer prints upon execution of each program.

3. 
```
10 LET N = 2
20 PRINT N
30 LET N = N + 2
40 IF N < = 20 THEN 20
50 END
```

4. 
```
10 LET X = 0
20 LET X = X + 1
30 LET N = X ∧ 2
40 PRINT X, N
50 IF X < 10 THEN 20
60 END
```

5. 
```
10 READ A, B
20 PRINT SQR (A ∧ 2 + B ∧ 2)
30 DATA 3,4,6,8,9,12
40 IF B = 12 THEN 60
50 GOTO 10
60 END
```

6. 
```
10 READ M
20 PRINT 1,6093*M
30 IF M = 3,0 THEN 60
40 GOTO 10
50 DATA 1,0, 1,5, 2,0, 3,0
60 END
```

7. Write a BASIC program for the flowchart in Fig. 18-3.

8. Use the INPUT statement and modify the program in Fig. 18-3 so that the computer will accept the input of three grades from the operator.

9. Use the READ and DATA statements to write a BASIC program that prints in columns, with headings, the diameter and circumference of a circle when the radius = 1, 2, 3, 4, and 5 cm.

10. Use the READ and DATA statements to write a program that prints in columns, with headings, the distance in kilometers and in miles for 20, 40, 60, 80, and 100 mi.

11. Use the INPUT statement to write a BASIC program that prints the solution of $ax + b = 0$, given values of $a$ and $b$.

12. Do the same as in problem 11 for $ax^2 + bx + c = 0$, given values of $a$, $b$, and $c$.

13. Write a BASIC program for the flowchart in Fig. 18-5.

14. Write a BASIC program that prints the sum of the first 100 whole numbers. (See Fig. 18-6.)

15. Write a BASIC program that prints the sum of the squares of the first 10 whole numbers.

16. Write a BASIC program that prints the sum of the first 10 terms of the geometric progression 2, 4, 8, . . . .

17. Given the price of an item, write a BASIC program that prints the final price of the item if it sells for one-third off the retail price and is subject to 5 percent sales tax.

18. Given the price of an automobile, write a BASIC program that prints the price of the new model after 2 years if the inflation rate is 12 percent.

19. Use the READ and DATA statements to write a BASIC program that prints the square roots of 1.2, 2.4, 3.8, 4.4, and 5.1.

20. Use the READ and DATA statements to write a program that prints $\sqrt{x^2 + y^2}$ for $(x, y)$ equal to (1, 0), (3, 3), (1, 2), and (4, 3).

21. Modify the program in Example 18-14 by using the INPUT statement to print the same results for a given input of $T$.

22. Modify the program in Example 18-17 by using the READ and DATA statements to print the same results for the following values of voltage and resistance:

| $E$ | 2.0 | 2.6 | 3.1 | 5.8 |
|---|---|---|---|---|
| $R$ | 5.0 | 8.0 | 12 | 16 |

**23–30:** Write a BASIC program for problems 23 to 30 in Exercise 18-3.

31. Given the circle $x^2 + y^2 = 25$ and a point $(x, y)$, write a BASIC program that tells whether the point lies inside $(x^2 + y^2 < 25)$, outside $(x^2 + y^2 > 25)$, or on the circle. Construct a flowchart first.

32. Use the READ and DATA statements to write a BASIC program that prints the largest number of $\sqrt{2}$, 1.414, and 18/13. Construct a flowchart first.

33. Given angles in the second quadrant of 120°, 135°, 150°, and 180°, write a BASIC program that prints a table of values like that shown in Example 18-18.

34. Given a right triangle with tan $A = 0.40$, write a BASIC program that prints $A$, sin $A$, and cos $A$ in three columns with headings. [*Hint:* Use ATN (X).]

35. Given $x = 0.8$, 1.1, 1.5, 2.2, and 2.5, write a BASIC program that prints $x$, $e^x$, ln $x$, and log $x$ in four columns with headings. [*Note:* log $x = 0.4343(\ln x)$.]

36. Modify the program in Example 18-19 to print the radius and the angle when $G = \pi/2$ or $3\pi/2$.

## 18-5 | *Review Questions*

1. Give an example of a mathematical procedure most effectively done by *(a)* a computer, *(b)* a hand calculator, and *(c)* the human mind.

2. Time how long it takes you to divide 37 into 23,680 by hand. How many times faster is a computer that can do this problem in 50 ns (1 ns $= 10^{-9}$ s)?

3. Suppose a millimeter of magnetic tape contains 80 characters. If 6 characters make up a ''word'' and a tape input reads 900 millimeters per minute, how many words per minute are processed by the tape input?

4. What will the video display show if you type the following?

```
GS←←HAB←VE A NIX←CR←E D←DAU←Y
```

5. What does the computer print when given the command LIST for the following?

```
10  TO ERR
40  IS DIVINE
30  TO FORGET,
40  DIVINE
30  TO FORGIVE,
20  IS HUMAN
```

6. What does the computer print upon execution of the following program?

```
10  LET A = 2
20  LET C = 4
30  LET A = 3
40  PRINT C ∧ 2 - A ∧ 2
50  END
```

7. Write in BASIC: $(x + 2y)^2$.

8. Write this BASIC expression in algebraic notation: $5 * A/(2 + B) - 3$.

What does the computer print upon execution of each program in problems 9 and 10?

9.
```
10  REM TRIG
20  LET A = 4.5
30  LET B = 6.0
40  LET C = 7.5
50  PRINT "SIN A", "COS A"
60  PRINT 4.5/7.5, 6.0/7.5
70  END
```

10.
```
10  PRINT "SERIES CIRCUIT"
20  LET C1 = 0.020
30  LET C2 = 0.030
40  LET C = C1 * C2/(C1 + C2)
50  PRINT "C1", "C2", "C TOTAL"
60  PRINT C1, C2, C
70  END
```

11. Construct a flowchart that shows the steps you go through each morning before breakfast.

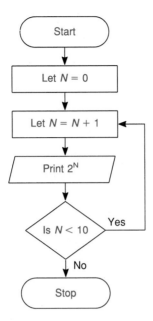

FIG. 18-13 *Flowchart, problems 12 and 21*

**12.** Explain what the flowchart in Fig. 18-13 does.

Construct a flowchart for each program in problems 13 to 16.

**13.** Print the surface area and volume of a cube whose side is 3.5 mm.

**14.** Given a speed in kilometers per hour, print the speed in miles per hour.

**15.** Print the first 10 even numbers, starting with $N = 0$.

**16.** The formula for the distance $s$ above the ground of a body thrown straight up with a velocity of 44 m/s is given by $s = 44t - \frac{1}{2}gt^2$. Given $g = 9.8$ m/s$^2$, print the values of time and distance for $t = 0.0, 1.0, 2.0, 2.5,$ and $3.0$ s in two columns with headings.

What does the computer print upon execution of each program in problems 17 and 18?

**17.**
```
10 LET N = 1
20 LET N = N + 1
30 PRINT N, N ∧ 2
40 IF N < 10 THEN 20
50 END
```

**18.**
```
10 READ R1, R2
20 DATA 1, 2, 4, 3, 3, 1
30 PRINT 1/(1/R1 + 1/R2)
40 IF R1 = 3 THEN 60
```

```
50 GOTO 10
60 END
```

**19.** Use the READ and DATA statements to write a BASIC program that prints the cube roots of 2, 3, 5, 6, 7, and 8.

**20.** Use the READ and DATA statements to write a BASIC program that prints the difference in volume between two cubes whose sides are *(a)* 1 and 2, *(b)* 2 and 3, *(c)* 3 and 4, and *(d)* 4 and 5.

**21.** Write a BASIC program for the flowchart in Fig. 18-13.

**22.** Use the INPUT statement to write a BASIC program that asks the operator to input the difference between 0.099 and 0.113 and prints CORRECT if the input is correct and INCORRECT, TRY AGAIN, if not.

**23.** Write a BASIC program that tells whether a given point lies on the line $y = 2x + 1$.

**24.** Given the values $a$ and $b$ of two legs of a right triangle, write a BASIC program that prints angle $A$, sin $A$, cos $A$, and tan $A$ in four columns with headings. [*Hint:* Use ATN (x).]

# C H A P T E R

# 19

# Limits and the Derivative

## 19-1 | *Limits*

Study the following example closely. It will give you some understanding of the basic concept of a limit. A more complete understanding will come as you study more examples and do many exercises.

**C**alculus is a powerful mathematical tool used in science and technology. It was discovered in the 17th century by both Isaac Newton and Gottfried Leibniz. There are two types of calculus: *Differential calculus* involves the concept of a derivative, which is the rate of change of one quantity with respect to another. *Integral calculus* involves the concept of an integral, which is an infinite sum. Evaluating an integral uses the inverse of finding a derivative. The integral and the derivative both depend upon the concept of a limit. This chapter therefore begins with limits and then studies the derivative and its application.

### Example 19-1

The distance $s$ in meters that a body falls from rest, from a height above the ground, is a function of the time $t$ in seconds given by:

$$s = f(t) = 4.90t^2$$

1. Find the *average velocity* $v_{av}$ during the first second, that is, from $t = 0$ to $t = 1$ s.

2. Find the *instantaneous velocity* $v$ at exactly $t = 1$ s.

3. Show algebraically that $v = \lim\limits_{t \to 1} v_{av} = 9.80$ m/s.

### Solution

1. Find the *average velocity* $v_{av}$ during the first second, that is, from $t = 0$ to $t = 1$ s.

   The average velocity $v_{av}$ is the change in distance divided by the change in time:

$$v_{av} = \frac{\text{change in distance}}{\text{change in time}} = \frac{\Delta s}{\Delta t}$$

where $\Delta$ (delta) means the difference or change. The table below shows the distances traveled by the body during the first 4 s, as calculated from the function $s = 4.90t^2$:

| $t$ | 0 | 1 | 2 | 3 | 4 |
|---|---|---|---|---|---|
| $s$ | 0 | 4.90 | 19.6 | 44.1 | 78.4 |

The average velocity from $t = 0$ to $t = 1$ s is then:

$$v_{av} = \frac{4.90 - 0}{1 - 0} = 4.90 \text{ m/s}$$

**2.** Find the *instantaneous velocity* $v$ at exactly $t = 1$ s.

The velocity of the body is continually increasing and at any instant cannot be calculated directly from the above formula for $v_{av}$. This is because $\Delta t = 0$ and $\Delta s = 0$ at any instant and the fraction 0/0 is undefined. The instantaneous velocity can be determined by using the concept of a limit and calculating the average velocity over smaller and smaller time intervals, using values of $t$ close to $t = 1$.

Let $t$ = some value of time between 0 and 1 s. Then $s = 4.90t^2$ is the corresponding value of the distance between 0 and 4.90 m. The average velocity between the time $t$ and 1 s is therefore:

$$v_{av} = \frac{4.90 - s}{1 - t} = \frac{4.90 - 4.90t^2}{1 - t}$$

Use this formula to compute $v_{av}$ for $t = 0.50$, 0.90, 0.99, 0.999, and 0.9999 s:

| $t$, s | 0.50 | 0.90 | 0.99 | 0.999 | 0.9999 |
|---|---|---|---|---|---|
| $v_{av}$, m/s | 7.35 | 9.31 | 9.75 | 9.795 | 9.7995 |

For example, when $t = 0.99$ s:

$$v_{av} = \frac{4.90 - 4.90(0.99)^2}{1 - 0.99} = \frac{0.0975}{0.01} = 9.75 \text{ m/s}$$

If you continue the table, you will find that as $t$ becomes closer and closer to 1, or as $t$ approaches 1 $(t \rightarrow 1)$, $v_{av}$ approaches 9.80 $(v_{av} \rightarrow 9.80)$.

This idea is expressed by using the concept of a limit:

$$\lim_{t \to 1} v_{av} = \lim_{t \to 1} \frac{4.90 - 4.90t^2}{1 - t} = 9.80 \text{ m/s}$$

It follows that the instantaneous velocity at exactly $t = 1$ s is:

$$v = \lim_{t \to 1} v_{av} = 9.80 \text{ m/s}$$

Note again that $v_{av} = 0/0$ and is undefined at exactly $t = 1$ s.

3. Show algebraically that $v = \lim_{t \to 1} v_{av} = 9.80$ m/s.

Simplify the formula for $v_{av}$ by factoring and reducing the fraction:

$$v = \lim_{t \to 1} v_{av} = \lim_{t \to 1} \frac{4.90 - 4.90t^2}{1 - t}$$

$$= \lim_{t \to 1} \frac{4.90(1 - t^2)}{1 - t}$$

$$= \lim_{t \to 1} \frac{4.90(1 + t)(1 - t)}{1 - t} = \lim_{t \to 1} 4.90(1 + t)$$

It is important to realize that the factor $1 - t$ can be divided out only if $t \neq 1$; otherwise, you are dividing by zero. The condition $t \to 1$ means $t$ can assume any value as close to 1 as you wish, but $t$ cannot equal 1.

As $t \to 1$, $4.90(1 + t) \to 4.90(1 + 1)$; thus, the limit can be found by substituting $t = 1$:

$$v = \lim_{t \to 1} 4.90(1 + 1) = 4.90(2) = 9.80 \text{ m/s}$$

The preceding example leads to the definition of the *limit of a function:*

If a function $f(x) \to L$ [$f(x)$ approaches $L$] as $x \to a$ ($x$ approaches $a$), that is, $f(x)$ becomes infinitesimally close to $L$ as $x$ becomes infinitesimally close to $a$, then:

$$\lim_{x \to a} f(x) = L \tag{19-1}$$

Study the following examples to better understand the concept of a limit.

**Example 19-2**

Find $\lim_{x \to -1} (2x + 5)$.

### Solution

The function $2x + 5$ is defined for $x = -1$. As $x \rightarrow -1$, $2x + 5 \rightarrow 2(-1) + 5$, and the limit can be found by substitution:

$$\lim_{x \to -1} (2x + 5) = 2(-1) + 5 = 3$$

### Example 19-3

Find $\lim_{x \to 0} [(x^2 + x)/x]$.

### Solution

The function $f(x) = (x^2 + x)/x$ is undefined for $x = 0$ since $f(0) = 0/0$. The limit is found by first simplifying the expression for $f(x)$:

$$\lim_{x \to 0} \frac{x^2 + x}{x} = \lim_{x \to 0} \frac{\cancel{x}(x + 1)}{\cancel{x}} = \lim_{x \to 0} (x + 1)$$

Note you can divide out $x$ since $x \neq 0$. As $x \rightarrow 0$, $x + 1 \rightarrow 0 + 1$; therefore:

$$\lim_{x \to 0} (x + 1) = 0 + 1 = 1$$

In elementary algebra the function $f(x) = (x^2 + x)/x$ is considered the same as the function $g(x) = x + 1$, since $f(x)$ can be reduced to $g(x)$. However, this is not exactly correct since $f(0) = 0/0$ and is undefined when $x = 0$, whereas $g(0) = 0 + 1 = 1$.

The two functions are identical for all values of $x$ except $x = 0$:

$$f(x) = \frac{x^2 + x}{x} = x + 1 = g(x) \qquad x \neq 0$$

The limits are the same when $x \rightarrow 0$:

$$\lim_{x \to 0} \frac{x^2 + x}{x} = \lim_{x \to 0} (x + 1) = 1$$

However, $f(0) \neq g(0)$.

In calculus it is important to make these distinctions. The graph of $g(x)$ is a straight line and is said to be continuous, while the graph of $f(x)$ is a straight line *with a point missing* and is discontinuous or not continuous (Fig. 19-1).

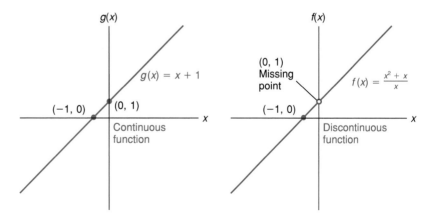

FIG. 19-1 *Continuous and discontinuous functions, Example 19-3*

The preceding example leads to the definition of the *continuity of a function:*

A function $f(x)$ is *continuous at a point* $x = a$ if $f(a)$ exists, $\lim\limits_{x \to a} f(x)$

exists, and

$$\lim\limits_{x \to a} f(x) = f(a) \tag{19-2}$$

A function is *continuous in an interval* if it is continuous at every point in the interval.

It follows from the definition of continuity (19-2) that when $f(x)$ is continuous at $x = a$, you can find $\lim\limits_{x \to a} f(x)$ by substituting $a$ for $x$ in $f(x)$, that is, by finding $f(a)$ as in Example 19-2.

### Example 19-4

Find $\lim\limits_{x \to 3} \sqrt{x^2 + 1}$.

### Solution

This function is continuous at $x = 3$, and you can find the limit by substitution:

$$\lim\limits_{x \to 3} \sqrt{x^2 + 1} = \sqrt{3^2 + 1} = \sqrt{10} \approx 3.16$$

### Example 19-5

Find $\lim\limits_{x \to -2} \dfrac{x^2 + 4x + 4}{x + 2}$.

## Solution

This function is not continuous at $x = -2$ since the denominator is zero at that $x$ value. Find the limit by simplifying the expression:

$$\lim_{x \to -2} \frac{x^2 + 4x + 4}{x + 2} = \lim_{x \to -2} \frac{(x + 2)^2}{x + 2}$$

$$= \lim_{x \to -2} (x + 2) = -2 + 2 = 0$$

Note that you can divide out the factor $x + 2$ because $x \neq -2$ and you are not dividing by zero.

### Example 19-6

Find $\lim\limits_{h \to 0} \dfrac{(x + h)^2 - x^2}{h}$.

## Solution

$$\lim_{h \to 0} \frac{(x + h)^2 - x^2}{h} = \lim_{h \to 0} \frac{x^2 + 2xh + h^2 - x^2}{h}$$

$$= \lim_{h \to 0} \frac{h(2x + h)}{h} = 2x + 0 = 2x$$

Limits where $x$ becomes infinitely large $(x \to \infty)$ are important in curve sketching and in integral calculus. Study the following examples.

### Example 19-7

Find $\lim\limits_{x \to \infty} \dfrac{5x + 1}{x^2}$.

## Solution

As $x \to \infty$, the numerator and denominator both become infinitely large, and the limit cannot be found with the fraction in this form. However, the limit can be found by separating the fraction into two fractions:

$$\lim_{x \to \infty} \frac{5x + 1}{x^2} = \lim_{x \to \infty} \left( \frac{5x}{x^2} + \frac{1}{x^2} \right)$$

As $x \to \infty$, note that $5/x \to 0$ and $1/x^2 \to 0$. Therefore:

$$\lim_{x \to \infty} \left( \frac{5}{x} + \frac{1}{x^2} \right) = 0 + 0 = 0$$

## Example 19-8

Find $\lim\limits_{x \to \infty} \dfrac{2x^2 + 1}{3x^2 - x}$.

## Solution

The numerator and denominator both become infinitely large as $x \to \infty$. The limit is found by dividing each term *by the largest power of x*, which is $x^2$:

$$\lim_{x \to \infty} \frac{2x^2 + 1}{3x^2 - x} = \lim_{x \to \infty} \frac{\dfrac{2x^2}{x^2} + \dfrac{1}{x^2}}{\dfrac{3x^2}{x^2} - \dfrac{x}{x^2}}$$

$$= \lim_{x \to \infty} \frac{2 + \dfrac{1}{x^2}}{3 - \dfrac{1}{x}} = \frac{2 + 0}{3 - 0} = \frac{2}{3}$$

**Exercise 19-1**

1. (a) Apply the method shown in Example 19-1(1), and find the average velocity of the body during the second second, that is, from $t = 1$ to $t = 2$ s. (b) Applying the method shown in Example 19-1(2), find the instantaneous velocity at exactly $t = 2$ s by computing the average velocity between $t = 2$ s and $t = 1.50$, 1.90, 1.99, 1.999, and 1.9999 s and finding the limit as $t \to 2$.

2. The height above the ground of a rocket shot straight up is given by the function $s = 160t - 16t^2$, where $s = $ distance in feet and $t = $ time in seconds. (a) Find the average velocity during the first second, that is, from $t = 0$ to $t = 1$ s. (b) Find the instantaneous velocity at exactly $t = 1$ s by computing the average velocity between $t = 1$ s and $t = 0.50$, 0.90, 0.99, 0.999, and 0.9999 s and finding the limit as $t \to 1$. (See Example 19-1.)

3. If:

$$f(x) = \frac{x - 4}{x^2 - 4x} \qquad \text{show that} \qquad \lim_{x \to 4} f(x) = 0.25$$

by computing $f(x)$ for $x = 3.50$, 3.90, 3.99, 3.999, and 3.9999.

4. If:

$$f(x) = \frac{x - 1}{\sqrt{x} - 1} \qquad \text{show that} \qquad \lim_{x \to 1} f(x) = 2$$

by computing $f(x)$ for $x = 1.50$, 1.10, 1.01, 1.001, and 1.0001.

In problems 5 to 32 find each limit.

5. $\lim\limits_{x \to 4} (x - 6)$             6. $\lim\limits_{x \to -3} (3x^2 + 3)$

**7.** $\lim\limits_{x \to 0} \dfrac{x}{x+2}$

**8.** $\lim\limits_{x \to 1/2} \dfrac{x-1}{x}$

**9.** $\lim\limits_{x \to 1/4} \sqrt{x}$

**10.** $\lim\limits_{x \to 0.6} \sqrt{1-x^2}$

**11.** $\lim\limits_{x \to 0} \dfrac{x-x^2}{x}$

**12.** $\lim\limits_{x \to 0} \dfrac{x}{2x^2+2x}$

**13.** $\lim\limits_{x \to 3} \dfrac{x^2-9}{x-3}$

**14.** $\lim\limits_{x \to -1} \dfrac{x^2+2x+1}{x+1}$

**15.** $\lim\limits_{x \to 2} \dfrac{x^2+x-6}{x-2}$

**16.** $\lim\limits_{x \to -4} \dfrac{x+4}{x^2-16}$

**17.** $\lim\limits_{k \to 0} \dfrac{2.5k+3.5k^2}{0.5k}$

**18.** $\lim\limits_{m \to 0.1} \dfrac{m^2-0.01}{m-0.1}$

**19.** $\lim\limits_{x \to 1} \dfrac{x^2-x}{x^2-1}$

**20.** $\lim\limits_{x \to -2} \dfrac{x^2+3x+2}{x^2+x-2}$

**21.** $\lim\limits_{h \to 0} \dfrac{(1+h)^2-1}{h}$

**22.** $\lim\limits_{t \to 1} \dfrac{1/t-1}{t-1}$

For problems 23 and 24 rationalize the denominator.

**23.** $\lim\limits_{r \to 0} \dfrac{r}{\sqrt{r}}$

**24.** $\lim\limits_{m \to 4} \dfrac{m-4}{\sqrt{m}-2}$

**25.** $\lim\limits_{v \to v_0} \dfrac{v^2-v_0^2}{v-v_0}$

**26.** $\lim\limits_{\Delta x \to 0} \dfrac{(x+\Delta x)^3-x^3}{\Delta x}$

**27.** $\lim\limits_{x \to \infty} \dfrac{2}{x}$

**28.** $\lim\limits_{x \to \infty} \dfrac{2+x}{x}$

**29.** $\lim\limits_{n \to \infty} \dfrac{n-1}{n^2}$

**30.** $\lim\limits_{k \to \infty} \dfrac{k+3}{2k+2}$

**31.** $\lim\limits_{y \to \infty} \dfrac{3y^2+2}{y^2-y}$

**32.** $\lim\limits_{n \to \infty} \dfrac{1-n^2}{n^2-n^3}$

**33.** Given

$$f(x) = \dfrac{x^2+2x}{x} \qquad \text{and} \qquad g(x) = x+2$$

*(a)* Show that $f(0) \neq g(0)$.

*(b)* Show that

$$\lim_{x \to 0} f(x) = \lim_{x \to 0} g(x)$$

*(c)* Draw the graphs of $f(x)$ and $g(x)$ and tell which function is not continuous and why. (See Example 19-3.)

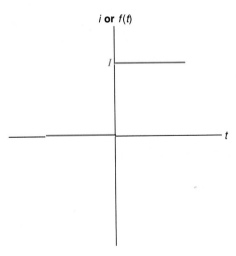

FIG. 19-2 *Step function, problem 34*

**34.** In a dc circuit containing a resistance, the current $i$ rises almost instantaneously as soon as the switch is closed. This can be described by a step function (Fig. 19-2)

$$i = \begin{cases} f(t) = 0 & t \le 0 \\ f(t) = I & t > 0 \end{cases}$$

This means that the current $i = 0$ when $t \le 0$ and $i = I$ when $t > 0$. Show that $\lim_{t \to 0} f(t)$ does not exist and $f(t)$ is not continuous at $t = 0$ by showing that

$$\lim_{t \to 0^+} f(t) \ne \lim_{t \to 0^-} f(t)$$

where $t \to 0^+$ means $t$ approaches zero from the positive direction and $t \to 0^-$ means $t$ approaches zero from the negative direction.

**35.** A ball is dropped from 10 m above the ground and bounces up to a height of 5 m. On each successive bounce the height is one-half the previous height. *(a)* If $h =$ height and $n =$ number of bounces, find $\lim_{n \to \infty} h$. *(b)* If the total distance traveled by the ball is $S = 10 + 5 + 2.5 + \cdots$, find $\lim_{n \to \infty} S$. (*Note:* $S$ is the sum of a geometric series.)

**36.** The perimeter of a polygon with $n$ sides inscribed in a circle of radius 1 unit is given by

$$P = 2n \sin \frac{\pi}{n}$$

where *the angle $\pi/n$ is in radians.* Calculate $P$ for $n = 10, 20, 50, 100,$ and $200$, and show that $\lim_{n \to \infty} P = 2\pi \approx 6.283$.

C **37.** The number $e$ is defined as $e = \lim\limits_{n \to 0} (1 + n)^{1/n}$. Calculate $(1 + n)^{1/n}$ for $n = 1, 0.1,$ $0.01, 0.001, 0.0001,$ and show that $e \approx 2.718$.

C **38.** In a dc circuit containing a resistance and an inductance, the current $t$ seconds after the switch is closed is given by $I = 2.5(1 - e^{-20t})$, where $e \approx 2.718$. Calculate values of $I$ for $t = 0.05, 0.10, 0.15, 0.20,$ and $0.25$ s, and find $\lim\limits_{t \to \infty} I$.

## 19-2 | *Slope and the Derivative*

The derivative of a function uses the concept of a limit. It is related to the slope of the tangent line at a point on the curve of the function. The following example shows how the slope of a tangent line is determined by finding a limit. This leads to the definition of the derivative.

### Example 19-9

**1.** Find the slope of the tangent line at the point $x = 1$ on the parabola $y = f(x) = x^2$.

**2.** Find a formula for the slope at any point $(x, y)$ on the curve $y = f(x) = x^2$.

### Solution

**1.** Find the slope of the tangent line at the point $x = 1$ on the parabola $y = f(x) = x^2$.

The parabola and the tangent line are shown in Fig. 19-3. The slope $m$ of a straight line is defined in Sec. 14-1 as:

$$m = \frac{\text{change in } y}{\text{change in } x} = \frac{y_2 - y_1}{x_2 - x_1} = \frac{\Delta y}{\Delta x} \tag{19-3}$$

The delta ($\Delta$) notation is a convenient one that is used in calculus. Formula (19-3) requires two points to evaluate the slope; however, only one point, $(1, 1)$, on the tangent line is known. The slope of the tangent line is found by first setting up an expression for the slope of a line passing through $(1, 1)$ and a variable point on the curve near $(1, 1)$. Then the limit of this expression is found as the variable point approaches $(1, 1)$. Figure 19-4 shows the portion of the curve containing the fixed point $(1, 1)$ and a variable point near $(1, 1)$. If $\Delta x$ is the difference in $x$ between the two points, then $1 + \Delta x$ is the $x$ coordinate of the variable point. Similarly, if $\Delta y$ is the difference in $y$ between the two points, then $1 + \Delta y$ is the $y$ coordinate of the variable point and is equal to $f(1 + \Delta x)$:

$$1 + \Delta y = f(1 + \Delta x) = (1 + \Delta x)^2$$
$$= 1 + 2\,\Delta x + (\Delta x)^2$$

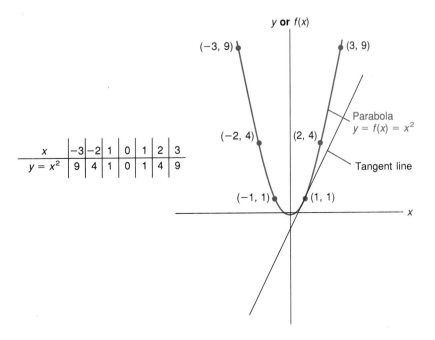

FIG. 19-3 *Parabola and tangent line, Example 19-9*

| x | -3 | -2 | 1 | 0 | 1 | 2 | 3 |
|---|----|----|----|----|----|----|----|
| y = x² | 9 | 4 | 1 | 0 | 1 | 4 | 9 |

Note that $\Delta x$ and $\Delta y$ are variables like $x$ and $y$. You cannot separate the $\Delta$ from the $x$ or $y$. Then:

$$\Delta y = 2\,\Delta x + (\Delta x)^2$$

and the slope between the two points in terms of $\Delta x$ is:

$$\frac{\Delta y}{\Delta x} = \frac{2\,\Delta x + (\Delta x)^2}{\Delta x} = \frac{\Delta x(2 + \Delta x)}{\Delta x} = 2 + \Delta x \qquad (\Delta x \neq 0)$$

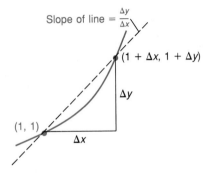

FIG. 19-4 *Portion of parabola $y = x^2$ near point (1, 1), Example 19-9(1)*

Notice you can divide out $\Delta x$, providing $\Delta x \neq 0$. As the variable point approaches (1, 1), $\Delta x \to 0$ and $\Delta y \to 0$. The slope of the tangent line is then:

$$\lim_{\Delta x \to 0} \frac{\Delta y}{\Delta x} = \lim_{\Delta x \to 0} (2 + \Delta x) = 2 + 0 = 2$$

Compare this to finding the instantaneous velocity in Example 19-1(3). The notation is different, but the process is the same.

2. Find a formula for the slope at any point $(x, y)$ on the curve $y = f(x) = x^2$.

The slope at a point means the slope of the tangent line at that point. Figure 19-5 shows a portion of the curve near a point $(x, y)$. The procedure is similar to that in (1). Choose a variable point near $(x, y)$. Let $\Delta x$ be the difference in $x$ and $\Delta y$ be the difference in $y$ between the two points. The coordinates of the variable point are then $(x + \Delta x, y + \Delta y)$. Express $\Delta y$ in terms of $x$ and $\Delta x$ as follows:

$$y + \Delta y = f(x + \Delta x)$$
$$\Delta y = f(x + \Delta x) - y$$

Since $y = f(x)$:

$$\Delta y = f(x + \Delta x) - f(x)$$

Use this formula to evaluate $\Delta y$:

$$\Delta y = (x + \Delta x)^2 - x^2$$
$$= x^2 + 2x\,\Delta x + (\Delta x)^2 - x^2 = 2x\,\Delta x + (\Delta x)^2$$

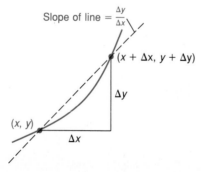

FIG. 19-5 *Portion of parabola* $y = x^2$ *near point* $(x, y)$, *Example 19-9(2)*

Find $\Delta y/\Delta x$:

$$\frac{\Delta y}{\Delta x} = \frac{2x\,\Delta x + (\Delta x)^2}{\Delta x} = \frac{\Delta x(2x + \Delta x)}{\Delta x}$$

$$= 2x + \Delta x \qquad (\Delta x \neq 0)$$

Then find $\lim\limits_{\Delta x \to 0} (\Delta y/\Delta x)$:

$$\lim_{\Delta x \to 0} \frac{\Delta y}{\Delta x} = \lim_{\Delta x \to 0} (2x + \Delta x) = 2x + 0 = 2x$$

This process is called the *delta process*. The formula obtained is the formula for the slope of any point $(x, y)$ on the parabola $y = x^2$. For example, at $x = 1$, slope $= 2(1) = 2$, which is the result obtained in (1) above. At $x = 2$, slope $= 2(2) = 4$; and at $x = 0$, slope $= 2(0) = 0$. These results agree with the shape of the curve in Fig. 19-3.

The formula obtained is called the *derivative of y with respect to x* and is written $f'(x) = 2x$. It is a formula that expresses the *instantaneous rate of change of y with respect to x*. The *derivative $f'(x)$ of a function $f(x)$* is defined as follows:

$$f'(x) = \lim_{\Delta x \to 0} \frac{\Delta y}{\Delta x} = \lim_{\Delta x \to 0} \frac{f(x + \Delta x) - f(x)}{\Delta x} \qquad (19\text{-}4)$$

Other notations used for the derivative in addition to $f'(x)$ (or just $f'$) are:

$$y' \qquad \frac{dy}{dx} \qquad D_x y$$

The notation $dy/dx$ is useful because it indicates the variable that the derivative is taken with respect to. The importance of this notation is shown in Sec. 19-6.

Note that *slope is a geometric model of the derivative* (Example 19-9) and *instantaneous velocity is a physical model* (Example 19-1). These ideas are discussed further below and in subsequent sections, particularly Sec. 19-4.

### Example 19-10

Find the derivative $f'(x)$ for $y = f(x) = 3x^2 + x$, using the delta process.

### Solution

Find $\Delta y$:

$$\Delta y = f(x + \Delta x) - f(x)$$

$$= [3(x + \Delta x)^2 + (x + \Delta x)] - (3x^2 + x)$$

$$= 3[x^2 + 2x\,\Delta x + (\Delta x)^2] + x + \Delta x - 3x^2 - x$$
$$= 3x^2 + 6x\,\Delta x + 3(\Delta x)^2 + x + \Delta x - 3x^2 - x$$
$$= 6x\,\Delta x + 3(\Delta x)^2 + \Delta x$$

Then find $\Delta y/\Delta x$ and $f'(x) = \lim\limits_{\Delta x \to 0} (\Delta y/\Delta x)$:

$$\frac{\Delta y}{\Delta x} = \frac{6x\,\Delta x + 3(\Delta x)^2 + \Delta x}{\Delta x}$$
$$= 6x + 3\,\Delta x + 1 \qquad\qquad (\Delta x \neq 0)$$

Observe that you can divide $\Delta x$ into each term in the numerator without factoring out $\Delta x$ first:

$$f'(x) = \lim\limits_{\Delta x \to 0} \frac{\Delta y}{\Delta x} = \lim\limits_{\Delta x \to 0} (6x + 3\,\Delta x + 1) = 6x + 1$$

The derivative $f'(x) = 6x + 1$ represents a formula for the slope of any point $x$ on the curve $f(x) = 3x^2 + x$. For example, at $x = -1$, the slope is:

$$f'(-1) = 6(-1) + 1 = -5$$

### Example 19-11
Find $y' = dy/dx$ for $y = f(x) = mx + b$, where $m$ and $b$ are constants.

### Solution
This is the slope-intercept form of the equation of a straight line given in Sec. 14-1. Since $m$ represents the slope of the line, $dy/dx$ should equal $m$. By using the delta process this is shown as follows:

Find $\Delta y$:

$$\Delta y = f(x + \Delta x) - f(x)$$
$$= [m(x + \Delta x) + b] - (mx + b)$$
$$= m\,\Delta x$$

Find $\Delta y/\Delta x$ and $dy/dx$:

$$\frac{\Delta y}{\Delta x} = \frac{m\,\Delta x}{\Delta x} = m$$

$$y' = \frac{dy}{dx} = \lim\limits_{\Delta x \to 0} \frac{\Delta y}{\Delta x} = \lim\limits_{\Delta x \to 0} m = m$$

Since $m$ is a constant and does not vary with $\Delta x$, the limit is simply $m$, as expected.

## Example 19-12

Find $f'(x)$ for $y = f(x) = x^3 - 3$.

## Solution

Find $\Delta y$:

$$\Delta y = f(x + \Delta x) - f(x)$$
$$= [(x + \Delta x)^3 - 3] - (x^3 - 3)$$

Multiply $x + \Delta x$ by itself 3 times and simplify:

$$\Delta y = \cancel{x^3} + 3x^2\,\Delta x + 3x(\Delta x)^2 + (\Delta x)^3 - \cancel{3} - \cancel{x^3} + \cancel{3}$$
$$= 3x^2\,\Delta x + 3x(\Delta x)^2 + (\Delta x)^3$$

Find $\Delta y/\Delta x$ and $f'(x)$:

$$\frac{\Delta y}{\Delta x} = \frac{3x^2\,\cancel{\Delta x} + 3x(\Delta x)^{\cancel{2}} + (\Delta x)^{\cancel{3}^{2}}}{\cancel{\Delta x}}$$
$$= 3x^2 + 3x\,\Delta x + (\Delta x)^2 \qquad\qquad (\Delta x \neq 0)$$
$$f'(x) = \lim_{\Delta x \to 0} \frac{\Delta y}{\Delta x} = \lim_{\Delta x \to 0}\,[3x^2 + 3x\,\Delta x + (\Delta x)^2] = 3x^2$$

The next example involves combining fractions to find a derivative.

## Example 19-13

Find $I' = dI/dR$ for $I = f(R) = 2/R$.

## Solution

In this example $R$ takes the place of $x$, and $I$ takes the place of $y$. Find $\Delta I$:

$$\Delta I = f(R + \Delta R) - f(R)$$
$$= \frac{2}{R + \Delta R} - \frac{2}{R}$$
$$= \frac{2R - 2(R + \Delta R)}{R(R + \Delta R)} = \frac{-2\,\Delta R}{R(R + \Delta R)}$$

Find $\Delta I/\Delta R$ and $dI/dR$:

$$\frac{\Delta I}{\Delta R} = \frac{-2\,\cancel{\Delta R}}{R(R + \Delta R)} \cdot \frac{1}{\cancel{\Delta R}} = \frac{-2}{R(R + \Delta R)}$$

$$I' = \frac{dI}{dR} = \lim_{\Delta R \to 0} \frac{\Delta I}{\Delta R} = \lim_{\Delta R \to 0} \frac{-2}{R(R + \Delta R)} = \frac{-2}{R^2}$$

**Exercise 19-2**

1. *(a)* Apply the method shown in Example 19-9(1), and find the slope of the tangent line at $x = 1$ on the curve $y = f(x) = 2x^2$. Sketch the curve and the tangent line. *(b)* Apply the method shown in Example 19-9(2), and find a formula for the slope at any point $(x, y)$ on $f(x)$.

2. Do the same as in problem 1 for the function $f(x) = x^3$.

3. *(a)* Using the method of Example 19-9(2), find a formula for the slope at any point $(x, y)$ for the function $f(x) = 1 - x^2$. *(b)* Use the formula in *(a)* to find the slope of the curve at $x = -2$, 0, and 2. That is, find $f'(-2)$, $f'(0)$, and $f'(2)$. Sketch the curve and the tangent lines.

4. Do the same as in problem 3 for $f(x) = x^3 - 4x$.

In problems 5 to 26, find the derivative of each function, using the delta process.

5. $y = f(x) = 5x + 1$

6. $y = f(x) = 2 - 2x$

7. $y = f(x) = 3x^2$

8. $y = f(x) = x^2 - 2$

9. $y = f(x) = 2x^2 + 3x + 1$

10. $d = f(x) = x - x^2$

11. $x = f(y) = 3y^2 - y + 3$

12. $v = f(t) = 4 - t^2$

13. $s = f(t) = t^3 + 1$

14. $y = f(x) = x - x^3$

15. $y = f(x) = \dfrac{1}{x}$

16. $y = f(x) = \dfrac{-2}{x}$

17. $F = f(C) = 1.8C + 32$

18. $P = f(I) = 5.5I^2 + 1.1$

19. $q = f(t) = 4.9t^2 - 6.2t$

20. $V = f(r) = \dfrac{1}{3}\pi r^3$

21. $y = f(x) = \dfrac{1}{x - 1}$

22. $y = f(x) = \dfrac{x}{x + 1}$

23. $y = f(x) = \dfrac{1}{x^3}$

24. $y = f(x) = x - \dfrac{1}{x}$

25. $y = f(x) = ax^2 + bx + c$ ($a$, $b$, $c$ constants)

26. $R = f(A) = \dfrac{\rho l}{A}$ ($\rho$, $l$ constants)

27. *(a)* Find the derivative of $y = f(x) = \sqrt{x}$. (Multiply top and bottom of $\Delta y/\Delta x$ by $\sqrt{x + \Delta x} + \sqrt{x}$ to help find the limit.) *(b)* Show that $f'(0)$ does not exist.

28. Do the same as in problem 27 for $y = f(x) = 1/\sqrt{x}$.

29. The distance $s$ in feet above the ground of a surface-to-air missile is given by $s = f(t) = -16t^2 + 320t - 20$, where $t =$ time in seconds. *(a)* Sketch $s$ vs. $t$ for $t = 0$, 5, 10, 15, and 20. *(b)* The derivative $s' = f'(t)$ is the instantaneous rate of change of distance with respect to time and therefore is a formula for the vertical velocity $v$. Find the formula for $v$. *(c)* Find $v$ when $t = 0$, 10, and 20 s.

30. The velocity $v$ (in meters per second) of a car during its travel around a curve is given by $v = f(t) = 3t^2 - 15t + 30$, where the time $t$ varies from $t = 0$ to $t = 5$ s. *(a)* Sketch $v$ vs. $t$ for $t = 0$, 1, 2, 3, 4, and 5 s. *(b)* The derivative $v' = f'(t)$ is the instantaneous rate of change of velocity with respect to time and therefore is a

formula for the acceleration $a$. Find the formula for $a$. *(c)* Find $a$ when $t = 0, 3$, and 5 s.

**31.** The formula for the area of a circle is $A = \pi r^2$. The derivative $A' = dA/dr$ is the instantaneous rate of change of area with respect to the radius. *(a)* Show that $A'$ is the formula for the circumference of a circle. *(b)* Find $A'$ when $r = 5$.

**32.** At a constant temperature, Boyle's law states that the pressure $P$ of a gas is inversely proportional to the volume $V$:

$$P = \frac{k}{V}$$

where $k$ is the constant of variation. The derivative $P' = dP/dV$ is the instantaneous rate of change of the pressure with respect to the volume. *(a)* Find the formula for $P'$. *(b)* If $k = 1300$, find $P'$ when $V = 2.5$.

## 19-3 | *Derivative of a Polynomial*

To better understand the concept of a derivative, it is useful to study many of its applications. Some are shown in Sec. 19-4, and more are studied in Chap. 20. However, before these applications are presented, this section develops the formulas for differentiating any polynomial. These formulas allow you to find the derivative of any polynomial without having to use the delta process each time. Therefore, with these formulas, you will be able to apply and understand the new ideas more effectively in subsequent sections.

### Example 19-14
Find the derivative of the *constant function* $y = f(x) = c$.

### Solution
The equation $y = c$ is the equation of a horizontal line whose slope is zero. Therefore the derivative should be zero as follows:

$$\Delta y = f(x + \Delta x) - f(x)$$
$$\Delta y = c - c = 0$$
$$\frac{\Delta y}{\Delta x} = \frac{0}{\Delta x} = 0$$
$$f'(x) = \lim_{\Delta x \to 0} \frac{\Delta y}{\Delta x} = \lim_{\Delta x \to 0} 0 = 0$$

Another way to think about this result is that since the instantaneous rate of change of a constant is zero, *the derivative of a constant is zero:*

$$\text{If } y = c, \text{ then } y' = 0. \qquad (c = \text{constant}) \qquad (19\text{-}5)$$

### Example 19-15
Find the derivative of $y = \frac{2}{3}$.

### Solution
Apply formula (19-5):

$$y' = 0$$

### Example 19-16
Find the derivative of the *power function* $y = f(x) = x^n$.

### Solution

$$\Delta y = f(x + \Delta x) - f(x) = (x + \Delta x)^n - x^n$$

To expand $(x + \Delta x)^n$, use the binomial formula (16-12) from Sec. 16-3:

$$\Delta y = \left[ \cancel{x^n} + nx^{n-1}\,\Delta x + \frac{n(n-1)}{2}x^{n-2}\,(\Delta x)^2 + \cdots + (\Delta x)^n \right] - \cancel{x^n}$$

$$= nx^{n-1}\,\Delta x + \frac{n(n-1)}{2}x^{n-2}\,(\Delta x)^2 + \cdots + (\Delta x)^n$$

Divide each term by $\Delta x$ to obtain:

$$\frac{\Delta y}{\Delta x} = nx^{n-1} + \frac{n(n-1)}{2}x^{n-2}\,\Delta x + \cdots + (\Delta x)^{n-1}$$

Only the first term above needs to be considered to find the derivative. All the other terms contain $\Delta x$ and approach zero when $\Delta x \to 0$:

$$f'(x) = \lim_{\Delta x \to 0} \frac{\Delta y}{\Delta x} = nx^{n-1}$$

This very useful formula is called the *power formula:*

$$\text{If } y = x^n, \text{ then } y' = nx^{n-1}. \qquad (n = \text{positive integer}) \qquad (19\text{-}6)$$

The power formula (19-6) is proved in Example 19-16 for $n$ a positive integer. The formula also works for $n = 0$, which is the constant function shown in Example 19-14. In Sec. 19-5 the formula is proved for $n$ a negative integer and in Sec. 19-6 for $n$ equal to any rational number.

### Example 19-17
Find $y'$ for $y = x$.

**Solution**

Apply formula (19-6) where $n = 1$:

$$y' = (1)x^{1-1} = 1x^0 = 1(1) = 1$$

This is to be expected since the slope of the line $y = x$ is 1.

### Example 19-18

Find $f'(x)$ for $f(x) = x^4$.

**Solution**

Apply formula (19-6):

$$f'(x) = 4x^{4-1} = 4x^3$$

Two basic rules are necessary before you can apply formula (19-6) to differentiate any polynomial. The first rule is:

The derivative of a function times a constant is equal to the constant times the derivative of the function:

$$\text{If } u \text{ is a function of } x \text{ and } y = c \cdot u, \text{ then } y' = c \cdot u'. \qquad \text{(19-7)}$$

### Example 19-19

Find $y'$ for $y = 5x^2$

**Solution**

Apply formulas (19-6) and (19-7) with $c = 5$ and $u = x^2$:

$$y' = 5u' = 5(2x) = 10x$$

## Proof of Formula (19-7)

$$y = f(x) = c \cdot u(x)$$
$$\Delta y = f(x + \Delta x) - f(x)$$
$$= c \cdot u(x + \Delta x) - c \cdot u(x)$$

Let $\Delta u = u(x + \Delta x) - u(x)$, where $\Delta u$ is determined by $\Delta x$:

$$\Delta y = c[u(x + \Delta x) - u(x)] = c \cdot \Delta u$$

Then:

$$\frac{\Delta y}{\Delta x} = c\left(\frac{\Delta u}{\Delta x}\right)$$

It can be shown that the limit of a constant times a function equals the constant times the limit of the function. Therefore:

$$y' = \lim_{\Delta x \to 0} \frac{\Delta y}{\Delta x} = \lim_{\Delta x \to 0} c\left(\frac{\Delta u}{\Delta x}\right) = c\left(\lim_{\Delta x \to 0} \frac{\Delta u}{\Delta x}\right)$$

The last limit is the definition of the derivative (19-4) applied to $u(x)$. Hence:

$$y' = c \cdot u'$$

The second rule is as follows:

The derivative of the sum of two functions equals the sum of the derivatives of the functions:

If $u$ and $v$ are functions of $x$, and $y = u + v$, then $y' = u' + v'$.  (19-8)

### Example 19-20
Find $f'(r)$ for $f(r) = 2r^3 - 3r^2$.

### Solution
Apply formulas (19-6), (19-7), and (19-8):

$$f'(r) = 2(3r^2) - 3(2r) = 6r^2 - 6r$$

## Proof of Formula (19-8)

$$y = f(x) = u(x) + v(x)$$
$$\Delta y = f(x + \Delta x) - f(x)$$
$$= [u(x + \Delta x) + v(x + \Delta x)] - [u(x) + v(x)]$$

Let $\Delta u = u(x + \Delta x) - u(x)$ and $\Delta v = v(x + \Delta x) - v(x)$, where $\Delta u$ and $\Delta v$ are determined by $\Delta x$. Then:

$$\Delta y = u(x + \Delta x) - u(x) + v(x + \Delta x) - v(x) = \Delta u + \Delta v$$
$$\frac{\Delta y}{\Delta x} = \frac{\Delta u}{\Delta x} + \frac{\Delta v}{\Delta x}$$

It can be shown that the limit of the sum of two functions equals the sum of the limits of each function. Therefore:

$$y' = \lim_{\Delta x \to 0} \frac{\Delta y}{\Delta x} = \lim_{\Delta x \to 0} \left(\frac{\Delta u}{\Delta x} + \frac{\Delta v}{\Delta x}\right) = \lim_{\Delta x \to 0} \frac{\Delta u}{\Delta x} + \lim_{\Delta x \to 0} \frac{\Delta v}{\Delta x}$$

The last two limits are the derivatives of $u(x)$ and $v(x)$ by formula (19-4). Hence:

$$y' = u' + v'$$

This rule can be extended to three or more functions, and in general:

The derivative of a sum of functions equals the sum of the derivatives of the functions.           (19-9)

Rules (19-5) through (19-9) enable you to differentiate any polynomial function:

$$y = f(x) = a_n x^n + a_{n-1} x^{n-1} + \cdots + a_1 x + a_0$$

Study the following examples, which apply all the above rules.

### Example 19-21
Find $y'$ for $y = 4x^3 - 5x^2 + 3x - 7$.

### Solution
Apply formulas (19-5) through (19-9):

$$y' = 4(3x^2) - 5(2x^1) + 3(x^0) - 0 = 12x^2 - 10x + 3$$

### Example 19-22
Find $dy/dx$ for $f(x) = ax^2 + bx + c$, where $a$, $b$, and $c$ are constants.

### Solution

$$\frac{dy}{dx} = a(2x) + b(1x^0) + 0 = 2ax + b$$

### Example 19-23
Given the parabola $y = f(x) = x^2 + 4x + 3$:

**1.** Find $f'(x)$.
**2.** Find the slope of the parabola at $x = -4$, $-2$, and 0.

### Solution
**1.** Find $f'(x)$.

$$f'(x) = 2x + 4$$

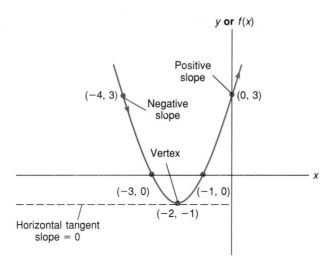

FIG. 19-6 *Parabola $y = x^2 + 4x + 3$, Example 19-23*

**2.** Find the slope of the parabola at $x = -4$, $-2$, and 0.

$$f'(-4) = 2(-4) + 4 = -4$$
$$f'(-2) = 2(-2) + 4 = 0$$
$$f'(0) = 2(0) + 4 = 4$$

The parabola has a negative slope at $x = -4$ because $f(x)$ is decreasing (Fig. 19-6). The curve has a zero slope (horizontal tangent) at $x = -2$. This is the vertex of the parabola. At $x = 0$, the curve has a positive slope since $f(x)$ is increasing.

### Example 19-24

The linear displacement (in feet) of a machine robot from its starting point is given by $s = t^3 - 6t^2$, where $t =$ time in seconds.

**1.** Find $s' = ds/dt$.

**2.** Find $v$ when $t = 0$, 2, 4, and 6 s.

### Solution

**1.** Find $s' = ds/dt$.

Since $s'$ is the instantaneous rate of change of distance with respect to time, it is a formula for the velocity $v$ of the robot:

$$v = s' = 3t^2 - 12t$$

**2.** Find $v$ when $t = 0, 2, 4,$ and 6 s.

$$t = 0: \qquad v = 3(0^2) - 12(0) = 0 \text{ ft/s}$$
$$t = 2: \qquad v = 3(2^2) - 12(2) = -12 \text{ ft/s}$$
$$t = 4: \qquad v = 3(4^2) - 12(4) = 0 \text{ ft/s}$$
$$t = 6: \qquad v = 3(6^2) - 12(6) = 36 \text{ ft/s}$$

At $t = 0$ the robot's velocity is zero, and at $t = 2$ the robot is moving in a negative direction (usually to the left). At $t = 4$ the robot has stopped momentarily and changed direction, and at $t = 6$ the robot is moving in a positive direction.

**Exercise 19-3**

In problems 1 to 22, find the derivative of each function.

**1.** $y = -3$

**2.** $y = x^5$

**3.** $f(x) = \dfrac{1}{3}x^3$

**4.** $f(x) = -1.5x^4$

**5.** $y = 2x^2 - 6x + 1$

**6.** $s = 3t^3 - 4t$

**7.** $f(x) = 2x^5 - 6x^3 + x$

**8.** $f(x) = 4x^4 + x^2 + 7$

**9.** $q = \dfrac{t - t^3}{3}$

**10.** $F = \dfrac{1}{2}(x^3 - x^2)$

**11.** $m = \dfrac{v^2(v^2 + 1)}{v}$

**12.** $H = \dfrac{1}{T}(T^3 - T)$

**13.** $M = 1.0d^4 - 2.3d^2 + 7.5$

**14.** $a = 5.5 - 3.3t^3$

**15.** $f(n) = \dfrac{3}{4}n^2 - \dfrac{1}{6}n^3$

**16.** $V = 2\pi r^2 + 5\pi r$

**17.** $f(y) = (y - 1)^3$ (*Hint:* Multiply out.)

**18.** $P = V(V - 2)(V + 4)$

**19.** $y = ax^3 + bx^2 + cx + d$
($a, b, c, d$ constants)

**20.** $h = 2m^2 - mp + mp^2$
($m$ constant)

**21.** $P = I^2 R$ ($R$ constant)

**22.** $E = \dfrac{CV^2}{2}$ ($C$ constant)

**23.** Given the parabola $y = x^2 + 2x - 8$: (*a*) Find $y'$. (*b*) Find the slope of the parabola at $x = -4, -1,$ and 2, and sketch the parabola, showing the three points. (See Example 19-23.)

**24.** Given the curve $y = x^3 - 3x$: (*a*) Find $y'$. (*b*) Find the slope of the curve at $x = -2, -1, 0, 1,$ and 2, and sketch the curve, showing the five points. (See Example 19-23.)

**25.** An electronic component moves in an assembly line according to the function $s = t^3 - 6t^2 + 9t$, where $s$ is the distance in meters from its starting point and $t$ is the time in seconds. (*a*) Find the formula for the velocity $v = s'$. (*b*) Find $v$ when $t = 0, 1, 2,$ and 3 s. (See Example 19-24.)

26. The velocity (in feet per second) of a machine part is given by $v = 8t^2 - t^4$, where $t$ is the time (in seconds). The instantaneous rate of change of velocity with respect to time is the acceleration $a$. *(a)* Find the formula for $a = v'$. *(b)* Find $a$ when $t = 0, 1, 2,$ and 3 s.

27. The power in an electric circuit is the instantaneous rate of change of energy consumed $W$ (or work done) with respect to time $t$, or the derivative of $W$ with respect to $t$. The energy used in joules (J) by an alternator charging a battery as a function of time $t$ (in seconds) is given by $W = 420t - 0.056t^2$. *(a)* Find the formula for the power $P = dW/dt$. *(b)* Find $P$ in watts when $t = 0$ and $t = 60$ s.

28. The current at a point in an electric circuit is the instantaneous rate of change of charge $q$ with respect to time $t$, or the derivative of $q$ with respect to $t$. The charge in coulombs (C) at a point in a circuit is given by $q = 8t - 1.5t^3$. *(a)* Find the formula for the current $I = dq/dt$. *(b)* Find $I$ in amperes (A) when $t = 0$ and $t = 1$ s.

29. The length $L$ of a metal rod as a function of temperature $T$ is given by

$$L = aL_0T + L_0$$

where $a$ = coefficient of linear expansion (assumed constant) and $L_0$ = length at 0°C. Find the expression for the instantaneous rate of change of $L$ with respect to $T$.

30. The formula for the volume of a sphere is $V = 4\pi r^3/3$. Show that $dV/dr$ is the formula for the surface area of a sphere.

31. The monthly profit $P$ made by a microcomputer manufacturer is a function of the number $n$ of computers sold, where $P = 300n - 10n^2$. *(a)* Find the formula for $P'$, the instantaneous rate of change of $P$ with respect to $n$. *(b)* Set $P' = 0$ and find the value of $n$ when $P' = 0$. This is when the profit no longer increases with the sales and begins to decrease.

32. The Stefan-Boltzmann law states that the rate of radiation $R$ of a perfect radiator is proportional to the fourth power of its temperature $T$ in kelvins. The constant of porportionality is $\sigma = 4.88 \times 10^{-8}$. *(a)* Express $R$ as a function of $T$. *(b)* Find the formula for $dR/dT$.

## 19-4 | *Understanding the Derivative*

Now that you can readily find the derivative of a polynomial by using the formulas in Sec. 19-3, this section further explores the concept of a derivative. Its meaning and some of its applications are studied in greater detail to help you better understand this fundamental idea.

The meaning of a derivative in words is:

Given a function $y = f(x)$, the *derivative of $y$ with respect to $x$ is a formula which expresses the instantaneous rate of change of the quantity $y$ with respect to the quantity $x$. When applied to the graph of $y = f(x)$, the geometric model of a derivative* is the formula for the *slope of the curve at* *any point.* When $y$ is replaced by $s$, the distance traveled by an object, and $x$ is replaced by $t$, the time elapsed, *the physical model of a derivative* is a formula for the *velocity of the object at any time*.

Important physical models of derivatives are illustrated in the next example.

### Example 19-25
List some important models of derivatives.

### Solution
1. *Slope:* When $y = f(x)$, then $y' = dy/dx = f'(x)$ is a formula for the slope of the curve at any point.

2. *Velocity:* When $s = f(t)$, where $s$ is the distance traveled by an object in a straight line and $t$ is the time, $v = s' = ds/dt$ is a formula for the velocity of the object at any time.

3. *Acceleration:* When $v = f(t)$, where $v$ is the velocity of an object traveling in a straight line and $t$ is the time, $a = v' = dv/dt$ is a formula for the acceleration of the object at any time.

4. *Current:* When $q = f(t)$, where $q$ is the charge in coulombs at a point in an electric current and $t$ is the time in seconds, $I = q' = dq/dt$ is a formula for the current in amperes at any time.

5. *Power:* When $W = f(t)$, where $W$ is the energy consumed or work done over a time interval $t$, then $P = W' = dW/dt$ is a formula for the power output at any time. When $W$ is in joules and $t$ is in seconds, $P$ is measured in watts.

6. *Temperature:* When $T = f(x)$, where $T$ is the temperature of a substance and $x$ is the distance in a certain direction, $T' = dT/dx = f'(x)$ is the temperature gradient, or the rate of change of temperature in the direction $x$.

7. *Voltage:* When $I = f(t)$, where $I$ is the current in amperes in a circuit containing an inductance $L$ (henrys) and $t$ is the time in seconds, $V = LI' = L\,(dI/dt)$ is the voltage across the inductance.

The following examples illustrate the use of derivatives in different applications.

### Example 19-26
Given the parabola $y = -x^2 + 6x - 8$:

1. Find the formula for the slope at any point $x$.
2. Find the coordinates of the vertex of the parabola.
3. Find the equation of the tangent to the parabola at $x = 2$.

### Solution
1. Find the formula for the slope at any point $x$.

Apply rules (19-5) to (19-9) to find $y'$:

$$y' = -2x + 6$$

2. Find the coordinates of the vertex of the parabola.

The *vertex* is where the *slope is zero*. Set $y' = 0$ and solve for $x$:

$$y' = -2x + 6 = 0$$
$$x = 3$$

and $\qquad\qquad\qquad y = -(3)^2 + 6(3) - 8 = 1$

The vertex is therefore $(3, 1)$. See Fig. 19-7.

3. Find the equation of the tangent to the parabola at $x = 2$.

Substitute into the formula for $y'$ to find the slope of the tangent:

$$y' = -2(2) + 6 = 2$$

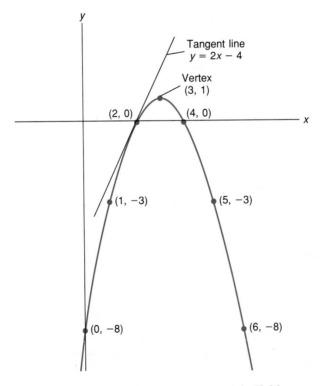

FIG. 19-7 *Parabola and tangent line, Example 19-26*

Find the corresponding $y$ value when $x = 2$:

$$y = -(2)^2 + 6(2) - 8 = 0$$

Use the point-slope form of the equation of a straight line (14-5), letting $m = 2$, $x_1 = 2$, and $y_1 = 0$:

$$y - y_1 = m(x - x_1)$$
$$y - 0 = 2(x - 2)$$
$$y = 2x - 4$$

### Example 19-27
A rocket is shot straight upward, and its distance $s$ (in meters) above the ground is given by $s = 245t - 4.90t^2$, where $t = $ time (in seconds).

1. Find the formula for the velocity.
2. After how many seconds does the rocket reach the highest point, and how high does it travel?
3. Find the formula for the acceleration at any time $t$.

### Solution
1. Find the formula for the velocity.

   Differentiate the formula for $s$:

   $$v = s' = 245 - 9.80t$$

2. After how many seconds does the rocket reach the highest point, and how high does it travel?

   At the highest point the velocity is zero. Set $v = 0$:

   $$245 - 9.80t = 0$$
   $$t = \frac{245}{9.80} = 25 \text{ s}$$

   Substitute $t = 25$ s to find the height:

   $$s = 245(25) - 4.90(25)^2 = 3063 \text{ m}$$

3. Find the formula for the acceleration at any time $t$.

   Apply Example 19-25(3) and differentiate the formula for $v$:

   $$a = v' = -9.80 \text{ m/s}^2$$

The acceleration is constant, as expected, since it is due to gravity, which is a constant force. The formula for $a$ is called the *second derivative* of $s$, written $s''$. Higher derivatives are discussed in Sec. 20-1.

### Example 19-28

In an electric circuit containing an inductance, the charge $q$ in coulombs is given by $q = 0.5t^2 - 0.01t^3$, where $t$ = time (in seconds).

1. Find the formula for the current at any time $t$.
2. Find the current when $t = 1$ and $t = 10$ s.
3. If the inductance $L = 0.040$ H (henry), find the formula for the voltage across the inductance.

### Solution

1. Find the formula for the current at any time $t$.

   Apply Example 19-25(4) and differentiate the formula for $q$:

   $$I = q' = 1.0t - 0.03t^2$$

2. Find the current when $t = 1$ and $t = 10$ s.

   $t = 1$ s:     $I = 1.0(1) - 0.03(1)^2 = 1.0 - 0.03 = 0.97$ A

   $t = 10$ s:     $I = 1.0(10) - 0.03(10)^2 = 10 - 3.0 = 7.0$ A

3. If the inductance $L = 0.040$ H (henry), find the formula for the voltage across the inductance.

   Apply Example 19-25(7) to the formula for $I$ in (1):

   $$V = LI' = (0.040)(1.0 - 0.06t) = 0.040 - 0.0024t$$

### Example 19-29

The mechanical work done in foot-pounds (ft · lb) by a wind generator over a time interval $t$ is given by $W = t^3 - 6t^2 + 12t$, where $t$ varies from 0 to 24 h.

1. Find the formula for the power output.
2. Find the power in foot-pounds per hour (ft · lb/h) when $t = 12$ and $t = 24$ h.
3. Find the answers to (2) in watts if 1 ft · lb/h $= 3.766 \times 10^{-4}$ W.

### Solution

1. Find the formula for the power output.

   Apply 19-25(5) and differentiate the formula for $W$:

   $$P = W' = 3t^2 - 12t + 12$$

2. Find the power in foot-pounds per hour (ft · lb/h) when $t = 12$ and $t = 24$ h.

$t = 12$ h:  $P = 3(12)^2 - 12(12) + 12 = 300$ ft · lb/h

$t = 24$ h:  $P = 3(24)^2 - 12(24) + 12 = 1452$ ft · lb/h

3. Find the answers to (2) in watts if 1 ft · lb/h $= 3.766 \times 10^{-4}$ W.

The answers, to three significant figures, are

$t = 12$ h:

$$300 \text{ ft} \cdot \text{lb/h} \left( \frac{3.766 \times 10^{-4} \text{ W}}{1 \text{ ft} \cdot \text{lb/h}} \right) = 1130 \times 10^{-4} \text{ W} = 0.113 \text{ W}$$

$t = 24$ h:

$$1452 \text{ ft} \cdot \text{lb/h} \left( \frac{3.766 \times 10^{-4} \text{ W}}{1 \text{ ft} \cdot \text{lb/h}} \right) = 5468 \times 10^{-4} \text{ W} = 0.547 \text{ W}$$

**Exercise 19-4**  In problems 1 to 4, for each curve find *(a)* the formula for the slope at any point $x$ and *(b)* the coordinates of any points where the slope is zero. (See Example 19-26.)

**1.** $y = 2x^2 + 2$    **2.** $y = x^2 + 4x - 5$

**3.** $y = 2x^3 - 6x$    **4.** $y = x^3 - 6x^2 + 9x + 1$

In problems 5 to 8, for each equation of motion in a straight line, find *(a)* the formula for the velocity and *(b)* the formula for the acceleration. (See Example 19-27.)

**5.** $s = t - 2t + 4$    **6.** $s = 64t - 16t^2$

**7.** $s = 10 + 2t^2 - t^3$    **8.** $s = \dfrac{t^3}{3} + \dfrac{t^2}{2} - t$

9. Given the parabola $y = x^2 + 2x - 3$: *(a)* Find the formula for the slope. *(b)* Find the coordinates of the vertex. *(c)* Find the equation of the tangent to the parabola at $x = 1$. (See Example 19-26.)

10. Given the curve $y = 2x^3 + 1$: *(a)* Find the formula for the slope. *(b)* Find the equation of the tangent to the curve at $x = -1$. (See Example 19-26.) *(c)* Find the equation of the *normal* to the curve at $x = -1$. The *normal* is the line perpendicular to the tangent, and its slope is the negative reciprocal of the slope of the tangent.

11. A ball is thrown straight up, and its distance $s$ (in meters) above the ground is given by $s = 19.6t - 4.90t^2$, where $t$ = time (in seconds). *(a)* Find the formula for the velocity. *(b)* How long does it take the ball to reach the highest point, and how high does it travel? (See Example 19-27.)

12. Upon reversing the engines, the distance traveled by a supertanker in a straight line before coming to a stop is given by $s = 27t - t^3$, where $s$ is the distance (in feet) and $t$ is the time (in minutes). *(a)* Find the formula for the velocity. *(b)* What is the velocity as soon as the engine is put in reverse ($t = 0$)? *(c)* How long does it take for the tanker to stop, and how far does it travel?

13. In an electric circuit containing an inductance, the charge (in coulombs) is given by $q = 1.0t^2 - 0.02t^3$, where $t$ = time (in seconds). (a) Find the formula for the current $I$. (b) Find $I$ when $t = 5$ and $t = 10$ s. (See Example 19-28.)

14. The inductance in the circuit of problem 13 is $L = 0.070$ H. (a) Find the formula for the voltage $V$ across the inductance. (b) Find the voltage when $t = 0$ and $t = 1$ s. (See Example 19-28.)

15. Across the insulation wall of a refrigerator the temperature as a function of the perpendicular distance $x$ from the interior is given by $T = 10x^2 - 10x - 30$, where $x$ varies from 0 to 4 in. (a) Find the formula for the temperature gradient. (b) Find the temperature gradient when $x = 1.5$ and $x = 2.5$ in. [See Example 19-25(6)]

16. In a chemical reaction, the rate of reaction is the rate of change of a substance with respect to time. In a certain reaction, the number of grams of a substance undergoing chemical change is given by $Q = 11.3t - 2.1t^2$, where $t$ = time (in minutes). (a) Find the formula for $Q'$, the rate of reaction. (b) Find $Q'$ when $t = 2$ min. (c) After how many minutes does the reaction end ($Q' = 0$)?

17. The electric energy consumed in a transmission line over a time interval $t$ (in hours) is given by $W = -t^4 + 12t^3 - 54t^2 + 108t$. (a) Find the formula for the power output. (See Example 19-29.) (b) Find the power output in watts when $t = 0$ and $t = 1$ h.

18. The work done (in foot-pounds) by a crane lifting a bridge support is given by $W = 100t^3$, where $t$ varies from 0 to 10 s. (a) Find the formula for the power output. (See Example 19-29.) (b) What is the final power output of the crane? (c) If $1$ ft $\cdot$ lb/s $= 0.0018$ hp, find the final horsepower output of the crane.

19. In a certain engine, the temperature in the cylinder is related to the volume by:

$$\frac{T}{V} = 0.0151$$

Find the formula for the rate of change of temperature with respect to volume.

20. Oil is being drained from a spherical tank 10 ft in diameter, and the volume of oil as a function of the height $h$ (in feet) is given by:

$$V = \pi\left(\frac{1}{3h^3} - 5h^2\right)$$

(a) Find the formula for the rate of decrease of volume with respect to the height. (b) Find the rate of decrease in cubic feet per foot when $h = 10$ and $h = 5$ ft.

21. The voltage across a capacitance in a circuit is given by $V = 40t^2 + 2$ for a short interval of time $t$ (in seconds). The current to the capacitance is given by $I = CV' = C\ dV/dt$, where the capacitance $C = 3.0 \times 10^{-4}$ F (farad). Find the formula for the current.

22. The voltage $V$ produced by a thermocouple is a function of the temperature $T$, given approximately by $V = a + bT + cT^2$, where $a$, $b$, and $c$ are constants. Find the rate of change of $V$ with respect to $T$, called the *thermoelectric power* of the couple.

23. A gear turns through an angle $\theta$ in time $t$ given by $\theta = 32t - 8t^2$. The *angular velocity* $\omega$ is the rate of change of $\theta$ with respect to $t$. (a) Find the formula for $\omega$. (b) Find $\omega$ in radians per second when $t = 1$ and $t = 3$ s.

**24.** A bridge cable hangs in a parabolic arc given approximately by $y = (1.7 \times 10^{-4})x^2$, where the vertex of the parabola is at the origin. *(a)* Find the slope of the cable at the right support, which is 1000 ft from the center. *(b)* Find the angle $\theta$ that the cable makes with the horizontal at the right support. *(Note:* Slope $= \tan \theta$.)

# 19-5 | *Derivatives of a Product and a Quotient*

This section develops formulas for differentiating the product of two functions and the quotient of two functions. These formulas are useful when you are working with polynomial functions and with the trigonometric and exponential functions introduced in Chap. 22.

## *Product Formula*

*The derivative of the product of two functions equals the first function times the derivative of the second plus the second times the derivative of the first:*

If $u$ and $v$ are functions of $x$ and $y = f(x) = u \cdot v$, then:

$$y' = f'(x) = uv' + vu' \qquad (19\text{-}10)$$

### Example 19-30
Find the derivative of $f(x) = (3x + 1)(x^2 - 4)$ by using the product formula (19-10).

### Solution

Let
$$u = 3x + 1 \qquad v = x^2 - 4$$
Then
$$u' = 3 \qquad v' = 2x$$

Apply formula (19-10):

$$
\begin{aligned}
f'(x) &= uv' + vu' \\
&= (3x + 1)(2x) + (x^2 - 4)(3) \\
&= 6x^2 + 2x + 3x^2 - 12 = 9x^2 + 2x - 12
\end{aligned}
$$

You can also find the derivative by first multiplying out the binomials and then differentiating:

$$
\begin{aligned}
f(x) &= (3x + 1)(x^2 - 4) = 3x^3 + x^2 - 12x - 4 \\
f'(x) &= 9x^2 + 2x - 12
\end{aligned}
$$

However, the product of two functions cannot always be easily multiplied together, in which case formula (19-10) is the more desirable way to find the derivative. (See Example 19-39.)

## Proof of the Product Formula (19-10)

$$y = f(x) = u(x) \cdot v(x)$$
$$\Delta y = f(x + \Delta x) - f(x)$$
$$= [u(x + \Delta x)] \cdot [v(x + \Delta x)] - u(x) \cdot v(x)$$

Let $u(x + \Delta x) = u(x) + \Delta u$ and $v(x + \Delta x) = v(x) + \Delta v$, where $\Delta u$ and $\Delta v$ are determined by $\Delta x$:

$$\Delta y = [u(x) + \Delta u][v(x) + \Delta v] - u(x) \cdot v(x)$$
$$= u(x)\,\Delta v + v(x)\,\Delta u + \Delta u\,\Delta v$$
$$\frac{\Delta y}{\Delta x} = u(x)\,\frac{\Delta v}{\Delta x} + v(x)\,\frac{\Delta u}{\Delta x} + \frac{\Delta u}{\Delta x}\,\Delta v$$
$$\lim_{\Delta x \to 0} \frac{\Delta y}{\Delta x} = u(x)\,\lim_{\Delta x \to 0} \frac{\Delta v}{\Delta x} + v(x)\,\lim_{\Delta x \to 0} \frac{\Delta u}{\Delta x} + \lim_{\Delta x \to 0} \frac{\Delta u}{\Delta x}\,\Delta v$$

As $\Delta x \to 0$, $u(x)$ and $v(x)$ do not change but $\Delta u \to 0$ and $\Delta v \to 0$:

$$y' = \lim_{\Delta x \to 0} \frac{\Delta y}{\Delta x} = u(x) \cdot v'(x) + v(x) \cdot u'(x) + u'(x)(0)$$

Therefore:

$$y' = uv' + vu'$$

## Example 19-31

Find the derivative of $y = (x^2 + x + 1)(x^2 - x + 1)$ by using the product formula (19-10).

## Solution

Let $\qquad$ $u = x^2 + x + 1$ $\qquad$ $v = x^2 - x + 1$

Then $\qquad$ $u' = 2x + 1$ $\qquad$ $v' = 2x - 1$

Apply formula (19-10):

$$y' = uv' + vu'$$
$$= (x^2 + x + 1)(2x - 1) + (x^2 - x + 1)(2x + 1)$$

$$= 2x^3 + \cancel{x^2} + x - \cancel{1} + 2x^3 - \cancel{x^2} + x + \cancel{1}$$
$$= 4x^3 + 2x$$

## Quotient Formula

*The derivative of the quotient of two functions is the denominator times the derivative of the numerator minus the numerator times the derivative of the denominator, all divided by the denominator squared:*

If $u$ and $v$ are functions of $x$ and if $y = f(x) = u/v$, then

$$y' = f'(x) = \frac{vu' - uv'}{v^2} \tag{19-11}$$

### Example 19-32

Find the derivative of $f(x) = \dfrac{x^2 + 1}{3x}$.

### Solution

Let $\qquad\qquad\qquad u = x^2 + 1 \qquad v = 3x$

Then $\qquad\qquad\quad u' = 2x \qquad\quad v' = 3$

Apply formula (19-11):

$$f'(x) = \frac{vu' - uv'}{v^2}$$

$$= \frac{(3x)(2x) - (x^2 + 1)(3)}{(3x)^2}$$

$$= \frac{3x^2 - 3}{9x^2} = \frac{x^2 - 1}{3x^2}$$

## Proof of the Quotient Formula (19-11)

$$y = f(x) = \frac{u(x)}{v(x)}$$

$$\Delta y = f(x + \Delta x) - f(x)$$

$$= \frac{u(x + \Delta x)}{v(x + \Delta x)} - \frac{u(v)}{v(v)}$$

Let $u(x + \Delta x) = u(x) + \Delta u$ and $v(x + \Delta x) = v(x) + \Delta v$, where **$\Delta u$ and $\Delta v$ are** determined by $\Delta x$:

$$\Delta y = \frac{u(x) + \Delta u}{v(x) + \Delta v} - \frac{u(x)}{v(x)}$$

$$= \frac{v(x)[u(x) + \Delta u] - u(x)[v(x) + \Delta v]}{v(x)[v(x) + \Delta v]}$$

$$= \frac{v(x)\, \Delta u - u(x)\, \Delta v}{v(x)[v(x) + \Delta v]}$$

$$\frac{\Delta y}{\Delta x} = \frac{v(x)(\Delta u/\Delta x) - u(x)(\Delta v/\Delta x)}{v(x)[v(x) + \Delta v]}$$

As $\Delta x \to 0$, $\Delta u \to 0$ and $\Delta v \to 0$:

$$y' = \lim_{\Delta x \to 0} \frac{\Delta y}{\Delta x} = \frac{v(x) \cdot u'(x) - u(x) \cdot v'(x)}{v(x)[v(x) + 0]}$$

Therefore
$$y' = \frac{vu' - uv'}{v^2}$$

### Example 19-33

Find the derivative of $y = \dfrac{3x^2 + 3}{2 - x}$.

### Solution

Let             $u = 3x^2 + 3 \qquad v = 2 - x$

Then           $u' = 6x \qquad\quad v' = -1$

Apply formula (19-11):

$$y' = \frac{vu' - uv'}{v^2}$$

$$= \frac{(2 - x)(6x) - (3x^2 + 3)(-1)}{(2 - x)^2}$$

$$= \frac{12x - 6x^2 + 3x^2 + 3}{(2 - x)^2}$$

$$= \frac{-3x^2 + 12x + 3}{(2 - x)^2} = \frac{-3(x^2 - 4x - 1)}{(2 - x)^2}$$

## *Extension of the Power Formula*

The power formula (19-6) can be shown to apply when $n$ = negative integer, by using the quotient formula (19-11) as follows:

Let $$y = x^n = x^{-m} = \frac{1}{x^m} \qquad (m = \text{positive integer})$$

then:

$$y' = \frac{x^m(0) - 1(mx^{m-1})}{(x^m)^2}$$

$$= \frac{-mx^{m-1}}{x^{2m}} = -mx^{(m-1)-2m}$$

$$= -mx^{-m-1}$$

$$= nx^{n-1} \qquad \text{since } n = -m$$

Therefore:

$$\text{If } y = x^n, \text{ then } y' = nx^{n-1}. \qquad (n = any \text{ integer}) \qquad (19\text{-}12)$$

### Example 19-34
Find the derivative of $s = 3/t - 2/t^2$.

### Solution
Write the function as $s = 3t^{-1} - 2t^{-2}$ and apply (19-12) and (19-17):

$$s' = (3)(-1)t^{-1-1} - 2(-2)t^{-2-1}$$

$$= -3t^{-2} + 4t^{-3}$$

or $$s' = \frac{4}{t^3} - \frac{3}{t^2}$$

Check that you get the same result for the derivative by using the quotient formula.

**Exercise 19-5**

In problems 1 to 10, find the derivative of each function in two ways: (a) using the product formula (19-10) and (b) multiplying out and then taking the derivative.

**1.** $f(x) = 2x(x^2 - 1)$          **2.** $f(x) = x^3(x - x^2)$

**3.** $y = (x - 1)(x + 2)$          **4.** $y = (3x + 1)(x^2 - 4)$

**5.** $f(v) = (v^3 - 0.8)(v + 0.2)$      **6.** $M = (5.0d + 1.0)(d^3 - 1.0d)$

**7.** $y = (x^2 - 2x + 1)(x + 3)$      **8.** $y = (x^2 + x - 2)(x^2 + x + 2)$

**9.** $y = (ax + b)(cx + d)$
$\quad$ ($a, b, c, d$ constants)

**10.** $V = 2\pi r(\pi r^2 + 2\pi rh)$
$\quad$ ($h$ constant)

In problems 11 to 22, find the derivative of each function, using the quotient formula (19-11) or the power formula (19-12).

**11.** $y = \dfrac{1}{x}$

**12.** $A = \dfrac{3}{2r^3}$

**13.** $v = \dfrac{3}{t^3} + \dfrac{2}{t^2}$

**14.** $d = \dfrac{1}{m} - \dfrac{1}{3m^3}$

**15.** $y = \dfrac{4}{x - 3}$

**16.** $y = \dfrac{x}{x + 2}$

**17.** $f(x) = \dfrac{x + 1}{2x^2}$

**18.** $f(x) = \dfrac{x^2 - 3}{x - 3}$

**19.** $Q = \dfrac{h + 5}{1 - h^2}$

**20.** $F(a) = \dfrac{a^2 - 2}{a^2 + a}$

**21.** $y = \dfrac{4}{2x^2 + x + 1}$

**22.** $y = \dfrac{2 - 3x}{3x^2 - 2x - 3}$

In problems 23 to 28, find the derivative of each function.

**23.** $a = 3v^3 + \dfrac{1}{v^2}$

**24.** $R = 5T^2 + (2T - 1)(T^2 - T + 3)$

**25.** $S = \dfrac{v + d}{v - d}$ $\quad$ ($d$ constant)

**26.** $p = p_0x + \dfrac{x}{2gx + 1}$ $\quad$ ($p_0, g$ constants)

**27.** $f(x) = \left(\dfrac{1}{x}\right)\left(\dfrac{1}{2 - x}\right)$

**28.** $F = \dfrac{(w + 1)(w - 2)}{w + 2}$

**29.** Given the curve $y = 1/(x + 1)$: *(a)* Find the slope formula. *(b)* Find the value of the slope when $x = -2, 0, 2$. Can the slope ever be positive?

**30.** Given the curve $y = x^2/(x - 1)$: *(a)* Find the slope formula. *(b)* Find the values of $x$ when the slope is zero.

☐ **31.** The displacement of an automobile chassis in an assembly line is given by

$$s = 10 - \dfrac{10}{t^2 + 1}$$

where the time $t$ varies from 0 to 2 min. *(a)* Find the formula for the velocity $v$ (feet per minute). *(b)* Find $v$ when $t = 0, 0.5, 1.0,$ and 1.5 min.

**32.** The angular displacement $\theta$ of a pulley is given by $\theta = 12/t^2$, where the time $t$

varies from 1 to 4 s. *(a)* Find the formula for the *angular velocity* $\omega = \theta'$. *(b)* Find the formula for the *angular acceleration* $\alpha = \omega'$.

**33.** The power in an electric circuit is given by

$$P = \frac{144R}{R^2 + 25}$$

where $R$ = resistance. *(a)* Find the formula for $P'$, the rate of change of $P$ with respect to $R$. *(b)* Find $R$ when $P' = 0$.

**34.** The voltage in a dc circuit is given by $E = 0.5t$, where $t$ = time. The resistance in the circuit is given by $R = 0.1t^2 + 1.0$. *(a)* If the current $I = E/R$, find the formula for $I'$. *(b)* Find $I'$ when $t = 2$ s.

**35.** The torsional stress on a circular shaft of radius $R$ is given by

$$s = \frac{220,000}{\pi R^3}$$

*(a)* Find the formula for $s'$. *(b)* Find $s'$ when $R = 10$.

**36.** The equivalent diameter of a rectangular pipe with cross section of length $l$ and width $w$ is given by

$$d = \frac{2lw}{l + w}$$

*(a)* Assuming the width $w$ is constant, find $d'$, the formula for the rate of change of $d$ with respect to $l$. *(b)* Find the value of $d'$ when $l = 2w$.

**37.** Find the derivative of

$$f(t) = \frac{2}{t + 1} - \frac{3}{t}$$

in two ways: *(a)* Differentiate each fraction separately and then combine. *(b)* Combine fractions and then differentiate.

**38.** Show that given a product of three functions $y = f(x) \cdot g(x) \cdot h(x)$, the derivative $y' = f(x) \cdot g(x) \cdot h'(x) + f(x) \cdot g'(x) \cdot h(x) + f'(x) \cdot g(x) \cdot h(x)$. [*Hint*: Apply the product formula (19-10), letting $u = f(x)$ and $v = g(x) \cdot h(x)$.]

# 19-6 | *Derivative of a Power Function*

This section extends the power formula (19-12) to the power of a function and to $n$ = any rational number. This will enable you to find the derivative of functions such as $y = (2x + 1)^4$ and $y = (x^2 + 3x + 1)^3$ without multiplying out the expressions. It will also enable you to find the derivative of radicals such as $y = 2\sqrt{x} = 2x^{1/2}$, $y = \sqrt{x-2} = (x-2)^{1/2}$, and $y = \sqrt[3]{x^2 - 1} = (x^2 - 1)^{1/3}$. It is first necessary to develop the *chain rule* for the derivative of a *composite function*. Consider the function $y = (2x + 1)^4$. If you let $u = 2x + 1$, then $y = u^4$. This is called a *composite function: y* is a function of $u$, and $u$ in turn is a function of $x$. So $y$ still depends on $x$ and is a composite function of $x$. This can be expressed as $y = f(u(x))$.

## Chain Rule

If $y$ is a function of $u$ and $u$ is a function of $x$, then the derivative of $y$ with respect to $x$ equals the derivative of $y$ with respect to $u$, times the derivative of $u$ with respect to $x$:

$$\frac{dy}{dx} = \frac{dy}{du}\frac{du}{dx} \tag{19-13}$$

Formula (19-13) uses the derivative notation introduced in Sec. 19-2, since this notation indicates which variable each derivative is taken with respect to.

### Example 19-35
Use the chain rule to find the derivative of $y = (2x + 1)^4$.

### Solution
Let $u = 2x + 1$. Then $y = u^4$ and:

$$\frac{dy}{du} = 4u^3 \qquad \frac{du}{dx} = 2$$

Apply formula (19-13):

$$\frac{dy}{dx} = \frac{dy}{du}\frac{du}{dx}$$

$$\frac{dy}{dx} = (4u^3)(2) = 4(2x + 1)^3(2) = 8(2x + 1)^3$$

## Proof of the Chain Rule (19-13)

Let $y = f(u(x))$. When $x$ changes by $\Delta x$, $u$ changes by $\Delta u$ and $y$ changes by $\Delta y$. The following identity is true:

$$\frac{\Delta y}{\Delta x} = \frac{\Delta y}{\Delta u}\frac{\Delta u}{\Delta x}$$

As $\Delta x \to 0$, $\Delta u \to 0$ and $\Delta y \to 0$. Then:

$$\frac{dy}{dx} = \lim_{\Delta x \to 0} \frac{\Delta y}{\Delta x} = \lim_{\Delta x \to 0} \left( \frac{\Delta y}{\Delta u} \frac{\Delta u}{\Delta x} \right) = \lim_{\Delta u \to 0} \frac{\Delta y}{\Delta u} \cdot \lim_{\Delta x \to 0} \frac{\Delta u}{\Delta x}$$

since it can be shown that the limit of a product equals the product of the limits. Therefore:

$$\frac{dy}{dx} = \frac{dy}{du} \frac{du}{dx}$$

When the chain rule is applied to the power of a function, as shown in Example 19-35, it leads to the *general power formula:*

If $y = u^n$ and $u$ is a function of $x$, then:

$$\frac{dy}{dx} = y' = nu^{n-1}(u') \qquad (n = \text{any rational number}) \qquad (19\text{-}14)$$

### Example 19-36
Find the derivative of $y = (x^2 + 3x + 1)^3$.

### Solution
Let $u = x^2 + 3x + 1$. Then $u' = 2x + 3$. Apply formula (19-14):

$$y' = nu^{n-1}(u') = 3(x^2 + 3x + 1)^2(2x + 3)$$

Do not forget the factor $u'$ when you are applying the general power formula (19-14). Note that Example 19-35 can also be done more directly by using formula (19-14).

## *Proof of the General Power Formula (19-14)*

*The general power formula (19-14) applies when $n$ = any rational number.* This is shown by using the chain rule (19-13) as follows:

Let $y = u^{p/q}$ where $p$ and $q$ are *integers* and $u$ is a function of $x$. Raise each side to the $q$th power:

$$y^q = (u^{p/q})^q$$
$$y^q = u^p$$

Apply the chain rule (19-13) and power formula (19-12) to each side, differentiating with respect to $x$:

$$qy^{q-1}(y') = pu^{p-1}(u')$$

$$y' = \frac{pu^{p-1}u'}{qy^{q-1}}$$

$$= \frac{pu^{p-1}u'}{q(u^{p/q})^{q-1}} = \frac{pu^{p-1}u'}{qu^{p-p/q}}$$

$$= \frac{p}{q}u^{p-1-(p-p/q)}u' = \frac{p}{q}u^{p/q-1}u'$$

The general power formula (19-14) includes the power formula (19-12) as a special case. Therefore (19-12) also applies when $n =$ any rational number.

Study the following examples, which illustrate the use of the power formulas.

### Example 19-37
Find $f'(x)$ for $f(x) = 2\sqrt{x}$.

### Solution
Express the function as $f(x) = 2x^{1/2}$. Apply formula (19-12):

$$f'(x) = 2\left(\frac{1}{2}x^{1/2-1}\right) = x^{-1/2} \qquad \text{or} \qquad f'(x) = \frac{1}{\sqrt{x}}$$

### Example 19-38
Find $y'$ for $y = \dfrac{1}{\sqrt{x^2+1}}$.

### Solution
Express the function as:

$$y = (x^2 + 1)^{-1/2}$$

Let $u = x^2 + 1$ and apply formula (19-14). Do not leave out $u' = 2x$:

$$y' = -\frac{1}{2}(x^2 + 1)^{-1/2-1}(2x)$$

$$y' = -x(x^2 + 1)^{-3/2} \qquad \text{or} \qquad y' = -\frac{x}{(\sqrt{x^2+1})^3}$$

The next example shows the use of the product formula (19-10) along with the power formula (19-14).

### Example 19-39
Find $f'(h)$ for $f(h) = h\sqrt{2 - 3h}$.

### Solution
Let:

$$u = h \qquad \text{and} \qquad v = \sqrt{2 - 3h} = (2 - 3h)^{1/2}$$

Apply formula (19-14):

$$u' = 1 \qquad \text{and} \qquad v' = \frac{1}{2}(2 - 3h)^{1/2-1}(-3) = -\frac{3}{2}(2 - 3h)^{-1/2}$$

Then apply the product formula (19-10):

$$f'(h) = uv' + vu'$$

$$f'(h) = h\left[ -\frac{3}{2}(2 - 3h)^{-1/2} \right] + (2 - 3h)^{1/2}(1)$$

$$= \frac{-3h}{2\sqrt{2 - 3h}} + \sqrt{2 - 3h}$$

Combine terms for a simpler expression:

$$f'(h) = \frac{-3h + 2(2 - 3h)}{2\sqrt{2 - 3h}} = \frac{4 - 9h}{2\sqrt{2 - 3h}}$$

The next two examples show the use of the quotient formula (19-11) combined with the power formula (19-14).

### Example 19-40
Find $I'$ for $I = \dfrac{(1.6 - t)^3}{4.2t}$.

### Solution
Apply (19-11) and (19-14):

$$I' = \frac{4.2t[3(1.6 - t)^2(-1)] - (1.6 - t)^3(4.2)}{(4.2t)^2}$$

Note that the derivative of the numerator requires the use of the power formula (19-14). The answer can be simplified by factoring out $4.2(1.6 - t)^2$ from the numerator and dividing out 4.2:

$$I' = \frac{\cancel{4.2}(1.6 - t)^2[-3t - (1.6 - t)]}{(\cancel{4.2})(4.2t^2)} = \frac{(1.6 - t)^2(-2t - 1.6)}{4.2t^2}$$

This example can also be done by using the product and power formulas and expressing the function as $I = (1.0 - t)^3(4.2t)^{-1}$. *It is important to know your algebra well when you apply the formulas of calculus.*

## Example 19-41

Find $w'$ for $w = \sqrt{\dfrac{p}{p-2}}$.

## Solution

Apply (19-11) and (19-14):

$$w = \left(\frac{p}{p-2}\right)^{1/2}$$

$$w' = \frac{1}{2}\left(\frac{p}{p-2}\right)^{-1/2}\frac{(p-2)(1) - (p)(1)}{(p-2)^2}$$

$$= \frac{1}{2}\left(\frac{p-2}{p}\right)^{1/2}\left[\frac{-2}{(p-2)^2}\right]$$

Note that rule (8-1) for exponents is applied above and the fraction is inverted by changing the sign of the exponent. Then simplifying:

$$w' = \frac{-\sqrt{p-2}}{(p-2)^2\sqrt{p}} \qquad \text{or} \qquad w' = \frac{-\sqrt{p(p-2)}}{p(p-2)^2}$$

## Table 19-1   IMPORTANT DERIVATIVE FORMULAS

| Formula | $y$ | $y'$ |
|---|---|---|
| Constant | $c$ | $0$ |
| Power | $x^n$ | $nx^{n-1}$ |
| Power | $u^n$ | $nu^{n-1}(u')$ |
| Product | $uv$ | $uv' + vu'$ |
| Quotient | $\dfrac{u}{v}$ | $\dfrac{vu' - uv'}{v^2}$ |

The answer may also be expressed in terms of exponents:

$$w' = -(p - 2)^{-3/2}p^{-1/2} \quad \text{or} \quad w' = \frac{-1}{(p - 2)^{3/2}p^{1/2}}$$

    Table 19.1 (on the previous page) shows the important derivative formulas introduced so far where $u$ and $v$ are functions of $x$, $n =$ any rational number and $y' = dy/dx$.

**Exercise 19-6**

In problems 1 to 28, find the derivative of each function.

**1.** $y = (3x - 1)^3$                              **2.** $y = 2(2x + 5)^4$

**3.** $f(d) = \dfrac{(3 - 3d)^4}{2}$             **4.** $f(w) = 0.4(w - w^2)^5$

**5.** $y = (2x^2 + 3x + 1)^3$          **6.** $A = (2 - \pi r - 2\pi r^2)^2$

**7.** $y = \sqrt[3]{x}$                                 **8.** $y = 4\sqrt{x}$

**9.** $s = \dfrac{2.6}{\sqrt{t}}$                      **10.** $h = \dfrac{0.316}{\sqrt[4]{R}}$

**11.** $y = x^{3/2}$                              **12.** $Q = 1.5H^{2/3}$

**13.** $f(x) = \dfrac{3}{(4x + 3)^2}$          **14.** $f(v) = \dfrac{1}{(v^2 + 1)^3}$

**15.** $y = (x + 1)^3(x - 1)^3$        **16.** $q = \left(\dfrac{m}{m + 1}\right)^2$

**17.** $f(t) = \dfrac{(t + 1)^3}{(t - 2)^2}$         **18.** $f(k) = \dfrac{(3.0 + k)^3}{1.5k}$

**19.** $a = \dfrac{-2}{\sqrt{2d - 1}}$           **20.** $f(n) = \dfrac{1.2}{\sqrt{1 - n^2}}$

**21.** $y = x\sqrt{x + 1}$                **22.** $y = \dfrac{1 + x}{\sqrt{x}}$

**23.** $y = \dfrac{\sqrt{2x + 1}}{3x - 2}$         **24.** $f(x) = \dfrac{2 - x}{\sqrt{1 - x}}$

**25.** $R = \sqrt{\dfrac{h + 5}{2h}}$           **26.** $w = \sqrt{\dfrac{p}{p + 3}}$

**27.** $y = \sqrt{r^2 - x^2}$  ($r$ constant)     **28.** $v = R\sqrt{a^2 + w^4}$  ($R$, $a$ constants)

**29.** Given the circle $x^2 + y^2 = 25$: (*a*) Solve for $y$ and find the slope formula. (*b*) Find the slope and the equation of the tangent line at the point (3, 4).

**30.** Given the curve $y^3 = x$: (*a*) Solve for $y$ and find the slope formula. (*b*) Find the slope and the equation of the tangent line at the point (8, 2).

**31.** A body moves a distance $s$ in a straight line given by $s = \frac{1}{4}(2t - 1)^3$, where $t =$ time. (*a*) Find the formula for the velocity. (*b*) Find the formula for the acceleration. (Acceleration is the derivative of the velocity.)

32. The velocity of a falling body is given by $v = \sqrt{v_0^2 + 2gs}$, where $v_0$ = initial velocity, $g$ = gravitational acceleration, and $s$ = distance. Find the formula for the rate of change of $v$ with respect to $s$.

C 33. The charge on a capacitor over a short interval of time $t$ s is given by $q = 3t^2 + \sqrt{2 - 4t}$. (a) Find the formula for the current $I = q'$. (b) Find the current when $t = 0.3$ s.

C 34. In an $RC$ circuit the current $I$ is given by:

$$I = \frac{V}{\sqrt{R^2 + X_C^2}}$$

where $V$ = voltage, $R$ = resistance, and $X_C$ = capacitance reactance. If $V = 120$ V and $R = 40$ $\Omega$, (a) find the formula for $I'$, the rate of change of $I$ with respect to $X_C$, and (b) find the value of $I'$ when $X_C = 30$ $\Omega$.

C 35. The discharge $Q$ of a triangular weir (open channel) is given by:

$$Q = 0.56\sqrt{2gH^5}$$

where $g$ = gravitational acceleration and $H$ = height of the triangular cross section. (a) Find $Q'$, the derivative of $Q$ with respect to $H$. (b) Find the value of $Q'$ when $H = 5.0$ m. Use $g = 9.8$ m/s$^2$.

C 36. In a certain adiabatic process (no heat added or removed from the system) the pressure is related to the volume by:

$$P = \frac{1.68 \times 10^5}{V^{1.407}}$$

(a) Find $P'$, the rate of change of pressure with respect to volume. (b) Find the value of $P'$ when $V = 9.08$.

37. When light passes from air into a transparent medium, the ratio of reflected light to incident light is given approximately by:

$$R = \left(\frac{n-1}{n+1}\right)^2$$

where $n$ = index of refraction for the medium. Find the derivative of $R$ with respect to $n$.

38. The angular velocity of a conical pendulum is given by:

$$\omega = \frac{\sqrt{g}}{(l^2 - r^2)^{1/4}}$$

where $l$ = length, $r$ = radius of the circle traveled by the end of the pendulum, and $g$ = gravitational acceleration. Assuming $r$ is constant, find the formula for the rate of change of $\omega$ with respect to $l$.

39. In an $RLC$ series circuit, the current is given by:

$$I = \frac{-0.030}{(3 + t)^{3/2}}$$

where $t$ = time. The voltage across the inductance $V = LI'$. Find the formula for $V$ if $L = 0.0080$ H.

**40.** If $u = ax^2 + bx + c$, where $a$, $b$, and $c$ are constants, and $y = 2\sqrt{u}$, show that:

$$y' = \frac{2ax + b}{\sqrt{u}}$$

# 19-7 | *Derivative of an Implicit Function*

Up to now, the functions you have been differentiating have been of the form $y = f(x)$. In this form $y$ is expressed directly in terms of $x$ and is an *explicit function* of $x$. When two variables are related by a formula or equation and one variable is not expressed directly in terms of the other, then one variable is defined as an *implicit function* of the other. An implicit function may sometimes be solved for one of the variables and expressed as an explicit function, but not always. Study the following example.

### Example 19-42
Give examples of implicit and explicit functions.

### Solution
**1.** The line $2x - y = 3$. In this form $y$ is defined as an implicit function of $x$. When this equation is solved for $y = 2x - 3$, $y$ is expressed as an explicit function of $x$.

**2.** The formula $IR = 12$. In this form $I$ is defined as an implicit function of $R$. When this equation is solved for $I = 12/R$, $I$ is an explicit function of $R$.

**3.** The circle $x^2 + y^2 = 16$. This equation defines $y$ as an implicit function of $x$. When this equation is solved for $y = \pm\sqrt{16 - x^2}$, it defines two explicit functions: $y = \sqrt{16 - x^2}$ and $y = -\sqrt{16 - x^2}$. This is because a function can have only one value of $y$ corresponding to each value of $x$.

**4.** The equation $s^3 + s^2t - t^3 = 0$. This equation defines $s$ as an implicit function of $t$. It is difficult to solve this equation for $s$ and to express $s$ as an explicit function of $t$. It is easier to work with this function in the implicit form. For complex implicit functions it is not always possible or desirable to change them to explicit functions.

The *derivative of an implicit function* is found by applying the following rule:

When a variable $y$ is defined implicitly in terms of another variable $x$, differentiate each term of the equation with respect to $x$ and solve for the derivative $y'$.    (19-15)

### Example 19-43
Find $y' = dy/dx$ implicitly for the line $2x - y = 3$.

### Solution
Apply rule (19-15). Differentiate each term with respect to $x$ and solve for $y'$:

$$2 - y' = 0$$
$$y' = 2$$

Note for the derivative of $y$ you just write $y'$. Check that you get the same answer when you solve for $y$ and differentiate explicitly.

### Example 19-44
Given the formula $IR = 12$.

**1.** Find $I' = dI/dR$ implicitly.

**2.** Find $I'$ explicitly.

**3.** Show that the answers to (1) and (2) are the same.

### Solution
**1.** Find $I' = dI/dR$ implicitly.

Differentiate $IR$ with respect to $R$, using the product formula (19-10):

$$I(1) + RI' = 0$$

Since you are differentiating with respect to $R$, $R' = 1$. Then solve for $I'$:

$$I' = \frac{-I}{R}$$

Notice that $I'$ is expressed in terms of $I$ and $R$.

**2.** Find $I'$ explicitly.

Solve for $I$ and differentiate:

$$I = \frac{12}{R} = 12R^{-1}$$

$$I' = -12R^{-2} = \frac{-12}{R^2}$$

**3.** Show that the answers to (1) and (2) are the same.

Substitute $12/R$ for $I$ in the answer to (1):

$$I' = \frac{-I}{R} = \frac{-12/R}{R} = \frac{-12}{R^2}$$

## Example 19-45

Find $y' = dy/dx$ implicitly for the circle:

$$x^2 + y^2 = 16$$

### Solution

Differentiate each term with respect to $x$ and solve for $y'$:

$$2x(1) + 2yy' = 0$$
$$2yy' = -2x$$
$$y' = \frac{-x}{y}$$

Notice that to find the derivative of the term $y^2$, you treat $y$ as a function of $x$, like $u$ in the power formula (19-14).

## Example 19-46

Find $s' = ds/dt$ for the function $s^3 + s^2t - t^3 = 0$.

### Solution

Differentiate each term with respect to $t$, applying the product formula to $s^2t$:

$$3s^2s' + [s^2(1) + t(2ss')] - 3t^2(1) = 0$$

Solve for $s'$ by factoring out $s'$ and putting the other terms on the opposite side:

$$s'(3s^2 + 2st) = 3t^2 - s^2$$
$$s' = \frac{3t^2 - s^2}{3s^2 + 2st}$$

**Exercise 19-7**

In problems 1 to 4, find the indicated derivative (a) implicitly and (b) explicitly. Show that the two answers agree.

**1.** $\dfrac{y}{x} = 1$; $y' = \dfrac{dy}{dx}$

**2.** $PV = 20$; $P' = \dfrac{dP}{dV}$

**3.** $at + 3 = 7t$; $a' = \dfrac{da}{dt}$

**4.** $mv^2 - 3.3 = 0$; $m' = \dfrac{dm}{dv}$

In problems 5 to 20, find the indicated derivative implicitly.

**5.** $3y + 2x = 6;\ \dfrac{dy}{dx}$                 **6.** $4x - 3y = 0;\ \dfrac{dy}{dx}$

**7.** $4t^2 + t - 3w = 4;\ \dfrac{dw}{dt}$         **8.** $v^2 = 3v - t;\ \dfrac{dv}{dt}$

**9.** $xy + x = 6;\ \dfrac{dy}{dx}$             **10.** $3x^2 + 2tx = 2;\ \dfrac{dx}{dt}$

**11.** $x^2 + y^2 = 25;\ \dfrac{dy}{dx}$          **12.** $2h^2 + r^2 = 9;\ \dfrac{dh}{dr}$

**13.** $Mt^2 - 1.5t + 3.1 = 0;\ \dfrac{dM}{dt}$      **14.** $F^2x - F = 4.3;\ \dfrac{dF}{dx}$

**15.** $(x - 2)^2 + (y - 1)^2 = 16;\ \dfrac{dy}{dx}$      **16.** $y^2 + 2y = x + 2;\ \dfrac{dy}{dx}$

**17.** $p^2 + pq + q^2 = 0;\ \dfrac{dp}{dq}$        **18.** $y^3 + xy^2 = 1;\ \dfrac{dy}{dx}$

**19.** $V = \dfrac{\pi r^2 h}{3}$  (V constant)$;\ \dfrac{dh}{dr}$      **20.** $\sqrt{x^2 + y^2} = r$  (r constant)$;\ \dfrac{dy}{dx}$

Solve problems 21 to 24 by differentiating implicitly.

**21.** Find the slope of the tangent to the hyperbola $x^2 - y^2 = 9$ at the point $(5, 4)$.

**22.** Find the slope of the tangent to the ellipse $2x^2 + y^2 = 9$ at the point $(2, 1)$.

**23.** The displacement of a body moving in a straight line is given by $3.2s - t^2 = 8.2t$. Find the velocity when $t = 0$.

**24.** The voltage as a function of time in an $RC$ parallel circuit is given by $v^2 = 2t^3 - 1$. Find $dv/dt$ in terms of $t$.

# 19-8 | *Review Questions*

In problems 1 to 6, find each limit.

**1.** $\lim\limits_{x \to -1} (2 - 3x)$            **2.** $\lim\limits_{x \to 2} \dfrac{x}{x^2 - 1}$

**3.** $\lim\limits_{h \to 1} \dfrac{h - 1}{h^2 + h - 2}$       **4.** $\lim\limits_{t \to 0.3} \dfrac{t^2 - 0.09}{t^2 - 0.3t}$

**5.** $\lim\limits_{n \to \infty} \dfrac{n^2 + n}{n^2 + 2}$        **6.** $\lim\limits_{a \to b} \dfrac{a^2 - ab}{a^2 - b^2}$

In problems 7 to 10, use the delta process to find the derivative of each function.

**7.** $y = 3x + 7$             **8.** $f(x) = 2 - x^2$

**9.** $s = t^3 - t$              **10.** $P = \dfrac{5}{V}$

In problems 11 to 26, find the derivative of each function.

**11.** $y = 6x^2 - 3x + 2$

**12.** $f(x) = \dfrac{x^4 + x^2 + 1}{4}$

**13.** $S = 3.6\pi r + 2.8\pi r^2$

**14.** $v = (t + 1)^2(t - 1)$

**15.** $m = (3k - 3)^3$

**16.** $f(h) = 5h - \dfrac{1}{h} + \dfrac{3}{h^2}$

**17.** $y = \dfrac{x^2}{2x + 1}$

**18.** $f(d) = \dfrac{2d^2 + 2}{d - 1}$

**19.** $s = \dfrac{M}{3I + 2} + 2I^3$   ($M$ constant)

**20.** $E = E_0\left(\dfrac{f - 3.2}{f + 5.1}\right)$   ($E_0$ constant)

**21.** $y = 4(2 - 3x)^4$

**22.** $q = (p + 3)^3(p - 2)^2$

**23.** $s = \dfrac{\sqrt{t}}{t - 1}$

**24.** $f(x) = \dfrac{2}{(x + 1)^{3/2}}$

**25.** $F = \sqrt{\dfrac{x}{x - 2}}$

**26.** $h = \sqrt{a^2 + r^2}$   ($a$ constant)

In problems 27 to 30, find the indicated derivative implicitly.

**27.** $xy^2 = 1$; $\dfrac{dy}{dx}$

**28.** $Mx - 5 = x^2$; $\dfrac{dM}{dx}$

**29.** $s^2 - st + t^2 = 0$; $\dfrac{ds}{dt}$

**30.** $ax^2 + by^2 = a^2b^2$   ($a$, $b$ constants); $\dfrac{dy}{dx}$

**31.** A voltage is applied to a dc circuit, causing the charge on a capacitance to increase to 0.004 C during the first tenth of a second. Every tenth of a second thereafter, the charge increases by one-half of the increase during the last tenth of a second. The total charge is then given by the infinite series $Q = 0.004 + 0.002 + 0.001 + \cdots$. If $t = $ time, find $\lim\limits_{t \to \infty} Q$.

**32.** The velocity of sound in air (in meters per second) at a temperature $t°C$ is given approximately by:

$$V = 331.5\sqrt{1 + \dfrac{t}{273}}$$

Find $\lim\limits_{t \to 0} V$.

**33.** Given the parabola $y = 2x^2 - x$: *(a)* Find the slope at the point $(2, 6)$. *(b)* Find the coordinates of the vertex where $y' = 0$.

**34.** Given the curve $y = -1/x$: *(a)* Find the slope at the point $(2, -\frac{1}{2})$. *(b)* Find the two points where the slope $y' = 1$.

**35.** The distance (in meters) traveled by a supersonic aircraft after landing is given approximately by $s = 410 - (4.5 - t)^4$, where $t = $ time (in seconds). *(a)* Find the velocity upon touchdown ($t = 0$). *(b)* Find the total distance traveled after landing.

**36.** The angular velocity (in radians per second) of a flywheel is given by:

$$\omega = \frac{10}{t^2 + 1}$$

where the time $t$ varies from 0 to 10 s. *(a)* Find the formula for the angular acceleration $\alpha = \omega'$. *(b)* Find $\alpha$ when $t = 5$ s.

37. The charge (in coulombs) in an electric circuit is given by $q = 1.6t^{3/2} + 2.0t$. Find the current (in amperes) at $t = 4$ s. (See Example 19-28.)

38. The induced voltage in a coil is given by:

$$V = N \frac{d\phi}{dt}$$

where $N$ = number of turns, $\phi$ = magnetic flux in webers (Wb), and $t$ = time in seconds. A coil of 20 turns is linked by a magnetic flux $\phi = 0.4t^3 + 0.7\sqrt{t}$. Find the induced voltage at $t = 0.5$ s.

39. The force on a piston is related to the distance $x$ between the piston and the end of the cylinder by the equation $Fx = 20$. Find $F' = dF/dx$.

40. The deflection $y$ of a cantilever beam with a load at the free end is given by the equation:

$$16y = 50{,}000\left(3x - \frac{x^2}{2}\right)$$

where $x$ is the distance from the fixed end. Find the rate of change of $y$ with respect to $x$.

41. The radius in centimeters of a solid circular shaft needed to transmit $H$ hp at $n$ r/min with a maximum torsional stress is given by:

$$r = \sqrt[3]{\frac{45{,}600H}{nS}}$$

Assuming $n$ and $S$ are constant, find $r' = dr/dH$.

42. In thermodynamics, in the study of flow through a nozzle, the following equation occurs:

$$y = \left(\frac{p}{p_0}\right)^{1.4} - \left(\frac{p}{p_0}\right)^{1.7}$$

Assuming $p_0$ = constant, find $y' = dy/dp$.

43. The power in a dc circuit containing a variable resistance $R$ is given by:

$$P = \frac{24R}{(2 + R)^2}$$

*(a)* Find the formula for $P'$. *(b)* Find the value of $R$ that makes $P' = 0$.

44. If $x^{1/3} + y^{1/3} = a^{1/3}$ with $a$ constant, show that:

$$y' = \frac{dy}{dx} = -\left(\frac{y}{x}\right)^{2/3}$$

# 20

# Applications of the Derivative

## 20-1 | *Velocity and Acceleration: Higher-Order Derivatives*

Section 19-4 introduced problems involving motion in a straight line and showed how derivatives are used to find velocity and acceleration. This section explores these ideas further and studies angular motion and motion along a curve which involve velocity components. The first example leads to the concept of higher-order derivatives.

### Example 20-1

A commuter train approaching a station applies its brakes. The distance traveled (in feet) from the moment its brakes are applied, is given by $s = 120t - t^2$, where the time $t$ varies from 0 to 60 s.

1. Find the initial and final velocities.
2. Find the initial and final acceleration.

### Solution

1. Find the initial and final velocities.

   Since velocity is the derivative of distance, find $s' = ds/dt$:

   $$v = s' = \frac{ds}{dt} = 120 - 2t$$

The initial velocity at $t = 0$ is:

$$v = 120 - 2(0) = 120 \text{ ft/s}$$

The final velocity at $t = 60$ s is:

$$v = 120 - 2(60) = 0 \text{ ft/s}$$

The train therefore comes to a stop after 60 s.

2. Find the initial and final acceleration.

Since acceleration $a$ is the derivative of velocity, find $v' = dv/dt$:

$$a = v' = \frac{dv}{dt} = -2 \text{ ft/s}^2$$

Notice the acceleration is constant and therefore the initial and final acceleration are the same.

Motion where the acceleration is constant is called *uniform accelerated motion*. A body falling in a gravitational field is an example of uniform accelerated motion where the constant acceleration is that due to gravity $g = 32$ ft/s$^2$ or $g = 9.8$ m/s$^2$. When the acceleration is zero and the velocity is constant, the motion is called *uniform motion*.

## Higher-Order Derivatives

Acceleration is also the *second derivative* of the distance. This is expressed as:

$$a = s'' = \frac{d^2s}{dt^2}$$

In general, if $y = f(x)$, the derivative of the derivative is called the *second derivative* and is expressed as:

$$y'' = f''(x) \quad \text{or} \quad \frac{d^2y}{dx^2} \quad \text{or} \quad D_x^2 y$$

Similarly, the *third derivative* of $y = f(x)$ is expressed as:

$$y''' = f'''(x) \quad \text{or} \quad \frac{d^3y}{dx^3} \quad \text{or} \quad D_x^3 y$$

Higher-order derivatives are expressed similarly.

The next example shows an application to angular motion.

### Example 20-2

A wheel turns through an angle $\theta$ (in radians) given by $\theta = 0.04t^{5/2} + 0.10t^2$, where $t = $ time (in seconds).

**1.** Find the angular velocity $\omega$ when $t = 4$ s.

**2.** Find the angular acceleration when $t = 4$ s.

### Solution

**1.** Find the angular velocity $\omega$ when $t = 4$ s.

The angular velocity $\omega$ is the derivative of the angular displacement $\theta$:

$$\omega = \theta' = \frac{d\theta}{dt} = 0.10t^{3/2} + 0.20t$$

At $t = 4$ s:

$$\omega = 0.10(4)^{3/2} + 0.20(4)$$
$$= 0.10(8) + 0.80 = 1.6 \text{ rad/s}$$

**2.** Find the angular acceleration when $t = 4$ s.

The angular acceleration $\alpha$ is the derivative of the angular velocity $\omega$ or the second derivative of $\theta$:

$$\alpha = \omega' = \frac{d\omega}{dt} = \frac{d^2\theta}{dt^2} = 0.15t^{1/2} + 0.20$$

At $t = 4$ s:

$$\alpha = 0.15(4)^{1/2} + 0.20$$
$$= 0.15(2) + 0.20 = 0.50 \text{ rad/s}^2$$

The next example illustrates motion along a curve. When a body moves along a curve, its velocity at any point is a vector which has components in the $x$ direction and the $y$ direction. Vectors and vector components are presented in Sec. 10-4. In describing the position of a body along a curve, *parametric equations* are usually used. The $x$ coordinate and the $y$ coordinate are each given in terms of a third variable, or parameter (Sec. 11-4).

### Example 20-3

A body moves in the path of a parabola given by the parametric equations $x = t$ and $y = 9 - t^2$ (Fig. 20-1).

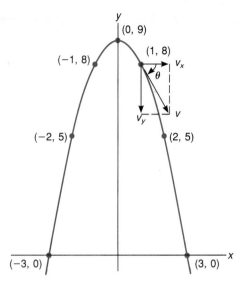

FIG. 20-1  *Velocity components along a curve, Example 20-3*

1. Find the velocity components at $t = 1$.
2. Find the resultant velocity at $t = 1$.

### Solution

1. Find the velocity components at $t = 1$.

   The velocity component in the $x$ direction is $v_x = dx/dt$, and the velocity component in the $y$ direction is $v_y = dy/dt$:

$$v_x = \frac{dx}{dt} = 1 \qquad v_y = \frac{dy}{dt} = -2t$$

   At $t = 1$:

$$v_x = 1 \qquad v_y = -2(1) = -2$$

   Notice that $v_x$ is constant along the curve.

2. Find the resultant velocity at $t = 1$.

   The magnitude of the resultant velocity $v = \sqrt{v_x^2 + v_y^2}$ by the pythagorean theorem:

$$v = \sqrt{(1)^2 + (-2)^2} = \sqrt{5} = 2.2$$

The negative angle $\theta$ that $v$ makes with the horizontal is:

$$\theta = \tan^{-1} \frac{-2}{1} = -63.4°$$

**Exercise 20-1**

In problems 1 to 4, for each equation of linear motion, find the velocity and acceleration for the given value of $t$. See Example 20-1.

1. $s = 4t^2 - 3$; $t = 3$

2. $s = 2t - \dfrac{1}{t}$; $t = -1$

3. $s = \dfrac{t}{t + 1}$; $t = -2$

4. $s = t + \sqrt{t}$; $t = 1$

In problems 5 to 8, for each equation of angular motion, find the angular velocity and angular acceleration for the given value of $t$. See Example 20-2.

5. $\theta = t - t^3$; $t = \dfrac{1}{2}$

6. $\theta = \dfrac{5}{t^2}$; $t = 1.5$

7. $\theta = \sqrt{2t + 3}$; $t = 3$

8. $\theta = 3t^{4/3} + 6t^{1/3}$; $t = 8$

In problems 9 to 12, for each pair of parametric equations describing the motion of a body along a curve, find (a) velocity components and (b) resultant velocity for the given value of $t$. See Example 20-3.

9. $x = 2t$, $y = t + 1$; $t = 3$

10. $x = t$, $y = t^2 + t$; $t = -2$

11. $x = t^2$, $y = \dfrac{1}{t}$; $t = 0.5$

12. $x = \dfrac{1}{\sqrt{t + 1}}$, $y = t$; $t = 0$

13. A weather balloon is released from the ground and rises in the atmosphere. Its distance above the ground (in meters) from the moment it is released is given by $s = 90t - 9t^2$, where the time $t$ varies from 0 to 4 min. (a) Find the initial and final velocities. (b) Find the initial and final accelerations.

14. A car accelerates from rest to a maximum velocity at $t = 30$ s and then decelerates to a stop at $t = 60$ s. The distance traveled (in meters) is given by $s = t^2 - t^3/90$. (a) Find the maximum velocity. (b) Find the initial and final acceleration. This type of motion where the acceleration is not constant is called *nonuniform accelerated motion*.

15. A gear turns through an angle $\theta$ (in radians) given by $\theta = 1.6t^2 + 2.0t^{3/2}$, where $t$ = time (in seconds). *(a)* Find the angular velocity when $t = 4$ s. *(b)* Find the angular acceleration when $t = 4$ s.

16. The angular displacement (in radians) of a flywheel is given by $\theta = t^2 - 4t^{3/2}$, where $t$ = time (in seconds) and $t > 0$. *(a)* Find the value of $t$ when the angular velocity is zero. *(b)* Find the angular acceleration when the wheel changes direction, that is, when the angular velocity is zero.

17. A body moves in the path of a parabola given by the parametric equations $x = t$ and $y = t^2 + 3$. *(a)* Sketch the parabola and find the velocity components at $t = 2$. *(b)* Find the resultant velocity at $t = 2$.

18. A projectile moves according to the parametric equations $x = 60t$ and $y = 120t - 16t^2$, where $x$ and $y$ are in feet and $t$ is in seconds. *(a)* Find the magnitude of the resultant velocity at $t = 1$ s. *(b)* Find the magnitude of the acceleration at $t = 1$ s. (*Hint:* Differentiate the velocity components.)

19. Show for the quadratic function $y = ax^2 + bx + c$ (*a, b, c* constants) that $d^3y/dx^3 = 0$.

20. Given $y' = 1/x$, show that $y''' = -2y''/x$.

## 20-2 | *Curve Sketching: Maximum, Minimum, and Inflection Points*

Chapter 19 shows that the geometric model of the derivative is a formula for the slope of a curve at any point. Because of this property, derivatives are very useful in sketching curves. When the first derivative or slope is positive, the curve is increasing. When the first derivative or slope is negative, the curve is decreasing. When a curve turns through a *relative maximum point* or a *relative minimum point*, the tangent is horizontal and the first derivative or slope $y' = f'(x) = 0$ at these points (Fig. 20-2). These points are simply referred to as maximum and minimum points. A point where $y' = 0$ is called a *critical point*.

The second derivative is the rate of change of the slope, and it gives information about the curvature or concavity of the curve. When the second derivative is positive, the slope is increasing and the curve is concave up (Fig. 20-2). When the second derivative is negative, the slope is decreasing and the curve is concave down. Where a curve twists and changes concavity, the curvature is zero at this point and the second derivative $y'' = f''(x) = 0$ at this point. This point is called an *inflection point*.

Carefully study the first example, which explains these basic properties of curves.

### Example 20-4
Given the function $y = x^3 - 6x^2 + 9x + 3$:

1. Find the maximum and minimum points of the function.

**2.** Find the inflection point of the curve of the function and sketch the graph.

**Solution**

**1.** Find the maximum and minimum points of the function.

At a maximum or minimum point, the first derivative equals zero. Therefore first find $y'$:

$$y' = 3x^2 - 12x + 9$$

Set $y' = 0$ and solve the equation for $x$:

$$3x^2 - 12x + 9 = 0$$
$$x^2 - 4x + 3 = 0 \qquad \text{(divide by 3)}$$
$$(x - 1)(x - 3) = 0$$
$$x = 1 \qquad x = 3$$

The values $x = 1$ and $x = 3$ where $y' = 0$ are called *critical values*. To determine whether a critical value is a maximum or minimum, find the values of $y$ and $y'$ at points on either side of the critical value. The following table shows the values of $y$ and $y'$ for $x = 0, 1, 2, 3, 4$:

| | | Maximum point ↓ | | | Minimum point ↓ | |
|---|---|---|---|---|---|---|
| $x$ | 0 | 1 | 2 | | 3 | 4 |
| $y$ | 3 | 7 | 5 | | 3 | 7 |
| $y'$ | +9 | 0 | −3 | | 0 | +9 |
| | | Sign change | | | Sign change | |

For example, when $x = 2$:

$$y = (2)^3 - 6(2)^2 + 9(2) + 3 = 5$$
and $\qquad y' = 3(2)^2 - 12(2) + 9 = -3$

The value $x = 1$ is a maximum point because the values of $y$ are less than 7 on either side of $x = 1$. The value $x = 3$ is a minimum point because the values of $y$ are greater than 3 on either side of $x = 3$. Note that these points are *relative* maxima and minima. The values of $y$ are greater than 7 when $x > 4$ and are less than 3 when $x < 0$ (Fig. 20-2).

Maxima and minima can also be determined by the values of the first derivative or slope on either side of the critical point. The *first-derivative test for maxima and minima* is:

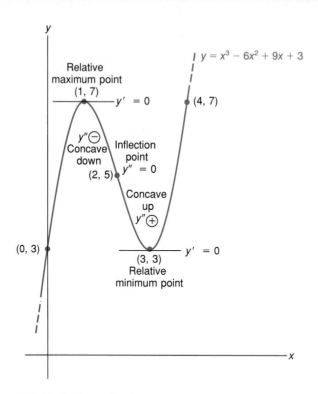

FIG. 20-2  *Curve sketching, Example 20-4*

If the first derivative changes from positive to negative as $x$ increases through a critical point, then the point is a relative maximum point.

If the first derivative changes from negative to positive as $x$ increases through a critical point, then the point is a relative minimum point.

(20-1)

Notice how the first-derivative test applies in the above table. The sign of $y'$ changes from ⊕ to ⊖ as $x$ increases through the maximum point. The sign of $y'$ changes from ⊖ to ⊕ as $x$ increases through the minimum point.

2. Find the inflection point of the curve of the function and sketch the graph.

At an inflection point the second derivative equals zero. Therefore first find $y''$:

$$y'' = 6x - 12$$

Set $y'' = 0$ and solve for $x$:

$$6x - 12 = 0$$
$$x = 2$$

The value $x = 2$ is where the rate of change of the slope is zero. The curve "flattens" out at this point, and the curvature is zero.

To test if the point where $y'' = 0$ is an inflection point, it is necessary to determine if the curvature or concavity changes. Find values of $y''$ on either side of $x = 2$.

At $x = 1$:         $y'' = 6(1) - 12 = -6$

sign change

At $x = 3$:         $y'' = 6(3) - 12 = +6$

The second derivative changes sign at $x = 2$. Therefore it is an inflection point. The sketch is shown in Fig. 20-2.

Notice in Fig. 20-2 that the slope is decreasing at the maximum point and increasing at the minimum point. This leads to the *second-derivative test for maxima and minima:*

If the second derivative is negative at a critical point, then the curve is concave down and the point is a relative maximum point. If the second derivative is positive at a critical point, then the curve is concave up and the point is a relative minimum point.                      (20-2)

From Example 20-4 the *test for an inflection point* is:

If the second derivative changes sign at a point on a curve, then the point is called an inflection point.                      (20-3)

Study the next example, which applies all the above ideas.

### Example 20-5
Sketch the graph of $f(x) = 2x^3 - 3x^2 - 12x + 6$, showing the maximum, minimum, and inflection points.

### Solution
Find $f'(x)$ and $f''(x)$. Set each equal to zero and solve for $x$:

$$f'(x) = 6x^2 - 6x - 12 \qquad\qquad f''(x) = 12x - 6$$

$$6x^2 - 6x - 12 = 0 \qquad\qquad 12x - 6 = 0$$

$$x^2 - x - 2 = 0 \quad \text{(divide by 6)} \qquad x = \frac{1}{2} \text{ or } 0.5$$

$$(x + 1)(x - 2) = 0$$

$$x = -1 \qquad x = 2$$

Set up a table for values of $x$, $f(x)$, $f'(x)$, and $f''(x)$. Choose values of $x$ which include values on either side of the critical points and on either side of the point where $f''(x) = 0$. For this example, choose $x = -2, -1, 0, \frac{1}{2}, 1, 2, 3$. Calculate values of $f(x)$ for all values of $x$. Calculate values of $f'(x)$ on either side of the critical points. Calculate values of $f''(x)$ on either side of the point where $f''(x) = 0$ and at the critical points. The table looks as follows:

| | | Maximum point ↓ | | Inflection point ↓ | | Minimum point ↓ | |
|---|---|---|---|---|---|---|---|
| $x$ | $-2$ | $-1$ | $0$ | $\frac{1}{2}$ | $1$ | $2$ | $3$ |
| $f(x)$ | $2$ | $13$ | $6$ | $-\frac{1}{2}$ | $-7$ | $-14$ | $-3$ |
| $f'(x)$ | $+24$ | $0$ | $-12$ | | $-12$ | $0$ | $+24$ |
| $f''(x)$ | | $-18$ | $-6$ | $0$ | $+6$ | $+18$ | |
| | | ↑ Concave down | | ↳ Sign ↲ change | | ↑ Concave up | |

For example, at $x = 1$:

$$f(1) = 2(1)^3 - 3(1)^2 - 12(1) + 6 = -7$$

$$f'(1) = 6(1)^2 - 6(1) - 12 = -12$$

$$f''(1) = 12(1) - 6 = +6$$

Study the table carefully and notice the following:

1. At the critical point $x = -1$, three things in the table show it is a maximum point:
   a. On either side of $x = -1$ the values of $f(x)$ are less than $f(-1) = 13$.
   b. Test (20-1) applies: The value of $f'(x)$ changes from positive to negative, $f'(-2) = +24$ and $f'(0) = -12$.
   c. Test (20-2) applies: The value of $f''(x)$ at $x = -1$ is negative. So $f''(-1) = -18$, and the curve is *concave down*.

2. At the critical point $x = 2$, three things in the table show it is a minimum point:

   a. On either side of $x = 2$ the values of $f(x)$ are greater than $f(2) = -14$.
   b. Test (20-1) applies: The value of $f'(x)$ changes from negative to positive, $f'(1) = -12$ and $f'(3) = +24$.
   c. Test (20-2) applies: The value of $f''(x)$ at $x = 2$ is positive. So $f''(2) = +18$, and the curve is *concave up*.

**3.** At the point $x = \frac{1}{2}$, $f''(x)$ changes sign: $f''(0) = -6$ and $f''(1) = +6$. The concavity changes at $x = \frac{1}{2}$, and, therefore, by test (20-3), it is an inflection point.

   Using the above information, you can sketch the curve, showing the important points. Notice in the sketch in Fig. 20-3 that the scales are different on each axis to make plotting the points easier. This curve can have at most two turning points. It increases indefinitely as $x$ increases above 2, and it decreases indefinitely as $x$ decreases below $-1$. If a more precise graph is necessary, more points can be plotted.

   Examples 20-4 and 20-5 are graphs of polynomials which have certain properties because of the nature of their derivatives. If $y = f(x)$ is a polynomial function of degree $n$, then the first derivative $y' = f'(x)$ is a polynomial function of degree $n - 1$ and the second derivative is a polynomial function of degree

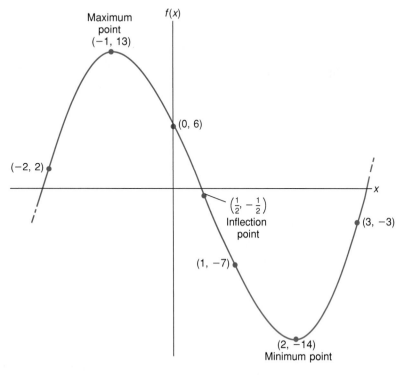

FIG. 20-3 *Sketch of* $y = 2x^3 - 3x^2 - 12x + 6$, *Example 20-5*

$n - 2$ (see Sec. 15-1). When $y'$ and $y''$ are set equal to zero, the polynomial equations that result can have *at most* $n - 1$ and $n - 2$ distinct real roots, respectively (see Sec. 15-4). Therefore, it follows that the graph of $y = f(x)$ can have *at most* $n - 1$ critical points and $n - 2$ inflection points. In Examples 20-4 and 20-5 the polynomial functions are both of degree 3, and each has two critical points and one inflection point. This is the greatest number of each type of point that they can have. Note that the graph of $f(x)$ *can have fewer* than $n - 1$ critical points and $n - 2$ inflection points. The next example illustrates such a case.

### Example 20-6

Sketch the graph of $y = x^4 - 4x^3 + 10$, showing the maximum, minimum, and inflection points.

### Solution

Find $y'$ and $y''$. Set each equal to zero and solve for $x$:

$$y' = 4x^3 - 12x^2 \qquad y'' = 12x^2 - 24x$$
$$4x^3 - 12x^2 = 0 \qquad 12x^2 - 24x = 0$$

Solve equations by factoring:

$$4x^2(x - 3) = 0 \qquad\qquad 12x(x - 2) = 0$$
$$4x^2 = 0 \qquad x - 3 = 0 \qquad 12x = 0 \qquad x - 2 = 0$$
$$x = 0 \qquad\quad x = 3 \qquad\quad x = 0 \qquad\quad x = 2$$

Construct a table of values of $x$, $y$, $y'$, and $y''$ that includes values of $x$ on either side of the points $x = 0$, $x = 2$, and $x = 3$. Calculate values of $y'$ and $y''$ on either side of the points where each is equal to zero:

|   |   | Inflection point ↓ |   |   | Inflection point ↓ | Minimum point ↓ |   |
|-----|------|------|------|------|------|------|------|
| $x$  | $-1$  | 0   | 1   | 2   | 3   | 4 |
| $y$  | 15   | 10  | 7   | $-6$  | $-17$ | 10 |
| $y'$ | $-16$ | 0   | $-8$  | $-16$ | 0   | $+64$ |
| $y''$| $+36$ | 0   | $-12$ | 0   | $+36$ | |

Sign change ↗↖          Sign change ↗

Study the table carefully. Notice that when $x = 0$, both $y' = 0$ and $y'' = 0$. When this happens, the second-derivative test for maxima and minima (20-2) cannot be used. When you apply the first-derivative test (20-1), you see that $y'$ does not change sign; therefore $x = 0$ is not a maximum or minimum point.

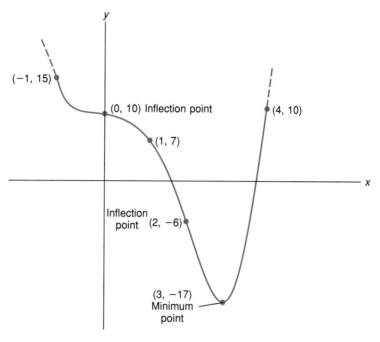

FIG. 20-4 *Sketch of* $y = x^4 - 4x^3 + 10$, *Example 20-6*

However, $y''$ does change sign at $x = 0$, and, therefore, by applying (20-3), it is an inflection point. The curve at $x = 0$ levels off and straightens out, creating a horizontal inflection point. Similarly, $x = 2$ is an inflection point, and $x = 3$ satisfies the tests for a minimum point. The sketch in Fig. 20-4 shows the concavity changing at each of the inflection points.

Table 20-1 summarizes the three cases encountered in applying derivatives to curve sketching and indicates the most efficient test for each case.

**Table 20-1 APPLICATION OF DERIVATIVES TO CURVE SKETCHING**

| Derivative | Explanation | Point Test |
|---|---|---|
| $y' = 0$ and $y'' \neq 0$ | Horizontal tangent | Maximum if $y''$ is $\ominus$<br>Minimum if $y''$ is $\oplus$ |
| $y'' = 0$ and $y' \neq 0$ | "Flat" spot | Inflection point if $y''$ changes sign |
| $y' = 0$ and $y'' = 0$ | Both of the above | Maximum if $y'$ changes from $\oplus$ to $\ominus$<br>Minimum if $y'$ changes from $\ominus$ to $\oplus$<br>Inflection point if $y''$ changes sign |

### Example 20-7

Sketch the graph of $y = 2x^4 - x - 2$.

### Solution

Set $y'$ and $y''$ equal to zero and solve for $x$:

$$y' = 8x^3 - 1 \qquad\qquad y'' = 24x^2$$
$$8x^3 - 1 = 0 \qquad\qquad 24x^2 = 0$$
$$x^3 = \frac{1}{8} \qquad\qquad\qquad x = 0$$
$$x = \frac{1}{2} = 0.5$$

The table of values showing $x$, $y$, $y'$, and $y''$ is:

| | | | | Minimum point ↓ | |
|---|---|---|---|---|---|
| $x$ | −1 | −0.5 | 0 | 0.5 | 1 |
| $y$ | +1 | −1.4 | −2 | −2.4 | −1 |
| $y'$ | | | −1 | 0 | +7 |
| $y''$ | | +6 | 0 | +6 | |

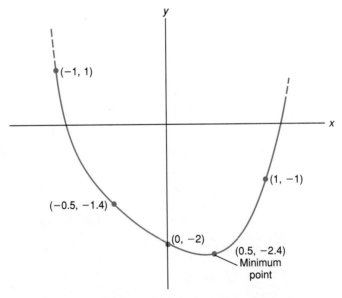

FIG. 20-5 *Sketch of $y = 2x^4 - x - 2$, Example 20-7*

The point $x = -0.5$ is included to aid in sketching the curve. Notice that $y''$ is always positive, and the curve is therefore always concave up. It has no inflection points and one minimum point. At $x = 0$, where $y'' = 0$, the curve straightens out, but the concavity does not change (Fig. 20-5).

**Exercise 20-2**

In problems 1 to 4, find the coordinates of any maximum or minimum points of the function.

**1.** $y = x^2 + 6x$                                         **2.** $y = x^3 + 9x^2$

**3.** $f(x) = 3x^3 - 6x^2 + 3x - 1$              **4.** $f(t) = 4t^2 - 2t^4$

In problems 5 to 8, find the coordinates of any inflection points of the function.

**5.** $y = 2x^3 - 12x^2 + 3x + 8$                 **6.** $y = 5x^3 - 3x^2 + x$

**7.** $f(x) = 2x^4 - 2x^3 + x$                        **8.** $f(t) = t - t^4$

In problems 9 to 30, sketch the graph of each function, showing all maximum, minimum, and inflection points.

**9.** $y = x^2 - 4x + 3$                                   **10.** $y = 5x - x^2$

**11.** $y = x^3 - 3x$                                        **12.** $y = 12x - x^3$

**13.** $f(x) = x^3 - 3x^2 - 9x + 10$             **14.** $f(x) = x^3 + 6x^2 + 9x$

**15.** $f(x) = x^3 - 2$                                     **16.** $f(x) = x^3 - 3x^2 + 3x - 1$

**17.** $y = 2x^3 + 3x^2 - 12x - 8$              **18.** $y = 2x^3 - 6x^2 + 16$

**19.** $s = \dfrac{t^3}{3} + 3t^2$                          ⚆ **20.** $s = 8t^3 - 9t^2 + 1$

**21.** $y = 6x^2 - x^4$                                    ⚆ **22.** $y = x^4 - 8x^3 + 10$

**23.** $f(x) = x^4 + 4x^3 + 12$                      **24.** $f(t) = 8t^3 - 2t^4$

**25.** $y = x^4 + 1$                                         **26.** $y = (x + 4)^4$

**27.** $w = x^4 + 4x$                                      **28.** $M = 16x - \dfrac{x^4}{2}$

**29.** $f(x) = x^5 - 5x$                                  ⚆ **30.** $f(t) = t^4 - \dfrac{t^5}{5}$

⚆ **31.** The electric energy supplied by a generator in joules (J) is given by $W = 2000(14t - 2t^2)$, where $t =$ time (in hours). Sketch the graph of $W$ vs. $t$, showing the point of maximum energy.

⚆ **32.** The power in an electric circuit is given by $P = I^2R$, where the resistance $R = 48\ \Omega$ and the current is given by $I = t^2 - t$. Sketch the curve of $P$ vs. $t$ from $t = 0.30$ to $t = 1.3$ s, showing the maximum and minimum power.

⚆ **33.** The deflection $y$ (in centimeters) of a beam is given by:

$$y = 0.01(2x^3 - 24x^2 + 72x - 40)$$

where $x =$ distance (in meters) from the left end. If the beam is 8.0 m long, sketch the curve of deflection, showing the points of maximum deflection.

$\boxed{\texttt{C}}$ **34.** The monthly profit of a company that manufactures microcomputers is given by:

$$P = 300n^2 - 10n^3$$

where $n$ = number of microcomputers manufactured each month. Sketch the profit curve from $n = 0$, showing the point of maximum profit.

**35.** Show that the function $y = ax^3$ ($a$ = constant) has no maximum or minimum points.

**36.** Show that the function $y = ax^4$ ($a$ = constant) has no inflection points.

## 20-3 | *Curve Sketching: Intercepts, Asymptotes, and Symmetry*

In addition to using derivatives to sketch the graph of a function, three properties of the function itself help to determine the curve. The first of these, the *intercepts,* are used in Chaps. 14 and 15 to sketch curves. The *y intercepts* are found by setting $x = 0$, and the *x intercepts* are found by setting $y = 0$ [or $f(x) = 0$]. The *asymptotes* of a curve are introduced in Sec. 14-5 on the hyperbola. An *asymptote* is a line that a curve approaches but never touches as either $x \to \infty$ or $y \to \infty$. For example, Fig. 14-29 in Sec. 14-5 shows the asymptotes of a hyperbola. The tests for horizontal and vertical asymptotes are explained in the first example.

The third property, *symmetry,* relates to the shape of the curve with respect to the axes or the origin. *Symmetry to an axis* (the $x$ axis or the $y$ axis) means that half the curve on one side of the axis is the mirror image of the other half on the opposite side of the axis. For example, the parabola shown in Fig. 14-13, Sec. 14-2, is symmetric to the $y$ axis, and the hyperbola shown in Fig. 14-31, Sec. 14-5, is symmetric to the $x$ axis. *Symmetry to the origin* means that for each point $(x, y)$ on the curve there is a symmetric point $(-x, -y)$ that is the same distance from the origin on the opposite side. For example, the third-degree polynomial function shown in Fig. 15-3, Sec. 15-1, is symmetric to the origin. The tests for symmetry are explained in the first example. Study this first example carefully to learn how to apply the above ideas to curve sketching.

### Example 20-8
Sketch the graph of $y = f(x) = \dfrac{1}{x^2}$.

### Solution
The solution applies the ideas of derivatives shown in the preceding section as well as the ideas of intercepts, asymptotes, and symmetry.

*Derivatives.* Set $y'$ and $y''$ equal to zero:

$$y' = \frac{-2}{x^3} \qquad y'' = \frac{6}{x^4}$$

$$\frac{-2}{x^3} = 0 \qquad \frac{6}{x^4} = 0$$

Both these equations have no solutions, and therefore the curve has no maximum, minimum, or inflection points. Because $x^4$ is always positive, the second derivative is always positive, and therefore the curve is concave up throughout (Fig. 20-6).

*Intercepts.* To find the $y$ intercepts, set $x = 0$:

$$y = \frac{1}{0}$$

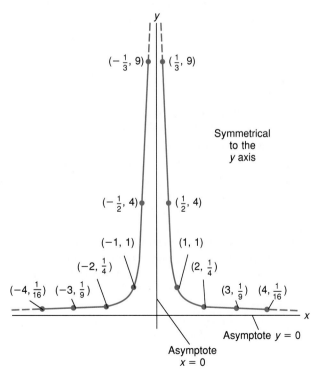

FIG. 20-6 *Sketch of $y = \dfrac{1}{x^2}$, Example 20-8*

But $y$ is not defined when $x = 0$, and the curve has no $y$ intercepts. To find the $x$ intercepts, set $y = 0$:

$$\frac{1}{x^2} = 0$$

This equation has no solution, and the curve has no $x$ intercepts.

*Asymptotes.*    To test for a *horizontal asymptote,* let $x \to \infty$ and examine the behavior of $y$.

As $x \to \infty$, $y = 1/x^2 \to 0$ and therefore the curve approaches the line $y = 0$, or the $x$ axis, as $x$ increases. The $x$ axis is therefore an asymptote to the curve.

To test for a *vertical asymptote,* examine the function for values of $x$ that make $y$ infinite. This can happen when *the denominator of a term approaches zero.* As $x \to 0$, $1/x^2 \to \infty$ and therefore the curve approaches the line $x = 0$, or the $y$ axis, as $y$ increases. The $y$ axis is therefore an asymptote to the curve.

*Symmetry.*    *A curve is symmetric to the y axis if when x is replaced by $-x$, the function does not change, that is, $f(x) = f(-x)$.*

Substitute $-x$ for $x$ in the function:

$$y = \frac{1}{(-x)^2} = \frac{1}{x^2}$$

Since the function does *not* change, the curve *is* symmetric to the $y$ axis.

*A curve is symmetric to the x axis if when y is replaced by $-y$, the equation does not change.*

Substitute $-y$ for $y$:

$$-y = \frac{1}{x^2} \qquad \text{or} \qquad y = \frac{-1}{x^2}$$

Since the equation or function *does* change, the curve is *not* symmetric to the $x$ axis. Note that *a function cannot be symmetric to the x axis* since only one value of $y$ can be defined for each value of $x$. For example, the equation of a circle, $x^2 + y^2 = r^2$, is *not* a function and is symmetric to the $x$ axis.

*A curve is symmetric to the origin if when x is replaced by $-x$ and y by $-y$ [or f(x) by $-f(x)$], the function does not change.*

Substitute $-x$ for $x$ and $-y$ for $y$:

$$-y = \frac{1}{(-x)^2} \qquad \text{or} \qquad y = \frac{-1}{x^2}$$

Since the function *does* change, the curve is *not* symmetric to the origin.

*Graph.* The curve therefore does not cross the $x$ or $y$ axis but approaches each axis as an asymptote. Furthermore, the half on the right side of the $y$ axis is symmetric to the half on the left side. To sketch the curve, you can therefore group positive and negative values of $x$, since they will have the same values for $y$:

| $x$ | $\pm\frac{1}{4}$ | $\pm\frac{1}{3}$ | $\pm\frac{1}{2}$ | $\pm 1$ | $\pm 2$ | $\pm 3$ | $\pm 4$ |
|---|---|---|---|---|---|---|---|
| $y$ | 16 | 9 | 4 | 1 | $\frac{1}{4}$ | $\frac{1}{9}$ | $\frac{1}{16}$ |

The graph is shown in Fig. 20-6, which uses different scales on the $x$ and $y$ axes to better show the shape of the curve.

## Example 20-9

Sketch the graph of $y = f(x) = \dfrac{1}{1 - x^2}$.

## Solution

*Derivatives.* Set $f'(x)$ and $f''(x)$ equal to zero and solve for $x$:

$$f'(x) = \frac{(1 - x^2)(0) - 1(-2x)}{(1 - x^2)^2} = \frac{2x}{(1 - x^2)^2} = 0$$

To find when a fraction is zero, set the numerator equal to zero:

$$2x = 0$$
$$x = 0$$

The function has one critical point at $x = 0$:

$$f''(x) = \frac{(1 - x^2)^2(2) - (2x)(2)(1 - x^2)(-2x)}{(1 - x^2)^{4^3}}$$

$$= \frac{(1 - x^2)(2) - (2x)(2)(-2x)}{(1 - x^2)^3}$$

$$= \frac{6x^2 + 2}{(1 - x^2)^3}$$

Set the numerator equal to zero:

$$6x^2 + 2 = 0$$

$$x^2 = -\frac{1}{3}$$

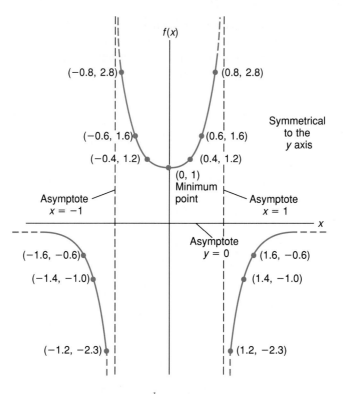

FIG. 20-7  *Sketch of* $y = \dfrac{1}{1 - x^2}$, *Example 20-9*

This equation has only imaginary roots, so the graph has no inflection points. When $x = 0$, $f''(0) = +2$, and therefore the critical point is a minimum point (Fig. 20-7).

*Intercepts.*   When $x = 0$, $f(x) = 1/(1 - 0) = 1$. The $y$ intercept $(0, 1)$ is therefore the minimum point.

When $f(x) = 0$, you have the equation $1/(1 - x^2) = 0$, which has no solutions for $x$, and therefore the curve does not cross the $x$ axis.

*Asymptotes.*   As $x \to \infty$, $f(x) \to 0$ and the curve approaches the $x$ axis as $x$ increases. The $x$ axis is therefore a horizontal asymptote.

The denominator of $f(x)$ will be zero when $1 - x^2 = 0$ or $x = \pm 1$. Therefore as $x \to +1$, or as $x \to -1$, $f(x) \to \infty$. The lines $x = 1$ and $x = -1$ are therefore vertical asymptotes.

*Symmetry.*   When $x$ is replaced by $-x$, the function does not change and the curve is symmetric to the $y$ axis.

Since a function can have only one value for each value of $x$, the curve is not symmetric to the $x$ axis.

When $x$ is replaced by $-x$ and $f(x)$ by $-f(x)$, the function is not the same, and the curve is not symmetric to the origin.

*Graph.* The table of values for the graph cannot include the asymptotes $x = \pm 1$. Choose several values on either side of $x = \pm 1$ to sketch the graph. The curve is symmetric to the $y$ axis, so positive and negative values can be grouped in the table:

| $x$ | 0 | ±0.4 | ±0.6 | ±0.8 | ±1.2 | ±1.4 | ±1.6 |
|---|---|---|---|---|---|---|---|
| $f(x)$ | 1 | 1.2 | 1.6 | 2.8 | −2.3 | −1.0 | −0.6 |

The graph is shown in Fig. 20-7, which uses different scales on the $x$ and $y$ axes to better show the shape of the curve.

## Example 20-10

Sketch the graph of $y = \dfrac{x}{x^2 + 1}$.

**Solution**
*Derivatives.* Set $y'$ and $y''$ equal to zero and solve for $x$:

$$y' = \frac{(x^2 + 1)(1) - x(2x)}{(x^2 + 1)^2} = \frac{1 - x^2}{(x^2 + 1)^2} = 0$$

$$1 - x^2 = 0$$

$$x = \pm 1$$

$$y'' = \frac{(x^2 + 1)^{2}(-2x) - (1 - x^2)(2)(x^2 + 1)(2x)}{(x^2 + 1)^{4}}$$

$$= \frac{(x^2 + 1)(-2x) - (1 - x^2)(2)(2x)}{(x^2 + 1)^3}$$

$$= \frac{2x^3 - 6x}{(x^2 + 1)^3} = 0$$

$$2x^3 - 6x = 0$$

$$x(2x^2 - 6) = 0$$

$$x = 0 \qquad x = \pm\sqrt{3} \text{ or } \pm 1.73$$

The nature of the points where $y'$ and $y''$ equal zero is shown in the table at the end of the example.

*Intercepts.* When $x = 0$, $y = 0$ and vice versa. Therefore there is only one intercept, (0, 0).

*Asymptotes.*     When $x \to \infty$, you need to find $\lim\limits_{x\to\infty} y$. This is found by dividing the top and bottom of the fraction by $x^2$:

$$\lim_{x\to\infty} y = \lim_{x\to\infty} \frac{x}{x^2 + 1} = \lim_{x\to\infty} \frac{x/x^2}{x^2/x^2 + 1/x^2}$$

$$= \lim_{x\to\infty} \frac{1/x}{1 + 1/x^2} = \frac{0}{1+0} = 0$$

The line $y = 0$, or the $x$ axis, is therefore a horizontal asymptote.

The denominator of $y$, $x^2 + 1$, can never be zero, and therefore the curve has no vertical asymptotes.

*Symmetry.*     The curve is not symmetric to the $x$ or $y$ axis. However, when $x$ is replaced by $-x$ and $y$ by $-y$, you obtain:

$$-y = \frac{-x}{(-x)^2 + 1} = \frac{-x}{x^2 + 1}$$

which is the same as:

$$y = \frac{x}{x^2 + 1}$$

The curve is therefore symmetric to the origin (Fig. 20-8).

*Graph.*     Choose values of $x$ on either side of the values where $y' = 0$ and $y'' = 0$, and calculate the derivatives to determine the nature of these points:

|      |      | Inflec-<br>tion<br>point<br>↓ | Mini-<br>mum<br>point<br>↓ |      |      | Inflec-<br>tion<br>point<br>↓ |      |      | Maxi-<br>mum<br>point<br>↓ | Inflec-<br>tion<br>point<br>↓ |      |
|------|------|------|------|------|------|------|------|------|------|------|------|
| $x$  | $-3$ | $-\sqrt{3} \approx -1.7$ | $-1$ | $-0.5$ | $0$ | $0.5$ | $1$ | $\sqrt{3} \approx 1.7$ | $3$ |
| $y$  | $-0.3$ | $-0.4$ | $-0.5$ | $-0.4$ | $0$ | $0.4$ | $0.5$ | $0.4$ | $0.3$ |
| $y'$ |      | $-0.1$ | $0$ | $+0.1$ |      | $+0.1$ | $0$ | $-0.1$ |      |
| $y''$ | $-0.04$ | $0$ | $+0.5$ | $+1.4$ | $0$ | $-1.4$ | $-0.5$ | $0$ | $+0.04$ |

The graph in Fig. 20-8 uses different scales on the $x$ and $y$ axes to better show the shape of the curve. Notice that for each point on one side of the origin, such as $(3, 0.3)$, there is a symmetric point $(-3, -0.3)$ on the opposite side of the origin.

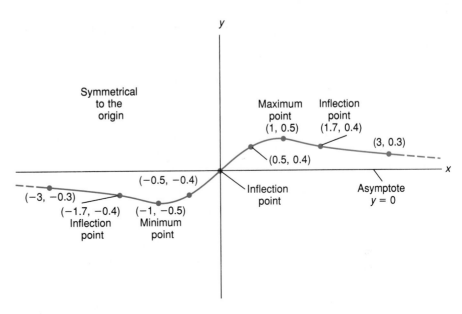

FIG. 20-8  *Sketch of* $y = \dfrac{x}{x^2 + 1}$, *Example 20-10*

**Exercise 20-3**   Sketch the graph of each function in problems 1 to 12, investigating the derivatives, intercepts, asymptotes, and symmetry for each.

**1.** $y = \dfrac{4}{x}$          **2.** $y = \dfrac{1}{x^3}$

**3.** $f(x) = \dfrac{2}{x^2}$          **4.** $f(x) = 1 + \dfrac{1}{x^2}$

**5.** $y = \dfrac{1}{x^2 - 1}$          **6.** $y = \dfrac{1}{x^2 + 2x}$

**7.** $f(x) = \dfrac{x}{x - 1}$          **8.** $f(x) = \dfrac{x}{3 - x}$

□ **9.** $y = \dfrac{4}{x^2 + 1}$          □ **10.** $y = \dfrac{6}{2 + x^2}$

**11.** $f(x) = \dfrac{x}{1 - x^2}$          □ **12.** $f(x) = \dfrac{2x}{1 + x^2}$

**13.** The current in a dc generator is given by:

$$I = \dfrac{V}{R + r}$$

where $V$ = voltage, $R$ = load resistance, and $r$ = internal resistance of the genera-tor. Sketch $I$ vs. $R$ for $R \geq 0$ when $E = 12$ V and $r = 2\ \Omega$.

14. In problem 13 the power in the load $P = I^2R$. By using the formula for $I$ this becomes:

$$P = \left(\frac{V}{R + r}\right)^2 R = \frac{V^2R}{(R + r)^2}$$

Using the values of $V$ and $r$ from problem 13, sketch the curve of $P$ vs. $R$ for $R \geq 0$.

15. The compressive load $F$ that a cylindrical column can withstand is directly proportional to the fourth power of its diameter and inversely proportional to the square of its length $l$:

$$F = \frac{kd^4}{l^2}$$

If $k = 10$ lb/ft$^2$ and $d = 1$ ft, sketch the curve of $F$ vs. $l$ for $F \geq 0$.

16. The speed of sound in a liquid is given by:

$$V = \sqrt{\frac{K}{\rho}}$$

where $K =$ bulk modulus and $\rho$ (rho) = mass density. If $K = 2.0 \times 10^9$, sketch the graph of $V$ vs. $\rho$. Use different scales for $V$ and $\rho$.

# 20-4 | *Applied Maxima and Minima Problems*

In science and technology there are many types of problems where the maximum or minimum value of a function is desired. When the function is known, the procedure is the same as that shown in Sec. 20-2 for finding a maximum or minimum point of a curve. The first example illustrates this basic procedure.

### Example 20-11
The velocity (in miles per second) of a satellite during a 2½-h period is given by:

$$v = 1.5 + 2.0t - \frac{t^3}{6.0}$$

where the time $t$ varies from 0.5 to 3.0 h. What is the maximum velocity attained by the satellite during this period?

### Solution
The maximum velocity occurs where the function achieves its maximum value and the derivative is zero. Find $v'$ and set it equal to zero:

$$v' = 2.0 - \frac{t^2}{2.0} = 0$$

$$t^2 = 4.0$$

$$t = \pm 2.0$$

The negative value of $t$ does not apply. Check that when $t = 2.0$, $v$ attains a maximum value, by finding the value of the second derivative when $t = 2.0$:

$$v'' = -t = -2.0$$

Since the second derivative is negative, the curve is concave down and $v$ attains a maximum value. Substitute $t = 2.0$ in the formula for $v$ to find the maximum value:

$$v = 1.5 + 2.0(2.0) - \frac{(2.0)^3}{6.0} = 4.2 \text{ mi/s}$$

The next example involves finding the minimum value of a function. The function, however, is not given and must be set up based on the conditions of the problem. This requires the techniques used to set up verbal problems that are shown in earlier chapters. Study the next example and the ones that follow to learn how to set up these functions.

### Example 20-12
Patricia wants to fence in two equal adjacent rectangular plots in her backyard, using an existing fence as one of the sides of the two plots (Fig. 20-9). What is the minimum amount of fencing she needs to enclose a total area of 1200 ft$^2$?

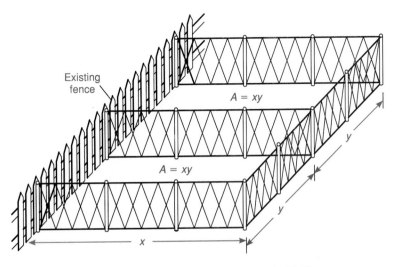

FIG. 20-9 *Minimizing the amount of fencing, Example 20-12*

## Solution

In this problem you want to minimize the amount of fencing. The amount of fencing depends on the dimensions of the rectangular plots. Let one dimension of each plot be $x$ and the other $y$, as shown in Fig. 20-9. The amount of fencing $F$ is then:

$$F = 3x + 2y$$

Before you can differentiate this function to find the minimum value, you need to express it in terms of one variable. You know that the total area is to be 1200 ft², and therefore the area of each plot is 600 ft²:

$$xy = 600 \quad \text{and} \quad y = \frac{600}{x}$$

Substitute this expression for $y$ in the function $F$:

$$F = 3x + 2\left(\frac{600}{x}\right) = 3x + \frac{1200}{x}$$

This expresses $F$ as a function of $x$ only. Now you can find $F' = dF/dx$:

$$F' = 3 - \frac{1200}{x^2}$$

Set $F' = 0$ to find the minimum value:

$$3 - \frac{1200}{x^2} = 0$$
$$3x^2 = 1200$$
$$x^2 = 400$$
$$x = \pm 20$$

The negative value of $x$ does not apply. Check the value of the second derivative when $x = 20$:

$$F'' = \frac{2400}{x^3} = \frac{2400}{(20)^3} = 0.30$$

The second derivative is positive, so $F$ attains a minimum value when $x = 20$. When $x = 20$, $y = 600/20 = 30$ and the minimum value of $F$ is:

$$F = 3(20) + 2(30) = 120 \text{ ft}$$

The preceding example illustrates the *basic procedure for solving applied maximum and minimum problems*. Study the following, which outlines this procedure:

1. Identify which variable is to be maximized or minimized.
2. Set up an equation which contains this variable.
3. Express the variable as a function of one independent variable. Use another equation if necessary, based on the given information.
4. Differentiate with respect to the independent variable, and set the derivative equal to zero.
5. Check that the critical value is the desired maximum or minimum.

The next example of a geometric problem involves finding the maximum value of a function for a specific solution and for a general solution.

### Example 20-13

A square piece of steel, $s$ cm on a side, is to be made into an equipment chassis by cutting equal squares out of the corners, folding up the sides, and welding the seams to form a pan (Fig. 20-10).

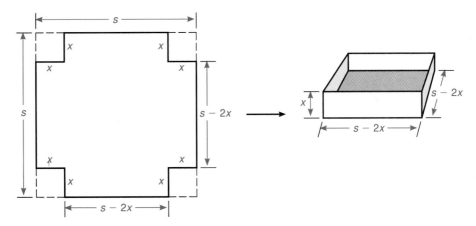

FIG. 20-10 *Maximizing the volume of a steel chassis, Example 20-13*

1. If $s = 60$ cm, what should the side of the cutout square be, to maximize the volume of the chassis?

2. In general, what is the relationship between the side of the steel square $s$ and the side of the cutout square $x$ that will produce the maximum volume of the chassis?

### Solution

1. If $s = 60$ cm, what should the side of the cutout square be, to maximize the volume of the chassis?

Let $x$ = side of the cutout square. Then the height of the chassis is $x$, and the sides of the base are each $s - 2x = 60 - 2x$.

The volume $V$ of the chassis is:

$$V = x(60 - 2x)^2 = 4x^3 - 240x^2 + 3600x$$

Find $V' = dV/dx$, and set it equal to zero:

$$V' = 12x^2 - 480x + 3600 = 0$$
$$x^2 - 40x + 300 = 0 \qquad \text{(divide by 12)}$$
$$(x - 10)(x - 30) = 0$$
$$x = 10 \qquad x = 30$$

Check the value of the second derivative when $x = 10$ and $x = 30$:

$$V'' = 24x - 480$$

When $x = 10$:  $V'' = 24(10) - 480 = -240$  Maximum
When $x = 30$:  $V'' = 24(30) - 480 = +240$  Minimum

Therefore when $x = 10$, the maximum value of $V$ is:

$$V = 10[60 - 2(10)]^2 = 16{,}000 \text{ cm}^3$$

When $x = 30$, it is not possible to construct the chassis, and $V = 0$. This is a minimum value of $V$.

2.  In general, what is the relationship between the side of the steel square $s$ and the side of the cutout square $x$ that will produce the maximum volume of the chassis?

Express the volume $V$ in terms of $x$ and $s$:

$$V = x(s - 2x)^2 = 4x^3 - 4x^2s + s^2x$$

Find $V' = dV/dx$ and set it equal to zero. Treat $s$ as a constant:

$$V' = 12x^2 - 8xs + s^2 = 0$$
$$(6x - s)(2x - s) = 0$$
$$6x - s = 0 \qquad 2x - s = 0$$
$$s = \frac{s}{6} \qquad x = \frac{s}{2}$$

When $x = s/6$, the volume is a maximum. When $x = s/2$, it is not possible to construct the chassis and $V = 0$. Therefore, the relationship between $s$

and $x$ for maximum volume is that the side of the cutout square should be one-sixth of the side of the steel square.

The next example illustrates an application to an electrical problem where the maximum power in a dc circuit is required.

### Example 20-14

A battery supplies a constant voltage of 8.0 V to a circuit containing a variable resistor $R$ in series with a fixed resistance of 4.0 $\Omega$ (Fig. 20-11). Find the value of $R$ that will produce the maximum power consumption in the variable resistor.

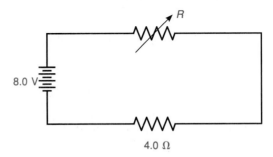

8.0 V

4.0 $\Omega$

FIG. 20-11 *Maximizing the power in a circuit resistance, Example 20-14*

### Solution

The power consumption $P$ in the variable resistor $R$ is given by $P = I^2R$, where $I$ is the current in the circuit. The expression for $I$ is found by using Ohm's law and dividing the voltage by the total resistance, which is $R + 4.0$:

$$I = \frac{8.0}{R + 4.0}$$

The power consumption $P$ is then:

$$P = \left(\frac{8}{R + 4}\right)^2 R = \frac{64R}{(R + 4)^2}$$

Find the maximum power by finding $P' = dP/dR$ and setting it equal to zero:

$$P' = \frac{(R + 4)^2(64) - 64R(2)(R + 4)}{(R + 4)^3} = \frac{256 - 64R}{(R + 4)^3}$$

$$256 - 64R = 0$$

$$R = 4.0 \ \Omega$$

In this example it is easier to use the first-derivative test (20-1) to determine if $R = 4\ \Omega$ produces a maximum value of $P$. When $R < 4$, $P'$ is positive; and when $R > 4$, $P'$ is negative. Therefore $R = 4.0\ \Omega$ produces the maximum power consumption in the variable resistor.

The next example shows an application to a mechanical problem where the minimum force to lift a weight is to be found.

### Example 20-15

A load of 500 lb is to be lifted by a steel bar weighing 4 lb/ft as shown in Fig. 20-12. What should be the length $x$ of the bar which will minimize the lifting force $F$, and what is the minimum force?

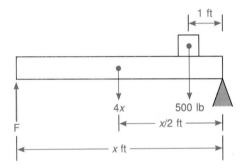

FIG. 20-12 *Minimizing the force to lift a load, Example 20-15*

### Solution

By the principle of the lever, the moment arm (force times distance) acting up must equal the sum of the moment arms acting down. [See Example 3-10(2), Sec. 3-3.] The moment arms acting down consist of that due to the load and that due to the weight of the bar, which acts at the center of gravity and which is $x/2$ ft from the fulcrum. The weight of the bar is $4x$ since the bar weighs 4 lb/ft and is $x$ ft long. Equating the moment arms, you have:

$$Fx = 4x\left(\frac{x}{2}\right) + 500(1)$$

The force $F$ expressed as a function of $x$ is then:

$$F = 2x + \frac{500}{x}$$

Find $F' = dF/dx$ and set it equal to zero:

$$F' = 2 - \frac{500}{x^2} = 0$$
$$x^2 = 250$$
$$x = 15.8 \text{ ft}$$

When $x = 15.8$ ft, the second derivative is positive. The minimum force is therefore:

$$F = 2(15.8) + \frac{500}{15.8} = 63.2 \text{ lb}$$

**Exercise 20-4**

1. The angular velocity (in radians per second) of a flywheel is given by:

$$\omega = 8t + \frac{8}{t}$$

where $t$ varies from 0.25 to 1.25 s. What is the minimum angular velocity of the flywheel?

2. The acceleration (meters per second per second) of a rocket during liftoff is given by $a = 0.9t - 0.001t^3$, where $t$ varies from 0 to 30 s. What is the maximum acceleration of the rocket?

3. The current (amperes) in a circuit containing an inductance is given by $I = 40t^2 - 160t^3$, where $t =$ time (seconds). What is the maximum current in the circuit?

4. In a circuit the current (amperes) in a constant resistance of 2 $\Omega$ is given by $I = t^2 - 4$, where $t$ varies from 0 to 4 s. Find the maximum and minimum power in the circuit. Use the formula for power $P = I^2R$.

5. The deflection $y$ (in inches) of a beam is given by $y = 0.001(x^3 - 9x^2 + 24x - 19)$, where $x =$ distance from one end. Find the maximum positive and *maximum negative* deflection of the beam. The maximum negative deflection is the minimum point on the curve of deflection.

6. The daily cost of production for a company that produces large solar panels is given by $c = n^3 - 30n^2 + 225n + 300$, where $n =$ number of solar panels produced each day. How many solar panels should be produced to minimize the cost?

7. A rectangular area is to be enclosed by 120 m of fencing. What should be the length and width of the rectangle to maximize the area? (See Example 20-12.)

8. If the rectangular area in problem 7 is to have a fence across the middle, dividing it into two equal rectangular areas, what should be the length and width of the large rectangle to maximize the area? (See Example 20-12.)

9. A rectangular piece of steel 8 by 5 in. is to be made into an equipment chassis by cutting equal squares out of the corners, folding up the sides, and welding the seams to form a pan. What should be the size of the cutout square to maximize the volume of the chassis? [See Example 20-13(1) and Fig. 20-10.]

**10.** Solve problem 9 if the dimensions of the rectangular piece are 20 by 10 cm.

**11.** A piece of copper $s$ m wide is to be bent into an open trough of rectangular cross section by bending up the sides as shown in Fig. 20-13. If $s = 4$ m, what should be the height $h$ to maximize the cross-sectional area?

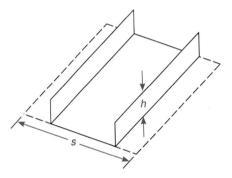

FIG. 20-13  *Copper trough, problems 11 and 12*

**12.** In general, for the piece of copper in problem 11, what should be the relationship between $s$ and $h$ to maximize the cross-sectional area? [See Example 20-13(2).]

**13.** A battery supplies a constant voltage of 6.0 V to a circuit containing a variable resistor $R$ in series with a fixed resistance of 2.0 $\Omega$. Find the value of $R$ that will produce the maximum power consumption in the variable resistor. (See Example 20-14.)

**14.** A dc generator supplies a constant voltage of 12 V to a circuit containing a variable resistor $R$ in parallel with a fixed resistance of 3.0 $\Omega$. What value of $R$ will produce the maximum power consumption in the variable resistor? The total resistance of two resistances $R_1$ and $R_2$ in parallel is given by $R_1 R_2/(R_1 + R_2)$. (See Example 20-14.)

**15.** If the load in Example 20-15 is placed 1.5 ft from the fulcrum, what should be the length $x$ of the bar to minimize the force $F$, and what is this minimum force?

**16.** Show that if the load in Example 20-15 weighs $W$ lb, the length of the bar to minimize the force $F$ should be $\sqrt{2W}/2$.

**17.** An electronics company finds that for each calculator over 50 sold to a retailer, the price can be reduced by $1 each. If the base price is $100 each, how many calculators must be sold to produce the maximum income?

**18.** The cost per hour of operating an oil tanker is proportional to the cube of its speed and is $100 per hour when the speed is 10 nmi/h. There is also a fixed operating expense of $150 per hour. At what speed should the tanker cruise to minimize the cost for a trip of 2000 nmi?

**19.** The current in an ac circuit containing a resistance $R$ and a reactance $X$ is given by:

$$I = \frac{V}{\sqrt{R^2 + X^2}}$$

Assuming $V$ and $X$ are constant, find the value of $R$ for maximum power $P$ in the resistance. (*Note:* $P = I^2 R$.)

**20.** The current in a capacitance of $C$ microfarads ($\mu$F) is given by:

$$I = C\,\frac{dV}{dt}$$

During a short time $t$, the voltage is given by $V = (t^2 - 1)^2$. At what value of $t$ is the *current* a minimum?

**21.** An open equipment box is to have a square bottom of side $x$ and height $h$. If the volume is to be 4 ft$^3$, what should $x$ and $h$ be to minimize the amount of material to construct the box, that is, the surface area of the bottom and four sides?

**22.** In general, for the box in problem 21, what should be the relationship between $x$ and $h$ to minimize the surface area? (*Hint:* Differentiate both the formulas for the volume and the surface area implicitly, treating the volume as a constant.)

**23.** The bending moment of a beam is given by:

$$y = \frac{2x^3}{3} - \frac{3Lx^2}{2} + L^2 x$$

where $L$ = length of the beam and $x$ = distance from one end. (*a*) Show that the points where the beam is most likely to fail are at $x = L/2$ and $x = L$. These are the points where the bending moment is a maximum. (*b*) Find the bending moment at each of these points, and show that it is greater at $x = L/2$.

**24.** The maximum force that a rectangular beam can withstand varies directly as the beam's width $w$ and the square of its height $h$: $F = kwh^2$, where $k$ is constant. Find the width and height of the strongest beam that can be cut from a circular log 1.5 m in diameter.

**25.** The cost of building a transmission line is proportional to the cross-sectional area $A$. At the same time the cost of operating the transmission line is inversely proportional to $A$, since the larger the cross section, the less the heating and power loss. This can be expressed by:

$$C = kA + \frac{K}{A}$$

where $k$ and $K$ are constants. Show that $C$ is a minimum when $kA = K/A$.

**26.** A ship is anchored 1 mi from a straight shore. A boat is to pick up a passenger on shore who is 4 mi from the point on the shore directly opposite the ship. If the boat can travel at 3 mi/h and the passenger can walk along the shore at 4 mi/h, at what point on the shore should the boat meet the passenger in order to make the pickup in the shortest time?

## 20-5 | *Related Rates*

When a relationship between two or more variables is known, you can differentiate the expression implicitly, as shown in Sec. 19-7, and obtain a relationship between the *rates of change of each variable* with respect to some specific variable, such as time. It is then possible to find the rate of change of a variable, given information about the other variables and their rates. This is called a *related rate problem,* and it occurs in many scientific and engineering applications. The following examples show some of the types of related rate problems you may encounter.

### Example 20-16

Boyle's law states that at a constant temperature the pressure $P$ of a gas is inversely proportional to the volume $V$: $P = k/V$. A gas contracts at a constant temperature while the pressure increases at the rate of 10 kPa/s. If $k = 1300$, at what rate is the volume changing when $V = 3.0$ m³?

### Solution

Differentiate the equation relating pressure and volume implicitly with respect to time to obtain an equation involving the rates of change:

$$\frac{dP}{dt} = \frac{-k}{V^2}\left(\frac{dV}{dt}\right)$$

or

$$P' = \frac{-kV'}{V^2}$$

Substitute $P' = 10$ kPa/s, $k = 1300$, and $V = 3.0$ m³, and solve for $V'$:

$$10 = \frac{-1300V'}{(3.0)^2}$$

$$V' = -\frac{90}{1300} = -0.069 \text{ m}^3/\text{s}$$

Therefore the volume is *decreasing* at 0.069 m³/s.

### Example 20-17

Two planes at the same altitude are flying toward an airport. The first plane is 40 mi south of the airport, and its speed is 300 mi/h. The second plane is 30 mi east of the airport, and its speed is 200 mi/h. How fast is the distance between them decreasing at this time?

### Solution

Figure 20-14 shows the distances and the speeds of the planes. The distances are continually changing, and it is necessary to represent each of them with variables, even though their values are known at a specific time. Let:

$x$ = distance of first plane from airport

$y$ = distance of second plane from airport

$z$ = distance between two planes

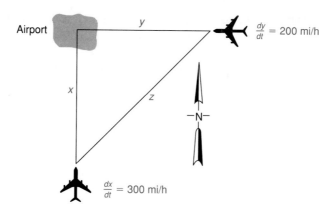

FIG. 20-14 *Planes flying toward an airport, Example 20-17*

Then, by the pythagorean theorem, the relationship between the distances of the planes is:

$$x^2 + y^2 = z^2$$

From this equation the value of z *at the given time* is:

$$z = \sqrt{x^2 + y^2} = \sqrt{40^2 + 30^2} = \sqrt{2500} = 50 \text{ mi}$$

From the distance equation we obtain the relationship between the distances and the speeds of the planes by differentiating each term implicitly with respect to the time t:

$$2x \frac{dx}{dt} + 2y \frac{dy}{dt} = 2z \frac{dz}{dt}$$

or
$$2xx' + 2yy' = 2zz'$$

which can be simplified to:

$$xx' + yy' = zz'$$

Notice that the general power formula (20-14) is used to obtain the derivatives. The meaning of each derivative is:

$$\frac{dx}{dt} = x' = \text{speed of first plane} = 300 \text{ mi/h}$$

$$\frac{dy}{dt} = y' = \text{speed of second plane} = 200 \text{ mi/h}$$

$$\frac{dz}{dt} = z' = \text{speed of planes approaching each other}$$

*At the given time,* $z'$ can be found from the above equation, since the value of each of the other variables is known or has been determined:

$$40(300) + 30(200) = 50z'$$

$$z' = \frac{12,000 + 6000}{50} = 360 \text{ mi/h}$$

Note that this solution is for a *specific time* and $z'$ continually changes depending on the distances of the planes.

## Example 20-18
A rectangular metal plate is heated, causing the area to increase. If the length is increasing at 0.2 mm/min and the width at 0.1 mm/min, how fast is the area increasing when the length is 40 cm and the width is 20 cm?

### Solution
The relationship between the area $A$, the length $l$, and the width $w$ is:

$$A = lw$$

We differentiate this equation implicitly with respect to time, using the product formula (19-10):

$$A' = lw' + wl'$$

At the given time:

$$l = 40 \text{ cm}$$
$$w = 20 \text{ cm}$$
$$l' = 0.2 \text{ mm/min} = 0.02 \text{ cm/min}$$
$$w' = 0.1 \text{ mm/min} = 0.01 \text{ cm/min}$$

We substitute these values into the above equation to find the rate of change of the area at the specific time:

$$A' = 40(0.01) + 20(0.02) = 0.8 \text{ cm}^2\text{/min}$$

## Example 20-19
The current $I$ through a constant resistance $R = 2.0 \ \Omega$ is increasing at the rate of 1.3 A/s. At what rate is the power changing (in watts per second) when $I = 0.35$ A?

## Solution

The power in the resistance is given by:

$$P = I^2R = I^2(2.0)$$

Differentiate $P$ with respect to time:

$$P' = 4.0I(I')$$

When the current is 0.35 A:

$$P' = 4.0(0.35)(1.3) = 1.82 \text{ W/s}$$

**Exercise 20-5**

1. The volume of a rising weather balloon is increasing at the rate of 250 cm³/s. At what rate is the radius $r$ increasing when $r = 30$ cm? [*Note:* The volume of a sphere is given by $V = (\frac{4}{3})\pi r^3$.]

2. Newton's second law of motion states that the force $F$ on a body is proportional to the acceleration $a$: $F = ma$. A body whose mass $m = 3$ kg experiences a changing force at the rate of 1 N/s (N = newton). At what rate is the acceleration changing (in meters per second cubed)?

3. In a dc circuit with a constant voltage of 9.0 V, a variable resistance is increased at the rate of 1.0 $\Omega$/min. Using Ohm's law, $R = V/I$, find the rate of change of the current when $I = 5.5$ A.

4. The velocity of a water jet at the bottom of a water tank is given by $v = 0.95\sqrt{2gh}$, where $g$ = gravitational acceleration and $h$ = height. Using $g = 32$ ft/s², find the rate of change of $v$ in feet per second per second when $h = 16$ ft and is *decreasing* at the rate of 2 ft/s.

5. Two ships leave the same port at the same time. One ship travels east at 9 nmi/h, and the other ship travels south at 12 nmi/h. How fast is the distance between them increasing after 2 h? (See Example 20-17.)

6. A plane 40 km west of an airport is *flying toward* it at 400 km/h. A second plane, at the same altitude, is 75 km north of the same airport and is *flying away* from it at 500 km/h. What is the rate of change of the distance between them at this time? (See Example 20-17.)

7. A hot rectangular steel plate is left to cool, causing the length to decrease at 0.03 in./min and the width to decrease at 0.005 in./min. At what rate is the area decreasing when the length is 10 in. and the width is 2 in.? (See Example 20-18.)

8. A metal equipment chassis base in the shape shown in Fig. 20-15 becomes warm and expands. If the dimension $x$ is increasing at 1 mm/min and the dimension $y$ is increasing at 2 mm/min, how fast is the area increasing when $x = 10$ cm and $y = 20$ cm? (See Example 20-18.)

9. The current $I$ through a constant resistance of 4.5 $\Omega$ is increasing at the rate of 0.5 A/s. At what rate is the power changing when $I = 2.0$ A? (See Example 20-19.)

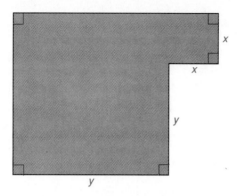

FIG. 20-15 *Metal equipment chassis base, problem 8*

**10.** The voltage $E$ in a dc circuit is decreasing at 0.2 V/s, and the current $I$ is decreasing at 0.1 A/s. How fast is the power $P$ changing when $E = 24$ V and $I = 11$ A? *Note: $P = EI$.*

**11.** A rectangular fuel tank with a square base 0.5 m on a side is being filled at the rate of 10 L/min. How fast is the level rising? (*Note:* 1 L $= 1000$ cm$^3$.)

**12.** A conical funnel whose height $h$ equals twice its radius $r$ is filled with oil. If the oil is pouring out at the rate of 30 in.$^3$/s, how fast is the level dropping when it is 8 in. deep? (*Note:* $V = \frac{1}{3}\pi r^2 h$.)

**13.** A ladder 15 ft long is resting against a vertical wall. If the foot of the ladder is pulled along the ground at 1 ft/s so the ladder slips down the wall, how fast is the top sliding when it is 9 ft from the ground?

**14.** A plane at an altitude of 1.5 mi is approaching a point directly above a military base on the ground. The radar unit at the base shows that the direct-line distance to the plane is 4 mi, and this distance is decreasing at the rate of 90 mi/h. How fast is the plane approaching the point above the base at this moment?

**15.** The impedance $Z$ of a circuit containing a resistance and an inductance in parallel is given by:

$$Z = \frac{RX}{R + X}$$

If $R$ is constant at 5.0 $\Omega$ and $X$ is decreasing at 2.0 $\Omega$/s, at what rate is $Z$ changing when $X = 3.0$ $\Omega$?

**16.** In problem 15, if $Z$ is increasing at 1.0 $\Omega$/s, find the rate at which $X$ is changing when $X = 3.0$ $\Omega$.

**17.** The total resistance $R$ of a circuit containing two resistances $R_1$ and $R_2$ in parallel is given by:

$$\frac{1}{R} = \frac{1}{R_1} + \frac{1}{R_2}$$

If $R_1$ is constant at 10 Ω and $R_2$ is decreasing at the rate of 5 Ω/min, at what rate is $R$ changing when $R_2 = 10$ Ω?

**18.** Show in problem 17 that:

$$\frac{R}{R_2} = \sqrt{\frac{R'}{R_2'}}$$

**19.** A rectangular block of ice with a square base is melting so that the edge of the base is decreasing at 2 in./h and the height is decreasing at 4 in./h. How fast is the volume decreasing in cubic feet per hour when the base edge is 20 in. and the height is 20 in.?

**20.** A trough 5 m long has a triangular cross section that is 1 m wide at the top and 1 m high. Water is poured into the trough at the rate of 0.5 m³/min. How fast is the height increasing when it is 0.5 m high?

**21.** Judy is approaching a streetlight at the rate of 4 ft/s. If Judy is 6 ft tall and the streetlight is 20 ft high, how fast is the tip of Judy's shadow moving?

**22.** A yacht is anchored in 22 ft of water. Milton, the skipper, is pulling in the anchor line at the rate of 2 ft/s, causing the yacht to move toward the point directly over the anchor. The anchor line passes over the bow, which is 6 ft above the water. How fast is the yacht moving when there is 100 ft of anchor line still out? Assume the anchor line is straight.

## 20-6 | *The Differential*

The differential is a very useful idea when you are working with small changes of a variable or function. It is of particular importance in the meaning of an integral and the process of integration, which are discussed in the next chapter.

Consider the function $y = f(x) = x^2$. The change in $y$ is defined as:

$$\Delta y = f(x + \Delta x) - f(x) = (x + \Delta x)^2 - x^2$$
$$= x^2 + 2x\,\Delta x + (\Delta x)^2 - x^2 = 2x\,\Delta x + (\Delta x)^2$$

If $x = 4$ and the change in $x$ is a small quantity, say $\Delta x = \frac{1}{4}$, then:

$$\Delta y = 2(4)\left(\frac{1}{4}\right) + \left(\frac{1}{4}\right)^2 = 2 + \frac{1}{16}$$

However, notice that the value of $\Delta y$ can be approximated by using the formula:

$$\Delta y \approx f'(x)\,\Delta x = 2x\,\Delta x = 2(4)\left(\frac{1}{4}\right) = 2$$

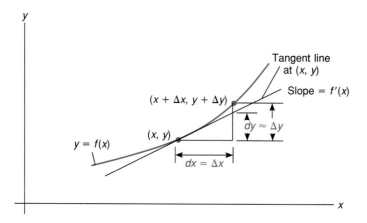

FIG. 20-16  *The differential dy compared with $\Delta y$*

The value $(\Delta x)^2 = \frac{1}{16}$ represents the amount of error in the approximation. As $\Delta x \to 0$, the error approaches zero and the approximation approaches the actual value. For this reason, the *differential of a function dy* is defined as follows:

$$dy = f'(x)\,\Delta x \qquad \text{or} \qquad dy = f'(x)\,dx \tag{20-4}$$

where the *differential of x, dx,* is defined equal to $\Delta x$.

Figure 20-16 shows the differentials $dy$ and $dx$, and how $dy$ compares closely with $\Delta y$. As $\Delta x \to 0$, $dy \to \Delta y$. Differentials $dy$ and $dx$ should be thought of as variables, like $\Delta x$ and $\Delta y$, that satisfy formula (20-4). Study the next example, which shows how to apply formula (20-4). The meaning of the differential is discussed further after this example.

### Example 20-20
Find the differential of each function.

1. $y = f(x) = 2x^4 + x^2$
2. $s = f(t) = (3t + 3)^3$
3. $y = f(x) = \dfrac{2x}{x - 1}$

### Solution
1. $y = f(x) = 2x^4 + x^2$

   Apply formula (20-4):

$$dy = f'(x)\,dx = (8x^3 + 2x)\,dx$$

**2.** $s = f(t) = (3t + 3)^3$:

$$ds = f'(t) \, dt = [3(3t + 3)^2(3)] \, dt = 9(3t + 3)^2 \, dt$$

**3.** $y = f(x) = \dfrac{2x}{x - 1}$:

$$dy = \frac{(x - 1)(2) - (2x)(1)}{(x - 1)^2} \, dx = \frac{-2}{(x - 1)^2} \, dx$$

Notice that formula (20-4) can be written:

$$\frac{dy}{dx} = f'(x)$$

The differentials are purposely defined this way so that the ratio of $dy$ to $dx$ is the derivative. This is the original concept used by Leibniz, one of the discoverers of the calculus. He visualized $dy$ and $dx$ as infinitesimal changes whose ratio was the slope of the curve.

The following examples show how differentials can be applied to approximate small changes.

### Example 20-21
The radius $r$ of a sphere is increased from 12 to 12.01 cm. Approximately how much does the volume $V$ increase?

### Solution
The formula for the volume of a sphere is $V = \frac{4}{3} \pi r^3$. Find the differential $dV$, which is approximately equal to $\Delta v$:

$$dV = V' \, dr = 4\pi r^2 \, dr$$

Let $r = 12$ and $dr = \Delta r = 0.01$. Then:

$$dV = 4\pi(12)^2(0.01) = 18.1 \text{ cm}^3$$

This agrees with the actual change $\Delta V$ to three figures:

$$\Delta V = \frac{4}{3} \pi(12.01)^3 - \frac{4}{3} \pi(12)^3 = 18.1$$

### Example 20-22
In a dc circuit, the current $I$ through a constant resistance $R$ of 4.0 $\Omega$ is measured to be 5.0 A. A more accurate measurement, made later, shows that $I$ is

closer to 5.1 A. Approximately how much error in the calculation of the power results from the first measurement?

**Solution**

The formula for power $P = I^2R = 4.0I^2$. Find $dP$ when $I = 5.0$ and $dI = 0.1$:

$$dP = P'\, dI = 8.0I\, dI = 8.0(5.0)(0.1) = 4.0 \text{ W}$$

The actual error calculated directly is:

$$\Delta P = (5.1)^2(4.0) - (5.0)^2(4.0) = 4.04 \text{ W}$$

**Exercise 20-6**

In problems 1 to 4, find $\Delta y$ and $dy$ for the given values of $x$ and $\Delta x$.

1. $y = f(x) = 2x^2$; $x = 3$, $\Delta x = \dfrac{1}{2}$

2. $y = f(x) = x^3$; $x = 1$, $\Delta x = 0.1$

[C]   3. $y = f(x) = \sqrt{x}$; $x = 4$, $\Delta x = 0.2$

[C]   4. $y = f(x) = x^{3/2}$; $x = 8$, $\Delta x = 0.05$

In problems 5 to 12, find the differential of each function.

5. $y = f(x) = 3x^3 - 4$

6. $y = f(x) = 4x^4 - 2x^2$

7. $s = f(t) = (t^2 + t)^3$

8. $I = f(t) = \dfrac{1}{(t - 1)^2}$

9. $M = f(x) = \dfrac{x}{2 - x}$

10. $v = f(t) = \dfrac{t}{\sqrt{t^2 + 1}}$

11. $Z = \sqrt{R^2 + X^2}$   ($X$ constant)

12. $f = \dfrac{1}{2\pi\sqrt{LC}}$   ($C$ constant)

In problems 13 to 20, use differentials to solve each problem.

13. The side $s$ of a cube of ice decreases from 10 to 9.99 cm. Approximately how much does the volume decrease?

14. The radius $r$ of a circle increases from 1.50 to 1.52 ft. Approximately how much does the area increase?

15. In a dc circuit, the voltage drop $E$ across a constant resistance $R$ of 3.0 $\Omega$ is measured to be 9.0 V. A more accurate measurement, made later, shows that the voltage drop is closer to 8.9 V. Approximately how much error in the calculation of the power results from the first measurement? (*Note:* $P = E^2/R$.)

[C] **16.** The resistance in a circuit increases from 8.00 to 8.03 $\Omega$. If the voltage is constant at 120 V, how much does the current $I$ change? Use Ohm's law, $I = E/R$.

[C] **17.** A spherical satellite of diameter 15 m becomes coated with ice to a thickness of 1.5 cm. How much does the volume increase? (*Note:* Volume of a sphere $V = \frac{4}{3}\pi r^3$.)

**18.** The relationship between Fahrenheit and Celsius degrees is $F = 1.8C + 32$. If a Celsius thermometer can be read to the nearest 0.5°, what is the corresponding accuracy in Fahrenheit degrees?

[C] **19.** The impedance of an ac circuit is given by $Z = \sqrt{R^2 + 4.00}$. If the resistance $R$ is measured to be $7.00 \pm 0.01$ $\Omega$, what is the corresponding range for the impedance?

**20.** When a gas expands adiabatically (no heat gain or loss), the pressure and volume are related by $PV^{1.4} = k$. If $k = 1100$ and $V = 1.5$ m³, how much does the pressure change, in kilopascals, when $V$ increases by 0.001 m³?

# 20-7 | *Antiderivatives*

In science and engineering, very often information is known about the rate of change of a quantity, that is, its derivative or differential. The information is then used to determine the original function of which the derivative is known. This process is called *antidifferentiation*. Consider the following example, which illustrates this situation.

## Example 20-23
A stone is dropped from a cliff 100 m high. Using the fact that the acceleration due to gravity is $g = -9.8$ m/s² (negative for down):

**1.** Find the formula for the velocity as a function of time.

**2.** Find the formula for the distance above the ground as a function of $t$. This is called the *equation of motion*.

## Solution
**1.** Find the formula for the velocity as a function of time.

Acceleration is the rate of change of velocity $v$ with respect to time $t$, or $dv/dt$. Therefore you can write:

$$\frac{dv}{dt} = g = -9.8$$

You now need to find the function $f(t)$ such that $-9.8$ is the derivative of it. Recall how the power formula (19-14) works (Sec. 19-6). The exponent is multiplied by the coefficient, and the power is reduced by 1. *One* antiderivative of the above is then:

$$v = -9.8t$$

Since the derivative of a constant is zero, the derivative of $-9.8t$ plus *any* constant equals $-9.8$. Therefore the formula for *any* antiderivative must be written with a constant $C$:

$$v = -9.8t + C$$

You can verify this by taking the derivative and checking that you get $-9.8$. The value of the constant can be determined in this example from the given information. Since the stone is dropped, its velocity $v = 0$ when $t = 0$. Substitute these values into the formula to find $C$:

$$0 = -9.8(0) + C$$
$$C = 0$$

The formula for velocity is therefore:

$$v = -9.8t$$

2. Find the formula for the distance above the ground as a function of $t$. This is called the *equation of motion*.

Velocity is the rate of change of distance $s$ with respect to time, or $ds/dt$. You can then write:

$$\frac{ds}{dt} = -9.8t$$

Apply the power formula (19-14) again in reverse. The antiderivative must come from $t^2$, and its coefficient must be one-half of 9.8 since it is multiplied by 2 when you differentiate. A constant must also be added. The formula for the antiderivative is then:

$$s = -4.9t^2 + C$$

Verify this by taking the derivative and checking that you get $-9.8t$. Find the constant $C$ by substituting $s = 100$ when $t = 0$:

$$100 = -4.9(0)^2 + C$$
$$C = 100$$

Therefore the equation of motion is:

$$s = -4.9t^2 + 100$$

The preceding example leads to the *power formula for the antiderivative:*

If $f'(x) = kx^n$, then

$$f(x) = \frac{kx^{n+1}}{n+1} + C \qquad (n \neq -1) \tag{20-5}$$

In formula (20-5), $k$ = coefficient and $C$ = constant. The constant $C$ can be determined when certain additional information is known about $f'(x)$, as in Example 20-23. Otherwise $C$ is left as an unknown constant.

Formula (20-5) applies when $n$ = any rational number except $-1$, since then you would be dividing by zero. You can verify the formula by differentiating $f(x)$ and checking that you get $f'(x)$. The next example shows how to use this formula. The importance of antiderivatives is discussed further after this example.

### Example 20-24
Find the antiderivative of each derivative function.

**1.** $f'(x) = 4x^3 - 6x^2 + 4$

**2.** $y' = 2x - \dfrac{1}{x^2}$

**3.** $f'(x) = \dfrac{3}{\sqrt{x}}$

**Solution**
**1.** $f'(x) = 4x^3 - 6x^2 + 4$

Since the derivative of a sum equals the sum of the derivatives, you can apply the power formula (20-5) separately to each term. Multiply each result by the coefficient as you do when you differentiate:

$$f(x) = \frac{4x^{3+1}}{3+1} - \frac{6x^{2+1}}{2+1} + \frac{4x^{0+1}}{0+1} + C$$

$$= x^4 - 2x^3 + 4x + C$$

The constant is left as $C$ since no other information is given about $f(x)$. Check this result by differentiation.

**2.** $y' = 2x - \dfrac{1}{x^2}$

Apply formula (20-5), using $n = -2$ for the term $1/x^2$:

$$y = \frac{2x^{1+1}}{1+1} - \frac{x^{-2+1}}{-2+1} + C$$

$$= x^2 + x^{-1} + C = x^2 + \frac{1}{x} + C$$

Check this result by differentiation.

**3.** $f'(x) = \dfrac{3}{\sqrt{x}}$

Apply formula (20-5), using $n = -\frac{1}{2}$ for $1/\sqrt{x}$:

$$f(x) = 3\left(\frac{x^{-1/2+1}}{-\frac{1}{2}+1}\right) + C = 6\sqrt{x} + C$$

Check this result by differentiation.

The antiderivative is important in the process of integration, which is discussed in the next few chapters. It is also important in solving differential equations, which are equations containing derivatives; these are discussed in Chap. 25. The two equations in Example 20-23 which contain $dv/dt$ and $ds/dt$ are examples of elementary differential equations. The next example shows another application of the antiderivative to a differential equation.

### Example 20-25
In a circuit containing an inductance $L = 0.04$ H, the voltage across the inductance is given by $V = 0.06t^2 + 0.08$ for a short interval of time $t$. Given that $V = L \, dI/dt$, find the current $I$ as a function of $t$ if $I = 0$ A when $t = 0$ s.

### Solution
Equate the two expressions for $V$ to obtain the differential equation:

$$L \frac{dI}{dt} = 0.06t^2 + 0.08$$

Substitute $L = 4.0 \times 10^{-2}$:

$$(4.0 \times 10^{-2}) \frac{dI}{dt} = 0.06t^2 + 0.08$$

Solve for $dI/dt$ and take the antiderivative:

$$\frac{dI}{dt} = 1.5t^2 + 2.0$$

$$I = 1.5\left(\frac{t^3}{3}\right) + 2.0t + C$$

$$= 0.50t^3 + 2.0t + C$$

Find $C$ by substituting $I = 0$ when $t = 0$:

$$0 = 0.50(0)^3 + 2.0(0) + C$$
$$C = 0$$

The current $I$ as a function of $t$ is then:

$$I = 0.5t^3 + 2.0t$$

**Exercise 20-7**

In problems 1 to 18, find the antiderivative of each derivative function.

**1.** $f'(x) = 4x^3$

**2.** $f'(x) = 2x - 1$

**3.** $y' = 6x^2 + 2x - 3$

**4.** $y' = 10x^5 - 6x^3 + 3x$

**5.** $f'(x) = \dfrac{x^3 - 2x^2}{x^2}$

**6.** $y' = (1 - x)^2$

**7.** $f'(x) = 3 - \dfrac{3}{x^2}$

**8.** $f'(x) = \dfrac{1}{x^3} - \dfrac{1}{x^4}$

**9.** $y' = \dfrac{1}{4x^3}$

**10.** $y' = 3x^2 - \dfrac{1}{3x^2}$

**11.** $f'(x) = \sqrt{x}$

**12.** $f'(x) = \dfrac{1}{2\sqrt{x}}$

**13.** $y' = \dfrac{x^{3/2}}{2} - \dfrac{x^{1/2}}{2}$

**14.** $y' = 4\sqrt[3]{x}$

**15.** $\dfrac{ds}{dt} = 0.60t^3 - 1.5t^2 + 2.0t$

**16.** $\dfrac{dV}{dt} = 0.5t\sqrt{t} - 0.3\sqrt{t}$

**17.** $\dfrac{dy}{dx} = 2ax + 2b$    ($a$, $b$ constants)

**18.** $\dfrac{dP}{dI} = 2IR + 2Ir$    ($R$, $r$ constants)

**19.** A ball is dropped from a window 50 ft high. Using $g = -32$ ft/s², (*a*) find the formula for the velocity as a function of time and (*b*) find the equation of motion. (See Example 20-23.)

20. A ball is thrown upward from the ground with a velocity $v = 5$ m/s. Using $g = -9.8$ m/s$^2$, find (a) the velocity as a function of time and (b) the equation of motion. (See Example 20-23.)

21. Find the equation of the curve whose slope is given by $y' = 2x + 2$ and passes through the point (1, 4).

22. The *rate of change of slope* of a curve is given by $f''(x) = 6x - 2$. If $f'(0) = f(0) = 0$, find the equation of the curve.

23. The current through a capacitance $C = 2.0 \times 10^{-2}$ F is given by $I = 2t + 1$ for a short time $t$. Given that $I = C\ dV/dt$, find the voltage $V$ as a function of $t$ if $V = 0$ V when $t = 0$ s. (See Example 20-25.)

24. The voltage across an inductance $L = 3 \times 10^{-3}$ H is given by $V = (9 \times 10^{-3})t^2 + (3 \times 10^{-3})t$ for a short interval $t$. Given that $V = L\ dI/dt$, find the current $I$ as a function of $t$ if $I = 0.5$ A when $t = 1.0$ s. (See Example 20-25.)

25. A ship traveling at 20 nmi/h shuts off its engines and slows down. If the deceleration due to the water resistance is given by:

$$\frac{dv}{dt} = \frac{-10}{\sqrt{t}}$$

how long will it take for the ship to stop?

26. The differential of a function is given by:

$$dy = \frac{x^2 - 2}{x^2}\ dx$$

Find the function.

## 20-8 | *Review Questions*

1. The equation of motion for an electronic part in an assembly line is given by:

$$s = \frac{20}{t + 1}$$

where $s$ = distance (in centimeters). Find the velocity and acceleration when $t = 1$ s.

2. A pulley turns through an angle $\theta$ (in radians) given by $\theta = 1.5t^3 - 0.5t$, where $t$ = time. Find the angular velocity $\omega$ and angular acceleration $\alpha$ when $t = 5$ s.

3. A particle moves along a curve according to the parametric equations $x = 2t$ and $y = t^2 + 1$. When $t = 2$, find (a) the velocity components and (b) the resultant velocity.

4. A rocket's coordinates (in meters) at any time $t$ are given by $x = 10\sqrt{t}$ and $y = 2t^3$. What is the resultant velocity when $t = 2.5$ s?

In problems 5 to 12, sketch the graph of each function, showing intercepts, asymptotes, symmetry, and maximum, minimum, and inflection points.

**5.** $y = x^3 - 3x^2$

**6.** $f(x) = 9x - 3x^3$

**7.** $f(x) = x^4 - 4x^2$

ⓒ **8.** $y = t^4 - 2t^3$

**9.** $y = \dfrac{1}{x^2} - 4$

**10.** $y = x + \dfrac{1}{x}$

**11.** $y = \dfrac{2}{1 - x^2}$

ⓒ **12.** $f(x) = \dfrac{1}{1 + 2x^2}$

In problems 13 and 14, find the differential of each function.

**13.** $y = f(x) = 3x^2 + 4x^4$

**14.** $I = f(t) = \sqrt{t} - 1/t$

In problems 15 to 18, find the antiderivative of each derivative.

**15.** $y' = 9x^2 + 5$

**16.** $f'(x) = \dfrac{x^4 + x^3}{x}$

**17.** $f'(x) = 5x - \dfrac{6}{x^2}$

**18.** $\dfrac{dI}{dt} = \sqrt{t} + \dfrac{2}{\sqrt{t}}$

ⓒ **19.** The deflection $y$ (in inches) of a beam is given by:

$$y = 0.01x^3 - 0.08x^2 + 0.20$$

where $x$ = distance (in feet) from one end. If the beam is 8 ft long, sketch the curve of deflection, showing the points of maximum deflection.

**20.** A resistance $R_1$ = 4.0 $\Omega$ is placed in parallel with a variable resistance $R_2$. The total resistance of the circuit is given by:

$$R = \dfrac{R_1 R_2}{R_1 + R_2}$$

Sketch the curve of $R$ vs. $R_2$ for $R_2 \geq 0$.

**21.** The coordinates of a projectile are given by the parametric equations $x = t$ and $y = 10t - t^2$. Find the coordinates of the point where $y$ is a maximum.

22. The voltage across a capacitance is given by $V = t + 4/t$ from $t = 0.5$ to $t = 8$ s. What is the maximum voltage during this time interval?

23. Joel wants to fence in three sides of a rectangular area for his livestock, using a stream as the fourth side. What is the maximum area he can enclose with 300 m of fencing?

24. A solar cell supplies a constant voltage of 1.5 V to a circuit containing a variable resistance $R$ in series with a fixed resistance of 1.0 $\Omega$. Find the value of $R$ that will produce the maximum power consumption in $R$. (See Example 20-14.)

25. The bending moment of a beam is given by:

$$y = 3.0x^3 - 4.5Lx^2 + 2.0L^2x$$

where $L$ = length of the beam and $x$ = distance from one end. Find the maximum bending moment of the beam in terms of $L$ for $x < L/2$.

26. Using the formulas in Table 4-5 for the surface area and volume of a closed right circular cylinder (Sec. 4-6), show that, for a given surface area, the volume is a maximum when the height $h$ equals twice the radius $r$. (*Hint:* Differentiate both formulas implicitly with respect to $r$ and eliminate $h'$.)

🖳 27. A volume $V = 4.50$ m³ of nitrogen at a pressure $P = 20$ kPa is being compressed adiabatically (without gain or loss of heat) at the rate of 1.00 kPa/s. The relationship between $P$ and $V$ is given by $PV^{1.41} = 167$. What is the rate of change of $V$ at this time, in cubic meters per second?

28. A variable resistance in a circuit is decreasing at the rate of 2.0 $\Omega$/s. If the voltage across the resistance is constant at 120 V, at what rate is the current $I$ changing when $I = 3.0$ A?

29. A car leaves Los Angeles at noon traveling north at 75 mi/h. At 1 P.M. a train leaves Los Angeles traveling east at 80 mi/h. How fast is the distance between them increasing at 2 P.M.?

30. Ice is forming on a spherical satellite of radius 3.0 m such that the radius is increasing at the rate of 2.0 cm/s. If the ice weighs 1.0 g/cm³, at what rate is the satellite's weight increasing, in kilograms per second, when the thickness of the ice is 10 cm?

31. The height of the water in a cylindrical rain barrel increases from 5.0 to 5.1 in. If the radius of the barrel is 12 in., use differentials to find the approximate increase in the volume of the water.

32. A car accelerates from rest at the constant rate of 2.5 m/s². (*a*) Find the formula for the velocity as a function of time. (*b*) Find the equation of motion. (See Example 20-23.)

# 21

# The Integral and Its Application

## 21-1 | *The Indefinite Integral*

The concept of an integral combines the ideas of the differential and the antiderivative. Given a function $F(x)$, its differential is $F'(x)\,dx$. If you let $f(x) = F'(x)$, then the *indefinite integral of* $f(x)$ is defined as follows:

$$\int f(x)\,dx = F(x) + C$$
$$C = \text{constant} \qquad (21\text{-}1)$$

The *integral sign* $\int$ is an elongated S and comes from the word *sum*. The reason for this is shown in the next section. The differential $dx$ must be included in the integral and indicates that the integral is "with respect to $x$." The *constant of integration* $C$ is an arbitrary constant and can be determined when certain additional information about $F(x)$ is known.

> This chapter introduces the concept of the integral which is the basis for integral calculus. The process of integration, or evaluating an integral, is essentially the same as that of antidifferentiation (Sec. 20-7). Integration is shown in Sec. 21-1, which defines the indefinite integral and shows you some of its applications. Section 21-2 discusses the definite integral and its geometric model as an area under a curve. The remainder of the chapter explores several applications of integrals and shows you the importance of integration in solving various problems in engineering and technology.

From the power formula for the antiderivative (20-5), it follows from formula (21-1) that for $n$ = any rational number except $-1$:

$$\int kx^n\,dx = \frac{kx^{n+1}}{n+1} + C \qquad (21\text{-}2)$$

where $k$ is a constant and $f(x) = F'(x) = kx^n$.

### Example 21-1

Find each indefinite integral.

**1.** $\int 3x^2 \, dx$

**2.** $\int 5 \, dx$

**3.** $\int \dfrac{dt}{\sqrt{t}}$

### Solution

**1.** $\int 3x^2 \, dx$

Applying formula (21-1) to this integral yields $f(x) = F'(x) = 3x^2$. Apply formula (21-2) to obtain:

$$\int 3x^2 \, dx = \frac{3x^3}{3} + C = x^3 + C$$

**2.** $\int 5 \, dx$

Applying formula (21-1) to this integral gives $f(x) = 5x^0$. Apply formula (21-2):

$$\int 5 \, dx = \frac{5x^{0+1}}{0+1} + C = 5x + C$$

**3.** $\int \dfrac{dt}{\sqrt{t}}$

In this integral $f(t) = 1/\sqrt{t} = t^{-1/2}$. Apply formula (21-2):

$$\int \frac{dt}{\sqrt{t}} = \int t^{-1/2} \, dt = \frac{t^{-1/2+1}}{-\frac{1}{2}+1} + C$$

$$= \frac{t^{1/2}}{\frac{1}{2}} + C = 2\sqrt{t} + C$$

As a direct result of properties (19-7) and (19-8) for derivatives (Sec. 19-3), the following two properties apply to integrals.

The integral of a constant times a function is equal to the constant times the integral of the function:

If $u$ is a function of $x$ and $k$ is a constant, then:

$$\int ku \, dx = k \int u \, dx \qquad (21\text{-}3)$$

The integral of a sum of functions is equal to the sum of the integrals of the functions:
  If $u$ and $v$ are functions of $x$, then:

$$\int (u + v)\,dx = \int u\,dx + \int v\,dx \qquad (21\text{-}4)$$

The above two properties allow you to integrate certain functions, as the next example shows.

**Example 21-2**
Find each indefinite integral.

**1.** $\int (8x^3 + 3x^2 + 2x)\,dx$

**2.** $\int \dfrac{x^2 - 3}{x^2}\,dx$

**Solution**
**1.** $\int (8x^3 + 3x^2 + 2x)\,dx$:

$$\int (8x^3 + 3x^2 + 2x)\,dx = 8\int x^3\,dx + 3\int x^2\,dx + 2\int x\,dx$$

$$= 8\left(\frac{x^4}{4}\right) + 3\left(\frac{x^3}{3}\right) + 2\left(\frac{x^2}{2}\right) + C$$

$$= 2x^4 + x^3 + x^2 + C$$

Note that the constant $C$ represents the sum of the constants from each integral.

**2.** $\int \dfrac{x^2 - 3}{x^2}\,dx$:

$$\int \frac{x^2 - 3}{x^2}\,dx = \int \frac{x^2\,dx}{x^2} - \int 3\,\frac{dx}{x^2}$$

$$= \int dx - 3\int \frac{dx}{x^2} = x - 3\left(\frac{x^{-2+1}}{-2+1}\right) + C$$

$$= x + \frac{3}{x} + C \quad \text{or} \quad \frac{x^2 + 3}{x} + C$$

Notice in the $\int dx$ the 1 is understood and $\int dx = x + C$.

The above two examples show how to integrate powers of $x$. However, to integrate a power function such as $\int (2x + 1)^2 \, dx$ or $\int \sqrt{3x - 1} \, dx$, a more general formula than (21-2) is needed. From the general power formula for derivatives (19-14), Sec. 19-6, it follows that for $n =$ any rational number except $-1$:

If $u$ is a function of $x$, then:

$$\int u^n (u') \, dx = \frac{u^{n+1}}{n + 1} + C \qquad (n \neq -1) \tag{21-5}$$

Since $du = u' \, dx$, this formula can also be written:

$$\int u^n \, du = \frac{u^{n+1}}{n + 1} + C \qquad (n \neq -1) \tag{21-5a}$$

To apply formula (21-5), it is necessary to have the correct form and to identify the function $u$ and its derivative $u'$. Study the next example carefully.

### Example 21-3
Find each indefinite integral.

1. $\int (2x + 1)^2 \, 2 \, dx$
2. $\int (2x + 1)^2 \, dx$
3. $\int (2x^2 + 3)^3 \, x \, dx$
4. $\int \sqrt{3r - 1} \, dr$

### Solution
1. $\int (2x + 1)^2 \, 2 \, dx$

This integral is in the form $\int u^2 (u') \, dx$, where $u = 2x + 1$ and $u' = 2$ or $du = 2 \, dx$. You can apply formula (21-5) or (21-5a) to obtain:

$$\int (2x + 1)^2 \, 2 \, dx = \int u^2 \, du = \frac{u^3}{3} + C$$

$$= \frac{(2x + 1)^3}{3} + C$$

The 2 does not appear in the result but is a necessary "ingredient" that allows you to do the integration. If you apply formula (19-14) and differentiate the result, the 2 will appear in the derivative.

**2.** $\int (2x + 1)^2 \, dx$

Compare this integral to (1) above. To apply formula (21-5), you need $u' = 2$ in the integral. Because of property (21-3) you can multiply 2 into the integral and "balance" it by multiplying the integral by $\frac{1}{2}$:

$$\int (2x + 1)^2 \, dx = \int \frac{1}{2}(2x + 1)^2(2) \, dx = \frac{1}{2} \int (2x + 1)^2(2) \, dx$$

Note that *an integral can only be multiplied by a constant in this way*. A function that is not a constant cannot be multiplied into an integral because it will change the function in the integral.

The integral is now in the form $\int u^2(u') \, dx$ and can be integrated by using formula (21-5):

$$\frac{1}{2} \int (2x + 1)^2(2) \, dx = \frac{1}{2} \int u^2 \, du = \frac{1}{2}\left(\frac{u^3}{3}\right) + C$$

$$= \frac{1}{2}\left[\frac{(2x + 1)^3}{3}\right] + C = \frac{(2x + 1)^3}{6} + C$$

The $\frac{1}{2}$ *does* appear in the result, but the 2 *does not*. Check this result by differentiating and seeing that you get the original function.

This integral can also be done by multiplying out $(2x + 1)^2$:

$$\int (2x + 1)^2 \, dx = \int (4x^2 + 4x + 1) \, dx = \frac{4x^3}{3} + 2x^2 + x + C_1$$

This answer is the same as the one above except that $C_1 = C + \frac{1}{6}$. The $\frac{1}{6}$ is the last term when you multiply out $(2x + 1)^3/6$ in the first answer.

**3.** $\int (2x^2 + 3)^3 \, x \, dx$

In this integral $u = 2x^2 + 3$ and $u' = 4x$. Put the integral into the necessary form by multiplying by 4 on the inside and $\frac{1}{4}$ on the outside. Then apply formula (21-5):

$$\int (2x^2 + 3)^3 \, x \, dx = \frac{1}{4} \int (2x^2 + 3)^3(4)x \, dx$$

$$= \frac{1}{4}\left[\frac{(2x^2 + 3)^4}{4}\right] + C = \frac{(2x^2 + 3)^4}{16} + C$$

**4.** $\int \sqrt{3r - 1} \, dr$

Since $\sqrt{3r - 1} = (3r - 1)^{1/2}$, $u = 3r - 1$, $n = \frac{1}{2}$, and $u' = 3$. Multiply by 3 on the inside and $\frac{1}{3}$ on the outside, and apply formula (21-5):

$$\int \sqrt{3r-1}\ dr = \frac{1}{3}\int (3r-1)^{1/2}(3)\ dr$$

$$= \frac{1}{3}\left[\frac{(3r-1)^{1/2+1}}{\frac{1}{2}+1}\right] + C = \frac{2(3r-1)^{3/2}}{9} + C$$

The next three examples show applications of indefinite integrals where information is provided that allows you to find the constant of integration.

### Example 21-4
Find the equation of the curve whose slope $y' = dy/dx = 2x$ and which passes through the point (2, 5).

### Solution
Write the equation $dy/dx = 2x$ in differential notation, $dy = 2x\ dx$. Then integrate to find the equation of the curve:

$$\int dy = \int 2x\ dx$$

$$y = \frac{2x^2}{2} + C = x^2 + C$$

Determine the constant by substituting the coordinates of the point (2, 5) into the equation $y = x^2 + C$:

$$5 = (2)^2 + C$$
$$C = 1$$

The equation of the curve is therefore $y = x^2 + 1$, and it is a parabola.

### Example 21-5
The velocity of an object is given by $v = 3t^2 + 2$. If the distance traveled $s = 6$ ft when $t = 1$ s, find $s$ as a function of $t$, that is, the equation of motion.

### Solution
Since $ds/dt = v$, you can write the equation in differential notation as $ds = v\ dt$ and integrate to find $s$:

$$\int ds = \int v\ dt = \int (3t^2 + 2)\ dt$$

$$s = \frac{3t^3}{3} + 2t + C = t^3 + 2t + C$$

Find $C$ by substituting $s = 6$ when $t = 1$:

$$6 = (1)^3 + 2(1) + C$$

$$C = 3$$

The equation of motion is then:

$$s = t^3 + 2t + 3$$

As shown in Example 21-5 for velocity and distance, the derivative formulas for motion each have an equivalent integral form. Table 21-1 lists these formulas in their derivative and integral forms.

**Table 21-1**  MOTION FORMULAS

| Description | Derivative Form | Integral Form |
|---|---|---|
| Distance $s$ and velocity $v$ | $v = \dfrac{ds}{dt}$ | $s = \displaystyle\int v\, dt$ |
| Velocity $v$ and acceleration $a$ | $a = \dfrac{dv}{dt}$ | $v = \displaystyle\int a\, dt$ |
| Angular displacement $\theta$ and angular velocity $\omega$ | $\omega = \dfrac{d\theta}{dt}$ | $\theta = \displaystyle\int \omega\, dt$ |
| Angular velocity $\omega$ and angular acceleration $\alpha$ | $\alpha = \dfrac{d\omega}{dt}$ | $\omega = \displaystyle\int \alpha\, dt$ |

## Application to Electric Circuits

The derivative formulas for electric circuits each have an equivalent integral form. Three basic formulas are shown in Table 21-2 in their derivative and integral forms.

**Table 21-2**  ELECTRICAL FORMULAS

| Description | Derivative Form | Integral Form |
|---|---|---|
| Current $I$ and charge $q$ | $I = \dfrac{dq}{dt}$ | $q = \displaystyle\int I\, dt$ |
| Current $I$ to and voltage $V$ across a capacitance $C$ | $I = C\dfrac{dV}{dt}$ | $V = \dfrac{1}{C}\displaystyle\int I\, dt$ |
| Voltage $V$ across and current $I$ in an inductance $L$ | $V = L\dfrac{dI}{dt}$ | $I = \dfrac{1}{L}\displaystyle\int V\, dt$ |

### Example 21-6

The current to a capacitance $C = 0.03$ F in an electric circuit is given by $I = 0.3 - 3t^2$ during an interval $t$. If the voltage across the capacitance is zero when $t = 0$ s, find the voltage across the capacitance when $t = 0.3$ s.

### Solution

From Table 21-2 the voltage across the capacitance is given by:

$$V = \frac{1}{C} \int I\, dt$$

Substitute the formula for the current $I$ and the value for the capacitance $C$ and integrate:

$$V = \frac{1}{0.03} \int (0.3 - 3t^2)\, dt$$

$$= \frac{1}{0.03} \left( 0.3t - \frac{3t^3}{3} \right) + C = 10t - 33t^3 + C$$

To find $C$, substitute $V = 0$ when $t = 0$:

$$0 = 10(0) - 33(0)^3 + C$$
$$C = 0$$

Then $V = 10t - 33t^3$, and when $t = 0.3$ s:

$$V = 10(0.3) - 33(0.3)^3 = 2.1 \text{ V}$$

### Exercise 21-1

In problems 1 to 32, find each indefinite integral.

1. $\displaystyle\int 6x^2\, dx$    2. $\displaystyle\int 4x\, dx$

3. $\displaystyle\int 3\, dx$    4. $\displaystyle\int dx$

5. $\displaystyle\int \frac{dx}{x^2}$    6. $\displaystyle\int \frac{5\, dx}{x^3}$

7. $\displaystyle\int 3\sqrt{t}\, dt$    8. $\displaystyle\int \frac{dt}{2\sqrt{t}}$

9. $\displaystyle\int (3x^2 + 2x + 1)\, dx$    10. $\displaystyle\int (x^4 - x^2)\, dx$

11. $\displaystyle\int x^2\left(1 - \frac{2}{x}\right) dx$    12. $\displaystyle\int \frac{1}{x}(x^3 + 5x)\, dx$

**13.** $\int 2\pi(r-1)\,dr$

**14.** $\int \sqrt{t}(t+2)\,dt$

**15.** $\int \dfrac{3x^2+8}{x^2}\,dx$

**16.** $\int \dfrac{4-x}{x^3}\,dx$

**17.** $\int (3x-1)^2\,3\,dx$

**18.** $\int (2-2x)^3(2)\,dx$

**19.** $\int (2x-3)^3\,dx$

**20.** $\int (5x+1)^4\,dx$

**21.** $\int (x^2+4)^4 x\,dx$

**22.** $\int (3-x^2)^3 x\,dx$

**23.** $\int (4x^2+1)^3(2x)\,dx$

**24.** $\int (1-2x^2)^5(2x)\,dx$

**25.** $\int \dfrac{x\,dx}{(2x-1)^2}$

**26.** $\int \dfrac{6x^2\,dx}{(x^3+1)^2}$

**27.** $\int \sqrt{t+1}\,dt$

**28.** $\int 4\sqrt{3-2t}\,dt$

**29.** $\int \dfrac{x\,dx}{\sqrt{x^2-4}}$

**30.** $\int \dfrac{2x^2\,dx}{\sqrt{x^3+1}}$

**31.** $\int r\sqrt{\pi r^2+0.5}\,dr$

**32.** $\int (ax+b)^3\,dx$   (*a, b* constants)

**33.** Find the equation of the curve whose slope $dy/dx = 4x$ and which passes through the point $(-1, -1)$.

**34.** Find the equation of the curve whose slope is $dy/dx = 4x^3 - 8x$ and which passes through the origin.

**35.** Find the equation of the curve whose slope $dy/dx = -12/x^2$ and which passes through the point $(4, 3)$.

**36.** Find the equation of the curve whose *rate of change of slope* $d^2y/dx^2 = 6x$ and which is tangent to the *x* axis at $x = 1$.

**37.** The velocity of an object is given by $v = 5 - 6t^2$. If the distance traveled $s = 0$ when $t = 0$ s, find the equation of motion. (See Example 21-5.)

**38.** A ball is thrown straight up from the ground with a velocity of 10 ft/s. The gravitational acceleration is constant at $-32$ ft/s². *(a)* Find the formula for the velocity. *(b)* Find the equation of motion. (See Table 21-1.)

**39.** The deceleration of a ship is given approximately by $a = -1/(4\sqrt{t})$ for $t \geq 1$ s. The velocity $v = 15$ m/s when $t = 4$ s. *(a)* Find the formula for the velocity. (See Table 21-1.) *(b)* How many *minutes* will it take for the ship to stop?

**40.** The angular acceleration of a flywheel is given approximately by $\alpha = -1.5/t^2$ for $t \geq 0.5$ s. *(a)* If the angular velocity $\omega = 3$ rad/s when $t = 0.5$ s, find the formula for the angular velocity. (See Table 21-1.) *(b)* When will $\omega = 1$ rad/s?

**41.** At $t = 0$ s, the current to a discharged capacitor $I = 0.003$ A. If the current remains constant, what is the charge in coulombs on the capacitor at $t = 4$ s? (See Table 21-2.)

**42.** If the current in problem 41 is given by $I = 0.002t$, what is the charge at $t = 4$ s?

**43.** The current to a capacitance $C = 0.002$ F in an electric circuit is given by $I = 0.04t - 0.03t^2$ during an interval $t$. If the voltage across the capacitance is zero at $t = 0$ s, find the voltage across the capacitance when $t = 1$ s. (See Example 21-6.)

$\boxed{\text{C}}$ **44.** A current $I = 0.001\sqrt{t}$ flows to a discharged capacitor for 5 s. If the capacitance $C = 0.040$ F, what is the final voltage across the capacitor? (See Example 21-6.)

$\boxed{\text{C}}$ **45.** The voltage across an inductance $L = 0.20$ H in an electric circuit is given by $V = 0.03(t^2 - \sqrt{t})$ for an interval of time $t$. If the initial current is 3.0 A, find the current at $t = 5$ s. (See Table 21-2.)

**46.** The voltage across an inductance $L = 0.10$ H in an electric circuit is given by $V = 0.1/\sqrt{t + 1}$ for an interval $t$. If the initial current is zero, at what time is the current 2 A? (See Table 21-2.)

**47.** The work done by a variable force $F$ over a distance $x$ is given by $W = \int F\, dx$. Find the formula for the work done if $F = (x - 2)^3$ and $W = 10$ ft $\cdot$ lb when $x = 0$ ft.

**48.** The deflection $\delta$ of a beam at a distance $x$ meters from a *fixed end* is given by:

$$50\delta = \int 0.0020\, dx + \int 0.050x^2\, dx$$

Find $\delta$ when $x = 0.25$ m.

## 21-2 | *The Definite Integral: Area under a Curve*

The geometric model of the integral is the area under a curve, in the same way that the geometric model of a derivative is the slope of a curve. The methods for finding the area under a curve were developed separately from the methods for finding the slope of a curve. The discovery that the two methods represent inverse operations marked the discovery of the calculus by Newton and Leibniz. The first example shows how the area under a curve can be determined by finding the limit of the sum of an infinite number of terms. This leads to the meaning of the definite integral and its definition in terms of antiderivatives.

### Example 21-7
Determine the area under the parabola $y = f(x) = x^2$, which is bounded by the $x$ axis and the lines $x = 0$ and $x = 1$, by finding the limit of the sum of an infinite number of rectangular areas.

### Solution
Figure 21-1 shows the area to be found. It is bounded by the $x$ axis, the line $x = 1$, and a segment of the parabola. Figure 21-2 shows the same area in more detail. The area is approximated by inscribing $n$ rectangles of equal width under the curve. Study the figure carefully. The distance along the $x$ axis is 1 unit and is divided into $n$ equal intervals. *The width of each rectangle $\Delta x$ is therefore*

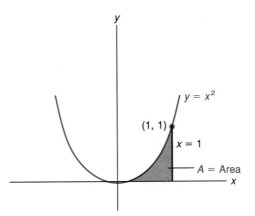

FIG. 21-1 *Area under the parabola $y = x^2$, Example 21-7*

$1/n$. For example, if $n = 100$, there would be 100 rectangles and each rectangle would have a width $\Delta x = 1/100$ of a unit.

*The height of each rectangle is determined by the y value of the point on the curve that each rectangle touches.* The first rectangle touches the curve at $x = 0$, and therefore its height is $y = 0$. The second rectangle touches the curve

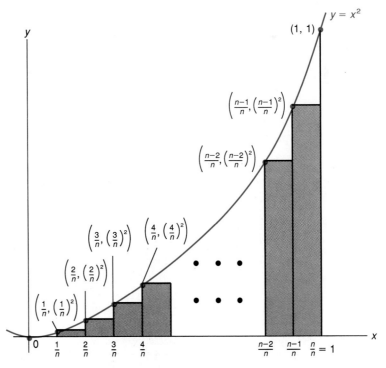

FIG. 21-2 *Approximation of area under the parabola $y = x^2$, Example 21-7*

at $x = 1/n$. The height of the second rectangle is therefore $y = x^2 = (1/n)^2$. The height of the third rectangle is therefore $y = (2/n)^2$. By continuing in this way, the height of the last, or $n$th, rectangle is $[(n-1)/n]^2$. The *area* of each rectangle is its width $1/n$ times its height, as follows:

$$\text{Area of 1st rectangle} = \frac{1}{n}(0) = 0$$

$$\text{Area of 2d rectangle} = \frac{1}{n}\left(\frac{1}{n}\right)^2 = \frac{1}{n^3}$$

$$\text{Area of 3d rectangle} = \frac{1}{n}\left(\frac{2}{n}\right)^2 = \frac{4}{n^3}$$

$$\text{Area of 4th rectangle} = \frac{1}{n}\left(\frac{3}{n}\right)^2 = \frac{9}{n^3}$$

$$\vdots$$

$$\text{Area of } n\text{th rectangle} = \frac{1}{n}\left(\frac{n-1}{n}\right)^2 = \frac{(n-1)^2}{n^3}$$

The total area of the $n$ rectangles is therefore:

$$A_n = 0 + \frac{1}{n^3} + \frac{4}{n^3} + \frac{9}{n^3} + \cdots + \frac{(n-2)^2}{n^3} + \frac{(n-1)^2}{n^3}$$

or $$A_n = \frac{1}{n^3}[1 + 4 + 9 + \cdots + (n-2)^2 + (n-1)^2]$$

The quantity in the brackets is the sum of the squares of the first $n-1$ whole numbers. This sum can be shown to be equal to $n^3/3 - n^2/2 + n/6$. For example, if $n - 1 = 9$, $n = 10$, and:

$$1 + 4 + 9 + \cdots + 81 = \frac{10^3}{3} - \frac{10^2}{2} + \frac{10}{6} = 285$$

therefore $A_n$ can be written:

$$A_n = \frac{1}{n^3}\left(\frac{n^3}{3} - \frac{n^2}{2} + \frac{n}{6}\right) = \frac{1}{3} - \frac{1}{2n} + \frac{1}{6n^2}$$

This formula can be used to calculate the total area of the rectangles for a given value of $n$. For example, if $n = 10$:

$$A_{10} = \frac{1}{3} - \frac{1}{2(10)} + \frac{1}{6(10)^2} = \frac{1}{3} - \frac{1}{20} + \frac{1}{600} = 0.285$$

If $n = 100$:

$$A_{100} = \frac{1}{3} - \frac{1}{2(100)} + \frac{1}{6(100)^2} = \frac{1}{3} - \frac{1}{200} + \frac{1}{60,000} = 0.3284$$

As $n \to \infty$, $A_n$ approaches the area $A$ under the curve shown in Fig. 21-1. The reason for this can be seen in Fig. 21-3, which shows how the area is more closely approximated when $n$ is doubled from 5 to 10. The area $A$ under the curve is then:

$$A = \lim_{n \to \infty} A_n = \lim_{n \to \infty} \left( \frac{1}{3} - \frac{1}{2n} + \frac{1}{6n^2} \right) = \frac{1}{3} \text{ square unit}$$

The last two fractions approach zero as $n \to \infty$. Note that the area $A$ is *exactly* $\frac{1}{3}$ of a square unit.

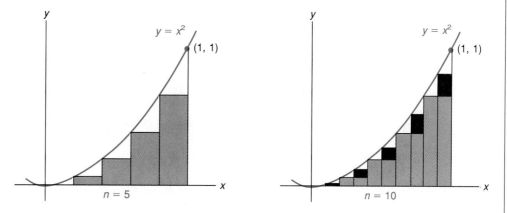

FIG. 21-3 *Effect of increasing the number of rectangles, Example 21-7*

The area under a curve $y = f(x)$ and the summation process shown in Example 21-7 are related to $\int y \, dx$ or $\int f(x) \, dx$. This relationship can be seen as follows.

Consider the area under a curve $y = f(x)$, bounded by the $x$ axis and the lines $x = a$ and $x = b$ (Fig. 21-4). Let $f(x) > 0$ throughout the interval $x = a$ to $x = b$. Let the function $A = A(x)$ represent the area under the curve from $x = a$ to a value of $x$ such that $a \le x \le b$. Let $\Delta A$ represent the change in $A$ between $x$ and $x + \Delta x$. The area $\Delta A$ is less than or equal to the large rectangle shown above the curve and greater than or equal to the small rectangle shown below the curve:

$$(y)(\Delta x) \le \Delta A \le (y + \Delta y)(\Delta x)$$

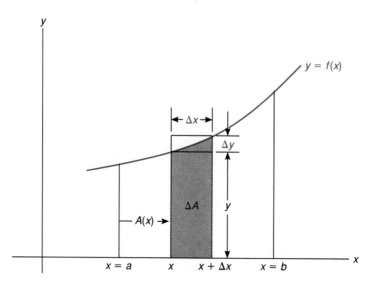

FIG. 21-4 *Area under a curve* $y = f(x)$

This inequality can be divided by $\Delta x$:

$$y \leq \frac{\Delta A}{\Delta x} \leq y + \Delta y$$

The inequality says that between the two points $x$ and $x + \Delta x$, the average rate of change of the area $A$ with respect to $x$ is between $y$ and $y + \Delta y$. If you take the limit as $\Delta x \to 0$, $\Delta y \to 0$ and:

$$\lim_{\Delta x \to 0} \frac{\Delta A}{\Delta x} = \frac{dA}{dx}$$

Therefore:

$$y \leq \frac{dA}{dx} \leq y$$

which means:

$$\frac{dA}{dx} = y \qquad\qquad (21\text{-}6)$$

It follows that $dA = v\, dx$ or:

$$A = \int y\, dx = \int f(x)\, dx \qquad\qquad (21\text{-}7)$$

Evaluating this integral, you obtain:

$$A = F(x) + C$$

where $f(x) = F'(x)$. When $x = a$, $A = 0$, therefore:

$$0 = F(a) + C$$
$$C = -F(a)$$

and the area function is:

$$A = F(x) - F(a)$$

When $x = b$, the area under the curve between $x = a$ and $x = b$ is then:

$$A_{a,b} = F(b) - F(a) \tag{21-8}$$

The area under a curve can thus be obtained by using integration and applying the above ideas. The next example shows how the area in Fig. 21-1 is obtained in this way.

### Example 21-8
Determine the area in Example 21-7 by integration.

### Solution
Apply formula (21-7):

$$A = \int y \, dx = \int x^2 \, dx = \frac{x^3}{3} + C$$

Then apply formula (21-8) to find the area under the curve between $x = 0$ and $x = 1$:

$$A_{0,1} = F(1) - F(0) = \frac{1^3}{3} - \frac{0^3}{3} = \frac{1}{3} \text{ square unit}$$

This result agrees with the limit found in Example 21-7.

Example 21-7 uses the limit of the sum of an infinite number of rectangular areas to determine the same area which Example 21-8 finds by integration. Examples 21-7 and 21-8 illustrate the fact that the limit of a sum is a basic part of the meaning of an integral. This is the reason that the integral sign is an elongated S. The definition of the *definite integral* of a function $y = f(x)$ is:

$$\int_a^b y \, dx = \int_a^b f(x) \, dx = F(x) \Big|_a^b = F(b) - F(a) \tag{21-9}$$

where $f(x) = F'(x)$. Here $b$ is called the *upper limit,* and $a$ is called the *lower limit.* Notice that there is no constant of integration in a definite integral. The definite integral represents a definite value, unlike an indefinite integral, which represents a function. Study the next example, which shows how to evaluate definite integrals. The meaning of the definite integral is explained further after this example.

### Example 21-9
Evaluate each definite integral.

1. $\displaystyle\int_0^1 2x^3 \, dx$

2. $\displaystyle\int_2^3 (x^2 - 4x) \, dx$

3. $\displaystyle\int_1^2 5\sqrt{5x - 1} \, dx$

### Solution

1. $\displaystyle\int_0^1 2x^3 \, dx$

   Apply formula (21-9):

$$\int_0^1 2x^3 \, dx = \frac{x^4}{2}\bigg|_0^1 = \frac{1^4}{2} - \frac{0^4}{2} = \frac{1}{2} - 0 = \frac{1}{2}$$

   Notice that the limits are first written next to $F(x) = x^4/2$ with a vertical line and are then substituted to evaluate the integral.

2. $\displaystyle\int_2^3 (x^2 - 4x) \, dx$:

$$\int_2^3 (x^2 - 4x) \, dx = \left(\frac{x^3}{3} - 2x^2\right)\bigg|_2^3$$

$$= \left[\frac{(3)^3}{3} - 2(3)^2\right] - \left[\frac{(2)^3}{3} - 2(2)^2\right]$$

$$= (9 - 18) - \left(\frac{8}{3} - 8\right) = -9 + \frac{16}{3} = -\frac{11}{3}$$

   Notice the value of a definite integral can be negative since it represents the limit of a sum. The interpretation of this is explained further in Sec. 21-3.

**3.** $\displaystyle\int_{1}^{2} 5\sqrt{5x-1}\ dx$:

$$\int_{1}^{2} 5\sqrt{5x-1}\ dx = \frac{2(5x-1)^{3/2}}{3}\bigg|_{1}^{2}$$

$$= \frac{2(9)^{3/2}}{3} - \frac{2(4)^{3/2}}{3}$$

$$= \frac{2(27)}{3} - \frac{2(8)}{3} = \frac{38}{3}$$

Notice that the power formula (21-5) is used to integrate the function where $u = 5x - 1$, $u' = 5$, and $n = \frac{1}{2}$.

When $y = f(x)$ *is positive* throughout the interval $x = a$ to $x = b$, that is, when the curve is above the $x$ axis, *the geometric model of the definite integral (21-9) is the area bounded by* $y = f(x)$, *the $x$ axis, and the lines $x = a$ and $x = b$.* Figure 21-5 shows the area $A$ and its interpretation as a definite integral. A rectangle whose width is the differential of $x$, $dx$, and whose height is $y$ or $f(x)$, represents an *element of area dA:*

$$dA = y\ dx = f(x)\ dx$$

From this it can be shown that the area $A$ is the definite integral:

$$A = \int_{a}^{b} y\ dx = \int_{a}^{b} f(x)\ dx \qquad b > a \qquad (21\text{-}10)$$

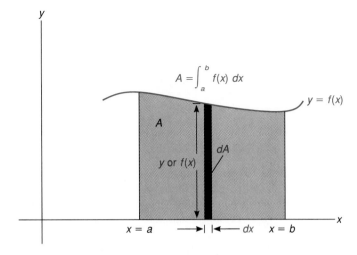

FIG. 21-5 *The definite integral and the area under a curve*

This definite integral represents the limit of the sum of an infinite number of rectangular areas $dA$ which are summed in the positive $x$ direction, similar to that shown in Example 21-7.

Study the next two examples, which show how to apply formula (21-10).

### Example 21-10

Find the area bounded by the parabola $f(x) = x^2 + 2$, $y = 0$, $x = 1$, and $x = 2$.

### Solution

The area is shown in Fig. 21-6. Note that $y = 0$ is the $x$ axis. You can apply formula (21-10) since $f(x)$ is positive throughout the interval $x = 1$ to $x = 2$:

$$A = \int_1^2 (x^2 + 2)\, dx = \left(\frac{x^3}{3} + 2x\right)\Bigg|_1^2$$

$$= \left[\frac{(2)^3}{3} + 2(2)\right] - \left[\frac{(1)^3}{3} + 2(1)\right]$$

$$= \left(\frac{8}{3} + 4\right) - \left(\frac{1}{3} + 2\right) = \frac{20}{3} - \frac{7}{3} = \frac{13}{3} \quad \text{or} \quad 4.3 \text{ square units}$$

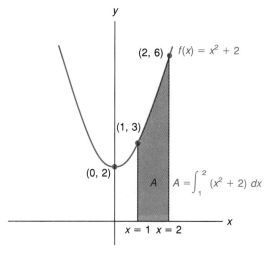

FIG. 21-6  *Area under the parabola $f(x) = x^2 + 2$, Example 21-10*

### Example 21-11

Find the area in the first quadrant bounded by the curve $y = 4x - x^3$ and the $x$ axis.

### Solution

You must sketch the curve to determine the area required (Fig. 21-7). *The limits of the integral are the $x$ intercepts.* Find these by setting $y = 0$:

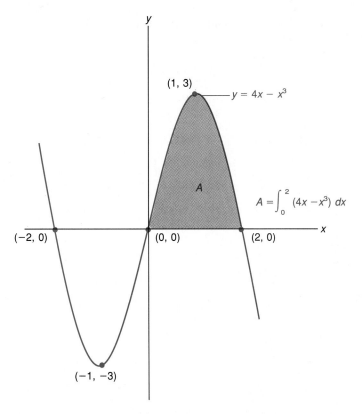

$$A = \int_0^2 (4x - x^3)\, dx$$

FIG. 21-7 *Area under the curve* $y = 4x - x^3$, *Example 21-11*

$$4x - x^3 = 0$$
$$x(4 - x^2) = 0$$
$$x = 0 \qquad 4 - x^2 = 0$$
$$x = \pm 2$$

The limits of the integral are therefore 0 and 2. Apply formula (21-10) to find the area:

$$A = \int_0^2 (4x - x^3)\, dx = \left(2x^2 - \frac{x^4}{4}\right)\Bigg|_0^2$$
$$= (8 - 4) - 0 = 4 \text{ square units}$$

**Exercise 21-2** In problems 1 to 6, determine the area bounded by the given curves and the $x$ axis by finding the limit of the sum of an infinite number of rectangular areas. Follow the procedure of Example 21-7. For problems 1 and 2, note that $1 + 2 + 3 + \cdots + (n - 1) = (n)(n - 1)/2$. Check problems 1 and 2, using the formula for the area of a triangle.

1. $y = x$, $x = 0$, $x = 1$
2. $y = 2x$, $x = 0$, $x = 1$
3. $y = 2x^2$, $x = 0$, $x = 1$
4. $y = 3x^2$, $x = 0$, $x = 1$
5. $f(x) = x^2 + 1$, $x = 0$, $x = 1$
6. $f(x) = 1 - x^2$, $x = 0$, $x = 1$

In problems 7 to 22, evaluate each definite integral.

7. $\displaystyle\int_0^2 x \, dx$

8. $\displaystyle\int_0^1 2x^2 \, dx$

9. $\displaystyle\int_0^1 (x^3 - 1) \, dx$

10. $\displaystyle\int_0^2 (1 - x^2) \, dx$

11. $\displaystyle\int_2^3 (x^2 + 3x) \, dx$

12. $\displaystyle\int_0^1 (5x^4 + 4x^3) \, dx$

13. $\displaystyle\int_1^2 \frac{dt}{t^2}$

14. $\displaystyle\int_1^4 6\sqrt{t} \, dt$

15. $\displaystyle\int_1^4 (4\sqrt{x} + x) \, dx$ $\boxed{\texttt{C}}$

16. $\displaystyle\int_1^{1.5} \frac{dr}{\sqrt[3]{r}}$

17. $\displaystyle\int_{-1}^0 (3x + 3)^3 \, dx$

18. $\displaystyle\int_{-2}^2 (x^2 - 4)^4 x \, dx$

19. $\displaystyle\int_0^4 2\sqrt{2x + 1} \, dx$

20. $\displaystyle\int_0^3 \frac{dx}{\sqrt{x + 1}}$

21. $\displaystyle\int_{1/2}^{3/2} \frac{x \, dx}{(2x^2 + 1)^2}$

22. $\displaystyle\int_{-1}^2 \frac{x \, dx}{(2 - x^2)^2}$

In problems 23 to 28, find the area bounded by each curve and the $x$ axis, using formula (21-10). Check problems 23 and 24 by using the formula for the area of a triangle or trapezoid.

23. $f(x) = x + 1$, $x = 0$, $x = 1$
24. $f(x) = 2 - 2x$, $x = 0$, $x = 1$
25. $y = 2x^2$, $x = 0$, $x = 1$   (Compare result with problem 3.)
26. $y = x^2 + 1$, $x = 0$, $x = 1$   (Compare result with problem 5.)
27. $y = x^3$, $x = 1$, $x = 2$
28. $y = 4x^3 - 3x^2$, $x = 2$, $x = 3$

In problems 29 to 32, find the area in the first quadrant bounded by each curve and the $x$ axis. (See Example 21-11.)

29. $y = 4 - x^2$
30. $y = 3x - x^2$
31. $y = x - x^3$
32. $y = 2x^3 - x^4$

**33.** The charge on a capacitor in millicoulombs (mC) is given by:

$$Q = \int_1^3 (0.6t^2 - 0.2t^3)\, dt$$

Evaluate this integral and determine the charge.

**34.** The force on a charge in an electric field is given by:

$$F = \pi \int_{0.5}^{1.0} \frac{r^2 - 1}{r^2}\, dr$$

Evaluate this integral.

☐ **35.** The moment of the force on a dam in foot-pounds (ft · lb), about a point at the base of the dam, is given by:

$$M = \int_0^{100} 62.4L\left(50y - \frac{y^2}{2}\right) dy$$

where $L$ = length of the dam. Evaluate this integral in terms of the constant $L$.

**36.** The work done in pumping water out of a reservoir is given by:

$$W = \int_0^{8.5} 8.8\pi(9.7y^2 - y^3)\, dy$$

Evaluate this integral in terms of $\pi$.

# 21-3 | *Areas between Curves*

Section 21-2 shows that the area under a curve can be determined by the limit of the sum of an infinite number of rectangular areas, and that this sum is basic to the concept of an integral. This section shows how to interpret and apply these ideas to find the area between two curves, by using integration.

Consider the two curves $y_2 = f_2(x)$ and $y_1 = f_1(x)$ as shown in Fig. 21-8. The area $A$ between the two curves is that bounded by $y_2 = f_2(x)$, $y_1 = f_1(x)$, and the lines $x = a$ and $x = b$. A rectangle whose width is the differential of $x$, $dx$, and whose height is $y_2 - y_1$, or $f_2(x) - f_1(x)$, represents an element of area $dA$:

$$dA = (y_2 - y_1)\, dx = [f_2(x) - f_1(x)]\, dx$$

From this it can be shown that *the area $A$ between the two curves $y_2 = f_2(x)$ and $y_1 = f_1(x)$ is the definite integral:*

$$A = \int_a^b (y_2 - y_1)\, dx = \int_a^b [f_2(x) - f_1(x)]\, dx \qquad (21\text{-}11)$$

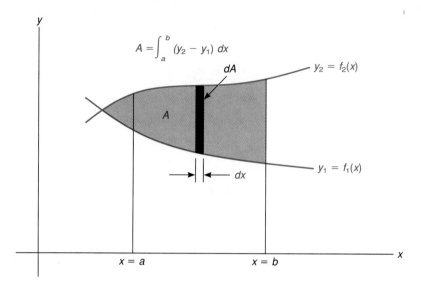

FIG. 21-8  *Area between two curves summed in x direction*

where $b > a$. This definite integral represents the sum of an infinite number of rectangular areas $dA$ which are summed in the positive $x$ direction, similar to what is shown in Example 21-7 and Fig. 21-2. Formula (21-11) applies whether the curves are above *or below* the $x$ axis. When the lower curve is subtracted from the upper curve, $y_2 - y_1$ will always be positive and the value of the definite integral will be positive. Study the following examples.

### Example 21-12
Find the area bounded by the parabola $y = 5x - x^2$ and the line $y = 4$.

### Solution
The area is shown in Fig. 21-9. You need to solve the two equations simultaneously to find the points of intersection. These points are the limits of the integral. Substitute $y = 4$ into the first equation:

$$5x - x^2 = 4$$
$$x^2 - 5x + 4 = 0$$
$$(x - 1)(x - 4) = 0$$
$$x = 1 \qquad x = 4$$

Apply (21-11) to find the area. Let $y_2 = 5x - x^2$ and $y_1 = 4$:

$$A = \int_1^4 (y_2 - y_1)\, dx = \int_1^4 [(5x - x^2) - (4)]\, dx$$

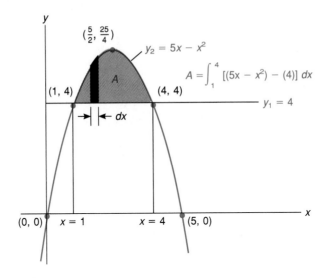

FIG. 21-9  *Area bounded by parabola and line, Example 21-12*

$$= \frac{5x^2}{2} - \frac{x^3}{3} - 4x \Big|_1^4$$

$$= \left[ \frac{5(4)^2}{2} - \frac{(4)^3}{3} - 4(4) \right] - \left[ \frac{5(1)^2}{2} - \frac{(1)^3}{3} - 4(1) \right]$$

$$= \frac{8}{3} - \left( -\frac{11}{6} \right) = \frac{9}{2} \text{ square units}$$

## Example 21-13
Find the area bounded by $y = x^3$, $y = -x$, and $x = 1$.

### Solution
The area is shown in Fig. 21-10. From the figure it can be seen that the limits of the integral are $x = 0$ to $x = 1$. Even though part of the area is below the $x$ axis, if you let the upper curve be $y_2 = x^3$ and the lower curve be $y_1 = -x$, formula (21-11) will produce the correct area:

$$A = \int_0^1 (y_2 - y_1) \, dx$$

$$= \int_0^1 [x^3 - (-x)] \, dx = \left( \frac{x^4}{4} + \frac{x^2}{2} \right) \Big|_0^1$$

$$= \left( \frac{1}{4} + \frac{1}{2} \right) - 0 = \frac{3}{4} \text{ square unit}$$

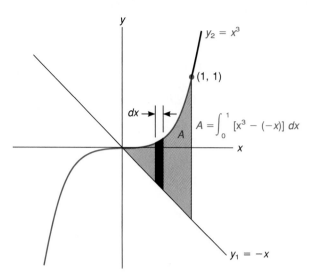

FIG. 21-10  *Area bounded by* $y = x^3$, $y = -x$, *and* $x = 1$, *Example 21-13*

The next example explains the relation between an area below the *x* axis and a negative definite integral.

### Example 21-14
Find the area bounded by the parabola $y = x^2 - 4$ and the *x* axis.

### Solution
It is important to sketch the curve and the area, which Fig. 21-11 shows is below the *x* axis. The intercepts are found by setting $x^2 - 4 = 0$. This yields $x = \pm 2$. If the *x* axis is considered the upper curve, $y_2 = 0$ and $y_1 = x^2 - 4$. The area is then:

$$A = \int_{-2}^{2} [0 - (x^2 - 4)] \, dx = \int_{-2}^{2} (4 - x^2) \, dx$$

$$= \left(4x - \frac{x^3}{3}\right)\Bigg|_{-2}^{2} = \frac{16}{3} - \left(-\frac{16}{3}\right) = \frac{32}{3} \text{ square units}$$

From the above example, it can be seen that:

$$A = \int_{-2}^{2} (0 - y_1) \, dx = -\int_{-2}^{2} y_1 \, dx$$

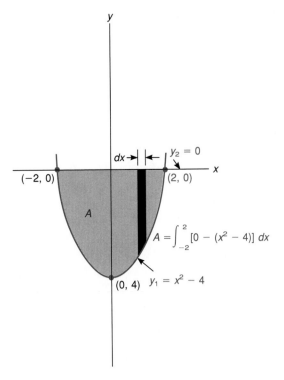

FIG. 21-11 *Area bounded by parabola and x axis, Example 21-14*

Therefore *if* $y = f(x)$ *is negative* throughout the interval $x = a$ to $x = b$, then *the area bounded by* $y = f(x)$, *the x axis, and the lines* $x = a$ *and* $x = b$ *is:*

$$A = -\int_a^b y \, dx - \int_a^b f(x) \, dx \qquad (21\text{-}12)$$

### Example 20-15
Find the area bounded by $y = x^3 - 9x$ and the $x$ axis.

### Solution
The area $A$ is shown shaded in Fig. 21-12. The $x$ intercepts of the curve are found by setting $y = 0$, which yields $x = 0$ and $x = \pm 3$. Since the area to the right of $x = 0$ is below the $x$ axis, apply formula (21-12) to find that area and formula (21-10) to find the area to the left of $x = 0$:

$$A = \int_{-3}^0 (x^3 - 9x) \, dx - \int_0^3 (x^3 - 9x) \, dx$$

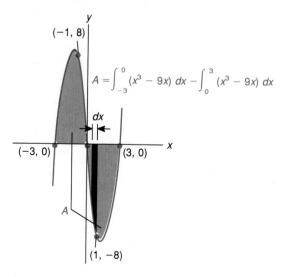

$$A = \int_{-3}^{0} (x^3 - 9x)\, dx - \int_{0}^{3} (x^3 - 9x)\, dx$$

FIG. 21-12  *Area bounded by* $y = x^3 - 9x$ *and the x axis, Example 21-15*

$$= \left( \frac{x^4}{4} - \frac{9x^2}{2} \right)\Big|_{-3}^{0} - \left( \frac{x^4}{4} - \frac{9x^2}{2} \right)\Big|_{0}^{3}$$

$$= \frac{81}{4} + \frac{81}{4} = 40.5 \text{ square units}$$

Observe that since the curve is symmetric to the origin, the area can also be found by doubling the area from $x = -3$ to $x = 0$. That is:

$$A = 2 \int_{-3}^{0} (x^3 - 9x)\, dx = 40.5 \text{ square units}$$

Note also that if you integrate from $x = -3$ to $x = 3$, you obtain:

$$\int_{-3}^{3} (x^3 - 9x)\, dx = 0$$

In the preceding examples the definite integral (21-11) that is used to find each area uses an element of area whose width is $dx$, and these areas are summed in the positive $x$ direction. In a similar way, an element of area of width $dy$ can be summed in the positive $y$ direction to find an area between two curves. Figure 21-13 illustrates this situation. Each curve must be solved for $x$

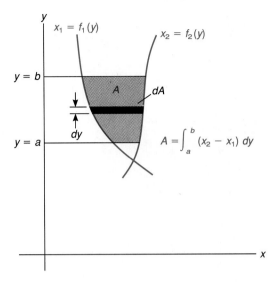

FIG. 21-13 *Area between two curves summed in y direction*

and expressed as a function of **y**. *The area summed in the positive y direction* is then given by:

$$A = \int_a^b (x_2 - x_1)\, dy \tag{21-13}$$

Study the following example, which shows how to apply formula (21-13).

### Example 21-16

Find the area bounded by $y = x^3$, $y = -x$, and $y = 2$.

### Solution

The area is shown in Fig. 21-14. If elements of area are summed in the $x$ direction, it is necessary to set up more than one definite integral since the area to the left of the $y$ axis is bounded by $y = 2$ and $y = -x$, whereas the area to the right of the $y$ axis is bounded by $y = 2$ and $y = x^3$. If you use an element in the $y$ direction, you can find the area with only one definite integral. Solve each of the equations for $x$ to obtain $x = y^{1/3}$ and $x = -y$. Apply (21-13), letting $x_2 = y^{1/3}$ (because it is to the right) and $x_1 = -y$, and sum from $y = 0$ to $y = 2$:

$$A = \int_0^2 (x_2 - x_1)\, dy = \int_0^2 [y^{1/3} - (-y)]\, dy$$

$$= \left( \frac{3y^{4/3}}{4} + \frac{y^2}{2} \right) \Bigg|_0^2 = \frac{3(2)^{4/3}}{4} + \frac{2^2}{2} = 3.9 \text{ square units}$$

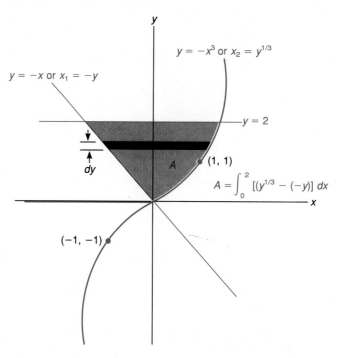

FIG. 21-14 *Area bounded by $y = x^3$, $y = -x$, and $y = 2$, Example 21-16*

**Exercise 21-3**    In problems 1 to 22, sketch and find the area bounded by the given curves.

1. $y = x$, $y = 2x$, $x = 2$

2. $y = x$, $y = \dfrac{x}{2}$, $x = \dfrac{3}{2}$

3. $y = 4x - x^2$, $y = 3$

4. $y = 2 - x^2$, $x = 0$, $y = 1$

5. $y = x^2 - 2x + 1$, $y = 1$

6. $y = x^2 - 4x + 5$, $y = 5$

7. $y = x^2$, $y = x$

8. $y = x^2$, $y = x + 2$

9. $y = x^3$, $y = x$

10. $y = \dfrac{1}{x^2}$, $y = x$, $x = 2$ $\left(\text{area above } y = \dfrac{1}{x^2}\right)$

11. $y = x^3$, $y = x^2$

12. $y = x^2$, $y = 8 - x^2$

13. $y = x^2 - 1$, $y = 0$

14. $y = x^2 - 4x$, $y = 0$

15. $y = \sqrt{x}$, $y = x$

[C] 16. $y = 2\sqrt{x}$, $y = x$, $x = 1$, $x = 3$

17. $y = 4x^3 - x$, $y = 0$

18. $y = x^3 - 3x$, $y = 0$

[C] 19. $y = x^2$, $y = 1$, $y = 2$

20. $y = 4 - x^2$, $y = 0$, $y = 3$

[C] 21. $y = x^3$, $y = -x$, $y = 3$

22. $y = x^4$, $y = 2x$, $y = 1$

23. The current to a capacitor is given by $I = 0.4t^2 - 0.1t^3$. The amount of charge from $t = t_1$ to $t = t_2$ is equal to the area bounded by the curve of $I$ vs. $t$ and the lines $I = 0$, $t = t_1$, and $t = t_2$. Find the amount of charge in coulombs from $t_1 = 1$ to $t_2 = 2$ s.

24. The current through a resistance $R = 3.0$ $\Omega$ is given by $I = 2.0\sqrt{t}$. The energy dissipated from $t = t_1$ to $t = t_2$ is equal to the area bounded by the power curve of $P$ vs. $t$ and the lines $P = 0$, $t = t_1$, and $t = t_2$. Using $P = I^2R$, find the energy dissipated in joules (J) from $t_1 = 0.5$ to $t_2 = 1.5$ s.

25. The force required to stretch a certain spring is given by $F = 1.5x$, where $x$ is the distance the spring moves from its unstretched position. The work done in stretching the spring from $x = a$ to $x = b$ is equal to the area bounded by the curve of $F$ vs. $x$, and the lines $F = 0$, $x = a$, and $x = b$. Find the work done in foot-pounds when the spring is stretched from $x = 0.1$ to $x = 0.3$ ft.

[C] 26. The pressure of a gas under compression is given by $P = 100/V^{1.3}$, where $V = $ volume. The work required to compress the gas from $V = V_1$ to $V = V_2$ is equal to the area bounded by the curve of $P$ vs. $V$ and the lines $P = 0$, $V = V_1$, and $V = V_2$. Find the work needed in foot-pounds to compress the gas from $V_1 = 2$ ft$^3$ to $V_2 = 1$ ft$^3$.

27. Find the area between the parabola $y = x^2$ and the line which passes through the points (0, 0) and (1, 2).

28. Find the area between the parabola $y = 9 - x^2$ and the line which passes through the points (1, 8) and (2, 5).

## 21-4 | *Physical Applications*

Consider a constant force $F$ which moves through a distance $x$. The work done $W$ is the product of the force times the distance: $W = Fx$. The work can be interpreted as the area of a rectangle whose height is $F$ and whose width is $x$. Suppose now the force $F$ is not constant, but varies as a function of $x$, that is, $F = f(x)$. The area under the graph of $F$ vs. $x$ between $x = a$ and $x = b$ represents the limit of the sum of an infinite number of rectangular areas whose height is $F$ and whose width is the differential of $x$, $dx$ (Fig. 21-15). Each of these areas can be thought of as an element of work $dW = F\,dx$. The total work

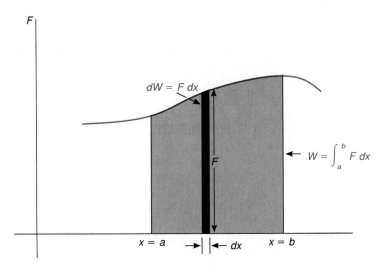

FIG. 21-15 *Work as an area under the force curve*

done from $x = a$ to $x = b$ is then the area under the graph, which is the definite integral:

$$W = \int_a^b F \; dx$$

In a similar way, the amount of charge $q$ and the energy used $E$, during an interval of time $t_1$ to $t_2$ in an electric circuit, are defined in terms of definite integrals, as shown in Table 21-3.

## Table 21-3   PHYSICAL APPLICATIONS OF INTEGRALS

| Description | Integral |
|---|---|
| Work $W$ and force $F$ as a function of distance $x$ | $W = \int_a^b F \; dx$ |
| Charge $q$ and current $I$ as a function of time $t$ | $q = \int_{t_1}^{t_2} I \; dt$ |
| Energy $E$ and power $P$ as a function of time $t$ | $E = \int_{t_1}^{t_2} P \; dt$ |

The following two examples show how to apply the formulas in Table 21-3.

## Example 21-17

A weight of 10 lb stretches a spring 2 in. beyond its natural length. How much work is done, *in foot-pounds,* when the spring is stretched from 2 to 3 in. beyond its natural length? (The work done is equal to the potential energy stored in the spring.)

### Solution

According to Hooke's law, *the force F required to stretch a spring is proportional to the distance x it is stretched:* $F = kx$. First find $k$ by letting $F = 10$ and $x = 2$:

$$10 = k(2)$$
$$k = 5 \text{ lb/in.}$$

Therefore $F = 5x$. Then apply the formula for work in Table 21-3, letting $a = 2$ and $b = 3$:

$$W = \int_a^b F \, dx = \int_2^3 5x \, dx$$
$$= 5\left(\frac{x^2}{2}\right)\Big|_2^3 = \frac{45}{2} - \frac{20}{2} = 12.5 \text{ in.} \cdot \text{lb}$$

The work in foot-pounds is then $W = 12.5/12 = 1.04 \text{ ft} \cdot \text{lb}$.

## Example 21-18

The current in milliamperes flowing to a capacitor during a short time interval is given by $I = 0.3t(2 + t)$. If the initial charge at $t = 0$ s is 0.2 mC, find the charge at $t = 3$ s.

### Solution

The *accumulation* of charge from $t = 0$ to $t = 3$ s is found by applying the formula in Table 21-3, letting $t_1 = 0$ and $t_2 = 3$:

$$q = \int_{t_1}^{t_2} I \, dt = \int_0^3 [0.3t(2 + t)] \, dt$$
$$= \int_0^3 (0.6t + 0.3t^2) \, dt$$
$$= (0.3t^2 + 0.1t^3)\Big|_0^3 = (2.7 + 2.7) - 0 = 5.4 \text{ mC}$$

Since the initial charge is 0.2 mC, the total charge at $t = 3$ s is $5.4 + 0.2 = 5.6$ mC.

There are many other physical applications of integrals besides those shown in Table 21-3. Some appear in the problems at the end of this section. In addition, the following concept has application to certain mechanical and electrical problems.

## Average or Mean Value

Consider an electric circuit where the current and therefore the power vary with time. In this situation, the average current or average power is more useful than the current or power at any instant. This is because the amount of energy used over an interval of time is of more interest than the energy being used at any instant. For example, consider a room being warmed by a heater. The amount of heat energy released during 1 h is more important than the heat energy being released at any given moment. Similarly, if a force varies over a distance, the average force is more important than the force at any instant.

Given a function $y = f(x)$, the *average* or *mean value* of $y$ over an interval from $x = a$ to $x = b$ is defined as:

$$y_{av} = \frac{\int_a^b y \, dx}{b - a} \qquad (21\text{-}14)$$

That is, the mean value of a function over an interval is the area under the curve for this interval divided by the width of the interval.

### ▣ Example 21-19
The current in a circuit containing a resistance of 2.0 Ω is given by $I = 3.0t^{3/2}$ from $t = 0$ to $t = 2$ s.

1. Find the average current during this interval.
2. Find the average power in the resistance during this interval.

**Solution**
1. Find the average current during this interval.

Apply formula (21-14) where $y = I$ and $x = t$:

$$I_{av} = \frac{\int_0^2 I \, dt}{2 - 0} = \frac{\int_0^2 3.0t^{3/2} \, dt}{2}$$

$$= \frac{3.0}{2} \left( \frac{2}{5} t^{5/2} \right) \Big|_0^2 = \frac{3}{5}(2)^{5/2} - 0 = 3.4 \text{ A}$$

2. Find the average power in the resistance during this interval.

The power $P = I^2R = (3.0t^{3/2})^2(2.0) = 18t^3$. Apply formula (21-14) where $y = P$ and $x = t$:

$$P_{av} = \frac{\displaystyle\int_0^2 18t^3 \, dt}{2 - 0} = 9\left(\frac{t^4}{4}\right)\Big|_0^2 = 36 \text{ W}$$

**Exercise 21-4**

1. A weight of 5 lb stretches a spring 3 in. beyond its natural length. How much work is done in *foot-pounds* in stretching the spring from 2 to 4 in. beyond its natural length?

2. How much work is done in *foot-pounds* in stretching the spring of problem 1 from 1.0 to 1.5 ft beyond its natural length?

3. An automobile spring is 1.0 m long. A force of 80 kgf stretches it to 1.2 m. How much work is done when the spring is stretched from 1.2 to 1.3 m?

4. How much work is done in compressing the automobile spring in problem 3 to a length of 0.9 m?

5. An anchor chain weighs 20 lb/ft and is hanging vertically. If the force necessary to lift it is proportional to its length, find the work done in hauling up a 100-ft length of chain.

6. If the anchor chain in problem 5 has a 10-ft length of heavy chain weighing 40 lb/ft added to its end, find the work done in hauling up the chain.

C 7. Two charged particles repel each other with a force that is inversely proportional to the distance $r$ between them: $F = k/r^2$. When they are 1.0 m apart, the force of repulsion is $10^{-6}$ N. Find the work done in joules when the distance between the charges increases from 1.0 to 100 m.

C 8. The gravitational force of attraction between two bodies is inversely proportional to the distance $r$ between them: $F = k/r^2$. If $k = 4.0 \times 10^{14}$ between two heavenly bodies, find the work done in joules when the distance between the bodies changes from $2.0 \times 10^8$ to $1.0 \times 10^8$ m.

9. The current in milliamperes flowing to a capacitor during a short interval $t$ is given by $I = 0.6t^2 + 0.2t$. If the capacitor is discharged at $t = 0$ s, find the charge at $t = 2$ s. (See Example 21-18.)

10. Find the accumulation of charge for the capacitor in problem 9 from $t = 3$ to $t = 4$ s.

11. A capacitor with a charge of 8.2 mC discharges for 1.5 s. The current is given by $I = 6 - t^3$ during that interval. How much charge is lost during the first second?

C 12. The current in milliamperes which flows to a capacitor is given by $I = 0.4t^2(3 - t)$ from $t = 0$ to $t = 4$ s. The maximum charge on the capacitor occurs when $I = dq/dt = 0$. If the capacitor is discharged at $t = 0$ s, find the maximum charge.

C 13. The power in a circuit is given by $P = 0.5\sqrt{t}$ for a short time $t$. Find the energy used in joules from $t = 0.5$ to $t = 1.0$ s. (See Table 21-3.)

14. The current in a circuit is given by $I = 1.5t$ from $t = 0$ to $t = 1.5$ s. Find the energy used (in joules) in a resistance $R = 4.0\ \Omega$ during that interval. (*Note*: $P = I^2R$.)

**15.** Apply (21-14) to find the average value of the function $y = \sqrt{2/x}$ from $x = 1$ to $x = 4$.

**16.** The force in a nonlinear system is given by $F = (x - 10)^3 + 20$. Find the average force in pounds from $x = 10$ to $x = 20$.

[C] **17.** The current in a circuit is given by $I = 3\sqrt{t}$ from $t = 0.25$ to $t = 2.25$ s. Find the average current in amperes during that interval.

[C] **18.** In problem 13 find the average power in the circuit from $t = 0$ to $t = 1.5$ s.

**19.** In problem 14 find the average power in the resistance from $t = 1.0$ to $t = 2.0$ s.

**20.** If the circuit in problem 17 contains a resistance of 2.5 $\Omega$, find the average power in the resistance during the given interval.

**21.** The voltage in a circuit is given by $V = 1/\sqrt{t + 1}$ over an interval of time $t$. Find the average voltage from $t = 3$ to $t = 8$ s.

**22.** In the circuit of problem 21, the current is given by $I = 10t\sqrt{t + 1}$ over the same interval $t$. Find the average power in the circuit from $t = 3$ to $t = 8$ s. (*Note: P = VI.*)

[C] **23.** The rate of flow of water from a reservoir in liters per minute is given by $F = 10^6 t^{0.40}$, where $t$ is the time in minutes. (*a*) Find the amount of water flowing out of the reservoir during the first hour. (*Note:* The amount of water is the area under the curve of $F$ vs. $t$.) (*b*) Find the average rate of flow during the first hour.

[C] **24.** A rocket weighing 5000 lb at the earth's surface is fired vertically. The force of gravity, that is, the weight of the rocket, varies inversely as the square of its distance from the earth's center. (*a*) Find the work done or energy required, in foot-pounds, to lift the rocket to a height of 1000 mi. (*Note:* The earth's radius = 3960 mi.) (*b*) Find the average weight of the rocket over the distance of 1000 mi.

**25.** A rectangular tank with a square base 10 by 10 ft and a height of 20 ft is full of fuel. If the fuel weighs 58 lb/ft³, how much work is done pumping all the fuel out of the top of the tank? See Fig. 21-16. [The element of force $dF$ necessary to lift each volume element of width $dx$ is the weight of the volume element =

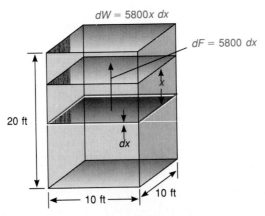

FIG. 21-16 *Lifting fuel from a tank, problem 25*

58(10)(10) $dx$ = 5800 $dx$. Each volume element must be lifted $x$ ft, and therefore the element of work done $dW$ = 5800$x$ $dx$.]

26. A cylindrical tank 8 ft in diameter and 10 ft high is full of water. If the water weighs 62 lb/ft$^3$, find the work done in pumping all the water out of the top of the tank. (See problem 25.)

27. The capacitor in problem 12 reaches a maximum charge and then discharges. The capacitor is completely discharged at time $t$ when:

$$\int_0^t I \, dt = 0$$

Find the time $t$ when the capacitor is discharged.

28. The potential difference or voltage between two points $a$ and $b$ in an electric field is given by:

$$V_{ab} = \int_a^b E \, dr$$

where $E$ = electric field intensity. If $E = k/r^2$, show that:

$$V_{ab} = \frac{k(b - a)}{ab}$$

# 21-5 | *Volumes of Revolution*

Integration can be used to find certain volumes that are obtained by revolving an area bounded by a known curve about the $x$ or $y$ axis. There are two basic methods, which are explained in this section: the disk method and the shell method.

## *Disk Method*

Consider the area bounded by the curve $y = f(x)$, the $x$ axis, and the lines $x = a$ and $x = b$. When this area is revolved about the $x$ axis, it generates the volume $V$ shown in Fig. 21-17. An element of volume $dV$ is a disk whose width is the differential of $x$, $dx$. The disk is actually a right circular cylinder of radius $y$ and height $dx$. The element of volume is then:

$$dV = \pi y^2 \, dx \qquad (21\text{-}15)$$

Similar to the area under a curve, the total volume $V$ is the limit of the sum of an infinite number of circular disks $dV$ summed in the positive $x$ direction from $x = a$ to $x = b$:

$$V = \pi \int_a^b y^2 \, dx \qquad (21\text{-}16)$$

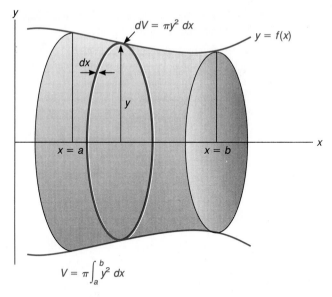

FIG. 21-17  *Volume of revolution: Disk method*

The following example illustrates the application of formula (21-16).

### Example 21-20

Find the volume generated by revolving the area bounded by $y = x^2$, $y = 0$, $x = 1$, and $x = 2$ about the $x$ axis.

### Solution

Figure 21-18 shows the volume of revolution. Apply formula (21-16) to find the volume, where $a = 1$, $b = 2$, and $y = x^2$:

$$V = \pi \int_a^b y^2 \, dx = \pi \int_1^2 (x^2)^2 \, dx = \pi \left( \frac{x^5}{5} \right) \Big|_1^2$$

$$= \pi \left( \frac{32}{5} - \frac{1}{5} \right) = \frac{31\pi}{5} \text{ or } 19.5 \text{ cubic units}$$

Consider now the area bounded by the curve $y = f(x)$, the $y$ axis, and the lines $y = a$ and $y = b$. A volume of revolution is generated by revolving this area about the $y$ axis. An element of volume $dV$ is a disk whose width is the differential of $y$, $dy$, and whose radius is $x$ (Fig. 21-19). The element of volume is then $dV = \pi x^2 \, dy$, and the total volume $V$ is:

$$V = \pi \int_a^b x^2 \, dy \qquad\qquad (21\text{-}17)$$

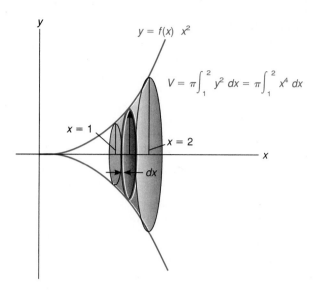

FIG. 21-18 *Volume generated by revolving a parabola about the x axis, Example 21-20*

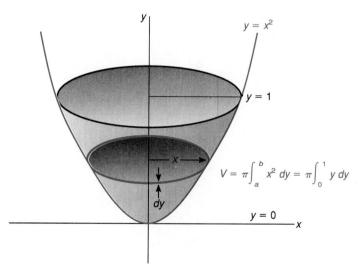

FIG. 21-19 *Volume generated by revolving a parabola about the y axis, Example 21-21*

### Example 21-21

Find the volume generated by revolving the area bounded by $y = x^2$, $x = 0$, and $y = 1$ about the $y$ axis. When a parabola is revolved about its axis of symmetry, the volume generated is called a *paraboloid of revolution*.

### Solution

Figure 21-19 shows the paraboloid of revolution. Apply formula (21-17) to find the volume where $a = 0$, $b = 1$, and $x^2 = y$:

$$V = \pi \int_a^b x^2 \, dy = \pi \int_0^1 y \, dy$$

$$= \pi \left( \frac{y^2}{2} \right) \Big|_0^1$$

$$= \pi \left( \frac{1}{2} - 0 \right) = \frac{\pi}{2} \text{ or } 1.57 \text{ cubic units}$$

## Shell Method

A volume of revolution can also be found by using an element of volume $dV$ which is a shell instead of a disk. Figure 21-20 shows the same volume of revolution as that of Fig. 21-19, but using a shell instead of a disk as the element of volume. The volume of the shell can be determined by thinking of it as cut along the height and laid out flat, forming a rectangular sheet. The length of the sheet is equal to the circumference of the shell, $2\pi r$, where $r$ = radius of

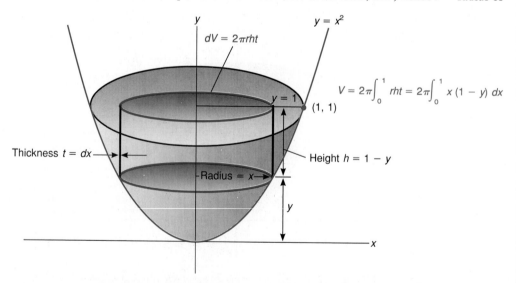

FIG. 21-20 *Volume of revolution: Shell method, Example 21-22*

the shell. The height $h$ of the sheet depends on the curve and the area being revolved, and the thickness $t$ is either $dx$ or $dy$ depending on the axis that the area is revolved about. The element of volume is then given by:

$$dV = 2\pi rht \qquad\qquad (21\text{-}18)$$

The radius, height, and thickness depend on the area being revolved and the axis of revolution. They must be determined for each particular problem. Study the next example, which shows how to apply formula (21-18) to find the same volume as Example 21-21, by using shells instead of disks.

### Example 21-22

Use the shell method to find the volume of the paraboloid of revolution in Example 21-21.

### Solution

The volume is shown in Fig. 21-20. The radius of the shell $r = x$, the height $h = 1 - y$, and the thickness $t = dx$. Apply formula (21-18) to find the element of volume, substituting $x^2$ for $y$:

$$dV = 2\pi rht = 2\pi(x)(1 - y)\ dx = 2\pi(x)(1 - x^2)\ dx$$

The total volume $V$ is found by summing the volumes of the shells in the $x$ direction from $x = 0$ to $x = 1$, because when $y = 1$, $x = 1$ on the parabola $y = x^2$:

$$V = \int_0^1 dV = 2\pi \int_0^1 (x)(1 - x^2)\ dx$$

$$= 2\pi \int_0^1 (x - x^3)\ dx$$

$$= 2\pi \left(\frac{x^2}{2} - \frac{x^4}{4}\right)\Bigg|_0^1 = 2\pi\left(\frac{1}{2} - \frac{1}{4}\right) = \frac{\pi}{2} \text{ or } 1.57 \text{ cubic units}$$

The next example shows a volume of revolution which lends itself to the use of the shell method.

### Example 21-23

Use the shell method to find the volume generated by revolving the area bounded by $y = \sqrt{x}$, $y = 2$, and $x = 0$ about the $x$ axis.

## Solution

The volume is shown in Fig. 21-21. The radius of the shell $r = y$, the height $h = x$, and the thickness $t = dy$. Apply formula (21-18) to find the element of volume:

$$dV = 2\pi rht = 2\pi(y)(x)\ dy$$

Since $y = \sqrt{x}$, $x = y^2$ and therefore:

$$dV = 2\pi(y)(y^2)\ dy = 2\pi y^3\ dy$$

Find the total volume $V$ by summing the volumes of the shells in the $y$ direction from $y = 0$ to $y = 2$:

$$V = \int_0^2 dV = 2\pi \int_0^2 y^3\ dy$$

$$= 2\pi \left(\frac{y^4}{4}\right)\bigg|_0^2 = 8\pi \text{ or } 25.1 \text{ cubic units}$$

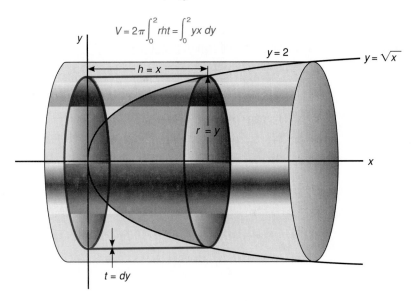

FIG. 21-21   *Volume of revolution: Shell method, Example 21-23*

Note that if Example 21-23 were to be done by using the disk method, the disks would have holes in them and a modification of formula (21-16) would be necessary.

The following example compares the use of both methods to find the same volume.

### Example 21-24

The area bounded by the parabola $y = 1 - 4x^2$, $x = 0$, and $y = 0$ is revolved about the $y$ axis, generating a volume of revolution.

**1.** Find the volume, using the disk method.

**2.** Find the volume, using the shell method.

### Solution

**1.** Find the volume, using the disk method.

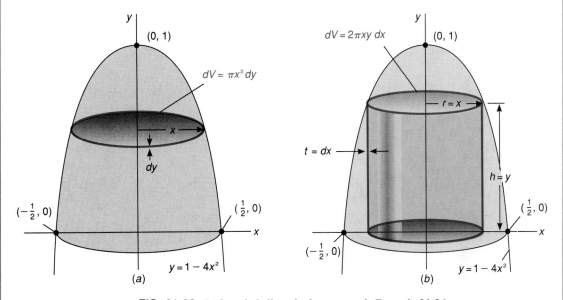

FIG. 21-22 *Disk and shell methods compared, Example 21-24*

The volume is shown in Fig. 21-22a. Apply formula (21-17) to find the volume, substituting $(1 - y)/4$ for $x^2$:

$$V = \pi \int_0^1 x^2 \, dy = \pi \int_0^1 \frac{1 - y}{4} \, dy = \frac{\pi}{4} \left( y - \frac{y^2}{2} \right) \Big|_0^1$$

$$= \frac{\pi}{8} \text{ or } 0.393 \text{ cubic unit}$$

2. Find the volume, using the shell method.

The volume is shown in Fig. 21-22b. The radius of the shell $r = x$, the height $h = y$, and the thickness $t = dx$. Apply formula (21-18) to find the element of volume:

$$dV = 2\pi xy \, dx = 2\pi x(1 - 4x^2) \, dx = 2\pi(x - 4x^3) \, dx$$

Find the upper limit of the integral by finding the *positive* value of $x$ that corresponds to $y = 0$:

$$1 - 4x^2 = 0$$

$$x^2 = \frac{1}{4}$$

$$x = \frac{1}{2}$$

The volume is then:

$$V = 2\pi \int_0^{1/2} (x - 4x^3) \, dx = 2\pi \left( \frac{x^2}{2} - x^4 \right) \Big|_0^{1/2}$$

$$= \frac{\pi}{8} \text{ or } 0.393 \text{ cubic unit}$$

**Exercise 21-5**

In problems 1 to 24, the areas bounded by the given curves are revolved about the axis indicated, generating a volume of revolution. In problems 1 to 8, find the volume generated by the disk method.

1. $y = x^2$, $y = 0$, $x = 1$   ($x$ axis)
2. $y = 2x^2$, $y = 0$, $x = 1$, $x = 2$   ($x$ axis)
3. $y = x^3$, $y = 0$, $x = 1$   ($x$ axis)
4. $y = \sqrt{x}$, $y = 0$, $x = 2$, $x = 4$   ($x$ axis)
5. $y = x^2$, $x = 0$, $y = 1$, $y = 3$   ($y$ axis)
6. $y = \dfrac{x^2}{2}$, $x = 0$, $y = 2$   ($y$ axis)

**7.** $y = x^2 - 1$, $x = 0$, $y = 0$, $y = 1$   ($y$ axis)

**8.** $y = 4 - x^2$, $x = 0$, $y = 0$   ($y$ axis)

In problems 9 to 16, find the volume generated by the shell method.

**9.** $y = x^2$, $y = 4$, $x = 0$   ($x$ axis)

**10.** $y = \dfrac{x^2}{4}$, $y = 1$, $x = 0$   ($x$ axis)

**11.** $y = \sqrt{x}$, $y = 0$, $x = 4$   ($x$ axis)

**12.** $y = x^3$, $y = 8$, $x = 0$   ($x$ axis)

**13.** $y = 3x^2$, $y = 3$, $x = 0$   ($y$ axis)

**14.** $y = x^2 + 1$, $y = 0$, $x = 1$, $x = 2$   ($y$ axis)

**15.** $y = (x - 1)^2$, $y = 0$, $x = 0$   ($y$ axis)

**16.** $y = 2x - x^2$, $y = 0$, $x = 1$, $x = 2$   ($y$ axis)

In problems 17 to 24, find the volume generated by either method. Note that one method may be easier in some problems.

**17.** $y = \dfrac{x^3}{2}$, $y = 0$, $x = 2$   ($y$ axis)

**18.** $y = x^4$, $y = 0$, $x = 1$   ($x$ axis)

**19.** $y = x - x^2$, $y = 0$   ($x$ axis)

**20.** $y = 9 - x^2$, $y = 0$, $x = 1$, $x = 2$   ($y$ axis)

**21.** $y = \dfrac{1}{x}$, $x = 1$, $x = 3$, $y = 0$   ($x$ axis)

**22.** $y = \dfrac{1}{2x}$, $x = 1$, $x = 2$, $y = 0$   ($y$ axis)

$\boxed{\text{C}}$ **23.** $y = \sqrt{x}$, $x = 3$, $y = 0$   ($y$ axis)

$\boxed{\text{C}}$ **24.** $y = x^3$, $y = 2$, $x = 0$   ($x$ axis)

**25.** The area bounded by the lines $y = x$, $y = 2$, and $x = 0$ is revolved about the $y$ axis, generating a right circular cone. *(a)* Find the volume of the cone, using integration. *(b)* Find the volume of the cone, using the geometric formula $V = \frac{1}{3}\pi r^2 h$, where $r =$ radius of the base and $h =$ height. Check that both answers agree.

**26.** Derive the formula for the volume of a sphere by finding the volume generated by revolving a quadrant of the circle $x^2 + y^2 = r^2$ about the $x$ or $y$ axis and doubling the expression.

**27.** The area bounded by the parabola $y = 1 - x^2/4$, $x = 0$, and $y = 0$ is revolved about the $y$ axis. *(a)* Find the volume generated, using the disk method. *(b)* Find the volume generated, using the shell method. Check that both answers agree.

**28.** The area bounded by the curve $y = x^{1/3}$, $y = 0$, and $x = 1$ is revolved about the $x$ axis. *(a)* Find the volume generated, using the disk method. *(b)* Find the volume generated, using the shell method. Check that both answers agree.

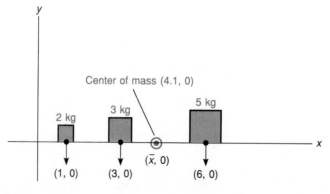

**29.** The reflector of an antenna is in the shape of a paraboloid of revolution. Its cross section is the parabola $y = 0.30x^2$. If the reflector is 12 m deep in the center, find the volume of the reflector.

**30.** A loudspeaker horn is formed by revolving the parabola $y = 0.05x^2$ about the $x$ axis from $x = 2$ to $x = 8$ in. Find the volume of the horn.

## 21-6 | *Center of Mass and Centroids*

The *center of mass* or center of gravity of a body is the balance point of the body. A body acts as if its total mass were concentrated at its center of mass. For example, if a board with weights on either end is supported by a fulcrum at its center of mass, the board will balance in a level position. A *moment* is defined as mass times distance. The sum of the moments about a point equals the moment of the center of mass. The following example illustrates the concepts of moment and center of mass, which are important in the study of mechanics.

### Example 21-25
A board placed on the $x$ axis contains a mass of 2 kg at (1, 0), 3 kg at (3, 0), and 5 kg at (6, 0). Neglecting the mass of the board, find the center of mass of the system along the $x$ axis.

### Solution
The system is shown in Fig. 21-23. Assume the center of mass is at the point $(\bar{x}, 0)$. Use the origin as the point about which the moments are calculated. For example, the moment for the mass of 3 kg is 3(3) = 9. (*Note:* If each unit in the $x$ direction were 1 m, the units for the moment would be 9 kg · m.) The moment for the center of mass is the total mass times the distance to the origin:

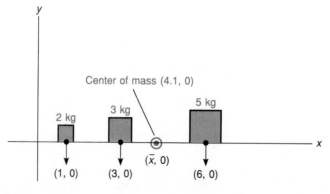

FIG. 21-23 *Center of mass along a line, Example 21-25*

$(2 + 3 + 5)\bar{x} = 10\bar{x}$. Find the center of mass by *equating the sum of the moments to the moment of the center of mass:*

$$2(1) + 3(3) + 5(6) = 10\bar{x}$$

$$\bar{x} = \frac{41}{10} \quad \text{or} \quad 4.1$$

The board would therefore balance at the point $(4.1, 0)$.

## *Plane Areas*

When mass is distributed *uniformly* throughout a plane area, the center of mass of the area is called the *centroid*. The centroid of a plane area is located by two coordinates $(\bar{x}, \bar{y})$. In Example 21-25, the center of mass along a line is found by dividing the sum of the moments by the total mass. Area replaces mass when you are finding the centroid, because if the mass is distributed uniformly, the area is proportional to the mass.

The *x* coordinate $\bar{x}$ of the centroid of an area is equal to the sum of the moments about the *y* axis, $M_y$, divided by the total area, and the *y* coordinate $\bar{y}$ is equal to the sum of the moments about the *x* axis, $M_x$, divided by the total area:

$$\bar{x} = \frac{M_y}{A} \qquad \bar{y} = \frac{M_x}{A} \tag{21-19}$$

Study the next example, which shows how to apply formula (21-19).

### Example 21-26
Find the centroid of the area in Fig. 21-24.

### Solution
Divide the area into rectangles as shown by the dotted lines. The centroid, or center of mass, of each rectangle is located at its geometric center. For example, the centroid of the middle rectangle is:

$$\bar{x} = 2 + \frac{1}{2} = \frac{5}{2} \qquad \bar{y} = \frac{4}{2} = 2$$

*The moment of the middle rectangle about the y axis is the area times the distance of its centroid to the y axis:* $1(4)(5/2) = 10$. In this way calculate the

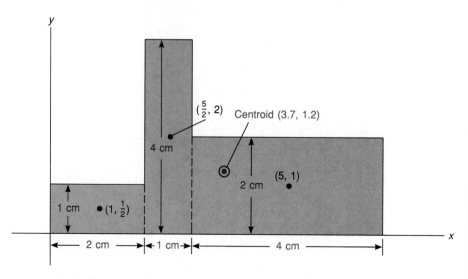

FIG. 21-24 *Centroid of a plane area, Example 21-26*

sum of the moments of the rectangles about the *y* axis. Then apply formula (21-19) and divide by the total area to find the *x* coordinate of the centroid:

$$\bar{x} = \frac{M_y}{A} = \frac{2(1)(1) + 1(4)(5/2) + 4(2)(5)}{2(1) + 1(4) + 4(2)} = 3.7$$

Similarly, find the *y* coordinate by dividing the sum of the moments about the *x* axis by the total area:

$$\bar{y} = \frac{M_x}{A} = \frac{2(1)(1/2) + 1(4)(2) + 4(2)(1)}{2(1) + 1(4) + 4(2)} = 1.2$$

The centroid or center of mass is then $(\bar{x}, \bar{y}) = (3.7, 1.2)$.

Consider now an area bounded by the curve $y = f(x)$, $x = a$, $x = b$, and the *x* axis (Fig. 21-25). As shown in Example 21-26, the centroid of the area depends only on the shape of the area when the mass is uniformly distributed. The area is divided into elements of area $dA = y\, dx$. The moment of $dA$ about the *y* axis is then $x\, dA = x(y, dx)$, and the moment of $dA$ about the *x* axis is $(y/2)\, dA = (y/2)(y\, dx)$. The sum of the moments about each axis is then the limit of the sum of an infinite number of moments and can be expressed as an integral, in

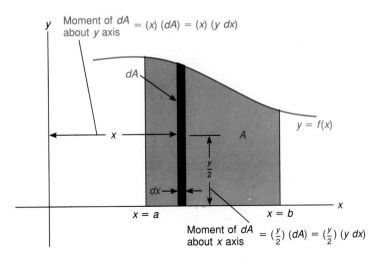

FIG. 21-25 *Centroid of area under a curve*

the same way as the area is expressed as an integral. By applying (21-19), the coordinates of the centroid are thus given by:

$$\bar{x} = \frac{M_y}{A} = \frac{\displaystyle\int_a^b xy \, dx}{\displaystyle\int_a^b y \, dx} \qquad \bar{y} = \frac{M_x}{A} = \frac{\displaystyle\int_a^b \frac{y^2}{2} \, dx}{\displaystyle\int_a^b y \, dx} \qquad (21\text{-}20)$$

### Example 21-27
Find the centroid of the area bounded by $y = x^2$, $x = 0$, $x = 2$, and the $x$ axis.

### Solution
The area is shown in Fig. 21-26. Apply formula (21-20) to find the coordinates of the centroid:

$$\bar{x} = \frac{\displaystyle\int_a^b xy \, dx}{\displaystyle\int_a^b y \, dx} = \frac{\displaystyle\int_0^2 x(x^2) \, dx}{\displaystyle\int_0^2 x^2 \, dx} = \frac{4}{8/3} = \frac{3}{2} \quad \text{or} \quad 1.5$$

$$\bar{y} = \frac{\displaystyle\int_a^b y^2/2 \, dx}{\displaystyle\int_a^b y \, dx} = \frac{\displaystyle\int_0^2 (x^2)^2/2 \, dx}{\displaystyle\int_0^2 x^2 \, dx} = \frac{16/5}{8/3} = \frac{6}{5} \quad \text{or} \quad 1.2$$

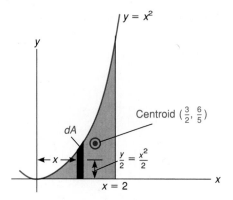

FIG. 21-26 *Centroid of area under a parabola, Example 21-27*

The centroid is therefore ($\frac{3}{2}$, $\frac{6}{5}$). Note that the area $A$ need only be calculated once for both coordinates.

### Example 21-28
Find the centroid of an isosceles triangle whose base is 2 and whose height is 1.

### Solution
Place the base of the triangle on the $x$ axis with the vertex on the $y$ axis, as shown in Fig. 21-27. Because of symmetry, the $x$ coordinate of the centroid is on the $y$ axis. Also because of symmetry, *the $y$ coordinate of the centroid of the right half of the triangle* (which is an isosceles right triangle) *is the same as that of the entire triangle.* Consider then the right triangle. The area of the right

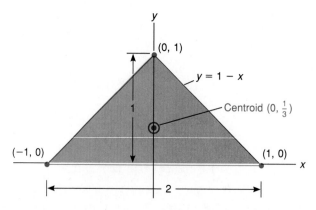

FIG. 21-27 *Centroid of an isosceles triangle, Example 21-28*

triangle can be computed directly: $A = \frac{1}{2}(1)(1) = \frac{1}{2}$. The equation of the hypotenuse is $y = 1 - x$. Apply (21-20) to find $\bar{y}$:

$$\bar{y} = \frac{M_x}{A} = \frac{\displaystyle\int_0^1 \frac{y^2}{2}\,dx}{1/2} = \frac{\displaystyle\int_0^1 \frac{(1-x)^2}{2}\,dx}{1/2} = \frac{1}{3}$$

The centroid is therefore $(0, \frac{1}{3})$, or two-thirds of the distance from the vertex angle to the base.

## Solids of Revolution

The centroid of a solid of revolution is found in a similar way to that of an area. Assuming that the mass is uniformly distributed, the centroid depends only on the shape of the solid. Because of symmetry the centroid will lie on the axis of rotation. It is therefore only necessary to find one coordinate on the axis of rotation.

When an area bounded by the curve $y = f(x)$, $x = a$, $x = b$, and the $x$ axis is revolved about the $x$ axis, the $x$ coordinate of the centroid of the solid formed is equal to the sum of the moments of the elements of volume about the $y$ axis divided by the total volume. By using disks as elements of volume (Fig. 21-28), this is expressed by:

$$\bar{x} = \frac{\displaystyle\int_a^b xy^2\,dx}{\displaystyle\int_a^b y^2\,dx} \qquad (21\text{-}21)$$

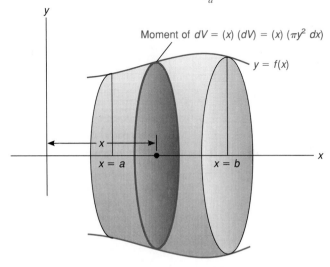

FIG. 21-28 *Centroid of a solid of revolution*

Notice in formula 21-21 that the $\pi$ divides out and is not shown.

### Example 21-29

Find the centroid of the solid of revolution formed by revolving the area bounded by $y = x^2$, $y = 0$, and $x = 2$ about the $x$ axis.

### Solution

The volume is shown in Fig. 21-29. Apply formula (21-21) to find $\bar{x}$:

$$\bar{x} = \frac{\displaystyle\int_a^b xy^2 \, dx}{\displaystyle\int_a^b y^2 \, dx} = \frac{\displaystyle\int_0^2 x(x^2)^2 \, dx}{\displaystyle\int_0^2 (x^2)^2 \, dx} = \frac{\left.\dfrac{x^6}{6}\right|_0^2}{\left.\dfrac{x^5}{5}\right|_0^2} = \frac{5}{3} \quad \text{or} \quad 1.7$$

The centroid is therefore at $(\tfrac{5}{3}, 0)$ or $(1.7, 0)$.

Similarly, when an area bounded by the curve $y = f(x)$, $y = a$, $y = b$, and the $y$ axis is revolved about the $y$ axis, the $y$ coordinate of the centroid of the solid formed is given by:

$$\bar{y} = \frac{\displaystyle\int_a^b yx^2 \, dx}{\displaystyle\int_a^b x^2 \, dx} \tag{21-22}$$

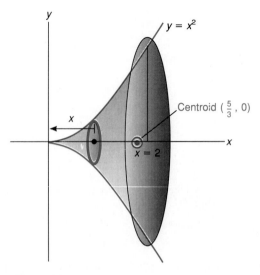

FIG. 21-29 *Centroid of a solid of revolution about the x axis, Example 21-29*

### Example 21-30
Find the centroid of a hemisphere of radius $r$.

### Solution
A hemisphere can be generated by revolving a quadrant of a circle about the $y$ axis (Fig. 21-30). The equation of the quadrant is $x^2 + y^2 = r^2$, where $x$ and $y$ are both positive. Apply formula (21-22) to find $\bar{y}$:

$$\bar{y} = \frac{\displaystyle\int_a^b yx^2 \, dy}{\displaystyle\int_a^b x^2 \, dy} = \frac{\displaystyle\int_0^r y(r^2 - y^2) \, dy}{\displaystyle\int_0^r (r^2 - y^2) \, dy}$$

$$= \frac{\left(\dfrac{r^2y^2}{2} - \dfrac{y^4}{4}\right)\Big|_0^r}{\left(r^2y - \dfrac{y^3}{3}\right)\Big|_0^r} = \frac{r^4/4}{2r^3/3} = \frac{3r}{8}$$

The centroid is therefore $(0, 3r/8)$.

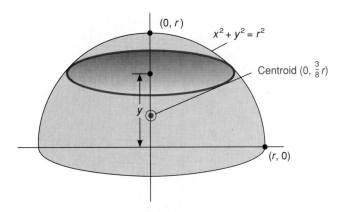

FIG. 21-30 *Centroid of a hemisphere*

**Exercise 21-6** In problems 1 to 4, find the center of mass for each system containing the given masses at the given points on the $x$ axis. (See Example 21-25.)

    **1.** 3 kg at (1, 0), 5 kg at (4, 0)

    **2.** 4 lb at (3, 0), 2 lb at (5, 0)

    **3.** 8 lb at (0, 0), 3 lb at (3, 0), 1 lb at (4, 0)

    **4.** 5 kg at (−1, 0), 1 kg at (1, 0), 4 kg at (7, 0)

In problems 5 to 8, find the centroid of each plane area. (See Example 21-26.)

5.

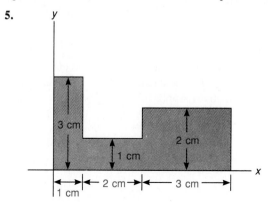

FIG. 21-31

6.

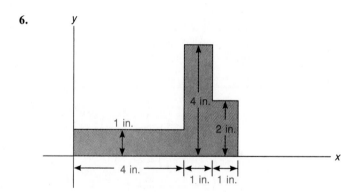

FIG. 21-32

7.

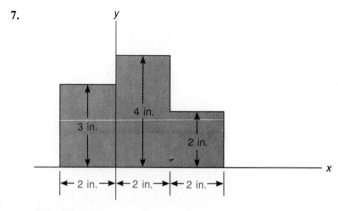

FIG. 21-33

**8.**

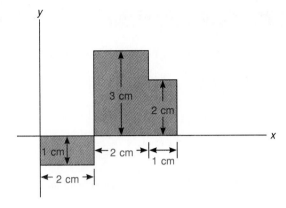

FIG. 21-34

In problems 9 to 22, find the centroid of the given figures.

**9.** The area bounded by $y = x^2$, $y = 0$, $x = 1$.

**10.** The area bounded by $y = x^3$, $y = 0$, $x = 2$.

**11.** The area bounded by $y = x^2 + 2$, $y = 0$, $x = 0$, $x = 1$.

**12.** The area bounded by $y = \sqrt{x}$, $y = 0$, $x = 1$.

**13.** A right triangle whose legs are 1 and 2.

**14.** A semicircle of radius $r$.

**15.** The solid formed by revolving the area bounded by $y = x^2$, $y = 0$, and $x = 1$ about the $x$ axis.

**16.** The solid formed by revolving the area bounded by $y = \sqrt{x}$, $y = 0$, and $x = 1$ about the $x$ axis.

**17.** The solid formed by revolving the area bounded by $y = x^2$, $x = 0$, and $y = 2$ about the $y$ axis.

**18.** The solid formed by revolving the area bounded by $y = x^3$, $x = 0$, and $y = 1$ about the $y$ axis.

**19.** A cone whose height is 1 and radius of the base is 1. (*Hint:* Revolve the line $y = 1 - x$ about the $x$ or $y$ axis.)

**20.** Any cone whose height is $h$ and radius of the base is $r$. (See problem 19.)

**21.** The area bounded by $y_1 = x^2$ and $y_2 = x$. [Use an element of area $dA = (y_2 - y_1)\, dx$, and modify formulas (21-20).]

**22.** The area bounded by $y = x^2$, $x = 0$, $y = 1$. [*Note:* Use a horizontal element of area $dA = x\, dy$, and modify formulas (21-20).]

# 21-7 | *Review Questions*

In problems 1 to 6, find each indefinite integral.

1. $\displaystyle\int 5x^3\,dx$

2. $\displaystyle\int \left(x^2 + \frac{1}{x^2}\right) dx$

3. $\displaystyle\int (2x + 1)^3(2)\,dx$

4. $\displaystyle\int (2 - x^2)^2 x\,dx$

5. $\displaystyle\int 3(6x + 1)^3\,dx$

6. $\displaystyle\int \sqrt{t + 3}\,dt$

In problems 7 to 12, evaluate each definite integral.

7. $\displaystyle\int_0^2 (1 - x^4)\,dx$

8. $\displaystyle\int_1^3 \frac{dt}{t^3}$

9. $\displaystyle\int_4^9 \sqrt{x}\,dx$

10. $\displaystyle\int_1^4 3\sqrt{3x - 3}\,dx$

11. $\displaystyle\int_{-1}^1 \frac{4\,dx}{(4x + 3)^2}$

12. $\displaystyle\int_{0.5}^{1.5} (3 - 2x)^3\,dx$

In problems 13 to 20, find the area bounded by the given curves.

13. $y = \dfrac{x^2}{2}$, $y = 0$, $x = 1$

14. $y = 9 - x^2$, $y = 0$

15. $y = 3x^2$, $y = 4 - x^2$

16. $y = x^2 - 2x$, $y = 0$

17. $y = 9x - x^3$, $y = 0$

18. $y = x^2$, $y = 3$

In problems 19 to 24, find the volume generated when the area bounded by the given curves is revolved about the axis indicated.

19. $y = \dfrac{x^2}{2}$, $y = 0$, $x = 1$   (x axis)

20. $y = x^2 - 4$, $x = 0$, $y = 0$   (y axis)

☐ 21. $y = x^2$, $y = 2$, $x = 0$   (x axis)

22. $y = x^2 + 2$, $y = 0$, $x = 1$, $x = 2$   (y axis)

23. $y = \dfrac{1}{x}$, $y = 1$, $y = 2$, $x = 0$   (y axis)

24. $y = 4\sqrt{x}$, $y = 4$, $x = 0$   (x axis)

In problems 25 to 28, find the centroid of the given figure.

25. The area bounded by $y = x^3/3$, $y = 0$, $x = 2$.

26. The area bounded by $y = 1 - x^2$, $y = 0$.

27. The solid formed by revolving the area bounded by $y = 2x^2$, $y = 0$, $x = 2$ about the x axis.

28. The solid formed by revolving the area bounded by $y = \sqrt{x}$, $y = 2$, and $x = 0$ about the y axis.

29. The angular velocity of a gas turbine is given by $\omega = 4.2(16t - t^5)$, where $t$ varies from $t = 0$ to $t = 1$ s. *(a)* If the angular displacement $\theta = 0$ rad/s when $t = 0$ s, find the formula for the angular displacement. *(b)* Find $\theta$ when $t = 1$ s.

30. A capacitor has an initial charge of 0.020 C at $t = 1$ s. A current flows to the capacitor for 1 s, given by $I = 0.030/t^2$. *(a)* Find the formula for the charge on the capacitor during this interval. (See Table 21-2.) *(b)* What is the charge on the capacitor at $t = 2$ s?

31. A strong spring in a plane's landing gear has a natural length of 15 ft. If a force of 100 lb compresses it to 10 ft, how much work is required to compress it from 15 to 5 ft?

□ 32. The power in a circuit is given by $P = 5t^{3/2}$ for a short time interval $t$. Find the energy used in joules from $t = 1.0$ to $t = 1.2$ s. (See Table 21-3.)

□ 33. The total force exerted on a dam 10 ft high is given by:

$$F = \int_0^{10} 120x\sqrt{x^2 + 44}\ dx$$

where $x = $ depth. Find the total force (in pounds) on the dam.

34. The voltage in a circuit is given by $V = 12 - t^2$ for a short time interval $t$. *(a)* Find the average voltage from $t = 0$ to $t = 2$ s. *(b)* If the circuit contains a resistance of 3 $\Omega$, find the average power in the resistance from $t = 0$ to $t = 2$ s. (*Note*: $P = V^2/R$.)

# C H A P T E R

# 22

# Transcendental Functions

## 22-1 | *Derivatives of Trigonometric Functions*

### *Sine and Cosine Functions*

The process of differentiation is introduced in Chap. 19, and the inverse process, integration, is introduced in Chap. 21. In these chapters the derivatives and integrals of certain algebraic functions are studied. This chapter studies the derivatives and integrals of trigonometric, exponential, and logarithmic functions. These functions are not algebraic functions and are called *transcendental functions*. Transcendental functions are very important in engineering and science. Applications of these functions, using calculus, are shown in each section of this chapter as the concepts are presented.

The derivative of the sine function provides the basis for the derivatives of the five other trigonometric functions. The derivative of $y = \sin x$ is shown in what follows to be $y' = \cos x$. To prove this, it is first necessary to determine the $\lim_{\theta \to 0} (\sin \theta/\theta)$ where $\theta$ *is in radians*. To help see if this limit exists, you can use a calculator to make a table of values of $\theta$, $\sin \theta$, and $\sin \theta/\theta$ for values of $\theta$ *in radians* as $\theta \to 0$, as shown at the bottom of this page.

It appears from the table that as $\theta \to 0$, $\sin \theta \to 0$ and $\sin \theta/\theta \to 1$. To prove that this is the actual limit, consider Fig. 22-1, which shows the angle $\theta$ in a unit circle of radius $r = 1$. The following is true of the areas of the two triangles and the sector of the circle shown

| $\theta$ | 1.00 | 0.100 | 0.0100 | 0.00100 | $\to 0$ |
|---|---|---|---|---|---|
| $\sin \theta$ | 0.8414710 | 0.09983342 | 0.009999833 | 0.0009999998 | $\to 0$ |
| $\dfrac{\sin \theta}{\theta}$ | 0.8414710 | 0.9983342 | 0.9999833 | 0.9999998 | $\to 1$ |

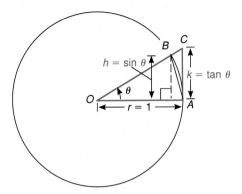

FIG. 22-1 *Relationship between* sin $\theta$, $\theta$, *and* tan $\theta$

in the figure when $\theta > 0$:

$$\text{Area } \triangle AOB < \text{area of sector } AOB < \text{area } \triangle AOC$$

Apply the formulas for the area of a triangle and the area of a sector [formula (4-5), Sec. 4-5] to obtain:

$$\frac{1}{2}rh < \frac{1}{2}r^2\theta < \frac{1}{2}rk$$

Since $r = 1$ and the inequality may be divided by $\frac{1}{2}$:

$$h < \theta < k$$

From the figure, sin $\theta = h/1 = h$ and tan $\theta = k/1 = k$; therefore:

$$\sin \theta < \theta < \tan \theta$$

By dividing by sin $\theta$, simplifying, and taking reciprocals, which reverses the inequality, you obtain:

$$1 < \frac{\theta}{\sin \theta} < \frac{\tan \theta}{\sin \theta}$$

$$1 < \frac{\theta}{\sin \theta} < \frac{1}{\cos \theta}$$

$$\cos \theta < \frac{\sin \theta}{\theta} < 1$$

As $\theta \to 0$, $\cos \theta \to 1$, and therefore $\sin \theta / \theta$ must approach 1. Hence:

$$\lim_{\theta \to 0} \frac{\sin \theta}{\theta} = 1 \qquad (22\text{-}1)$$

Now to find the derivative of $y = \sin x$, use the delta process (Sec. 19-2):

$$y = \sin x$$
$$\Delta y = \sin (x + \Delta x) - \sin x$$

Apply the sum and product formula shown in Example 12-9(2):

$$\sin X - \sin Y = 2 \cos \frac{1}{2} (X + Y) \ \sin \frac{1}{2} (X - Y)$$

to the right side of the equation for $\Delta y$. The equation can then be written (where $X = x + \Delta x$ and $Y = x$):

$$\Delta y = 2 \cos \tfrac{1}{2} (x + \Delta x + x) \sin \tfrac{1}{2} (x + \Delta x - x)$$
$$= 2 \cos \left(x + \frac{\Delta x}{2}\right) \sin \frac{\Delta x}{2}$$

Then:

$$\frac{\Delta y}{\Delta x} = \frac{2 \cos (x + \Delta x / 2) \sin (\Delta x / 2)}{\Delta x}$$
$$= \cos \left(x + \frac{\Delta x}{2}\right) \left[\frac{\sin (\Delta x / 2)}{\Delta x / 2}\right]$$

and:

$$\frac{dy}{dx} = y' = \lim_{\Delta x \to 0} \frac{\Delta y}{\Delta x}$$
$$= \lim_{\Delta x \to 0} \left\{\cos \left(x + \frac{\Delta x}{2}\right) \left[\frac{\sin (\Delta x / 2)}{\Delta x / 2}\right]\right\}$$

Apply formula (22-1), where $\theta = \Delta x / 2$:

$$y' = \cos (x + 0)(1) = \cos x$$

Therefore:

$$\text{If } y = \sin x, \text{ then } y' = \cos x. \qquad (22\text{-}2)$$

**Example 22-1**

Find the derivative of $y = 2 \sin x + 1$.

**Solution**

Apply formula (22-2) and the formulas for derivatives:

$$y' = 2 \cos x + 0 = 2 \cos x$$

To find the derivative of $y = \sin u$, where $u$ *is a function of x,* apply the chain rule (19-13):

$$\text{If } y = \sin u, \text{ then } y' = (\cos u)(u'). \tag{22-3}$$

**Example 22-2**

Find the derivative of each function.

**1.** $y = 3 \sin 2x$
**2.** $y = \sin^2 x$

**Solution**

**1.** $y = 3 \sin 2x$

Apply formula (22-3), where $u = 2x$ and $u' = 2$:

$$y' = 3 \cos u(u') = 3(\cos 2x)(2) = 6 \cos 2x$$

**2.** $y = \sin^2 x$

Apply the general power formula (19-14) where $n = 2$, $u = \sin x$ and $u' = \cos x$:

$$y' = nu^{n-1}(u') = 2(\sin x)(\cos x) = 2 \sin x \cos x \qquad \text{or} \qquad y' = \sin 2x$$

To find the derivative of $y = \cos u$, where $u$ is a function of $x$, let:

$$y = \cos u = \sin \left( \frac{\pi}{2} - u \right)$$

Then $\qquad y' = \cos \left( \frac{\pi}{2} - u \right)(-u') = -(\sin u)(u')$

Therefore, when $u$ is a function of $x$:

$$\text{If } y = \cos u, \text{ then } y' = -(\sin u)(u'). \tag{22-4}$$

### Example 22-3
Find the derivative of each function.

1. $y = 2 \cos (3x + \pi)$
2. $y = \sin t \cos t$

### Solution
1. $y = 2 \cos (3x + \pi)$

Apply formula (22-4), where $u = 3x + \pi$ and $u' = 3$:

$$y' = -2 \sin u(u') = -2[\sin (3x + \pi)](3) = -6 \sin (3x + \pi)$$

2. $y = \sin t \cos t$

Apply the product formula (19-10), where $u = \sin t$ and $v = \cos t$, and formulas (22-3) and (22-4):

$$y' = uv' + vu'$$
$$y' = (\sin t)(-\sin t) + (\cos t)(\cos t)$$
$$= \cos^2 t - \sin^2 t \quad \text{or} \quad y' = \cos 2t$$

## Other Trigonometric Functions

To find the derivative of $y = \tan u$, where $u$ is a function of $x$, let:

$$y = \tan u = \frac{\sin u}{\cos u}$$

Then apply the quotient formula (19-11), where $u = \sin u$ and $v = \cos u$:

$$y' = \frac{(\cos u)(\cos u)(u') - (\sin u)(-\sin u)(u')}{\cos^2 u}$$
$$= \frac{(\sin^2 u + \cos^2 u)(u')}{\cos^2 u} = \frac{1}{\cos^2 u}(u') = (\sec^2 u)(u')$$

Therefore, when $u$ is a function of $x$:

$$\text{If } y = \tan u, \text{ then } y' = (\sec^2 u)(u'). \tag{22-5}$$

In a similar way, by letting $\cot u = 1/\tan u$, $\sec u = 1/\cos u$, and $\csc u = 1/\sin u$, the following formulas are derived, where $u$ is a function of $x$:

$$\text{If } y = \cot u, \text{ then } y' = -(\csc^2 u)(u'). \tag{22-6}$$
$$\text{If } y = \sec u, \text{ then } y' = (\sec u \tan u)(u') \tag{22-7}$$

$$\text{If } y = \csc u, \text{ then } y' = -(\csc u \cot u)(u') \qquad (22\text{-}8)$$

The derivations of formulas (22-6) and (22-7) are left as problems 65 and 66, respectively, at the end of this section.

### Example 22-4
Find the derivative of each function.

**1.** $y = 2 \tan \dfrac{x}{2}$

**2.** $y = \cot^2 x + 1$

**3.** $y = 4 \sec 2x$

**4.** $y = \sqrt{\csc t}$

### Solution

**1.** $y = 2 \tan \dfrac{x}{2}$

Apply formula (22-5):

$$y' = 2\left(\sec^2 \frac{x}{2}\right)\left(\frac{1}{2}\right) = \sec^2 \frac{x}{2}$$

**2.** $y = \cot^2 x + 1$

Apply the power formula (19-14) and formula (22-6):

$$y' = (2 \cot x)(-\csc^2 x) = -2 \cot x \csc^2 x$$

**3.** $y = 4 \sec 2x$

Apply formula (22-7):

$$y' = (4 \sec 2x \tan 2x)(2) = 8 \sec 2x \tan 2x$$

**4.** $y = \sqrt{\csc t}$

Apply the power formula (19-14) and formula (22-8):

$$y' = \frac{1}{2}(\csc t)^{-1/2}(-\csc t \cot t)$$

This can be simplified to:

$$y' = \frac{-\csc t \cot t}{2\sqrt{\csc t}} = \frac{-\cot t \sqrt{\csc t}}{2}$$

## Applications

The applications of derivatives shown in Chap. 20 also apply to trigonometric functions. The following examples illustrate some of these applications.

### Example 22-5

A weight vibrates on a spring in simple harmonic motion given by $y = 8 \cos (3t/2)$, where $y$ is the displacement from its initial position in centimeters and $t$ is the time in seconds.

1. Find the formulas for the velocity and acceleration.
2. Find the acceleration at the time of maximum displacement.

### Solution

1. Find the formulas for the velocity and acceleration.

   The velocity $v$ is the derivative of the displacement:

$$v = y' = -8 \left( \sin \frac{3t}{2} \right) \left( \frac{3}{2} \right) = -12 \sin \frac{3t}{2}$$

   The acceleration $a$ is the derivative of the velocity:

$$a = v' = -12 \left( \cos \frac{3t}{2} \right) \left( \frac{3}{2} \right) = -18 \cos \frac{3t}{2}$$

2. Find the acceleration at the time of maximum displacement.

   Similar to the maximum point on a curve, the time of maximum displacement occurs when the first derivative or velocity is zero:

$$v = -12 \sin \frac{3t}{2} = 0$$

   It is only necessary to consider values of the angle between 0 and $2\pi$ since the values repeat:

$$\frac{3t}{2} = 0 \qquad \frac{3t}{2} = \pi$$

$$t = 0 \qquad t = \frac{2\pi}{3}$$

   The values of $t$ can also be determined directly from the cosine curve $y = 8 \cos (3t/2)$ by locating the turning points.
   The acceleration is then:

$$t = 0: \qquad a = -\left[ 18 \cos \left( \frac{3}{2} \right) \right](0) = -18(1) \stackrel{\bullet}{=} -18 \text{ cm/s}^2$$

$$t = \pi: \qquad a = -18 \cos\left[\left(\frac{3}{2}\right)\left(\frac{2\pi}{3}\right)\right] = -18(-1) = 18 \text{ cm/s}^2$$

### Example 22-6

The current in an ac circuit is given by $I = \sin t + 2 \cos t$.

1. Find the maximum and minimum values of the current for $0 \le t < 2\pi$.
2. Sketch 1 cycle of the curve from $t = 0$, showing the maximum and minimum values.

### Solution

1. Find the maximum and minimum values of the current for $0 \le t < 2\pi$.

   The maximum and minimum values occur when $I' = 0$:

   $$I' = \cos t - 2 \sin t = 0$$

   $$\frac{2 \sin t}{\cos t} = 1$$

   $$\tan t = \frac{1}{2} = 0.5$$

   $$t = 0.464, \ 3.61 \text{ rad}$$

   When $t = 0.464$ rad, the maximum value is $I = 2.2$ A; and when $t = 3.61$ rad, the minimum value is $I = -2.2$ A.

2. Sketch 1 cycle of the curve from $t = 0$, showing the maximum and minimum values.

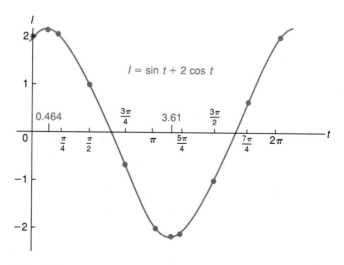

FIG. 22-2 *Sinusoidal current in an ac circuit, Example 22-6*

Using the maximum and minimum values from (1), construct a table of values and sketch the curve (Fig. 22-2). The curve repeats starting at $t = 2\pi$:

| $t$ | 0 | 0.464 | $\dfrac{\pi}{4}$ | $\dfrac{\pi}{2}$ | $\dfrac{3\pi}{4}$ | $\pi$ | 3.61 | $\dfrac{5\pi}{4}$ | $\dfrac{3\pi}{2}$ | $\dfrac{7\pi}{4}$ | $2\pi$ |
|---|---|---|---|---|---|---|---|---|---|---|---|
| $I = \sin t + 2 \cos t$ | 2 | 2.2 | 2.1 | 1 | $-0.7$ | $-2$ | $-2.2$ | $-2.1$ | $-1$ | 0.7 | 2 |

### Example 22-7

A searchlight rotates at the rate of 0.10 rad/s, illuminating a straight wall 500 ft from the searchlight. How fast is the light moving along the wall at a point 600 ft from the searchlight?

### Solution

This is a problem in related rates. The situation is shown in Fig. 22-3. The equation relating the angle $\theta$ and the distance $x$ is:

$$\tan \theta = \frac{x}{500}$$

Differentiate this equation implicitly with respect to time:

$$\sec^2 \theta \left( \frac{d\theta}{dt} \right) = \frac{1}{500} \frac{dx}{dt}$$

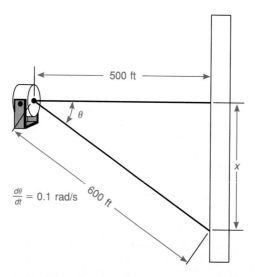

500 ft

$\dfrac{d\theta}{dt} = 0.1$ rad/s

600 ft

FIG. 22-3 *Searchlight moving along a wall, Example 22-7*

It is necessary to find $dx/dt$. At the given instant $\sec \theta = 600/500 = 1.2$ and $d\theta/dt = 0.10$ rad/s. Therefore:

$$(1.2)^2(0.10) = \frac{1}{500} \frac{dx}{dt}$$

$$\frac{dx}{dt} = 72 \text{ ft/s}$$

**Exercise 22-1**

In problems 1 to 40, find the derivative of each function.

**1.** $y = 3 \sin x$

**2.** $y = \dfrac{\sin x}{2}$

**3.** $y = \sin 3x$

**4.** $y = 2 \sin 2x$

**5.** $y = 3 \cos 2x$

**6.** $y = 4 \cos (x + 3)$

**7.** $y = \dfrac{1}{2} \cos^2 x$

**8.** $y = 3 - \cos^2 x$

**9.** $y = \sin 2x + 2 \cos x$

**10.** $y = 2 \cos 3x - \sin 2x$

**11.** $y = x \cos x$

**12.** $y = x \sin^2 x$

**13.** $y = \dfrac{\sin x}{x}$

**14.** $y = \dfrac{\cos 3x}{3x}$

**15.** $y = 2 \sin t \cos t$

**16.** $y = \sin 2t \cos t$

**17.** $y = \sqrt{1 + \cos^2 x}$

**18.** $y = \sin^2 x \cos x$

**19.** $y = \sin 60\pi t + \cos 60\pi t$

**20.** $y = \sin^2 120\pi t$

**21.** $y = 4 \tan 2x$

**22.** $y = \tan \dfrac{x}{2} - x$

**23.** $y = 3 \cot (\pi - x)$

**24.** $y = 1.5 \cot 3x$

**25.** $y = 2 \sec (x + 2)$

**26.** $y = 3x - \csc x$

**27.** $y = \tan^2 x - 1$

**28.** $y = (1 - \cot x)^3$

**29.** $y = \csc^3 t$

**30.** $y = 2\sqrt{\sec t}$

**31.** $y = 3x \tan x$

**32.** $y = \sec x \tan x$

**33.** $y = \dfrac{\cot 2x}{x}$

**34.** $y = \dfrac{1 - \csc x}{2x}$

**35.** $y = \sin 2x \tan x$

**36.** $y = \csc x \sqrt{\cos x}$

**37.** $y = \dfrac{\cot x}{1 - \cos x}$

**38.** $y = \dfrac{\sin x}{\tan x + 1}$

**39.** $y = a \sin (bx + \phi)$   $(a, b, \phi$ constants)

**40.** $y = A \sin \pi t + B \cos \pi t$   $(A, B$ constants)

In problems 41 to 44, find the differential of each function.

**41.** $y = 4 \sin^2 3x$

**42.** $y = \sqrt{\cos 2x}$

**43.** $y = \cos x \cot x$

**44.** $y = \dfrac{\sec x}{x}$

**45.** In simple harmonic motion, the force, and therefore the acceleration, is proportional to the displacement $y$:

$$y'' = \frac{d^2y}{dt^2} = -ky$$

Show that the above is true for the equation of simple harmonic motion: $y = A \sin (t\sqrt{k})$, where $A =$ constant.

**46.** Using the fact that $\lim\limits_{\theta \to 0} (\sin \theta/\theta) = 1$, show that:

$$\lim_{\theta \to 0} \frac{1 - \cos \theta}{\theta} = 0$$

(*Hint:* Multiply top and bottom by $1 + \cos \theta$.)

**47.** Find the slope of the line tangent to the curve $y = 2 \cos 3x$ at the point where $x = \pi/3$.

**48.** Find the slope of the line tangent to the curve $y = \tan (2x + \pi)$ at the point where $x = \pi/8$.

**49.** The equation of motion of a machine part is given by $s = 1 - 6 \cos (t/3)$. Find the velocity in feet per second when $t = 3.14$ s.

☐ **50.** A missile moves in the path of a parabola whose $x$ and $y$ coordinates are given by $x = \sin t$ and $y = \cos 2t$. (*a*) Find the formulas for the $x$ and $y$ components of the velocity. (*b*) Find the resultant velocity when $t = \pi/6$. (See Example 20-3.)

**51.** The current in a circuit containing an inductance $L = 0.050$ H (henry) is given by $I = 4.0 \cos 60t$. Using the fact that the voltage across the inductance $V = L \, dI/dt$, find the formula for $V$.

**52.** In an ac circuit containing an inductor and a variable resistor, the current in the resistor is given by $I = I_0 \cot \theta$, where $I_0 =$ inductor current and $\theta =$ phase angle, $0 < \theta \le \pi/2$. Assuming $I_0$ is constant, find $dI/d\theta$.

**53.** A weight vibrates on a spring in simple harmonic motion given by $y = 4 \sin 0.5t$, where $y =$ displacement in centimeters and $t =$ time in seconds. (*a*) Find the formulas for the velocity and acceleration. (*b*) Find the acceleration when the displacement is a maximum. (See Example 22-5.)

☐ **54.** The motion of an engine piston approximates simple harmonic motion given by $y = 1.5 \cos 50t$, where $y =$ displacement in inches and $t =$ time in seconds. Find the velocity and acceleration when $t = 0.01$ s. (See Example 22-5.)

**55.** Sketch one cycle of the graph of $y = 2 \sin x - \cos x$, showing the maximum and minimum points. (See Example 22-6.)

**56.** Sketch one cycle of the graph of $y = \cos 2x + 2 \cos x$, showing the maximum and minimum points. (See Example 22-6.)

$\boxed{c}$ **57.** The current in an electric circuit is given by $I = 2 \sin t + \cos t$. *(a)* Find the maximum and minimum values of the current for $0 \le t < 2\pi$. *(b)* Sketch one cycle of the curve from $t = 0$, showing the maximum and minimum values. (See Example 22-6.)

**58.** The power in an ac circuit containing an inductance is given by $P = VI$, where the voltage $V = 120 \sin 10t$ and the current $I = 5 \cos 10t$. *(a)* Find the maximum and minimum values of the power during one cycle. *(b)* Sketch one cycle of the power curve from $t = 0$, showing the maximum and minimum points.

**59.** A radar antenna is on a ship which is 3.0 mi from a straight shore. If the antenna rotates at the rate of 0.20 rad/s, how fast is the radar beam moving along the shore at a point 4.0 mi from the ship? (See Example 22-7.)

**60.** A motorboat leaves a dock on a straight shore and travels perpendicular to the shore. Maria, who is standing on shore 100 ft from the dock, observes that the angle between the shore and her line of sight to the boat is changing at 0.05 rad/s, at the instant when the angle is 1.0 rad. How fast is the boat moving at this instant?

**61.** The current in a 10-$\Omega$ resistor is given by $I = 2 \sin 30t$. Find the expression for the rate of change of the power at any instant. (*Note:* $P = I^2R$.)

**62.** When a flat surface, which makes an angle $\theta$ with the horizontal, is moved through the air, the lifting power of the air resistance is given by:

$$E = k \sin^2 \theta \cos \theta$$

where $k$ is a constant of proportion. Find the expression for the rate of change of $E$ with respect to $\theta$.

**63.** The sides of a triangular trough are both $s$ units long. What should the angle between the sides be for maximum capacity? (*Note:* Area of a triangle = $\frac{1}{2} ab \sin C$.)

**64.** At a point 50 m from the base of a building, the angle of elevation is 60°. Using differentials, find the approximate error in the height of the building due to an error of 0.010 rad in the measurement of the angle. (See Sec. 20-6.)

**65.** Prove formula (22-6), using the proof of formula (22-5) as a guide.

**66.** Prove formula (22-7), using the proof of formula (22-5) as a guide.

# 22-2 | *Integrals Involving Trigonometric Functions*

Since an indefinite integral is an antiderivative, the following integral formulas follow from the derivatives of the trigonometric functions, where $u$ is a function of $x$:

$$\int (\sin u)(u') \, dx = \int \sin u \, du = -\cos u + C \qquad (22\text{-}9)$$

$$\int (\cos u)(u') \, dx = \int \cos u \, du = \sin u + C \qquad (22\text{-}10)$$

$$\int (\sec^2 u)(u') \, dx = \int \sec^2 u \, du = \tan u + C \qquad (22\text{-}11)$$

$$\int (\csc^2 u)(u') \, dx = \int \csc^2 u \, du = -\cot u + C \qquad (22\text{-}12)$$

$$\int (\sec u \tan u)(u') \, dx = \int \sec u \tan u \, du = \sec u + C \qquad (22\text{-}13)$$

$$\int (\csc u \cot u)(u') \, dx = \int \csc u \cot u \, du = -\csc u + C \qquad (22\text{-}14)$$

The formulas for the integrals of tan $u$, cot $u$, sec $u$, and csc $u$ involve the logarithmic function and are presented in Chap. 23 after logarithmic functions are discussed. Notice in formulas (22-9) to (22-14) that when $u$ is a function of $x$, then $du = u' \, dx$. Study the following examples, which show the application of these formulas.

### Example 22-8
Find each indefinite integral.

1. $\int 3 \sin 3x \, dx$
2. $\int \cos 2x \, dx$
3. $\int 4x \sec x^2 \tan x^2 \, dx$

### Solution
1. $\int 3 \sin 3x \, dx$

This integral is in the form $\sin u \, du$, where $u = 3x$ and $u' = 3$, or $du = 3 \, dx$. Apply formula (22-9):

$$\int 3 \sin 3x \, dx = -\cos 3x + C$$

2. $\int \cos 2x \, dx$

In this integral $u = 2x$ and $u' = 2$. You need to multiply by 2 on the inside and $\frac{1}{2}$ on the outside and apply formula (22-10):

$$\frac{1}{2} \int (\cos 2x)(2) \, dx = \frac{1}{2} \sin 2x + C$$

3. $\int 4x \sec x^2 \tan x^2 \, dx$

In this integral $u = x^2$ and $u' = 2x$. Move a factor of 2 to the outside and apply formula (22-13):

$$2 \int 2x \sec x^2 \tan x^2 \, dx = 2 \sec x^2 + C$$

Certain integrals can be evaluated by using the trigonometric identities from Chap. 12 as follows.

### Example 22-9
Find the indefinite integrals.

1. $\int \cot^2 x \, dx$
2. $\int \sin^2 x \, dx$

### Solution
1. $\int \cot^2 x \, dx$

None of the above formulas can be used directly to evaluate this integral. Use formula (12-3) to express $\cot^2 x$ as $\csc^2 x - 1$ and apply (22-12):

$$\int \cot^2 x \, dx = \int (\csc^2 x - 1) \, dx$$
$$= \int \csc^2 x \, dx - \int dx = -\cot x - x + C$$

2. $\int \sin^2 x \, dx$

Use formula (12-8) for $\cos 2x = 1 - 2 \sin^2 x$, and solve for $\sin^2 x = \frac{1}{2}(1 - \cos 2x)$. Then apply (22-10):

$$\int \sin^2 x \, dx = \frac{1}{2} \int (1 - \cos 2x) \, dx$$
$$= \frac{1}{2} \int dx - \frac{1}{2}\left(\frac{1}{2}\right) \int (\cos 2x)(2) \, dx$$
$$= \frac{1}{2}x - \frac{1}{4} \sin 2x + C$$

Some trigonometric integrals can be evaluated by using the power formula (21-5) and may or may not use the above formulas.

### Example 22-10
Find each indefinite integral.

1. $\int \sin^2 x \cos x \, dx$

2. $\int \dfrac{\cot x}{\sin x} \, dx$

## Solution

**1.** $\int \sin^2 x \cos x\, dx$

This integral is in the form $\int u^n(u')\, dx$, where $u = \sin x$, $u' = \cos x$, and $n = 2$. Apply the power formula (21-5):

$$\int \sin^2 x \cos x\, dx = \frac{\sin^3 x}{3} + C$$

**2.** $\int \dfrac{\cot x}{\sin x}\, dx$

This integral can be evaluated in two ways. The first method uses formula (22-14):

$$\int \frac{\cot x}{\sin x}\, dx = \int \cot x\left(\frac{1}{\sin x}\right) dx$$

$$= \int \cot x \csc x\, dx = -\csc x + C$$

The second method uses the power formula (21-5):

$$\int \frac{\cot x}{\sin x} = \int \frac{\cos x/\sin x}{\sin x}\, dx$$

$$= \int (\sin x)^{-2} \cos x\, dx = -\frac{1}{\sin x} + C$$

$$= -\csc x + C$$

## Applications

The next two examples show applications of trigonometric integrals to areas and electric circuits.

### Example 22-11

Find the area bounded by the curve $y = \sin x$ and the $x$ axis from $x = 0$ to $x = \pi/2$.

## Solution

The area is shown in Fig. 22-4. By applying formula (22-10), it is equal to:

$$\int_0^{\pi/2} \sin x\, dx = -\cos x \,\Big|_0^{\pi/2}$$

$$= -\cos \frac{\pi}{2} + \cos 0 = 0 + 1 = 1 \text{ square unit}$$

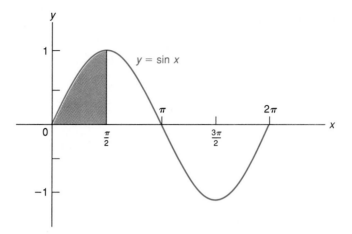

FIG. 22-4  *Area under y = sin x, Example 22-11*

In Sec. 21-4 the average or mean value of a function over an interval is defined as the area under the curve for this interval divided by the width of the interval. Consider an alternating current $I = I_{max} \sin 2\pi ft$, where $f =$ frequency. The charge flowing in one direction is equal to the charge flowing in the other direction during one cycle. *The mean value of the current over one cycle is therefore zero* since the area above the $x$ axis is equal to the area below the $x$ axis for the sine curve (Fig. 22-4). Since this current does produce power, a more useful measure of the current is the root mean square, or effective, value. The *root mean square* (rms) value of a function $y = f(x)$ over an interval from $x = a$ to $x = b$ is defined as:

$$y_{rms} = \sqrt{\frac{1}{b-a} \int_a^b y^2 \, dx} \qquad (22\text{-}15)$$

Note the difference of this formula and formula (21-14). The power of a current $I$ flowing through a resistance $R$ is $I^2R$, and the rms, or effective, value of an alternating current is that value of direct current which produces the same amount of power or heating effect.

### Example 22-12
Find the rms value of the current $I = 2 \sin 2\pi t$ over one cycle of the current.

### Solution
Apply formula (22-15), where $y = I$, $x = t$, $a = 0$, and $b = 1$ since the period of the current $= 1/f$ where $f = 1$:

$$I_{rms} = \sqrt{\frac{1}{1-0} \int_0^1 (2 \sin 2\pi t)^2 \, dt}$$

First evaluate the integral. Use formula (12-8) as shown in Example 22-9(2):

$$\int_0^1 (2 \sin 2\pi t)^2 \, dt = 4 \int_0^1 (\sin 2\pi t)^2 \, dt$$

$$= 4\left(\frac{1}{2}\right) \int_0^1 (1 - \cos 4\pi t) \, dt$$

$$= 2 \int_0^1 dt - 2\left(\frac{1}{4\pi}\right) \int_0^1 (\cos 4\pi t)(4\pi) \, dt$$

$$= 2t\Big|_0^1 - \left(\frac{1}{2\pi}\right) \sin 4\pi t\Big|_0^1 = 2 - 0 = 2$$

The rms value of the current is then:

$$I_{\text{rms}} = \sqrt{1(2)} = 1.41 \text{ A}$$

Therefore, a direct current of $I = 1.41$ A will produce the same heating effect as an alternating current where $I_{\text{max}} = 2$ A. It can be shown that $I_{\text{rms}} = I_{\text{max}}/\sqrt{2}$. (See Exercise 22-2, problem 39.)

**Exercise 22-2**

In problems 1 to 28, find each indefinite integral.

1. $\int 2 \sin 2x \, dx$

2. $\int 6 \cos 3x \, dx$

3. $\int \cos (2x + 1) \, dx$

4. $\int \sin \frac{x}{2} \, dx$

5. $\int \sec^2 2x \, dx$

6. $\int \frac{\csc^2 x}{3} \, dx$

7. $\int x \csc x^2 \cot x^2 \, dx$

8. $\int \sec 3x \tan 3x \, dx$

9. $\int \tan^2 x \, dx$

10. $\int (\cot^2 x + 1) \, dx$

11. $\int (\cos^2 x - \sin^2 x) \, dx$

12. $\int \frac{dx}{\cos^2 x}$

13. $\int \cos^2 x \, dx$

14. $\int 2 \sin^2 x \, dx$

15. $\int \cot^2 2x \, dx$

16. $\int (\tan^2 2x - 1) \, dx$

17. $\int \sin x \cos x \, dx$

18. $\int \cos^2 x \sin x \, dx$

19. $\int \sin^2 3t \cos 3t \, dt$

20. $\int \cos t \sqrt{\sin t} \, dt$

21. $\int \tan^2 x \sec^2 x \, dx$

22. $\int \sec^3 x \tan x \, dx$

23. $\int \frac{\tan x}{\cos x} \, dx$

24. $\int \frac{\cot 2x}{\sin 2x} \, dx$

**25.** $\displaystyle\int \frac{1 - \sin x}{\cos^2 x}\, dx$      **26.** $\displaystyle\int \frac{1 + \cos x}{1 - \cos^2 x}\, dx$

**27.** $\displaystyle\int 3 \sin 2\pi t\, dt$      **28.** $\displaystyle\int \cos (4\pi t + 1.5)\, dt$

In problems 29 to 32, evaluate each definite integral.

**29.** $\displaystyle\int_0^{\pi/2} (2 \sin x + 3 \cos x)\, dx$

**30.** $\displaystyle\int_{\pi/4}^{\pi/2} \frac{dx}{\sin^2 x}$

**31.** $\displaystyle\int_0^{\pi/4} \sec^2 x \tan x\, dx$

**32.** $\displaystyle\int_0^{0.01} \sin 100\pi t\, dt$

**33.** Find the area bounded by $y = 2 \cos x$ and the $x$ axis from $x = 0$ to $x = \pi/2$.

**34.** Find the area bounded by $y = \sin 2x$ and the $x$ axis from $x = \pi/2$ to $x = \pi$.

**35.** Find the area bounded by $y = \sin x$, $y = \cos x$, and the $y$ axis for $x \geq 0$.

**36.** Find the average value of the function $y = 3 \sin^2 x$ from $x = 0$ to $x = \pi$. [See formula (21-14), Sec. 21-4.]

**37.** Find the rms value of the current $I = 5 \sin \pi t$ over one cycle of the current.

**38.** Find the rms value of the voltage $V = 120 \cos 100\pi t$ over one cycle of the voltage.

**39.** Given an alternating current $I = I_{max} \sin 2\pi f t$, show that $I_{rms} = I_{max}/\sqrt{2}$.

**40.** The current flowing to a capacitor is given by $I = 10 \cos 120\pi t$. Find the formula for the charge $q$ on the capacitor as a function of the time $t$ if $q = 0$ when $t = 0$. (See Table 21-2, Sec. 21-1.)

**41.** The following integral occurs in a mechanical problem involving forced vibrations:

$$\int (A\omega \sin^2 \omega t - A\omega \cos^2 \omega t)\, dt$$

where $A$ and $\omega$ are constants. Find this indefinite integral.

**42.** The plate power in a rectifier circuit is given by:

$$P = \frac{V}{2\pi R} \int_0^\pi (\sin \theta)(\sin \theta - V_p)\, d\theta$$

where $V$, $R$, and $V_p$ are constants. Evaluate this integral.

**43.** The acceleration of a body in simple harmonic motion is given by $a = -48 \cos 4t$. *(a)* Find the formula for the velocity as a function of time $t$ if $v = 0$ when $t = 0$. *(b)* Find the formula for the displacement as a function of time $t$ if $s = 0$ when $t = 0$.

**44.** Find $\int \tan x \sec^2 x\, dx$ in two ways, using the power formula: *(a)* Let $u = \tan x$ and $u' = \sec^2 x$. *(b)* Let $u = \sec x$ and $u' = \sec x \tan x$. *(c)* Show that both answers are equivalent even though they do not appear the same.

# 22-3 | *Inverse Trigonometric Functions*

## *Derivatives*

The derivative of $y = \sin^{-1} u$, where $u$ is a function of $x$, is found by first expressing the function in the form:

$$u = \sin y \qquad \text{where} \quad -\frac{\pi}{2} \le y \le \frac{\pi}{2}$$

Then taking derivatives implicitly with respect to $x$ and solving for $y'$ yields:

$$u' = (\cos y)(y') \qquad y' = \frac{u'}{\cos y}$$

Now since $-\pi/2 \le y \le \pi/2$, $\cos y$ is positive and $\cos y = +\sqrt{1 - \sin^2 y} = \sqrt{1 - u^2}$. Therefore, when $u$ is a function of $x$:

$$\text{If } y = \sin^{-1} u, \text{ then } y' = \frac{u'}{\sqrt{1 - u^2}}. \tag{22-16}$$

Similarly, it can be shown that when $u$ is a function of $x$:

$$\text{If } y = \cos^{-1} u, \text{ then } y' = \frac{-u'}{\sqrt{1 - u^2}}. \tag{22-17}$$

The proof of formula (22-17) is left as problem 37.

The derivative of $y = \tan^{-1} u$, where $u$ is a function of $x$, is found as follows:

$$u = \tan y \qquad u' = \sec^2 y \, (y')$$

$$y' = \frac{u'}{\sec^2 y} = \frac{u'}{1 + \tan^2 y}$$

Therefore, when $u$ is a function of $x$:

$$\text{If } y = \tan^{-1} u, \text{ then } y' = \frac{u'}{1 + u^2}. \tag{22-18}$$

The functions $\cot^{-1} x$, $\sec^{-1} x$, and $\csc^{-1} x$ are the inverse functions of the cotangent, secant, and cosecant, respectively. They are defined for the following angles:

$$0 < \cot^{-1} x < \pi$$

$$0 \le \sec^{-1} x \le \pi \tag{22-19}$$

$$-\frac{\pi}{2} \le \csc^{-1} x \le \frac{\pi}{2}$$

The derivatives of the above functions are found in a similar way as those for $\sin^{-1} x$, $\cos^{-1} x$, and $\tan^{-1} x$. They are as follows, where $u$ is a function of $x$:

$$\text{If } y = \cot^{-1} u, \text{ then } y' = \frac{-u'}{1 + u^2}. \tag{22-20}$$

$$\text{If } y = \sec^{-1} u, \text{ then } y' = \frac{u'}{u\sqrt{u^2 - 1}}. \tag{22-21}$$

$$\text{If } y = \csc^{-1} u, \text{ then } y' = \frac{-u'}{u\sqrt{u^2 - 1}}. \tag{22-22}$$

Notice that the derivative of $\cos^{-1} u$ is the negative of the derivative of $\sin^{-1} x$ and similarly for $\tan^{-1} u$ and $\cot^{-1} u$, and $\sec^{-1} u$ and $\csc^{-1} u$. The proof of formula (22-20) is left as problem 38.

### Example 22-13
Find the derivative of each function.

**1.** $y = 3 \sin^{-1} 2x$

**2.** $y = \dfrac{1}{2} \tan^{-1} \dfrac{x}{2}$

**3.** $y = \sec^{-1} x + \sqrt{x^2 - 1}$

**4.** $y = x \cos^{-1} 3x$

### Solution
**1.** $y = 3 \sin^{-1} 2x$

Apply formula (22-16), where $u = 2x$ and $u' = 2$:

$$y' = 3\frac{2}{\sqrt{1 - (2x)^2}} = \frac{6}{\sqrt{1 - 4x^2}}$$

**2.** $y = \dfrac{1}{2} \tan^{-1} \dfrac{x}{2}$

Apply formula (22-18), where $u = x/2$ and $u' = \frac{1}{2}$:

$$y' = \frac{1}{2}\left[\frac{1/2}{1 + (x/2)^2}\right] = \frac{1}{4(1 + x^2/4)} = \frac{1}{4 + x^2}$$

**3.** $y = \sec^{-1} x + \sqrt{x^2 - 1}$

Apply formula (22-21) and the power formula (19-14) and then combine fractions:

$$y' = \frac{1}{x\sqrt{x^2 - 1}} + \frac{x}{\sqrt{x^2 - 1}} = \frac{1 + x^2}{x\sqrt{x^2 - 1}}$$

**4.** $y = x \cos^{-1} 3x$

Apply the product formula (19-10) and formula (22-17), where $u = 3x$ and $u' = 3$:

$$y' = x\frac{-3}{\sqrt{1 - (3x)^2}} + (\cos^{-1} 3x)(1)$$

$$= \frac{-3x}{\sqrt{1 - 9x^2}} + \cos^{-1} 3x$$

## Integrals

The derivatives of the inverse trigonometric functions are all algebraic functions. As a result, they yield useful integration formulas. It is only necessary to consider $\sin^{-1} x$, $\tan^{-1} x$, and $\sec^{-1} x$, since the derivatives of the other three inverse functions are the negatives of the derivatives of these three. The following three useful integration formulas result from the derivative formulas, where $u$ is a function of $x$ and $a$ is a constant:

$$\int \frac{u' \, dx}{\sqrt{a^2 - u^2}} = \int \frac{du}{\sqrt{a^2 - u^2}} = \sin^{-1} \frac{u}{a} + C \qquad (22\text{-}23)$$

$$\int \frac{u' \, dx}{a^2 + u^2} = \int \frac{du}{a^2 + u^2} = \frac{1}{a} \tan^{-1} \frac{u}{a} + C \qquad (22\text{-}24)$$

$$\int \frac{u' \, dx}{u\sqrt{u^2 - a^2}} = \int \frac{du}{u\sqrt{u^2 - a^2}} = \frac{1}{a} \sec^{-1} \frac{u}{a} + C \qquad (22\text{-}25)$$

These three formulas can easily be verified by taking derivatives. For example, consider formula (22-24):

If $$y = \frac{1}{a} \tan^{-1} \frac{u}{a} + C$$

then $$y' = \frac{1}{a} \left( \frac{u'/a}{1 + (u/a)^2} \right) = \frac{u'}{a^2 + u^2}$$

Verification of formulas (22-23) and (22-25) is left as problems 41 and 42, respectively.

### Example 22-14
Find each indefinite integral.

1. $\displaystyle\int \frac{2\,dx}{\sqrt{9 - 4x^2}}$

2. $\displaystyle\int \frac{dx}{1 + 16x^2}$

3. $\displaystyle\int \frac{5\,dx}{x\sqrt{x^2 - 25}}$

4. $\displaystyle\int \frac{dx}{x^2 - 4x + 8}$

### Solution

1. $\displaystyle\int \frac{2\,dx}{\sqrt{9 - 4x^2}}$

This integral fits formula (22-23), where $u = 2x$, $u' = 2$, and $a = 3$:

$$\int \frac{2\,dx}{\sqrt{(3)^2 - (2x)^2}} = \sin^{-1} \frac{2x}{3} + C$$

2. $\displaystyle\int \frac{dx}{1 + 16x^2}$

This integral fits formula (22-24), where $u = 4x$, $u' = 4$, and $a = 1$. Multiply by 4 on the inside and by ¼ on the outside:

$$\int \frac{dx}{1 + 16x^2} = \frac{1}{4} \int \frac{4\,dx}{1 + 16x^2} = \frac{1}{4}\,\tan^{-1} 4x + C$$

3. $\displaystyle\int \frac{5\,dx}{x\sqrt{x^2 - 25}}$

Apply formula (22-25), where $u = x$, $u' = 1$, and $a = 5$:

$$\int \frac{5\,dx}{x\sqrt{x^2 - 25}} = 5 \int \frac{dx}{x\sqrt{x^2 - 25}} = (5)\left(\frac{1}{5}\right) \sec^{-1} \frac{x}{5} + C$$

$$= \sec^{-1} \frac{x}{5} + C$$

4. $\displaystyle\int \frac{dx}{x^2 - 4x + 8}$

This integral can be made to fit formula (22-24) by writing the denominator as the sum of two squares: $x^2 - 4x + 8 = x^2 - 4x + 4 + 4 = (x - 2)^2 + 2^2$. Then $u = x - 2$, $u' = 1$, and $a = 2$:

$$\int \frac{dx}{x^2 - 4x + 8} = \int \frac{dx}{(x - 2)^2 + 2^2} = \frac{1}{2} \tan^{-1} \frac{x - 2}{2} + C$$

To apply formulas (22-23), (22-24), and (22-25), you must be sure that the integral clearly fits one of these formulas. For example, the integral:

$$\int \frac{2x \, dx}{\sqrt{1 - x^2}}$$

does not fit any of the formulas. It can be integrated by using the power formula, where $u = 1 - x^2$, $u' = -2x$, and $n = -\frac{1}{2}$:

$$\int \frac{2x \, dx}{\sqrt{1 - x^2}} = -\int (1 - x^2)^{-1/2}(-2x) \, dx = -2\sqrt{1 - x^2} + C$$

### Example 22-15
Find the area bounded by:

$$y = \frac{1}{x\sqrt{x^2 - 1}}, \qquad y = 0, \, x = 1, \text{ and } x = 2$$

### Solution
The area below the curve between $x = 1$ and $x = 2$ is the following definite integral. Apply (22-25) and the definition of $\sec^{-1} x$ to evaluate the integral:

$$\int_1^2 \frac{dx}{x\sqrt{x^2 - 1}} = \sec^{-1} x \Big|_1^2$$

$$= \sec^{-1} 2 - \sec^{-1} 1 = \frac{\pi}{3} - 0 = \frac{\pi}{3}$$

**Exercise 22-3**

In problems 1 to 16, find the derivative of each function.

1. $y = \sin^{-1} 2x$

2. $y = \cos^{-1} (x - 1)$

3. $y = \frac{1}{3} \tan^{-1} \frac{x}{3}$

4. $y = \cot^{-1} 3x^2$

5. $y = \cos^{-1} \frac{x}{2}$

6. $y = \sin^{-1} \sqrt{x}$

7. $y = 4 \sec^{-1} 4x$

8. $y = \csc^{-1} x - x$

9. $y = \sin^{-1} x - \sqrt{1 - x^2}$

10. $y = \sqrt{1 - x^2} - \cos^{-1} x$

**11.** $y = \cot^{-1} \dfrac{1}{t}$        **12.** $y = t \tan^{-1} t$

**13.** $y = \dfrac{1}{4} \csc^{-1} \dfrac{x + 1}{4}$        **14.** $y = \sec^{-1} 2x + \sqrt{4x^2 - 1}$

**15.** $y = \tan^{-1} x$        **16.** $y = \sin^{-1} \dfrac{1}{1 - x}$

In problems 17 to 32, find each indefinite integral.

**17.** $\displaystyle\int \dfrac{5 \, dx}{\sqrt{1 - x^2}}$        **18.** $\displaystyle\int \dfrac{3 \, dx}{\sqrt{1 - 9x^2}}$

**19.** $\displaystyle\int \dfrac{dx}{4 + x^2}$        **20.** $\displaystyle\int \dfrac{dx}{1 + 4x^2}$

**21.** $\displaystyle\int \dfrac{3 \, dx}{x\sqrt{x^2 - 9}}$        **22.** $\displaystyle\int \dfrac{dx}{3x\sqrt{9x^2 - 1}}$

**23.** $\displaystyle\int \dfrac{dx}{4 + 25x^2}$        **24.** $\displaystyle\int \dfrac{dx}{\sqrt{1 - 16x^2}}$

**25.** $\displaystyle\int \dfrac{4 \, dx}{\sqrt{25 - 4x^2}}$        **26.** $\displaystyle\int \dfrac{x \, dx}{4 + x^4}$

**27.** $\displaystyle\int \dfrac{dx}{x\sqrt{16x^2 - 1}}$        **28.** $\displaystyle\int \dfrac{dx}{(x + 1)^2 + 4}$

**29.** $\displaystyle\int \dfrac{dx}{x^2 - 2x + 2}$        **30.** $\displaystyle\int \dfrac{dx}{x^2 + 4x + 5}$

**31.** $\displaystyle\int \dfrac{dx}{\sqrt{2x - x^2}}$        **32.** $\displaystyle\int \dfrac{dx}{(x + 1)\sqrt{x^2 + 2x}}$

In problems 33 to 36, evaluate each definite integral.

**33.** $\displaystyle\int_0^1 \dfrac{2 \, dx}{\sqrt{1 - x^2}}$        **34.** $\displaystyle\int_0^{1/2} \dfrac{dx}{\frac{1}{4} + x^2}$

**35.** $\displaystyle\int_{\sqrt{2}}^2 \dfrac{2 \, dx}{x\sqrt{x^2 - 1}}$        $\boxed{\text{C}}$ **36.** $\displaystyle\int_{1.0}^{1.5} \dfrac{dx}{\sqrt{4 - x^2}}$

**37.** Following the proof of formula (22-16), show that formula (22-17) is true.

**38.** Following the proof of formula (22-18), show that formula (22-20) is true.

**39.** The following expression arises from a problem in mechanics: $p = \cos(\sin^{-1} t)$. Find $dp/dt$.

**40.** Show that if $y = \tan^{-1} x + \tan^{-1}(1/x)$, then $y' = 0$.

**41.** Verify formula (22-23) by showing that if $y = \sin^{-1}(u/a) + C$, then:

$$y' = \dfrac{u'}{\sqrt{a^2 - u^2}}$$

**42.** Verify formula (22-25) by showing that if $y = (1/a)\sec^{-1}(u/a) + C$, then:

$$y' = \frac{u'}{u\sqrt{u^2 - a^2}}$$

**43.** Find the area bounded by $y = 1/(1 + x^2)$, $y = 0$, $x = 0$, and $x = 1$.

**44.** Find the area bounded by $y = 1/\sqrt{1 - x^2}$, $y = 0$, $x = 0$, and $x = \frac{1}{2}$.

**45.** The current flowing to a capacitor is given by:

$$I = \frac{0.50}{0.25 + t^2}$$

Find the formula for the charge $q$ on the capacitor as a function of the time $t$ if $q = 0$ when $t = 0$. (See Table 21-2, Sec. 21-1.)

**46.** The velocity of a particle is given by:

$$v = \frac{0.01}{\sqrt{0.01 - t^2}} \qquad 0 \le t \le 0.1$$

Find the equation of motion if the distance $s = 0.10$ mm when the time $t = 0$ s.

## 22-4 | *Derivatives of Logarithmic and Exponential Functions*

The derivatives of logarithmic and exponential functions depend on the irrational number $e$, which is introduced in Sec. 9-1 and is the base of natural logarithms, as shown in Sec. 9-2. The number $e$ is defined as follows:

$$e = \lim_{n \to \infty} \left(1 + \frac{1}{n}\right)^n \tag{22-26}$$

You can approximate this limit with a calculator by choosing large values of $n$ and calculating $(1 + 1/n)^n$:

| $n$ | $10^3$ | $10^4$ | $10^5$ | $10^6 \to \infty$ |
|-----|--------|--------|--------|-------------------|
| $\left(1 + \dfrac{1}{n}\right)^n$ | 2.71692 | 2.71815 | 2.71827 | $2.71828 \to e$ |

The last entry in the table, 2.71828, is the value of $e$ accurate to six figures.

## Logarithmic Function

The derivative of $y = \log_b x$ is found by applying the delta process:

$$y = \log_b x$$
$$\Delta y = \log_b (x + \Delta x) - \log_b x$$

By applying the subtraction rule for logarithms (9-4), $\Delta y$ can be written:

$$\Delta y = \log_b \frac{x + \Delta x}{x} = \log_b \left(1 + \frac{\Delta x}{x}\right)$$

and:

$$\frac{\Delta y}{\Delta x} = \frac{\log_b (1 + \Delta x/x)}{\Delta x} = \left(\frac{x}{x}\right) \frac{1}{\Delta x} \log_b \left(1 + \frac{\Delta x}{x}\right)$$

The factor $x/x$ is multiplied into the equation so that the limit can be evaluated as follows. Apply the multiplication rule for logarithms (9-5):

$$\frac{\Delta y}{\Delta x} = \frac{1}{x} \frac{x}{\Delta x} \log_b \left(1 + \frac{\Delta x}{x}\right) = \frac{1}{x} \log_b \left(1 + \frac{\Delta x}{x}\right)^{x/\Delta x}$$

Then:

$$\frac{dy}{dx} = y' = \lim_{\Delta x \to 0} \frac{\Delta y}{\Delta x} = \lim_{\Delta x \to 0} \frac{1}{x} \log_b \left(1 + \frac{\Delta x}{x}\right)^{x/\Delta x}$$

To find the limit, let $\Delta x/x = 1/n$ or $x/\Delta x = n$. Then as $\Delta x \to 0$, $n \to \infty$ and:

$$\lim_{\Delta x \to 0} \left(1 + \frac{\Delta x}{x}\right)^{x/\Delta x} = \lim_{n \to \infty} \left(1 + \frac{1}{n}\right)^n = e$$

Therefore:

$$\text{If } y = \log_b x, \text{ then } y' = \frac{1}{x} \log_b e. \qquad (22\text{-}27)$$

### Example 22-16
Find the derivative of $y = 4 \log_2 x$.

## Solution
Apply formula (22-27):

$$y' = (4)\frac{1}{x}\log_2 e = \frac{4}{x}\log_2 e$$

Formula (22-27) is much simpler when the base is $e$ since $\log_e e = \ln e = 1$. For this reason, logarithms with base $e$ are the most natural ones to use in calculus and hence are called *natural logarithms* (written ln). Therefore:

$$\text{If } y = \ln x, \text{ then } y' = \frac{1}{x}. \tag{22-28}$$

## Example 22-17
Find the derivative of $y = \ln x + 1/x$.

## Solution
Apply formula (22-28) and the power formula (19-14):

$$y' = \frac{1}{x} - \frac{1}{x^2} = \frac{x-1}{x^2}$$

Formulas (22-27) and (22-28) are extended to $u$, where $u$ is a function of $x$, by applying the chain rule (19-13):

$$\text{If } y = \log_b u, \text{ then } y' = \frac{u'}{u}\log_b e. \tag{22-29}$$

$$\text{If } y = \ln u, \text{ then } y' = \frac{u'}{u}. \tag{22-30}$$

## Example 22-18
Find the derivative of each function.

**1.** $y = \log (3x^2 + 1)$

**2.** $y = t \ln (1 - t)$

**3.** $y = \ln (\cos x)$

**4.** $y = \ln \dfrac{2x}{x^2 - 1}$

## Solution
**1.** $y = \log (3x^2 + 1)$

Notice in this function the base is 10 (common logarithm) since it is not written. Apply formula (22-29), where $u = 3x^2 + 1$ and $u' = 6x$:

$$y' = \frac{6x}{3x^2 + 1} \log e$$

**2.** $y = t \ln (1 - t)$

Apply formula (22-30) and the product formula (19-10):

$$y' = t\left(\frac{-1}{1 - t}\right) + [\ln (1 - t)](1) = \frac{t}{t - 1} + \ln (1 - t)$$

**3.** $y = \ln (\cos x)$

Apply formula (22-30), when $u = \cos x$ and $u' = -\sin x$:

$$y' = \frac{-\sin x}{\cos x} = -\tan x$$

**4.** $y = \ln \dfrac{2x}{x^2 - 1}$

The derivative is more easily found by applying the subtraction rule for logarithms (9-4) and then differentiating:

$$y = \ln 2x - \ln (x^2 - 1)$$
$$y' = \frac{2}{2x} - \frac{2x}{x^2 - 1} = \frac{1}{x} - \frac{2x}{x^2 - 1}$$
$$= \frac{-x^2 - 1}{x(x^2 - 1)} \quad \text{or} \quad \frac{1 + x^2}{x(1 - x^2)}$$

## Exponential Function

The derivative of the exponential function $y = e^x$ is found by writing the equation in logarithmic form and differentiating implicitly with respect to $x$:

$$y = e^x$$
$$\ln y = x$$
$$\frac{y'}{y} = 1$$
$$y' = y = e^x$$

Therefore the derivative of $e^x$ is $e^x$. That makes this function very useful and important in applications of calculus to science and engineering. See Examples 22-21 and 22-24.

When $y = b^x$, then:

$$\ln y = x \ln b$$

$$\frac{y'}{y} = \ln b$$

$$y' = y \ln b = b^x \ln b$$

The above results are extended to $u$, where $u$ is a function of $x$, by applying the chain rule (19-13):

$$\text{If } y = e^u, \text{ then } y' = e^u \, (u'). \tag{22-31}$$

$$\text{If } y = b^u, \text{ then } y' = b^u \, (u') \ln b. \tag{22-32}$$

### Example 22-19

Find the derivative of each function.

**1.** $y = xe^x$

**2.** $y = 2^{2x-1}$

**3.** $y = e^{\sin t}$

**4.** $y = 2\sqrt{e^x}$

### Solution

**1.** $y = xe^x$

Apply formula (22-31) and the product formula (19-10):

$$y' = x(e^x) + e^x(1) = e^x(x + 1)$$

**2.** $y = 2^{2x-1}$

Apply formula (22-32), where $u = 2x - 1$ and $u' = 2$:

$$y' = 2^{2x-1}(2) \ln 2 = 2^{2x} \ln 2 = (2^2)^x \ln 2 = 4^x \ln 2$$

Notice that the addition rule (8-3) and the multiplication rule (8-5) for exponents are used to simplify the result.

**3.** $y = e^{\sin t}$

Apply formula (22-31), where $u = \sin t$ and $u' = \cos t$:

$$y' = e^{\sin t} \cos t$$

**4.** $y = 2\sqrt{e^x}$

You can apply the multiplication rule for exponents (8-5) and write:

$$y = 2(e^x)^{1/2} = 2e^{x/2}$$

Then:

$$y' = 2e^{x/2}\left(\frac{1}{2}\right) = e^{x/2} = \sqrt{e^x}$$

You can also use the power formula (19-14) to find $y'$:

$$y = 2(e^x)^{1/2}$$

$$y' = 2\left(\frac{1}{2}\right)(e^x)^{-1/2}(e)^x = (e^x)^{1/2} = \sqrt{e^x}$$

At this point, all the necessary derivative formulas for algebraic and transcendental functions have been introduced. A summary of these formulas is given in App. B-7.

## *Applications*

Exponential and logarithmic functions occur often in science and engineering. The first example shows the use of derivatives in curve sketching.

C **Example 22-20**
Sketch the graph of $y = (\ln x)/x$.

**Solution**
This example applies the elements of curve sketching shown in Secs. 20-2 and 20-3. First note that $x > 0$, since $\ln x$ is not defined for $x \le 0$.

*Derivatives.* Set $y'$ and $y''$ equal to zero, and solve for $x$ to find any maximum, minimum, or inflection points:

$$y' = \frac{x(1/x) - (\ln x)(1)}{x^2} = \frac{1 - \ln x}{x^2} = 0$$

$$1 - \ln x = 0$$

$$\ln x = 1$$

$$x = e \approx 2.7$$

$$y'' = \frac{x^2(-1/x) - (1 - \ln x)(2x)}{x^4} = \frac{2\ln x - 3}{x^3} = 0$$

$$2 \ln x - 3 = 0$$

$$\ln x = \frac{3}{2}$$

$$x = e^{3/2} \approx 4.5$$

The above values of $x$ found where $y'$ and $y''$ equal zero are studied in the table below.

*Intercepts.*    When $x = 0$, $y$ is undefined; and when $y = 0$, $\ln x = 0$ and $x = e^0 = 1$. Therefore, there is one intercept on the $x$ axis, $(1, 0)$.

*Asymptotes.*    When the denominator $x \to 0$, $y \to \infty$ and therefore $x = 0$ is a vertical asymptote. When $x \to \infty$, $\ln x \to \infty$ and $\lim\limits_{x \to \infty} y$ cannot be evaluated with the methods presented so far. However, when you choose values of $x$, including large values, and construct a table, you obtain the following:

| $x$ | 0.5 | 1 | Maximum point ↓ $e \approx 2.7$ | Inflection point ↓ $e^{3/2} \approx 4.5$ | 10 | 100 |
|---|---|---|---|---|---|---|
| $y$ | $-1.4$ | 0 | $\dfrac{1}{e} \approx 0.4$ | $\dfrac{3}{2e^{3/2}} \approx 0.3$ | 0.2 | 0.04 |
| $y'$ | | $+1$ | 0 | $-0.02$ | | |
| $y''$ | | | $-0.05$ | 0 | $+0.002$ | |

*Graph.*    The graph is shown in Fig. 22-5. Examine the table carefully. When $x = e$, there is a maximum point since $y''$ is negative, or because $y'$ changes sign from positive to negative. When $x = e^{3/2}$, there is an inflection point since $y''$ changes sign. As $x$ becomes larger, $y$ becomes smaller, and since the curve has no more critical points to the right of $x = e$, the curve cannot become horizontal and therefore cannot increase. It can be shown by more advanced methods that $y = 0$ is an asymptote. However, the preceding analysis provides a good sketch of the curve without more advanced techniques.

The next example illustrates an important property of the exponential function $y = e^{kx}$, where $k$ is a constant. The derivative of this function $y' = ke^{kx} = ky$. This means that *the rate of change of the function is proportional to the function itself.* When $k > 0$, $y$ is called an *exponential growth function.* When $k < 0$, $y$ is called an *exponential decay function.* Many natural growth and decay processes, such as population growth and radioactive decay, behave as exponential functions. Exponential growth and decay functions are also important in the theory of electric circuits. Some of these applications are discussed

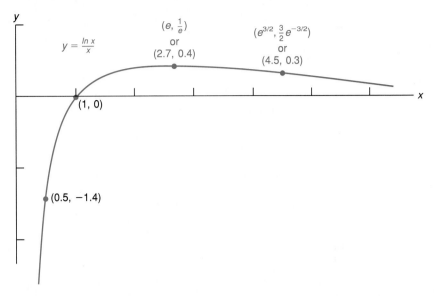

FIG. 22-5 *Sketch of y = (ln x)/x, Example 22-20*

in Chap. 9. The next example shows an application of exponential decay to electric circuits, using calculus.

### Example 22-21

In a circuit containing a resistance $R$, an inductance $L$ ($RL$ circuit), and a voltage source $V$, the current is given by $i = (V/R)(1 - e^{-Rt/L})$. Show that this formula for $i$ satisfies the differential equation $Ri + L\,di/dt = V$.

### Solution

A *differential equation* is an equation containing derivatives or differentials. Find $di/dt$ from the formula for $i$:

$$\frac{di}{dt} = \left(\frac{V}{R}\right)\left(\frac{R}{L}\right) e^{-Rt/L} = \frac{V}{L}\, e^{-Rt/L}$$

Substitute into the left side of the differential equation and show it equals the right side:

$$R\left(\frac{V}{R}\right)(1 - e^{-Rt/L}) + L\left(\frac{V}{L}\right)(e^{-Rt/L}) = V(1 - e^{-Rt/L} + e^{-Rt/L}) = V$$

**Exercise 22-4**

In problems 1 to 36, find the derivative of each function.

**1.** $y = 3 \log_2 x$        **2.** $y = \log x^2$

**3.** $y = \ln x - x$        **4.** $y = \ln x + 2x$

**5.** $y = \log\sqrt{x}$        **6.** $y = \log_3 (2x + 2)$

**7.** $y = x \ln x$        **8.** $y = \dfrac{\ln x}{x}$

**9.** $y = \ln (\sin x)$        **10.** $y = \ln (2 \tan x)$

**11.** $y = 1.5\sqrt{\ln x}$        **12.** $y = 0.5(\ln x)^2$

**13.** $V = \ln (p - \sqrt{p})$        **14.** $s = t^2 \ln (t^2 + 1)$

**15.** $y = \ln \dfrac{x}{x^2 + 1}$        **16.** $y = \ln (2x\sqrt{x - 1})$

**17.** $y = \ln (x \cos x)$        **18.** $y = \ln \dfrac{x + 1}{x - 1}$

**19.** $y = 2e^{3x}$        **20.** $y = e^{x^2+1}$

**21.** $y = 3^{1+x}$        **22.** $y = 10^{x/2}$

**23.** $y = xe^{2x}$        **24.** $y = \dfrac{e^x}{x}$

**25.** $y = \dfrac{e^x}{x + 1}$        **26.** $y = e^x\sqrt{x}$

**27.** $y = e^x \sin x$        **28.** $y = \dfrac{e^x}{\cos x}$

**29.** $y = \dfrac{e^x}{e^x + 1}$        **30.** $y = \sqrt{2e^x + 1}$

**31.** $I = e^{-t/2} \sin 2t$        **32.** $r = e^{\cos \theta}$

**33.** $y = \ln (e^x + x)$        **34.** $y = \dfrac{\ln x}{e^x}$

**35.** $y = \ln (\sin e^x)$        **36.** $y = e^{\ln (\cos x)}$

In problems 37 to 40, sketch the graph of each function. See Example 22-20.

**37.** $y = \ln x - x$

**38.** $y = x \ln x$

**39.** $y = e^{-x^2}$

**40.** $y = xe^x$

In problems 41 and 42, the functions are examples of damped vibration, which is important in electrical and mechanical applications. Sketch the graph of each function for $0 \leq t \leq 2\pi$.

**41.** $y = e^{-t} \sin t$

**42.** $y = e^{-t/2} \sin t$

**43.** The speed of the signal $s$ in a telecommunications line is given by:

$$s = -kt^2 \ln t$$

where $k$ is a constant and the variable $t$ is a function of the thickness of the line. *(a)* Find $s' = ds/dt$. *(b)* Find the maximum value of $s$ in terms of $k$.

**44.** The temperature difference $\Delta T$ between the inner and outer walls of a steam pipe is given by:

$$\Delta T = \frac{1}{2\pi} \ln \frac{R_1}{R_2}$$

where $R_1$ = outer radius and $R_2$ = inner radius. Find the rate of change of $\Delta T$ with respect to $R_2$ if $R_1$ is constant.

**45.** In a circuit containing a resistance $R$, an inductance $L$, and a capacitance $C$ (*RLC* circuit), the current is given by the damped vibration function:

$$I = (A \sin bt + \cos bt)e^{-at}$$

where $A$, $b$, and $a$ are constants. Find $I' = dI/dt$.

**46.** The displacement of a vehicle undergoing a damped vibration is given by:

$$s = e^{-at} \cos bt$$

where $a$ and $b$ are constants. Show that $s$ is a maximum or minimum when $\tan bt = -a/b$.

**47.** The population growth in an underdeveloped country is given approximately by $P = 10^6 e^{0.01t}$, where $t$ is the time in years. Find $P'' = d^2P/dt^2$.

**48.** An overheated computer at 60°C is allowed to cool in a room whose temperature is 15°C. By Newton's law of cooling, which states that the rate of cooling is proportional to the temperature difference, the temperature of the computer is given by:

$$T = 15 + 40e^{-0.20t}$$

where $t$ = time in minutes. Show that $T' = dT/dt = 3.0 - 0.20T$.

**49.** In a circuit containing a resistance $R$, a capacitance $C$ (*RC* circuit), and a voltage source $V$, the charge $q$ on the capacitor after closing the switch is given by $q = CV(1 - e^{-t\,(RC)})$. Show that this formula satisfies the differential equation:

$$R \frac{dq}{dt} + \frac{q}{C} = V$$

**50.** In an electric circuit containing a resistance $R$ and a capacitance $C$, the voltage $V$ across the capacitance satisfies the differential equation:

$$i - C \frac{dV}{dt} = \frac{V}{R}$$

where $i$ = current. Show that if $R = 5.0$ Ω, $C = 4 \times 10^{-3}$ F, and $i = 3.0$ A, the formula for the voltage $V = 15(1 - e^{-50t})$ satisfies the differential equation.

**51.** Show that the derivative of $y = x^x$ is $y' = x^x(1 + \ln x)$ by first taking logarithms of each side and then differentiating implicitly. This is known as *logarithmic differentiation*. [*Note:* The power formula (19-14) cannot be used directly because the exponent is a variable, and the exponential formula (22-32) cannot be used directly because the base is a variable.]

**52.** Two important functions in engineering and science are the *hyperbolic sine* and the *hyperbolic cosine,* defined as follows:

$$\sinh x = \frac{1}{2}(e^x - e^{-x}) \qquad \cosh x = \frac{1}{2}(e^x + e^{-x})$$

Show that *(a)* if $y = \sinh x$, then $y' = \cosh x$; *(b)* if $y = \cosh x$, then $y' = \sinh x$.

## 22-5 | *Integrals Involving Logarithmic and Exponential Functions*

Corresponding to formula (22-30) for the derivative of ln $u$ is the following integral formula, where $u$ is a function of $x$:

$$\int \frac{u'\, dx}{u} = \int \frac{du}{u} = \ln u + C \qquad (22\text{-}33)$$

However, ln $u$ is only defined when $u > 0$. If $u < 0$, it is possible to alter formula (22-33) so it still applies. When $u < 0$, $-u > 0$, and the differential of $-u$ is $-du$. Formula (22-33) is then used as follows:

$$\int \frac{-du}{-u} = \int \frac{du}{u} = \ln(-u) + C \qquad (22\text{-}34)$$

The two integrals (22-33) and (22-34) are combined into one formula by using the absolute, or positive, value of $u = |u|$. If $u$ is a function of $x$, then:

$$\int \frac{u'\, dx}{u} = \int \frac{du}{u} = \ln |u| + C \qquad (22\text{-}35)$$

Note that formula (22-35) allows you to integrate the power of a function $u^n$ when $n = -1$. This is the one case that does not apply with the power formula (21-5).

### Example 22-22
Find each indefinite integral.

**1.** $\displaystyle\int \frac{x\, dx}{x^2 - 1}$

**2.** $\int \tan x\, dx$

**3.** $\displaystyle\int \frac{6e^x}{3e^x + 2}\, dx$

**4.** $\displaystyle\int \frac{x^2 + 2x + 1}{x - 1}\, dx$

**Solution**

**1.** $\displaystyle\int \frac{x\, dx}{x^2 - 1}$

Apply formula (22-35), where $u = x^2 - 1$ and $u' = 2x$. Multiply by 2 on the inside and by ½ on the outside:

$$\int \frac{x\, dx}{x^2 - 1} = \frac{1}{2} \int \frac{(2)x\, dx}{x^2 - 1} = \frac{1}{2}\ln|x^2 - 1| + C$$

**2.** $\displaystyle\int \tan x\, dx$

Express $\tan x$ as $\sin x/\cos x$. Apply (22-35), where $u = \cos x$ and $u' = -\sin x$. Multiply on the inside and outside by $-1$:

$$\int \tan x\, dx = \int \frac{\sin x}{\cos x}\, dx = -\int \frac{-\sin x}{\cos x}\, dx$$

$$= -\ln|\cos x| + C$$

Note that formula (22-35) allows you to integrate $\cot x$, $\sec x$, and $\csc x$ also. These integrals are shown in Sec. 23-1.

**3.** $\displaystyle\int \frac{6e^x}{3e^x + 2}\, dx$

Apply formula (22-35), where $u = 3e^x + 2$ and $u' = 3e^x$. Factor the 6 and put 3 on the inside and 2 on the outside:

$$\int \frac{6e^x}{3e^x + 2}\, dx = 2\int \frac{(3)e^x}{3e^x + 2}\, dx = 2\ln|3e^x + 2| + C$$

**4.** $\displaystyle\int \frac{x^2 + 2x + 1}{x - 1}\, dx$

Divide $x - 1$ into $x^2 + 2x + 1$ and integrate each term. Apply (22-35) to the remainder, where $u = x - 1$ and $u' = 1$:

$$\int \frac{x^2 + 2x + 1}{x - 1}\, dx = \int \left(x + 3 + \frac{4}{x - 1}\right) dx$$

$$= \frac{x^2}{2} + 3x + 4\ln|x - 1| + C$$

Corresponding to the formula (22-31) for the derivative of $e^u$ is the following integral formula, where $u$ is a function of $x$:

$$\int e^u \, (u') \, dx = \int e^u \, du = e^u + C \qquad \text{(22-36)}$$

### Example 22-23
Find each indefinite integral.

1. $\int 10e^{5x} \, dx$
2. $\int 2xe^{x^2+1} \, dx$
3. $\int \sqrt{e^t} \, dt$

### Solution
1. $\int 10e^{5x} \, dx$

   Apply formula (22-36), where $u = 5x$ and $u' = 5$. Factor the 10 and put 2 on the inside and 5 on the outside:

   $$\int 10e^{5x} \, dx = 2 \int e^{5x}(5) \, dx = 2e^{5x} + C$$

2. $\int 2xe^{x^2+1} \, dx$

   Apply formula (22-36), where $u = x^2 + 1$ and $u' = 2x$:

   $$\int 2xe^{x^2+1} \, dx = \int e^{x^2+1}(2x) \, dx = e^{x^2+1} + C$$

3. $\int \sqrt{e^t} \, dt$

   Express $\sqrt{e^t}$ as $(e^t)^{1/2} = e^{t/2}$ and apply formula (22-36), where $u = t/2$ and $u' = 1/2$:

   $$\int \sqrt{e^t} \, dt = \int e^{t/2} \, dt = 2 \int e^{t/2} \left(\frac{1}{2}\right) dt$$

   $$= 2e^{t/2} + C$$

In Sec. 22-4 it is shown that the rate of change of the exponential function $y = e^{kx}$ is proportional to the function itself, that is, $y' = dy/dx = ky$. The next example shows an application of integration which leads to an exponential decay function where $k < 0$.

### Example 22-24
A radioactive substance decays exponentially; that is, it decays at a rate that is proportional to the amount of the substance present at any time. If the initial mass of a radioactive isotope of $^{237}$U is $m_0$ and the half-life of $^{237}$U is 6.6 days, find the formula for the radioactive decay of $^{237}$U.

## Solution

Since the rate of change of the mass $m$ is proportional to the mass present at any time $t$, you can write the differential equation:

$$\frac{dm}{dt} = km$$

Express this equation in differential form and integrate:

$$\frac{dm}{m} = k\,dt$$

$$\int \frac{dm}{m} = \int k\,dt$$

$$\ln m = kt + C$$

Express the preceding equation in exponential form:

$$e^{kt+C} = m$$

or $\qquad m = e^{kt} \cdot e^C = ce^{kt} \qquad (c = e^C)$

This is the general formula for the decay of a radioactive substance. (See Example 9-3.) It is necessary to determine the constants $c$ and $k$ for $^{237}U$. At $t = 0$, $m = m_0$:

$$m_0 = ce^{k(0)} = c$$

At $t = 6.6$, half the original substance has decayed and $m/m_0 = \frac{1}{2}$, or $m = m_0/2$:

$$m = m_0 e^{kt}$$

$$\frac{m_0}{2} = m_0 e^{k(6.6)}$$

$$e^{6.6k} = \frac{1}{2}$$

Take the logarithm of each side:

$$6.6k = \ln \frac{1}{2}$$

$$k = -0.105$$

Therefore, the formula for radioactive decay is:

$$m = m_0 e^{-0.105t}$$

where $t$ is in days.

**Exercise 22-5**    In problems 1 to 30, find each indefinite integral.

1. $\displaystyle\int \frac{2\ dx}{2x + 1}$

2. $\displaystyle\int \frac{dx}{3 - 2x}$

3. $\displaystyle\int \frac{2x\ dx}{2x^2 + 5}$

4. $\displaystyle\int \frac{3x^3\ dx}{x^4 - 1}$

5. $\displaystyle\int \cot x\ dx$

6. $\displaystyle\int \frac{\sec^2 x\ dx}{\tan x + 1}$

7. $\displaystyle\int \frac{e^x - e^{-x}}{e^x + e^{-x}}\ dx$

8. $\displaystyle\int \frac{e^{2x}\ dx}{e^{2x} + 1}$

9. $\displaystyle\int \frac{x^2 + 1}{x}\ dx$

10. $\displaystyle\int \frac{3x - 1}{x^2}\ dx$

11. $\displaystyle\int \frac{x^2 + 3x + 1}{x + 1}\ dx$

12. $\displaystyle\int \frac{2x^2 + x + 1}{x - 1}\ dx$

13. $\displaystyle\int \frac{\sin 2t}{\sin^2 t + 1}\ dt$

14. $\displaystyle\int \frac{2 \sin 2t}{2 - \cos^2 t}\ dt$

15. $\displaystyle\int 6e^{2x}\ dx$

16. $\displaystyle\int (e^x - e^{-x})\ dx$

17. $\displaystyle\int e^{x/3}\ dx$

18. $\displaystyle\int e^x \cdot e^{x-1}\ dx$

19. $\displaystyle\int xe^{x^2}\ dx$

20. $\displaystyle\int 2xe^{3x^2}\ dx$

21. $\displaystyle\int (e^x + 1)^2\ dx$

22. $\displaystyle\int \frac{e^{2x} + 1}{e^x}\ dx$

23. $\displaystyle\int \sqrt{e^{3t}}\ dt$

24. $\displaystyle\int \frac{dt}{\sqrt{e^t}}$

25. $\displaystyle\int e^{\sin t} \cos t\ dt$

26. $\displaystyle\int e^{\cos 2t} \sin 2t\ dt$

27. $\displaystyle\int \frac{e^{1/x}}{x^2}\ dx$

28. $\displaystyle\int \frac{dx}{e^{2x}}$

29. $\displaystyle\int 1 - e^{-kt}\ dt$   (k constant)

30. $\displaystyle\int \frac{dp}{pV \ln p}$   (V constant)

In problems 31 to 34, evaluate each definite integral.

31. $\displaystyle\int_1^2 \frac{3\ dx}{3x - 2}$

32. $\displaystyle\int_{\pi/2}^{\pi/3} \frac{\sin x}{1 - \cos x}\ dx$

33. $\displaystyle\int_0^1 (e^x + e^{-x})\ dx$

34. $\displaystyle\int_1^2 \frac{dx}{2e^x}$

35. Find the area bounded by $y = 1/x$, $y = 0$, $x = 1$, and $x = 2$.

**36.** Find the area bounded by $y = 1/x$ and the line $2x + 2y = 5$.

**37.** Find the area bounded by $y = e^x$, $y = 0$, $x = 0$, and $x = 1$.

**38.** Find the area bounded by $y = 1 - e^x$, $y = 0$, and $x = 1$.

**39.** The following integral occurs in a problem concerning electric circuits:

$$\int \frac{dI}{E - IR}$$

where $E$ and $R$ are constants. Find this indefinite integral.

**40.** The following integral occurs in a problem involving an $RL$ circuit:

$$\int_0^{L/R} e^{-Rt/L} \, dt$$

where $L$ and $R$ are constants. Evaluate this definite integral in terms of $e$, $R$, and $L$.

**41.** The following integral occurs in a problem involving the compression of a gas:

$$\int_{30}^{90} P \, dV$$

If $PV = 1000$, evaluate this definite integral.

**42.** The following integral occurs in a problem about heat transfer:

$$\int_0^{10} (T - T_0)e^{-0.10t} \, dt$$

Evaluate this definite integral, where $T$ and $T_0$ are constants.

**43.** The current flowing to a capacitor in an electric circuit is given by $I = 5e^{-10t}$. If the capacitor is discharged at $t = 0$, find the charge on the capacitor at $t = 0.20$ s.

**44.** The current in an electric circuit is given by $I = 1 - e^{-0.40t}$. Find the rms value of the current from $t = 0$ to $t = 5.0$. [See formula (22-15).]

**45.** A radioactive isotope of radium ($^{228}$Ra) decays exponentially, that is, at a rate proportional to the amount present at any time. The half-life of $^{228}$Ra is 6.7 years. If the initial mass is 10.0 g, find the formula for the radioactive decay of $^{228}$Ra. (See Example 22-24.)

**46.** The population of a certain virus increases exponentially. If the number of viruses increases from $10^6$ to $10^7$ in 10 min, find the formula for the population $P$ at any time $t$ min. Assume $P = 10^6$ when $t = 0$ min.

**47.** The velocity of a particle is given by:

$$v = \frac{t}{t^2 + 1}$$

for $t \geq 0$. Find the equation of motion if the displacement $s = 0$ when $t = 0$.

**48.** A telecommunications line hangs in a curve called a *catenary* between two poles 200 m apart. Its equation is given by:

$$y = 10(e^{x/20} + e^{-x/20})$$

where the middle of the catenary is on the *y* axis at $x = 0$. The length of the line is given by the integral:

$$2 \int_0^{100} \sqrt{1 + (y')^2} \, dx$$

Find the length of the telecommunications line. {*Note:* $\frac{1}{4}(e^{x/10} + 2 + e^{-x/10}) = [\frac{1}{2}(e^{x/20} + e^{-x/20})]^2$.}

## 22-6 | *Review Questions*

In problems 1 to 20, find the derivative of each function.

**1.** $y = 3 \sin \dfrac{x}{2}$  **2.** $y = 2 \cos^2 x$

**3.** $y = x^2 \cos x$  **4.** $y = \dfrac{\sin 2x}{x}$

**5.** $y = \dfrac{1}{3} \tan 3x$  **6.** $y = \tan \sqrt{x}$

**7.** $y = t\sqrt{\sec t}$  **8.** $y = \csc t \cot t$

**9.** $y = 2 \sin^{-1} 3x$  **10.** $y = \cos^{-1} \dfrac{x}{3}$

**11.** $y = \tan^{-1}\sqrt{2x}$  **12.** $y = \sec^{-1} x - \sqrt{x^2 - 1}$

**13.** $y = 5 \log (4x - 3)$  **14.** $y = \dfrac{\ln x}{x}$

**15.** $y = \ln (\cos t)$  **16.** $y = \sqrt{t} \ln t$

**17.** $y = \dfrac{e^{2x}}{2}$  **18.** $y = 2xe^x$

**19.** $y = \sqrt{e^{3x}}$  **20.** $y = e^{-x}(\sin x - \cos x)$

In problems 21 to 40, find each indefinite integral.

**21.** $\displaystyle\int 3 \cos 2x \, dx$  **22.** $\displaystyle\int x \sin x^2 \, dx$

**23.** $\displaystyle\int \dfrac{\sec^2 x}{2} \, dx$  **24.** $\displaystyle\int \cot^2 x \, dx$

**25.** $\displaystyle\int 2 \cos^2 x \, dx$  **26.** $\displaystyle\int 2 \sin x \cos x \, dx$

**27.** $\displaystyle\int \sec^2 x \tan x \, dx$  **28.** $\displaystyle\int \dfrac{\cot x}{\sin x} \, dx$

**29.** $\displaystyle\int \frac{x\,dx}{\sqrt{4 - x^2}}$      **30.** $\displaystyle\int \frac{dx}{x\sqrt{x^2 - 1}}$

**31.** $\displaystyle\int \frac{6\,dx}{1 + 9x^2}$      **32.** $\displaystyle\int \frac{dx}{9 + 4x^2}$

**33.** $\displaystyle\int \frac{2\,dx}{x - 2}$      **34.** $\displaystyle\int \frac{x\,dx}{x^2 + 1}$

**35.** $\displaystyle\int \frac{2 \sin 2x}{\cos 2x}\,dx$      **36.** $\displaystyle\int \frac{1 - x^2}{2x}\,dx$

**37.** $\displaystyle\int (e^{2x} + e^{-2x})\,dx$      **38.** $\displaystyle\int e^x(1 - e^{-x})\,dx$

**39.** $\displaystyle\int 3\sqrt{e^t}\,dt$      **40.** $\displaystyle\int \frac{t + e^t}{te^t}\,dt$

In problems 41 to 44, evaluate each definite integral.

**41.** $\displaystyle\int_{\pi/6}^{\pi/4} 3 \csc^2 x\,dx$      **42.** $\displaystyle\int_0^{1/2} \frac{2\,dx}{\sqrt{1 - 4x^2}}$

**43.** $\displaystyle\int_0^1 \frac{dx}{2x + 2}$      **44.** $\displaystyle\int_{-1}^0 \frac{4\,dt}{e^{2t}}$

**45.** Sketch one cycle of the graph of $y = \sin x - 2 \cos x$, showing the maximum and minimum points.

**46.** Sketch the graph of $y = e^x - x$.

**47.** Find the area bounded by $y = \sin x + \cos x$, $y = 0$, $x = 0$, and $x = \pi/2$.

**48.** Find the area bounded by $y = 2/(1 + x^2)$, $y = 0$, $x = 0$, and $x = 2$.

**49.** Find the area bounded by $y = 2/x$, $y = 0$, $x = 2$, and $x = 4$.

**50.** Show that the equation of simple harmonic motion, $y = a \sin (bx + \phi)$, satisfies the differential equation $y'' + b^2 y = 0$.

**51.** The displacement of a point on a spring is given by $y = 3 \sin t + 5 \cos t$, where $y$ is in centimeters and $t$ is in seconds. *(a)* Find the formulas for the velocity and acceleration. *(b)* Find the acceleration when the displacement is a maximum for $0 \le t \le 2\pi$.

**52.** The angular velocity of a pendulum rotating in a circle (conical pendulum) is given by:

$$\omega = \sqrt{\frac{g \tan \phi}{R}}$$

where $g$ = gravitational acceleration, $R$ = radius of the circle, and $\phi$ = angle the pendulum makes with the vertical. If $R$ is constant, find $d\omega/d\phi$ in terms of $R$, $g$, and $\phi$.

**53.** The voltage across a 4.0-$\Omega$ resistor is given by $V = 12 \cos 3t$. Find the rate of change of power in watts per second when $t = 0.50$ s.

**54.** A weather balloon 100 ft high is rising at the rate of 7.0 ft/s. How fast is its angle of elevation increasing from a point on the ground where the angle is 60°?

55. The current flowing to a capacitor of 0.0020 F is given by $I = -5.0 \sin 120\pi t$. Find the formula for the voltage across the capacitor if $V = 0$ when $t = 0$. (See Table 21-2, Sec. 21-1.)

56. In a certain amplifier, the angle of current flow is given by $\theta = 2 \cos^{-1} (V/10)$, where $V$ is the voltage. Find $d\theta/dV$.

57. The insulation resistance of a high-voltage power line is given by:

$$R = \frac{\rho}{2\pi} \ln \frac{r}{0.30}$$

where $\rho$ = resistivity of the insulation material and $r$ = outer radius of the power line. Assuming $\rho$ is constant, find $dR/dr$.

☐ 58. A beam of a bridge undergoes damped vibration given by:

$$s = 0.010e^{-2t} \sin t$$

where $s$ = displacement in meters and $t$ = time in seconds, $t \geq 0$. Find the first maximum value of $s$. (This is the absolute maximum since the amplitude of the curve decreases as $t$ increases.)

59. The power supplied to a capacitor in a circuit is given by $P = 4e^{-2t} - 4e^{-4t}$. Find the energy in joules stored in the circuit from $t = 0$ to $t = 1$ s. (See Table 21-3, Sec. 21-4.)

60. The energy supplied by the piston of a certain engine moving from $x = a$ to $x = b$ is given by:

$$E = \int_a^b \frac{50}{x} \, dx$$

Find $E$ in terms of $a$ and $b$.

# CHAPTER

# 23

# Methods of Integration

## 23-1 | *Trigonometric Integrals*

Section 22-2 introduces the formulas for the integrals of the sine and cosine but not for the other four basic trigonometric functions. The derivative of the logarithmic function is needed for these other four, and it is introduced in Sec. 22-4. The integral of tan $u$, where $u$ is a function of $x$, is derived as follows by applying formula (22-35):

$$\int (\tan u)(u')\, dx = \int \frac{\sin u}{\cos u}(u')\, dx$$

$$= -\int \frac{-\sin u}{\cos u}(u')\, dx$$

$$= -\ln |\cos u| + C$$

The integral of cot $u$ is derived similarly. The formula (23-2), is shown below and its proof is left as problem 47.

The integral of sec $u$, where $u$ is a function of $x$, is derived by multiplying by the fraction shown below and applying formula (22-35):

$$\int (\sec u)(u')\, dx = \int \sec u \left( \frac{\sec u + \tan u}{\sec u + \tan u} \right)(u')\, dx$$

$$= \int \frac{\sec u \tan u + \sec^2 u}{\sec u + \tan u}(u')\, dx = \ln |\sec u + \tan u| + C$$

Notice that the top of the function becomes the derivative of the bottom, and therefore the fraction is of the form $du/u$.

The integral of csc $u$ is derived similarly to that for sec $u$. The formula, (23-4), is shown below, and its proof is left as problem 48.

The formulas for the other four basic trigonometric functions are then as follows, where $u$ is a function of $x$:

$$\int \tan u \,(u')\, dx = \int \tan u \, du = -\ln |\cos u| + C \qquad (23\text{-}1)$$

$$\int \cot u \,(u')\, dx = \int \cot u \, du = \ln |\sin u| + C \qquad (23\text{-}2)$$

$$\int \sec u \,(u')\, dx = \int \sec u \, du = \ln |\sec u + \tan u| + C \qquad (23\text{-}3)$$

$$\int \csc u \,(u')\, dx = \int \csc u \, du = \ln |\csc u - \cot u| + C \qquad (23\text{-}4)$$

### Example 23-1
Find each indefinite integral.

1. $\int 2 \tan 3x \, dx$
2. $\int (\sec \theta + 1)^2 \, d\theta$
3. $\displaystyle\int \frac{4 \cos 2t - 2}{\sin 2t} \, dt$

### Solution
1. $\int 2 \tan 3x \, dx$

Apply formula (23-1), where $u = 3x$ and $u' = 3$. Multiply by 3 on the inside and by ⅓ on the outside:

$$\int 2 \tan 3x \, dx = 2 \left(\frac{1}{3}\right) \int (\tan 3x)(3) \, dx = -\frac{2}{3}\ln |\cos 3x| + C$$

2. $\int (\sec \theta + 1)^2 \, d\theta$

Expand and integrate each term. Apply formula (22-11) to the first term and formula (23-3) to the second term:

$$\int (\sec \theta + 1)^2 \, d\theta = \int (\sec^2 \theta + 2 \sec \theta + 1) \, d\theta$$

$$= \int \sec^2 \theta \, d\theta + 2 \int \sec \theta \, d\theta + \int d\theta$$

$$= \tan \theta + 2 \ln |\sec \theta + \tan \theta| + \theta + C$$

**3.** $\displaystyle\int \frac{4 \cos 2t - 2}{\sin 2t}\,dt$

Write the fraction as two fractions, and apply formula (23-2) to the first fraction and formula (23-4) to the second:

$$\int \frac{4 \cos 2t - 2}{\sin 2t}\,dt = \int \frac{4 \cos 2t}{\sin 2t}\,dt - \int \frac{2}{\sin 2t}\,dt$$
$$= 2 \int (\cot 2t)(2)\,dt - \int (\csc 2t)(2)\,dt$$
$$= 2 \ln |\sin 2t| - \ln |\csc 2t - \cot 2t| + C$$

## Trigonometric Forms

Certain trigonometric integrals assume forms for which there is a prescribed method of integration. The first form is an integral of the following type, where $u$ is a function of $x$:

$$\int \sin^m u \, \cos^n u \, du \qquad (23\text{-}5)$$

When *either m or n is odd* in (23-5), the function can be integrated by use of the pythagorean identity formula (12-3): $\sin^2 u + \cos^2 u = 1$.

Study the following examples.

### Example 23-2

Integrate $\int \sin^2 x \cos^3 x \, dx$.

### Solution

Change the odd power and express as powers of $\sin x$ times $\cos x$ by letting $\cos^2 x = 1 - \sin^2 x$:

$$\int \sin^2 x \cos^3 x \, dx = \int \sin^2 x \, (\cos^2 x)(\cos x) \, dx$$
$$= \int \sin^2 x \, (1 - \sin^2 x)(\cos x) \, dx$$
$$= \int \sin^2 x \cos x \, dx - \int \sin^4 x \cos x \, dx$$

Both integrals can now be found by applying the power formula, (21-5):

$$\int \sin^2 x \cos x \, dx - \int \sin^4 x \, dx = \frac{\sin^3 x}{3} - \frac{\sin^5 x}{5} + C$$

### Example 23-3

Integrate $\int \sin^3 2t \, dt$.

## Solution

Express as powers of cos $2t$ times sin $2t$ by letting $\sin^2 2t = 1 - \cos^2 2t$:

$$\int \sin^3 2t \, dt = \int (\sin^2 2t)(\sin 2t) \, dt$$

$$= \int (1 - \cos^2 2t)(\sin 2t) \, dt = \int \sin 2t \, dt - \int \cos^2 2t \sin 2t \, dt$$

$$= \left(\frac{1}{2}\right)\int \sin 2t(2) \, dt + \left(\frac{1}{2}\right)\int \cos^2 2t \, -\sin 2t(2) \, dt$$

$$= -\frac{\cos 2t}{2} + \frac{\cos^3 2t}{6} + C$$

When *both m and n are even* in formula (23-5), the function can be integrated by use of the double-angle formula (12-8):

$$\cos 2u = 2\cos^2 u - 1 \qquad \text{or} \qquad \cos 2u = 1 - 2\sin^2 u$$

One illustration of the use of these formulas is shown in Example 22-9(2) (Sec. 22-2). The following example shows another illustration.

## Example 23-4

Integrate $\int \sin^2 x \cos^2 x \, dx$.

## Solution

Express $\sin^2 x$ as $(1 - \cos 2x)/2$ and $\cos^2 x$ as $(1 + \cos 2x)/2$:

$$\int \sin^2 x \, dx \cos^2 x \, dx = \int \frac{1 - \cos 2x}{2} \frac{1 + \cos 2x}{2} \, dx$$

$$= \int \frac{1 - \cos^2 2x}{4} \, dx$$

Apply formula (12-8) again and express $\cos^2 2x$ as $(1 + \cos 4x)/2$:

$$\int \frac{1 - \cos^2 2x}{4} \, dx = \frac{1}{4}\int \left(1 - \frac{1 + \cos 4x}{2}\right) dx$$

$$= \frac{1}{8}\int (1 - \cos 4x) \, dx$$

$$= \frac{1}{8}\int dx - \frac{1}{8}\left(\frac{1}{4}\right)\int (\cos 4x)(4) \, dx = \frac{x}{8} - \frac{\sin 4x}{32} + C$$

The second trigonometric form is an integral of either of the following two types, where $u$ is a function of $x$:

$$\int \sec^m u \tan^n u \, du \quad \text{or} \quad \int \csc^m u \cot^n u \, du \qquad (23\text{-}6)$$

Study the following examples, which show how to apply formula (23-6).

## Example 23-5

Integrate $\int \sec^2 x \tan^2 x \, dx$.

### Solution

When $m = 2$ in (23-6), you can use the power formula (21-5), where $u = \tan x$ and $u' = \sec^2 x$ (or $u = \cot x$ and $u' = -\csc^2 x$):

$$\int \tan^2 x \sec^2 x \, dx = \frac{\tan^3 x}{3} + C$$

When $m > 2$ *and is even, or n is odd* in (23-6), the function can be integrated by using the pythagorean identities (12-3):

$$\tan^2 u + 1 = \sec^2 u \quad \text{or} \quad \cot^2 u + 1 = \csc^2 u$$

## Example 23-6

Integrate $\int \csc^4 x \, dx$.

### Solution

When $m = 4$ in (23-6), use the pythagorean identity (12-3), as follows:

$$\int \csc^4 x \, dx = \int \csc^2 x \, (\csc^2 x) \, dx$$
$$= \int \csc^2 x \, (\cot^2 x + 1) \, dx$$
$$= \int \csc^2 x \cot^2 x \, dx + \int \csc^2 x \, dx$$

The first integral is found by using the power formula (21-5), where $u = \cot x$ and $u' = -\csc^2 x$, and the second by applying (22-12):

$$\int \csc^2 x \cot^2 x \, dx + \int \csc^2 x \, dx = -\int \cot^2 x \, (-\csc^2 x) \, dx + \int \csc^2 x \, dx$$
$$= -\frac{\cot^3 x}{3} - \cot x + C$$

## Example 23-7

Integrate $\int \sec x \tan^3 x \, dx$.

### Solution

When $n$ is odd in (23-6), use the pythagorean identity (12-3) as follows:

$$\int \sec x \tan^3 x \, dx = \int \sec x \tan x \, (\tan^2 x) \, dx$$

$$= \int \sec x \tan x \, (\sec^2 x - 1) \, dx$$

$$= \int \sec^2 x \, (\sec x \tan x) \, dx - \int \sec x \tan x \, dx$$

The first integral is found by using the power formula (21-5), where $u = \sec x$ and $u' = \sec x \tan x$, and the second by applying (22-13):

$$\int \sec x \tan^3 x \, dx = \frac{\sec^3 x}{3} - \sec x + C$$

### Example 23-8

Integrate $\int \cot^3 x \, dx$.

### Solution

Use the pythagorean identity (12-3) as follows:

$$\int \cot^3 x \, dx = \int \cot x \, (\cot^2 x) \, dx$$

$$= \int \cot x \, (\csc^2 x - 1) \, dx = \int (\cot x \csc^2 x - \cot x) \, dx$$

$$= -\int \cot x \, (-\csc^2 x \, dx) \, dx - \int \cot x \, dx$$

$$= \frac{-\cot^2 x}{2} - \ln |\sin x| + C$$

A third trigonometric form is an integral containing only even powers of the tangent or cotangent and is done as follows.

### Example 23-9

Evaluate the integral $\displaystyle\int_0^{\pi/4} \tan^4 x \, dx$.

### Solution

Apply the pythagorean identity (12-3) twice:

$$\int_0^{\pi/4} \tan^4 x \, dx = \int_0^{\pi/4} \tan^2 x \, (\tan^2 x) \, dx$$

$$= \int_0^{\pi/4} \tan^2 x \, (\sec^2 x - 1) \, dx$$

$$= \int_0^{\pi/4} (\tan^2 x \sec^2 x - \tan^2 x)\, dx$$

$$= \int_0^{\pi/4} [\tan^2 x \sec^2 x - (\sec^2 x - 1)]\, dx$$

$$= \int_0^{\pi/4} \tan^2 x \sec^2 x - \int_0^{\pi/4} \sec^2 x\, dx + \int_0^{\pi/4} dx$$

$$= \left( \frac{\tan^3 x}{3} - \tan x + x \right)\Big|_0^{\pi/4} = \left( \frac{1}{3} - 1 + \frac{\pi}{4} \right) - 0$$

$$= \frac{\pi}{4} - \frac{2}{3} = 0.119$$

**Exercise 23-1**    In problems 1 to 34, find each indefinite integral.

1. $\displaystyle\int 2 \tan 2x\, dx$    2. $\displaystyle\int \cot \frac{x}{2}\, dx$

3. $\displaystyle\int 6 \sec 3x\, dx$    4. $\displaystyle\int 2 \csc 4x\, dx$

5. $\displaystyle\int \cot (2t - 1)\, dt$    6. $\displaystyle\int \tan \pi t\, dt$

7. $\displaystyle\int (1 - \csc \theta)^2\, d\theta$    8. $\displaystyle\int (\sec \theta + 2)^2\, d\theta$

9. $\displaystyle\int (\tan x + 1)^2\, dx$    10. $\displaystyle\int (\cot 2x - 1)^2\, dx$

11. $\displaystyle\int \frac{1 + \cos x}{\sin x}\, dx$    12. $\displaystyle\int \frac{\sin x - 1}{\cos x}\, dx$

13. $\displaystyle\int \frac{1 - \sin 2t}{2 \cos 2t}\, dt$    14. $\displaystyle\int \frac{\sin t + \cos t}{\sin t \cos t}\, dt$

15. $\displaystyle\int \frac{\sin x + \cot x}{\cos x}\, dx$    16. $\displaystyle\int \frac{1 - \sin x}{\tan x}\, dx$

17. $\displaystyle\int \sin 2x \cos^2 2x\, dx$    18. $\displaystyle\int \sin^3 x \cos x\, dx$

19. $\displaystyle\int \sin^3 x \cos^2 x\, dx$    20. $\displaystyle\int 2 \sin^2 2x \cos^3 2x\, dx$

21. $\displaystyle\int 3 \cos^3 2t\, dt$    22. $\displaystyle\int \sin^3 2\pi t\, dt$

23. $\displaystyle\int 4 \sin^2 x \cos^2 x\, dx$    24. $\displaystyle\int \sin^2 2x \cos^2 2x\, dx$

25. $\displaystyle\int \cos^4 x\, dx$    26. $\displaystyle\int \sin^4 x \cos x\, dx$

27. $\displaystyle\int \sec^2 x \tan^3 x\, dx$    28. $\displaystyle\int \csc^2 x \cot^2 x\, dx$

29. $\displaystyle\int 4 \csc^4 2x\, dx$    30. $\displaystyle\int 2 \sec^4 x\, dx$

31. $\displaystyle\int (\tan^3 x - 1)\, dx$    32. $\displaystyle\int \csc x \cot^3 x\, dx$

33. $\displaystyle\int (\tan^2 x + 1)^2\, dx$    34. $\displaystyle\int 3 \cot^4 2x\, dx$

In problems 35 to 38, evaluate each definite integral.

C **35.** $\displaystyle\int_{\pi/6}^{\pi/2} (\cot\theta + \cos\theta)\, d\theta$    C **36.** $\displaystyle\int_{0}^{\pi/6} \sec 2r\, dr$

**37.** $\displaystyle\int_{0}^{\pi/2} \sin^3 t\, dt$    C **38.** $\displaystyle\int_{0}^{\pi/4} \tan^3 \phi\, d\phi$

C **39.** Find the area bounded by $y = \tan x$, $x = \pi/4$, and $y = 0$.

C **40.** Find the area bounded by $y = \csc x$, $y = 0$, $x = \pi/6$, and $x = \pi/3$.

**41.** The following integral comes from a problem involving vibration:

$$2\pi \int_{0}^{0.25} \sin^2 2\pi t \cos^3 2\pi t\, dt$$

Evaluate this integral.

**42.** The velocity of an object is given by:

$$v = \frac{1 - \sin t}{\cos t}$$

where $t$ = time. If the distance $s = 0$ when $t = 0$, find the equation of motion.

**43.** The current in a circuit is given by $I = 2.5 \cos^3 t$, where $I$ is in amperes and $t$ is in seconds. Find the charge transmitted in coulombs from $t = 0$ to $t = \pi/2$.

**44.** A current given by $I = 2 \tan t$ flows in a 10-$\Omega$ resistor, where $I$ is in amperes and $t$ is in seconds. Find the energy dissipated in joules from $t = 0$ to $t = \pi/4$. (See Table 21-3, Sec. 21-4.)

**45.** Find the rms value of the current $I = 4 \sin t \cos t$ from $t = 0$ to $t = \pi$. (See Example 22-12, Sec. 22-2.)

**46.** Find the rms value of the current $I = \sin t + \cos t$ from $t = 0$ to $t = 2\pi$. (See Example 22-12, Sec. 22-2.)

**47.** Prove formula (23-2), using the proof of (23-1) as a guide.

**48.** Prove formula (23-4), using the proof of (23-3) as a guide.

## 23-2 | *Trigonometric Substitution*

One of the useful applications of trigonometric integrals is to integrate certain algebraic functions by using trigonometric substitution. When an integral contains the square root of a quadratic expression, such as $\sqrt{1 - x^2}$, trigonometric substitution can be used. The first example illustrates the procedure. The three possible situations are then presented.

## Example 23-10

Integrate $\int \sqrt{1 - x^2}\, dx$.

### Solution
This integral cannot be found by any of the methods presented so far. If you let $x = \sin\theta$, the differential $dx = \cos\theta\, d\theta$, and $\sqrt{1 - x^2} = \sqrt{1 - \sin^2\theta} = \sqrt{\cos^2\theta} = \cos\theta$. The integral can then be written:

$$\int (\cos\theta)(\cos\theta\, d\theta) = \int \cos^2\theta\, d\theta$$

This trigonometric function can be integrated by using the double-angle identity, as shown in Example 23-4:

$$\int \cos^2\theta\, d\theta = \int \frac{1 + \cos 2\theta}{2} d\theta = \frac{\theta}{2} + \frac{\sin 2\theta}{4} + C$$

The result, however, should be expressed in terms of $x$. If $x = \sin\theta$, then $\theta = \sin^{-1} x$, $\cos\theta = \sqrt{1 - x^2}$, and $\sin 2\theta = 2\sin\theta\cos\theta = 2x\sqrt{1 - x^2}$. Therefore:

$$\int \sqrt{1 - x^2}\, dx = \frac{\sin^{-1} x}{2} + \frac{x\sqrt{1 - x^2}}{2} + C$$

The three functions that lend to trigonometric substitution, and the substitutions that apply, are as follows:

| If the integral contains: | Let: | Then | |
|---|---|---|---|
| $\sqrt{a^2 - x^2}$ | $x = a\sin\theta$ | $dx = a\cos\theta\, d\theta$ | |
| $\sqrt{a^2 + x^2}$ | $x = a\tan\theta$ | $dx = a\sec^2\theta\, d\theta$ | (23-7) |
| $\sqrt{x^2 - a^2}$ | $x = a\sec\theta$ | $dx = a\sec\theta\tan\theta\, d\theta$ | |

## Example 23-11

Integrate $\int \dfrac{dx}{x\sqrt{4 + x^2}}$.

### Solution
Apply formula (23-7), where $a = 2$, and let $x = 2\tan\theta$. Then $dx = 2\sec^2\theta\, d\theta$ and:

$$\sqrt{4 + x^2} = \sqrt{4 + 4\tan^2\theta} = \sqrt{4(1 + \tan^2\theta)}$$
$$= \sqrt{4\sec^2\theta} = 2\sec\theta$$

The integral therefore becomes:

$$\int \frac{dx}{x\sqrt{4 + x^2}} = \int \frac{2\sec^2\theta \, d\theta}{(2\tan\theta)(2\sec\theta)}$$

$$= \int \frac{\sec\theta \, d\theta}{2\tan\theta} = \frac{1}{2}\int \csc\theta \, d\theta$$

$$= \frac{1}{2}\ln|\csc\theta - \cot\theta| + C$$

To express the result in terms of $x$, it is helpful to construct a right triangle where $x = 2\tan\theta$ or $\tan\theta = x/2$ (Fig. 23-1). By using the pythagorean theorem, the hypotenuse is $\sqrt{4 + x^2}$. Then $\csc\theta = \sqrt{4 + x^2}/x$ and $\cot\theta = 2/x$.

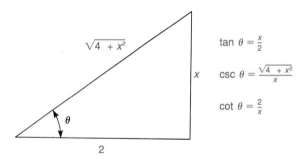

FIG. 23-1 *Trigonometric substitution, Example 23-11*

Therefore the integral is:

$$\int \frac{dx}{x\sqrt{4 + x^2}} = \frac{1}{2}\ln\left|\frac{\sqrt{4 + x^2} - 2}{x}\right| + C$$

By using the properties of logarithms, the answer can be expressed as:

$$\int \frac{dx}{x\sqrt{4 + x^2}} = -\frac{1}{2}\ln\left|\frac{x}{\sqrt{4 + x^2} - 2}\right| + C$$

and simplified to:

$$\int \frac{dx}{x\sqrt{4 + x^2}} = -\frac{1}{2}\ln\left|\frac{x}{\sqrt{4 + x^2} - 2}\left(\frac{\sqrt{4 + x^2} + 2}{\sqrt{4 + x^2} + 2}\right)\right| + C$$

$$= -\frac{1}{2}\ln\left|\frac{x(\sqrt{4 + x^2} + 2)}{4 + x^2 - 4}\right| + C$$

$$= -\frac{1}{2}\ln \left| \frac{\sqrt{4 + x^2} + 2}{x} \right| + C$$

This is how the answer appears in the table of integrals (form 29, App. B-8).

### Example 23-12

Integrate $\displaystyle\int \frac{dx}{\sqrt{x^2 - 1}}$.

### Solution

Apply formula (23-7), where $a = 1$, and let $x = \sec \theta$. Then $dx = \sec \theta \tan \theta \, d\theta$ and $\sqrt{x^2 - 1} = \sqrt{\sec^2 \theta - 1} = \sqrt{\tan^2 \theta} = \tan \theta$. The integral becomes:

$$\int \frac{dx}{\sqrt{x^2 - 1}} = \int \frac{\sec \theta \tan \theta \, d\theta}{\tan \theta} = \int \sec \theta \, d\theta$$

$$= \ln |\sec \theta + \tan \theta| + C$$

By substituting $x = \sec \theta$ and $\sqrt{x^2 - 1} = \tan \theta$, the integral in terms of $x$ is:

$$\int \frac{dx}{\sqrt{x^2 - 1}} = \ln |x + \sqrt{x^2 - 1}| + C$$

**Exercise 23-2**  In problems 1 to 16, find each indefinite integral.

1. $\displaystyle\int \frac{dx}{\sqrt{1 - x^2}}$

2. $\displaystyle\int \frac{dx}{x^2\sqrt{4 - x^2}}$

3. $\displaystyle\int \frac{dx}{x^2\sqrt{9 + x^2}}$

4. $\displaystyle\int \frac{dx}{\sqrt{1 + x^2}}$

5. $\displaystyle\int \frac{\sqrt{x^2 - 1}}{x}dx$

6. $\displaystyle\int \frac{dx}{\sqrt{x^2 - 4}}$

7. $\displaystyle\int \sqrt{16 - x^2}\, dx$

8. $\displaystyle\int \frac{\sqrt{9 - x^2}}{2x^2}\, dx$

9. $\displaystyle\int 5x^3\sqrt{x^2 + 4}\, dx$

10. $\displaystyle\int \frac{dx}{(x^2 + 1)^{3/2}}$

11. $\displaystyle\int \frac{dx}{x^3\sqrt{x^2 - 1}}$

12. $\displaystyle\int \frac{\sqrt{x^2 - 9}}{x}\, dx$

13. $\displaystyle\int \frac{3\, dt}{\sqrt{9t^2 - 1}}$  (Let $x = 3t$.)

14. $\displaystyle\int \frac{2\, dt}{2t\sqrt{1 - 4t^2}}$  (Let $x = 2t$.)

15. $\displaystyle\int \frac{x^3\, dx}{\sqrt{x^2 - r^2}}$  ($r$ constant)

16. $\displaystyle\int \frac{2\, dx}{4x^2\sqrt{9 + 4x^2}}$

C 17. Find the area bounded by $y = 1/\sqrt{x^2 + 1}$, $y = 0$, $x = 0$, and $x = 1$.

**18.** Using integration, find the area of the quadrant bounded by the circle $x^2 + 2y^2 = 1$, $y = 0$, and $x = 0$. (Check the answer by geometry.)

**19.** The current in a circuit in amperes is given by $I = 1/(1 + t^2)^{3/2}$. Find the charge transmitted from $t = 0$ to $t = 0.5$ s.

**20.** The primary current $I$ in a transformer is given by the integral:

$$I = 0.25 \int \frac{\sqrt{t^2 + 1}}{t^2} \, dt$$

Find the formula for $I$ if $I = 0$ when $t = \pi/2$.

## 23-3 | *Integration by Parts*

One of the most useful methods of integration is integration by parts. It is based on the product formula for derivatives. If $y = uv$, where $u$ and $v$ are functions of $x$, the product formula states that:

$$y' = (uv)' = uv' + vu'$$

By integrating this equation term by term with respect to $x$, it becomes:

$$uv = \int uv' \, dx + \int vu' \, dx$$

This is then written in the useful form:

$$\int uv' \, dx = uv - \int vu' \, dx \qquad \text{or} \qquad \int u \, dv = uv - \int v \, du \quad (23\text{-}8)$$

Formula (23-8) allows you to change one integral into another one that may be easier to integrate. Study the following examples.

### Example 23-13
Integrate $\int x \cos x \, dx$.

### Solution
This integral cannot be found by any of the methods shown so far. Let $u = x$ and $v' = \cos x$ or $dv = \cos x \, dx$ for the left side of formula (23-8). Then $u' = 1$ or $du = dx$, and $v = \sin x + C'$. It is convenient to list this as follows:

$$u = x \qquad dv = \cos x \, dx$$
$$du = dx \qquad v = \sin x + C'$$

Substitute these functions into formula (23-8) to obtain:

$$\int x \cos x \, dx = x(\sin x + C') - \int (\sin x + C') \, dx$$
$$= x \sin x + C'x - (-\cos x + C'x) + C$$
$$= x \sin x + \cos x + C$$

Notice that the constant associated with $v$, $C'$, cancels. This will always happen, and therefore it is not necessary to include it when you are doing integration by parts.

How do you decide which function to call $u$ and which to call $dv$? In most cases, *choose $u$ so that $du$ is of simpler form, and choose $dv$ to be the more complicated function that can be integrated.*

### Example 23-14

Integrate $\int xe^x \, dx$.

### Solution

Let $u = x$ and $dv = e^x \, dx$. Then:

$$u = x \qquad dv = e^x \, dx$$
$$du = dx \qquad v = e^x$$

Apply formula (23-8):

$$\int xe^x \, dx = xe^x - \int e^x \, dx = xe^x - e^x + C$$

Notice what would happen if you let $u = e^x$ and $v' = x$. Formula (23-8) would become:

$$\int xe^x \, dx = \frac{x^2 e^x}{2} - \int \frac{x^2 e^x}{2} \, dx$$

This is clearly more difficult to integrate.

### Example 23-15

Integrate $\int \ln x \, dx$.

### Solution

The integral of $\ln x$ has not been shown so far. Integration by parts allows it to be determined. Let $u = \ln x$ and $dv = dx$. Then:

$$u = \ln x \qquad dv = dx$$
$$du = \frac{1}{x} \, dx \qquad v = x$$

Apply formula (23-8):

$$\int \ln x \, dx = x \ln x - \int x \left(\frac{1}{x}\right) dx = x \ln x - x + C$$

Sometimes it is necessary to integrate by parts more than once, as the next example shows.

## Example 23-16
Integrate $\int x^2 \sin x \, dx$.

**Solution**
Let:

$$u = x^2 \qquad dv = \sin x \, dx$$
$$du = 2x \, dx \qquad v = -\cos x$$

Apply formula (23-8) once to obtain:

$$\int x^2 \sin x \, dx = -x^2 \cos x - \int (-\cos x)(2x) \, dx$$
$$= -x^2 \cos x + 2 \int x \cos x \, dx$$

The last integral is the one shown in Example 23-13. Integrate by parts again to obtain:

$$u = x \qquad dv = \cos x \, dx$$
$$du = dx \qquad v = \sin x$$

Then:

$$-x^2 \cos x + 2 \int x \cos x \, dx = -x^2 \cos x + 2 \left(x \sin x - \int \sin x \, dx\right)$$

Therefore the final result is:

$$\int x^2 \sin x \, dx = -x^2 \cos x + 2x \sin x + 2 \cos x + C$$

The next example shows an integral where two applications of integration by parts lead back to the original integral. By an algebraic technique the original integral can be found as follows.

## Example 23-17
Integrate $\int e^x \cos x \, dx$.

**Solution**

Let:

$$u = e^x \qquad dv = \cos x\, dx$$
$$du = e^x\, dx \qquad v = \sin x$$

Then:

$$\int e^x \cos x\, dx = e^x \sin x - \int e^x \sin x\, dx$$

For the last integral let:

$$u = e^x \qquad dv = \sin x\, dx$$
$$du = e^x\, dx \qquad v = -\cos x$$

Therefore:

$$\int e^x \cos x\, dx = e^x \sin x - \int e^x \sin x\, dx$$
$$= e^x \sin x + e^x \cos x - \int e^x \cos x\, dx$$

Notice $\int e^x \cos x\, dx$ appears on both sides of the equation. Add this integral to both sides to obtain the final result:

$$2 \int e^x \cos x\, dx = e^x \sin x + e^x \cos x$$
$$\int e^x \cos x\, dx = \frac{e^x}{2}\,(\sin x + \cos x) + C$$

**Exercise 23-3**

In problems 1 to 18, find each indefinite integral.

1. $\displaystyle\int x \sin x\, dx$   2. $\displaystyle\int x \cos (x + 1)\, dx$

3. $\displaystyle\int 2x e^{2x}\, dx$   4. $\displaystyle\int x e^{-x}\, dx$

5. $\displaystyle\int \ln \sqrt{t}\, dt$   6. $\displaystyle\int t \ln t\, dt$

7. $\displaystyle\int \sin^{-1} x\, dx$   8. $\displaystyle\int \cos^{-1} x\, dx$

9. $\displaystyle\int x \sin^2 x\, dx$   10. $\displaystyle\int x^2 \cos x\, dx$

11. $\displaystyle\int x^2 e^x\, dx$   12. $\displaystyle\int x e^{2x}\, dx$

13. $\displaystyle\int \ln (t + 1)\, dt$   14. $\displaystyle\int t^2 \ln t\, dt$

15. $\displaystyle\int e^x \sin x\, dx$   16. $\displaystyle\int e^{2x} \cos x\, dx$

17. $\displaystyle\int 5x\sqrt{x + 1}\, dx$   18. $\displaystyle\int 3x\sqrt{1 - x}\, dx$

In problems 19 to 22, evaluate each definite integral.

**19.** $\displaystyle\int_0^{\pi/4} 2\theta \cos 2\theta \; d\theta$        **20.** $\displaystyle\int_0^2 te^{t/2} \; dt$

C **21.** $\displaystyle\int_0^1 \tan^{-1} x \; dx$        **22.** $\displaystyle\int_0^\pi e^x \cos 2x \; dx$

**23.** Find the area bounded by $y = \ln x$, $y = 0$, and the line $x = e$.

C **24.** Find the area bounded by the damped vibration curve $y = e^{-x} \sin x$ and the $x$ axis during the first half-cycle from $x = 0$ to $x = \pi/2$.

**25.** The current in a circuit containing an inductance is given by $I = 4.0e^{-t} \cos t$. Find the formula for the charge $q$ transmitted at any time $t$, if $q = 0$ when $t = 0$.

C **26.** Find the rms value of the voltage $V = \sqrt{t} \cos t$ from $t = 0$ to $t = \pi/2$. (See Example 22-12, Sec. 22-2.)

**27.** The velocity of a particle in millimeters is given by $v = t/\sqrt{t + 1}$. Find the displacement of the particle from $t = 0$ to $t = 3$ s.

**28.** The following integral arises in a problem involving the work done by a variable magnetic force:

$$W = \int_0^e (r - \ln r) \; dr$$

Evaluate this definite integral in terms of $e$.

## 23-4 | *Use of Integral Tables*

Tables of integrals are necessary because of the many integral forms and integration techniques that exist. Appendix B-8 contains a short table of integrals, and more extensive tables exist in reference sources containing mathematical formulas. The following examples illustrate the use of some of the formulas in App. B-8 and indicate how to recognize certain forms.

### Example 23-18
Integrate $\displaystyle\int \frac{x \; dx}{3 + 2x}$.

### Solution
This integral fits form 19 in App. B-8:

$$\int \frac{u \; du}{a + bu} = \frac{1}{b^2} (a + bu - a \ln |a + bu|) + C$$

where $u = x$, $a = 3$, $b = 2$, and $du = dx$:

$$\int \frac{x \, dx}{3 + 2x} = \frac{1}{4}(3 + 2x - 3 \ln |3 + 2x|) + C$$

## Example 23-19

Integrate $\int \sqrt{4x^2 + 1} \, dx$.

### Solution
This integral fits form 26 in App. B-8:

$$\int \sqrt{u^2 + a^2} \, du = \frac{u}{2}\sqrt{u^2 + a^2} + \frac{a^2}{2}\ln |u + \sqrt{u^2 + a^2}| + C$$

where $u = 2x$, $a = 1$, and $du = 2 \, dx$:

$$\int \sqrt{4x^2 + 1} \, dx = \frac{1}{2} \int \sqrt{(2x)^2 + 1^2} \, (2) \, dx$$
$$= \frac{1}{4}(2x\sqrt{4x^2 + 1} + \ln |2x + \sqrt{4x^2 + 1}|) + C$$

## ⊡ Example 23-20

Evaluate $\int_0^{\pi/4} \sec^3 \theta \, d\theta$.

### Solution
This integral fits form 48 in App. B-8:

$$\int \sec^n u \, du = \frac{1}{n - 1}\tan u \sec^{n-2} u + \frac{n - 2}{n - 1} \int \sec^{n-2} u \, du + C$$

where $n = 3$, $u = \theta$, and $du = d\theta$:

$$\int_0^{\pi/4} \sec^3 \theta \, d\theta = \frac{\sec \theta \tan \theta}{2}\Big|_0^{\pi/4} + \frac{1}{2} \int_0^{\pi/4} \sec \theta \, d\theta$$

This is an example of a reduction formula where the exponent is reduced by 2 with the first integration. Formula (23-3) is now applied to evaluate the last integral:

$$\int \sec^3 u \, du = \frac{\sec \theta \tan \theta}{2}\Big|_0^{\pi/4} + \frac{1}{2} \ln |\sec \theta + \tan \theta|\Big|_0^{\pi/4}$$

$$= \left[\frac{(\sqrt{2})(1)}{2} + \frac{1}{2} \ln |\sqrt{2} + 1|\right] - \left(0 + \frac{1}{2} \ln |1 + 0|\right)$$

$$= \frac{\sqrt{2}}{2} + \frac{1}{2} \ln |\sqrt{2} + 1| = 1.15$$

## Example 23-21

Integrate $\int xe^x \sin x \, dx$.

## Solution

This integral does not fit any form directly; however, part of it fits form 60 in App. B-8:

$$\int e^{au} \sin bu \, du = \frac{e^{au}(a \sin bu - b \cos bu)}{a^2 + b^2}$$

Use integration by parts to separate the integral. Let $u = x$ and $dv = e^x \sin x$. Then, by using form 60:

$$u = x \qquad dv = e^x \sin x$$

$$du = dx \qquad v = \frac{e^x (\sin x - \cos x)}{2}$$

Apply formula (23-8) for integration by parts:

$$\int xe^x \sin x \, dx = \frac{xe^x (\sin x - \cos x)}{2} - \int \frac{e^x (\sin x - \cos x)}{2} \, dx$$

Use forms 60 and 61 to find the last integral:

$$\int xe^x \sin x \, dx = \frac{xe^x (\sin x - \cos x)}{2} - \frac{1}{2} \left[ \frac{e^x (\sin x - \cos x)}{2} - \frac{e^x (\cos x + \sin x)}{2} \right]$$

$$= \frac{e^x}{2} (x \sin x - x \cos x + \cos x) + C$$

**Exercise 23-4**    In problems 1 to 30, find each indefinite integral.

1. $\displaystyle\int \frac{dx}{x(2 + 5x)}$    2. $\displaystyle\int \frac{dx}{x^2(2x + 1)}$

3. $\displaystyle\int 3x\sqrt{4 + 3x} \, dx$    4. $\displaystyle\int \frac{4x \, dx}{\sqrt{5 + 2x}}$

5. $\displaystyle\int \frac{dx}{\sqrt{1 - 4x^2}}$    6. $\displaystyle\int \frac{\sqrt{4x^2 - 9}}{x} \, dx$

7. $\displaystyle\int \cos^3 2t \, dt$    8. $\displaystyle\int \tan^3 \theta \, d\theta$

9. $\displaystyle\int \tan^{-1} x \, dx$    10. $\displaystyle\int \sin^{-1} \frac{x}{2} \, dx$

11. $\displaystyle\int \frac{dx}{\sqrt{(x^2 + 1)^3}}$    12. $\displaystyle\int \frac{dx}{\sqrt{(1 - x^2)^3}}$

**13.** $\displaystyle\int x^2 \ln 3x$

**14.** $\displaystyle\int e^{2x} \sin 2x$

**15.** $\displaystyle\int \frac{8dx}{16 - x^2}$

**16.** $\displaystyle\int \frac{dx}{x\sqrt{4 - 9x^2}}$

**17.** $\displaystyle\int \frac{x \, dx}{x^2 + 2x + 1}$    (*Hint:* Factor denominator.)

**18.** $\displaystyle\int \frac{dx}{2x + 3x^2}$    (*Hint:* Factor denominator.)

**19.** $\displaystyle\int \sin 2x \cos 3x \, dx$

**20.** $\displaystyle\int \sin^3 x \cos^2 x \, dx$

**21.** $\displaystyle\int \frac{x + 1}{x + 2} \, dx$    (*Hint:* Separate fractions.)

**22.** $\displaystyle\int \frac{x - 1}{\sqrt{x + 1}} \, dx$    (*Hint:* Separate fractions.)

**23.** $\displaystyle\int \frac{\sqrt{t^2 + 0.04}}{t} \, dt$

**24.** $\displaystyle\int \frac{\sqrt{1.0 - 0.01t^2}}{0.1t} \, dt$

**25.** $\displaystyle\int \sin^4 x \, dx$

**26.** $\displaystyle\int \csc^3 2x \, dx$

**27.** $\displaystyle\int (x^2 - 4)^{3/2} \, dx$

**28.** $\displaystyle\int 2\sqrt{2x - x^2} \, dx$    (*Hint:* Complete the square.)

**29.** $\displaystyle\int xe^x \cos x \, dx$    (See Example 23-21.)

**30.** $\displaystyle\int x \sin^{-1} x \, dx$    (See Example 23-21.)

In problems 31 to 34, evaluate each definite integral.

**C 31.** $\displaystyle\int_{-1}^{0} \frac{x \, dx}{2 + x}$

**C 32.** $\displaystyle\int_{1}^{2} \frac{2 \, dx}{4x^2 - 1}$

**33.** $\displaystyle\int_{0}^{\pi/4} \sin 3x \sin x \, dx$

**34.** $\displaystyle\int_{3}^{6} \frac{\sqrt{x^2 - 9}}{x} \, dx$

**35.** Find the area bounded by $y = x^2 e^x$, $y = 0$, $x = 0$, and $x = 1$.

**36.** In calculating the total force on a structure, the following integral arises:

$$\int_{0}^{2L} \frac{L \, dx}{\sqrt{L^2 x^2 + 4}}$$

Evaluate this definite integral in terms of the constant $L$.

$\boxed{\text{c}}$ **37.** The current in a circuit is given by $I = t/(1 + t)$. Find the charge in coulombs transmitted from $t = 0.5$ to $t = 1.5$ s.

**38.** The current to a capacitance $C = 0.01$ F in an electric circuit is given by:

$$I = \frac{1}{t^2 + t}$$

where $t \geq 1$. If the voltage across the capacitance is zero when $t = 1$ s, find the formula for the voltage across the capacitance at any time $t$. (See Example 21-6.)

## 23-5 | *Approximate Integration*

The methods of integration introduced so far do not allow you to integrate all functions. Although additional methods can be found in other sources, there are some functions that cannot be integrated by any direct method, such as:

$$\int \frac{dx}{\sqrt{1 + x^3}}$$

For these integrals certain approximation methods are necessary. One of these is shown in this section, and one is shown in Chap. 24, Sec. 24-2.

The definite integral $\int_a^b y \, dx$ represents the area under the curve $y = f(x)$ from $x = a$ to $x = b$ (Fig. 23-2). This area can be approximated geometrically

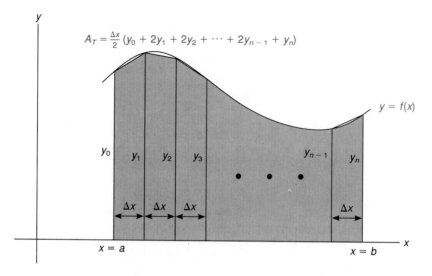

$$A_T = \frac{\Delta x}{2} (y_0 + 2y_1 + 2y_2 + \cdots + 2y_{n-1} + y_n)$$

FIG. 23-2 *Trapezoidal rule*

by a method similar to that shown in Sec. 21-2, but by using trapezoids instead of rectangles because they provide a better approximation. The area is divided into $n$ trapezoids each of width $\Delta x = (b - a)/n$. The top of each trapezoid is a chord of the curve, that is, a line connecting two points on the curve. The area of each trapezoid equals one-half the height times the sum of the bases. The height of each trapezoid is $\Delta x$, and its bases are the two values of $y$ corresponding to the points on the curve. The area $A_T$ in Fig. 23-2 represents the sum of the areas of the $n$ trapezoids and is therefore given by:

$$A_T = \frac{\Delta x}{2}\,(y_0 + y_1) + \frac{\Delta x}{2}\,(y_1 + y_2)$$

$$+ \frac{\Delta x}{2}\,(y_2 + y_3) + \cdots + \frac{\Delta x}{2}\,(y_{n-1} + y_n) \qquad (23\text{-}9)$$

$$= \frac{\Delta x}{2}\,(y_0 + 2y_1 + \cdots + 2y_{n-1} + y_n)$$

The definite integral is then approximately equal to $A_T$:

$$\int_a^b y\,dx = \int_a^b f(x)\,dx \approx A_T \qquad (23\text{-}10)$$

Formula (23-10) is called the *trapezoidal rule*. Study the following examples, which show how to apply the trapezoidal rule.

### C Example 23-22

Approximate $\displaystyle\int_0^1 \sqrt{x}\,dx$, using the trapezoidal rule and $n = 4$.

### Solution

The width of each trapezoid is $\Delta x = (1 - 0)/4 = \frac{1}{4} = 0.250$. Each $y$ coordinate is calculated from each $x$ coordinate, by using the function $y = \sqrt{x}$, as follows:

$$
\begin{aligned}
x_0 &= 0 & y_0 &= \sqrt{0} = 0 \\
x_1 &= x_0 + \Delta x = 0.2500 & y_1 &= \sqrt{0.2500} = 0.5000 \\
x_2 &= x_1 + \Delta x = 0.5000 & y_2 &= \sqrt{0.5000} = 0.7071 \\
x_3 &= x_2 + \Delta x = 0.7500 & y_3 &= \sqrt{0.7500} = 0.8660 \\
x_4 &= x_3 + \Delta x = 1.000 & y_4 &= \sqrt{1.000} = 1.000
\end{aligned}
$$

Apply formulas (23-9) and (23-10) to approximate the integral:

$$\int_0^1 \sqrt{x}\, dx \approx A_T$$

$$= \frac{0.250}{2}\, [0 + 2(0.5000) + 2(0.7071) + 2(0.8660) + 1.000]$$

$$= 0.6433$$

This integral may be evaluated directly:

$$\int_0^1 \sqrt{x}\, dx = \left. \tfrac{2}{3}(x)^{3/2} \right|_0^1 = 0.6667$$

The error is therefore $0.6667 - 0.6433 = 0.0234$. The error can be made as small as desired by increasing the number of trapezoids. The trapezoidal rule can be readily programmed on a calculator or a computer.

☐ **Example 23-23**
Approximate

$$\int_2^4 \frac{dx}{\sqrt{1 + x^3}}$$

by using the trapezoidal rule and $n = 5$.

**Solution**
This integral cannot be evaluated by any direct method. The width of each trapezoid is $\Delta x = (4 - 2)/5 = 0.400$. The $x$ and $y$ coordinates are:

$$
\begin{aligned}
x_0 &= 2.000 & y_0 &= 0.3333 \\
x_1 &= 2.400 & y_1 &= 0.2597 \\
x_2 &= 2.800 & y_2 &= 0.2087 \\
x_3 &= 3.200 & y_3 &= 0.1721 \\
x_4 &= 3.600 & y_4 &= 0.1449 \\
x_5 &= 4.000 & y_5 &= 0.1240
\end{aligned}
$$

Apply formulas (23-9) and (23-10) to approximate the integral:

$$\int_2^4 \frac{dx}{\sqrt{1 + x^3}} \approx A_T$$

$$= \frac{0.400}{2}[0.3333 + 2(0.2597) + 2(0.2087) + 2(0.1721)$$

$$+ 2(0.1449) + 0.1240] = 0.4056$$

□ **Example 23-24**

Approximate $\displaystyle\int_0^1 \sin x^2 \, dx$ by using the trapezoidal rule and $n = 10$.

**Solution**

This integral cannot be evaluated by any direct method. The width of each trapezoid is $\Delta x = (1 - 0)/10 = 0.100$. The $x$ and $y$ coordinates are, where $x$ is in radians, as follows:

$$
\begin{aligned}
x_0 &= 0 & y_0 &= \sin 0 = 0 \\
x_1 &= 0.1000 & y_1 &= \sin 0.01000 = 0.01000 \\
x_2 &= 0.2000 & y_2 &= \sin 0.04000 = 0.03999 \\
x_3 &= 0.3000 & y_3 &= \sin 0.09000 = 0.08988 \\
x_4 &= 0.4000 & y_4 &= \sin 0.1600 = 0.1593 \\
x_5 &= 0.5000 & y_5 &= \sin 0.2500 = 0.2474 \\
x_6 &= 0.6000 & y_6 &= \sin 0.3600 = 0.3523 \\
x_7 &= 0.7000 & y_7 &= \sin 0.4900 = 0.4706 \\
x_8 &= 0.8000 & y_8 &= \sin 0.6400 = 0.5972 \\
x_9 &= 0.9000 & y_9 &= \sin 0.8100 = 0.7243 \\
x_{10} &= 1.000 & y_{10} &= \sin 1.000 = 0.8415
\end{aligned}
$$

Apply formulas (23-9) and (23-10) to approximate the integral:

$$\int_0^1 \sin x^2 \, dx \approx A_T$$

$$= \frac{0.10}{2}[0 + 2(0.01000) + 2(0.03999)$$

$$+ \cdots + 2(0.7243) + 0.8415]$$

$$= 0.3112$$

**Exercise 23-5**    In problems 1 to 6, *(a)* approximate each definite integral, using the trapezoidal rule
▬▬▬▬▬    (23-10) and the given value of *n*; *(b)* compare the approximate answer to the answer
□    obtained by direct integration.

1. $\displaystyle\int_0^1 x^2 \, dx; \; n = 4$    2. $\displaystyle\int_0^1 x^3 \, dx; \; n = 4$

3. $\displaystyle\int_1^2 \sqrt{x} \, dx; \; n = 5$    4. $\displaystyle\int_0^2 \sqrt{x + 1} \, dx; \; n = 5$

5. $\displaystyle\int_1^2 \frac{2}{x} \, dx; \; n = 6$    6. $\displaystyle\int_2^3 \frac{dx}{x - 1}; \; n = 6$

$\boxed{C}$ In problems 7 to 16, approximate each definite integral by using the trapezoidal rule (23-10) and the given value of $n$. These integrals cannot be evaluated by any direct method.

**7.** $\int_0^1 \sqrt{2 + x^3}\, dx;\ n = 5$

**8.** $\int_0^1 \dfrac{dx}{\sqrt{1 + x^3}};\ n = 6$

**9.** $\int_2^4 \dfrac{dx}{\sqrt{3 + x^3}};\ n = 5$

**10.** $\int_1^3 \sqrt{3 + 2x^3}\, dx;\ n = 6$

**11.** $\int_0^1 x \tan x\, dx;\ n = 8$

**12.** $\int_{0.1}^{1.1} \dfrac{\sin t}{t}\, dt;\ n = 10$

**13.** $\int_{0.5}^{1.0} \sin x^2\, dx;\ n = 5$

**14.** $\int_0^{1/2} \cos (x^2 + 1)\, dx;\ n = 5$

**15.** $\int_0^2 e^{-x^2}\, dx;\ n = 10$

**16.** $\int_2^4 \dfrac{dx}{\ln x};\ n = 10$

## 23-6 | *Review Questions*

In problems 1 to 16, find each indefinite integral without using the integral table.

**1.** $\int \tan 3x\, dx$

**2.** $\int \sec (2x + 1)\, dx$

**3.** $\int (2 \cot \theta + 1)^2\, d\theta$

**4.** $\int \dfrac{1 + \cos \phi}{\cot \phi}\, d\phi$

**5.** $\int \cos^3 x \sin x\, dx$

**6.** $\int \sin^3 3x\, dx$

**7.** $\int \sec^4 x \tan^2 x\, dx$

**8.** $\int \csc^4 2x\, dx$

**9.** $\int \dfrac{dx}{x^2\sqrt{1 - x^2}}$

**10.** $\int \dfrac{dx}{\sqrt{x^2 + 4}}$

**11.** $\int x\sqrt{x^2 - 9}\, dx$

**12.** $\int \dfrac{2\, dt}{\sqrt{1 - 4t^2}}$

**13.** $\int x \sec^2 x\, dx$

**14.** $\int xe^{-2x}\, dx$

**15.** $\int 2x^2 \sin 2x\, dx$

**16.** $\int \dfrac{\ln t}{t^2}\, dt$

In problems 17 to 22, find each indefinite integral by using the table of integrals, App. B-8.

**17.** $\int \dfrac{4x\, dx}{5 + 2x}$

**18.** $\int \dfrac{dx}{\sqrt{4x^2 + 1}}$

**19.** $\int \dfrac{\sqrt{9x^2 - 4}}{x}\, dx$

**20.** $\int \sin 3\theta \cos 2\theta\, d\theta$

**21.** $\int e^{3t} \cos 2t\, dt$

**22.** $\int \dfrac{dx}{\sqrt{x^3 + x^2}}$

In problems 23 to 26, evaluate each definite integral.

**23.** $\displaystyle\int_0^{\pi/3} \cot\left(\theta + \frac{\pi}{2}\right) d\theta$     [C] **24.** $\displaystyle\int_0^{0.5} 2 \cos^{-1} t \, dt$

**25.** $\displaystyle\int_0^{\pi/2} \sin^3 \theta \, d\theta$     **26.** $\displaystyle\int_0^4 \sqrt{16 - x^2} \, dx$

In problems 27 and 28, approximate each definite integral by using the trapezoidal rule (23-10) and the given value of $n$.

[C] **27.** $\displaystyle\int_0^1 \sqrt{1 + x^4} \, dx; \ n = 5$     [C] **28.** $\displaystyle\int_{1/2}^1 \frac{\sin 2x}{x} \, dx; \ n = 10$

[C] **29.** Find the area bounded by $y = \sec x$, $y = 2$, $x = 0$, and $x = \pi/4$.

[C] **30.** Find the area above the $x$ axis bounded by the hyperbola $x^2 - y^2 = 1$, $y = 0$, and $x = 2$.

[C] **31.** The current in a circuit is given by $I = 3.0(1 + \cos^3 t)$, where $I$ is in amperes and $t$ is in seconds. Find the charge transmitted from $t = 0$ to $t = 1.0$ s.

[C] **32.** The voltage across an inductance $L = 0.10$ H in an electric circuit is given by $V = 0.10/\sqrt{1 + t^2}$ from $t = 0$ to $t = 3.0$ s. If $I = 0$ A when $t = 0$ s, find the current when $t = 3.0$ s. (See Table 21-2, Sec. 21-1.)

**33.** The acceleration of a particle is given by $a = t/\sqrt{t + 4}$. Find the formula for the velocity $v$ of the particle if $v = 0$ when $t = 0$.

**34.** The following integral comes from a problem involving wave motion:

$$\int e^{-at} \cos at \, dt$$

where $a$ is a constant. Find this integral in terms of $a$.

# C H A P T E R

# 24

# Infinite Series

---

## 24-1 | *The Maclaurin Series*

The binomial series (16-16) when $b = -x$ and $n = -1$ becomes:

$$(1 - x)^{-1} = \frac{1}{1 - x} = 1 + x + x^2 + x^3 + \cdots \qquad (24\text{-}1)$$

**E**xamples of infinite series are presented in Chap. 16 in the form of infinite geometric series (Sec. 16-2) and binomial series (Sec. 16-3). Many functions, including transcendental functions, can be expressed in terms of infinite series and can then be evaluated in terms of these series. This chapter explores three types of series and their application to functions.

The above series is a true representation of the function $f(x) = 1/(1 - x)$ when $-1 < x < 1$. For example, when $x = \frac{1}{2}$:

$$f\left(\frac{1}{2}\right) = \frac{1}{1 - \frac{1}{2}} = 2$$

and:

$$1 + x + x^2 + x^3 + \cdots =$$
$$1 + \frac{1}{2} + \frac{1}{4} + \frac{1}{8} + \cdots$$

The last series is an infinite geometric series where the first term $a_1 = 1$ and the ratio $r = \frac{1}{2}$. Applying the formula for the infinite sum $S_\infty$ (16-11) yields:

$$S_\infty = \frac{a_1}{1 - r} = \frac{1}{1 - \frac{1}{2}} = 2$$

The results on both sides of (24-1) therefore agree when $x = \frac{1}{2}$. The series (24-1) is an example of a *power series* which takes the following form for a function $f(x)$:

$$f(x) = a_0 + a_1 x + a_2 x^2 + a_3 x^3 + \cdots \qquad (24\text{-}2)$$

Assuming that a function can be represented by the power series (24-2), the coefficients can be determined as follows by taking successive derivatives:

$$f'(x) = a_1 + 2a_2x + 3a_3x^2 + 4a_4x^3 + \cdots$$
$$f''(x) = 2a_2 + 2(3)a_3x + 3(4)a_4x^2 + 4(5)a_5x^3 + \cdots$$
$$f'''(x) = 2(3)a_3 + 2(3)(4)a_4x + 3(4)(5)a_5x^2 + 4(5)(6)a_6x^3 + \cdots$$

Assuming now that the series (24-2) and the above derivatives are defined for $x = 0$:

$$f(0) = a_0, \quad f'(0) = a_1, \quad f''(0) = 2a_2, \quad f'''(0) = 2(3)a_3, \quad \ldots$$

or $\quad a_0 = f(0), \quad a_1 = f'(0), \quad a_2 = \dfrac{f''(0)}{2!}, \quad a_3 = \dfrac{f'''(0)}{3!}, \quad \ldots$

where $2! = 2 \cdot 1$, $3! = 3 \cdot 2 \cdot 1$, and in general $n! = n(n-1)(n-2) \ldots (1)$. By substituting these expressions for the coefficients, the series (24-2) becomes the *Maclaurin series:*

$$f(x) = f(0) + f'(0)x + \frac{f''(0)}{2!}x^2 + \frac{f'''(0)}{3!}x^3 + \cdots \qquad (24\text{-}3)$$

In the examples that follow, the only functions considered will be those that can be represented by a Maclaurin series (24-3) for certain values of $x$.

### Example 24-1

Find the first four nonzero terms of the Maclaurin series for $f(x) = 1/(1 + x) = (1 + x)^{-1}$.

### Solution

Apply (24-3) and find the coefficients:

$$f(x) = (1 + x)^{-1} \qquad a_0 = f(0) = (1 + 0)^{-1} = 1$$
$$f'(x) = -(1 + x)^{-2} \qquad a_1 = f'(0) = -(1 + 0)^{-2} = -1$$
$$f''(x) = 2(1 + x)^{-3} \qquad a_2 = \frac{f''(0)}{2!} = \frac{2(1 + 0)^{-3}}{2} = 1$$
$$f'''(x) = -6(1 + x)^{-4} \qquad a_3 = \frac{f'''(0)}{3!} = \frac{-6(1 + 0)^{-4}}{6} = -1$$

The Maclaurin series is then:

$$\frac{1}{1 + x} = 1 - x + x^2 - x^3 + \cdots$$

The above series is an example of an *alternating series*. This series represents the function when $-1 < x < 1$. It requires more advanced methods to determine the range of values for which the series is valid. It will therefore be assumed that a series is valid for all real values of $x$ unless otherwise specified.

## Example 24-2
Find the first four nonzero terms of the Maclaurin series for $f(x) = e^x$.

## Solution
Apply series (24-3) and find the coefficients:

$$f(x) = e^x \qquad a_0 = f(0) = e^0 = 1$$
$$f'(x) = e^x \qquad a_1 = f'(0) = e^0 = 1$$
$$f''(x) = e^x \qquad a_2 = \frac{f''(0)}{2!} = \frac{e^0}{2!} = \frac{1}{2!}$$
$$f'''(x) = e^x \qquad a_3 = \frac{f'''(0)}{3!} = \frac{e^0}{3!} = \frac{1}{3!}$$

The Maclaurin series is then:

$$e^x = 1 + x + \frac{x^2}{2!} + \frac{x^3}{3!} + \cdots$$

This series is valid for all real values of $x$.

## Example 24-3
Find the first four nonzero terms of the Maclaurin series for $f(x) = \cos x$.

## Solution
Apply series (24-3). *It is necessary to find seven coefficients* to obtain the first four nonzero terms since three coefficients are zero:

$$f(x) = \cos x \qquad a_0 = f(0) = 1$$
$$f'(x) = -\sin x \qquad a_1 = f'(0) = 0$$
$$f''(x) = -\cos x \qquad a_2 = \frac{f''(0)}{2!} = -\frac{1}{2!}$$
$$f'''(x) = \sin x \qquad a_3 = \frac{f'''(0)}{3!} = 0$$
$$f^4(x) = \cos x \qquad a_4 = \frac{f^4(0)}{4!} = \frac{1}{4!}$$

$$f^5(x) = -\sin x \qquad a_5 = \frac{f^5(0)}{5!} = 0$$

$$f^6(x) = -\cos x \qquad a_6 = \frac{f^6(0)}{6!} = -\frac{1}{6!}$$

The Maclaurin series is then:

$$\cos x = 1 - \frac{x^2}{2!} + \frac{x^4}{4!} - \frac{x^6}{6!} + \cdots$$

This series is valid for all real values of $x$.

### Example 24-4

Find the first four nonzero terms of the Maclaurin series for $f(x) = \ln (1 + x)$.

### Solution

Apply series (24-3):

$$f(x) = \ln (1 + x) \qquad a_0 = f(0) = \ln 1 = 0$$
$$f'(x) = (1 + x)^{-1} \qquad a_1 = f'(0) = 1$$

$$f''(x) = -(1 + x)^{-2} \qquad a_2 = \frac{f''(0)}{2!} = -\frac{1}{2}$$

$$f'''(x) = 2(1 + x)^{-3} \qquad a_3 = \frac{f'''(0)}{3!} = \frac{1}{3}$$

$$f^4(x) = -6(1 + x)^{-4} \qquad a_4 = \frac{f^4(0)}{4!} = -\frac{1}{4}$$

The Maclaurin series is then:

$$\ln (1 + x) = x - \frac{x^2}{2} + \frac{x^3}{3} - \frac{x^4}{4} + \cdots$$

This series is valid for $-1 < x < 1$.

Examples 24-2, 24-3, and 24-4 and the series for $\sin x$ are important basic series which are useful in evaluating transcendental functions. They are listed below for reference. Verification of the series for $\sin x$ is left as problem 5:

$$e^x = 1 + x + \frac{x^2}{2!} + \frac{x^3}{3!} + \cdots \qquad (24\text{-}4)$$

$$\sin x = x - \frac{x^3}{3!} + \frac{x^5}{5!} - \frac{x^7}{7!} + \cdots \qquad (24\text{-}5)$$

$$\cos x = 1 - \frac{x^2}{2!} + \frac{x^4}{4!} - \frac{x^6}{6!} + \cdots \qquad (24\text{-}6)$$

$$\ln(1+x) = x - \frac{x^2}{2} + \frac{x^3}{3} - \frac{x^4}{4} + \cdots \qquad (-1 < x < 1) \qquad (24\text{-}7)$$

The first few terms of a Maclaurin series provide a good approximation to a function for small values of $x$. The series is therefore said to *converge rapidly* for small values of $x$. The next example illustrates how the series is used to evaluate transcendental functions and can be used to construct tables or program a calculator or a computer.

### C Example 24-5

Approximate the value of $e$ by evaluating the first five terms of the Maclaurin series for $e^x$.

**Solution**

Let $x = 1$ in (24-4):

$$e \approx 1 + 1 + \frac{1}{2!} + \frac{1}{3!} + \frac{1}{4!}$$

$$\approx 1 + 1 + \frac{1}{2} + \frac{1}{6} + \frac{1}{24}$$

$$\approx 2.71$$

Greater accuracy can, of course, be obtained by using more terms of the series.

### Exercise 24-1

In problems 1 to 18, find the first four nonzero terms of the Maclaurin series for each function. Each series is valid for all real values of $x$ unless indicated otherwise.

1. $f(x) = e^{2x}$                              2. $f(x) = e^{-x}$

3. $f(x) = \dfrac{1}{(1+x)^2}$    $(-1 < x < 1)$       4. $f(x) = \dfrac{1}{2-x}$    $(-2 < x < 2)$

5. $f(x) = \sin x$                           6. $f(x) = \cos 2x$

7. $f(x) = \ln(1-x)$    $(-1 < x < 1)$       8. $f(x) = \ln(e+x)$    $(-e < x < e)$

9. $f(x) = \tan^{-1} x$    $(-1 < x < 1)$       10. $f(x) = \sin^{-1} x$    $(-1 < x < 1)$

11. $f(x) = \frac{1}{2}(e^x - e^{-x})$                 12. $f(x) = \frac{1}{2}(e^x + e^{-x})$

13. $f(x) = xe^x$                             14. $f(x) = e^x \cos x$

15. $f(x) = \ln \cos x$    $(-\pi/4 < x < \pi/4)$     16. $f(x) = \sqrt{1+x}$    $(-1 < x < 1)$

17. $f(x) = e^{\sin x}$                          18. $f(x) = \tan x$    $(-\pi/4 < x < \pi/4)$

C In problems 19 to 22, approximate the value of each function by evaluating the first five terms of its Maclaurin series.

**19.** $\sqrt{e}$

**20.** $\ln 1.1$

**21.** $\sin \dfrac{\pi}{20}$

**22.** $\cos 4°$   (Change angle to radians.)

**23.** Explain why $f(x) = 1/x$ cannot be represented by a Maclaurin series.

**24.** Explain why $f(x) = \ln x$ cannot be represented by a Maclaurin series.

**25.** Obtain the Maclaurin series for $1/(1 + x)$ (Example 24-1) by dividing $1 + x$ into 1, using polynomial division (see Sec. 15-2).

C **26.** Using the first four terms of the series for:

$$\sin^{-1} x = x + \frac{x^3}{6} + \frac{1}{2}\left(\frac{3}{4}\right)\left(\frac{x^5}{5}\right) + \frac{1}{2}\left(\frac{3}{4}\right)\left(\frac{5}{6}\right)\left(\frac{x^7}{7}\right)$$

approximate the value of $\pi$ by letting $x = 0.5$.

## 24-2 | *Operations with Series*

This section studies various operations one can perform with known series to find new series. Some other applications of series are also studied. The first example illustrates how a new series is found by a change of variable in one of the basic series (24-4) to (24-7) given in the last section.

### Example 24-6
Find the Maclaurin series for $\sin 2x$ by using the series for $\sin x$.

### Solution
Using the series for $\sin x$, (24-5), you can change the variable and write the series for $\sin u$:

$$\sin u = u - \frac{u^3}{3!} + \frac{u^5}{5!} - \frac{u^7}{7!} + \cdots$$

Let $u = 2x$ to obtain the series for $\sin 2x$:

$$\sin 2x = 2x - \frac{(2x)^3}{3!} + \frac{(2x)^5}{5!} - \frac{(2x)^7}{7!} + \cdots$$

or        $$\sin 2x = 2x - \frac{4x^3}{3} + \frac{4x^5}{3 \cdot 5} - \frac{8x^7}{3 \cdot 3 \cdot 5 \cdot 7} + \cdots$$

This example can also be done simply by replacing $x$ by $2x$ in the series for $\sin x$.

Series may be added, subtracted, multiplied, or divided to obtain new series, as shown in the next example.

### Example 24-7
Find the Maclaurin series for $e^x \sin x$ by using the series for $e^x$ and $\sin x$.

### Solution
Multiply the series for $e^x$, (24-4), and $\sin x$, (24-5), as follows:

$$e^x \sin x = \left(1 + x + \frac{x^2}{2!} + \frac{x^3}{3!} + \frac{x^4}{4!} + \cdots\right)\left(x - \frac{x^3}{3!} + \frac{x^5}{5!} - \frac{x^7}{7!} + \cdots\right)$$

Choose the combinations of products that give you the first four nonzero terms (up to $x^5$ since there is no term for $x^4$):

$$e^x \sin x = 1(x) + x(x)$$

$$+ \left[\frac{x^2}{2!}(x) + 1\left(-\frac{x^3}{3!}\right)\right]$$

$$+ \left[x\left(-\frac{x^3}{3!}\right) + \frac{x^3}{3!}(x)\right]$$

$$+ \left[1\left(\frac{x^5}{5!}\right) + \frac{x^2}{2!}\left(-\frac{x^3}{3!}\right) + \frac{x^4}{4!}(x)\right] + \cdots$$

Therefore:

$$e^x \sin x = x + x^2 + \frac{x^3}{3} - \frac{x^5}{30} + \cdots$$

The following example shows an application of series to a problem in electricity.

### Example 24-8
In a circuit the voltage induced in a coil is given by $V = (e^{t/2} - 1)/t$ for an interval of time $t$.

**1.** Express $V$ in terms of a Maclaurin series.
**2.** Using the Maclaurin series, find $V$ when $t = 0.01$ s.

## Solution

1. Express $V$ in terms of a Maclaurin series.

Use the series for $e^x$ to obtain:

$$V = \frac{[1 + t/2 + (t/2)^2/2! + (t/2)^3/3! + (t/2)^4/4!] - 1}{t}$$

$$= \frac{1}{2} + \frac{t}{8} + \frac{t^2}{48} + \frac{t^3}{384} + \cdots$$

2. Using the Maclaurin series, find $V$ when $t = 0.01$ s:

$$V = \frac{1}{2} + \frac{0.01}{8} + \frac{(0.01)^2}{48} + \frac{(0.01)^3}{384} + \cdots = 0.5013 \text{ V}$$

(4 significant figures)

Series can be differentiated or integrated term by term to obtain new series. Integration of series is also used to evaluate integrals that are difficult or cannot be evaluated by any direct method. The next three examples illustrate these techniques.

### Example 24-9

Find the Maclaurin series for $1/(1 + x)$ by differentiating the series for $\ln (1 + x)$ term by term.

### Solution

The series for $\ln (1 + x)$, (24-7), is:

$$\ln (1 + x) = x - \frac{x^2}{2} + \frac{x^3}{3} - \frac{x^4}{4} + \cdots$$

Differentiate $\ln (1 + x)$ and the series term by term to obtain:

$$\frac{1}{1 + x} = 1 - x + x^2 - x^3 + \cdots$$

###  Example 24-10

Evaluate $\displaystyle\int_0^\pi x \cos x \, dx$ by expressing the function as a Maclaurin series.

### Solution

Multiply the series for cos $x$, (24-6), by $x$ to obtain the series for $x \cos x$. Then integrate the series term by term:

$$\int_0^\pi x \cos x \, dx = \int_0^\pi x \left( 1 - \frac{x^2}{2!} + \frac{x^4}{4!} - \frac{x^6}{6!} + \cdots \right) dx$$

$$= \int_0^\pi \left( x - \frac{x^3}{2!} + \frac{x^5}{4!} - \frac{x^7}{6!} + \cdots \right) dx$$

$$= \left( \frac{x^2}{2} - \frac{x^4}{8} + \frac{x^6}{144} - \frac{x^8}{5760} + \cdots \right) \Big|_0^\pi$$

$$= \frac{\pi^2}{2} - \frac{\pi^4}{8} + \frac{\pi^6}{144} - \frac{\pi^8}{5760} + \cdots$$

$$= -2.001 \qquad \text{(4 significant figures)}$$

### $\boxed{\text{C}}$ Example 24-11

Evaluate $\displaystyle\int_0^{0.5} e^{-x^2} \, dx$ by expressing the function as a Maclaurin series.

### Solution

This integral cannot be evaluated by any direct method. Substitute $-x^2$ for $x$ in the series for $e^x$, (24-4), and integrate term by term:

$$\int_0^{0.5} e^{-x^2} \, dx = \int_0^{0.5} \left( 1 - x^2 + \frac{x^4}{2!} - \frac{x^6}{3!} + \cdots \right) dx$$

$$= \left( x - \frac{x^3}{3} + \frac{x^5}{10} - \frac{x^7}{42} + \cdots \right) \Big|_0^{0.5} = 0.4613 \qquad \text{(4 significant figures)}$$

## The Exponential $e^{j\theta}$

In Chap. 13, Sec. 13-5, the exponential form of a complex number is introduced:

$$e^{j\theta} = \cos \theta + j \sin \theta \qquad (24\text{-}8)$$

With the aid of the series expansions for $e^x$, $\sin x$, and $\cos x$, the reason for this definition can now be seen. In more advanced courses, the series for $e^x$, $\sin x$, and $\cos x$ are shown to be true for complex numbers. You can then write:

$$e^{j\theta} = 1 + j\theta + \frac{(j\theta)^2}{2!} + \frac{(j\theta)^3}{3!} + \frac{(j\theta)^4}{4!} + \frac{(j\theta)^5}{5!} + \cdots$$

Grouping the real and the imaginary terms, you obtain the two series:

$$e^{j\theta} = \left(1 - \frac{\theta^2}{2!} + \frac{\theta^4}{4!} - \cdots\right) + j\left(\theta - \frac{\theta^3}{3!} + \frac{\theta^5}{5!} - \cdots\right)$$

Therefore $e^{j\theta} = \cos\theta + j\sin\theta$. Note when $\theta = \pi$ in (24-8), the remarkable mathematical relationship between $e$, $\pi$, and $j$ results:

$$e^{j\pi} = -1 \qquad\qquad (24\text{-}9)$$

**Exercise 24-2**

In problems 1 to 6, find the first four nonzero terms of the Maclaurin series for each function by changing the variable in one of the basic series (24-4) to (24-7).

1. $\cos(x/2)$                    2. $\sin x^2$

3. $e^{4x}$                        4. $e^{-2t}$

5. $\ln(1 + 2x)$                  6. $\ln(1 - x)$

In problems 7 to 12, find the first four nonzero terms of the Maclaurin series for each function by performing the indicated operations on the basic series.

7. $x \sin x$                     8. $\dfrac{\ln(1 + x)}{x}$

9. $e^x \cos x$                   10. $e^{-x} \sin x$

11. $\sin^2 x$                    12. $te^{t/2}$

In problems 13 to 18, evaluate each integral by using the first four terms of the Maclaurin series for each function.

13. $\displaystyle\int_0^{\pi/2} \frac{\sin x}{x}\, dx$     14. $\displaystyle\int_0^1 \cos x^2\, dx$

15. $\displaystyle\int_0^{0.5} xe^x\, dx$                   16. $\displaystyle\int_0^{0.2} e^{-t^2}\, dt$

17. $\displaystyle\int_0^{0.3} x \ln(1 - x)\, dx$          18. $\displaystyle\int_0^{0.5} \frac{1 - \cos x}{x}\, dx$

19. Show that by differentiating the series for $\sin x$ term by term the series for $\cos x$ is obtained.

20. Find the series for $2x/(1 + x^2)$ by differentiating the series for $\ln(1 + x^2)$ term by term.

21. Show that by integrating the series for $e^x$ term by term the result is also the series for $e^x$. What is the value of the constant of integration?

22. Find the first three nonzero terms of the series for $\tan x$ by dividing the series for $\sin x$ by the series for $\cos x$. Apply polynomial division to the first three terms of each series.

23. The voltage in a circuit is given by $V = 0.3e^{0.3t}$. Express $V$ in terms of a Maclaurin series.

24. The current in an $RC$ circuit is given by $I = 0.4e^{-0.1t}$. *(a)* Express $I$ in terms of a Maclaurin series. *(b)* Using the first four terms of the series, find $I$ when $t = 0.2$ s.

25. The charge on a capacitor in millicoulombs (mC) is given by the integral:

$$Q = \int_0^{1.0} (1.0 - e^{0.1t^2})\, dt$$

Evaluate this integral, using the first four terms of the Maclaurin series for this function.

26. The current in milliamperes (mA) flowing to a capacitor during a short time interval $t$ is given by $I = e^{-0.2t^2}$. Using a series expansion, find the accumulation of charge on the capacitor from $t = 1$ to $t = 2$ s. (See Example 21-18, Sec. 21-4.)

27. *(a)* The *hyperbolic sine,* a function important in engineering, is given by $\sinh x = \frac{1}{2}(e^x - e^{-x})$. Find the Maclaurin series for this function by combining two series. *(b)* Do the same for the *hyperbolic cosine* $\cosh x = \frac{1}{2}(e^x + e^{-x})$. (See problem 52, Sec. 22-4.)

28. *(a)* Using the formula for the exponential $e^{j\theta}$, (24-8), show that $\sin y = (1/2j)(e^{jy} - e^{-jy})$. *(b)* Do the same for $\cos y = \frac{1}{2}(e^{jy} + e^{-jy})$.

## 24-3 | *The Taylor Series*

The Taylor series is a more general type of power series and includes the Maclaurin series as a special case. The Maclaurin series uses zero as a point of reference and converges rapidly for values of $x$ near $x = 0$. The Taylor series uses any value $a$ as a point of reference and converges rapidly for values of $x$ near $x = a$. The *Taylor series* assumes that a function $f(x)$ can be represented by a power series of the form:

$$f(x) = a_0 + a_1(x - a) + a_2(x - a)^2 + a_3(x - a)^3 + \cdots \quad (24\text{-}10)$$

By taking successive derivatives of (24-10) and evaluating them at $x = a$ (similar to the procedure shown for the Maclaurin series), the coefficients can be found. The coefficients resemble those of the Maclaurin series where 0 is replaced by $a$:

$$f(x) = f(a) + f'(a)(x - a) + \frac{f''(a)}{2!}(x - a)^2 + \frac{f'''(a)}{3!}(x - a)^3 + \cdots \quad (24\text{-}11)$$

The Taylor series (24-11) can be used to express certain functions that cannot be expressed by a Maclaurin series. It can also be used to evaluate functions at values for which the Maclaurin series converges too slowly. The following examples illustrate some applications.

### Example 24-12

**1.** Expand $f(x) = \ln x$ in a Taylor series about $a = 1$.

**2.** Approximate the value of $\ln 0.8$ by evaluating the first three terms of the Taylor series.

### Solution

**1.** Expand $f(x) = \ln x$ in a Taylor series about $a = 1$.

Apply series (24-11):

$$f(x) = \ln x \qquad a_0 = f(1) = \ln 1 = 0$$

$$f'(x) = \frac{1}{x} \qquad a_1 = f'(1) = 1$$

$$f''(x) = -\frac{1}{x^2} \qquad a_2 = \frac{f''(1)}{2!} = -\frac{1}{2}$$

$$f'''(x) = \frac{2}{x^3} \qquad a_3 = \frac{f'''(1)}{3!} = \frac{2}{6} = \frac{1}{3}$$

The Taylor series is then:

$$\ln x = (x - 1) - \frac{(x-1)^2}{2} + \frac{(x-1)^3}{3} - \cdots$$

This series represents the function, that is, converges, when $0 < x \le 2$. Note that $\ln x$ cannot be expanded in a Maclaurin series because $f'(0) = 1/0$ is not defined.

**2.** Approximate the value of $\ln 0.8$ by evaluating the first three terms of the Taylor series.

Let $x = 0.8$ in the above series:

$$\ln 0.8 \approx (0.8 - 1) - \frac{(0.8-1)^2}{2} + \frac{(0.8-1)^3}{3}$$

$$\approx -0.2 - 0.02 - 0.0027 = -0.223$$

Notice that the terms in the series decrease rapidly since $0.8$ is close to $1$.

### Example 24-13

**1.** Expand $\sin x$ in a Taylor series about any point $x = a$.

**2.** Approximate the value of $\sin 62°$ to four significant figures by evaluating the first four terms of the Taylor series.

## Solution

**1.** Expand $\sin x$ in a Taylor series about any point $x = a$.

Apply series (24-11):

$$f(x) = \sin x \qquad a_0 = f(a) = \sin a$$
$$f'(x) = \cos x \qquad a_1 = f'(a) = \cos a$$
$$f''(x) = -\sin x \qquad a_2 = \frac{f''(a)}{2!} = \frac{-\sin a}{2!}$$
$$f'''(x) = -\cos x \qquad a_3 = \frac{f'''(a)}{3!} = \frac{-\cos a}{3!}$$

The Taylor series is then:

$$\sin x = \sin a + (\cos a)(x - a) - \frac{(\sin a)(x - a)^2}{2!} - \frac{(\cos a)(x - a)^3}{3!} + \cdots$$

**2.** Approximate the value of $\sin 62°$ to four significant figures by evaluating the first four terms of the Taylor series.

Choose the reference point $a = 60°$, which is close to $x = 62°$ and is one of the special angles for which the sine and cosine are found without a table or calculator. Note that $a$ must be in radians and $x - a = 62° - 60° = 2° = \pi/90$ rad. Then:

$$\sin 62° \approx \sin \frac{\pi}{3} + \left(\cos \frac{\pi}{3}\right)\left(\frac{\pi}{90}\right) - \frac{[\sin (\pi/3)](\pi/90)^2}{2} - \frac{[\cos (\pi/3)](\pi/90)^3}{6}$$
$$= 0.86603 + 0.01745 - 0.00053 - 0.000004 = 0.8829$$

**Exercise 24-3**

In problems 1 to 8, expand each function in a Taylor series about the given value of $a$ by finding the first three nonzero terms.

**1.** $1/x$, $a = 1$         **2.** $\sqrt{x}$, $a = 4$

**3.** $e^x$, $a = 1$         **4.** $e^{-x}$, $a = 1$

**5.** $\cos x$, $a = \pi/3$     **6.** $\sin x$, $a = \pi/6$

**7.** $\ln x$, $a = 2$         **8.** $\tan x$, $a = \pi/4$

In problems 9 to 16, use the first three terms of a Taylor series developed in the text or in problems 1 to 8 to approximate the given value of each function.

**9.** $\ln 1.5$             **10.** $\sqrt{3.9}$

**11.** $e^{0.9}$            **12.** $e^{-1.2}$

**13.** $\cos 58°$          **14.** $\sin 31°$

**15.** $\sin 59°$          **16.** $\tan 42°$

**17.** *(a)* Expand $\cos x$ in a Taylor series about any point $x = a$. *(b)* Approximate the value of $\cos 47°$ by evaluating the first four terms of the series.

**18.** *(a)* Expand ln $x$ in a Taylor series about any point $x = a$. *(b)* Approximate ln 3 by evaluating the first four terms of the series. (*Hint*: Let $a = e \approx 2.718$.)

## 24-4 | *The Fourier Series*

The Fourier series and Fourier analysis are a remarkable result of mathematical research and very useful tools in wave analysis. They are named for J. B. Fourier, a French mathematician, who published his results in 1812.

The Fourier series is a trigonometric series which finds application in engineering problems involving the analysis of waves, such as mechanical vibrations, and electrical voltages and currents. The functions represented by a Fourier series are periodic functions; that is, they repeat after a certain interval, as do the trigonometric functions. A *Fourier series* for a periodic function $f(x)$ takes the following form:

$$f(x) = a_0 + (a_1 \cos x + a_2 \cos 2x + \cdots)$$
$$+ (b_1 \sin x + b_2 \sin 2x + \cdots) \qquad (24\text{-}12)$$

The coefficients are equal to the integrals:

$$a_0 = \frac{1}{2\pi} \int_0^{2\pi} f(x)\, dx$$

$$a_n = \frac{1}{\pi} \int_0^{2\pi} f(x) \cos nx\, dx \qquad n = 1, 2, 3, \ldots \qquad (24\text{-}13)$$

$$b_n = \frac{1}{\pi} \int_0^{2\pi} f(x) \sin nx\, dx \qquad n = 1, 2, 3, \ldots$$

The coefficients are found by using methods of integration applied to the series (24-12). Their derivation is omitted for the sake of brevity.

The following two examples show how to find the Fourier series for a function. The second is an application of Fourier analysis to a problem in electronics.

### Example 24-14

Given the periodic wave function:

$$f(x) = \begin{cases} 1 & \text{for } 0 \le x < \pi \\ -1 & \text{for } \pi \le x < 2\pi \end{cases}$$

and

$$f(x + 2\pi) = f(x)$$

find the Fourier series for this function.

## Solution

This function is a discontinuous function and has the value 1 when $0 \leq x < \pi$ and the value $-1$ when $\pi \leq x < 2\pi$. It is periodic and repeats every $2\pi$ rad. The graph of three periods of this function is shown in Fig. 24-1. Functions of this type are found in square wave voltages and currents. Apply (24-13) to find the coefficients. It is necessary to evaluate two integrals for each coefficient because the function is defined differently for the first half of the cycle than for the second half of the cycle:

$$a_0 = \frac{1}{2\pi}\left[\int_0^\pi (1)\, dx + \int_\pi^{2\pi} (-1)\, dx\right] = \frac{1}{2\pi}(\pi - \pi) = 0$$

$$a_1 = \frac{1}{\pi}\left(\int_0^\pi \cos x\, dx + \int_\pi^{2\pi} -\cos x\, dx\right)$$

$$= \frac{1}{\pi}\left(\sin x\Big|_0^\pi - \sin x\Big|_\pi^{2\pi}\right) = 0$$

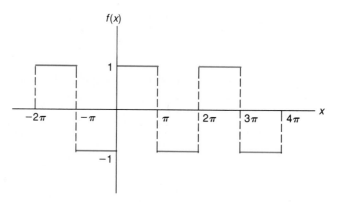

FIG. 24-1  *Square wave function, Example 24-14*

Similarly, all the $a_n$'s equal zero:

$$a_n = \frac{1}{\pi}\left(\int_0^\pi \cos nx\, dx + \int_\pi^{2\pi} -\cos nx\, dx\right)$$

$$= \frac{1}{\pi}\left(\frac{\sin nx}{n}\Big|_0^\pi - \frac{\sin nx}{n}\Big|_\pi^{2\pi}\right) = 0$$

Therefore the Fourier series has no cosine terms. The $b_n$'s are:

$$b_n = \frac{1}{\pi}\left(\int_0^\pi \sin nx\, dx + \int_\pi^{2\pi} - \sin nx\, dx\right)$$

$$= \frac{1}{\pi}\left(-\frac{\cos nx}{n}\Big|_0^\pi + \frac{\cos nx}{n}\Big|_\pi^{2\pi}\right)$$

When $n$ is odd $= 1, 3, 5, \ldots$ :

$$b_n = \frac{1}{\pi}\left[-\left(-\frac{1}{n} - \frac{1}{n}\right) + \left(\frac{1}{n} + \frac{1}{n}\right)\right] = \frac{4}{n\pi}$$

When $n$ is even $= 2, 4, 6, \ldots$ :

$$b_n = \frac{1}{\pi}\left[-\left(\frac{1}{n} - \frac{1}{n}\right) + \left(\frac{1}{n} - \frac{1}{n}\right)\right] = 0$$

The Fourier series for $f(x)$ is therefore:

$$f(x) = \frac{4}{\pi}\left(\sin x + \frac{\sin 3x}{3} + \frac{\sin 5x}{5} + \cdots\right)$$

The graph of the first three terms of the series is shown in Fig. 24-2. With only three terms of the series considered, the graph approaches the shape of a square wave.

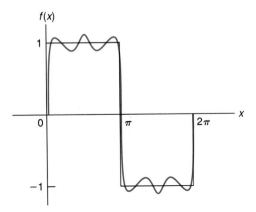

FIG. 24-2 *Synthesis of square wave function, Example 24-14*

## Fourier Analysis

*Fourier analysis* involves finding the following relative amounts of the components of a waveform:

1. The *average, or dc, component* equals the coefficient $a_0$ in the Fourier series (24-12).

2. The *fundamental sinusoidal component* equals the sum of the terms whose coefficients are $a_1$ and $b_1$ in the Fourier series:

$$a_1 \cos x + b_1 \sin x$$

**3.** The infinite series of *sinusoidal harmonics* equals the remaining terms of the Fourier series:

$$(a_2 \cos 2x + a_3 \cos 3x + \cdots) + (b_2 \sin 2x + b_3 \sin 3x + \cdots)$$

Thus, finding the Fourier series for a waveform is equivalent to the Fourier analysis of the waveform. The next example illustrates this analysis.

### Example 24-15

In a certain electronic circuit, the current takes the form of a sawtooth wave:

$$i = f(t) = t \qquad 0 \le t < 2\pi \qquad \text{and} \qquad f(t + 2\pi) = f(t)$$

Perform a Fourier analysis of this waveform.

### Solution

The sawtooth wave is shown in Fig. 24-3. The average, or dc, component of the wave is:

$$a_0 = \frac{1}{2\pi} \int_0^{2\pi} t \, dt = \pi$$

The fundamental sinusoidal component is computed as follows:

$$a_1 = \frac{1}{\pi} \int_0^{2\pi} t \cos t \, dt = \frac{1}{\pi}(\cos t + t \sin t) \Big|_0^{2\pi} = 0$$

$$b_1 = \frac{1}{\pi} \int_0^{2\pi} t \sin t \, dt = \frac{1}{\pi}(\sin t - t \cos t) \Big|_0^{2\pi} = -2$$

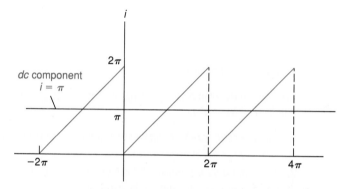

FIG. 24-3 *Sawtooth wave with dc component of $\pi$, Example 24-15*

These integrals are evaluated by using integration by parts or formulas 50 and 51 in App. B-8. The fundamental sinusoidal component is then:

$$b_1 \sin t = -2 \sin t$$

The sinusoidal harmonics are computed as follows:

$$a_n = \frac{1}{\pi} \int_0^{2\pi} t \cos nt \, dt = \frac{1}{\pi} \left( \frac{\cos nt}{n^2} + \frac{t \sin nt}{n} \right) \Big|_0^{2\pi} = 0$$

$$b_n = \frac{1}{\pi} \int_0^{2\pi} t \sin nt \, dt = \frac{1}{\pi} \left( \frac{\sin nt}{n^2} - \frac{t \cos nt}{n} \right) \Big|_0^{2\pi} = -\frac{2}{n}$$

These integrals are evaluated in a similar way as those for $a_1$ and $b_1$. The sinusoidal harmonics are then:

$$b_2 \sin 2t + b_3 \sin 3t + b_4 \sin 4t + \cdots = -\sin 2t - \frac{2}{3}\sin 3t - \frac{1}{2}\sin 4t - \cdots$$

The Fourier series for the current $i$ is therefore:

$$i = \pi - 2 \sin t - \sin 2t - \frac{2}{3}\sin 3t - \frac{1}{2}\sin 4t - \cdots$$

or
$$i = \pi - 2 \left( \sin t + \frac{1}{2}\sin 2t + \frac{1}{3}\sin 3t + \frac{1}{4}\sin 4t + \cdots \right)$$

## Exercise 24-4

In problems 1 to 6, find the Fourier series for each periodic function where $f(x + 2\pi) = f(x)$, and sketch the graph of each function.

**1.** $f(x) = \begin{cases} 2 & 0 \leq x < \pi \\ -2 & \pi \leq x < 2\pi \end{cases}$
    **2.** $f(x) = \begin{cases} 1 & 0 \leq x < \pi \\ 0 & \pi \leq x < 2\pi \end{cases}$

**3.** $f(x) = \dfrac{x}{2} \quad 0 \leq x < 2\pi$
    **4.** $f(x) = x - \pi \quad 0 \leq x < 2\pi$

**5.** $f(x) = x^2 \quad 0 \leq x < 2\pi$
    **6.** $f(x) = e^x \quad 0 \leq x < 2\pi$

In problems 7 to 10 perform a Fourier analysis of each waveform and sketch the graph of each function, where $f(t + 2\pi) = f(t)$.

**7.** $f(t) = \begin{cases} \pi & 0 \leq t < \pi \\ 0 & \pi \leq t < 2\pi \end{cases}$
    **8.** $f(t) = \begin{cases} \pi & 0 \leq t < \pi/2 \\ -\pi & \pi/2 \leq t < 3\pi/2 \\ \pi & 3\pi/2 \leq t < 2\pi \end{cases}$

**9.** $f(t) = 2t \quad 0 \leq t < 2\pi$
    **10.** $f(t) = \begin{cases} t & 0 \leq t < \pi \\ 0 & \pi \leq t < 2\pi \end{cases}$

**11.** A *half-wave rectifier* in an electronic circuit produces the following sinusoidal current wave:

$$i = f(t) = \begin{cases} \sin t & 0 \le t < \pi \\ 0 & \pi \le t < 2\pi \end{cases}$$

Sketch this function, where $f(t + 2\pi) = f(t)$, and perform a Fourier analysis of the waveform.

**12.** Do the same as problem 11 for the sinusoidal current wave produced by a *full-wave rectifier*, where $f(t + 2\pi) = f(t)$ and:

$$i = f(t) = \begin{cases} \sin t & 0 \le t < \pi \\ -\sin t & \pi \le t < 2\pi \end{cases}$$

## 24-5 | *Review Questions*

In problems 1 to 8, find the first four nonzero terms of the Maclaurin series for each function.

**1.** $e^{3x}$

**2.** $1/(1 - x)^2$

**3.** $\sin 3x$

**4.** $\cos t^2$

**5.** $\ln (1 + \sqrt{x})$

**6.** $xe^x$

**7.** $e^{-t} \cos t$

**8.** $\cos^2 x$

In problems 9 and 10, expand each function in a Taylor series about the given value of $a$ by finding the first three nonzero terms.

**9.** $\ln x$, $a = \frac{1}{2}$

**10.** $\sin x$, $a = \pi/4$

In problems 11 and 12, find the Fourier series for each periodic function, where $f(x + 2\pi) = f(x)$, and sketch the graph of each function.

**11.** $f(x) = \begin{cases} 2 & 0 \le x < \pi \\ -1 & \pi \le x < 2\pi \end{cases}$

**12.** $f(x) = \begin{cases} 2x & 0 \le x < \pi \\ 0 & \pi \le x < 2\pi \end{cases}$

In problems 13 and 14, approximate the given value of each function, using the first three terms of an appropriate series.

**13.** $\ln 0.9$

**14.** $\sin 44°$

In problems 15 and 16, evaluate each integral by using the first four terms of the series for each function.

**15.** $\displaystyle\int_0^{\pi/2} x \cos x \, dx$

**16.** $\displaystyle\int_0^1 \frac{e^x - 1}{x} \, dx$

**17.** The voltage induced in a coil is given by $V = t(e^{2t} - 1)$ for an interval of time $t$. *(a)* Express $V$ in terms of a Maclaurin series. *(b)* Using the first four terms of the series, find $V$ when $t = 0.5$ s.

**18.** Show that by integrating the Maclaurin series for $\cos x$ the series for $\sin x$ is obtained.

# CHAPTER

# 25

# Introduction to Differential Equations

## 25-1 | *Differential Equations*

The *order* of a differential equation is the order of the highest derivative occurring in the equation. The *degree* of a differential equation is the power of the highest-order derivative in the equation. Study the following example.

**A** *differential equation* **is an equation which contains derivatives or differentials. Differential equations are important in engineering and science because they occur in many applications. This chapter shows how to solve some basic types of differential equations and examines some of their applications.**
**The first section begins with some of the fundamental ideas concerning differential equations and their solutions.**

**Example 25-1**
List some examples of differential equations.

**1.** $\dfrac{dy}{dx} + 2 = x^2$

This equation is a *first-order* differential equation of the *first degree*, or a *linear differential equation of the first order*.

**2.** $x - \dfrac{d^2y}{dx^2} = y$

This equation is a *second-order* differential equation of the *first degree*, or a *linear differential equation of the second order*.

**3.** $L \dfrac{d^2q}{dt^2} + R \dfrac{dq}{dt} = 5$

This equation is a *second-order* differential equation of the *first degree*, or a *linear differential equation of the second order*. A second-order equation may also contain first-order derivatives.

**4.** $(y'')^2 - y' = 0$

This equation is a *second-order* differential equation of the *second degree*.

**5.** $y''' + (y')^2 = y + 3$

This equation is a *third-order* differential equation of the *first degree*. The second power of the first derivative does not affect the degree of the equation since it is not the highest-order derivative.

This chapter studies linear differential equations of the first and second orders.

A *solution* of a differential equation is a relation between the variables that does not contain any derivatives or differentials and satisfies the differential equation, as shown in the following examples.

## Example 25-2

Show that $y = x^2 + cx$, where $c$ is an arbitrary constant, is a solution of the differential equation:

$$xy' - x^2 = y$$

**Solution**

If $y = x^2 + cx$, then $y' = 2x + c$. Substituting into the differential equation yields:

$$x(2x + c) - x^2 = x^2 + cx$$
$$x^2 + cx = x^2 + cx$$

Therefore $y = x^2 + cx$ is a solution of the differential equation. It is called a *general solution* because it satisfies the following:

A *general solution* of a differential equation of the *n*th order is a solution which contains *n* arbitrary constants.

## Example 25-3

Show that $y = 2 \sin x$ is a solution of the differential equation:

$$\frac{d^2y}{dx^2} + y = 0$$

**Solution**

If $y = 2 \sin x$, then $dy/dx = 2 \cos x$ and $d^2y/dx^2 = -2 \sin x$. Then substituting into the differential equation yields:

$$-2 \sin x + 2 \sin x = 0$$

Therefore $y = 2 \sin x$ is a solution of the differential equation. This is not a general solution because it does not contain any constants. It is called a *particular solution*.

> A *particular solution* of a differential equation is a solution which can be obtained from the general solution by assigning one or more values to the constants.

For example, the general solution to Example 25-3 is $y = c_1 \sin (x + c_2)$, where $c_1$ and $c_2$ are arbitrary constants. The particular solution shown above, $y = 2 \sin x$, can be obtained from the general solution by letting $c_1 = 2$ and $c_2 = 0$.

### Example 25-4
Show that $y = ce^x - x - 1$ is the general solution to the differential equation $y' = x + y$.

### Solution
If $y = ce^x - x - 1$, then $y' = ce^x - 1$. Substituting into the differential equation yields:

$$y' = x + y$$
$$ce^x - 1 = x + (ce^x - x - 1)$$
$$ce^x - 1 = ce^x - 1$$

The solution $y = ce^x - x - 1$ is the general solution because it contains one constant, and the differential equation is of the first order.

### Example 25-5
Show that $y = c_2 x - \ln x$ is a solution of the differential equation:

$$\frac{d^2y}{dx^2} = \frac{1}{x^2}$$

### Solution:

$$\frac{dy}{dx} = c_2 - \frac{1}{x} \quad \text{and} \quad \frac{d^2y}{dx^2} = \frac{1}{x^2}$$

The solution is a particular solution because it contains one constant and the differential equation is of the second order.

### Example 25-6
Show that $x^2 + y^2 = c$ is a solution of the differential equation $yy' + x = 0$.

**Solution**

Differentiate the given solution implicitly:

$$2x + 2yy' = 0$$
$$yy' + x = 0$$

The solution is a general solution because it contains one constant and the differential equation is of the first order.

**Exercise 25-1**

For each differential equation in problems 1 to 20, show that the given function is a solution. Tell whether the solution is a general solution or a particular solution.

1. $\dfrac{dy}{dx} = 2;\ y = 2x + 5$

2. $\dfrac{dy}{dx} = -4x;\ y = 2 - 2x^2$

3. $\dfrac{dy}{dx} = -\dfrac{y}{x};\ y = \dfrac{c}{x}$

4. $\dfrac{dy}{dx} = \dfrac{y}{x};\ y = cx$

5. $y' - y = 0;\ y = e^{x+c}$

6. $y' = 2xy;\ y = ce^{x^2}$

7. $\dfrac{d^2y}{dx^2} + y = 0;\ y = \cos x$

8. $\dfrac{d^2y}{dx^2} = x;\ y = \dfrac{x^3}{6} + 3x$

9. $xy'' = y';\ y = c_1x^2 + c_2$

10. $xy'' = 2y';\ y = c_1x^3 + c_2$

11. $y' + y = e^{-x};\ y = xe^{-x} + e^{-x}$

12. $y' - e^x = 2y;\ y = -e^x + e^{2x}$

13. $\dfrac{dy}{dx} = \dfrac{y}{x} = 2;\ y = x + \dfrac{c}{x}$

14. $\dfrac{dy}{dx} + \dfrac{2y}{x} = x;\ y = \dfrac{x^2}{4} + \dfrac{c}{x^2}$

15. $yy' = -x;\ x^2 + y^2 = 3$

16. $y^3y' = x^2;\ 3y^4 - 4x^3 = 0$

17. $xy'' + y' = 0;\ y = \ln x + c$

18. $yy'' + 2y' = (y')^2;\ 3y = ce^{3x} - 2$

19. $y'' + 4y = 0;\ y = c_1 \sin 2x + c_2 \cos 2x$

20. $xy''' + 3y'' = 0;\ y = c_1x + \dfrac{c_2}{x} + c_3$

**21.** In an electric circuit containing a constant resistance $R$ and a constant inductance $L$, the current $i$ at any time $t$ satisfies the differential equation:

$$L\frac{di}{dt} + Ri = V$$

where $V$ = constant voltage. Show that $i = V/R + Ae^{-Rt/L}$ is a general solution of the differential equation, where $A$ = constant.

**22.** In an electric circuit containing a constant resistance $R$ and a constant capacitance $C$, the charge $q$ at any time $t$ satisfies the differential equation:

$$R\frac{dq}{dt} + \frac{q}{C} = V$$

where $V$ = constant voltage. Show that $q = CV + Ae^{-t/(RC)}$ is a general solution of the differential equation, where $A$ = constant.

## 25-2 | *Separation of Variables*

This section and the following section study the solution of differential equations of the first order and first degree. Consider the first example.

### Example 25-7
Solve the differential equation:

$$\frac{dy}{dx} = 3x^2y$$

**Solution**
This equation is solved by first writing it in the *differential form:*

$$dy = 3x^2y\ dx$$

Dividing by $y$ to *separate the variables* yields:

$$\frac{dy}{y} = 3x^2\ dx$$

The general solution is now found by integrating each term separately:

$$\ln y = x^3 + c$$

The solution can be left in this form or changed to the exponential form:

$$y = e^{x^3+c} = e^c e^{x^3} = c_1 e^{x^3}$$

When a differential equation of the first order and first degree can be put in the form:

$$M \, dx + N \, dy = 0 \qquad\qquad (25\text{-}1)$$

where $M$ is a function of $x$ only and $N$ is a function of $y$ only, the equation can be solved by integrating each term separately. This method of solution is called *separation of variables*. Study the following examples.

### Example 25-8
Solve the differential equation:

$$2x^2(1 + y^2) + x \, \frac{dy}{dx} = 0$$

### Solution
Write the equation in differential form. This can be thought of as multiplying through by $dx$ (though it is not mathematically precise):

$$2x^2(1 + y^2) \, dx + x \, dy = 0$$

Then separate the variables by dividing by $x$ and then $(1 + y^2)$:

$$2x(1 + y^2) \, dx + dy = 0$$

$$2x \, dx + \frac{dy}{1 + y^2} = 0$$

Integrate each term to find the general solution:

$$x^2 + \tan^{-1} y = c$$

This solution can also be written in the form:

$$\tan^{-1} y = c - x^2$$
$$y = \tan (c - x^2)$$

When certain values of the variables are known, then a particular solution of a differential equation can be found, as the next example illustrates.

### Example 25-9
**1.** Solve the differential equation:

$$e^x \, dy + y \, dx = 0$$

**2.** Find the particular solution of the differential equation if $y = 1$ when $x = -1$.

**Solution**

**1.** Solve the differential equation:

$$e^x \, dy + y \, dx = 0$$

Divide by $y$ and $e^x$ to separate the variables:

$$\frac{dy}{y} + \frac{dx}{e^x} = 0$$

Integrate each term to find the general solution:

$$\ln y - e^{-x} = c$$
$$\ln y = e^{-x} + c$$

**2.** Find the particular solution of the differential equation if $y = 1$ when $x = -1$.

Substitute in the general solution to find the constant:

$$\ln 1 = e^{-(-1)} + c$$
$$0 = e + c'$$
$$c = -e$$

The particular solution is then:

$$\ln y = e^{-x} - e$$

The solution of a differential equation may take more than one form, as Examples 25-7 and 25-8 illustrate. Sometimes one form is preferred to another, as the next example shows.

**Example 25-10**

Solve the differential equation:

$$2xy \, dy = (1 - y^2) \, dx$$

**Solution**

Separate the variables and integrate:

$$\frac{y \, dy}{1 - y^2} = \frac{dx}{2x}$$

$$-\tfrac{1}{2} \ln (1 - y^2) = \tfrac{1}{2} \ln x + c_1$$

$$-\ln (1 - y^2) = \ln x + 2c_1$$

This is the general solution; however, it can be put into a better form, as follows. Let $2c_1 = \ln c$ and apply the rules for logarithms (9-3) and (9-5):

$$-\ln (1 - y^2) = \ln x + \ln c$$

$$\ln (1 - y^2)^{-1} = \ln cx$$

Then take the antilog of both sides:

$$\frac{1}{1 - y^2} = cx$$

or
$$cx \, (1 - y^2) = 1$$

The next example shows an application of differential equations to a problem involving an electric circuit.

### Example 25-11

**1.** In an electric circuit containing a constant resistance $R$ and a constant capacitance $C$, the charge $q$ at any time $t$ satisfies the differential equation:

$$R \frac{dq}{dt} + \frac{q}{C} = V$$

where $V$ = constant voltage. Solve this differential equation for $q$ in terms of $t$.

**2.** If the initial charge $q = 0$, find the particular solution for this condition.

### Solution

**1.** In an electric circuit containing a constant resistance $R$ and a constant capacitance $C$, the charge $q$ at any time $t$ satisfies the differential equation:

$$R \frac{dq}{dt} + \frac{q}{C} = V$$

where $V$ = constant voltage. Solve this differential equation for $q$ in terms of $t$.

Clear the fraction and isolate the derivative:

$$RC \frac{dq}{dt} + q = CV$$

$$RC \frac{dq}{dt} = CV - q$$

Write the equation in differential form:

$$RC \, dq = (CV - q) \, dt$$

Divide by $RC$ and $CV - q$ to separate the variables:

$$\frac{dq}{CV - q} = \frac{dt}{RC}$$

Integrate to find the general solution:

$$-\ln (CV - q) = \frac{t}{RC} + K \qquad K = \text{constant}$$

To solve for $q$ in terms of $t$, multiply by $-1$ and let $-K = \ln A$, where $A = \text{constant}$:

$$\ln (CV - q) = -\frac{t}{RC} - K = \ln A - \frac{t}{RC} = \ln Ae^{-t/(RC)}$$

Write the equation in exponential form by taking the antilog of both sides:

$$CV - q = Ae^{-t/(RC)}$$

Solving for $q$ puts the equation in the final form:

$$q = CV - Ae^{-t/(RC)}$$

*Note:* See problem 22, Sec. 25-1.

2. If the initial charge $q = 0$, find the particular solution for this condition.

The given condition is that $q = 0$ when $t = 0$:

$$0 = CV - Ae^0 = CV - A$$

Therefore $A = CV$, and the particular solution is:

$$q = CV(1 - e^{-t/(RC)})$$

**Exercise 25-2**    In problems 1 to 20, solve each differential equation.

**1.** $\dfrac{dy}{dx} = xy$             **2.** $\dfrac{dy}{dx} = \dfrac{x}{y}$

**3.** $dx + x^2 \, dy = 0$          **4.** $y^2 \, dx + x \, dy = 0$

**5.** $xy^2 \, dx + dy = 0$         **6.** $dx + x^2 y^2 \, dy = 0$

**7.** $(4 + x^2)y' = y$           **8.** $x^2 y' = 1 + y^2$

**9.** $\dfrac{dy}{dt} = \dfrac{2e^t}{y}$        **10.** $R\,\dfrac{dq}{dt} + q = 0$     $(R = \text{constant})$

**11.** $x \, dx + y(x^2 + 1) \, dy = 0$     **12.** $(1 + y^2) \, dx - xy \, dy = 0$

**13.** $ye^x \, dx + dy = 0$         **14.** $y^2 \, dx + e^{2x} \, dy = 0$

**15.** $dx + \cos^2 x \, dy = 0$      **16.** $\cos^2 y \, dx + \sec x \, dy = 0$

**17.** $xyy' + \sqrt{1 + y^2} = 0$      **18.** $xy' - 2x^2 y = 3y$

**19.** $xy \, dx + \ln y \, dy = 0$      **20.** $(3x + 1) \, dx + e^{x+y} \, dy = 0$

In problems 21 to 24, find the particular solution of each differential equation for the given conditions.

**21.** $\dfrac{dy}{dx} = \dfrac{y}{x}$; $y = 2$, $x = 1$

**22.** $x^2 y \, dx + y^2 \, dy = 0$; $y = 1$, $x = 1$

**23.** $dx - \sqrt{1 - x^2} \, dy = 0$; $y = 3$, $x = 0$

**24.** $(x + 1) \, dx - 2xy \, dy = 0$; $y = 2$, $x = 1$

**25.** The following differential equation appears in a problem involving the motion of a charged particle:

$$m\,\frac{dv}{dt} + kv = F$$

where $m$ = mass of the particle, $v$ = velocity, $t$ = time, $F$ = constant force, and $k$ = constant. *(a)* If $k = 0.10$ and $m = 1.0$ g, solve this differential equation for $v$ in terms of $t$. *(b)* If the initial velocity $v = 0$, find the particular solution for these conditions. (See Example 25-11.)

**26.** In an electric circuit containing a constant resistance $R$ and a constant inductance $L$, the current $i$ at any time $t$ satisfies the differential equation:

$$L\,\frac{di}{dt} + Ri = V$$

where $V$ = constant voltage. *(a)* Solve this differential equation for $i$ in terms of $t$. *(b)* If the initial current $i = 0$, find the particular solution for these conditions. (See Example 25-11.)

**27.** The slope of a certain curve $y' = 3x^2 y$. Find the equation of the curve if it passes through the point $(1, 1)$.

28. Newton's law of cooling states that the rate of decrease in the temperature of a warm object is proportional to the difference in temperature between the object and its surroundings at any time $t$. (a) If $T$ = temperature of the object and $T_0$ = temperature of the surroundings, express this law as a differential equation. (b) Solve this differential equation for $T$ in terms of $t$.

29. In a simple dc circuit containing a battery and a constant resistance $R$, Ohm's law states that $I = E/R$, where $I$ = current and $E$ = constant voltage. Given that $I = dq/dt$, where $q$ = charge flowing past a point in the circuit and $t$ = time, find the equation that expresses $q$ as a function of $t$.

30. A resistance $R$ is connected in parallel with a capacitance $C$. Over a certain interval a constant current $I$ is sent through the combination. The voltage drop $V$ across the combination at time $t$ is given by the differential equation:

$$V = IR - RC \frac{dV}{dt}$$

If, at $t = 0$, $V = 0$, solve this differential equation for $V$ in terms of $t$. (See Example 25-11.)

## 25-3 | *Linear Differential Equations of the First Order*

A *linear differential equation of the first order* is an equation of the form:

$$\frac{dy}{dx} + Py = Q \tag{25-2}$$

where $P$ and $Q$ are functions of $x$ or constants. This equation is solved by multiplying each term by the *integrating factor* $e^{\int P\,dx}$:

$$(e^{\int P\,dx}) \frac{dy}{dx} + (e^{\int P\,dx})\, Py = Q\,(e^{\int P\,dx}) \tag{25-3}$$

This makes the integral of the left side of Eq. (25-3) exactly equal to $ye^{\int P\,dx}$ because if:

$$u = ye^{\int P\,dx}$$

then $$u' = (e^{\int P\,dx}) \frac{dy}{dx} + y(e^{\int P\,dx})(P)$$

Notice that the derivative of $\int P\,dx = P$ because differentiation is the inverse of integration.

Therefore the *general solution of Eq. (25-2)* is obtained by integrating Eq. (25-3):

$$ye^{\int P\,dx} = \int Qe^{\int P\,dx}\,dx + C \qquad (25\text{-}4)$$

The following examples show the application of Eq. (25-4).

### Example 25-12
Solve the differential equation:

$$\frac{dy}{dx} + y = 2x$$

### Solution
This equation is a linear differential equation of the first order of the form (25-2), where $P = 1$ and $Q = 2x$. First find the integrating factor $e^{\int P\,dx}$:

$$e^{\int P\,dx} = e^{\int dx} = e^x$$

Then apply (25-4) and formula 57 in App. B-8 to find the integral:

$$ye^x = \int 2xe^x\,dx + C = 2e^x(x - 1) + C$$

The general solution for $y$ is then:

$$y = ce^{-x} + 2x - 2$$

### Example 25-13
Solve the differential equation:

$$xy' + y = x^2$$

### Solution
Put the equation in the form (25-2) by dividing by $x$:

$$y' + \frac{1}{x}y = x$$

Then $P = 1/x$, $Q = x$, and:

$$e^{\int P\,dx} = e^{\int dx/x} = e^{\ln x} = x$$

Note that $e^{\ln x} = x$, since if $P = \ln x$, then, when the equation is written in exponential form, $e^P = x$. In general, $e^{\ln u} = u$, where $u$ is a function of $x$ or another variable. This is an important idea that occurs often in the solution of linear differential equations of the first order. By applying (25-4) the general solution is then found to be:

$$yx = \int x(x)\, dx + C$$

$$yx = \frac{x^3}{3} + C$$

$$y = \frac{x^2}{3} + \frac{C}{x}$$

### Example 25-14
**1.** Solve the differential equation:

$$x\, dy + 2y\, dx = \cos x\, dx$$

**2.** Find the particular solution of the differential equation if $y = 0$ when $x = \pi$.

### Solution
**1.** Solve the differential equation:

$$x\, dy + 2y\, dx = \cos x\, dx$$

Put the equation in the form (25-2) by dividing by $x$ and $dx$:

$$\frac{dy}{dx} + \frac{2}{x}y = \frac{\cos x}{x}$$

Then $P = 2/x$, $Q = (\cos x)/x$, and:

$$e^{\int P\, dx} = e^{\int 2\, dx/x} = e^{2 \ln x} = e^{\ln x^2} = x^2$$

Apply (25-4) and formula 51 in App. B-8 to find the integral and the general solution:

$$yx^2 = \int \frac{\cos x}{x}(x^2)\, dx + C = \int x \cos x\, dx + C$$

$$= \cos x + x \sin x + C$$

**2.** Find the particular solution of the differential equation if $y = 0$ when $x = \pi$.

Substitute in the general solution to find the constant:

$$(0)(\pi)^2 = \cos \pi + \pi \sin \pi + C$$

$$0 = -1 + 0 + C$$

$$C = 1$$

The particular solution is then:

$$yx^2 = \cos x + x \sin x + 1$$

The next example shows an important application of a linear differential equation to a problem in electric circuits.

### Example 25-15

In a series $RL$ circuit which contains a constant resistance $R$ and a constant inductance $L$, the current $i$ at any time $t$ satisfies the differential equation:

$$L\frac{di}{dt} + Ri = v$$

where the voltage $v$ is a function of the time $t$.

1. Solve this differential equation for $i$ in terms of $v$ and $t$.
2. If the voltage in the circuit $v = V$, where $V$ is a constant, find the solution of the differential equation.

### Solution

1. Solve this differential equation for $i$ in terms of $v$ and $t$.

   Divide by $L$ to put the equation in the form (25-2), where $y = i$ and $x = t$:

$$\frac{di}{dt} + \frac{R}{L}i = \frac{1}{L}v$$

   Then $P = R/L$, $Q = (1/L)v$, and:

$$e^{\int P\,dt} = e^{\int (R/L)\,dt} = e^{Rt/L}$$

   Apply (25-4) to find the general solution:

$$ie^{Rt/L} = \frac{1}{L}\int v e^{Rt/L}\,dt + A$$

   where $A$ is a constant.
   When the voltage $v$ is specified, the integral may be evaluated as follows.

2. If the voltage in the circuit $v = V$, where $V$ is a constant, find the solution of the differential equation.

   Substitute in the general solution and integrate:

$$ie^{Rt/L} = \frac{1}{L} \int Ve^{Rt/L} \, dt + A$$

$$ie^{Rt/L} = \frac{V}{L}e^{Rt/L} \left(\frac{L}{R}\right) + A$$

$$i = \frac{V}{R} + Ae^{-Rt/L}$$

This solution can also be found by separation of variables. (See problem 26, Sec. 25-2.)

**Exercise 25-3**

In problems 1 to 20, solve each differential equation.

**1.** $\dfrac{dy}{dx} + y = 1$           **2.** $\dfrac{dy}{dx} - y = x$

**3.** $\dfrac{dy}{dx} - y = 2e^x$       **4.** $\dfrac{dy}{dx} + 2y = e^{-x}$

**5.** $xy' + y = 3x^2$          **6.** $xy' + y = 2x$

**7.** $xy' + y = xe^x$          **8.** $xy' + 2y = 4x^2$

**9.** $2x \, dy + 4y \, dx = x^3 \, dx$     **10.** $3x \, dy - y \, dx = 6x^2 \, dx$

**11.** $x \, dy + y \, dx = \sin x \, dx$     **12.** $x^2 \, dy + 2xy \, dx = \cos x \, dx$

**13.** $\dfrac{dy}{dt} = \sin t - y$       **14.** $\dfrac{dy}{dt} = \cos t + y$

**15.** $\dfrac{dy}{dx} = xe^{2x} + 2y$      **16.** $\dfrac{dq}{dt} = 3e^{3t} + 3q$

**17.** $y' \sin x - y \cos x = \cot x$     **18.** $y' \cot x - y = \cos x$

**19.** $y' + x^2y = (x^2 + 1)e^x$      **20.** $xy' + y = (x + 1)^2$

In problems 21 to 24, find the particular solution of each differential equation for the given conditions.

**21.** $\dfrac{dy}{dx} - y = e^x; \; y = 0, \; x = -1$

**22.** $x\dfrac{dy}{dx} - y = x; \; y = 2, \; x = 1$

**23.** $xy' + y = \sec^2 x; \; y = 0, \; x = \pi/4$

**24.** $xy' + (1 + x)y = e^{-x}; \; y = 0, \; x = 1$

**25.** In a series $RC$ circuit which contains a constant resistance $R$ and a constant capacitance $C$, the charge $q$ at any time $t$ satisfies the differential equation:

$$R\frac{dq}{dt} + \frac{q}{C} = v$$

where the voltage $v$ is a function of the time $t$. *(a)* Solve this differential equation for $q$ in terms of $v$ and $t$. (See Example 25-15.) *(b)* If the voltage $v = V$, where $V$ is a constant, find the solution of the differential equation.

26. For the circuit of problem 25 the current $i$ satisfies the differential equation:

$$R\frac{di}{dt} + \frac{i}{C} = \frac{dv}{dt}$$

Solve this equation for $i$ in terms of $t$ given that the voltage $v = V$, where $V$ is a constant.

27. Under certain conditions when a ship slows to a stop, its speed $v$ is given by the differential equation:

$$\frac{dv}{dt} + t = kv$$

where $k$ = constant. Solve this differential equation for $v$ in terms of $t$.

28. The following equation arises in aerodynamic theory:

$$\frac{ds}{dt} + ks = g \cos \alpha$$

where $k$, $g$, and $\alpha$ are constants. Solve this differential equation for $s$ in terms of $t$.

29. Refer to the circuit of Example 25-15. *(a)* Find the particular solution to the differential equation if $i = 0$ when $t = 0$. *(b)* Show that:

$$\lim_{t \to \infty} i = \frac{V}{R}$$

30. In the circuit of Example 25-15 find the solution to the differential equation if the voltage is given by the sinusoidal wave $v = V \sin \omega t$. (Note: Use formula 60 in App. B-8.)

## 25-4 | *Homogeneous Equations of the Second Order*

This section studies linear (or first-degree) differential equations of the second order. A *homogeneous linear equation of the second order* is an equation of the form:

$$\frac{d^2y}{dx^2} + p\frac{dy}{dx} + qy = 0 \qquad (25\text{-}5)$$

where $p$ and $q$ are functions of $x$ or constants. (When the right-hand side is not zero but is a function of $x$, the equation is called *nonhomogeneous*.) Only equations of the form (25-5) *where p and q are constants* will be considered, since these occur most often in applications. The *general solution of Eq. (25-5)* with *constant coefficients* can be shown to be:

$$y = c_1 e^{m_1 x} + c_2 e^{m_2 x} \qquad (25\text{-}6)$$

where $m_1$ and $m_2$ are the solutions of the *auxiliary equation:*

$$m^2 + pm + q = 0 \qquad (25\text{-}7)$$

Study the following examples, which show how to apply Eqs. (25-6) and (25-7).

### Example 25-16

Solve and check the differential equation:

$$\frac{d^2y}{dx^2} + 2\frac{dy}{dx} - 3y = 0$$

### Solution

Set up the auxiliary equation (25-7), with $p = 2$ and $q = -3$, and obtain the two solutions:

$$m^2 + 2m - 3 = 0$$
$$(m - 1)(m + 3) = 0$$
$$m_1 = 1 \qquad m_2 = -3$$

Apply Eq. (25-6) to obtain the general solution:

$$y = c_1 e^x + c_2 e^{-3x}$$

To check that this is a general solution, find $dy/dx$ and $d^2y/dx^2$ and substitute them into the differential equation:

$$\frac{dy}{dx} = c_1 e^x - 3c_2 e^{-3x} \qquad \frac{d^2y}{dx^2} = c_1 e^x + 9c_2 e^{-3x}$$

$$\frac{d^2y}{dx^2} + 2\frac{dy}{dx} - 3y = (c_1 e^x + 9c_2 e^{-3x})$$

$$+ 2(c_1 e^x - 3c_2 e^{-3x}) - 3(c_1 e^x + c_2 e^{-3x})$$

$$= 3c_1 e^x - 3c_1 e^x + 9c_2 e^{-3x} - 9c_2 e^{-3x} = 0$$

The solution therefore checks, and it verifies that Eq. (25-6) provides a general solution to the homogeneous linear equation.

## Example 25-17

Solve the differential equation $3y'' - y' = 0$.

### Solution

Write the equation in the form (25-5):

$$y'' - \frac{1}{3} y' = 0$$

The auxiliary equation then has $p = -\frac{1}{3}$ and $q = 0$. Find the two roots of the auxiliary equation:

$$m^2 - \tfrac{1}{3}m = 0$$
$$3m^2 - m = 0$$
$$m(3m - 1) = 0$$
$$m_1 = 0 \qquad m_2 = \tfrac{1}{3}$$

Apply Eq. (25-6) to find the general solution:

$$y = c_1 e^{0x} + c_2 e^{x/3} = c_1 + c_2 e^{x/3}$$

## Example 25-18

Find the particular solution of the differential equation $y'' = y' + 2y$ which satisfies the conditions that $y = 2$ and $y' = 4$ when $x = 0$.

### Solution

It is necessary to give two conditions because there are two constants to be determined. First find the general solution for $y$ and $y'$ by applying Eqs. (25-6) and (25-7):

$$y'' - y' - 2y = 0$$
$$m^2 - m - 2 = 0$$
$$(m - 2)(m + 1) = 0$$
$$m_1 = 2 \qquad m_2 = -1$$

Then
$$y = c_1 e^{2x} + c_2 e^{-x}$$
and
$$y' = 2c_1 e^{2x} - c_2 e^{-x}$$

Substitute the given conditions into $y$ and $y'$ to find $c_1$ and $c_2$:

$$2 = c_1 + c_2$$
$$4 = 2c_1 - c_2$$

The solution of this linear system yields $c_1 = 2$ and $c_2 = 0$. The particular solution is therefore:

$$y = 2e^{2x}$$

## Equal and Complex Roots

When the roots of the auxiliary equation (25-7) are equal or complex, Eq. (25-6) does not give the general solution to the homogeneous linear equation (25-5).

*When the roots of the auxiliary equation (25-7) are equal,* the general solution of (25-5) is given by:

$$y = (c_1 + c_2 x)e^{mx} \tag{25-8}$$

where $m$ is the equal root of the auxiliary equation.

### Example 25-19
Solve the differential equation:

$$\frac{d^2y}{dx^2} - 4\frac{dy}{dx} + 4y = 0$$

### Solution
The auxiliary equation and roots are:

$$m^2 - 4m + 4 = 0$$
$$(m - 2)(m - 2) = 0$$
$$m = 2 \qquad m = 2$$

By applying Eq. (25-8) the general solution is:

$$y = (c_1 + c_2 x)e^{2x}$$

*When the roots of the auxiliary equation (25-7) are complex,* the general solution of (25-5) is:

$$y = e^{ax}(c_1 \sin bx + c_2 \cos bx) \tag{25-9}$$

where $a \pm bj$ are the complex roots of the auxiliary equation.

### Example 25-20
**1.** Solve the differential equation:

$$y'' - 2y' + 5y = 0$$

**2.** Find the particular solution of the given differential equation which satisfies the conditions $y = 2$ and $y' = 0$ when $x = 0$.

**Solution**

**1.** Solve the differential equation:

$$y'' - 2y' + 5y = 0$$

The auxiliary equation and roots are:

$$m^2 - 2m + 5 = 0$$

$$m = \frac{2 \pm \sqrt{4 - 20}}{2} = \frac{2 \pm \sqrt{-16}}{2} = \frac{2 \pm 4j}{2} = 1 \pm 2j$$

Then $a = 1$, $b = 2$, and, by applying Eq. (25-9), the general solution is:

$$y = e^x(c_1 \sin 2x + c_2 \cos 2x)$$

**2.** Find the particular solution of the given differential equation which satisfies the conditions $y = 2$ and $y' = 0$ when $x = 0$.

First find $y'$:

$$y' = e^x(c_1 \sin 2x + c_2 \cos 2x + 2c_1 \cos 2x - 2c_2 \sin 2x)$$

Then substitute the conditions into $y'$ and $y$ to find $c_1$ and $c_2$:

$$0 = e^0(c_1 \sin 0 + c_2 \cos 0 + 2c_1 \cos 0 - 2c_2 \sin 0)$$
$$0 = c_2 + 2c_1$$
$$y = e^x(c_1 \sin 2x + c_2 \cos 2x)$$
$$2 = e^0(c_1 \sin 0 + c_2 \cos 0)$$
$$2 = c_2$$

The solution of this system of equations is then $c_1 = -1$ and $c_2 = 2$. The particular solution is therefore:

$$y = e^x(2 \cos 2x - \sin 2x)$$

The next example shows an application of a second-order homogeneous differential equation to a basic electric circuit.

### Example 25-21

Given is a series $RLC$ circuit which contains a constant resistance $R$, a constant inductance $L$, and a capacitance $C$ (Fig. 25-1). When the voltage $V$ is constant,

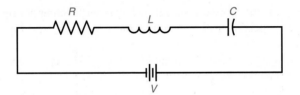

FIG. 25-1 *Series RLC circuit, Example 25-21*

the current $i$ at any time $t$ satisfies the differential equation:

$$\frac{d^2i}{dt^2} + \frac{R}{L}\frac{di}{dt} + \frac{1}{LC}i = 0$$

Solve this equation for $i$ in terms of $t$ if $R^2 < 4L/C$.

### Solution
This is a second-order homogeneous differential equation, where $p = R/L$ and $q = 1/(LC)$. The auxiliary equation is:

$$m^2 + \frac{R}{L}m + \frac{1}{LC} = 0$$

or

$$Lm^2 + Rm + \frac{1}{C} = 0$$

The roots of the auxiliary equation are:

$$m = \frac{-R \pm \sqrt{R^2 - 4L/C}}{2L}$$

When $R^2 < 4L/C$, the roots are complex and can be written as:

$$m = -\frac{R}{2L} \pm \frac{j}{2L}\sqrt{\frac{4L}{C} - R^2}$$

By applying Eq. (25-9) and letting $a = R/(2L)$ and $b = \sqrt{(4L/C) - R^2}/(2L)$, the solution can be written as:

$$i = e^{-at}(c_1 \sin bt + c_2 \cos bt)$$

This function represents an *underdamped oscillation*, and $e^{-at}$ is called the *damping factor*. The current is a sinusoidal oscillation which decreases in amplitude and approaches zero. This type of function not only is found in electrical applications but also is important in other engineering applications involving vibration.

Homogeneous differential equations of higher order can be solved by using techniques similar to those shown in this section. The auxiliary equations are of higher degree and are solved by using methods shown in Chap. 15.

**Exercise 25-4**

In problems 1 to 20, solve each differential equation.

1. $\dfrac{d^2y}{dx^2} - 3\dfrac{dy}{dx} + 2y = 0$     2. $\dfrac{d^2y}{dx^2} - 2\dfrac{dy}{dx} - 3y = 0$

3. $\dfrac{d^2y}{dx^2} - \dfrac{dy}{dx} - 6y = 0$     4. $2\dfrac{d^2y}{dx^2} + \dfrac{dy}{dx} - y = 0$

5. $y'' + 2y' = 0$     6. $2y'' - y' = 0$

7. $y'' = 4y$     8. $y'' = y$

9. $3y'' - 2y' - y = 0$     10. $2y'' - 5y' + 2y = 0$

11. $y'' - 1.4y = 0$     12. $y'' - 2y' = y$

13. $\dfrac{d^2y}{dx^2} - 2\dfrac{dy}{dx} + y = 0$     14. $\dfrac{d^2y}{dx^2} + 6\dfrac{dy}{dx} + 9y = 0$

15. $4y'' + 4y' + y = 0$     16. $9y'' - 6y' + y = 0$

17. $\dfrac{d^2y}{dx^2} + 4y = 0$     18. $\dfrac{d^2y}{dx^2} + y = 0$

19. $y'' + 2y' + 5y = 0$     20. $y'' - 2y' + 2y = 0$

In problems 21 to 24, find the particular solution of each differential equation for the given conditions.

21. $y'' - 4y' + 3y = 0$; $y = 2$, $y' = 4$, $x = 0$

22. $2y'' = y'$; $y = 3$, $y' = 2$, $x = 0$

23. $y'' + 4y' + 4y = 0$; $y = 0$, $y' = 1$, $x = 0$

24. $y'' + 4y' + 8y = 0$; $y = 2$, $y' = 4$, $x = 0$

25. In simple harmonic motion, the force, and therefore the acceleration, is proportional to the displacement $x$. This is expressed by the differential equation:

$$\frac{d^2x}{dt^2} = -\omega^2 x$$

where $\omega$ = angular velocity (see Sec. 11-1). Solve this equation for $x$ in terms of $t$.

26. Find the particular solution to the differential equation of problem 25, given the initial conditions $x = a$ and $dx/dt = 0$ when $t = 0$.

27. The current in a certain electric circuit satisfies the differential equation:

$$L\frac{d^2i}{dt^2} + R\frac{di}{dt} = 0$$

where $R$ = constant resistance and $L$ = constant inductance. Solve this equation for $i$ in terms of $t$.

28. When a mechanical vibration is damped, the displacement $x$ is given by the differential equation:

$$\frac{d^2x}{dt^2} + 2p\,\frac{dx}{dt} + k^2x = 0$$

where $p$ and $k$ are constants. Solve this equation for $x$ in terms of $t$ for the *overdamped vibration* given by $p^2 > k^2$. (See Example 25-21.)

29. Solve the differential equation of Example 25-21 for the *critical oscillation* given by $R^2 = 4L/C$.

30. Solve the differential equation of Example 25-21 for the *overdamped oscillation* given by $R^2 = 8L/C$.

## 25-5 | *Review Questions*

In problems 1 to 20, solve each differential equation.

1. $\dfrac{dy}{dx} = 2x^2y$

2. $2y^2\,dx + x^2\,dy = 0$

3. $ye^{2x}\,dy + dx = 0$

4. $x^2\dfrac{dy}{dx} + x(1 + y^2) = 0$

5. $dx + \sin^2 x\,dy = 0$

6. $xy\,dy = (1 + y^2)\,dx$

7. $\dfrac{dq}{dt} + 2q = 2$

8. $\dfrac{dy}{dx} + y = 3x$

9. $xy' + y = 3x$

10. $xy' + y = e^x$

11. $x\,dy + y\,dx + \cos x\,dx = 0$

12. $\dfrac{di}{dt} - 2e^{2t} = 3i$

13. $\dfrac{d^2y}{dx^2} + 4\dfrac{dy}{dx} + 3y = 0$

14. $\dfrac{d^2y}{dx^2} - \dfrac{dy}{dx} - 6y = 0$

15. $y'' - 2y' = 0$

16. $y'' = 9y$

17. $\dfrac{d^2y}{dx^2} + 2\dfrac{dy}{dx} + y = 0$

18. $4y'' - 4y' + 1 = 0$

19. $4y'' + y = 0$

20. $y'' + 2y' + 2y = 0$

In problems 21 to 24, find the particular solution of each differential equation for the given conditions.

21. $xy^2\,dx - y\,dy = 0;\ y = 1,\ x = 2$

22. $(1 + y)\,dx + x\,dy = 0;\ y = 1,\ x = 1$

23. $xy' + y + 4x^3 = 0;\ y = 0,\ x = 1$

24. $2y'' + 3y' + y = 0;\ y = 2,\ y' = 1,\ x = 0$

25. The radioactive decay of an element satisfies the differential equation:

$$\frac{dm}{dt} - km = 0$$

where $m$ = mass, $t$ = time, and $k$ = constant rate of decay. Solve this equation for $m$ in terms of $t$ if $m = m_0$ when $t = 0$.

26. The motion of an atomic particle in the $y$ direction is given by the differential equation:

$$\frac{dy}{dt} + gt = v_0 \sin \alpha$$

where $t$ = time, $g$ = gravitational force, $v_0$ = initial velocity, and $\alpha$ = angle between the $x$ axis and the direction of motion. Solve this equation for $y$ in terms of $t$ if $y = 0$ when $t = 0$.

27. In the circuit of Example 25-15, find the solution to the differential equation if $v = 2t$.

28. Solve the differential equation of Example 25-21 for the underdamped oscillation given by $R^2 = L/C$.

# Metric Units (SI) and Conversions

Chapter 4 (Secs 4.1 and 4.2) contains information and tables for the metric system or the more complete International System (SI, for Système International) and the U.S. Customary System. Below are more complete tables for the SI and U.S. systems.

## SI Prefixes

| Power of 10 | Prefix | Symbol |
|---|---|---|
| $10^{-18}$ | atto | a |
| $10^{-15}$ | femto | f |
| $10^{-12}$ | pico | p |
| $10^{-9}$ | nano | n |
| $10^{-6}$ | micro | $\mu$ |
| $10^{-3}$ | milli | m |
| $10^{-2}$ | centi | c |
| $10^{-1}$ | deci | d |
| $10^{0}$ | (base unit) | |
| $10^{1}$ | deka | da |
| $10^{2}$ | hecto | h |
| $10^{3}$ | kilo | k |
| $10^{6}$ | mega | M |
| $10^{9}$ | giga | G |
| $10^{12}$ | tera | T |
| $10^{15}$ | peta | P |
| $10^{18}$ | exa | E |

## Conversion Factors and Relations

### Length

| | |
|---|---|
| 1 centimeter (cm) | = 0.3937 inch (in.) |
| 1 foot (ft) | = 0.3048 meter (m) |
| 1 inch (in.) | = 2.540 centimeters (cm) |
| 1 kilometer (km) | = 0.6214 statute mile (mi) |

1 meter (m) = 3.281 feet (ft)
1 statute mile (mi) = 1.609 kilometers (km)
1 nautical mile (nmi) = 1.151 statute miles (mi)

## Mass

1 gram (g) = 0.03527 ounce avoirdupois (oz)
1 kilogram (kg) = 2.205 pounds avoirdupois (lb)
1 ounce avoirdupois (oz) = 28.35 grams (g)
1 pound avoirdupois (lb) = 0.4536 kilogram (kg)

## Force

1 newton (N) = 0.2248 pound-force (lb)
= 0.1020 kilogram-force (kgf)
1 pound-force (lb) = 4.448 newtons (N)
1 kilogram-force (kgf) = 9.807 newtons (N)

## Temperature

$$1 \text{ Celsius degree} = \frac{9}{5} \text{ Fahrenheit degrees}$$

$$C = \frac{5}{9}(F - 32)$$

$$1 \text{ Fahrenheit degree} = \frac{5}{9} \text{ Celsius degree}$$

$$F = \frac{9}{5}C + 32$$

$$1 \text{ kelvin} = 1 \text{ Celsius degree}$$
$$K = C + 273.16$$

## Capacity

1 gallon (gal) = 3.785 liters (L)
1 liter (L) = 0.2642 gallon (gal)
= 1.057 quarts (qt)
1 quart (qt) = 0.9464 liter (L)

## Angle

$$1° = \frac{\pi}{180} \text{ or } 0.01745 \text{ radian (rad)}$$

$$1 \text{ radian (rad)} = \frac{180}{\pi} \text{ or } 57.30°$$

## Area

| | |
|---|---|
| 1 square foot ($ft^2$) | = 0.09290 square meter ($m^2$) |
| 1 square kilometer ($km^2$) | = 0.3861 square mile ($mi^2$) |
| 1 square meter ($m^2$) | = 10.76 square feet ($ft^2$) |
| 1 square mile ($mi^2$) | = 2.590 square kilometers ($km^2$) |

## Volume

| | |
|---|---|
| 1 cubic foot ($ft^3$) | = 0.02832 cubic meter ($m^3$) |
| 1 cubic meter ($m^3$) | = 35.31 cubic feet ($ft^3$) |

## Velocity

| | |
|---|---|
| 1 foot per second (ft/s) | = 0.6818 mile per hour (mi/h) |
| | = 1.097 kilometers per hour (km/h) |
| 1 kilometer per hour (km/h) | = 0.9113 feet per second (ft/s) |
| 1 meter per second (m/s) | = 2.237 miles per hour (mi/h) |
| 1 mile per hour (mi/h) | = 1.467 feet per second (ft/s) |
| | = 0.4470 meter per second (m/s) |

## Pressure

| | |
|---|---|
| 1 pound per square inch (lb/$in^2$) | = 6.873 kilopascals (kPa) |
| 1 kilopascal (kPa) | = 0.1455 pound per square inch (lb/$in^2$) |

## Energy

| | |
|---|---|
| 1 British thermal unit (Btu) | = 1054 joules (J) |
| 1 foot-pound-force (ft · lb) | = 1.355 joules (J) |
| 1 joule (J) | = $9.485 \times 10^{-4}$ British thermal units (Btu) |
| | = 0.7380 foot-pound-force (ft · lb) |
| | = $2.778 \times 10^{-7}$ kilowatthours (kWh) |
| 1 kilowatthour (kWh) | = $3.600 \times 10^6$ joules (J) |

## Power

| | |
|---|---|
| 1 horsepower (hp) | = 0.7460 kilowatt (kW) |
| 1 kilowatt (kW) | = 1.341 horsepower (hp) |

# Tables

Table B-1 TRIGONOMETRIC FUNCTIONS IN DECIMAL DEGREES AND RADIANS

| Deg | Rad | Sin | Cos | Tan | Cot | | |
|-----|-----|-----|-----|-----|-----|---|---|
| 0.0 | 0.0000 | 0.00000 | 1.0000 | 0.00000 | — | 1.5708 | 90.0 |
| .1 | 0.0017 | 0.00175 | 1.0000 | 0.00175 | 573.0 | 1.5691 | .9 |
| .2 | 0.0035 | 0.00349 | 1.0000 | 0.00349 | 286.5 | 1.5673 | .8 |
| .3 | 0.0052 | 0.00524 | 1.0000 | 0.00524 | 191.0 | 1.5656 | .7 |
| .4 | 0.0070 | 0.00698 | 1.0000 | 0.00698 | 143.2 | 1.5638 | .6 |
| .5 | 0.0087 | 0.00873 | 1.0000 | 0.00873 | 114.6 | 1.5621 | .5 |
| .6 | 0.0105 | 0.01047 | 0.9999 | 0.01047 | 95.49 | 1.5603 | .4 |
| .7 | 0.0122 | 0.01222 | 0.9999 | 0.01222 | 81.85 | 1.5586 | .3 |
| .8 | 0.0139 | 0.01396 | 0.9999 | 0.01396 | 71.62 | 1.5568 | .2 |
| .9 | 0.0157 | 0.01571 | 0.9999 | 0.01571 | 63.66 | 1.5551 | .1 |
| 1.0 | 0.0175 | 0.01745 | 0.9998 | 0.01746 | 57.29 | 1.5533 | 89.0 |
| | | Cos | Sin | Cot | Tan | Rad | Deg |

*Table B-1 (continued)*

| Deg | Rad | Sin | Cos | Tan | Cot | | |
|---|---|---|---|---|---|---|---|
| 1.0 | 0.0175 | 0.01745 | 0.9998 | 0.01746 | 57.29 | 1.5533 | 89.0 |
| .1 | 0.0192 | 0.01920 | 0.9998 | 0.01920 | 52.08 | 1.5516 | .9 |
| .2 | 0.0209 | 0.02094 | 0.9998 | 0.02095 | 47.74 | 1.5499 | .8 |
| .3 | 0.0227 | 0.02269 | 0.9997 | 0.02269 | 44.07 | 1.5481 | .7 |
| .4 | 0.0244 | 0.02443 | 0.9997 | 0.02444 | 40.92 | 1.5464 | .6 |
| .5 | 0.0262 | 0.02618 | 0.9997 | 0.02619 | 38.19 | 1.5446 | .5 |
| .6 | 0.0279 | 0.02792 | 0.9996 | 0.02793 | 35.80 | 1.5429 | .4 |
| .7 | 0.0297 | 0.02967 | 0.0996 | 0.02968 | 33.69 | 1.5411 | .3 |
| .8 | 0.0314 | 0.03141 | 0.9995 | 0.03143 | 31.82 | 1.5394 | .2 |
| .9 | 0.0332 | 0.03316 | 0.9995 | 0.03317 | 30.14 | 1.5376 | .1 |
| 2.0 | 0.0349 | 0.03490 | 0.9994 | 0.03492 | 28.64 | 1.5359 | 88.0 |
| .1 | 0.0367 | 0.03664 | 0.9993 | 0.03667 | 27.27 | 1.5341 | .9 |
| .2 | 0.0384 | 0.03839 | 0.9993 | 0.03842 | 26.03 | 1.5324 | .8 |
| .3 | 0.0401 | 0.04013 | 0.9992 | 0.04016 | 24.90 | 1.5307 | .7 |
| .4 | 0.0419 | 0.04188 | 0.9991 | 0.04191 | 23.86 | 1.5289 | .6 |
| .5 | 0.0436 | 0.04362 | 0.9990 | 0.04366 | 22.90 | 1.5272 | .5 |
| .6 | 0.0454 | 0.04536 | 0.9990 | 0.04541 | 22.02 | 1.5254 | .4 |
| .7 | 0.0471 | 0.04711 | 0.9989 | 0.04716 | 21.20 | 1.5237 | .3 |
| .8 | 0.0489 | 0.04885 | 0.9988 | 0.04891 | 20.45 | 1.5219 | .2 |
| .9 | 0.0506 | 0.05059 | 0.9987 | 0.05066 | 19.74 | 1.5202 | .1 |
| 3.0 | 0.0524 | 0.05234 | 0.9986 | 0.05241 | 19.08 | 1.5184 | 87.0 |
| .1 | 0.0541 | 0.05408 | 0.9985 | 0.05416 | 18.46 | 1.5167 | .9 |
| .2 | 0.0559 | 0.05582 | 0.9984 | 0.05591 | 17.89 | 1.5149 | .8 |
| .3 | 0.0576 | 0.05756 | 0.9983 | 0.05766 | 17.34 | 1.5132 | .7 |
| .4 | 0.0593 | 0.05931 | 0.9982 | 0.05941 | 16.83 | 1.5115 | .6 |
| .5 | 0.0611 | 0.06105 | 0.9981 | 0.06116 | 16.35 | 1.5097 | .5 |
| .6 | 0.0628 | 0.06279 | 0.9980 | 0.06291 | 15.90 | 1.5080 | .4 |
| .7 | 0.0645 | 0.06453 | 0.9979 | 0.06467 | 15.46 | 1.5062 | .3 |
| .8 | 0.0663 | 0.06627 | 0.9978 | 0.06642 | 15.06 | 1.5045 | .2 |
| .9 | 0.0681 | 0.06802 | 0.9977 | 0.06817 | 14.67 | 1.5027 | .1 |
| 4.0 | 0.0698 | 0.06976 | 0.9976 | 0.06993 | 14.30 | 1.5010 | 86.0 |
| .1 | 0.0716 | 0.07150 | 0.9974 | 0.07168 | 13.95 | 1.4992 | .9 |
| .2 | 0.0733 | 0.07324 | 0.9973 | 0.07344 | 13.62 | 1.4975 | .8 |
| .3 | 0.0750 | 0.07498 | 0.9972 | 0.07519 | 13.30 | 1.4957 | .7 |
| .4 | 0.0768 | 0.07672 | 0.9971 | 0.07695 | 13.00 | 1.4940 | .6 |
| .5 | 0.0785 | 0.07846 | 0.9969 | 0.07870 | 12.71 | 1.4923 | .5 |
| .6 | 0.0803 | 0.08020 | 0.9968 | 0.08046 | 12.43 | 1.4905 | .4 |
| .7 | 0.0820 | 0.08194 | 0.9966 | 0.08221 | 12.16 | 1.4888 | .3 |
| .8 | 0.0838 | 0.08368 | 0.9965 | 0.08397 | 11.91 | 1.4870 | .2 |
| .9 | 0.0855 | 0.08542 | 0.9963 | 0.08573 | 11.66 | 1.4853 | .1 |
| 5.0 | 0.0873 | 0.08716 | 0.9962 | 0.08749 | 11.43 | 1.4835 | 85.0 |
| | | **Cos** | **Sin** | **Cot** | **Tan** | **Rad** | **Deg** |

*Table B-1 (continued)*

| Deg | Rad | Sin | Cos | Tan | Cot | | |
|-----|-----|-----|-----|-----|-----|-----|-----|
| 5.0 | 0.0873 | 0.08716 | 0.9962 | 0.08749 | 11.43 | 1.4835 | 85.0 |
| .1 | 0.0890 | 0.08889 | 0.9960 | 0.08925 | 11.20 | 1.4818 | .9 |
| .2 | 0.0908 | 0.09063 | 0.9959 | 0.09101 | 10.99 | 1.4800 | .8 |
| .3 | 0.0925 | 0.09237 | 0.9957 | 0.09277 | 10.78 | 1.4783 | .7 |
| .4 | 0.0942 | 0.09411 | 0.9956 | 0.09453 | 10.58 | 1.4765 | .6 |
| .5 | 0.0960 | 0.09585 | 0.9954 | 0.09629 | 10.39 | 1.4748 | .5 |
| .6 | 0.0977 | 0.09758 | 0.9952 | 0.09805 | 10.20 | 1.4731 | .4 |
| .7 | 0.0995 | 0.09932 | 0.9951 | 0.09981 | 10.02 | 1.4713 | .3 |
| .8 | 0.1012 | 0.1011 | 0.9949 | 0.1016 | 9.845 | 1.4696 | .2 |
| .9 | 0.1030 | 0.1028 | 0.9947 | 0.1033 | 9.677 | 1.4678 | .1 |
| 6.0 | 0.1047 | 0.1045 | 0.9945 | 0.1051 | 9.914 | 1.4661 | 84.0 |
| .1 | 0.1065 | 0.1063 | 0.9943 | 0.1069 | 9.357 | 1.4643 | .9 |
| .2 | 0.1082 | 0.1080 | 0.9942 | 0.1086 | 9.205 | 1.4626 | .8 |
| .3 | 0.1100 | 0.1097 | 0.9940 | 0.1104 | 9.058 | 1.4608 | .7 |
| .4 | 0.1117 | 0.1115 | 0.9938 | 0.1122 | 8.915 | 1.4591 | .6 |
| .5 | 0.1134 | 0.1132 | 0.9936 | 0.1139 | 8.777 | 1.4574 | .5 |
| .6 | 0.1152 | 0.1149 | 0.9934 | 0.1157 | 8.643 | 1.4556 | .4 |
| .7 | 0.1169 | 0.1167 | 0.9932 | 0.1175 | 8.513 | 1.4539 | .3 |
| .8 | 0.1187 | 0.1184 | 0.9930 | 0.1192 | 8.386 | 1.4521 | .2 |
| .9 | 0.1204 | 0.1201 | 0.9928 | 0.1210 | 8.264 | 1.4504 | .1 |
| 7.0 | 0.1222 | 0.1219 | 0.9925 | 0.1228 | 8.144 | 1.4486 | 83.0 |
| .1 | 0.1239 | 0.1236 | 0.9923 | 0.1246 | 8.028 | 1.4469 | .9 |
| .2 | 0.1257 | 0.1253 | 0.9921 | 0.1263 | 7.916 | 1.4451 | .8 |
| .3 | 0.1274 | 0.1271 | 0.9919 | 0.1281 | 7.806 | 1.4434 | .7 |
| .4 | 0.1292 | 0.1288 | 0.9917 | 0.1299 | 7.700 | 1.4416 | .6 |
| .5 | 0.1309 | 0.1305 | 0.9914 | 0.1317 | 7.596 | 1.4399 | .5 |
| .6 | 0.1326 | 0.1323 | 0.9912 | 0.1334 | 7.495 | 1.4382 | .4 |
| .7 | 0.1344 | 0.1340 | 0.9910 | 0.1352 | 7.396 | 1.4364 | .3 |
| .8 | 0.1361 | 0.1357 | 0.9907 | 0.1370 | 7.300 | 1.4347 | .2 |
| .9 | 0.1379 | 0.1374 | 0.9905 | 0.1388 | 7.207 | 1.4329 | .1 |
| 8.0 | 0.1396 | 0.1392 | 0.9903 | 0.1405 | 7.115 | 1.4312 | 82.0 |
| .1 | 0.1414 | 0.1409 | 0.9900 | 0.1423 | 7.026 | 1.4294 | .9 |
| .2 | 0.1431 | 0.1426 | 0.9898 | 0.1441 | 6.940 | 1.4277 | .8 |
| .3 | 0.1449 | 0.1444 | 0.9895 | 0.1459 | 6.855 | 1.4259 | .7 |
| .4 | 0.1466 | 0.1461 | 0.9893 | 0.1477 | 6.772 | 1.4242 | .6 |
| .5 | 0.1484 | 0.1478 | 0.9890 | 0.1495 | 6.691 | 1.4224 | .5 |
| .6 | 0.1501 | 0.1495 | 0.9888 | 0.1512 | 6.612 | 1.4207 | .4 |
| .7 | 0.1518 | 0.1513 | 0.9885 | 1.1530 | 6.535 | 1.4190 | .3 |
| .8 | 0.1536 | 0.1530 | 0.9882 | 0.1548 | 6.460 | 1.4172 | .2 |
| .9 | 0.1553 | 0.1547 | 0.9880 | 0.1556 | 6.386 | 1.4155 | .1 |
| 9.0 | 0.1571 | 0.1564 | 0.9877 | 0.1584 | 6.314 | 1.4137 | 81.0 |
| | | **Cos** | **Sin** | **Cot** | **Tan** | **Rad** | **Deg** |

*Table B-1 (continued)*

| Deg | Rad | Sin | Cos | Tan | Cot | | |
|---|---|---|---|---|---|---|---|
| 9.0 | 0.1571 | 0.1564 | 0.9877 | 0.1584 | 6.314 | 1.4137 | 81.0 |
| .1 | 0.1588 | 0.1582 | 0.9874 | 0.1602 | 6.243 | 1.4120 | .9 |
| .2 | 0.1606 | 0.1599 | 0.9871 | 0.1620 | 6.174 | 1.4102 | .8 |
| .3 | 0.1623 | 0.1616 | 0.9869 | 0.1638 | 6.107 | 1.4085 | .7 |
| .4 | 0.1641 | 0.1633 | 0.9866 | 0.1655 | 6.041 | 1.4067 | .6 |
| .5 | 0.1658 | 0.1650 | 0.9863 | 0.1673 | 5.976 | 1.4050 | .5 |
| .6 | 0.1676 | 0.1668 | 0.9860 | 0.1691 | 5.912 | 1.4032 | .4 |
| .7 | 0.1693 | 0.1685 | 0.9857 | 0.1709 | 5.850 | 1.4015 | .3 |
| .8 | 0.1710 | 0.1702 | 0.9854 | 0.1727 | 5.789 | 1.3998 | .2 |
| .9 | 0.1728 | 0.1719 | 0.9851 | 0.1745 | 5.730 | 1.3980 | .1 |
| 10.0 | 0.1745 | 0.1736 | 0.9848 | 0.1763 | 5.671 | 1.3963 | 80.0 |
| .1 | 0.1763 | 0.1754 | 0.9845 | 0.1781 | 5.614 | 1.3945 | .9 |
| .2 | 0.1780 | 0.1771 | 0.9842 | 0.1799 | 5.558 | 1.3928 | .8 |
| .3 | 0.1798 | 0.1788 | 0.9839 | 0.1817 | 5.503 | 1.3910 | .7 |
| .4 | 0.1815 | 0.1805 | 0.9836 | 0.1835 | 5.449 | 1.3893 | .6 |
| .5 | 0.1833 | 0.1822 | 0.9833 | 0.1853 | 5.396 | 1.3875 | .5 |
| .6 | 0.1850 | 0.1840 | 0.9829 | 0.1871 | 5.343 | 1.3858 | .4 |
| .7 | 0.1868 | 0.1857 | 0.9826 | 0.1890 | 5.292 | 1.3840 | .3 |
| .8 | 0.1885 | 0.1874 | 0.9823 | 0.1908 | 5.242 | 1.3823 | .2 |
| .9 | 0.1902 | 0.1891 | 0.9820 | 0.1926 | 5.193 | 1.3806 | .1 |
| 11.0 | 0.1920 | 0.1908 | 0.9816 | 0.1944 | 5.145 | 1.3788 | 79.0 |
| .1 | 0.1937 | 0.1925 | 0.9813 | 0.1962 | 5.097 | 1.3771 | .9 |
| .2 | 0.1955 | 0.1942 | 0.9810 | 0.1980 | 5.050 | 1.3753 | .8 |
| .3 | 0.1972 | 0.1959 | 0.9806 | 0.1998 | 5.005 | 1.3736 | .7 |
| .4 | 0.1990 | 0.1977 | 0.9803 | 0.2016 | 4.959 | 1.3718 | .6 |
| .5 | 0.2007 | 0.1994 | 0.9799 | 0.2035 | 4.915 | 1.3701 | .5 |
| .6 | 0.2025 | 0.2011 | 0.9796 | 0.2053 | 4.872 | 1.3683 | .4 |
| .7 | 0.2042 | 0.2028 | 0.9792 | 0.2071 | 4.829 | 1.3666 | .3 |
| .8 | 0.2059 | 0.2045 | 0.9789 | 0.2089 | 4.787 | 1.3648 | .2 |
| .9 | 0.2077 | 0.2062 | 0.9785 | 0.2107 | 4.745 | 1.3641 | .1 |
| 12.0 | 0.2094 | 0.2079 | 0.9781 | 0.2126 | 4.705 | 1.3614 | 78.0 |
| .1 | 0.2112 | 0.2096 | 0.9778 | 0.2144 | 4.665 | 1.3596 | .9 |
| .2 | 0.2129 | 0.2113 | 0.9774 | 0.2162 | 4.625 | 1.3579 | .8 |
| .3 | 0.2147 | 0.2130 | 0.9770 | 0.2180 | 4.586 | 1.3561 | .7 |
| .4 | 0.2164 | 0.2147 | 0.9767 | 0.2199 | 4.548 | 1.3544 | .6 |
| .5 | 0.2182 | 0.2164 | 0.9763 | 0.2217 | 4.511 | 1.3526 | .5 |
| .6 | 0.2199 | 0.2181 | 0.9759 | 0.2235 | 4.474 | 1.3509 | .4 |
| .7 | 0.2217 | 0.2198 | 0.9755 | 0.2254 | 4.437 | 1.3491 | .3 |
| .8 | 0.2234 | 0.2215 | 0.9751 | 0.2272 | 4.402 | 1.3474 | .2 |
| .9 | 0.2251 | 0.2233 | 0.9748 | 0.2290 | 4.366 | 1.3456 | .1 |
| 13.0 | 0.2269 | 0.2250 | 0.9744 | 0.2309 | 4.331 | 1.3439 | 77.0 |
| | | **Cos** | **Sin** | **Cot** | **Tan** | **Rad** | **Deg** |

*Table B-1 (continued)*

| Deg | Rad | Sin | Cos | Tan | Cot | | |
|---|---|---|---|---|---|---|---|
| 13.0 | 0.2269 | 0.2250 | 0.9744 | 0.2309 | 4.331 | 1.3439 | 77.0 |
| .1 | 0.2286 | 0.2267 | 0.9740 | 0.2327 | 4.297 | 1.3422 | .9 |
| .2 | 0.2304 | 0.2284 | 0.9736 | 0.2345 | 4.264 | 1.3404 | .8 |
| .3 | 0.2321 | 0.2300 | 0.9732 | 0.2364 | 4.230 | 1.3387 | .7 |
| .4 | 0.2339 | 0.2317 | 0.9728 | 0.2382 | 4.198 | 1.3369 | .6 |
| .5 | 0.2356 | 0.2334 | 0.9724 | 0.2401 | 4.165 | 1.3352 | .5 |
| .6 | 0.2374 | 0.2351 | 0.9720 | 0.2419 | 4.134 | 1.3334 | .4 |
| .7 | 0.2391 | 0.2368 | 0.9715 | 0.2438 | 4.102 | 1.3317 | .3 |
| .8 | 0.2409 | 0.2385 | 0.9711 | 0.2456 | 4.071 | 1.3299 | .2 |
| .9 | 0.2426 | 0.2402 | 0.9707 | 0.2475 | 4.041 | 1.3282 | .1 |
| 14.0 | 0.2443 | 0.2419 | 0.9703 | 0.2493 | 4.011 | 1.3265 | 76.0 |
| .1 | 0.2461 | 0.2436 | 0.9699 | 0.2512 | 3.981 | 1.3247 | .9 |
| .2 | 0.2478 | 0.2453 | 0.9694 | 0.2530 | 3.952 | 1.3230 | .8 |
| .3 | 0.2496 | 0.2470 | 0.9690 | 0.2549 | 3.923 | 1.3212 | .7 |
| .4 | 0.2513 | 0.2487 | 0.9686 | 0.2568 | 3.895 | 1.3195 | .6 |
| .5 | 0.2531 | 0.2504 | 0.9681 | 0.2586 | 3.867 | 1.3177 | .5 |
| .6 | 0.2548 | 0.2521 | 0.9677 | 0.2605 | 3.839 | 1.3160 | .4 |
| .7 | 0.2566 | 0.2538 | 0.9673 | 0.2623 | 3.812 | 1.3142 | .3 |
| .8 | 0.2583 | 0.2554 | 0.9668 | 0.2642 | 3.785 | 1.3125 | .2 |
| .9 | 0.2600 | 0.2571 | 0.9664 | 0.2661 | 3.758 | 1.3107 | .1 |
| 15.0 | 0.2618 | 0.2588 | 0.9659 | 0.2679 | 3.732 | 1.3090 | 75.0 |
| .1 | 0.2635 | 0.2605 | 0.9655 | 0.2698 | 3.706 | 1.3073 | .9 |
| .2 | 0.2653 | 0.2622 | 0.9650 | 0.2717 | 3.681 | 1.3055 | .8 |
| .3 | 0.2670 | 0.2639 | 0.9646 | 0.2736 | 3.655 | 1.3038 | .7 |
| .4 | 0.2688 | 0.2656 | 0.9641 | 0.2754 | 3.630 | 1.3020 | .6 |
| .5 | 0.2705 | 0.2672 | 0.9636 | 0.2773 | 3.606 | 1.3003 | .5 |
| .6 | 0.2723 | 0.2689 | 0.9632 | 0.2792 | 3.582 | 1.2985 | .4 |
| .7 | 0.2740 | 0.2706 | 0.9627 | 0.2811 | 3.558 | 1.2968 | .3 |
| .8 | 0.2758 | 0.2723 | 0.9622 | 0.2830 | 3.534 | 1.2950 | .2 |
| .9 | 0.2775 | 0.2740 | 0.9617 | 0.2849 | 3.511 | 1.2933 | .1 |
| 16.0 | 0.2793 | 0.2756 | 0.9613 | 0.2867 | 3.487 | 1.2915 | 74.0 |
| .1 | 0.2810 | 0.2773 | 0.9608 | 0.2886 | 3.465 | 1.2898 | .9 |
| .2 | 0.2827 | 0.2790 | 0.9603 | 0.2905 | 3.442 | 1.2881 | .8 |
| .3 | 0.2845 | 0.2807 | 0.9598 | 0.2924 | 3.420 | 1.2863 | .7 |
| .4 | 0.2862 | 0.2823 | 0.9593 | 0.2943 | 3.398 | 1.2846 | .6 |
| .5 | 0.2880 | 0.2840 | 0.9588 | 0.2962 | 3.376 | 1.2828 | .5 |
| .6 | 0.2897 | 0.2857 | 0.9583 | 0.2981 | 3.354 | 1.2811 | .4 |
| .7 | 0.2915 | 0.2874 | 0.9578 | 0.3000 | 3.333 | 1.2793 | .3 |
| .8 | 0.2932 | 0.2890 | 0.9573 | 0.3019 | 3.312 | 1.2776 | .2 |
| .9 | 0.2950 | 0.2907 | 0.9568 | 0.3038 | 3.291 | 1.2758 | .1 |
| 17.0 | 0.2967 | 0.2924 | 0.9563 | 0.3057 | 3.271 | 1.2741 | 73.0 |
| | | Cos | Sin | Cot | Tan | Rad | Deg |

*Table B-1 (continued)*

| Deg | Rad | Sin | Cos | Tan | Cot | | |
|---|---|---|---|---|---|---|---|
| 17.0 | 0.2967 | 0.2924 | 0.9563 | 0.3057 | 3.271 | 1.2741 | 73.0 |
| .1 | 0.2985 | 0.2940 | 0.9558 | 0.3076 | 3.251 | 1.2723 | .9 |
| .2 | 0.3002 | 0.2957 | 0.9553 | 0.3096 | 3.230 | 1.2706 | .8 |
| .3 | 0.3019 | 0.2974 | 0.9548 | 0.3115 | 3.211 | 1.2689 | .7 |
| .4 | 0.3037 | 0.2990 | 0.9542 | 0.3134 | 3.191 | 1.2671 | .6 |
| .5 | 0.3054 | 0.3007 | 0.9537 | 0.3153 | 3.172 | 1.2654 | .5 |
| .6 | 0.3072 | 0.3024 | 0.9532 | 0.3172 | 3.152 | 1.2636 | .4 |
| .7 | 0.3089 | 0.3040 | 0.9527 | 0.3191 | 3.133 | 1.2619 | .3 |
| .8 | 0.3107 | 0.3057 | 0.9521 | 0.3211 | 3.115 | 1.2601 | .2 |
| .9 | 0.3124 | 0.3074 | 0.9516 | 0.3230 | 3.096 | 1.2584 | .1 |
| 18.0 | 0.3142 | 0.3090 | 0.9511 | 0.3249 | 3.078 | 1.2566 | 72.0 |
| .1 | 0.3159 | 0.3107 | 0.9505 | 0.3269 | 3.060 | 1.2549 | .9 |
| .2 | 0.3177 | 0.3123 | 0.9500 | 0.3288 | 3.042 | 1.2531 | .8 |
| .3 | 0.3194 | 0.3140 | 0.9494 | 0.3307 | 3.024 | 1.2514 | .7 |
| .4 | 0.3211 | 0.3156 | 0.9489 | 0.3327 | 3.006 | 1.2497 | .6 |
| .5 | 0.3229 | 0.3173 | 0.9483 | 0.3346 | 2.989 | 1.2479 | .5 |
| .6 | 0.3246 | 0.3190 | 0.9478 | 0.3365 | 2.971 | 1.2462 | .4 |
| .7 | 0.3264 | 0.3206 | 0.9472 | 0.3385 | 2.954 | 1.2444 | .3 |
| .8 | 0.3281 | 0.3223 | 0.9466 | 0.3404 | 2.937 | 1.2427 | .2 |
| .9 | 0.3299 | 0.3239 | 0.9461 | 0.3424 | 2.921 | 1.2409 | .1 |
| 19.0 | 0.3316 | 0.3256 | 0.9455 | 0.3443 | 2.904 | 1.2392 | 71.0 |
| .1 | 0.3334 | 0.3272 | 0.9449 | 0.3463 | 2.888 | 1.2374 | .9 |
| .2 | 0.3351 | 0.3289 | 0.9444 | 0.3482 | 2.872 | 1.2357 | .8 |
| .3 | 0.3368 | 0.3305 | 0.9438 | 0.3502 | 2.856 | 1.2339 | .7 |
| .4 | 0.3386 | 0.3322 | 0.9432 | 0.3522 | 2.840 | 1.2322 | .6 |
| .5 | 0.3403 | 0.3338 | 0.9426 | 0.3541 | 2.824 | 1.2305 | .5 |
| .6 | 0.3421 | 0.3355 | 0.9421 | 0.3561 | 2.808 | 1.2287 | .4 |
| .7 | 0.3438 | 0.3371 | 0.9415 | 0.3581 | 2.793 | 1.2270 | .3 |
| .8 | 0.3456 | 0.3387 | 0.9409 | 0.3600 | 2.778 | 1.2252 | .2 |
| .9 | 0.3473 | 0.3404 | 0.9403 | 0.3620 | 2.762 | 1.2235 | .1 |
| 20.0 | 0.3491 | 0.3420 | 0.9397 | 0.3640 | 2.747 | 1.2217 | 70.0 |
| .1 | 0.3508 | 0.3437 | 0.9391 | 0.3659 | 2.733 | 1.2200 | .9 |
| .2 | 0.3526 | 0.3453 | 0.9385 | 0.3679 | 2.718 | 1.2182 | .8 |
| .3 | 0.3543 | 0.3469 | 0.9379 | 0.3699 | 2.703 | 1.2165 | .7 |
| .4 | 0.3560 | 0.3486 | 0.9373 | 0.3719 | 2.689 | 1.2147 | .6 |
| .5 | 0.3578 | 0.3502 | 0.9367 | 0.3739 | 2.675 | 1.2130 | .5 |
| .6 | 0.3595 | 0.3518 | 0.9361 | 0.3759 | 2.660 | 1.2113 | .4 |
| .7 | 0.3613 | 0.3535 | 0.9354 | 0.3779 | 2.646 | 1.2095 | .3 |
| .8 | 0.3630 | 0.3551 | 0.9348 | 0.3799 | 2.633 | 1.2078 | .2 |
| .9 | 0.3648 | 0.3567 | 0.9342 | 0.3819 | 2.619 | 1.2060 | .1 |
| 21.0 | 0.3665 | 0.3584 | 0.9336 | 0.3839 | 2.605 | 1.2043 | 69.0 |
| | | **Cos** | **Sin** | **Cot** | **Tan** | **Rad** | **Deg** |

Table B-1 (continued)

| Deg | Rad | Sin | Cos | Tan | Cot | | |
|------|--------|--------|--------|--------|-------|--------|------|
| 21.0 | 0.3665 | 0.3584 | 0.9336 | 0.3839 | 2.605 | 1.2043 | 69.0 |
| .1 | 0.3683 | 0.3600 | 0.9330 | 0.3859 | 2.592 | 1.2025 | .9 |
| .2 | 0.3700 | 0.3616 | 0.9323 | 0.3879 | 2.578 | 1.2008 | .8 |
| .3 | 0.3718 | 0.3633 | 0.9317 | 0.3899 | 2.565 | 1.1990 | .7 |
| .4 | 0.3735 | 0.3649 | 0.9311 | 0.3919 | 2.552 | 1.1973 | .6 |
| .5 | 0.3752 | 0.3665 | 0.9304 | 0.3939 | 2.539 | 1.1956 | .5 |
| .6 | 0.3770 | 0.3681 | 0.9298 | 0.3959 | 2.526 | 1.1938 | .4 |
| .7 | 0.3787 | 0.3697 | 0.9291 | 0.3979 | 2.513 | 1.1921 | .3 |
| .8 | 0.3805 | 0.3714 | 0.9285 | 0.4000 | 2.500 | 1.1903 | .2 |
| .9 | 0.3822 | 0.3730 | 0.9278 | 0.4020 | 2.488 | 1.1886 | .1 |
| 22.0 | 0.3840 | 0.3746 | 0.9272 | 0.4040 | 2.475 | 1.1868 | 68.0 |
| .1 | 0.3857 | 0.3762 | 0.9265 | 0.4061 | 2.463 | 1.1851 | .9 |
| .2 | 0.3875 | 0.3778 | 0.9259 | 0.4081 | 2.450 | 1.1833 | .8 |
| .3 | 0.3892 | 0.3795 | 0.9252 | 0.4101 | 2.438 | 1.1816 | .7 |
| .4 | 0.3910 | 0.3811 | 0.9245 | 0.4122 | 2.426 | 1.1798 | .6 |
| .5 | 0.3927 | 0.3827 | 0.9239 | 0.4142 | 2.414 | 1.1781 | .5 |
| .6 | 0.3944 | 0.3843 | 0.9232 | 0.4163 | 2.402 | 1.1764 | .4 |
| .7 | 0.3962 | 0.3859 | 0.9225 | 0.4183 | 2.391 | 1.1746 | .3 |
| .8 | 0.3979 | 0.3875 | 0.9219 | 0.4204 | 2.379 | 1.1729 | .2 |
| .9 | 0.3997 | 0.3891 | 0.9212 | 0.4224 | 2.367 | 1.1711 | .1 |
| 23.0 | 0.4014 | 0.3907 | 0.9205 | 0.4245 | 2.356 | 1.1694 | 67.0 |
| .1 | 0.4032 | 0.3923 | 0.9198 | 0.4265 | 2.344 | 1.1676 | .9 |
| .2 | 0.4049 | 0.3939 | 0.9191 | 0.4286 | 2.333 | 1.1659 | .8 |
| .3 | 0.4067 | 0.3955 | 0.9184 | 0.4307 | 2.322 | 1.1641 | .7 |
| .4 | 0.4084 | 0.3971 | 0.9178 | 0.4327 | 2.311 | 1.1624 | .6 |
| .5 | 0.4102 | 0.3987 | 0.9171 | 0.4348 | 2.300 | 1.1606 | .5 |
| .6 | 0.4119 | 0.4003 | 0.9164 | 0.4369 | 2.289 | 1.1589 | .4 |
| .7 | 0.4136 | 0.4019 | 0.9157 | 0.4390 | 2.278 | 1.1572 | .3 |
| .8 | 0.4154 | 0.4035 | 0.9150 | 0.4411 | 2.267 | 1.1554 | .2 |
| .9 | 0.4171 | 0.4051 | 0.9143 | 0.4431 | 2.257 | 1.1537 | .1 |
| 24.0 | 0.4189 | 0.4067 | 0.9135 | 0.4452 | 2.246 | 1.1519 | 66.0 |
| .1 | 0.4206 | 0.4083 | 0.9128 | 0.4473 | 2.236 | 1.1502 | .9 |
| .2 | 0.4224 | 0.4099 | 0.9121 | 0.4494 | 2.225 | 1.1484 | .8 |
| .3 | 0.4241 | 0.4115 | 0.9114 | 0.4515 | 2.215 | 1.1467 | .7 |
| .4 | 0.4259 | 0.4131 | 0.9107 | 0.4536 | 2.204 | 1.1449 | .6 |
| .5 | 0.4276 | 0.4147 | 0.9100 | 0.4557 | 2.194 | 1.1432 | .5 |
| .6 | 0.4294 | 0.4163 | 0.9092 | 0.4578 | 2.184 | 1.1414 | .4 |
| .7 | 0.4311 | 0.4179 | 0.9085 | 0.4599 | 2.174 | 1.1397 | .3 |
| .8 | 0.4328 | 0.4195 | 0.9078 | 0.4621 | 2.164 | 1.1379 | .2 |
| .9 | 0.4346 | 0.4210 | 0.9070 | 0.4642 | 2.154 | 1.1362 | .1 |
| 25.0 | 0.4363 | 0.4226 | 0.9063 | 0.4663 | 2.145 | 1.1345 | 65.0 |
| | | Cos | Sin | Cot | Tan | Rad | Deg |

*Table B-1 (continued)*

| Deg | Rad | Sin | Cos | Tan | Cot | | |
|---|---|---|---|---|---|---|---|
| 25.0 | 0.4363 | 0.4226 | 0.9063 | 0.4663 | 2.145 | 1.1345 | 65.0 |
| .1 | 0.4381 | 0.4242 | 0.9056 | 0.4684 | 2.135 | 1.1327 | .9 |
| .2 | 0.4398 | 0.4258 | 0.9048 | 0.4706 | 2.125 | 1.1310 | .8 |
| .3 | 0.4416 | 0.4274 | 0.9041 | 0.4727 | 2.116 | 1.1292 | .7 |
| .4 | 0.4433 | 0.4289 | 0.9033 | 0.4748 | 2.106 | 1.1275 | .6 |
| .5 | 0.4451 | 0.4305 | 0.9026 | 0.4770 | 2.097 | 1.1257 | .5 |
| .6 | 0.4468 | 0.4321 | 0.9018 | 0.4791 | 2.087 | 1.1240 | .4 |
| .7 | 0.4485 | 0.4337 | 0.9011 | 0.4813 | 2.078 | 1.1222 | .3 |
| .8 | 0.4503 | 0.4352 | 0.9003 | 0.4834 | 2.069 | 1.1205 | .2 |
| .9 | 0.4520 | 0.4368 | 0.8996 | 0.4856 | 2.059 | 1.1188 | .1 |
| 26.0 | 0.4538 | 0.4384 | 0.8988 | 0.4877 | 2.050 | 1.1170 | 64.0 |
| .1 | 0.4555 | 0.4399 | 0.8980 | 0.4899 | 2.041 | 1.1153 | .9 |
| .2 | 0.4573 | 0.4415 | 0.8973 | 0.4921 | 2.032 | 1.1135 | .8 |
| .3 | 0.4590 | 0.4431 | 0.8965 | 0.4942 | 2.023 | 1.1118 | .7 |
| .4 | 0.4608 | 0.4446 | 0.8957 | 0.4964 | 2.014 | 1.1100 | .6 |
| .5 | 0.4625 | 0.4462 | 0.8949 | 0.4986 | 2.006 | 1.1083 | .5 |
| .6 | 0.4643 | 0.4478 | 0.8942 | 0.5008 | 1.997 | 1.1065 | .4 |
| .7 | 0.4660 | 0.4493 | 0.8934 | 0.5029 | 1.988 | 1.1048 | .3 |
| .8 | 0.4677 | 0.4509 | 0.8926 | 0.5051 | 1.980 | 1.1030 | .2 |
| .9 | 0.4695 | 0.4524 | 0.8918 | 0.5073 | 1.971 | 1.1013 | .1 |
| 27.0 | 0.4712 | 0.4540 | 0.8910 | 0.5095 | 1.963 | 1.0996 | 63.0 |
| .1 | 0.4730 | 0.4555 | 0.8902 | 0.5117 | 1.954 | 1.0978 | .9 |
| .2 | 0.4747 | 0.4571 | 0.8894 | 0.5139 | 1.946 | 1.0961 | .8 |
| .3 | 0.4765 | 0.4586 | 0.8886 | 0.5161 | 1.937 | 1.0943 | .7 |
| .4 | 0.4782 | 0.4602 | 0.8878 | 0.5184 | 1.929 | 1.0926 | .6 |
| .5 | 0.4800 | 0.4617 | 0.8870 | 0.5206 | 1.921 | 1.0908 | .5 |
| .6 | 0.4817 | 0.4633 | 0.8862 | 0.5228 | 1.913 | 1.0891 | .4 |
| .7 | 0.4835 | 0.4648 | 0.8854 | 0.5250 | 1.905 | 1.0873 | .3 |
| .8 | 0.4852 | 0.4664 | 0.8846 | 0.5272 | 1.897 | 1.0856 | .2 |
| .9 | 0.4869 | 0.4679 | 0.8838 | 0.5295 | 1.889 | 1.0838 | .1 |
| 28.0 | 0.4887 | 0.4695 | 0.8829 | 0.5317 | 1.881 | 1.0821 | 62.0 |
| .1 | 0.4904 | 0.4710 | 0.8821 | 0.5340 | 1.873 | 1.0804 | .9 |
| .2 | 0.4922 | 0.4726 | 0.8813 | 0.5362 | 1.865 | 1.0786 | .8 |
| .3 | 0.4939 | 0.4741 | 0.8805 | 0.5384 | 1.857 | 1.0769 | .7 |
| .4 | 0.4957 | 0.4756 | 0.8796 | 0.5407 | 1.849 | 1.0751 | .6 |
| .5 | 0.4974 | 0.4772 | 0.8788 | 0.5430 | 1.842 | 1.0734 | .5 |
| .6 | 0.4992 | 0.4787 | 0.8780 | 0.5452 | 1.834 | 1.0716 | .4 |
| .7 | 0.5009 | 0.4802 | 0.8771 | 0.5475 | 1.827 | 1.0699 | .3 |
| .8 | 0.5027 | 0.4818 | 0.8763 | 0.5498 | 1.819 | 1.0681 | .2 |
| .9 | 0.5044 | 0.4833 | 0.8755 | 0.5520 | 1.811 | 1.0664 | .1 |
| 29.0 | 0.5061 | 0.4848 | 0.8746 | 0.5543 | 1.804 | 1.0647 | 61.0 |
| | | Cos | Sin | Cot | Tan | Rad | Deg |

Table B-1 *(continued)*

| Deg | Rad | Sin | Cos | Tan | Cot | | |
|---|---|---|---|---|---|---|---|
| 29.0 | 0.5061 | 0.4848 | 0.8746 | 0.5543 | 1.804 | 1.0647 | 61.0 |
| .1 | 0.5079 | 0.4863 | 0.8738 | 0.5566 | 1.797 | 1.0629 | .9 |
| .2 | 0.5096 | 0.4879 | 0.8729 | 0.5589 | 1.789 | 1.0612 | .8 |
| .3 | 0.5114 | 0.4894 | 0.8721 | 0.5612 | 1.782 | 1.0594 | .7 |
| .4 | 0.5131 | 0.4909 | 0.8712 | 0.5635 | 1.775 | 1.0577 | .6 |
| .5 | 0.5149 | 0.4924 | 0.8704 | 0.5658 | 1.767 | 1.0559 | .5 |
| .6 | 0.5166 | 0.4939 | 0.8695 | 0.5681 | 1.760 | 1.0542 | .4 |
| .7 | 0.5184 | 0.4955 | 0.8686 | 0.5704 | 1.753 | 1.0524 | .3 |
| .8 | 0.5201 | 0.4970 | 0.8678 | 0.5727 | 1.746 | 1.0507 | .2 |
| .9 | 0.5219 | 0.4985 | 0.8669 | 0.5750 | 1.739 | 1.0489 | .1 |
| 30.0 | 0.5236 | 0.5000 | 0.8660 | 0.5774 | 1.732 | 1.0472 | 60.0 |
| .1 | 0.5253 | 0.5015 | 0.8652 | 0.5797 | 1.725 | 1.0455 | .9 |
| .2 | 0.5271 | 0.5030 | 0.8643 | 0.5820 | 1.718 | 1.0437 | .8 |
| .3 | 0.5288 | 0.5045 | 0.8634 | 0.5844 | 1.711 | 1.0420 | .7 |
| .4 | 0.5306 | 0.5060 | 0.8625 | 0.5867 | 1.704 | 1.0402 | .6 |
| .5 | 0.5323 | 0.5075 | 0.8616 | 0.5890 | 1.698 | 1.0385 | .5 |
| .6 | 0.5341 | 0.5090 | 0.8607 | 0.5914 | 1.691 | 1.0367 | .4 |
| .7 | 0.5358 | 0.5105 | 0.8599 | 0.5938 | 1.684 | 1.0350 | .3 |
| .8 | 0.5376 | 0.5120 | 0.8590 | 0.5961 | 1.678 | 1.0332 | .2 |
| .9 | 0.5393 | 0.5135 | 0.8581 | 0.5985 | 1.671 | 1.0315 | .1 |
| 31.0 | 0.5411 | 0.5150 | 0.8572 | 0.6009 | 1.664 | 1.0297 | 59.0 |
| .1 | 0.5428 | 0.5165 | 0.8563 | 0.6032 | 1.658 | 1.0280 | .9 |
| .2 | 0.5445 | 0.5180 | 0.8554 | 0.6056 | 1.651 | 1.0263 | .8 |
| .3 | 0.5463 | 0.5195 | 0.8545 | 0.6080 | 1.645 | 1.0245 | .7 |
| .4 | 0.5480 | 0.5210 | 0.8536 | 0.6104 | 1.638 | 1.0228 | .6 |
| .5 | 0.5498 | 0.5225 | 0.8526 | 0.6128 | 1.632 | 1.0210 | .5 |
| .6 | 0.5515 | 0.5240 | 0.8517 | 0.6152 | 1.625 | 1.0193 | .4 |
| .7 | 0.5533 | 0.5255 | 0.8508 | 0.6176 | 1.619 | 1.0175 | .3 |
| .8 | 0.5550 | 0.5270 | 0.8499 | 0.6200 | 1.613 | 1.0158 | .2 |
| .9 | 0.5568 | 0.5284 | 0.8490 | 0.6224 | 1.607 | 1.0140 | .1 |
| 32.0 | 0.5585 | 0.5299 | 0.8480 | 0.6249 | 1.600 | 1.0123 | 58.0 |
| .1 | 0.5603 | 0.5314 | 0.8471 | 0.6273 | 1.594 | 1.0105 | .9 |
| .2 | 0.5620 | 0.5329 | 0.8462 | 0.6297 | 1.588 | 1.0088 | .8 |
| .3 | 0.5637 | 0.5344 | 0.8453 | 0.6322 | 1.582 | 1.0071 | .7 |
| .4 | 0.5655 | 0.5358 | 0.8443 | 0.6346 | 1.576 | 1.0053 | .6 |
| .5 | 0.5672 | 0.5373 | 0.8434 | 0.6371 | 1.570 | 1.0036 | .5 |
| .6 | 0.5690 | 0.5388 | 0.8425 | 0.6395 | 1.564 | 1.0018 | .4 |
| .7 | 0.5707 | 0.5402 | 0.8415 | 0.6420 | 1.558 | 1.0000 | .3 |
| .8 | 0.5725 | 0.5417 | 0.8406 | 0.6445 | 1.552 | 0.9983 | .2 |
| .9 | 0.5742 | 0.5432 | 0.8396 | 0.6469 | 1.546 | 0.9966 | .1 |
| 33.0 | 0.5760 | 0.5446 | 0.8387 | 0.6494 | 1.540 | 0.9948 | 57.0 |
| | | Cos | Sin | Cot | Tan | Rad | Deg |

*Table B-1 (continued)*

| Deg | Rad | Sin | Cos | Tan | Cot | | |
|-----|-----|-----|-----|-----|-----|-----|-----|
| 33.0 | 0.5760 | 0.5446 | 0.8387 | 0.6494 | 1.540 | 0.9948 | 57.0 |
| .1 | 0.5777 | 0.5461 | 0.8377 | 0.6519 | 1.534 | 0.9931 | .9 |
| .2 | 0.5794 | 0.5476 | 0.8368 | 0.6544 | 1.528 | 0.9913 | .8 |
| .3 | 0.5812 | 0.5490 | 0.8358 | 0.6569 | 1.522 | 0.9896 | .7 |
| .4 | 0.5829 | 0.5505 | 0.8348 | 0.6594 | 1.517 | 0.9879 | .6 |
| .5 | 0.5847 | 0.5519 | 0.8339 | 0.6619 | 1.511 | 0.9861 | .5 |
| .6 | 0.5864 | 0.5534 | 0.8329 | 0.6644 | 1.505 | 0.9844 | .4 |
| .7 | 0.5882 | 0.5548 | 0.8320 | 0.6669 | 1.499 | 0.9826 | .3 |
| .8 | 0.5899 | 0.5563 | 0.8310 | 0.6694 | 1.494 | 0.9809 | .2 |
| .9 | 0.5917 | 0.5577 | 0.8300 | 0.6720 | 1.488 | 0.9791 | .1 |
| 34.0 | 0.5934 | 0.5592 | 0.8290 | 0.6745 | 1.483 | 0.9774 | 56.0 |
| .1 | 0.5952 | 0.5606 | 0.8281 | 0.6771 | 1.477 | 0.9756 | .9 |
| .2 | 0.5969 | 0.5621 | 0.8271 | 0.6796 | 1.471 | 0.9739 | .8 |
| .3 | 0.5986 | 0.5635 | 0.8261 | 0.6822 | 1.466 | 0.9721 | .7 |
| .4 | 0.6004 | 0.5650 | 0.8251 | 0.6847 | 1.460 | 0.9704 | .6 |
| .5 | 0.6021 | 0.5664 | 0.8241 | 0.6873 | 1.455 | 0.9687 | .5 |
| .6 | 0.6039 | 0.5678 | 0.8231 | 0.6899 | 1.450 | 0.9669 | .4 |
| .7 | 0.6056 | 0.5693 | 0.8221 | 0.6924 | 1.444 | 0.9652 | .3 |
| .8 | 0.6074 | 0.5707 | 0.8211 | 0.6950 | 1.439 | 0.9634 | .2 |
| .9 | 0.6091 | 0.5721 | 0.8202 | 0.6976 | 1.433 | 0.9617 | .1 |
| 35.0 | 0.6109 | 0.5736 | 0.8192 | 0.7002 | 1.428 | 0.9599 | 55.0 |
| .1 | 0.6126 | 0.5750 | 0.8181 | 0.7028 | 1.423 | 0.9582 | .9 |
| .2 | 0.6144 | 0.5764 | 0.8171 | 0.7054 | 1.418 | 0.9564 | .8 |
| .3 | 0.6161 | 0.5779 | 0.8161 | 0.7080 | 1.412 | 0.9547 | .7 |
| .4 | 0.6178 | 0.5793 | 0.8151 | 0.7107 | 1.407 | 0.9530 | .6 |
| .5 | 0.6196 | 0.5807 | 0.8141 | 0.7133 | 1.402 | 0.9512 | .5 |
| .6 | 0.6213 | 0.5821 | 0.8131 | 0.7159 | 1.397 | 0.9495 | .4 |
| .7 | 0.6231 | 0.5835 | 0.8121 | 0.7186 | 1.392 | 0.9477 | .3 |
| .8 | 0.6248 | 0.5850 | 0.8111 | 0.7212 | 1.387 | 0.9460 | .2 |
| .9 | 0.6266 | 0.5864 | 0.8100 | 0.7239 | 1.381 | 0.9442 | .1 |
| 36.0 | 0.6283 | 0.5878 | 0.8090 | 0.7265 | 1.376 | 0.9425 | 54.0 |
| .1 | 0.6301 | 0.5892 | 0.8080 | 0.7292 | 1.371 | 0.9407 | .9 |
| .2 | 0.6318 | 0.5906 | 0.8070 | 0.7319 | 1.366 | 0.9390 | .8 |
| .3 | 0.6336 | 0.5920 | 0.8059 | 0.7346 | 1.361 | 0.9372 | .7 |
| .4 | 0.6353 | 0.5934 | 0.8049 | 0.7373 | 1.356 | 0.9355 | .6 |
| .5 | 0.6370 | 0.5948 | 0.8039 | 0.7400 | 1.351 | 0.9338 | .5 |
| .6 | 0.6388 | 0.5962 | 0.8028 | 0.7427 | 1.347 | 0.9320 | .4 |
| .7 | 0.6405 | 0.5976 | 0.8018 | 0.7454 | 1.342 | 0.9303 | .3 |
| .8 | 0.6423 | 0.5990 | 0.8007 | 0.7481 | 1.337 | 0.9285 | .2 |
| .9 | 0.6440 | 0.6004 | 0.7997 | 0.7508 | 1.332 | 0.9268 | .1 |
| 37.0 | 0.6458 | 0.6018 | 0.7986 | 0.7536 | 1.327 | 0.9250 | 53.0 |
| | | Cos | Sin | Cot | Tan | Rad | Deg |

Table B-1 (continued)

| Deg | Rad | Sin | Cos | Tan | Cot | | |
|---|---|---|---|---|---|---|---|
| 37.0 | 0.6458 | 0.6018 | 0.7986 | 0.7536 | 1.327 | 0.9250 | 53.0 |
| .1 | 0.6475 | 0.6032 | 0.7976 | 0.7563 | 1.322 | 0.9233 | .9 |
| .2 | 0.6493 | 0.6046 | 0.7965 | 0.7590 | 1.317 | 0.9215 | .8 |
| .3 | 0.6510 | 0.6060 | 0.7955 | 0.7618 | 1.313 | 0.9198 | .7 |
| .4 | 0.6528 | 0.6074 | 0.7944 | 0.7646 | 1.308 | 0.9180 | .6 |
| .5 | 0.6545 | 0.6088 | 0.7934 | 0.7673 | 1.303 | 0.9163 | .5 |
| .6 | 0.6562 | 0.6101 | 0.7923 | 0.7701 | 1.299 | 0.9146 | .4 |
| .7 | 0.6580 | 0.6115 | 0.7912 | 0.7729 | 1.294 | 0.9128 | .3 |
| .8 | 0.6597 | 0.6129 | 0.7902 | 0.7757 | 1.289 | 0.9111 | .2 |
| .9 | 0.6615 | 0.6143 | 0.7891 | 0.7785 | 1.285 | 0.9093 | .1 |
| 38.0 | 0.6632 | 0.6157 | 0.7880 | 0.7813 | 1.280 | 0.9076 | 52.0 |
| .1 | 0.6650 | 0.6170 | 0.7869 | 0.7841 | 1.275 | 0.9058 | .9 |
| .2 | 0.6667 | 0.6184 | 0.7859 | 0.7869 | 1.271 | 0.9041 | .8 |
| .3 | 0.6685 | 0.6198 | 0.7848 | 0.7898 | 1.266 | 0.9023 | .7 |
| .4 | 0.6702 | 0.6211 | 0.7837 | 0.7926 | 1.262 | 0.9006 | .6 |
| .5 | 0.6720 | 0.6225 | 0.7826 | 0.7954 | 1.257 | 0.8988 | .5 |
| .6 | 0.6737 | 0.6239 | 0.7815 | 0.7983 | 1.253 | 0.8971 | .4 |
| .7 | 0.6754 | 0.6252 | 0.7804 | 0.8012 | 1.248 | 0.8954 | .3 |
| .8 | 0.6772 | 0.6266 | 0.7793 | 0.8040 | 1.244 | 0.8936 | .2 |
| .9 | 0.6789 | 0.6280 | 0.7782 | 0.8069 | 1.239 | 0.8919 | .1 |
| 39.0 | 0.6807 | 0.6293 | 0.7771 | 0.8098 | 1.235 | 0.8901 | 51.0 |
| .1 | 0.6824 | 0.6307 | 0.7760 | 0.8129 | 1.230 | 0.8884 | .9 |
| .2 | 0.6842 | 0.6320 | 0.7749 | 0.8156 | 1.226 | 0.8866 | .8 |
| .3 | 0.6859 | 0.6334 | 0.7738 | 0.8185 | 1.222 | 0.8849 | .7 |
| .4 | 0.6877 | 0.6347 | 0.7727 | 0.8214 | 1.217 | 0.8831 | .6 |
| .5 | 0.6894 | 0.6361 | 0.7716 | 0.8243 | 1.213 | 0.8814 | .5 |
| .6 | 0.6912 | 0.6374 | 0.7705 | 0.8273 | 1.209 | 0.8796 | .4 |
| .7 | 0.6929 | 0.6388 | 0.7694 | 0.8302 | 1.205 | 0.8779 | .3 |
| .8 | 0.6946 | 0.6401 | 0.7683 | 0.8332 | 1.200 | 0.8762 | .2 |
| .9 | 0.6964 | 0.6414 | 0.7672 | 0.8361 | 1.196 | 0.8744 | .1 |
| 40.0 | 0.6981 | 0.6428 | 0.7660 | 0.8391 | 1.192 | 0.8727 | 50.0 |
| .1 | 0.6999 | 0.6441 | 0.7649 | 0.8421 | 1.188 | 0.8709 | .9 |
| .2 | 0.7016 | 0.6455 | 0.7638 | 0.8451 | 1.183 | 0.8691 | .8 |
| .3 | 0.7034 | 0.6468 | 0.7627 | 0.8481 | 1.179 | 0.8674 | .7 |
| .4 | 0.7051 | 0.6481 | 0.7615 | 0.8511 | 1.175 | 0.8656 | .6 |
| .5 | 0.7069 | 0.6494 | 0.7604 | 0.8541 | 1.171 | 0.8639 | .5 |
| .6 | 0.7086 | 0.6508 | 0.7593 | 0.8571 | 1.167 | 0.8622 | .4 |
| .7 | 0.7103 | 0.6521 | 0.7581 | 0.8601 | 1.163 | 0.8604 | .3 |
| .8 | 0.7121 | 0.6534 | 0.7570 | 0.8632 | 1.159 | 0.8587 | .2 |
| .9 | 0.7138 | 0.6547 | 0.7559 | 0.8662 | 1.154 | 0.8570 | .1 |
| 41.0 | 0.7156 | 0.6561 | 0.7547 | 0.8693 | 1.150 | 0.8552 | 49.0 |
| | | Cos | Sin | Cot | Tan | Rad | Deg |

*Table B-1 (continued)*

| Deg | Rad | Sin | Cos | Tan | Cot | | |
|---|---|---|---|---|---|---|---|
| 41.0 | 0.7156 | 0.6561 | 0.7547 | 0.8693 | 1.150 | 0.8552 | 49.0 |
| .1 | 0.7173 | 0.6574 | 0.7536 | 0.8724 | 1.146 | 0.8535 | .9 |
| .2 | 0.7191 | 0.6587 | 0.7524 | 0.8754 | 1.142 | 0.8517 | .8 |
| .3 | 0.7208 | 0.6600 | 0.7513 | 0.8785 | 1.138 | 0.8500 | .7 |
| .4 | 0.7226 | 0.6613 | 0.7501 | 0.8816 | 1.134 | 0.8482 | .6 |
| .5 | 0.7243 | 0.6626 | 0.7490 | 0.8847 | 1.130 | 0.8465 | .5 |
| .6 | 0.7261 | 0.6639 | 0.7478 | 0.8878 | 1.126 | 0.8447 | .4 |
| .7 | 0.7278 | 0.6652 | 0.7466 | 0.8910 | 1.122 | 0.8430 | .3 |
| .8 | 0.7295 | 0.6665 | 0.7455 | 0.8941 | 1.118 | 0.8412 | .2 |
| .9 | 0.7313 | 0.6678 | 0.7443 | 0.8972 | 1.115 | 0.8395 | .1 |
| 42.0 | 0.7330 | 0.6691 | 0.7431 | 0.9004 | 1.111 | 0.8378 | 48.0 |
| .1 | 0.7348 | 0.6704 | 0.7420 | 0.9036 | 1.107 | 0.8360 | .9 |
| .2 | 0.7365 | 0.6717 | 0.7408 | 0.9067 | 1.103 | 0.8343 | .8 |
| .3 | 0.7383 | 0.6730 | 0.7396 | 0.9099 | 1.099 | 0.8325 | .7 |
| .4 | 0.7400 | 0.6743 | 0.7385 | 0.9131 | 1.095 | 0.8308 | .6 |
| .5 | 0.7418 | 0.6756 | 0.7373 | 0.9163 | 1.091 | 0.8290 | .5 |
| .6 | 0.7435 | 0.6769 | 0.7361 | 0.9195 | 1.087 | 0.8273 | .4 |
| .7 | 0.7453 | 0.6782 | 0.7349 | 0.9228 | 1.084 | 0.8255 | .3 |
| .8 | 0.7470 | 0.6794 | 0.7337 | 0.9260 | 1.080 | 0.8238 | .2 |
| .9 | 0.7487 | 0.6807 | 0.7325 | 0.9293 | 1.076 | 0.8221 | .1 |
| 43.0 | 0.7505 | 0.6820 | 0.7314 | 0.9325 | 1.072 | 0.8203 | 47.0 |
| .1 | 0.7522 | 0.6833 | 0.7302 | 0.9358 | 1.069 | 0.8186 | .9 |
| .2 | 0.7540 | 0.6845 | 0.7290 | 0.9391 | 1.065 | 0.8168 | .8 |
| .3 | 0.7558 | 0.6858 | 0.7278 | 0.9424 | 1.061 | 0.8151 | .7 |
| .4 | 0.7575 | 0.6871 | 0.7266 | 0.9457 | 1.057 | 0.8133 | .6 |
| .5 | 0.7592 | 0.6884 | 0.7254 | 0.9490 | 1.054 | 0.8116 | .5 |
| .6 | 0.7610 | 0.6896 | 0.7242 | 0.9523 | 1.050 | 0.8098 | .4 |
| .7 | 0.7627 | 0.6909 | 0.7230 | 0.9556 | 1.046 | 0.8081 | .3 |
| .8 | 0.7645 | 0.6921 | 0.7218 | 0.9590 | 1.043 | 0.8063 | .2 |
| .9 | 0.7662 | 0.6934 | 0.7206 | 0.9623 | 1.039 | 0.8046 | .1 |
| 44.0 | 0.7679 | 0.6947 | 0.7193 | 0.9657 | 1.036 | 0.8029 | 46.0 |
| .1 | 0.7697 | 0.6959 | 0.7181 | 0.9691 | 1.032 | 0.8011 | .9 |
| .2 | 0.7714 | 0.6972 | 0.7169 | 0.9725 | 1.028 | 0.7994 | .8 |
| .3 | 0.7732 | 0.6984 | 0.7157 | 0.9759 | 1.025 | 0.7976 | .7 |
| .4 | 0.7749 | 0.6997 | 0.7145 | 0.9793 | 1.021 | 0.7959 | .6 |
| .5 | 0.7767 | 0.7009 | 0.7133 | 0.9827 | 1.018 | 0.7941 | .5 |
| .6 | 0.7784 | 0.7022 | 0.7120 | 0.9861 | 1.014 | 0.7924 | .4 |
| .7 | 0.7802 | 0.7034 | 0.7108 | 0.9896 | 1.011 | 0.7907 | .3 |
| .8 | 0.7819 | 0.7046 | 0.7096 | 0.9930 | 1.007 | 0.7889 | .2 |
| .9 | 0.7837 | 0.7059 | 0.7083 | 0.9965 | 1.003 | 0.7871 | .1 |
| 45.0 | 0.7854 | 0.7071 | 0.7071 | 1.0000 | 1.000 | 0.7854 | 45.0 |
| | | **Cos** | **Sin** | **Cot** | **Tan** | **Rad** | **Deg** |

**Table B-2**   TRIGONOMETRIC FUNCTIONS
IN DEGREES AND MINUTES

| Degrees | Sin | Cos | Tan | Cot | |
|---|---|---|---|---|---|
| 0°00′ | 0.0000 | 1.0000 | 0.0000 | — | 90°00′ |
| 10 | 0.0029 | 1.0000 | 0.0029 | 343.8 | 50 |
| 20 | 0.0058 | 1.0000 | 0.0058 | 171.9 | 40 |
| 30 | 0.0087 | 1.0000 | 0.0087 | 114.6 | 30 |
| 40 | 0.0116 | 0.9999 | 0.0116 | 85.94 | 20 |
| 50 | 0.0145 | 0.9999 | 0.0145 | 68.75 | 10 |
| 1°00′ | 0.0175 | 0.9998 | 0.0175 | 57.29 | 89°00′ |
| 10 | 0.0204 | 0.9998 | 0.0204 | 49.10 | 50 |
| 20 | 0.0233 | 0.9997 | 0.0233 | 42.96 | 40 |
| 30 | 0.0262 | 0.9997 | 0.0262 | 38.19 | 30 |
| 40 | 0.0291 | 0.9996 | 0.0291 | 34.37 | 20 |
| 50 | 0.0320 | 0.9995 | 0.0320 | 31.24 | 10 |
| 2°00′ | 0.0349 | 0.9994 | 0.0349 | 28.64 | 88°00′ |
| 10 | 0.0378 | 0.9993 | 0.0378 | 26.43 | 50 |
| 20 | 0.0407 | 0.9992 | 0.0407 | 24.54 | 40 |
| 30 | 0.0436 | 0.9990 | 0.0437 | 22.90 | 30 |
| 40 | 0.0465 | 0.9989 | 0.0466 | 21.47 | 20 |
| 50 | 0.0494 | 0.9988 | 0.0495 | 20.21 | 10 |
| 3°00′ | 0.0523 | 0.9986 | 0.0524 | 19.08 | 87°00′ |
| 10 | 0.0552 | 0.9985 | 0.0533 | 18.07 | 50 |
| 20 | 0.0581 | 0.9983 | 0.0582 | 17.17 | 40 |
| 30 | 0.0610 | 0.9981 | 0.0612 | 16.35 | 30 |
| 40 | 0.0640 | 0.9980 | 0.0641 | 15.60 | 20 |
| 50 | 0.0669 | 0.9978 | 0.0670 | 14.92 | 10 |
| 4°00′ | 0.0698 | 0.9976 | 0.0699 | 14.30 | 86°00′ |
| 10 | 0.0727 | 0.9974 | 0.0729 | 13.73 | 50 |
| 20 | 0.0756 | 0.9971 | 0.0758 | 13.20 | 40 |
| 30 | 0.0785 | 0.9969 | 0.0787 | 12.71 | 30 |
| 40 | 0.0814 | 0.9967 | 0.0816 | 12.25 | 20 |
| 50 | 0.0843 | 0.9964 | 0.0846 | 11.83 | 10 |
| 5°00′ | 0.0872 | 0.9962 | 0.0875 | 11.43 | 85°00′ |
| 10 | 0.0901 | 0.9959 | 0.0904 | 11.06 | 50 |
| 20 | 0.0929 | 0.9957 | 0.0934 | 10.71 | 40 |
| 30 | 0.0958 | 0.9954 | 0.0963 | 10.39 | 30 |
| 40 | 0.0987 | 0.9951 | 0.0992 | 10.08 | 20 |
| 50 | 0.1016 | 0.9948 | 0.1022 | 9.788 | 10 |
| 6°00′ | 0.1045 | 0.9945 | 0.1051 | 9.514 | 84°00′ |
| 10 | 0.1074 | 0.9942 | 0.1080 | 9.255 | 50 |
| 20 | 0.1103 | 0.9939 | 0.1110 | 9.010 | 40 |
| 30 | 0.1132 | 0.9936 | 0.1139 | 8.777 | 30 |
| 40 | 0.1161 | 0.9932 | 0.1169 | 8.556 | 20 |
| 50 | 0.1190 | 0.9929 | 0.1198 | 8.345 | 10 |
| 7°00′ | 0.1219 | 0.9925 | 0.1228 | 8.144 | 83°00′ |
| | Cos | Sin | Cot | Tan | Degrees |

*Table B-2 (continued)*

| Degrees | Sin | Cos | Tan | Cot | |
|---|---|---|---|---|---|
| 7°00′ | 0.1219 | 0.9925 | 0.1228 | 8.144 | 83°00′ |
| 10 | 0.1248 | 0.9922 | 0.1257 | 7.953 | 50 |
| 20 | 0.1276 | 0.9918 | 0.1287 | 7.770 | 40 |
| 30 | 0.1305 | 0.9914 | 0.1317 | 7.596 | 30 |
| 40 | 0.1334 | 0.9911 | 0.1346 | 7.429 | 20 |
| 50 | 0.1363 | 0.9907 | 0.1376 | 7.269 | 10 |
| 8°00′ | 0.1392 | 0.9903 | 0.1405 | 7.115 | 82°00′ |
| 10 | 0.1421 | 0.9899 | 0.1435 | 6.968 | 50 |
| 20 | 0.1449 | 0.9894 | 0.1465 | 6.827 | 40 |
| 30 | 0.1478 | 0.9890 | 0.1495 | 6.691 | 30 |
| 40 | 0.1507 | 0.9886 | 0.1524 | 6.561 | 20 |
| 50 | 0.1536 | 0.9881 | 0.1554 | 6.435 | 10 |
| 9°00′ | 0.1564 | 0.9877 | 0.1584 | 6.314 | 81°00′ |
| 10 | 0.1593 | 0.9872 | 0.1614 | 6.197 | 50 |
| 20 | 0.1622 | 0.9868 | 0.1644 | 6.084 | 40 |
| 30 | 0.1650 | 0.9863 | 0.1673 | 5.976 | 30 |
| 40 | 0.1679 | 0.9858 | 0.1703 | 5.871 | 20 |
| 50 | 0.1708 | 0.9853 | 0.1733 | 5.769 | 10 |
| 10°00′ | 0.1736 | 0.9848 | 0.1763 | 5.671 | 80°00′ |
| 10 | 0.1765 | 0.9843 | 0.1793 | 5.576 | 50 |
| 20 | 0.1794 | 0.9838 | 0.1823 | 5.485 | 40 |
| 30 | 0.1822 | 0.9833 | 0.1853 | 5.396 | 30 |
| 40 | 0.1851 | 0.9827 | 0.1883 | 5.309 | 20 |
| 50 | 0.1880 | 0.9822 | 0.1914 | 5.226 | 10 |
| 11°00′ | 0.1908 | 0.9816 | 0.1944 | 5.145 | 79°00′ |
| 10 | 0.1937 | 0.9811 | 0.1974 | 5.066 | 50 |
| 20 | 0.1965 | 0.9805 | 0.2004 | 4.989 | 40 |
| 30 | 0.1994 | 0.9799 | 0.2035 | 4.915 | 30 |
| 40 | 0.2022 | 0.9793 | 0.2065 | 4.843 | 20 |
| 50 | 0.2051 | 0.9787 | 0.2095 | 4.773 | 10 |
| 12°00′ | 0.2079 | 0.9781 | 0.2126 | 4.705 | 78°00′ |
| 10 | 0.2108 | 0.9775 | 0.2156 | 4.638 | 50 |
| 20 | 0.2136 | 0.9769 | 0.2186 | 4.574 | 40 |
| 30 | 0.2164 | 0.9763 | 0.2217 | 4.511 | 30 |
| 40 | 0.2193 | 0.9757 | 0.2247 | 4.449 | 20 |
| 50 | 0.2221 | 0.9750 | 0.2278 | 4.390 | 10 |
| 13°00′ | 0.2250 | 0.9744 | 0.2309 | 4.331 | 77°00′ |
| 10 | 0.2278 | 0.9737 | 0.2339 | 4.275 | 50 |
| 20 | 0.2306 | 0.9730 | 0.2370 | 4.219 | 40 |
| 30 | 0.2334 | 0.9724 | 0.2401 | 4.165 | 30 |
| 40 | 0.2363 | 0.9717 | 0.2432 | 4.113 | 20 |
| 50 | 0.2391 | 0.9710 | 0.2462 | 4.061 | 10 |
| 14°00′ | 0.2419 | 0.9703 | 0.2493 | 4.011 | 76°00′ |
| | **Cos** | **Sin** | **Cot** | **Tan** | **Degrees** |

*Table B-2 (continued)*

| Degrees | Sin | Cos | Tan | Cot | |
|---------|------|------|------|------|--------|
| 14°00′ | 0.2419 | 0.9703 | 0.2493 | 4.011 | 76°00′ |
| 10 | 0.2447 | 0.9696 | 0.2524 | 3.962 | 50 |
| 20 | 0.2476 | 0.9689 | 0.2555 | 3.914 | 40 |
| 30 | 0.2504 | 0.9681 | 0.2586 | 3.867 | 30 |
| 20 | 0.2532 | 0.9674 | 0.2617 | 3.821 | 20 |
| 50 | 0.2560 | 0.9667 | 0.2648 | 3.776 | 10 |
| 15°00′ | 0.2588 | 0.9659 | 0.2679 | 3.732 | 75°00′ |
| 10 | 0.2616 | 0.9652 | 0.2711 | 3.689 | 50 |
| 20 | 0.2644 | 0.9644 | 0.2742 | 3.647 | 40 |
| 30 | 0.2672 | 0.9636 | 0.2773 | 3.606 | 30 |
| 40 | 0.2700 | 0.9628 | 0.2805 | 3.566 | 20 |
| 50 | 0.2728 | 0.9621 | 0.2836 | 3.526 | 10 |
| 16°00′ | 0.2756 | 0.9613 | 0.2867 | 3.487 | 74°00′ |
| 10 | 0.2784 | 0.9605 | 0.2899 | 3.450 | 50 |
| 20 | 0.2812 | 0.9596 | 0.2931 | 3.412 | 40 |
| 30 | 0.2840 | 0.9588 | 0.2962 | 3.376 | 30 |
| 40 | 0.2868 | 0.9580 | 0.2994 | 3.340 | 20 |
| 50 | 0.2896 | 0.9572 | 0.3026 | 3.305 | 10 |
| 17°00′ | 0.2924 | 0.9563 | 0.3057 | 3.271 | 73°00′ |
| 10 | 0.2952 | 0.9555 | 0.3089 | 3.237 | 50 |
| 20 | 0.2979 | 0.9546 | 0.3121 | 3.204 | 40 |
| 30 | 0.3007 | 0.9537 | 0.3153 | 3.172 | 30 |
| 40 | 0.3035 | 0.9528 | 0.3185 | 3.140 | 20 |
| 50 | 0.3062 | 0.9520 | 0.3217 | 3.108 | 10 |
| 18°00′ | 0.3090 | 0.9511 | 0.3249 | 3.078 | 72°00′ |
| 10 | 0.3118 | 0.9502 | 0.3281 | 3.047 | 50 |
| 20 | 0.3145 | 0.9492 | 0.3314 | 3.018 | 40 |
| 30 | 0.3173 | 0.9483 | 0.3346 | 2.989 | 30 |
| 40 | 0.3201 | 0.9474 | 0.3378 | 2.960 | 20 |
| 50 | 0.3228 | 0.9465 | 0.3411 | 2.932 | 10 |
| 19°00′ | 0.3256 | 0.9455 | 0.3443 | 2.904 | 71°00′ |
| 10 | 0.3283 | 0.9446 | 0.3476 | 2.877 | 50 |
| 20 | 0.3311 | 0.9436 | 0.3508 | 2.850 | 40 |
| 30 | 0.3338 | 0.9426 | 0.3541 | 2.824 | 30 |
| 40 | 0.3365 | 0.9417 | 0.3574 | 2.798 | 20 |
| 50 | 0.3393 | 0.9407 | 0.3607 | 2.773 | 10 |
| 20°00′ | 0.3420 | 0.9397 | 0.3640 | 2.747 | 70°00′ |
| 10 | 0.3448 | 0.9387 | 0.3673 | 2.723 | 50 |
| 20 | 0.3475 | 0.9377 | 0.3706 | 2.699 | 40 |
| 30 | 0.3502 | 0.9367 | 0.3739 | 2.675 | 30 |
| 40 | 0.3529 | 0.9356 | 0.3772 | 2.651 | 20 |
| 50 | 0.3557 | 0.9346 | 0.3805 | 2.628 | 10 |
| 21°00′ | 0.3584 | 0.9336 | 0.3839 | 2.605 | 69°00′ |
| | **Cos** | **Sin** | **Cot** | **Tan** | **Degrees** |

*Table B-2 (continued)*

| Degrees | Sin | Cos | Tan | Cot | |
|---------|------|------|------|------|------|
| 21°00′ | 0.3584 | 0.9336 | 0.3839 | 2.605 | 69°00′ |
| 10 | 0.3611 | 0.9325 | 0.3872 | 2.583 | 50 |
| 20 | 0.3638 | 0.9315 | 0.3906 | 2.560 | 40 |
| 30 | 0.3665 | 0.9304 | 0.3939 | 2.539 | 30 |
| 40 | 0.3692 | 0.9293 | 0.3973 | 2.517 | 20 |
| 50 | 0.3719 | 0.9283 | 0.4006 | 2.496 | 10 |
| 22°00′ | 0.3746 | 0.9272 | 0.4040 | 2.475 | 68°00′ |
| 10 | 0.3773 | 0.9261 | 0.4074 | 2.455 | 50 |
| 20 | 0.3800 | 0.9250 | 0.4108 | 2.434 | 40 |
| 30 | 0.3827 | 0.9239 | 0.4142 | 2.414 | 30 |
| 40 | 0.3854 | 0.9228 | 0.4176 | 2.394 | 20 |
| 50 | 0.3881 | 0.9216 | 0.4210 | 2.375 | 10 |
| 23°00′ | 0.3907 | 0.9205 | 0.4245 | 2.356 | 67°00′ |
| 10 | 0.3934 | 0.9194 | 0.4279 | 2.337 | 50 |
| 20 | 0.3961 | 0.9182 | 0.4314 | 2.318 | 40 |
| 30 | 0.3987 | 0.9171 | 0.4348 | 2.300 | 30 |
| 40 | 0.4014 | 0.9159 | 0.4383 | 2.282 | 20 |
| 50 | 0.4041 | 0.9147 | 0.4417 | 2.264 | 10 |
| 24°00′ | 0.4067 | 0.9135 | 0.4452 | 2.246 | 66°00′ |
| 10 | 0.4094 | 0.9124 | 0.4487 | 2.229 | 50 |
| 20 | 0.4120 | 0.9112 | 0.4522 | 2.211 | 40 |
| 30 | 0.4147 | 0.9100 | 0.4557 | 2.194 | 30 |
| 40 | 0.4173 | 0.9088 | 0.4592 | 2.177 | 20 |
| 50 | 0.4200 | 0.9075 | 0.4628 | 2.161 | 10 |
| 25°00′ | 0.4226 | 0.9063 | 0.4663 | 2.145 | 65°00′ |
| 10 | 0.4253 | 0.9051 | 0.4699 | 2.128 | 50 |
| 20 | 0.4279 | 0.9038 | 0.4734 | 2.112 | 40 |
| 30 | 0.4305 | 0.9026 | 0.4770 | 2.097 | 30 |
| 40 | 0.4331 | 0.9013 | 0.4806 | 2.081 | 20 |
| 50 | 0.4358 | 0.9001 | 0.4841 | 2.066 | 10 |
| 26°00′ | 0.4384 | 0.8988 | 0.4877 | 2.050 | 64°00′ |
| 10 | 0.4410 | 0.8975 | 0.4913 | 2.035 | 50 |
| 20 | 0.4436 | 0.8962 | 0.4950 | 2.020 | 40 |
| 30 | 0.4462 | 0.8949 | 0.4986 | 2.006 | 30 |
| 40 | 0.4488 | 0.8936 | 0.5022 | 1.991 | 20 |
| 50 | 0.4514 | 0.8923 | 0.5059 | 1.977 | 10 |
| 27°00′ | 0.4540 | 0.8910 | 0.5095 | 1.963 | 63°00′ |
| 10 | 0.4566 | 0.8897 | 0.5132 | 1.949 | 50 |
| 20 | 0.4592 | 0.8884 | 0.5169 | 1.935 | 40 |
| 30 | 0.4617 | 0.8870 | 0.5206 | 1.921 | 30 |
| 40 | 0.4643 | 0.8857 | 0.5243 | 1.907 | 20 |
| 50 | 0.4669 | 0.8843 | 0.5280 | 1.894 | 10 |
| 28°00′ | 0.4695 | 0.8829 | 0.5317 | 1.881 | 62°00′ |
| | **Cos** | **Sin** | **Cot** | **Tan** | **Degrees** |

*Table B-2 (continued)*

| Degrees | Sin | Cos | Tan | Cot | |
|---|---|---|---|---|---|
| 28°00′ | 0.4695 | 0.8829 | 0.5317 | 1.881 | 62°00′ |
| 10 | 0.4720 | 0.8816 | 0.5354 | 1.868 | 50 |
| 20 | 0.4746 | 0.8802 | 0.5392 | 1.855 | 40 |
| 30 | 0.4772 | 0.8788 | 0.5430 | 1.842 | 30 |
| 40 | 0.4797 | 0.8774 | 0.5467 | 1.829 | 20 |
| 50 | 0.4823 | 0.8760 | 0.5505 | 1.816 | 10 |
| 29°00′ | 0.4848 | 0.8746 | 0.5543 | 1.804 | 61°00′ |
| 10 | 0.4874 | 0.8732 | 0.5581 | 1.792 | 50 |
| 20 | 0.4899 | 0.8718 | 0.5619 | 1.780 | 40 |
| 30 | 0.4924 | 0.8704 | 0.5658 | 1.767 | 30 |
| 40 | 0.4950 | 0.8689 | 0.5696 | 1.756 | 20 |
| 50 | 0.4975 | 0.8675 | 0.5735 | 1.744 | 10 |
| 30°00′ | 0.5000 | 0.8660 | 0.5774 | 1.732 | 60°00′ |
| 10 | 0.5025 | 0.8646 | 0.5812 | 1.720 | 50 |
| 20 | 0.5050 | 0.8631 | 0.5851 | 1.709 | 40 |
| 30 | 0.5075 | 0.8616 | 0.5890 | 1.698 | 30 |
| 40 | 0.5100 | 0.8601 | 0.5930 | 1.686 | 20 |
| 50 | 0.5125 | 0.8587 | 0.5969 | 1.675 | 10 |
| 31°00′ | 0.5150 | 0.8572 | 0.6009 | 1.664 | 59°00′ |
| 10 | 0.5175 | 0.8557 | 0.6048 | 1.653 | 50 |
| 20 | 0.5200 | 0.8542 | 0.6088 | 1.643 | 40 |
| 30 | 0.5225 | 0.8526 | 0.6128 | 1.632 | 30 |
| 40 | 0.5250 | 0.8511 | 0.6168 | 1.621 | 20 |
| 50 | 0.5275 | 0.8496 | 0.6208 | 1.611 | 10 |
| 32°00′ | 0.5299 | 0.8480 | 0.6249 | 1.600 | 58°00′ |
| 10 | 0.5324 | 0.8465 | 0.6289 | 1.590 | 50 |
| 20 | 0.5348 | 0.8450 | 0.6330 | 1.580 | 40 |
| 30 | 0.5373 | 0.8434 | 0.6371 | 1.570 | 30 |
| 40 | 0.5398 | 0.8418 | 0.6412 | 1.560 | 20 |
| 50 | 0.5422 | 0.8403 | 0.6453 | 1.550 | 10 |
| 33°00′ | 0.5446 | 0.8387 | 0.6494 | 1.540 | 57°00′ |
| 10 | 0.5471 | 0.8371 | 0.6536 | 1.530 | 50 |
| 20 | 0.5495 | 0.8355 | 0.6577 | 1.520 | 40 |
| 30 | 0.5519 | 0.8339 | 0.6619 | 1.511 | 30 |
| 40 | 0.5544 | 0.8323 | 0.6661 | 1.501 | 20 |
| 50 | 0.5568 | 0.8307 | 0.6703 | 1.492 | 10 |
| 34°00′ | 0.5592 | 0.8290 | 0.6745 | 1.483 | 56°00′ |
| 10 | 0.5616 | 0.8274 | 0.6787 | 1.473 | 50 |
| 20 | 0.5640 | 0.8258 | 0.6830 | 1.464 | 40 |
| 30 | 0.5664 | 0.8241 | 0.6873 | 1.455 | 30 |
| 40 | 0.5688 | 0.8225 | 0.6916 | 1.446 | 20 |
| 50 | 0.5712 | 0.8208 | 0.6959 | 1.437 | 10 |
| 35°00′ | 0.5736 | 0.8192 | 0.7002 | 1.428 | 55°00′ |
| | Cos | Sin | Cot | Tan | Degrees |

*Table B-2 (continued)*

| Degrees | Sin | Cos | Tan | Cot | |
|---|---|---|---|---|---|
| 35°00′ | 0.5736 | 0.8192 | 0.7002 | 1.428 | 55°00′ |
| 10 | 0.5760 | 0.8175 | 0.7046 | 1.419 | 50 |
| 20 | 0.5783 | 0.8158 | 0.7089 | 1.411 | 40 |
| 30 | 0.5807 | 0.8141 | 0.7133 | 1.402 | 30 |
| 40 | 0.5831 | 0.8124 | 0.7177 | 1.393 | 20 |
| 50 | 0.5854 | 0.8107 | 0.7221 | 1.385 | 10 |
| 36°00′ | 0.5878 | 0.8090 | 0.7265 | 1.376 | 54°00′ |
| 10 | 0.5901 | 0.8073 | 0.7310 | 1.368 | 50 |
| 20 | 0.5925 | 0.8056 | 0.7355 | 1.360 | 40 |
| 30 | 0.5948 | 0.8039 | 0.7400 | 1.351 | 30 |
| 40 | 0.5972 | 0.8021 | 0.7445 | 1.343 | 20 |
| 50 | 0.5995 | 0.8004 | 0.7490 | 1.335 | 10 |
| 37°00′ | 0.6018 | 0.7986 | 0.7536 | 1.327 | 53°00′ |
| 10 | 0.6041 | 0.7969 | 0.7581 | 1.319 | 50 |
| 20 | 0.6065 | 0.7951 | 0.7627 | 1.311 | 40 |
| 30 | 0.6088 | 0.7934 | 0.7673 | 1.303 | 30 |
| 40 | 0.6111 | 0.7916 | 0.7720 | 1.295 | 20 |
| 50 | 0.6134 | 0.7898 | 0.7766 | 1.288 | 10 |
| 38°00′ | 0.6157 | 0.7880 | 0.7813 | 1.280 | 52°00′ |
| 10 | 0.6180 | 0.7862 | 0.7860 | 1.272 | 50 |
| 20 | 0.6202 | 0.7844 | 0.7907 | 1.265 | 40 |
| 30 | 0.6225 | 0.7826 | 0.7954 | 1.257 | 30 |
| 40 | 0.6248 | 0.7808 | 0.8002 | 1.250 | 20 |
| 50 | 0.6271 | 0.7790 | 0.8050 | 1.242 | 10 |
| 39°00′ | 0.6293 | 0.7771 | 0.8098 | 1.235 | 51°00′ |
| 10 | 0.6316 | 0.7753 | 0.8146 | 1.228 | 50 |
| 20 | 0.6338 | 0.7735 | 0.8195 | 1.220 | 40 |
| 30 | 0.6361 | 0.7716 | 0.8243 | 1.213 | 30 |
| 20 | 0.6383 | 0.7698 | 0.8292 | 1.206 | 20 |
| 50 | 0.6406 | 0.7679 | 0.8342 | 1.199 | 10 |
| 40°00′ | 0.6428 | 0.7660 | 0.8391 | 1.192 | 50°00′ |
| 10 | 0.6450 | 0.7642 | 0.8441 | 1.185 | 50 |
| 20 | 0.6472 | 0.7623 | 0.8491 | 1.178 | 40 |
| 30 | 0.6494 | 0.7604 | 0.8541 | 1.171 | 30 |
| 40 | 0.6517 | 0.7585 | 0.8591 | 1.164 | 20 |
| 50 | 0.6539 | 0.7566 | 0.8642 | 1.157 | 10 |
| 41°00′ | 0.6561 | 0.7547 | 0.8693 | 1.150 | 49°00′ |
| 10 | 0.6583 | 0.7528 | 0.8744 | 1.144 | 50 |
| 20 | 0.6604 | 0.7509 | 0.8796 | 1.137 | 40 |
| 30 | 0.6626 | 0.7490 | 0.8847 | 1.130 | 30 |
| 40 | 0.6648 | 0.7470 | 0.8899 | 1.124 | 20 |
| 50 | 0.6670 | 0.7451 | 0.8952 | 1.117 | 10 |
| 42°00′ | 0.6691 | 0.7431 | 0.9004 | 1.111 | 48°00′ |
| | Cos | Sin | Cot | Tan | Degrees |

*Table B-2 (continued)*

| Degrees | Sin | Cos | Tan | Cot | |
|---------|-----|-----|-----|-----|---|
| 42°00' | 0.6691 | 0.7431 | 0.9004 | 1.111 | 48°00' |
| 10 | 0.6713 | 0.7412 | 0.9057 | 1.104 | 50 |
| 20 | 0.6734 | 0.7392 | 0.9110 | 1.098 | 40 |
| 30 | 0.6756 | 0.7373 | 0.9163 | 1.091 | 30 |
| 40 | 0.6777 | 0.7353 | 0.9217 | 1.805 | 20 |
| 50 | 0.6799 | 0.7333 | 0.9271 | 1.079 | 10 |
| 43°00' | 0.6820 | 0.7314 | 0.9325 | 1.072 | 47°00' |
| 10 | 0.6841 | 0.7294 | 0.9380 | 1.066 | 50 |
| 20 | 0.6862 | 0.7274 | 0.9435 | 1.060 | 40 |
| 30 | 0.6884 | 0.7254 | 0.9490 | 1.054 | 30 |
| 40 | 0.6905 | 0.7234 | 0.9545 | 1.048 | 20 |
| 50 | 0.6926 | 0.7214 | 0.9601 | 1.042 | 10 |
| 44°00' | 0.6947 | 0.7193 | 0.9657 | 1.036 | 46°00' |
| 10 | 0.6967 | 0.7173 | 0.9713 | 1.030 | 50 |
| 20 | 0.6988 | 0.7153 | 0.9770 | 1.024 | 40 |
| 30 | 0.7009 | 0.7133 | 0.9827 | 1.018 | 30 |
| 40 | 0.7030 | 0.7112 | 0.9884 | 1.012 | 20 |
| 50 | 0.7050 | 0.7092 | 0.9942 | 1.006 | 10 |
| 45°00' | 0.7071 | 0.7071 | 1.000 | 1.000 | 45°00' |
| | Cos | Sin | Cot | Tan | Degrees |

## Table B-3    VALUES OF $e^x$

| $x$ | $e^x$ | $e^{-x}$ | $x$ | $e^x$ | $e^{-x}$ |
|---|---|---|---|---|---|
| 0.00 | 1.0000 | 1.0000 | 2.5 | 12.182 | 0.0821 |
| 0.10 | 1.1052 | 0.9048 | 2.6 | 13.464 | 0.0743 |
| 0.20 | 1.2214 | 0.8187 | 2.7 | 14.880 | 0.0672 |
| 0.30 | 1.3499 | 0.7408 | 2.8 | 16.445 | 0.0608 |
| 0.40 | 1.4918 | 0.6703 | 2.9 | 18.174 | 0.0550 |
| 0.50 | 1.6487 | 0.6065 | 3.0 | 20.086 | 0.0498 |
| 0.60 | 1.8221 | 0.5488 | 3.1 | 22.198 | 0.0450 |
| 0.70 | 2.0138 | 0.4966 | 3.2 | 24.533 | 0.0408 |
| 0.80 | 2.2255 | 0.4493 | 3.3 | 27.113 | 0.0369 |
| 0.90 | 2.4596 | 0.4066 | 3.4 | 29.964 | 0.0334 |
| 1.0 | 2.7183 | 0.3679 | 3.5 | 33.115 | 0.0302 |
| 1.1 | 3.0042 | 0.3329 | 3.6 | 36.598 | 0.0273 |
| 1.2 | 3.3201 | 0.3012 | 3.7 | 40.447 | 0.0247 |
| 1.3 | 3.6693 | 0.2725 | 3.8 | 44.701 | 0.0224 |
| 1.4 | 4.0552 | 0.2466 | 3.9 | 49.402 | 0.0202 |
| 1.5 | 4.4817 | 0.2231 | 4.0 | 54.598 | 0.0183 |
| 1.6 | 4.9530 | 0.2019 | 4.1 | 60.340 | 0.0166 |
| 1.7 | 5.4739 | 0.1827 | 4.2 | 66.686 | 0.0150 |
| 1.8 | 6.0496 | 0.1653 | 4.3 | 73.700 | 0.0136 |
| 1.9 | 6.6859 | 0.1496 | 4.4 | 81.451 | 0.0123 |
| 2.0 | 7.3891 | 0.1353 | 4.5 | 90.017 | 0.0111 |
| 2.1 | 8.1662 | 0.1225 | 4.6 | 99.484 | 0.0091 |
| 2.2 | 9.0250 | 0.1108 | 4.7 | 109.95 | 0.0091 |
| 2.3 | 9.9742 | 0.1003 | 4.8 | 121.51 | 0.0082 |
| 2.4 | 11.023 | 0.0907 | 4.9 | 134.29 | 0.0074 |
|  |  |  | 5.0 | 148.41 | 0.0067 |

## Table B-4 COMMON LOGARITHMS

| N | 0 | 1 | 2 | 3 | 4 | 5 | 6 | 7 | 8 | 9 |
|---|---|---|---|---|---|---|---|---|---|---|
| 10 | 0000 | 0043 | 0086 | 0128 | 0170 | 0212 | 0253 | 0294 | 0334 | 0374 |
| 11 | 0414 | 0453 | 0492 | 0531 | 0569 | 0607 | 0645 | 0682 | 0719 | 0755 |
| 12 | 0792 | 0828 | 0864 | 0899 | 0934 | 0969 | 1004 | 1038 | 1072 | 1106 |
| 13 | 1139 | 1173 | 1206 | 1239 | 1271 | 1303 | 1335 | 1367 | 1399 | 1430 |
| 14 | 1461 | 1492 | 1523 | 1553 | 1584 | 1614 | 1644 | 1673 | 1703 | 1732 |
| 15 | 1761 | 1790 | 1818 | 1847 | 1875 | 1903 | 1931 | 1959 | 1987 | 2014 |
| 16 | 2041 | 2068 | 2095 | 2122 | 2148 | 2175 | 2201 | 2227 | 2253 | 2279 |
| 17 | 2304 | 2330 | 2355 | 2380 | 2405 | 2430 | 2455 | 2480 | 2504 | 2529 |
| 18 | 2553 | 2577 | 2601 | 2625 | 2648 | 2672 | 2695 | 2718 | 2742 | 2765 |
| 19 | 2788 | 2810 | 2833 | 2856 | 2878 | 2900 | 2923 | 2945 | 2967 | 2989 |
| 20 | 3010 | 3032 | 3054 | 3075 | 3096 | 3118 | 3139 | 3160 | 3181 | 3201 |
| 21 | 3222 | 3243 | 3263 | 3284 | 3304 | 3324 | 3345 | 3365 | 3385 | 3404 |
| 22 | 3424 | 3444 | 3464 | 3483 | 3502 | 3522 | 3541 | 3560 | 3579 | 3598 |
| 23 | 3617 | 3636 | 3655 | 3674 | 3692 | 3711 | 3729 | 3747 | 3766 | 3784 |
| 24 | 3802 | 3820 | 3838 | 3856 | 3874 | 3892 | 3909 | 3927 | 3945 | 3962 |
| 25 | 3979 | 3997 | 4014 | 4031 | 4048 | 4065 | 4082 | 4099 | 4116 | 4133 |
| 26 | 4150 | 4166 | 4183 | 4200 | 4216 | 4232 | 4249 | 4265 | 4281 | 4298 |
| 27 | 4314 | 4330 | 4346 | 4362 | 4378 | 4393 | 4409 | 4425 | 4440 | 4456 |
| 28 | 4472 | 4487 | 4502 | 4518 | 4533 | 4548 | 4564 | 4579 | 4594 | 4609 |
| 29 | 4624 | 4639 | 4654 | 4669 | 4683 | 4698 | 4713 | 4728 | 4742 | 4757 |
| 30 | 4771 | 4786 | 4800 | 4814 | 4829 | 4843 | 4857 | 4871 | 4886 | 4900 |
| 31 | 4914 | 4928 | 4942 | 4955 | 4969 | 4983 | 4997 | 5011 | 5024 | 5038 |
| 32 | 5051 | 5065 | 5079 | 5092 | 5105 | 5119 | 5132 | 5145 | 5159 | 5172 |
| 33 | 5185 | 5198 | 5211 | 5224 | 5237 | 5250 | 5263 | 5276 | 5289 | 5302 |
| 34 | 5315 | 5328 | 5340 | 5353 | 5366 | 5378 | 5391 | 5403 | 5416 | 5428 |
| 35 | 5441 | 5453 | 5465 | 5478 | 5490 | 5502 | 5514 | 5527 | 5539 | 5551 |
| 36 | 5563 | 5575 | 5587 | 5599 | 5611 | 5623 | 5635 | 5647 | 5658 | 5670 |
| 37 | 5682 | 5694 | 5705 | 5717 | 5729 | 5740 | 5752 | 5763 | 5775 | 5786 |
| 38 | 5798 | 5809 | 5821 | 5832 | 5843 | 5855 | 5866 | 5877 | 5888 | 5899 |
| 39 | 5911 | 5922 | 5933 | 5944 | 5955 | 5966 | 5977 | 5988 | 5999 | 6010 |
| 40 | 6021 | 6031 | 6042 | 6053 | 6064 | 6075 | 6085 | 6096 | 6107 | 6117 |
| 41 | 6128 | 6138 | 6149 | 6160 | 6170 | 6180 | 6191 | 6201 | 6212 | 6222 |
| 42 | 6232 | 6243 | 6253 | 6263 | 6274 | 6284 | 6294 | 6304 | 6314 | 6325 |
| 43 | 6335 | 6345 | 6355 | 6365 | 6375 | 6385 | 6395 | 6405 | 6415 | 6425 |
| 44 | 6435 | 6444 | 6454 | 6464 | 6474 | 6484 | 6493 | 6503 | 6513 | 6522 |
| 45 | 6532 | 6542 | 6551 | 6561 | 6571 | 6580 | 6590 | 6599 | 6609 | 6618 |
| 46 | 6628 | 6637 | 6646 | 6656 | 6665 | 6675 | 6684 | 6693 | 6702 | 6712 |
| 47 | 6721 | 6730 | 6739 | 6749 | 6758 | 6767 | 6776 | 6785 | 6794 | 6803 |
| 48 | 6812 | 6821 | 6830 | 6839 | 6848 | 6857 | 6866 | 6875 | 6884 | 6893 |
| 49 | 6902 | 6911 | 6920 | 6928 | 6937 | 6946 | 6955 | 6964 | 6972 | 6981 |
| 50 | 6990 | 6998 | 7007 | 7016 | 7024 | 7033 | 7042 | 7050 | 7059 | 7067 |
| 51 | 7076 | 7084 | 7093 | 7101 | 7110 | 7118 | 7126 | 7135 | 7143 | 7152 |
| 52 | 7160 | 7168 | 7177 | 7185 | 7193 | 7202 | 7210 | 7218 | 7226 | 7235 |
| 53 | 7243 | 7251 | 7259 | 7267 | 7275 | 7284 | 7292 | 7300 | 7308 | 7316 |
| 54 | 7324 | 7332 | 7340 | 7348 | 7356 | 7364 | 7372 | 7380 | 7388 | 7396 |

*Table B-4 (continued)*

| N | 0 | 1 | 2 | 3 | 4 | 5 | 6 | 7 | 8 | 9 |
|---|---|---|---|---|---|---|---|---|---|---|
| 55 | 7404 | 7412 | 7419 | 7427 | 7435 | 7443 | 7451 | 7459 | 7466 | 7474 |
| 56 | 7482 | 7490 | 7497 | 7505 | 7513 | 7520 | 7528 | 7536 | 7543 | 7551 |
| 57 | 7559 | 7566 | 7574 | 7582 | 7589 | 7597 | 7604 | 7612 | 7619 | 7627 |
| 58 | 7634 | 7642 | 7649 | 7657 | 7664 | 7672 | 7679 | 7686 | 7694 | 7701 |
| 59 | 7709 | 7716 | 7723 | 7731 | 7738 | 7745 | 7752 | 7760 | 7767 | 7774 |
| 60 | 7782 | 7789 | 7796 | 7803 | 7810 | 7818 | 7825 | 7832 | 7839 | 7846 |
| 61 | 7853 | 7860 | 7868 | 7875 | 7882 | 7889 | 7896 | 7903 | 7910 | 7917 |
| 62 | 7924 | 7931 | 7938 | 7945 | 7952 | 7959 | 7966 | 7973 | 7980 | 7987 |
| 63 | 7993 | 8000 | 8007 | 8014 | 8021 | 8028 | 8035 | 8041 | 8048 | 8055 |
| 64 | 8062 | 8069 | 8075 | 8082 | 8089 | 8096 | 8102 | 8109 | 8116 | 8122 |
| 65 | 8129 | 8136 | 8142 | 8149 | 8156 | 8162 | 8169 | 8176 | 8182 | 8189 |
| 66 | 8195 | 8202 | 8209 | 8215 | 8222 | 8228 | 8235 | 8241 | 8248 | 8254 |
| 67 | 8261 | 8267 | 8274 | 8280 | 8287 | 8293 | 8299 | 8306 | 8312 | 8319 |
| 68 | 8325 | 8331 | 8338 | 8344 | 8351 | 8357 | 8363 | 8370 | 8376 | 8382 |
| 69 | 8388 | 8395 | 8401 | 8407 | 8414 | 8420 | 8426 | 8432 | 8439 | 8445 |
| 70 | 8451 | 8457 | 8463 | 8470 | 8476 | 8482 | 8488 | 8494 | 8500 | 8506 |
| 71 | 8513 | 8519 | 8525 | 8531 | 8537 | 8543 | 8549 | 8555 | 8561 | 8567 |
| 72 | 8573 | 8579 | 8585 | 8591 | 8597 | 8603 | 8609 | 8615 | 8621 | 8627 |
| 73 | 8633 | 8639 | 8645 | 8651 | 8657 | 8663 | 8669 | 8675 | 8681 | 8686 |
| 74 | 8692 | 8698 | 8704 | 8710 | 8716 | 8722 | 8727 | 8733 | 8739 | 8745 |
| 75 | 8751 | 8756 | 8762 | 8768 | 8774 | 8779 | 8785 | 8791 | 8797 | 8802 |
| 76 | 8808 | 8814 | 8820 | 8825 | 8831 | 8837 | 8842 | 8848 | 8854 | 8859 |
| 77 | 8865 | 8871 | 8876 | 8882 | 8887 | 8893 | 8899 | 8904 | 8910 | 8915 |
| 78 | 8921 | 8927 | 8932 | 8938 | 8943 | 8949 | 8954 | 8960 | 8965 | 8971 |
| 79 | 8976 | 8982 | 8987 | 8993 | 8998 | 9004 | 9009 | 9015 | 9020 | 9025 |
| 80 | 9031 | 9036 | 9042 | 9047 | 9053 | 9058 | 9063 | 9069 | 9074 | 9079 |
| 81 | 9085 | 9090 | 9096 | 9101 | 9106 | 9112 | 9117 | 9122 | 9128 | 9133 |
| 82 | 9138 | 9143 | 9149 | 9154 | 9159 | 9165 | 9170 | 9175 | 9180 | 9186 |
| 83 | 9191 | 9196 | 9201 | 9206 | 9212 | 9217 | 9222 | 9227 | 9232 | 9238 |
| 84 | 9243 | 9248 | 9253 | 9258 | 9263 | 9269 | 9274 | 9279 | 9284 | 9289 |
| 85 | 9294 | 9299 | 9304 | 9309 | 9315 | 9320 | 9325 | 9330 | 9335 | 9340 |
| 86 | 9345 | 9350 | 9355 | 9360 | 9365 | 9370 | 9375 | 9380 | 9385 | 9390 |
| 87 | 9395 | 9400 | 9405 | 9410 | 9415 | 9420 | 9425 | 9430 | 9435 | 9440 |
| 88 | 9445 | 9450 | 9455 | 9460 | 9465 | 9469 | 9474 | 9479 | 9484 | 9489 |
| 89 | 9494 | 9499 | 9504 | 9509 | 9513 | 9518 | 9523 | 9528 | 9533 | 9538 |
| 90 | 9542 | 9547 | 9552 | 9557 | 9562 | 9566 | 9571 | 9576 | 9581 | 9586 |
| 91 | 9590 | 9595 | 9600 | 9605 | 9609 | 9614 | 9619 | 9624 | 9628 | 9633 |
| 92 | 9638 | 9643 | 9647 | 9652 | 9657 | 9661 | 9666 | 9671 | 9675 | 9680 |
| 93 | 9685 | 9689 | 9694 | 9699 | 9703 | 9708 | 9713 | 9717 | 9722 | 9727 |
| 94 | 9731 | 9736 | 9741 | 9745 | 9750 | 9754 | 9759 | 9763 | 9768 | 9773 |
| 95 | 9777 | 9782 | 9786 | 9791 | 9795 | 9800 | 9805 | 9809 | 9814 | 9818 |
| 96 | 9823 | 9827 | 9832 | 9836 | 9841 | 9845 | 9850 | 9854 | 9859 | 9863 |
| 97 | 9868 | 9872 | 9877 | 9881 | 9886 | 9890 | 9894 | 9899 | 9903 | 9908 |
| 98 | 9912 | 9917 | 9921 | 9926 | 9930 | 9934 | 9939 | 9943 | 9948 | 9952 |
| 99 | 9956 | 9961 | 9965 | 9969 | 9974 | 9978 | 9983 | 9987 | 9991 | 9996 |

## Table B-5   NATURAL LOGARITHMS

| $n$ | $\ln n$ | $n$ | $\ln n$ | $n$ | $\ln n$ |
|-----|---------|-----|---------|-----|---------|
| 1.0 | 0.0000 | 4.0 | 1.3863 | 7.0 | 1.9459 |
| 1.1 | 0.0953 | 4.1 | 1.4110 | 7.1 | 1.9601 |
| 1.2 | 0.1823 | 4.2 | 1.4351 | 7.2 | 1.9741 |
| 1.3 | 0.2624 | 4.3 | 1.4586 | 7.3 | 1.9879 |
| 1.4 | 0.3365 | 4.4 | 1.4816 | 7.4 | 2.0015 |
| 1.5 | 0.4055 | 4.5 | 1.5041 | 7.5 | 2.0149 |
| 1.6 | 0.4700 | 4.6 | 1.5261 | 7.6 | 2.0281 |
| 1.7 | 0.5306 | 4.7 | 1.5476 | 7.7 | 2.0412 |
| 1.8 | 0.5878 | 4.8 | 1.5686 | 7.8 | 2.0541 |
| 1.9 | 0.6419 | 4.9 | 1.5892 | 7.9 | 2.0669 |
| 2.0 | 0.6931 | 5.0 | 1.6094 | 8.0 | 2.0794 |
| 2.1 | 0.7419 | 5.1 | 1.6292 | 8.1 | 2.0919 |
| 2.2 | 0.7885 | 5.2 | 1.6487 | 8.2 | 2.1041 |
| 2.3 | 0.8329 | 5.3 | 1.6677 | 8.3 | 2.1163 |
| 2.4 | 0.8755 | 5.4 | 1.6864 | 8.4 | 2.1282 |
| 2.5 | 0.9163 | 5.5 | 1.7047 | 8.5 | 2.1401 |
| 2.6 | 0.9555 | 5.6 | 1.7228 | 8.6 | 2.1518 |
| 2.7 | 0.9933 | 5.7 | 1.7405 | 8.7 | 2.1633 |
| 2.8 | 1.0296 | 5.8 | 1.7579 | 8.8 | 2.1748 |
| 2.9 | 1.0647 | 5.9 | 1.7750 | 8.9 | 2.1861 |
| 3.0 | 1.0986 | 6.0 | 1.7918 | 9.0 | 2.1972 |
| 3.1 | 1.1314 | 6.1 | 1.8083 | 9.1 | 2.2083 |
| 3.2 | 1.1632 | 6.2 | 1.8245 | 9.2 | 2.2192 |
| 3.3 | 1.1939 | 6.3 | 1.8405 | 9.3 | 2.2300 |
| 3.4 | 1.2238 | 6.4 | 1.8563 | 9.4 | 2.2407 |
| 3.5 | 1.2528 | 6.5 | 1.8718 | 9.5 | 2.2513 |
| 3.6 | 1.2809 | 6.6 | 1.8871 | 9.6 | 2.2618 |
| 3.7 | 1.3083 | 6.7 | 1.9021 | 9.7 | 2.2721 |
| 3.8 | 1.3350 | 6.8 | 1.9169 | 9.8 | 2.2824 |
| 3.9 | 1.3610 | 6.9 | 1.9315 | 9.9 | 2.2925 |

$$\ln 10 = 2.3026$$
$$2 \ln 10 = 4.6052$$
$$3 \ln 10 = 6.9078$$
$$4 \ln 10 = 9.2103$$

## Table B-6 POWERS AND ROOTS

| No. | Sq. | Sq. Root | Cube | Cube Root | No. | Sq. | Sq. Root | Cube | Cube Root |
|---|---|---|---|---|---|---|---|---|---|
| 1 | 1 | 1.000 | 1 | 1.000 | 51 | 2,601 | 7.141 | 132,651 | 3.708 |
| 2 | 4 | 1.414 | 8 | 1.260 | 52 | 2,704 | 7.211 | 140,608 | 3.733 |
| 3 | 9 | 1.732 | 27 | 1.442 | 53 | 2,809 | 7.280 | 148,877 | 3.756 |
| 4 | 16 | 2.000 | 64 | 1.587 | 54 | 2,916 | 7.348 | 157,464 | 3.780 |
| 5 | 25 | 2.236 | 125 | 1.710 | 55 | 3,025 | 7.416 | 166,375 | 3.803 |
| 6 | 36 | 2.449 | 216 | 1.817 | 56 | 3,136 | 7.483 | 175,616 | 3.826 |
| 7 | 49 | 2.646 | 343 | 1.913 | 57 | 3,249 | 7.550 | 185,193 | 3.849 |
| 8 | 64 | 2.828 | 512 | 2.000 | 58 | 3,364 | 7.616 | 195,112 | 3.871 |
| 9 | 81 | 3.000 | 729 | 2.080 | 59 | 3,481 | 7.681 | 205,379 | 3.893 |
| 10 | 100 | 3.162 | 1,000 | 2.154 | 60 | 3,600 | 7.746 | 216,000 | 3.915 |
| 11 | 121 | 3.317 | 1,331 | 2.224 | 61 | 3,721 | 7.810 | 226,981 | 3.936 |
| 12 | 144 | 3.464 | 1,728 | 2.289 | 62 | 3,844 | 7.874 | 238,328 | 3.958 |
| 13 | 169 | 3.606 | 2,197 | 2.351 | 63 | 3,969 | 7.937 | 250,047 | 3.979 |
| 14 | 196 | 3.742 | 2,744 | 2.410 | 64 | 4,096 | 8.000 | 262,144 | 4.000 |
| 15 | 225 | 3.873 | 3,375 | 2.466 | 65 | 4,225 | 8.062 | 274,625 | 4.021 |
| 16 | 256 | 4.000 | 4,096 | 2.520 | 66 | 4,356 | 8.124 | 287,496 | 4.041 |
| 17 | 289 | 4.123 | 4,913 | 2.571 | 67 | 4,489 | 8.185 | 300,763 | 4.062 |
| 18 | 324 | 4.243 | 5,832 | 2.621 | 68 | 4,624 | 8.246 | 314,432 | 4.082 |
| 19 | 361 | 4.359 | 6,859 | 2.668 | 69 | 4,761 | 8.307 | 328,509 | 4.102 |
| 20 | 400 | 4.472 | 8,000 | 2.714 | 70 | 4,900 | 8.367 | 343,000 | 4.121 |
| 21 | 441 | 4.583 | 9,261 | 2.759 | 71 | 5,041 | 8.426 | 357,911 | 4.141 |
| 22 | 484 | 4.690 | 10,648 | 2.802 | 72 | 5,184 | 8.485 | 373,248 | 4.160 |
| 23 | 529 | 4.796 | 12,167 | 2.844 | 73 | 5,329 | 8.544 | 389,017 | 4.179 |
| 24 | 576 | 4.899 | 13,824 | 2.884 | 74 | 5,476 | 8.602 | 405,224 | 4.198 |
| 25 | 625 | 5.000 | 15,625 | 2.924 | 75 | 5,625 | 8.660 | 421,875 | 4.217 |
| 26 | 676 | 5.099 | 17,576 | 2.962 | 76 | 5,776 | 8.718 | 438,976 | 4.236 |
| 27 | 729 | 5.196 | 19,683 | 3.000 | 77 | 5,929 | 8.775 | 456,533 | 4.254 |
| 28 | 784 | 5.292 | 21,952 | 3.037 | 78 | 6,084 | 8.832 | 474,552 | 4.273 |
| 29 | 841 | 5.385 | 24,389 | 3.072 | 79 | 6,241 | 8.888 | 493,039 | 4.291 |
| 30 | 900 | 5.477 | 27,000 | 3.107 | 80 | 6,400 | 8.944 | 512,000 | 4.309 |
| 31 | 961 | 5.568 | 29,791 | 3.141 | 81 | 6,561 | 9.000 | 531,441 | 4.327 |
| 32 | 1,024 | 5.657 | 32,768 | 3.175 | 82 | 6,724 | 9.055 | 551,368 | 4.344 |
| 33 | 1,089 | 5.745 | 35,937 | 3.208 | 83 | 6,889 | 9.110 | 571,787 | 4.362 |
| 34 | 1,156 | 5.831 | 39,304 | 3.240 | 84 | 7,056 | 9.165 | 592,704 | 4.380 |
| 35 | 1,225 | 5.916 | 42,875 | 3.271 | 85 | 7,225 | 9.220 | 614,125 | 4.397 |
| 36 | 1,296 | 6.000 | 46,656 | 3.302 | 86 | 7,396 | 9.274 | 636,056 | 4.414 |
| 37 | 1,369 | 6.083 | 50,653 | 3.332 | 87 | 7,569 | 9.327 | 658,503 | 4.431 |
| 38 | 1,444 | 6.164 | 54,872 | 3.362 | 88 | 7,744 | 9.381 | 681,472 | 4.448 |
| 39 | 1,521 | 6.245 | 59,319 | 3.391 | 89 | 7,921 | 9.434 | 704,969 | 4.465 |
| 40 | 1,600 | 6.325 | 64,000 | 3.420 | 90 | 8,100 | 9.487 | 729,000 | 4.481 |
| 41 | 1,681 | 6.403 | 68,921 | 3.448 | 91 | 8,281 | 9.539 | 753,571 | 4.498 |
| 42 | 1,764 | 6.481 | 74,088 | 3.476 | 92 | 8,464 | 9.592 | 778,688 | 4.514 |
| 43 | 1,849 | 6.557 | 79,507 | 3.503 | 93 | 8,649 | 9.644 | 804,357 | 4.531 |
| 44 | 1,936 | 6.633 | 85,184 | 3.530 | 94 | 8,836 | 9.695 | 830,584 | 4.547 |
| 45 | 2,025 | 6.708 | 91,125 | 3.557 | 95 | 9,025 | 9.747 | 857,375 | 4.563 |
| 46 | 2,116 | 6.782 | 97,336 | 3.583 | 96 | 9,216 | 9.798 | 884,736 | 4.579 |
| 47 | 2,209 | 6.856 | 103,823 | 3.609 | 97 | 9,409 | 9.849 | 912,673 | 4.595 |
| 48 | 2,304 | 6.928 | 110,592 | 3.634 | 98 | 9,604 | 9.899 | 941,192 | 4.610 |
| 49 | 2,401 | 7.000 | 117,649 | 3.659 | 99 | 9,801 | 9.950 | 970,299 | 4.626 |
| 50 | 2,500 | 7.071 | 125,000 | 3.684 | 100 | 10,000 | 10.000 | 1,000,000 | 4.642 |

## Table B-7    DERIVATIVES

The table lists the important derivative formulas shown in the text. The functions $y$, $u$, and $v$ are functions of $x$ and $y' = dy/dx$.

| | $y$ | $y'$ |
|---|---|---|
| 1. | $c$ (constant) | $0$ |
| 2. | $u^n$ | $nu^{n-1}(u')$ |
| 3. | $uv$ | $uv' + vu'$ |
| 4. | $\dfrac{u}{v}$ | $\dfrac{vu' - uv'}{v^2}$ |
| 5. | $\sin u$ | $(\cos u)(u')$ |
| 6. | $\cos u$ | $-(\sin u)(u')$ |
| 7. | $\tan u$ | $(\sec^2 u)(u')$ |
| 8. | $\cot u$ | $-(\csc^2 u)(u')$ |
| 9. | $\sec u$ | $(\sec u \tan u)(u')$ |
| 10. | $\csc u$ | $-(\csc u \cot u)(u')$ |
| 11. | $\sin^{-1} u$ | $\dfrac{u'}{\sqrt{1 - u^2}}$ |
| 12. | $\cos^{-1} u$ | $\dfrac{-u'}{\sqrt{1 - u^2}}$ |
| 13. | $\tan^{-1} u$ | $\dfrac{u'}{1 + u^2}$ |
| 14. | $\cot^{-1} u$ | $\dfrac{-u'}{1 + u^2}$ |
| 15. | $\sec^{-1} u$ | $\dfrac{u'}{u\sqrt{u^2 - 1}}$ |
| 16. | $\csc^{-1} u$ | $\dfrac{-u'}{u\sqrt{u^2 - 1}}$ |
| 17. | $\log_b u$ | $\dfrac{1}{u}\left(\log_b e\right)(u')$ |
| 18. | $\ln u$ | $\dfrac{u'}{u}$ |
| 19. | $e^u$ | $e^u\,(u')$ |

## Table B-8 INTEGRALS

The first 18 formulas are basic formulas shown in the text. The remainder are a selection of useful integral formulas. The functions $u$ and $v$ are considered to be functions of $x$. The constant of integration is omitted.

## Basic Formulas

**1.** $\displaystyle\int u^n \, du = \frac{u^{n+1}}{n+1}$

**2.** $\displaystyle\int \sin u \, du = -\cos u$

**3.** $\displaystyle\int \cos u \, du = \sin u$

**4.** $\displaystyle\int \tan u \, du = -\ln \cos u$

**5.** $\displaystyle\int \cot u \, du = \ln \sin u$

**6.** $\displaystyle\int \sec u \, du = \ln |\sec u + \tan u|$

**7.** $\displaystyle\int \csc u \, du = \ln |\csc u - \cot u|$

**8.** $\displaystyle\int \sec^2 u \, du = \tan u$

**9.** $\displaystyle\int \csc^2 u \, du = -\cot u$

**10.** $\displaystyle\int \sec u \tan u \, du = \sec u$

**11.** $\displaystyle\int \csc u \tan u \, du = -\csc u$

**12.** $\displaystyle\int \frac{du}{\sqrt{a^2 - u^2}} = \sin^{-1} \frac{u}{a}$

**13.** $\displaystyle\int \frac{du}{a^2 + u^2} = \frac{1}{a} \tan^{-1} \frac{u}{a}$

**14.** $\displaystyle\int \frac{du}{u\sqrt{u^2 - a^2}} = \frac{1}{a} \sec^{-1} \frac{u}{a}$

**15.** $\displaystyle\int \frac{du}{u} = \ln |u|$

**16.** $\displaystyle\int e^u \, du = e^u$

**17.** $\displaystyle\int \ln u \, du = u \ln u - u$

**18.** $\displaystyle\int u \, dv = uv - \int v \, du$

### Forms Containing $a + bu$ and $\sqrt{a + bu}$

**19.** $\displaystyle\int \frac{u \, du}{a + bu} = \frac{1}{b^2} (a + bu - a \ln |a + bu|)$

**20.** $\displaystyle\int \frac{u \, du}{(a + bu)^2} = \frac{a}{b^2(a + bu)} + \frac{1}{b^2} \ln |a + bu|$

**21.** $\displaystyle\int \frac{du}{u(a + bu)} = -\frac{1}{a} \ln \left| \frac{a + bu}{u} \right|$

**22.** $\displaystyle\int \frac{du}{u^2(a + bu)} = -\frac{1}{au} + \frac{b}{a^2} \ln \left| \frac{a + bu}{u} \right|$

**23.** $\displaystyle\int u\sqrt{a + bu} \, du = \frac{2}{15b^2} (3bu - 2a)(a + bu)^{3/2}$

**24.** $\displaystyle\int \frac{u\,du}{\sqrt{a + bu}} = \frac{2}{3b^2}(bu - 2a)\sqrt{a + bu}$

**25.** $\displaystyle\int \frac{du}{u\sqrt{a + bu}} = \frac{1}{\sqrt{a}}\ln\left|\frac{\sqrt{a + bu} - \sqrt{a}}{\sqrt{a + bu} + \sqrt{a}}\right|, \qquad a > 0$

Forms Containing $\sqrt{u^2 \pm a^2}$ and $\sqrt{a^2 - u^2}$

**26.** $\displaystyle\int \sqrt{u^2 \pm a^2}\,du = \frac{1}{2}\left(u\sqrt{u^2 \pm a^2} \pm a^2\ln\left|u + \sqrt{u^2 \pm a^2}\right|\right)$

**27.** $\displaystyle\int \sqrt{a^2 - u^2}\,du = \frac{1}{2}\left(u\sqrt{a^2 - u^2} + a^2\sin^{-1}\frac{u}{a}\right)$

**28.** $\displaystyle\int \frac{du}{\sqrt{u^2 \pm a^2}} = \ln\left|u + \sqrt{u^2 \pm a^2}\right|$

**29.** $\displaystyle\int \frac{du}{u\sqrt{u^2 + a^2}} = -\frac{1}{a}\ln\left|\frac{a + \sqrt{u^2 + a^2}}{u}\right|$

**30.** $\displaystyle\int \frac{du}{u\sqrt{a^2 - u^2}} = -\frac{1}{a}\ln\left|\frac{a + \sqrt{a^2 - u^2}}{u}\right|$

**31.** $\displaystyle\int \frac{\sqrt{u^2 + a^2}}{u}\,du = \sqrt{u^2 + a^2} - a\ln\left|\frac{a + \sqrt{u^2 + a^2}}{u}\right|$

**32.** $\displaystyle\int \frac{\sqrt{u^2 - a^2}}{u}\,du = \sqrt{u^2 - a^2} - a\sec^{-1}\frac{u}{a} + C$

**33.** $\displaystyle\int \frac{\sqrt{a^2 - u^2}}{u}\,du = \sqrt{a^2 - u^2} - a\ln\left|\frac{a + \sqrt{a^2 - u^2}}{u}\right|$

**34.** $\displaystyle\int \frac{u^2\,du}{\sqrt{a^2 - u^2}} = -\frac{u}{2}\sqrt{a^2 - u^2} + \frac{a^2}{2}\sin^{-1}\frac{u}{a}$

**35.** $\displaystyle\int \frac{du}{u^2\sqrt{a^2 - u^2}} = -\frac{\sqrt{a^2 - u^2}}{a^2 u}$

**36.** $\displaystyle\int (u^2 \pm a^2)^{3/2}\,du = \frac{u}{8}(2u^2 \pm 5a^2)\sqrt{u^2 + a^2} + \frac{3a^4}{8}\ln\left|u + \sqrt{u^2 \pm a^2}\right|$

**37.** $\displaystyle\int (a^2 - u^2)^{3/2}\,du = \frac{u}{8}(5a^2 - 2u^2)\sqrt{a^2 - u^2} + \frac{3a^4}{8}\sin^{-1}\frac{u}{a}$

**38.** $\displaystyle\int \frac{du}{(u^2 \pm a^2)^{3/2}} = \pm\frac{u}{a^2\sqrt{u^2 \pm a^2}}$

**39.** $\displaystyle\int \frac{du}{(a^2 - u^2)^{3/2}} = \frac{u}{a^2\sqrt{a^2 - u^2}}$

## Trigonometric Forms

**40.** $\displaystyle\int \sin^2 u\ du = \frac{u}{2} - \frac{1}{2}\sin u \cos u$

**41.** $\displaystyle\int \cos^2 u\ du = \frac{u}{2} + \frac{1}{2}\sin u \cos u$

**42.** $\displaystyle\int \sin^3 u\ du = -\cos u + \frac{1}{3}\cos^3 u$

**43.** $\displaystyle\int \cos^3 u\ du = \sin u - \frac{1}{-}\sin^3 u$

**44.** $\displaystyle\int \sin^n u\ du = -\frac{1}{n}\sin^{n-1} u \cos u + \frac{n-1}{n}\int \sin^{n-2} u\ du$

**45.** $\displaystyle\int \cos^n u\ du = \frac{1}{n}\cos^{n-1} u \sin u + \frac{n-1}{n}\int \cos^{n-2} u\ du$

**46.** $\displaystyle\int \tan^n u\ du = \frac{\tan^{n-1} u}{n-1} - \int \tan^{n-2} u\ du$

**47.** $\displaystyle\int \cot^n u\ du = -\frac{\cot^{n-1} u}{n-1} - \int \cot^{n-2} u\ du$

**48.** $\displaystyle\int \sec^n u\ du = \frac{\sec^{n-2} u \tan u}{n-1} + \frac{n-2}{n-1}\int \sec^{n-2} u\ du$

**49.** $\displaystyle\int \csc^n u\ du = -\frac{\csc^{n-2} u \cot u}{n-1} + \frac{n-2}{n-1}\int \csc^{n-2} u\ du$

**50.** $\displaystyle\int u \sin u\ du = \sin u - u \cos u$

**51.** $\displaystyle\int u \cos u\ du = \cos u + u \sin u$

**52.** $\displaystyle\int \sin mu \sin nu\ du = \frac{\sin (m-n)u}{2(m-n)} - \frac{\sin (m+n)u}{2(m+n)}$

**53.** $\displaystyle\int \cos mu \cos nu\ du = \frac{\sin (m-n)u}{2(m-n)} + \frac{\sin (m+n)u}{2(m+n)}$

**54.** $\displaystyle\int \sin mu \cos nu\ du = -\frac{\cos (m-n)u}{2(m-n)} - \frac{\cos (m+n)u}{2(m+n)}$

**55.** $\displaystyle\int \sin^m u \cos^n u\ du = \frac{\sin^{m+1} u \cos^{n-1} u}{m+n} + \frac{n-1}{m+n}\int \sin^m u \cos^{n-2} u\ du$

$\displaystyle\qquad\qquad = -\frac{\sin^{m-1} u \cos^{n+1} u}{m+n} + \frac{m-1}{m+n}\int \sin^{m-2} u \cos^n u\ du$

## Other Forms

**56.** $\displaystyle\int \frac{du}{u^2 - a^2} = \frac{1}{2a} \ln \left| \frac{u - a}{u + a} \right|$    **57.** $\displaystyle\int ue^{au} \, du = \frac{e^{au}(au - 1)}{a^2}$

**58.** $\displaystyle\int u^n e^{au} \, du = \frac{u^n e^{au}}{a} - \frac{n}{a} \int u^{n-1} e^{au} \, du$

**59.** $\displaystyle\int u^n \ln u \, du = \frac{u^{n+1}}{n + 1} \left( \ln u - \frac{1}{n + 1} \right)$

**60.** $\displaystyle\int e^{au} \sin bu \, du = \frac{e^{au}(a \sin bu - b \cos bu)}{a^2 + b^2}$

**61.** $\displaystyle\int e^{au} \cos bu \, du = \frac{e^{au}(a \cos bu + b \sin bu)}{a^2 + b^2}$

**62.** $\displaystyle\int \sin^{-1} u \, du = u \sin^{-1} u + \sqrt{1 - u^2}$

**63.** $\displaystyle\int \cos^{-1} u \, du = u \cos^{-1} u - \sqrt{1 - u^2}$

**64.** $\displaystyle\int \tan^{-1} u \, du = u \tan^{-1} u - \frac{1}{2} \ln (1 + u^2)$

# Answers to Odd-Numbered Exercises

## Chapter 1

### Exercise 1-1   *(Page 3)*
**1.** 30
**3.** 120
**5.** 41
**7.** 30
**9.** 26
**11.** 5
**13.** 6
**15.** 8
**17.** 2
**19.** 57,600,000 mi
**21.** 78
**23.** 19

### Exercise 1-2   *(Page 7)*
**1.** 3/5
**3.** 4/5
**5.** 3/4
**7.** 2/15
**9.** 4/3
**11.** 2
**13.** 12
**15.** 5/8
**17.** 19/20
**19.** 85/24
**21.** 41/36
**23.** 39/10
**25.** 5
**27.** $5000
**29.** 6
**31.** 4

### Exercise 1-3   *(Page 12)*
**1.** 10.09
**3.** 8.91
**5.** 30
**7.** 0.08
**9.** 12.3
**11.** 4
**13.** 7
**15.** $1/5 = 0.20 = 20\%$
**17.** $17/100 = 0.17 = 17\%$
**19.** $7/125 = 0.056 = 5.6\%$
**21.** $1/250 = 0.004 = 0.4\%$
**23.** $1/4 = 0.25 = 25\%$
**25.** $6/5 = 1.2 = 120\%$
**27.** 40.12 V
**29.** 30 g
**31.** *(a)* $21.85
      *(b)* $25.13
**33.** *(a)* 5%
      *(b)* 12.6 V
**35.** 1902 Btu/h
**37.** 27

### Exercise 1-4   *(Page 17)*
**1.** 81
**3.** 1/4
**5.** 0.343
**7.** 1/9
**9.** 11/250
**11.** 1/20
**13.** 5/64
**15.** 4
**17.** 5/2

### Exercise 1-3 *(continued)*
**19.** 0.8
**21.** 4
**23.** 5
**25.** 1/2
**27.** 0.2
**29.** 7
**31.** $2\sqrt{2}$
**33.** $10\sqrt{2}$
**35.** 14
**37.** 2
**39.** $\sqrt{7}/4$
**41.** 4
**43.** 5/3
**45.** $2\sqrt{2}$
**47.** 1.728
**49.** 1.5 Ω
**51.** 2

### Exercise 1-5   *(Page 26)*
**1.** 2
**3.** 0.5
**5.** 2.4
**7.** 21.1
**9.** 0.68
**11.** 0.0143
**13.** 58.1
**15.** 0.0101
**17.** 0.655
**19.** 119
**21.** 0.242
**23.** 57
**25.** 15.0
**27.** 10.1

**29.** 57.4
**31.** 3.60
**33.** 2.74
**35.** 8.67
**37.** 23.0 kg, 0.00173 m³
**39.** 4.67% electrical, 19.1%
electronic, 5.97% civil,
7.83% mechanical
**41.** 78,900
**43.** 341 m/s
**45.** SOHO

**Exercise 1-6**   *(Page 31)*
**1.** 26.88
**3.** 1.9
**5.** 0.00825
**7.** 0.02
**9.** 0.0026
**11.** 0.0699
**13.** 0.031
**15.** 0.45
**17.** 224°
**19.** *(a)* 0.54 gm
   *(b)* 0.52, 3.7%
**21.** 8.19 min
**23.** 27.1 ft/s, 18.5 mi/h

**Review Questions 1-7**
*(Page 33)*
**1.** 22
**3.** 27
**5.** 30
**7.** 7/12
**9.** 4.04
**11.** 20
**13.** 4
**15.** 9
**17.** 0.2
**19.** $5\sqrt{3}$
**21.** $\dfrac{2\sqrt{6}}{3}$
**23.** 33
**25.** 0.95
**27.** 277
**29.** 0.136
**31.** 34.9
**33.** 67
**35.** 1.65

**37.** $43.16
**39.** 53.8 ft
**41.** 3.9 mm
**43.** 0.05

# Chapter 2

**Exercise 2-1**   *(Page 38)*
(For 1–13: variables; constants)
**1.** *P, l, w*; 2, 2
**3.** *W, F, d*; none
**5.** *P, V*; $P_o$, $V_o$
**7.** *C, C₁, C₂*; none
**9.** *R, I; E, r*
**11.** *V, t*; 331.5, 0.607
**13.** *l, T*; $l_o$, *a*, $T_o$
**15.** *(a)* 20 ft
   *(b)* 16 m
**17.** *(a)* 12.4 V,
   *(b)* 19.2 V
**19.** 172°F

**Exercise 2-2**   *(Page 46)*
**1.** *(a)* Time before event −,
   time after +
   *(b)* Up or right +, down or
   left −
(For 3–19: R = rational,
I = irrational)
**3.** +, yes, R
**5.** −, yes, R
**7.** +, no, I
**9.** +, yes, R
**11.** −, no, R
**13.** +, no, R
**15.** +, yes, R
**17.** +, no, I
**19.** +, yes, R
**21.** $-9/3 \quad -1 \qquad 5/4\ 2\sqrt{6}$

**23.** 8 > 4
**25.** −5 < −3
**27.** $\sqrt{5}$ < 3
**29.** −1.01 < −0.99

**31.** 400 < 10³
**33.** 8 > 7.21 > −11
**35.** 3
**37.** −10
**39.** 4
**41.** 5
**43.** −1/6
**45.** 5.3
**47.** 15
**49.** 7.8
**51.** −40
**53.** 3
**55.** −7
**57.** 1/2
**59.** 48
**61.** 0.06
**63.** 6/7
**65.** 0
**67.** 30
**69.** 12
**71.** −5/2
**73.** −9/5
**75.** −16
**77.** 3
**79.** 23/6
**81.** −4
**83.** −0.7
**85.** −5
**87.** 8848 − (−11,034) =
19,882 m, 29,028 −
(36,201) = 65,229 ft
**89.** −20° + 45° − 35° = −10°
**91.** −91 ft/s
**93.** +

**Exercise 2-3**   *(Page 50)*
**1.** −2*a*
**3.** 11*y* − 6*x*
**5.** 2*np* + 16*p*
**7.** 3*a*² − *a*²*t*
**9.** −1.5$R_1$ −2.3$R_2$
**11.** 3*ax* − 3
**13.** 5*r*
**15.** *x* + *y*
**17.** 5*x*² − *x* − 7
**19.** 2*u* + 6*v* − 4*w* + 1
**21.** 2/5 $F_y$ − 1/3 $F_x$
**23.** 5*d* − 2*x*

**25.** $xy^2 - 2$
**27.** $-n^2 + 3n$
**29.** $2g + 1$
**31.** $3x - y + 3$
**33.** $3\pi r^2 h + \pi r$
**35.** $\dfrac{W}{12} + 2$
**37.** $0.48IR - 0.06$

**Exercise 2-4** *(Page 57)*
**1.** 64
**3.** $-0.09$
**5.** $x^6$
**7.** $p^2$
**9.** $1/a^2$
**11.** 1
**13.** $w^9$
**15.** $81m^4n^4$
**17.** $\dfrac{a^3b^3}{8x^3}$
**19.** 8.8
**21.** 1/25
**23.** $-9/2$
**25.** 1
**27.** 64
**29.** $9/t^2$
**31.** $b^3$
**33.** $0.0016v^6$
**35.** $-12x^5$
**37.** $10I^2R$
**39.** $-3.6a^4b^3$
**41.** $3y^3$
**43.** $2x$
**45.** $2ht^2$
**47.** $2x^3 - 2x^2 + 6x$
**49.** $px^3 - p^3x$
**51.** $a^2b^2c^3 - ab^3c^3 + ab^2c^4$
**53.** $8ax^2y - 14bxy^2$
**55.** $x^3y + xy^3$
**57.** $3x + 6$
**59.** $-2b^2 + 3ab - 2a^2$
**61.** $C_1 + C_2$
**63.** $x - y$
**65.** $2.28t + 0.0152t^2$
**67.** 2
**69.** $2 * X \wedge 3 - 3 * X \wedge 2 + 5 * X$
**71.** $x^{a-b} = \dfrac{1}{x^{-(a-b)}} = \dfrac{1}{x^{b-a}}$

**Exercise 2-5** *(Page 61)*
**1.** $4.26 \times 10^{10}$
**3.** $1.178 \times 10^{11}$
**5.** $3.001 \times 10^{\circ}$
**7.** $1 \times 10^{-6}$
**9.** $5.64 \times 10^6$
**11.** $1 \times 10^4$
**13.** 26,400,000,000
**15.** 0.00000000009306
**17.** 0.0000320
**19.** 44,400,000
**21.** $2.72 \times 10^{13}$
**23.** 0.00942
**25.** $8.99 \times 10^7$
**27.** 0.0533
**29.** $4.00 \times 10^9$
**31.** 5.55%
**33.** $6.0 \times 10^{-5} \, m$
**35.** $1.09 \times 10^{11}$
**37.** $5.46 \times 10^8$
**39.** $4.39 \times 10^{11} \, kWh$

**Exercise 2-6** *(Page 67)*
**1.** $x^2 - 2x - 8$
**3.** $3a^2 - 10a + 3$
**5.** $5t^2 + 7t + 2$
**7.** $6d^2 + 2d - 20$
**9.** $20m^2 + mn - n^2$
**11.** $6x^2 - 13xy + 6y^2$
**13.** $4a^2b^2 - 13ab - 35$
**15.** $R^2 + R + 0.24$
**17.** $a^2 + 6a + 9$
**19.** $9x^2 - 24xy + 16y^2$
**21.** $3v^2t^2 + 6avt + 3a^2$
**23.** $y^2 - 9$
**25.** $25a^2 - 4x^2$
**27.** $4\pi r^2 - 12\pi rh + 8\pi h^2$
**29.** $x - 1$
**31.** $a - b$
**33.** $x^2 + ax + bx + ab$
**35.** $x^3 + 3x^2 - 2x - 6$
**37.** $x^3 - x^2 - 5x + 2$
**39.** $a^3 + b^3$
**41.** $x^3 + 6x^2 + 12x + 8$
**43.** $1 - 3j^2 - 2j^3$
**45.** $v_1^3 - v_1^2v_2 - v_1v_2^2 + v_2^3$
**47.** $pv - pb + a/v - ab/v^2$
**49.** $1.05Rt^2 + 324.8Rt - 8400R$

**Exercise 2-7** *(Page 73)*
**1.** 5
**3.** 1
**5.** $-4$
**7.** $-1$
**9.** 4
**11.** 6
**13.** $-1$
**15.** $-8.0$
**17.** 1.5
**19.** 3
**21.** 1/2
**23.** 2
**25.** 2.3
**27.** 2.4
**29.** 1/5
**31.** $-1/2$
**33.** $59^\circ$
**35.** 2.1
**37.** Not consistent

**Exercise 2-8** *(Page 80)*
**1.** $13,500; $18,500
**3.** $18,000
**5.** $6000
**7.** $1.60
**9.** 90
**11.** 8 A, 12 A, 6 A
**13.** 1050
**15.** 1.5 cal/min
**17.** 12.0 cm, 9.00 cm; 4.72 in., 3.54 in.
**19.** 2 qt
**21.** 12,000 mi

**Exercise 2-9** *(Page 85)*
**1.** $a = F/m$
**3.** $y = mx + b$
**5.** $R = \dfrac{E - Ir}{I}$
**7.** $v_0 = \dfrac{s - s_o + 4.9t^2}{t}$
**9.** 120 V
**11.** *(a)* $h = 550(212 - T)$
   *(b)* 1100 ft
**13.** 20 A
**15.** $136^\circ C$, $277^\circ F$

**17.** (a) $t = \dfrac{v - v_0}{g}$

    (b) 1.44 s

**19.** 1.2 kn, 1.38 mi/h

**21.** $D_0 = \dfrac{D(N + 2)}{N}$

    (b) −0.6 cm

**23.** (a) West

    (b) 9 mi/h

**Exercise 2-10**   *(Page 90)*

  **1.** 256

  **3.** 0.216

  **5.** −1.33

  **7.** −0.044

  **9.** 1.55

**11.** 1.76

**13.** 171

**15.** 539

**17.** $2.50 \times 10^8$

**19.** 0.215

**Review Questions 2-11**
*(Page 90)*

  **1.** 24

  **3.** −16

  **5.** −3/10

  **7.** 5/4

  **9.** $3x - 2x^2$

**11.** $2r - 6t$

**13.** $a - b$

**15.** $4x^6$

**17.** $n/2$

**19.** $90y^7$

**21.** $3mv$

**23.** $b^2y^4 - b^4y^2$

**25.** $2x^2 + 3xy - y$

**27.** $x^2 - 2x - 8$

**29.** $15a^2 - 2ab - b^2$

**31.** $4p^2x^2 - 1$

**33.** $x^3 - 5x + 2$

**35.** 6

**37.** −3/5

**39.** 1/2

**41.** $R = D/T$

**43.** $r = \dfrac{A - P}{Pt}$

**45.** 4230

**47.** 2.00

**49.** 10.1

**51.** $-100°F$

**53.** $0.4Lx^2 + 10Lx$

**55.** $1.834 \times 10^9$

**57.** $6.9 \times 10^{-4}$ *kWh*

**59.** $75.00

**61.** 40.5

**63.** −0.04 A

**65.** 2.98 cm

# Chapter 3

**Exercise 3-1**   *(Page 96)*

  **1.** 40 mi/h

  **3.** 1000

  **5.** 1/3

  **7.** 300

  **9.** $21/5 = 4.2$

**11.** 12.84

**13.** $5.9 \times 10^{-6}$

**15.** Virginia

**17.** 3.7

**19.** 200 g, 40 g, 10 g

**21.** 0.82

**23.** 300

**25.** 0.375 or 3/8 pt

**27.** 6.5 km

**29.** 146 h

**31.** 0.170 Ω

**33.** 7.63 cd

**35.** $20.6 \times 10^8$ J

**Exercise 3-2**   *(Page 102)*

  **1.** (a) $D = RT$

    (b) 57 mi/h

  **3.** (a) $V = \dfrac{4}{3}\pi r^3$

    (b) 8/1

  **5.** (a) 15 kg

    (b) 225 N

  **7.** (a) $200/min

    (b) $9000

  **9.** 0.247

**11.** (a) 2.5

    (b)

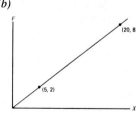

**13.** (a) 0.5

    (b)

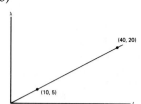

**15.** 265 ft

**17.** 10.5 m/s

**19.** 0.077 Ω

**21.** 27.9 A/m

**Exercise 3-3**   *(Page 108)*

  **1.** $xy = 45$

  **3.** $F = 0.36\ m_1m_2$

  **5.** $S = \dfrac{9.0w^3}{d}$

  **7.** 50

  **9.** 68

**11.** 2.43

**13.** (a) 5000 ft-lb

    (b) 250 lb

**15.** $1.6 \times 10^{-6}$ Ω · cm

**17.** $6000

**19.** (a) $E_g = 9.8$ mh

    (b) 245,000 m·kg

**21.** 3700 lb

**23.** 272 kPa

**25.** $T' = 16T$

**Review Questions 3-4**
*(Page 109)*

  **1.** 25.2

  **3.** 0.75

  **5.** 240 V

7. *(a)* $m = W/g$
   *(b)* 82.9 kg
9. $18.7 \times 10^8$ J/h
11. 112
13. *(a)* $F = k \dfrac{p_1 p_2}{r^2}$
   *(b)* $k = 10^{-7}$ N/A²
15. 78%

## Chapter 4

### Exercise 4-1   *(Page 116)*
1. 2300 g
3. 0.60 MV
5. 6.0 cm²
7. 5700 mL
9. $3.1 \times 10^3$ Ω
11. 373 K
13. $5.98 \times 10^{21}$ tons
15. $6.21 \times 10^6$ MJ
17. 53.9
19. 297 m
21. 5.81 m/s
23. Answers vary

### Exercise 4-2   *(Page 119)*
1. 54.4 kg
3. 13.5 cm
5. $9.49 \times 10^3$ Btu
7. 7.03 m³
9. 100 kPa
11. 190.5 cm
13. 35°C
15. 11 km/h
17. 0.063 in.
19. 3220 mi/h
21. *(a)* 0.01297 in.²
   *(b)* 0.6282 ohms/1000 ft
23. 48.2 km/s
25. 400 g for $1.35
27. $9 \times 10^{10}$ J

### Exercise 4-3   *(Page 124)*
1. $\pi/3$
3. $\pi/9$
5. 0.262 rad

7. 0.707 rad
9. 1.17 rad
11. 31°
13. 20°
15. 120°
17. 67.0°
19. 50.4°
21. *(a)* 19°40′
   *(b)* 19.7°
   *(c)* 0.344 rad
23. $\angle 1 = \angle 3 = 20°$, $\angle 2 = 160°$
25. $\angle 1 = 30°$, $\angle 2 = 90°$,
   $\angle 3 = 60°$
27. $\angle 1 = \angle 2 = 40°20′$
   $\angle 3 = 139°40′$
29. $\angle 1 = 70.1°$, $\angle 2 = 70.1°$,
   $\angle 3 = 71.5°$
31. 10
33. 30
35. *(a)* 15°
   *(b)* 75°
37. 135°
39. *(a)* $120\pi$ rad/s
   *(b)* 21,600°/s

### Exercise 4-4   *(Page 133)*
1. $x = 60°$
3. $x = y = 60°$
5. $x = 29.5°$, $y = 121°$
7. $x = 6$, $y = 4$
9. $x = 1.0$ cm
11. $\angle 1 = 40°$, $\angle 4 = 80°$
13. 50 m
15. $BC = 20$ ft, $CD = 10$ ft
17. 0.50 mm

### Exercise 4-5   *(Page 140)*
1. $A = 12$ in², $P = 16$ in.
3. $x = 60°$, 11 ft, 3.4 m
5. $C = 270$ cm, $A = 5810$ cm²
7. 8.9 m, 29 ft, 4.6 m², 50 ft²
9. 2.8 cm, 5.6 cm²
11. 6.3 ft
13. 4.54 cm²
15. 27 cm, 708 cm²
17. 0.14 in.², 89.4 mm²
19. 1.5 ha

21. 12 in.²
23. 1.20 m
25. 10.8 mm
27. 159 mi

### Exercise 4-6   *(Page 148)*
1. 1560 ft², 145 m², 3600 ft³,
   102 m³
3. *(a)* 12.6 m², 4.19 m³
   *(b)* 15.6 m²
5. $100
7. *(a)* 64 cm³
   *(b)* 67.5 cm³
9. 8.0 m
11. 0.86 cm³
13. $1.95 \times 10^8$ ft³,
   $5.51 \times 10^6$ m³
15. $1.3 \times 10^9$
17. 6.4 cm
19. 85 cm

### Exercise 4-7   *(Page 154)*
1. 5
3. 6
5. $\sqrt{5}$
7. 0.65
9. 5/6
11. 50 mi
13. 106 ft²
15. 6.21, 9.32
17. 67 m, 45 m
19. 7.5 Ω
21. Yes
23. 26 ft
25. 51 m

### Exercise 4-8   *(Page 164)*
(For 1–9: sin $A$, cos $A$, tan $A$)
1. 8/17, 15/17, 8/15
3. 5/13, 12/13, 5/12
5. 4/5, 3/5, 4/3
7. 0.60, 0.80, 0.75
9. 0.87, 0.50, 1.7
11. 0.9866
13. 0.8342
15. 0.1822
17. 0.0419

**19.** 18.5°
**21.** 64.9°
**23.** 48.2°
**25.** $B = 40°$, $a = 2.3$, $b = 1.9$
**27.** $b = 6.2$, $A = 38.7°$,
    $B = 51.3°$
**29.** $A = 17°$, $b = 6.5$, $c = 6.8$
**31.** $A = B = 45°$, $b = 2.0$,
    $c = 2.8$
**33.** 87 m
**35.** 119 m
**37.** 4.3°
**39.** $F = 46$ kg, $\Theta = 52°$
**41.** 76.0, 206 ft
**43.** (a) 72°
    (b) 19 min
**45.** 0.190 rad
**47.** 156 V
**49.** 2155 mi
**51.** $\cos A = 0.866$,
    $\tan A = 0.577$,
    $\sec A = 1.15$,
    $\csc A = 2.00$,
    $\cot A = 1.73$
**53.**

**Exercise 4-9**   *(Page 171)*
**1.** 85.3°
**3.** 50.013°
**5.** 6.1056°
**7.** 0.1236
**9.** 1.1538
**11.** 11.18°
**13.** 135.0°
**15.** 0.9493
**17.** 0.6573
**19.** 0.7151
**21.** 73.67°
**23.** 89.37°
**25.** No solution

**Review Questions 4-10**
*(Page 171)*
 **1.** 51 mm²
 **3.** 51 lb

**5.** 106 km/h
**7.** 0.876
**9.** 18°
**11.** $x = 49.7°$, $y = 139.7°$
**13.** $x = 40°$, $y = 40°$
**15.** $\sin A = 4/5$, $\cos A = 3/5$,
    $\tan A = 4/3$
**17.** $\sin A = 5/13$, $\cos A = 12/13$,
    $\tan A = 5/12$
**19.** 0.7159
**21.** 0.7490
**23.** 8.5°
**25.** 70.4°
**27.** $6.37 \times 10^3$ km
**29.** 62 lb/ft³
**31.** 40 m
**33.** 37.7 cm, 113 cm²
**35.** 21 cm
**37.** 120 L
**39.** 922 mi
**41.** 75 m
**43.** 25.6°
**45.** 3.4°

# Chapter 5

**Exercise 5-1**   *(Page 180)*
**1 through 9:**

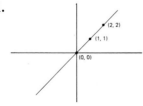

**11.**

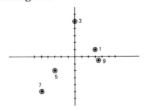

**13.**

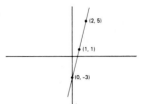

**15.**

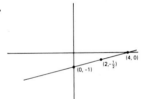

**17.**

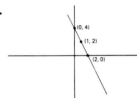

**19.**

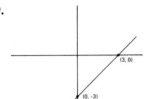

**21.**

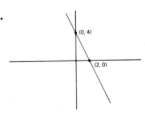

**23.**

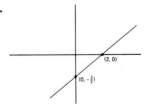

**25.**

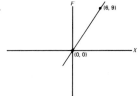

**27.**

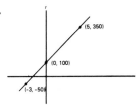

**29.**

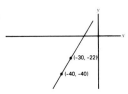

**31.**

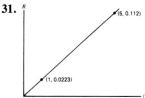

**Exercise 5-2** *(Page 185)*
 1. 3, 2
 3. 2, 1
 5. −2/3, 1/2
 7. −3, −3
 9. 0.8, 0.2
11. 0.8, 0.3
13. 2, 5
15. 3, −1
17. −14/5, −16/5
19. −12/5, 26/5
21. 3.0 A, 2.0 A
23. Same line

**Exercise 5-3** *(Page 191)*
 1. 1750 m², 2250 m²
 3. 80 lb, 80 lb, 100 lb

 5. 10.0 m, 32.8 ft
 7. 40 g (40%), 20 g (10%)
 9. *a* = 5.0 mi/h/s,
    *v* = 35 mi/h
11. 4.0 Ω, 32 V
13. 10-min ride, 20-min jog
15. $300, $200
17. 980°C

**Exercise 5-4** *(Page 198)*
 1. 1, 2, 3
 3. 4, −1, 3
 5. 0.1, 0.5, −0.3
 7. 3, −5, −1
 9. 1/2, 2/3, 3/5
11. 1, −2, −1, 2
13. 50, 200, 150
15. 1/2 gal, 1/3 gal, 1/6 gal;
    1.9 L, 1.3 L, 0.6 L
17. 71.4, 214, 214
19. 1.5 A, 0.83 A, 2.3 A
21. 40 km/h, 60 km/h, 80 km/h

**Review Questions 5-5**
*(Page 200)*
 1.

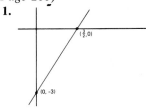

 3.

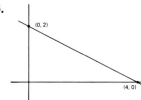

 5. −1, 2
 7. 0.4, 0.9
 9. 3, 1
11. 3, −1, 1
13. 0.8, 0.4, −0.2

**15.**

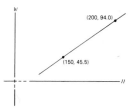

17. 500 mL, 500 mL
19. 2 mi/h/s, 20 mi/h
21. 300 g, 500 g, 200 g

# Chapter 6

**Exercise 6-1** *(Page 209)*
 1. 5
 3. 21
 5. −7.0
 7. 2
 9. 1
11. 18.0
13. 3, 2
15. −2/3, 1/2
17. −3, −3
19. 0.8, 0.3
21. 1, 2, 3
23. 0.1, 0.5, −0.3
25. 3, −5, −1
27. 1/2, 2/3, 3/5
29. 2.3, 0.39
31. 1.6, −9.2
33. 5, 6, 7
35. 0.1, −0.2, 0.3
37. 4.26, 7.39
39. 0.37 s, 370 m, 1210 ft
41. denominator = 0
43. $3500, $3500, $6000
45. 43% (14 carat), 14% (20
    carat), 43% (coinage)
47. 1.0 A, 0.4 A, 1.5 A

**Exercise 6-2** *(Page 212)*
 1. *(a)* row 3, column 2;
    *(b)* row 4, column 1;
    *(c)* row 5, column 5;
    *(d)* row *i*, column *j*

**3.** $\begin{pmatrix} 5 & 7 & 7 \\ -6 & 9 & 10 \\ 3 & 12 & 11 \end{pmatrix}$

**5.** $\begin{pmatrix} \frac{3}{4} & \frac{1}{2} \\ 0 & -1 \end{pmatrix}$

**7.** $\begin{pmatrix} 8 & 11 & -3 \\ -3 & 12 & 19 \end{pmatrix}$

**9.** $\begin{pmatrix} 1 & -8 & -12 \\ 3 & 6 & 11 \end{pmatrix}$

**11.** Cannot add

**13.** $2(A + B) =$
$\begin{pmatrix} 10 & 20 & 4 \\ -6 & 12 & 18 \end{pmatrix}$
and $2A + 2B =$
$\begin{pmatrix} 10 & 20 & 4 \\ -6 & 12 & 18 \end{pmatrix}$

### Exercise 6-3   (Page 218)

**1.** $\begin{pmatrix} 1 \\ 2 \end{pmatrix}$

**3.** $\begin{pmatrix} 28 & -11 \\ 27 & 5 \\ 11 & 8 \end{pmatrix}$

**5.** $\begin{pmatrix} 6 & 4 & -5 \\ -2 & 4 & -7 \end{pmatrix}$

**7.** $\begin{pmatrix} 0 & -1 & 15 \\ 9 & 3 & 18 \\ 12 & 6 & 13 \end{pmatrix}$

**9.** $\begin{pmatrix} 2 & 22 \\ 4 & 4 \end{pmatrix} \neq$
$\begin{pmatrix} 5 & 5 & 0 \\ 7 & 5 & 2 \\ 11 & 15 & -4 \end{pmatrix}$

**11.** $\begin{pmatrix} -2 & 3 \\ 1 & -1 \end{pmatrix}$

**13.** $\begin{pmatrix} \frac{1}{2} & \frac{3}{2} \\ 0 & -\frac{1}{2} \end{pmatrix}$

**15.** $\begin{pmatrix} \frac{4}{3} & -\frac{1}{3} \\ -3 & 1 \end{pmatrix}$

**17.** $\begin{pmatrix} 1 & -1 & -3 \\ 0 & -1 & -2 \\ 2 & 1 & -1 \end{pmatrix}$

**19.** $\begin{pmatrix} \frac{3}{4} & \frac{1}{4} & -\frac{1}{4} \\ -\frac{1}{2} & -\frac{1}{2} & 0 \\ 1 & 1 & \frac{1}{2} \end{pmatrix}$

**21.** $(70{,}000 \quad 130{,}000)$

**23.** $\begin{pmatrix} 2 & 3 \\ 3 & 4 \end{pmatrix}\begin{pmatrix} 1 \\ 2 \end{pmatrix} = \begin{pmatrix} 8 \\ 11 \end{pmatrix}$

### Exercise 6-4   (Page 224)

**1.** $\begin{pmatrix} -2 \\ 7 \end{pmatrix}$

**3.** $\begin{pmatrix} 3 \\ -1 \end{pmatrix}$

**5.** $\begin{pmatrix} \frac{1}{2} \\ 1 \end{pmatrix}$

**7.** $\begin{pmatrix} -1 \\ 2 \\ 1 \end{pmatrix}$

**9.** $\begin{pmatrix} 1 \\ 2 \\ 3 \end{pmatrix}$

**11.** $\begin{pmatrix} 2 \\ 2 \\ 4 \end{pmatrix}$

### Review Questions 6-5
*(Page 225)*

**1.** $-2$

**3.** 95

**5.** $\begin{pmatrix} 6 & 1 \\ 0 & -5 \end{pmatrix}$

**7.** $\begin{pmatrix} 6 & 3 \\ -8 & -7 \end{pmatrix}$

**9.** $\begin{pmatrix} 3 & -7 & 8 \\ 2 & 4 & -4 \end{pmatrix}$

**11.** $\begin{pmatrix} \frac{1}{3} & 0 \\ \frac{1}{3} & -\frac{1}{2} \end{pmatrix}$

**13.** $\begin{pmatrix} -9 & -\frac{13}{2} & 5 \\ 4 & \frac{5}{2} & -1 \end{pmatrix}$

**15.** 4, 3

**17.** 2, 2

**19.** 2, 1, 3

**21.** reg = 5 s, bold = 6 s

**23.** 1 A, 1 A, 2 A

# Chapter 7

### Exercise 7-1   (Page 229)

**1.** $2 \cdot 2 \cdot 3 \cdot 3$

**3.** $2 \cdot 7 \cdot 7 \cdot x \cdot y \cdot y$

**5.** $-3 \cdot 13 \cdot F_x \cdot F_v$

**7.** $3y(2 + y)$

**9.** $10b(a - 2b)$

**11.** $0.2k(3T_1 + T_2)$

**13.** $5(2x^2 + 4x - 3)$

**15.** $xyz(x^2 - x + 1)$

**17.** $x(m_0 - 5m^2)$

**19.** $\frac{1}{2}M(\theta_1 - \frac{1}{2}\theta_2)$

**21.** $4a^2b^2(9a^2 + 4ab - 3b^2)$

**23.** $x = \dfrac{b}{2a + 4}$

**25.** $C = \dfrac{10Q}{T_1 - T_2}$

**27.** $1.4R(4I_1{}^2 + 3I_2{}^2)$

**29.** $\pi h^2(r - h/3)$

### Exercise 7-2   (Page 233)

**1.** $(a + 2b)(a - 2b)$

**3.** $(10m + 11m_o)$
    $(10m - 11m_o)$

**5.** $(1 + K^2)(1 + K)(1 - K)$

**7.** prime

**9.** $(p + 0.3)(p - 0.3)$

**11.** $(0.9x + 0.4y)(0.9x - 0.4y)$

**13.** $(y + 1)(y + 5)$

**15.** $(w - 1)(w - 3)$

**17.** $(k + 5)(k - 2)$

**19.** prime

**21.** $(m + 5)^2$

**23.** $(3x + 1)(x + 1)$

**25.** $(5a + 6b)(a - b)$

**27.** $(3v - 4)(3v - 1)$

**29.** $(a + 1/2)^2$

**31.** $(p - 0.6)(2.0p - 0.3)$

**33.** $(4e + 1)(3e - 1)$

**35.** $6(h + 2)(h - 2)$

**37.** $4(2x - 1)(x + 4)$

**39.** $5a(2x + 5)(x - 1)$

**41.** $(x^2 + 1)^2$

**43.** $(f + 10^3)^2$

**45.** $2(4x^4 + 1)(2x^2 + 1)$
    $(2x^2 - 1)$

**47.** $(2a^2 + 1)(a^2 - 3)$

**49.** $(x + y)(x^2 - xy + y^2)$
**51.** $(2r - R)(4r^2 + 2rR + R^2)$
**53.** $V = w(2w + 1)(w - 2)$
**55.** $k(T^2 + T_o^2)(T + T_o)(T - T_o)$
**57.** $QH(Q + H)(Q^2 - QH + H^2)$

**Exercise 7-3**   *(Page 236)*
**1.** $2, -2$
**3.** $5, -2$
**5.** $2, 2/3$
**7.** $0.3, -0.3$
**9.** $0, 1/7$
**11.** $0, -0.433$
**13.** $1/2, -1$
**15.** $-0.6, -0.6$
**17.** $1/2, -3$
**19.** $2/3, -1$
**21.** $5, -8$
**23.** $1/4, 1/2$
**25.** $-0.5, 0.67$
**27.** $0, s$
**29.** $h/4, h/4$
**31.** $0, \dfrac{2v_o}{g}$
**33.** $t = 40\ s$
**35.** $w = 50\ cm,\ l = 200\ cm$
**37.** $0, \dfrac{V}{R_1 + R_2}$
**39.** $1.6, 3.0, 3.4$

**Exercise 7-4**   *(Page 241)*
**1.** $\dfrac{7y^2}{9}$
**3.** $\dfrac{c}{b + d}$
**5.** $\dfrac{c - d}{5}$
**7.** $\dfrac{a + 3b}{3a - b}$
**9.** Not reducible
**11.** $\dfrac{7t_1 - t_2}{9t_1 - 4t_2}$
**13.** $-2$
**15.** $5N$
**17.** $\dfrac{a^2 - ab + b^2}{a - b}$
**19.** $4r^2 - 2rR + R^2$

**21.** $6/x$
**23.** $5t$
**25.** $7/uv$
**27.** $\dfrac{t - h}{t}$
**29.** $\dfrac{2(x + 1)}{3}$
**31.** $5(2d - 1)$
**33.** $\dfrac{Q_h + 1}{Q_h}$
**35.** $\dfrac{2(a - 2x)}{(a - 2)(a + 2x)}$
**37.** $\dfrac{R - 1}{R + 1}$
**39.** $1$
**41.** $1$
**43.** $-(a^2 + 1)$
**45.** $-\dfrac{r}{R}$
**47.** $\dfrac{p + q}{p - q}$
**49.** $2(F_x + F_y)$
**51.** $0.5t^2$
**53.** $\dfrac{a^2 + aa' + a'^2}{a(a + a')}$

**Exercise 7-5**   *(Page 249)*
**1.** $23/24$
**3.** $37x/45$
**5.** $\dfrac{13}{3 \times 10^4}$
**7.** $\dfrac{3x^2 + 2y^2}{xy}$
**9.** $\dfrac{18 - 8b}{15ab^2}$
**11.** $\dfrac{2d^2 + 6cd - 5c^2}{2d^2}$
**13.** $\dfrac{7 - 3a}{a^2 - 1}$
**15.** $\dfrac{8.1 + 2.3k}{T(1 + k)}$
**17.** $2/3$
**19.** $\dfrac{1}{2x(x - 1)}$
**21.** $\dfrac{1}{1 - y}$

**23.** $\dfrac{a - b}{2(a + b)}$
**25.** $\dfrac{3x^2 + 4x - 3}{3(3x + 1)(x + 1)}$
**27.** $\dfrac{-2R - 13x}{(R + x)(R - x)(R + 2x)}$
**29.** $\dfrac{7I_1^2 + 9I_1I_2 - I_2^2}{I_1I_2(I_1 + I_2)}$
**31.** $\dfrac{6x^2 + 3x - 1}{(x - 1)^2(x + 1)}$
**33.** $\dfrac{7a^2 + 2}{(a^2 + 1)(a^2 - 1)}$
**35.** $\dfrac{x + y}{x^2 - xy + y^2}$
**37.** $\dfrac{1}{x + 1}$
**39.** $1 - x$
**41.** $2/m$
**43.** (a) $2r_1r_2/(r_2 + r_1)$
    (b) $24\ mi/h$
**45.** $\dfrac{R_1R_2R_3}{R_2R_3 + R_1R_3 + R_1R_2}$
**47.** $\dfrac{1}{G} = \dfrac{1}{\dfrac{1}{G_1} + \dfrac{1}{G_2}} = \dfrac{G_1G_2}{G_1 + G_2}$
**49.** $\dfrac{l^3(fl + 8F)}{384EI}$
**51.** $\dfrac{2\left(\dfrac{qf}{q - f}\right)q}{\left(\dfrac{qf}{q - f}\right) + q} = \dfrac{\dfrac{2q^2f}{q - f}}{\dfrac{q^2}{q - f}} = 2f$

**Exercise 7-6**   *(Page 255)*
**1.** $5$
**3.** $12$
**5.** $4$
**7.** $2/5$
**9.** $4.5$
**11.** $30$
**13.** $3$
**15.** $1/2$
**17.** $7$
**19.** $1/2$
**21.** $1$
**23.** $-2$

**25.** 2

**27.** $-1/5$

**29.** $(1, 1)$

**31.** $(1/2, 1/3)$

**33.** $\dfrac{5ab}{a + 3b}$

**35.** $\dfrac{5 + 3t}{10t}$

**37.** $\dfrac{kR_1R_2}{9 \times 10^9(R_1 - R_2)}$

**39.** $\dfrac{8h}{6 + \pi}$

**41.** $\dfrac{R_1R_2}{(\mu m - 1)(R_1 + R_2)}$

**43.** $4.8 \ \Omega$

**45.** $2.4$ h

**47.** 2 ms

**49.** 125, 75, 60

**51.** (a) $\dfrac{CC_2}{C_2 - C}$

(b) $C = 5.0 \ \mu F$

## Exercise 7-7 *(Page 266)*

**1.** $\pm 2$

**3.** $\pm 2.34$

**5.** $9, -5$

**7.** $-1 \pm \sqrt{5}; 1.24, 3.24$

**9.** $5/2, 1/2$

**11.** $-1, -3$

**13.** $2, -5/3$

**15.** $4, 1/2$

**17.** $1 \pm \sqrt{2}; 2.41, -0.41$

**19.** $\dfrac{5 \pm 3\sqrt{5}}{10}; 1.17, -0.17$

**21.** $1 \pm \sqrt{7}; 3.65, -1.65$

**23.** $\dfrac{3 \pm \sqrt{5}}{4}; 1.31, 0.19$

**25.** $1/3, -1$

**27.** $\dfrac{-2 \pm \sqrt{-4}}{2}$

**29.** $\dfrac{1 \pm \sqrt{-7}}{4}$

**31.** $0, -3$

**33.** $-4/3, 3/4$

**35.** $\pm 3/2$

**37.** $3/2, -2$

**39.** $0.70, 0.39$

**41.** $\pm \sqrt{-1}$

**43.** $2 \pm 2\sqrt{3}; 5.46, -1.46$

**45.** $\dfrac{m \pm m\sqrt{5}}{2}$

**47.** $\dfrac{1 \pm \sqrt{3}}{q}$

**49.** $\dfrac{-v_o \pm \sqrt{v_o^2 + 2gs}}{g}$

**51.** $7, 10, \sqrt{149} = 12.2$

**53.** $16

**55.** $3.9$ h, $6.9$ h

**57.** $-26, -3.8$

**59.** $100

**61.** 24

## Review Questions 7-8
*(Page 268)*

**1.** $4xy (2x - 3y)$

**3.** $3a (7a^2 + 5a - 9)$

**5.** $(3x + y)(3x - y)$

**7.** $(b + 7)(b - 3)$

**9.** $(3k - 2)(k + 4)$

**11.** $6 (t + 2)^2$

**13.** $(2Q^2 - 1)(Q^2 + 1)$

**15.** $\dfrac{x + 2y}{xy}$

**17.** $\dfrac{a(a + b)}{a - b}$

**19.** $9x/10$

**21.** $\dfrac{1}{c - d}$

**23.** 1

**25.** $\dfrac{29x}{24}$

**27.** $\dfrac{4d - 1}{d^2 - 1}$

**29.** $\dfrac{x^2 - 10x}{3(x + 2)^2}$

**31.** $\dfrac{p - q}{P}$

**33.** 4

**35.** $13/6$

**37.** $-1/2$

**39.** $5, -1$

**41.** $0, 7/8$

**43.** $1/3, -1$

**45.** $\dfrac{-3 \pm \sqrt{5}}{2}; -0.38, -2.62$

**47.** $\dfrac{-1 \pm \sqrt{19}}{2}; 1.68, -2.68$

**49.** $\dfrac{5 \pm \sqrt{97}}{6}; 2.47, -0.81$

**51.** $\dfrac{5}{t^2 - a^2}$

**53.** $\dfrac{m}{2} \left(v_1 + \dfrac{1}{2}v_2\right)\left(v_1 - \dfrac{1}{2}v_2\right)$

**55.** $4(h + 3)(h + 4)$

**57.** $\dfrac{2.2k^2 + 6.3}{K(1.3K + 2.1)}$

**59.** $10 \ \Omega, 15 \ \Omega$

**61.** $0.45$

## Chapter 8

### Exercise 8-1 *(Page 273)*

**1.** $x^4$

**3.** $a^4$

**5.** 2

**7.** $\dfrac{1}{3pq^4}$

**9.** $4.5r^3$

**11.** $0.9w$

**13.** $b^4/a^2$

**15.** $\dfrac{25L^2}{9}$

**17.** $\dfrac{-5}{a^2v^4}$

**19.** $\dfrac{54x^3}{25y^2}$

**21.** $\dfrac{m^3n^2 + 1}{m^2n^2}$

**23.** $\dfrac{qx}{pq + 1}$

**25.** $\dfrac{xy}{x + y}$

**27.** $2h^4 + 1$

**29.** $1 - e^{(a+b)t}$

**31.** $p^{2x} - \dfrac{1}{p^{2x}}$

**33.** $1 + \dfrac{1}{k^{2y}}$

**35.** $\dfrac{R_1(R_2 + R_3)}{R_1 + R_2 + R_3}$

**37.** $\dfrac{1}{e^{\gamma t}}$

**39.** $1 - \left(\dfrac{P_2}{P_1}\right)^3$

**Exercise 8-2** *(Page 277)*
**1.** 125
**3.** 81
**5.** 1/10
**7.** 1/27
**9.** 0.6
**11.** $32 \times 10^5$
**13.** 100
**15.** 343/64
**17.** 4
**19.** $1/x$
**21.** $-3ab^2$
**23.** $\dfrac{1}{1000w^3x^3}$
**25.** $\dfrac{1}{10^3 I}$
**27.** $1/p^2$
**29.** $108x^5$
**31.** $\dfrac{2ab^2}{3}$
**33.** $e^t$
**35.** $a - 1$
**37.** 3/2
**39.** $x^3 + 2x^2 + x$
**41.** 0.019
**43.** $(RC)^{0.4}$
**45.** 17.5 min

**Exercise 8-3** *(Page 281)*
**1.** $6x$
**3.** $0.2v$
**5.** $2x^2y$
**7.** $2\sqrt{3}$
**9.** $6\sqrt{3}$
**11.** $2\sqrt{2}$
**13.** $x^2y\sqrt{xy}$
**15.** $4rs^2\sqrt{2s}$
**17.** $2\sqrt[3]{3}$
**19.** $0.3v_o\sqrt{v_o}$
**21.** 15

**23.** 3
**25.** 0.2
**27.** $1.0h^2$
**29.** $200\sqrt{3}$
**31.** $1.2 \times 10^{-3}$
**33.** $x^n$
**35.** $a\sqrt[v]{a}$
**37.** $\sqrt{3}/3$
**39.** $\sqrt{30}/6$
**41.** $2\sqrt{3}/3$
**43.** $3\sqrt{2}/4$
**45.** $\sqrt[3]{2}/2$
**47.** $\sqrt{x}/x$
**49.** $\sqrt{2gh}/g$
**51.** $\sqrt[3]{6xy}/2y$
**53.** $\dfrac{\sqrt{R(R + 4)}}{2R}$
**55.** $\dfrac{10\sqrt[3]{q^2H^2}}{qH}$
**57.** $2\sqrt{5} + 2\sqrt{7}$
**59.** $7\sqrt{5}$
**61.** $8\sqrt{3}$
**63.** $-3\sqrt{5}$
**65.** $3\sqrt{2}$
**67.** $4\sqrt[3]{2}$
**69.** $5x\sqrt{2x}$
**71.** $0.4r\sqrt{\pi h}$
**73.** $3\sqrt{2}/4$
**75.** $17\sqrt{3}/6$
**77.** $\sqrt{10}$
**79.** $8.1t\sqrt{v}$
**81.** $2xy\sqrt[3]{xy}$
**83.** $23\sqrt{10}$
**85.** $5\sqrt{x - 1}$
**87.** $\dfrac{7\sqrt{2pq}}{4q}$
**89.** $\dfrac{2x\sqrt{x^2 - 1}}{x^2 - 1}$
**91.** $5x\sqrt{2}$
**93.** 77 Ω
**95.** $\dfrac{2\sqrt{3gh}}{3}$
**97.** $\dfrac{RX\sqrt{R^2 + X^2}}{R^2 + X^2}$

**Exercise 8-4** *(Page 286)*
**1.** $3\sqrt{2} - \sqrt{3}$

**3.** $10\sqrt{3} + 6\sqrt{10}$
**5.** $\dfrac{4\sqrt{2}}{3}$
**7.** $3\sqrt[3]{2} - 6$
**9.** $0.4x$
**11.** $a\sqrt{2b} + b\sqrt{3a}$
**13.** $\sqrt{3} - 2$
**15.** $2\sqrt{3}$
**17.** $24 + 20\sqrt{3}$
**19.** $11 - 6\sqrt{2}$
**21.** $-7$
**23.** 13
**25.** $4 - 6\sqrt{3}$
**27.** $xy - 1$
**29.** 9/8
**31.** $\sqrt{2} + 1$
**33.** $\sqrt{7} + \sqrt{2}$
**35.** $-7 - 4\sqrt{3}$
**37.** $\dfrac{7 + 2\sqrt{15}}{11}$
**39.** $\sqrt{2x} - \sqrt{x}$
**41.** $\dfrac{2p + 5\sqrt{pq} + 3q}{4p - 9q}$
**43.** $\pi ab(\sqrt{b} - \sqrt{a})$
**45.** $\dfrac{V_1 + \sqrt{V_1 V_2}}{V_1 - V_2}$

**Exercise 8-5** *(Page 290)*
**1.** 7
**3.** 9/20
**5.** 4/3
**7.** 1/8, 1/2
**9.** 0.034
**11.** 4
**13.** 1/4
**15.** 1/12
**17.** $3 - 2\sqrt{2}$
**19.** $\dfrac{\sqrt{R^2 - N^2}}{N}$
**21.** $s_0 = v_0 t - \frac{1}{2}gt^2$
**23.** 14 ft
**25.** $I_1 = 135\ cd,\ I_2 = 29\ cd$
**27.** $\dfrac{7.12 \times 10^5 d^3}{l}$
**29.** (a) $\dfrac{s}{s'} = \dfrac{\sqrt[3]{4}}{2}$
 (b) 2.9 cm

**Review Questions 8-6**
*(Page 291)*

1. $x$

3. $\dfrac{27p^6}{8}$

5. $\dfrac{12 + 4x}{3x^2}$

7. 9

9. $2ab^2$

11. $5/2\ K^4$

13. $3\sqrt{2}$

15. $10x^2\sqrt{3x}$

17. $1.1 \times 10^{-2}$

19. $\sqrt{2}/4$

21. $\sqrt{xy}/y^2$

23. $4\sqrt{6} - \sqrt{3}$

25. $5\sqrt[3]{2}$

27. $2.4\sqrt{x}$

29. $6\sqrt{6} - 12$

31. $2\sqrt{6} + 4$

33. $1/2$

35. $9/25$

37. $\dfrac{e^{2x} + 1}{2e^x}$

39. $1.44$

41. $\dfrac{C\sqrt{2g}(\sqrt{H} + \sqrt{h})}{2}$

43. *(a)* $\mu = \dfrac{\sqrt{3d^2 + d'^2}}{2d'}$

   *(b)* $\mu = 1.31$

## Chapter 9

**Exercise 9-1**   *(Page 301)*

1.
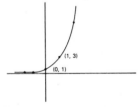

3.
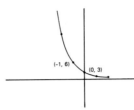

5.
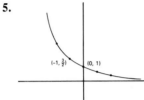

7.
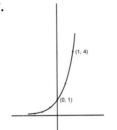

9.
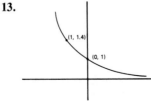

11.

13.

15.
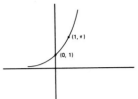

17.
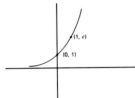

19.
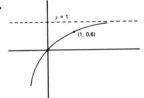

21.
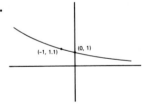

23.
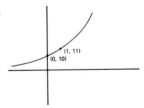

25. *(a)*
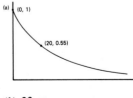

   *(b)* 23 yr

**27.** (*a*)

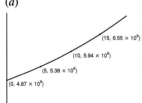

(*b*) During 1996

**29.**

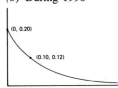

**Exercise 9-2**  (*Page 305*)
1. $2^4 = 16$
3. $5^3 = 125$
5. $4^{-2} = 1/16$
7. $10^3 = 1000$
9. $10^{0.4771} = 3.00$
11. $e^1 = e$
13. $e^{1.629} = 5.1$
15. $\log_2 32 = 5$
17. $\log 10,000 = 4$
19. $\log_4 8 = 3/2$
21. $\log 0.01 = -2$
23. $\ln 25.0 = 3.219$
25. $1/9$
27. $100,000$
29. $3'$
31. $1/2$
33. $-3$
35. $e^2$
37. $1$
39. $-1$
41. $1.2$
43. $2$
45.

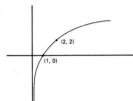

**47.**

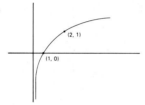

49. (*a*) $10^R = I$
   (*b*) $10$
51. $\ln (I/I_o) = -t/RC$
53. (*a*) $10^{\Delta L/10} = I_2/I_1$
   (*b*) $10^4$
55. $\log_2 (A/A_o) = 0.028t$

**Exercise 9-3**  (*Page 315*)
1. $2 + \log_2 x$
3. $3 + 2\log_3 p - \log_3 q$
5. $\log v - 1/2$
7. $3.7$
9. $\log_b a + x$
11. $1$
13. $3$
15. $2$
17. $\ln 3e^3$
19. $\log_b (y/t)$
21. $1.9345$
23. $-0.9788$ or $0.0212 - 1$
25. $3.7482$
27. $-2.2840$ or $0.7160 - 3$
29. $-2.2218$ or $0.7782 - 3$
31. $332$
33. $0.516$
35. $7.41$
37. $0.00632$
39. $2350$
41. $42.5$
43. $0.0370$
45. $7.96$
47. $1.3350$
49. $4.0943$
51. $-0.0834$
53. $10.1266$
55. $4.322$
57. $0.8614$
59. $-0.8340$
61. $\log P_2 + n\log V_2 - n\log V_1$

**63.** (*a*) 11.6 yr
   (*b*) 8.7 yr
**65.** 48 dB
**67.** (*a*) 8.1 s
   (*b*) 28,100 s or 7.8 h
**69.** $\log_b (X/Y) = x - y = \log_b X - \log_b Y$

**71.** $\log_a b = x = \dfrac{1}{1/x} = \dfrac{1}{\log_b a}$

**Exercise 9-4**  (*Page 321*)
1. $1$
3. $-1$
5. $-5/2$
7. $2.26$
9. $1.70$
11. $-0.276$
13. $0.307$
15. $0.178$
17. $6.03$
19. $3.74$
21. $99$
23. $3$
25. $1 + e^2$
27. $\dfrac{\log Q - 2}{\log 5}$
29. $t = (RC)\ln (E_0/E)$
31. $I_o \times 10^L$
33. $m_o e^{-kt}$
35. $13.3$ min
37. $11.9$ yr
39. $12.6$ yr
41. $0.14$ s
43. $P_2 = 60$ lb/ft$^2$
45. $1\ W/m^2$
47. $0.10$ mW
49. $6.1$ yr (in 1991)

**Exercise 9-5**  (*Page 326*)
1.

**3.**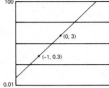

**5.**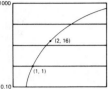

**7.**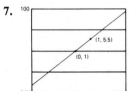

**9.**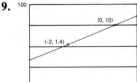

**11.**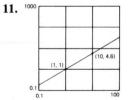

**13.**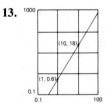

**15.**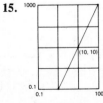

**17.**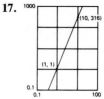

**19.**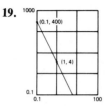

**21.**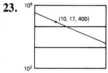

**23.**

**25.** (a)

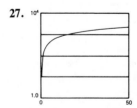

(b) 250 d

**27.**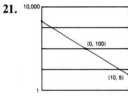

**Exercise 9-6** (Page 330)
1. 1.726
3. −1.287
5. 4.136

7. 0.0042286
9. 2.865
11. −0.1854
13. 4.572
15. 0.05663
17. 0.5771
19. 0.05569

**Review Questions 9-7**
(Page 330)

**1.**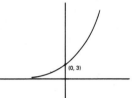

**3.**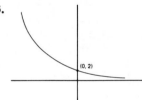

5. $3^2 = 9$
7. $\log_5 0.04 = -2$
9. 1/16
11. 3
13. $e$
15. 1/2
17. $\log_2 x - 2.5$
19. 3
21. 1.1206
23. 2570
25. 2.0149
27. 8
29. 34

**31.**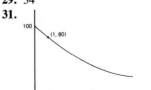

33. 5600 yr

**35.** (a) $e^{-Rt/L} = \dfrac{I_o - I}{I_o}$

$-Rt/L = \ln\left(\dfrac{I_o - I}{I_o}\right)$

$t = (L/R) \ln\left(\dfrac{I_o}{I_o - I}\right)$

(b) 0.080 s

**37.**

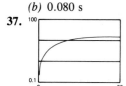

## Chapter 10

### Exercise 10-1 (Page 338)
(For 1–23: sin, cos, tan, csc, sec, cot)

**1.** 4/5, 3/5, 4/3, 5/4, 5/3, 3/4
**3.** −12/13, 5/13, −12/5, −13/12, 13/5, −5/12
**5.** −√2/2, −√2/2, 1, −√2, −√2, 1
**7.** 0.8, −0.6, −1.33, 1.25, −1.67, −0.75
**9.** −4/5, −3/5, 4/3, −5/4, −5/3, 3/4
**11.** 3/5, 4/5, 3/4, 5/3, 5/4, 4/3
**13.**

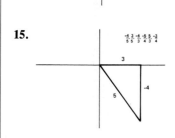

**15.**

**17.**

0.6, -0.8, -0.75, 1.7, -1.25, -1.3

**19.** $\frac{1}{3}, \frac{2\sqrt{2}}{3}, \frac{\sqrt{2}}{4}, 3, \frac{3\sqrt{2}}{4}, 2\sqrt{2},$

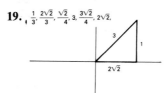

**21.**

$\frac{\sqrt{2}}{2}, -\frac{\sqrt{2}}{2}, -1, \sqrt{2}, -\sqrt{2}, -1$

**23.**

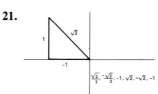

-0.866, 0.500, -1.73, -1.15, +2.00, -0.577

**25.**

| tan | cot |
|---|---|
| 0.7500, | 1.333, |
| sec | csc |
| 1.250, | 1.667 |

**27.**

| tan | cot |
|---|---|
| 0.5774, | 1.732, |
| sec | csc |
| −1.155, | −2.000 |

**29.**

| tan | cot |
|---|---|
| −3.078, | −0.3249, |
| sec | csc |
| −3.236, | 1.051 |

**31.** ±2
**33.** 0.88, −0.47

### Exercise 10-2 (Page 346)
**1.** sin 50° = 0.7660
**3.** −tan 30° = −√3/3
**5.** −cos 85° = −0.0872
**7.** −sec 60° = −2
**9.** sin π/4 = √2/2

**11.** cot 76° = 0.2493
**13.** cos 30° = √3/2
**15.** tan 0° = 0
**17.**

| | 30° | 45° | 60° |
|---|---|---|---|
| csc | 2 | √2 | $\frac{2\sqrt{3}}{3}$ |
| sec | $\frac{2\sqrt{3}}{3}$ | √2 | 2 |
| cot | √3 | 1 | $\frac{\sqrt{3}}{3}$ |

**19.**

| | 0 | 90 | 180 | 270 |
|---|---|---|---|---|
| csc | ∞ | 1 | ∞ | −1 |
| sec | 1 | ∞ | −1 | ∞ |
| cot | ∞ | 0 | ∞ | 0 |

**21.** (a) $(1/2)^2 + (-\sqrt{3}/2)^2 =$
  $1/4 + 3/4 = 1$
  (b) $(-2\sqrt{3}/3)^2 - (\sqrt{3}/3)^2 =$
  $4/3 - 1/3 = 1$
**23.** 0.7593
**25.** −0.7623
**27.** −1.19
**29.** −1.00
**31.** −1
**33.** 1.02
**35.** −√2
**37.** 0.492
**39.** −0.143
**41.** For $\theta = 30°$: $\sin(-30°) =$
  $-1/2 = -(1/2) =$
  $-\sin 30° \cos(-30°) =$
  $1/2 = \cos(30°)$
**43.** (a) $= \dfrac{0.87v_0^2}{g}$
  (b) $v_0^2/g$
  (c) $0.50v_0^2/g$
**45.** (a) 10 A
  (b) 8.7 A
**47.** 1.35
**49.** 110 mm²

## Exercise 10-3   (Page 351)

1. $60°$
3. $50°$
5. $8.5°$
7. $30°$
9. $\sqrt{3}/3$
11. $25.0°, 335.0°$
13. $139.8°, 319.8°$
15. $213.8°, 326.2°$
17. $42.2°, 222.2°$
19. $18.3°, 341.7°$
21. $135.0°$
23. $116.6°$
25. $208.4°$
27. $\pi/4, 5\pi/4$
29. $1.0472, 2.0944$
31. $1.009, 5.2738$
33. $\pi$
35. $0.5880, 3.7296$
37. $\pi/4, 3\pi/4, 5\pi/4, 7\pi/4$
39. $139.5°$
41. (a) $0.784, 2.36$
    (b) $0.0021$ s, $0.0063$ s
43. $36.9°, 143.1°, 216.9°,$
    $323.1°$

## Exercise 10-4   (Page 359)

1. $8.5, 14.7$
3. $-3.4, 9.4$
5. $-1.09, -0.29$
7. $0.020, -0.0059$
9. $34, 62°$
11. $19.5, 247°$
13. $586, 335°$
15. $1.89, 147°$
17. $27, 46°$
19. $3.03, 177°$
21. $0.587, 359°$
23. $42$ N, $123°$
25. $1.45$ A, $27.9°$
27. $174$ km/h, $985$ km/h
29. $35$ nmi
31. $60.5$ m, $129°$
33. $3.0$ mi/h, $71°$
35. $700$ kg$_f$, $1530$ lb
37. $666$ lb, $99°$
39. $9.7$ m/s$^2$, $-77°$

41. $9.8 \times 10^{-5}$ T, $15°$
43. $51$ N, $141°$
45. $27$ Ω, $48°$
47. (a) $4000\pi$ rad/min
    (b) $3140$ cm/s
49. $785$ cm/s

## Exercise 10-5   (Page 368)

1. $C = 40°, a = 7.5, c = 5.3$
3. $A = 41.9°, b = 21.2,$
   $c = 13.6$
5. $B = 42°, C = 81°, c = 5.9$
7. $A = 50.9°, B = 23.9°,$
   $b = 0.238$
9. $C = 0.44$ rad, $b = 56,$
   $c = 26$
11. $B = 38°50', a = 789,$
    $c = 429$
13. $B = 30.8°, A = 111.3°,$
    $a = 9.1 \times 10^{-3}$
15. $77°, 103°$
17. $AB = 1500$ ft, $AC = 1160$ ft
19. $1360$ N, $38.1°$
21. $0.65$ mi
23. $43.3$ A
25. $19.1$
27. $3.3$ mi
29. $a$ too short

## Exercise 10-6   (Page 375)

1. $c = 7.1, A = 42°, B = 68°$
3. $b = 0.42, A = 10.2°,$
   $C = 24.8°$
5. $C = 104°, A = 29°, B = 47°$
7. $C = 135.5°, A = 21.3°,$
   $B = 23.2°$
9. $a = 0.44, B = 54°,$
   $C = 110°$
11. $c = 3260, A = 7°10',$
    $B = 162°40'$
13. $b = 0.82, A = 0.32$ rad,
    $B = 0.52$ rad
15. $C = 112°, A = 31°, B = 37°$
17. $1220$ lb, $63°$
19. $5.9, 5.0°$
21. $x = 6.51$

23. $3.8$ mi/h, $81°$
25. $455$ m, $1492$ ft
27. $2000$ lb
29. $90°$

## Exercise 10-7   (Page 379)

1. $-0.4909$
3. $-0.1793$
5. $-0.7265$
7. $0.4067$
9. $-1.396$
11. $0.3827$
13. $7.185$
15. $191.7°, 348.3°$
17. $107.9°, 287.9°$
19. $114.5°, 245.5°$
21. No value

## Review Questions 10-8
(Page 379)
(For 1, 3: sin, cos, tan,
csc, sec, cot)

1. $12/13, -5/13, -12/5,$
   $13/12, -13/5, -5/12$
3. $-15/17, 8/17, -15/8,$
   $-17/15, 17/8, -8/15$
5. $-\sqrt{3}$ or $-1.732$
7. $-0.9833$
9. $-0.7660$
11. $-1$
13. $30°, 150°; \pi/6, 5\pi/6$
15. $136.0°, 316.0°; 2.37, 5.52$
17. $120°, 240°; 2\pi/3, 4\pi/3$
19. $20, 37°$
21. $13.5, 75.5°$
23. $C = 45°, b = 2.3, c = 2.1$
25. $A = 42°, B = 88°, c = 13$
27. $A = 26.4°, B = 36.3°, C = 117.3°$
29. (a) $540$ cm/s
    (b) $-722$ cm/s
31. $0.0573°$
33. $10$ lb, $212°$
35. $27$ mi/h, $5.0°$
37. $1027$ ft

# Chapter 11

## Exercise 11-1 *(Page 390)*

**1.**

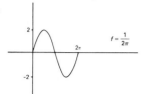

$f = \frac{1}{2\pi}$

**3.**

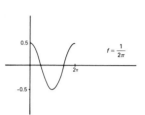

$f = \frac{1}{2\pi}$

**5.**

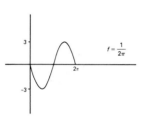

$f = \frac{1}{2\pi}$

**7.**

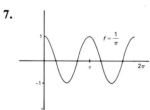

$f = \frac{1}{\pi}$

**9.**

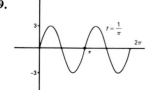

$f = \frac{1}{\pi}$

**11.**

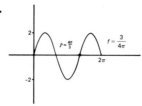

$P = \frac{4\pi}{3}$    $f = \frac{3}{4\pi}$

**13.**

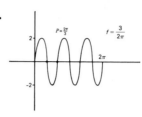

$P = \frac{2\pi}{3}$    $f = \frac{3}{2\pi}$

**15.**

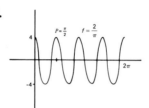

$P = \frac{\pi}{2}$    $f = \frac{2}{\pi}$

**17.**

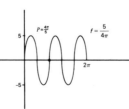

$P = \frac{2\pi}{3}$    $f = \frac{3}{2\pi}$

**19.**

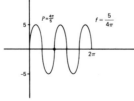

$P = \frac{4\pi}{5}$    $f = \frac{5}{4\pi}$

**21.**

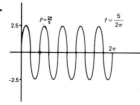

$P = \frac{2\pi}{5}$    $f = \frac{5}{2\pi}$

**23.**

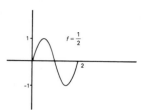

$f = \frac{1}{2}$

**25.**

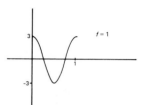

$f = 1$

**27.**

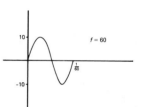

$f = 60$

**29.**

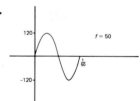

$f = 50$

**31.** $y = 12 \sin 2t$

**33.**

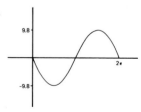

**35.** 8 A, 1/30, 30, $60\pi$

**37.** *(a)* $\dfrac{3}{2}$m, $6\pi$

*(b)*

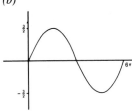

**39.** *(a)*

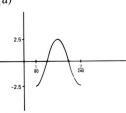

*(b)* 0.018 s, 0.024 s

**41.**

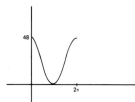

**Exercise 11-2**   *(Page 396)*
(For 1–19: amplitude, period, frequency, phase angle, phase shift)

**1.**

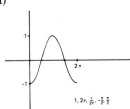

$1, 2\pi, \frac{1}{2\pi}, -\frac{\pi}{2}, \frac{\pi}{2}$

**3.**

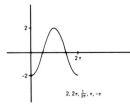

$2, 2\pi, \frac{1}{2\pi}, \pi, -\pi$

**5.**

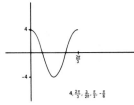

$4, \frac{2\pi}{3}, \frac{3}{2\pi}, \frac{\pi}{2}, -\frac{\pi}{6}$

**7.**

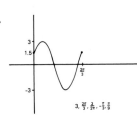

$3, \frac{2\pi}{3}, \frac{3}{2\pi}, -\frac{\pi}{3}, \frac{\pi}{9}$

**9.**

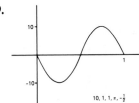

$10, 1, 1, \pi, -\frac{1}{2}$

**11.**

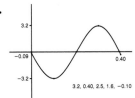

$3.2, 0.40, 2.5, 1.6, -0.10$

**13.**

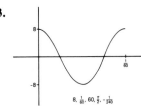

$8, \frac{1}{60}, 60, \frac{\pi}{2}, -\frac{1}{240}$

**15.**

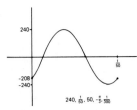

$240, \frac{1}{50}, 50, -\frac{\pi}{3}, \frac{1}{300}$

**17.**

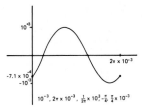

$10^{-3}, 2\pi \times 10^{-3}, \frac{1}{2\pi} \times 10^{3}, -\frac{\pi}{4}, \frac{\pi}{4} \times 10^{-3}$

**19.**

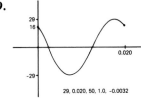

$29, 0.020, 50, 1.0, -0.0032$

**21.** *(a)* $y = 0.5 \sin (8\pi x - \pi/4)$
*(b)* $y = 10^{-4} \sin (2000\pi x + 0.80)$

**23.** *(a)*

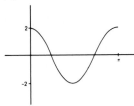

*(b)* $\alpha = 2 \cos 2t$

**25.**

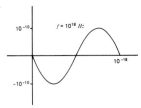

**27.** *(a)* $I = I_{max} \sin (2\pi ft + \pi/2)$
*(b)*

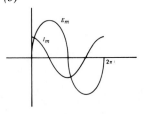

**29.**

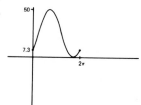

**9.**

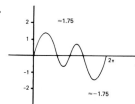

**21.**

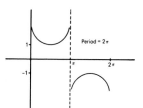

**31.** *(a)*

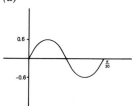

**11.**

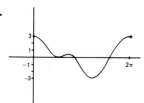

**23.**

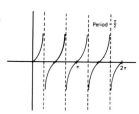

*(b)* emf $= 0.6 \sin 60t$

**Exercise 11-3** *(Page 402)*

**1.**

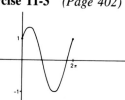

**13.**

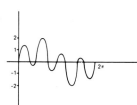

**25.**

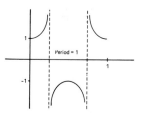

**27.** See Prob. 21

**3.**

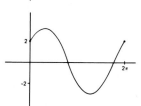

**15.**

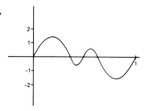

**29.**

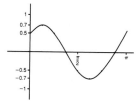

**5.**

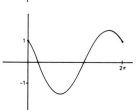

**17.**

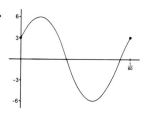

**31.**

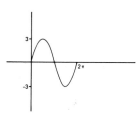

**7.**

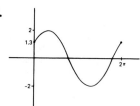

**19.**

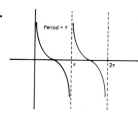

**33.**

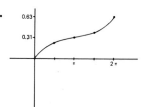

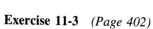

**35.** $\sqrt{5} \sin (x + 0.464)$

**37.**

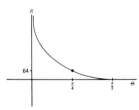

**Exercise 11-4**   *(Page 407)*

**1.**

**3.**

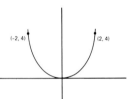

**5.**

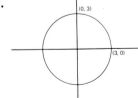

**7.**

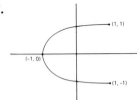

**9.**

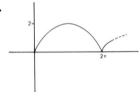

**11.**

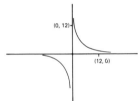

**13.**

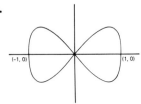

**15.**

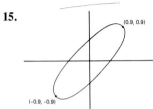

**17.**

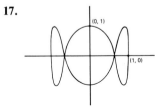

**19.**

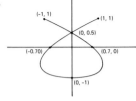

**21.**

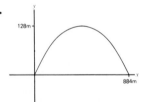

**23.**

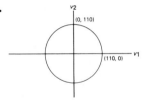

**Exercise 11-5**   *(Page 415)*
**1 through 9:**

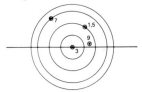

**11.** $(13, 113°)$
**13.** $(0.4, \pi)$
**15.** $(0.93, 2.9)$
**17.** $(0.5, 0.8)$
**19.**

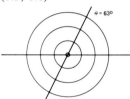

**21.**

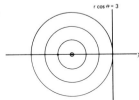

**23.**

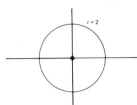

**25.**

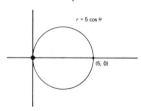

**27.**

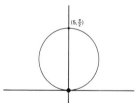

**29.**

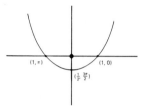

**31.**

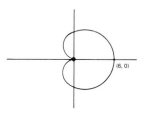

**33.**

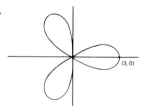

**35.**

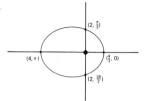

**37.**

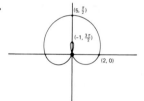

**39.**

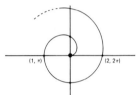

**41.**

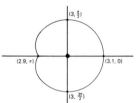

**43.** (9.6, 56.5°)

**45.** $A$(80°N, 45°E) $B$(60°N, 90°E) $C$(70°N, 135°E) $D$(50°N, 180°) $E$(60°N, 135°W) $F$(50°N, 0°)

**47.**

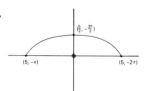

**49.**

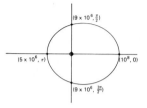

**51.**

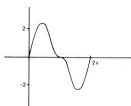

**53.**

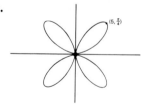

**Review Questions 11-6**
*(Page 418)*
(**For 1–5:** amplitude, period, frequency, phase angle, phase shift)

**1.**

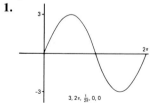

$3, 2\pi, \frac{1}{2\pi}, 0, 0$

**3.**

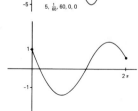

$2, 2\pi, \frac{1}{2\pi}, -\pi, +\pi$

**5.**

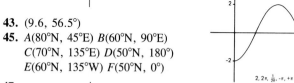

$5, \frac{1}{60}, 60, 0, 0$

**7.**

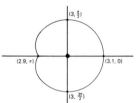

**9.**

**11.**

**13.**

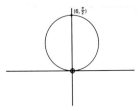

**15.** (a) $y = 3 \sin 2t$

(b)

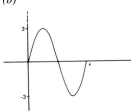

**17.**

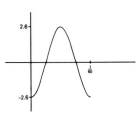

**19.** (a)

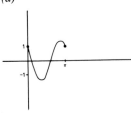

(b) $\pi$

**21.**

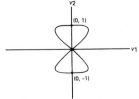

**23.**

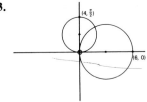

## Chapter 12

### Exercise 12-1   (Page 424)

**1.** $\dfrac{\sin^2 \theta}{\cos^2 \theta} + \dfrac{\cos^2 \theta}{\cos^2 \theta} =$

$\tan^2 \theta + 1 =$

$\sec^2 \theta = \dfrac{1}{\cos^2 \theta}$

(Answers to identities show suggested first changes to the left side unless indicated.)

**3.** $\cos \theta(1/\sin \theta)$

**5.** $\dfrac{(1/\cos \theta)}{(1/\sin \theta)}$

**7.** $(\sin x/\cos x)(\cos x)$
$(1/\sin x)$

**9.** $\cos^2 x - (1 - \cos^2 x)$

**11.** $\sin x \cos x \left( \dfrac{\cos x}{\sin x} + \dfrac{\sin x}{\cos x} \right)$

**13.** $\left( \dfrac{1}{\sin^2 A} \right) \cos^2 A$

**15.** $\sin^2 x \left( \dfrac{\cos^2 x}{(\sin^2 x)} \right)$

**17.** $(\sec^2 y - 1) + (\cot^2 y + 1)$

**19.** $\sin \theta(1/\cos \theta) + (\cos \theta/\sin \theta)$

**21.** $\dfrac{1 - \sin^2 A}{1 - \sin A}$

**23.** $\left( \dfrac{\tan x}{\tan x} \right)\left( \dfrac{\tan x}{\sec x + 1} \right)$

**25.** $(1 - \sin^2 \theta)(1 + \sin^2 \theta)$

**27.** $\dfrac{(\sin x/\cos x) - (\cos x/\sin x)}{\sin x \cos x}$

**29.** Right side:

$\left( \dfrac{\sin \phi}{\sin \phi} \right)\left( \dfrac{1 + \cos \phi}{\sin \phi} \right)$

**31.** (a) $0.9659 \neq 1.207$
(b) $1.000 \neq 0.8660$
(c) $0.5000 \neq 0.4142$

**33.** $\sqrt{1.0 - \cos^2 \alpha}$

**35.** Square both sides, change $\sin^2 \theta_i$ to $1 - \cos^2 \theta_i$ and solve for $\cos \theta_i$

**37.** $(4.0 \sin^2 \omega t + 4.0 \cos^2 \omega t)/2L$

**39.** $\dfrac{\cancel{a^2} \cos^2 t}{\cancel{a^2}} + \dfrac{\cancel{b^2} \sin^2 t}{\cancel{b^2}} = 1$

**41.** $\dfrac{\cos^2 u + \sin^2 u}{\cos^2 u} = \dfrac{1}{\cos^2 u} = \sec^2 u$

### Exercise 12-2   (Page 430)

**1.** (a) $-36/85$
(b) $77/85$
(c) $-36/77$
(d) IV

**3.** (a) $1.05$
(b) $A = 68.2°$, $B = 21.8°$

**5.** $\dfrac{\sqrt{6} + \sqrt{2}}{4}$, $0.966$

**7.** $-2 - \sqrt{3}$, $-3.73$

**9.** $\dfrac{\sqrt{2} - \sqrt{6}}{4}$, $-0.259$

**11.** $\sin \theta \cos (\pi/2) + \cos \theta \sin (\pi/2)$

**13.** $\cos \theta \cos \pi + \sin \theta \sin \pi$

**15.** $\dfrac{\tan \theta + \tan \pi}{1 - \tan \theta \tan \pi}$

**17.** $0.866 \neq 0.985$

**19.** (a) Expand right side
(b) Let $A + B = x$, $A - B = y$

**21.** Right side:

$\dfrac{\left( \dfrac{\sin A \cos B}{\cos A \cos B} \right) + \left( \dfrac{\cos A \sin B}{\cos A \cos B} \right)}{\left( \dfrac{\cos A \cos B}{\cos A \cos B} \right) - \left( \dfrac{\sin A \sin B}{\cos A \cos B} \right)}$

**23.** Expand both sides

**25.** $\sin A \cos A + \cos A \sin A$

**27.** $(\sin x \cos y + \cos x \sin y)\sin x +$

(cos $x$ cos $y$ −
sin $x$ sin $y$) cos $x$

**29.** $A_1$(cos $\omega t$ cos $\pi$ −
sin $\omega t$ sin $\pi$) +
$A_2$(cos $\omega t$ cos $\pi$ +
sin $\omega t$ sin $\pi$)

**31.** $I_{max}$ sin $120\pi t$

**33.** Right side: $t$(sin $\theta_i$ cos $\theta_r$ −
cos $\theta_i$ sin $\theta_r$)/cos $\theta_r$

**35.** 71.6°

### Exercise 12-3  *(Page 436)*

**1.** *(a)* 120/169
   *(b)* −119/169
   *(c)* −120/119
   *(d)* II, 67.4°
**3.** *(a)* −0.96
   *(b)* 0.28
   *(c)* −3.4
   *(d)* IV, 323°
**5.** *(a)* $\dfrac{\sqrt{10}}{10}$
   *(b)* $\dfrac{3\sqrt{10}}{10}$
   *(c)* 1/3
   *(d)* 36.9°
**7.** *(a)* 0.95
   *(b)* 0.31
   *(c)* 3.0
   *(d)* 143°
**9.** 336/625
**11.** cos 22.5° = cos(45°/2) =
$\dfrac{\sqrt{1 + \cos 45°}}{2}$ = 0.924

**13.** cot 75° = $\dfrac{1}{\tan(150°/2)}$ =
$\dfrac{1 + \cos 150°}{\sin 150°}$ = 0.268

**15.** csc 15° = $\dfrac{1}{\sin(30°/2)}$ =
$\sqrt{\dfrac{2}{1 - \cos 30°}}$ = 3.86

(Answers to identities show
suggested first changes to the left
side unless indicated.)

**17.** cos² $A$ − (1 − cos² $A$)

**19.** sin² $x$ − 2 sin $x$
cos $x$ + cos² $x$

**21.** $\dfrac{2 (\sin x/\cos x)}{1 - (\sin x/\cos x)^2}$

**23.** $\left(\dfrac{\sin x}{\cos x}\right)$ [1 + (1 − 2 sin² $x$)]

**25.** cos(2$x$ + $x$) = cos 2$x$ cos $x$ −
sin 2$x$ sin $x$

**27.** $\dfrac{1 - \cos x}{\sin x} + \dfrac{1 + \cos x}{\sin x}$

**29.** $\dfrac{\tan \theta - \tan(\pi/4)}{1 + \tan \theta \tan(\pi/4)} =$
$\dfrac{(\sin \theta/\cos \theta) - 1}{1 + (\sin \theta/\cos \theta)}$

**31.** Let 2$A$ = $x$ and solve for
cos ($x$/2)

**33.** Substitute values for $x$ and $y$

**35.** $P = V_{max}I_{max} \left(\dfrac{\sin 2\omega t}{2}\right)$

**37.** sin² $\theta_i$ = (1.5)²(1 − cos $\alpha$)/2

**39.** 2.22$C$(sin $\theta$)$H^{5/2}$ =
$Q$(1 + cos $\theta$)

**41.** 2 − $\sqrt{3}$

**43.** $\dfrac{-31\sqrt{2}}{50}$ or −0.877

### Exercise 12-4  *(Page 442)*

**1.** 2$\pi$/3, 4$\pi$/3
**3.** 59.0°, 239.0°
**5.** $\pi$/3, 2$\pi$/3, 4$\pi$/3, 5$\pi$/3
**7.** 26.6°, 153.4°, 206.6°,
   333.4°
**9.** 90°, 270°, 70.5°, 289.5°
**11.** 135°, 315°, 108.4°, 288.4°
**13.** $\pi$/4, 5$\pi$/4
**15.** $\pi$/2, $\pi$/6, 5$\pi$/6
**17.** 90°, 270°, 194.5°, 345.5°
**19.** 18.0°, 162.0°, 234.0°,
   306.0°
**21.** 96.8°, 263.2°
**23.** 30°, 120°, 60°, 150°
**25.** 126.9°
**27.** $\pi$/4
**29.** 0.72
**31.** 0, $\pi$

### Review Exercises 12-5
*(Page 443)*

**1.** cos $\theta$(sin $\theta$/cos $\theta$)
**3.** (1/cos² $\theta$) − (sin² $\theta$/cos² $\theta$)
**5.** $\left(\dfrac{1 - \cos x}{1 - \cos x}\right)\left(\dfrac{\sin x}{1 + \cos x}\right)$
**7.** cos $\pi$ cos $\theta$ + sin $\pi$ sin $\theta$
**9.** sin² $x$ + 2 sin $x$
cos $x$ + cos² $x$
**11.** 1/tan($x$/2)
**13.** *(a)* 56/65
   *(b)* 63/65
   *(c)* −56/33
**15.** 2 + $\sqrt{3}$ or 3.73
**17.** *(a)* 0.894
   *(b)* 0.447
**19.** $\pi$/6, 5$\pi$/6
**21.** $\pi$/3, 2$\pi$/3, 4$\pi$/3, 5$\pi$/3
**23.** 0, $\pi$
**25.** 2 − 2 cos $\theta$
**27.** $V_{max}$ cos 60$\pi t$
**29.** $\pi$/2, 7$\pi$/6, 11$\pi$/6

## Chapter 13

### Exercise 13-1  *(Page 448)*

**1.** 3$j$
**3.** 0.2$j$
**5.** −3$j\sqrt{2}$
**7.** 2$j\sqrt{10}$
**9.** $j$/2
**11.** −7
**13.** −10
**15.** 0.1$j$
**17.** 5$\sqrt{3}$
**19.** −4$j\sqrt{11}$
**21.** −12$j$
**23.** 15$\sqrt{15}$
**25.** 81
**27.** 1
**29.** 15 + 6$j$
**31.** −2 − 2$j$
**33.** 2 + $j\sqrt{3}$
**35.** 1 − $j$
**37.** 1 ± $j$
**39.** ±3$j$

**41.** $2/3 \pm 2j/3$
**43.** $1 \pm j\sqrt{2}$
**45.** $0.5 + 1.5j$
**47.** (a) Real
     (b) Pure imaginary

**Exercise 13-2**   *(Page 450)*
**1.** $4 + 2j$
**3.** $5j$
**5.** $8.4 + 2.7j$
**7.** $-4j$
**9.** $17 - 7j$
**11.** $8j$
**13.** $44.24 - 14.88j$
**15.** 61
**17.** 7
**19.** $39 + 54j$
**21.** $-54 - 54j$
**23.** $-1 + 3j$
**25.** $2/13 - 3j/13$
**27.** $1 - j$
**29.** $1 - 2j$
**31.** $16/29 - 11j/29$
**33.** $9/13 - 7j/13$
**35.** $1/3 - 2j\sqrt{2/3}$
**37.** $\sqrt{2} - j\sqrt{2}$
**39.** $0.20 + 0.05j$
**41.** $356.4 + 17.6j$
**43.** $\dfrac{(1 + 2j + j^2)}{2} =$

$\dfrac{2j}{2} = j$

**Exercise 13-3**   *(Page 458)*
**1.**

**3.**

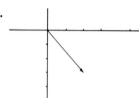

**5.**

**7.**

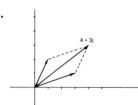

**9.**

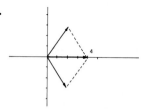

**11.**

**13.**

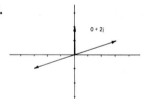

**15.**

**17.**

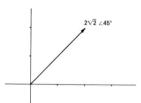

**19.**

**21.**

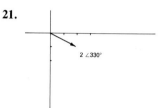

**23.**

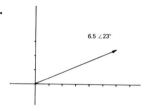

**25.**

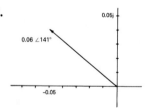

**27.**

**29.**

2.1 + 2.1j

**31.**

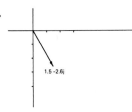

1.5 -2.6j

**33.**

-1.9 -5.2j

**35.**

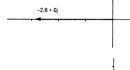

-2.8 + 0j

**37.**

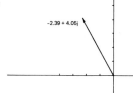

-2.39 + 4.05j

**39.**

-1+j

**41.**

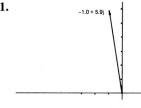

-1.0 + 5.9j

**43.** $6 \angle 100°$
**45.** $4.5 \angle 112°$
**47.** $1.5 \angle 80°$
**49.** $0.84 \angle 335.1°$
**51.** $96 \angle 3\pi/4$
**53.** $45 \angle 62°$
**55.** $4 + 0j = 4 \angle 0°$
**57.** $0 + \dfrac{2}{3}j = \dfrac{2}{3} \angle 90°$
**59.** $4 + 0j = 4 \angle 0°$
**61.** $6.77, 329°$
**63.** $531$
**65.** $\dfrac{r_1(\cos \theta_1 + j \sin \theta_1)}{r_2(\cos \theta_2 + j \sin \theta_2)} \dfrac{(\cos \theta_2 - j \sin \theta_2)}{(\cos \theta_2 - j \sin \theta_2)}$

$$= \dfrac{r_1[(\cos \theta_1 \theta_2 + \sin \theta_1 \theta_2) + j(\sin \theta_1 \cos \theta_2 - \cos \theta_1 \sin \theta_2)]}{r_2(\cos^2 \theta_2 + \sin^2 \theta_2)}$$

$$= \dfrac{r_1}{r_2} [\cos (\theta_1 - \theta_2) + j \sin (\theta_1 - \theta_2)]$$

**Exercise 13-4** *(Page 462)*
**1.** $0 + 25j$
**3.** $-8j$
**5.** $-2 + 2j$
**7.** $-18 - 26j$
**9.** $-33 - 21j$
**11.** $-64 + 0j$
**13.** $0 + 8j$
**15.** $5, -5$
**17.** $-1, \dfrac{1}{2} \pm \dfrac{\sqrt{3}}{2}j$
**19.** $1, -\dfrac{1}{2} \pm \dfrac{\sqrt{3}}{2}j$
**21.** $\pm\sqrt{2} \pm j\sqrt{2}$
**23.** $3, -\dfrac{3}{2} \pm \dfrac{3\sqrt{3}}{2}j$
**25.** $-0.29 + 1.1j, -0.79 - 0.79j, 1.1 - 0.29j$
**27.** $1.5 + 0.5j, -1.5 - 0.5j$
**29.** $1, j, -1, -j$
**31.** $0.16 \angle 254°$

**Exercise 13-5** *(Page 469)*
**1.** $1.1e^{j(\pi/6)}$
**3.** $0.32e^{0.70j}$
**5.** $3.4e^{5.50j}$
**7.** $1.56e^{1.18j}$
**9.** $0 + 6j$

**11.** $0 - 12.5j$
**13.** $0.92 - 0.38j$
**15.** $1.9 + 0.22j$
**17.** $0.21 \ \Omega, 3.0 \ \Omega$
**19.** $0.21 \ \Omega, -0.12 \ \Omega$
**21.** $0 \ \Omega, -5.6 \ \Omega$
**23.** $3.2 \ \Omega, 0.32 \ \text{rad}$
**25.** $35.0 \ \Omega, -0.644 \ \text{rad}$
**27.** $0.40 \ \Omega, -0.46 \ \text{rad}$

**29.**

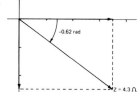

-0.62 rad

Z = 4.3 Ω

**31.**

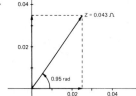

Z = 0.043 Ω

0.95 rad

**33.** $I = 0.5 + 0j$
**35.** $1.2 + 0.038j$
**37.** $6.0e^o$
**39.** $0.92 - 1.4j$

**41.** $0 - 24j$
**43.** $3 - 3j$
**45.** $X_L = 18.8 \ \Omega,$
   $X_c = 88.4 \ \Omega,$
   $Z = 0 - 69.6j$
**47.** $Z_l = j$ or $-j$

**Review Questions 13-6**
*(Page 471)*
**1.** $5j$
**3.** $10j\sqrt{2}$
**5.** $3 + 6j$
**7.** $9 - j$
**9.** $3 + 4j$
**11.** $0.28 - 0.96j$

**13.**

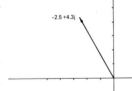

**15.**

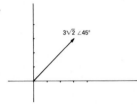

**17.**

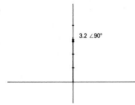

**19.**

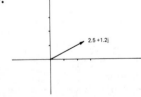

**21.** $-5 + 0j$
**23.** $32\sqrt{3} + 32j$
**25.** $2.6 + 1.5j, \ -2.6 - 1.5j$
**27.** $2, \ -1 + j\sqrt{3}, \ -1 - j\sqrt{3}$
**29.** $7.1e^{6.1j}$
**31.** $6.3, \ 11.8°$
**33.** $2.2, \ 0.46$ rad
**35.** $10e^{j(-\pi/4)}$

# Chapter 14

**Exercise 14-1**   *(Page 482)*
**1.** $5, \ 4/3, \ 53°$
**3.** $6.3, \ -1/3, \ 162°$
**5.** $8.7, \ -2.3, \ 114°$
**7.** $1, \ 0, \ 0°$
**9.** $2\sqrt{5} + 3\sqrt{5} = 5\sqrt{5}$
**11.** $m_{AC} = 5/4, \ m_{BC} = -4/5$
**13.** $m_{AB} = m_{CD} = -3/2,$
   $m_{BC} = m_{AD} = 2/3$
**15.** isosceles
**17.** $3, \ (0, \ -2)$
**19.** $2/3, \ (0, \ -2)$
**21.** $y = 2$
**23.** $y = -x + 3$
**25.** $y = 3x + 15$
**27.** $y = 2x/3 + 4/3$
**29.** $y = 3x/4 - 7/4$
**31.** $y = -4x/3 + 1/3$
**33.** $y = 5$
**35.** $y = -x$
**37.** $y = x - 1$
**39.** *(a)* $R = at + R_o$
   *(b)* $y = 0.00115t + 0.300$
**41.** $v = 0.60t + 332$
**43.** *(a)* $\theta = 0.00246$ s
   *(b)* $1.72°$

**Exercise 14-2**   *(Page 491)*
**1.**

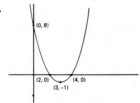

**3.**

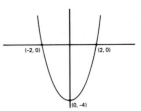

**5.**

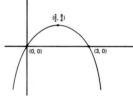

**7.**

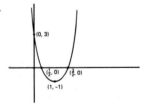

**9.**

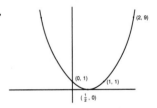

**11.**

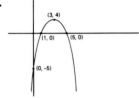

**13.**

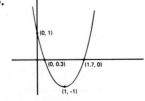

**15.**

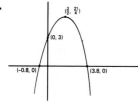

**17.**

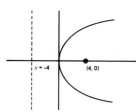

**19.**

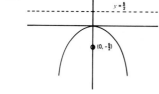

**21.**

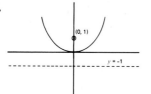

**23.**

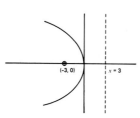

**25.**

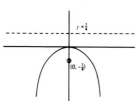

**27.** $x^2 = 12y$
**29.** $y^2 = 4x$

**31.** $x^2 = -2y$
**33.** $y^2 = -x$

**35.**

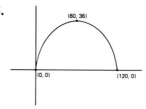

**37.** (a) 35 A, 245 W
   (b)

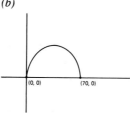

**39.** (a) 25 m
   (b) (50, 25)

**41.**

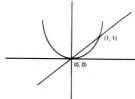

**43.**

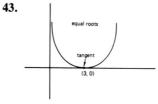

**45.** (a)

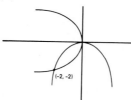

(b) $x^2 = -2y$,
   $y^2 = -2x$

**Exercise 14-3** *(Page 497)*
   **1.** $(x - 2)^2 + (y - 3)^2 = 9$
   **3.** $(x + 1)^2 + (y - 2)^2 = 1$
   **5.** $(x - 1/2)^2 + y^2 = 49/4$
   **7.** $x^2 + y^2 = 10$
   **9.** $x^2 + (y + 3)^2 = 9$
   **11.** $(x - 3)^2 + y^2 = 9$

**13.**

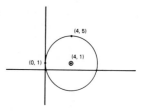

**15.**

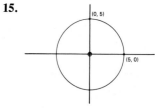

**17.**

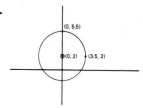

**19.**

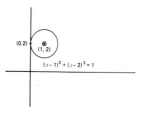

**21.**

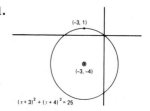

**23.**

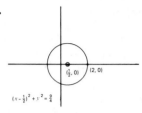

$(x - \frac{1}{2})^2 + y^2 = \frac{9}{4}$

$(\frac{1}{2}, 0)$  $(2, 0)$

**25.**

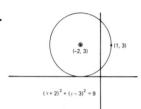

$(-2, 3)$  $(1, 3)$

$(x + 2)^2 + (y - 3)^2 = 9$

**27.**

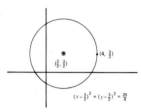

$(4, \frac{3}{2})$

$(\frac{3}{2}, \frac{3}{2})$

$(x - \frac{3}{2})^2 + (y - \frac{3}{2})^2 = \frac{25}{4}$

**29.** $x^2 + (y - 6)^2 = 17.64$
$(x - 8)^2 + y^2 = 4.41$
**31.** $(x - 4)^2 + (y - 5)^2 = 4$

**33.**

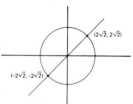

$(2\sqrt{2}, 2\sqrt{2})$

$(-2\sqrt{2}, -2\sqrt{2})$

**35.** $x^2 + y^2 + 2x + 2y - 23 = 0$

**Exercise 14-4**  *(Page 505)*
**1.**

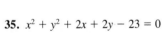

$(0, 3)$

$(5, 0)$
$(4, 0)$

**3.**

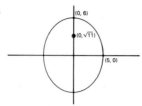

$(0, 6)$
$(0, \sqrt{11})$
$(5, 0)$

**5.**

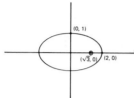

$(0, 1)$
$(\sqrt{3}, 0)$  $(2, 0)$

**7.**

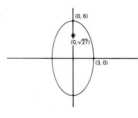

$(0, 6)$
$(0, \sqrt{27})$
$(3, 0)$

**9.**

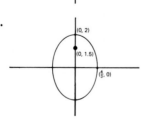

$(0, \sqrt{3})$
$(1, 0)$  $(2, 0)$

**11.**

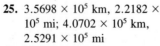

$(0, 2)$
$(0, 1.5)$
$(\frac{4}{3}, 0)$

**13.** $x^2/25 + y^2/16 = 1$
**15.** $x^2/36 + y^2/100 = 1$
**17.** $x^2/9 + y^2/25 = 1$
**19.** $x^2/6 + y^2/16 = 1$
**21.** $x^2/32 + y^2/16 = 1$
**23.** 20 cm, 10 cm
**25.** $3.5698 \times 10^5$ km, $2.2182 \times 10^5$ mi; $4.0702 \times 10^5$ km, $2.5291 \times 10^5$ mi
**27.** $x^2/676 + y^2/100 = 1$

**29.**

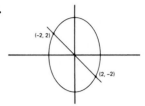

$(-2, 2)$
$(2, -2)$

**31.** $x^2/64 + y^2/100 = 1$

**Exercise 14-5**  *(Page 514)*
**1.**

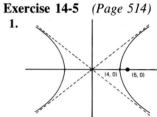

$(4, 0)$  $(5, 0)$

**3.**

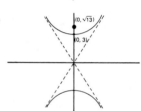

$(0, \sqrt{13})$
$(0, 3)$

**5.**

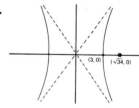

$(3, 0)$  $(\sqrt{34}, 0)$

**7.**

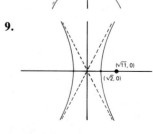

$(0, 7\sqrt{2})$
$(0, 7)$

**9.**

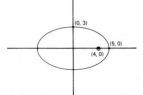

$(\sqrt{11}, 0)$
$(\sqrt{2}, 0)$

**11.**

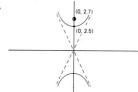

**13.** $x^2/16 - y^2/20 = 1$
**15.** $y^2/25 - x^2/11 = 1$
**17.** $3y^2 - 2x^2 = 6$
**19.** $x^2/4 - y^2/8 = 1$

**21.**

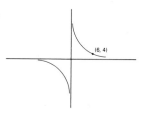

**23.**

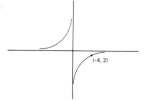

**25.**

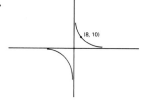

**27.** *(a)*

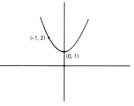

*(b)* $y^2 - 3x^2 = 1$

**29.**

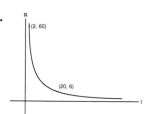

**31.**

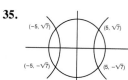

*PV = 2400*

**33.** $x^2 - y^2 = 9,$
$5y^2 - 3x^2 = 5$

**35.**

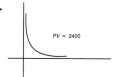

**Exercise 14-6** *(Page 520)*
**1.**

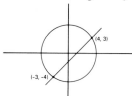

**3.**

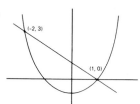

**5.**

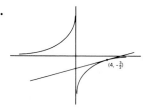

**7.**

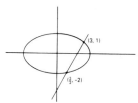

**9.**

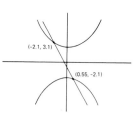

**11.**

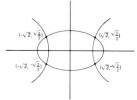

**13.**

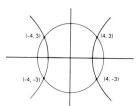

**15.**

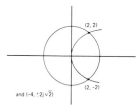

and $(-4, \pm 2j\sqrt{2})$

**17.**

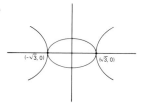

**19.**

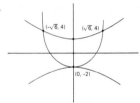

**21.** $p = 1$, $q = 3$; $p = 3$,
  $q = 1$
**23.** 5.83 Ω, 0.67 Ω
**25.** 150°C
**27.** 50 kg$f$, 3.0 m

**Exercise 14-7**  *(Page 528)*

**1.**

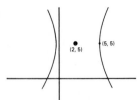

**3.**

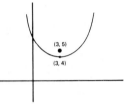

**5.**

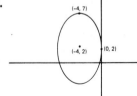

**7.**

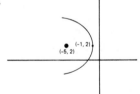

**9.** $(y - 2)^2 = 4(x - 2)$
**11.** $(x - 5)^2/25 +$
  $(y + 5)^2/16 = 1$
**13.** $(x - 5)^2/4 -$
  $(y + 2)^2/12 = 1$
**15.** $(x + 2)^2 = 12y$

**17.**

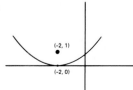

**19.**

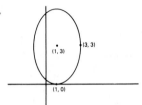

**21.**

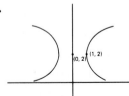

**23.**

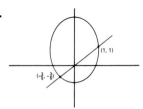

**25.**

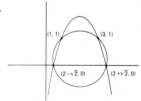

**27.** $x^2 = 2000(y + 125)$
**29.** *(a)* $x^2 = -100(y - 0.01)$

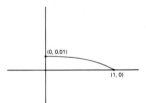

  *(b)* 1.0 cm
**31.** 6 mi/h, 2 mi/h
**33.** *(a)*

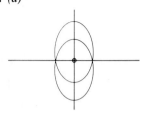

  *(b)* $(\pm 0.45, 0)$

**Review Questions 14-8**
*(Page 530)*
  **1.** $\sqrt{13}$, 3/2, 56°

  **3.** $y = \dfrac{1}{3}x - 2$

  **5.** $y = \dfrac{4}{3}x - \dfrac{7}{3}$

**7.**

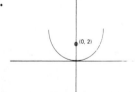

**9.**

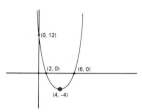

**11.**

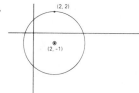

**13.**

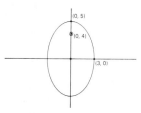

**15.**

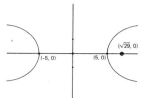

**17.**

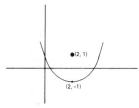

**19.**

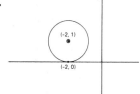

**21.**

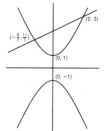

**23.** $y^2 = 12x$
**25.** $(x + 2)^2 + (y - 1)^2 = 9$
**27.** $x^2/25 + y^2/9 = 1$
**29.** $x^2 - y^2 = 9$
**31.** (a) $F = kx + 12$
    (b) 6.5
**33.** (a) parabola

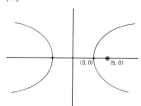

    (b) 2500 ft$^2$
**35.** $x^2/625 + y^2/225 = 1$
**37.** (a)

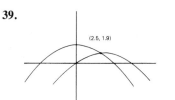

    (b) $x^2/9 + y^2/16 = 1$
**39.**

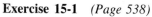

# Chapter 15

## Exercise 15-1   (Page 538)
  **1.** $-13, 0$
  **3.** $-9, 16$
  **5.** $-3, -x^5 - x - 1$
  **7.** $0, -1$
  **9.** $\sqrt{3}, \sqrt{6}$
 **11.** $-0.8, -0.4$
 **13.** $4x^2 + 4x + 1, x^2 + h^2 +$
    $2xh + 2x + 2h + 1$

**15.**

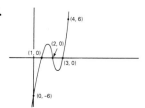

**17.**

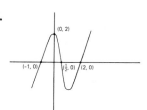

**19.**

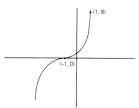

**21.**

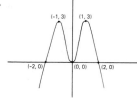

**23.**

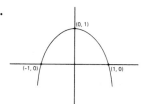

**25.**

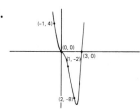

**27.**

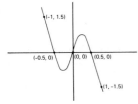

**29.**

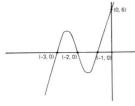

**31.**

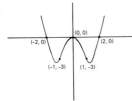

**33.**

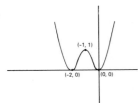

**35.**

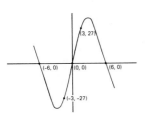

**37.**

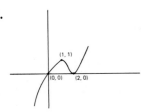

**Exercise 15-2**   *(Page 542)*
1. $x - 6,\ 21$
3. $x^2 + 3x - 1,\ 0$
5. $2x^2 - x - 1,\ 0$
7. $3h^3 - h^2 + h + 2,\ 1$
9. $x^2 + x - 4,\ 4$
11. $3x^2 + 2x + 1,\ 0$
13. $2y^3 + y + 2,\ 1$
15. $x^4 - x^3 + 2x^2 - 2x + 1,\ -1$
17. $2x^2 + 4x,\ 4$
19. $6w^3 - 3w^2 - 3w + 3,\ 0$
21. $4t^2 + 8t + 5,\ 8.5$
23. $x^4 + ax^3 + a^2x^2 + a^3x + a^4,$
    $0$
25. $(x - 2)(2x^2 + x + 3)$
27. no
29. $(y - 0.4)(5y^3 - 5)$
31. $f(3) = 43$
33. $x^2 + x + 1$

**Exercise 15-3**   *(Page 545)*
1. $15$
3. $10$
5. $78$
7. $1$
9. $-1$
11. $0.497$
13. $(x - 2)(3x^2 + x + 3)$
15. $(x + 1)(x^3 - x^2 - 3x + 3)$
17. $(x - 2)(x^3 + 2x^2 + 4x + 8)$
19. yes
21. no
23. no
25. no
27. $4, -3, 1$
29. $0, 1.1, -1.1$
31. $2, 2, -4$
33. $\left(x - \dfrac{1}{2}\right)(2x^2 - 2x - 2)$
35. $f(a) = a^n - a^n = 0$

**Exercise 15-4**   *(Page 550)*
1. $1, 1, -3$
3. $-2, -2, 3, -3$
5. $-\dfrac{1}{3}, 2, -1 \pm j\sqrt{3}$

7. $-2, -3$
9. $\dfrac{1 \pm \sqrt{5}}{2}$
11. $-1, -3$
13. $-2, -3$
15. $\dfrac{-1 \pm \sqrt{5}}{2}$
17. $-2j, 1, -1$
19. $1, -1, 0$
21. $-1, -1$
23. $0, L, L/3, 2L/3$
25. $4x^4 + 12x^3 + x^2 - 12x + 4 = 0$
27.
$$\begin{array}{r} 4 - 6d + 0\ \ + d^3 \\ 2d - 2d^2 - d^3 \\ \hline d/2\,\overline{)\,4 - 4d - 2d^2 + 0} \end{array}$$

**Exercise 15-5**   *(Page 557)*
1. $-1, -2, -3$
3. $1/2, 1/3, -1$
5. $4, \dfrac{-1 \pm \sqrt{17}}{4}$
7. $-3/4, \dfrac{-1 \pm j\sqrt{7}}{2}$
9. $2, -2, 1, 1$
11. $1, 2, 3, -2$
13. $1, 1/4, -1 \pm j\sqrt{2}$
15. $1, 1, -2, \pm\sqrt{2}$
17. $2, -2, 1/2, -1/2$
19. $1, 5; 4, 5/16$
21. $0, 4, 6, 8$
23. 2 in., 0.27 in.
25. 2 mm, 3 mm, 0.85 mm
27. $1, 1, -2$
29. $\pm j, \pm\sqrt{3}$
31. $\pi/2, 3\pi/2, 7\pi/6, 11\pi/6$

**Exercise 15-6**   *(Page 562)*
1. $1.4$
3. $-0.8$
5. $2.6$
7. $-1.90$
9. $0.86$
11. $1.68$
13. $-1, -1.35$

**15.** 2.618, 0.382
**17.** 0 m, 3.77 m
**19.** 0.47 in., 1.65 in.
**21.** 4.24 h
**23.** $\dfrac{\pm 3 \pm \sqrt{5}}{2}$
**25.** 0.98
**27.** 1.93
**29.** 0.28

**Review Questions 15-7**
*(Page 564)*
**1.** 22, 2

**3.**

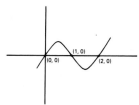

**5.**

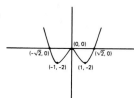

**7.** 10
**9.** 0
**11.** $(x + 2)(2x^2 + x - 2)$
**13.** no
**15.** $-3, -1, -1$
**17.** 3, 1, 1/2
**19.** $1, -1, 2 \pm \sqrt{3}$
**21.** 2.6

**23.**

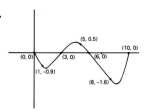

**25.** 1.0 Ω

# Chapter 16

**Exercise 16-1** *(Page 568)*
**1.** 4, 53
**3.** $-1, -14$
**5.** 3/5, 28/5
**7.** 0.3, 0.5
**9.** $x + 1, 9x + 9$
**11.** 315
**13.** $-270$
**15.** 33.75
**17.** 21.9
**19.** $-60bc$
**21.** $-3, 60$
**23.** 505
**25.** $\dfrac{n}{2}(1 + n) = \dfrac{n(n + 1)}{2}$
**27.** 11.4 mm
**29.** 220 ft
**31.** 9.769 m/s$^2$
**33.** 12
**35.** $6900

**Exercise 16-2** *(Page 576)*
**1.** 3, 729
**3.** 2/3, 16/3
**5.** $-2/5, -64/25$
**7.** 1.5, 18.225
**9.** $e^{-1}, e^{-2}$
**11.** $-1, -7$
**13.** 255
**15.** 3367
**17.** 34.1
**19.** $12.7\pi$
**21.** 9/2
**23.** 250/7
**25.** 1.00
**27.** 5/9
**29.** 1/11
**31.** 41/333
**33.** *(a)* $119.10
    *(b)* $119.41
**35.** 12%
**37.** $6227.80
**39.** 0.03125 or 1/32
**41.** 32.8 kPa
**43.** 16,384

**45.** 2059
**47.** 10 m
**49.** 100 yr
**51.** *(a)* Show 1/37 =
        0.027027 ⋯
    *(b)* Show 1/27 =
        0.037037 ⋯
**53.** 1/81

**Exercise 16-3** *(Page 583)*
**1.** $x^3 + 3x^2h + 3xh^2 + h^3$
**3.** $a^7 + 7a^6b + 21a^5b^2 + 35a^4b^3 + 35a^3b^4 + 21a^2b^5 + 7ab^6 + b^7$
**5.** $v^4 - 8v^3t + 24v^2t^2 - 32vt^3 + 16t^4$
**7.** $T^{10} + 20T^8 + 160T^6 + 640T^4 + 1280T^2 + 1024$
**9.** $\pi^6r^6 - 6\pi^5r^4 + 15\pi^4r^2 - 20\pi^3 + 15\pi^2/r^2 - 6\pi/r^4 + 1/r^6$
**11.** $x^2 + 4x\sqrt{xy} + 6xy + 4y\sqrt{xy} + y^2$
**13.** $I^3 + 7.5I^2 + 18.75I + 15.625$
**15.** $x^{12} + 12x^{11}y + 66x^{10}y^2 + 220x^9y^3$
**17.** $1 - 3y + 6y^2 - 10y^3$
**19.** 1.04
**21.** 0.958
**23.** $210t^4$
**25.** $792e^2$
**27.** $4\pi Rr(R^2 + r^2)$
**29.** 15,504
**31.** $5x^4 + 10x^3h + 10x^2h^2 + 5xh^3 + h^4$
**33.** let exponent $x = nx$.

**Review Questions 16-4**
*(Page 584)*
**1.** 3, 37, 246
**3.** 2/5, 3, 64/5
**5.** 2, 192, 381
**7.** 4, 96, 127.875
**9.** 12.5
**11.** $x^4 + 8x^3y + 24x^2y^2 + 32xy^3 + 16y^4$

**13.** $x^{12} + 6x^{10} + 15x^8 + 20x^6 +$
$15x^4 + 6x^2 + 1$

**15.** $1 + 1/2x - 1/8x^2 + 1/16x^3$

**17.** 32.5

**19.** *(a)* \$1102.50
*(b)* \$1103.81

**21.** *(a)* 0.4840
*(b)* 0.01285

# Chapter 17

**Exercise 17-1**   *(Page 590)*

**1.** $-1 > -2$
**3.** $2ab \leq a^2 + b^2$
**5.** $n + 1/n \geq 2$
**7.** $-1 < r < 1; |r| < 1$
**9.** $x < 2$
**11.** $x \leq -3$
**13.** $y > 1$
**15.** $t < 3.5$
**17.** $2 < q \leq 4$
**19.** $x < -1$ or $x > 3$
**21.** $-2 < e < 3$
**23.** Multiply by $a$
**25.** *(a)* $4.97 \leq r \leq 5.03$
*(b)* $9.94 \leq d \leq 10.06$
**27.** *(a)* $|\varepsilon| < 0.5$
*(b)* $6.5\ W < P < 7.5\ W$
**29.** $-23°C < T < 110°C$
**31.** $105 < V \leq 126.5$
**33.** $64\ \text{km/h} \leq s \leq 89\ \text{km/h}$

**Exercise 17-2**   *(Page 593)*

**1.** $1 < x < 4$
**3.** $x < -2$ or $x > 1/2$
**5.** $-10 \leq x \leq 0$
**7.** $y < -4$ or $y > 4; |y| > 4$
**9.** $|a| < 0.1$
**11.** $|v| \geq 12$
**13.** $x < 0$ or $x > 1$
**15.** $-1/3 \leq x < 4$
**17.** $-1 < h < 0$ or $h > 1$
**19.** $-1 < x < 0$ or $0 < x < 1$
**21.** Reduce to $(x + 1)^2 > 0$
**23.** Follows from
$a^2 - 2ab + b^2 > 0$
**25.** $x < 0.1$ or $x > 0.2$
**27.** $V > 2$

**Exercise 17-3**   *(Page 598)*

**1.**
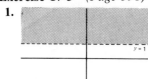

**3.**
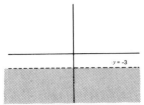

**5.**
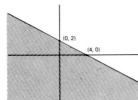

**7.**
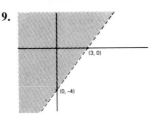

**9.**
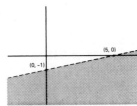

**11.**
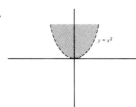

**13.**
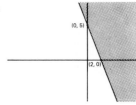

**15.**
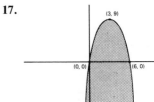

**17.**
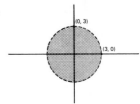

**19.**
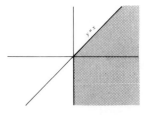

**21.**
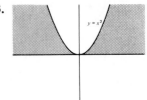

**23.**
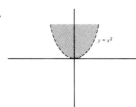

**25.**

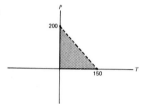

**27.** (a) $C \leq 0.70S$
(b)

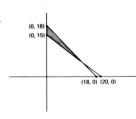

**Exercise 17-4** (Page 605)

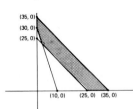

**3.**

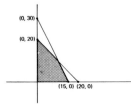

**5.**

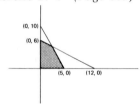

**7.**

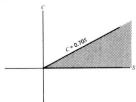

**9.**

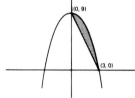

**13.** (a) 30 analog, 10 digital
(b) $1300
**15.** (a) $18D$, $10G$
(b) $8400
**17.** (a) Low: 75,000 $L$,
High: 25,000 $L$
(b) $22,500
**19.** (a) $X$: 160 mg,
$Y$: 180 mg
(b) $3.47

**Review Questions 17-5**
(Page 607)
**1.** $x \geq 1/3$
**3.** $4 < y < 6$
**5.** $r < -3$, $r > 5$
**7.** $x < -3$, $x > -1$
**9.** $|y| \leq 5$
**11.** $0.134 \leq x \leq 0.136$
**13.** $0 \leq v \leq 200$

**15.**

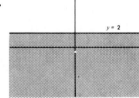

**17.**

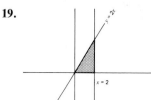

**19.**

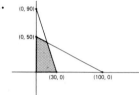

**21.**

**23.** (a) 3 hp: 10, 6 hp: 5
(b) $1750

# Chapter 18

**Exercise 18-1** (Page 612)
**1.** Answers vary
**3, 5.** See text
**7.** Input: body senses; Primary
storage: memory; Secondary
storage: written and recorded
material; Output: mouth,
hands
**9.** 26,000
**11.** (a) 393,216
(b) 12,288
**13.** MY NAME IS MARIA
**15.** 110, 111, 1000

**Exercise 18-2** (Page 619)
**1.** 10 THE SQUARE ROOT
OF 2

```
20  IS APPROXIMATELY
30  1.414214
```
**3.**
```
 1  LINCOLN
10  WITH MALICE
    TOWARD NONE
20  WITH CHARITY FOR
    ALL
```
**5.** See text

**7.** 0.72

**9.**
```
METRIC
2.54 CM = 1 IN
0.453 KG = 1 LB
```
**11.** $(A + 3)/(B - 2)$

**13.** $(F2 - F1) \wedge 3$

**15.** $SQR(2 * P) < = 6$

**17.** $y = ax^2 + bx + c$

**19.** $\dfrac{E_1 + E_2}{R} - I$

**21.** $9.9[R + (QH)^3]$

**23.**
```
AVERAGE
X     Y     A
89.5  76.8  83.15
```
**25.**
```
PARALLEL  CIRCUIT
R1   R2   R TOTAL
3.5  4.2  1.909
```
**27.**
```
SLOPE
(1, -2)  AND  (3, -1)
SLOPE = 0.5
```
**29.**
```
10  LET X = 2.78
20  LET Y = 3.12
30  LET Z = 5.01
40  LET A = (X + Y + Z)/3
50  PRINT A
```
**31.**
```
10  LET I = 2.40
20  LET R = 10.5
30  LET E = I*R
40  PRINT E
```

## Exercise 18-3    *(Page 628)*

**1.** See text

**3.**

**5.**

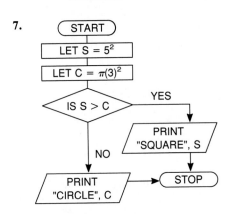

**7.**

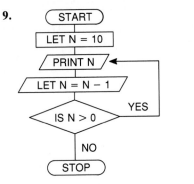

**9.**

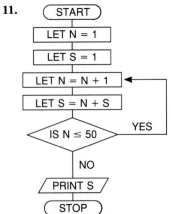

**11.**

**13.**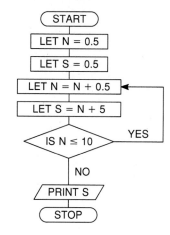

**21.**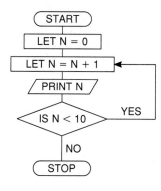

**15.**

**23.**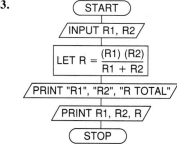

**17.**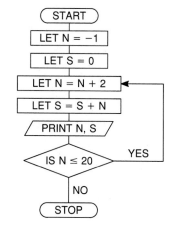

**25.**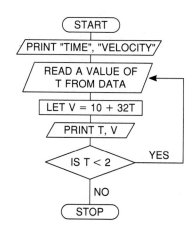

**19.** Given side *A* and hypotenuse *C* of right triangle, prints side *B*.

**27.**

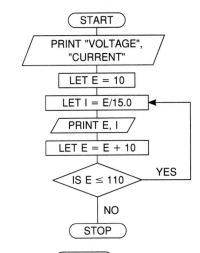

**29.**

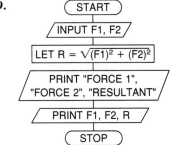

**Exercise 18-4** *(Page 635)*

**1.** See text

**3.** Even whole numbers from 2 to 20

**5.** 5, 10, 15

**7.**
```
10  LET A = (60 + 75 + 83)/3
20  IF A > = 70 THEN 50
30  PRINT 60, 75, 83, A, "FAIL"
40  GOTO 60
50  PRINT 60, 75, 83, A, "PASS"
60  END
```

**9.**
```
10  PRINT "DIAMETER",
    "CIRCUMFERENCE"
20  DATA 1, 2, 3, 4, 5
30  READ R
40  LET D = 2*R
50  LET C = 3.142*D
60  PRINT D, C
70  IF R < = 5 THEN 30
80  END
```

**11.**
```
10  INPUT A, B
```
```
20  LET X = - B/A
30  PRINT X
40  END
```

**13.**
```
10  LET N = 0
20  LET N = N + 1
30  PRINT N
40  IF N < 10 THEN 20
50  END
```

**15.**
```
10  LET N = 1
20  LET S = 1
30  LET N = N + 1
40  LET X = N2
50  LET S = S + X
60  IF N < 10 THEN 30
70  PRINT S
80  END
```

**17.**
```
10  INPUT P
20  LET F = (2/3)*P*1.05
30  PRINT "FINAL PRICE =" F
40  END
```

**19.**
```
10  DATA 1.2, 2.4, 3.8, 4.4, 5.1
20  READ X
30  PRINT SQR (X)
40  IF X < 5.1 THEN 20
50  END
```

**21.**
```
10  PRINT "TIME", "DISTANCE"
20  INPUT T
30  LET S = 4.9*T^2
40  PRINT T, S
50  END
```

**23.**
```
10  INPUT R1, R2
20  LET R = R1*R2/(R1 + R2)
30  PRINT "R1", "R2", "R TOTAL"
40  PRINT R1, R2, R
50  END
```

**25.**
```
10  PRINT "TIME", "VELOCITY"
20  DATA 0.2, 0.5, 0.9, 1.5
30  READ T
40  LET V = 10 + 32*T
50  PRINT T, V
60  IF T < 2 THEN 30
70  END
```

**27.**
```
10  PRINT "VOLTAGE", "CURRENT"
20  LET E = 10
30  LET I = E/15
40  PRINT E, I
50  LET E = E + 10
60  IF E < = 110 THEN 30
70  END
```

```
29. 10  INPUT F1, F2
    20  LET R = SQR(F1^2 + F2^2)
    30  PRINT "FORCE 1", "FORCE 2"
    40  PRINT "RESULTANT"
    50  PRINT F1, F2, R
    60  END
31. 10  INPUT X, Y
    20  IF X^2 + Y^2 = 25 THEN 80
    30  IF X^2 + Y^2 < 25 THEN 60
    40  PRINT "OUTSIDE THE CIRCLE"
    50  GOTO 90
    60  PRINT "INSIDE THE CIRCLE"
    70  GOTO 90
    80  PRINT "ON THE CIRCLE"
    90  END
33. 10  PRINT "ANGLE", "RADIAN",
        "SIN", "COS", "TAN"
    20  READ X
    30  DATA 120, 135, 150, 180
    40  LET R = 0.174529*X
    50  PRINT X, R, SIN(R), COS(R),
        TAN(R)
    60  IF X = 180 THEN 80
    70  GOTO 20
    80  END
35. 10  PRINT "X", "EXP X", "LN X",
        "LOG X"
    20  DATA 0.8, 1.1, 1.5, 2.2, 2.5
    30  PRINT X, EXP(X) LOG(X),
        0.434*LOG(X)
    40  IF X < 2.5 THEN 20
    50  END
```

**Review Questions 18-5**
*(Page 637)*

1. Answers vary
3. 12,000
5.
```
   10  TO ERR
   20  IS HUMAN
   30  TO FORGIVE
   40  DIVINE
```
7. $(X + 2Y)^2$
9.
| SIN A | COS A |
|-------|-------|
| 0.60  | 0.80  |
11. Answers vary

13.

```
START
LET S = 3.5
LET A = 6S²
LET V = S³
PRINT A, V
STOP
```

15.

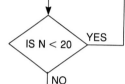

```
START
LET N = 0
PRINT N
LET N = N + 2
IS N < 20 ── YES
   │ NO
STOP
```

17.
| 2 | 4 |
|---|---|
| 3 | 9 |
| ⋮ | ⋮ |
| 10 | 100 |

19.
```
   10  DATA 2, 3, 5, 6, 7, 8
   20  READ X
   30  PRINT X^(1/3)
   40  IF X < 8 THEN 20
   50  END
```
21.
```
   10  LET N = 0
   20  LET N = N + 1
   30  PRINT 2N
   40  IF N < 10 THEN 20
   50  END
```
23.
```
   10  INPUT X, Y
   20  IF Y = 2*X + 1 THEN 50
   30  PRINT "NOT ON THE LINE Y =
       2X + 1"
   40  GO TO 60
   50  PRINT "ON THE LINE Y = 2X + 1"
   60  END
```

# Chapter 19

**Exercise 19-1**   *(Page 647)*
1. *(a)* 14.7 m/s
   *(b)* 19.6 m/s
3. 0.286, 0.256, 0.2506,
   0.25006, 0.250006
5. $-2$
7. 0
9. 1/2
11. 1
13. 6
15. 5
17. 5
19. 1/2
21. 2
23. 0
25. $2v_0$
27. 0
29. 0
31. 3
33. *(a)* $f(0)$ undefined; $g(0) = 2$
    *(b)* $\lim_{x \to 0} f(x) = 2 = \lim_{x \to 0} g(x)$
    *(c)*

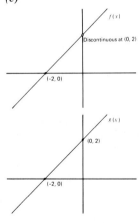

35. *(a)* 0
    *(b)* 20 m
37. 2, 2.59, 2.70, 2.717, 2.718

**Exercise 19-2**   *(Page 656)*
1. *(a)* 4

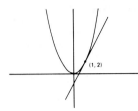

   *(b)* $4x$
3. *(a)* $-2x$
   *(b)* 4, 0, $-4$

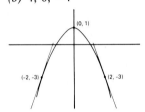

5. 5
7. $6x$
9. $4x + 3$
11. $6y - 1$
13. $3t^2$
15. $\dfrac{-1}{x^2}$
17. 1.8
19. $9.8t - 6.2$
21. $\dfrac{-1}{(x-1)^2}$
23. $\dfrac{-3}{x^4}$
25. $2ax + b$
27. $\dfrac{1}{2\sqrt{x}}$
29. *(a)*

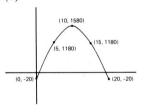

   *(b)* $v = -32t + 320$
   *(c)* 320, 0,
       $-320$ ft/s

31. *(a)* $\dfrac{dA}{dr} = 2\pi r$
    *(b)* $10\pi$ cm²/cm

**Exercise 19-3**   *(Page 663)*
1. 0
3. $x^2$
5. $4x - 6$
7. $10x^4 - 18x^2 + 1$
9. $\dfrac{1 - 3t^2}{3}$
11. $3v^2 + 1$
13. $4d^3 - 4.6d$
15. $\dfrac{3}{2}n - \dfrac{1}{2}n^2$
17. $3y^2 - 6y + 3$
19. $3ax^2 + 2bx + c$
21. $2IR$
23. *(a)* $y' = 2x + 2$
    *(b)* $-6, 0, 6$

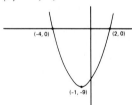

25. *(a)* $3t^2 - 12t + 9$
    *(b)* $9, 0, -3, 0$
27. *(a)* $420 - 0.112t$
    *(b)* 420 W, 413 W
29. $aL_0$
31. *(a)* $P' = 300 - 20n$
    *(b)* 15

**Exercise 19-4**   *(Page 669)*
1. *(a)* $4x$
   *(b)* $(0, 2)$
3. *(a)* $6x^2 - 6$
   *(b)* $(1, -4), (-1, 4)$
5. *(a)* $2t - 2$
   *(b)* 2
7. *(a)* $4t - 3t^2$
   *(b)* $4 - 6t$
9. *(a)* $2x + 2$
   *(b)* $(-1, -4)$

(c) $y = 4x - 4$

**11.** (a) $v = 19.6 - 9.80t$

(b) 2 s, 19.6 m

**13.** (a) $I = 2.0t - 0.06t^2$

(b) 8.5 A, 14 A

**15.** (a) $T' = 20x - 10$

(b) 20°/in., 40°/in.

**17.** (a) $P = -4t^3 +$
$36t^2 - 108t + 108$

(b) 108 W, 32 W

**19.** $T' = 0.0151$

**21.** $I = (2.4 \times 10^{-2})t$

**23.** (a) $\omega = 32 - 16t$

(b) 16 rad/s, $-16$ rad/s

**Exercise 19-5** *(Page 675)*

**1.** $6x^2 - 2$

**3.** $2x + 1$

**5.** $4v^3 + 0.6v^2 - 0.8$

**7.** $3x^2 - 2x - 5$

**9.** $2acx + ad + bc$

**11.** $\dfrac{-1}{x^2}$

**13.** $\dfrac{-9}{t^4} - \dfrac{4}{t^3}$

**15.** $\dfrac{-4}{(x - 3)^2}$

**17.** $\dfrac{-x - 2}{2x^3}$

**19.** $\dfrac{h^2 + 10h + 1}{(1 - h^2)^2}$

**21.** $\dfrac{-4(4x + 1)}{(2x^2 + x + 1)^2}$

**23.** $9v^2 - \dfrac{2}{v^3}$

**25.** $\dfrac{-2d}{(v - d)^2}$

**27.** $\dfrac{2x - 2}{x^2(2 - x)^2}$

**29.** (a) $\dfrac{-1}{(x + 1)^2}$

(b) $-1, -1, \dfrac{-1}{9}$, no

**31.** (a) $v = \dfrac{20t}{(t^2 + 1)^2}$

(b) 0, 6.4, 5.0, 2.8

**33.** (a) $P' = \dfrac{3600 - 144R^2}{(R^2 + 25)^2}$

(b) 5

**35.** (a) $s' = \dfrac{-660,000}{\pi R^4}$

(b) $\dfrac{-66}{\pi}$

**37.** $\dfrac{t^2 + 6t + 3}{t^2(t + 1)^2}$

**Exercise 19-6** *(Page 683)*

**1.** $9(3x - 1)^2$

**3.** $-6(3 - 3d)^3$

**5.** $3(4x + 3)(2x^2 + 3x + 1)^2$

**7.** $\dfrac{1}{3\sqrt[3]{x^2}}$

**9.** $\dfrac{-1.3}{t\sqrt{t}}$

**11.** $\dfrac{3x^{1/2}}{2}$

**13.** $\dfrac{-24}{(4x + 3)^3}$

**15.** $6x(x - 1)^2(x + 1)^2$

**17.** $\dfrac{(t - 8)(t + 1)^2}{(t - 2)^3}$

**19.** $2(2d - 1)^{-3/2}$

**21.** $\dfrac{3x + 2}{2\sqrt{x + 1}}$

**23.** $\dfrac{-3x - 5}{(3x - 2)^2\sqrt{2x + 1}}$

**25.** $-5(2h)^{-3/2}(h + 5)^{-1/2}$

**27.** $\dfrac{-x}{\sqrt{r^2 - x^2}}$

**29.** (a) $y = \sqrt{25 - x^2}$;
$y' = \dfrac{-x}{\sqrt{25 - x^2}}$

(b) $-3/4$; $3x + 4y = 25$

**31.** (a) $\dfrac{3(2t - 1)^2}{2}$

(b) $6(2t - 1)$

**33.** (a) $6t - \dfrac{2}{\sqrt{2 - 4t}}$

(b) $-0.44$ A

**35.** (a) $1.4\sqrt{2gH^3}$

(b) 69.3 m³/s

**37.** $\dfrac{4(n - 1)}{(n + 1)^3}$

**39.** $\dfrac{0.00036}{(3 + t)^{5/2}}$

**Exercise 19-7** *(Page 687)*

**1.** $\dfrac{y}{x} = 1$

**3.** $\dfrac{7 - a}{t} = \dfrac{3}{t^2}$

**5.** $\dfrac{-2}{3}$

**7.** $\dfrac{8t + 1}{3}$

**9.** $\dfrac{-y - 1}{x}$

**11.** $\dfrac{-x}{y}$

**13.** $\dfrac{1.5 - 2Mt}{t^2}$

**15.** $\dfrac{2 - x}{y - 1}$

**17.** $\dfrac{-p - 2q}{2p + q}$

**19.** $-2h/r$

**21.** $5/4$

**23.** $2.6$

**Review Questions 19-8**
*(Page 688)*

**1.** 5

**3.** 1/3

**5.** 1

**7.** 3

**9.** $3t^2 - 1$

**11.** $12x - 3$

**13.** $3.6\pi + 5.6\pi r$

**15.** $9(3k - 3)^2$

**17.** $\dfrac{2x^2 + 2x}{(2x + 1)^2}$

**19.** $\dfrac{-3M}{(3I + 2)^2} + 6I^2$

**21.** $-48(2 - 3x)^3$

**23.** $\dfrac{-\sqrt{t}(t+1)}{2t(t-1)^2}$

**25.** $\dfrac{-1}{\sqrt{x(x-2)^3}}$

**27.** $\dfrac{-y}{2x}$

**29.** $\dfrac{s-2t}{2s-t}$

**31.** 0.008

**33.** (a) 7
  (b) (1/4, −1/8)

**35.** (a) 365 m/s
  (b) 410 m

**37.** 6.8 A

**39.** $\dfrac{-20}{x^2}$

**41.** $\dfrac{11.9}{\sqrt[3]{nSH^2}}$

**43.** (a) $\dfrac{24(2-R)}{(2+R)^3}$
  (b) 2

# Chapter 20

**Exercise 20-1**  *(Page 695)*

**1.** 24, 8

**3.** 1/9, −2/27

**5.** 1/4, −3

**7.** 1/3, −1/27

**9.** (a) 2, 1
  (b) $\sqrt{5}$ or 2.24, 26.6°

**11.** (a) 1, −4
  (b) $\sqrt{17}$ or 4.12, −76.0°

**13.** (a) 90 m/min,
  18 m/min
  (b) −18 m/min², 
  −18 m/min²

**15.** (a) 18.8 rad/s
  (b) 3.95 rad/s²

**17.** (a) 1.4

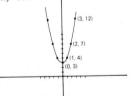

(b) $\sqrt{17}$ or 4.12, 76.0°

**19.** $y' = 2ax + b$,
$y'' = 2a$, $y''' = 0$

**Exercise 20-2**  *(Page 705)*

**1.** min: (−3, −9)

**3.** min: (1, −1), max: (1/3, −5/9)

**5.** (2, −18)

**7.** (0, 0), (1/2, 3/8)

**9.**

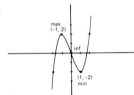

**11.**

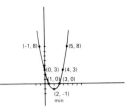

**13.**

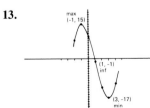

**15.**

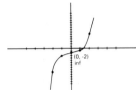

**17.**

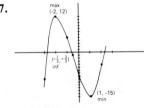

**19.**

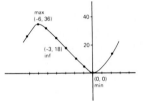

**21.**

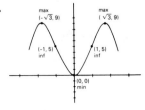

**23.**

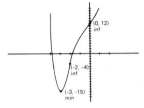

**25.**

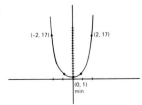

**27.**

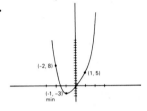

**29.**

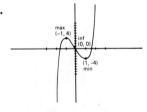

**31.**
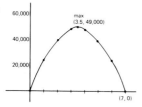

**33.**
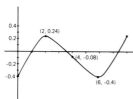

**35.** $y' = 3ax^2$, $y'$ is always positive

## Exercise 20-3   *(Page 713)*
**1.**
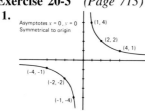

**3.**
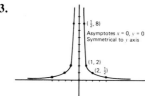

**5.**
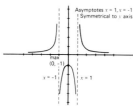

**7.**
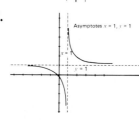

**9.**
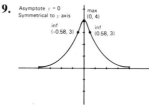

**11.**
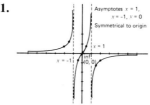

**13.**
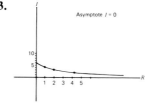

**15.**
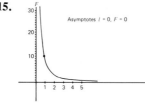

## Exercise 20-4   *(Page 721)*
**1.** 16 rad/s
**3.** 0.37 A
**5.** 0.001 in.; $-0.003$ in.
**7.** 30 m, 30 m
**9.** 1 in.
**11.** 1 m
**13.** 2.0 $\Omega$
**15.** 19.4 ft; 77.5 lb
**17.** 75
**19.** $R = X$
**21.** $x = 2$ ft, $h = 1$ ft
**23.** (a) $y' = 2x^2 - 3Lx + L^2 = 0$;
   $(2x - L)(x - L) = 0$,
   so $x = L/2$ and $x = L$
   (b) $\dfrac{5L^3}{24} > \dfrac{L^3}{6}$

**25.** $C' = k - K/A^2 = 0$,
   $k = K/A^2$ or $kA = K/A$

## Exercise 20-5   *(Page 727)*
**1.** 0.022 cm/s
**3.** $-3.4$ A/min
**5.** 15 nmi/h
**7.** $-0.11$ in.$^2$/min
**9.** 9 W/s
**11.** 4 cm/min
**13.** $-1.3$ ft/s
**15.** $-0.78$ $\Omega$/s
**17.** $-1.25$ $\Omega$/min
**19.** $-1.85$ ft$^3$/h
**21.** 5.7 ft/s

## Exercise 20-6   *(Page 732)*
**1.** $\Delta y = 6.5$, $dy = 6$
**3.** $\Delta y = 0.049$,
   $dy = 0.050$
**5.** $9x^2\,dx$
**7.** $3(t^2 + t)(2t + 1)\,dt$
**9.** $\dfrac{2\,dx}{(2 - x)^2}$
**11.** $\dfrac{R\,dR}{\sqrt{R^2 + X^2}}$
**13.** $-3$ cm$^3$
**15.** 0.6 W
**17.** 10.6 m$^3$
**19.** $7.28 \pm 0.01$ $\Omega$

## Exercise 20-7   *(Page 737)*
**1.** $x^4 + C$
**3.** $2x^3 + x^2 - 3x + C$
**5.** $\dfrac{x^2}{2} - 2x + C$
**7.** $3x + \dfrac{3}{x} + C$
**9.** $\dfrac{-1}{8x^2} + C$
**11.** $\dfrac{2x\sqrt{x}}{3} + C$
**13.** $\dfrac{x^{5/2}}{5} - \dfrac{x^{3/2}}{3} + C$
**15.** $0.15t^4 - 0.5t^3 + 1.0t^2 + C$
**17.** $ax^2 + bx + c$

**19.** (a) $v = -32t$
  (b) $s = -16t^2 + 50$
**21.** $y = x^2 + 2x + 1$
**23.** $V = 50(t^2 + t)$
**25.** 1 h

**Review Questions 20-8**
*(Page 738)*
**1.** $v = -5$ cm/s,
  $a = 5$ cm/s$^2$
**3.** (a) $v_x = 2$, $v_y = 4$
  (b) $v = 4.5$,
    $\theta = 63.4°$
**5.**

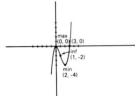

**7.**

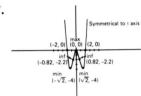

**9.**

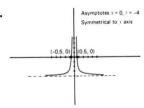

**11.**

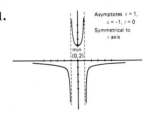

**13.** $dy = (6x + 16x^3)\,dx$
**15.** $y = 3x^3 + 5x + C$
**17.** $y = \dfrac{5x^2}{2} + \dfrac{6}{x} + C$

**19.**

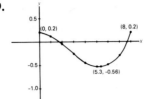

**21.** $(5, 25)$
**23.** $11{,}250$ m$^2$
**25.** $0.278$ $L^3$
**27.** $-0.160$ m$^3$/s
**29.** $104$ mi/h
**31.** $45$ in$^3$.

## Chapter 21

**Exercise 21-1**  *(Page 748)*
**1.** $2x^3 + C$
**3.** $3x + C$
**5.** $-\dfrac{1}{x} + C$
**7.** $2\sqrt{t^3} + C$
**9.** $x^3 + x^2 + x + C$
**11.** $\dfrac{x^3}{3} - x^2 + C$
**13.** $\pi(r^2 - 2r) + C$
**15.** $\dfrac{3x^2 - 8}{x} + C$
**17.** $\dfrac{(3x - 1)^3}{3} + C$
**19.** $\dfrac{(2x - 3)^4}{8} + C$
**21.** $\dfrac{(x^2 + 4)^5}{10} + C$
**23.** $\dfrac{(4x^2 + 1)^4}{16} + C$
**25.** $\dfrac{-1}{2(2x - 1)} + C$
**27.** $\dfrac{2(t + 1)^{3/2}}{3} + C$
**29.** $\sqrt{x^2 - 4} + C$
**31.** $\dfrac{(\pi r^2 + 0.5)^{3/2}}{3\pi} + C$
**33.** $y = 2x^2 - 3$

**35.** $xy = 12$
**37.** $s = 5t - 2t^3$
**39.** (a) $v = 16 - \dfrac{\sqrt{t}}{2}$
  (b) $1024$ s $= 17.1$ min
**41.** $0.012$ C
**43.** $5$ V
**45.** $8.1$ A
**47.** $W = \dfrac{(x - 2)^4}{4} + 6$

**Exercise 21-2**  *(Page 759)*
**1.** $A = \lim\limits_{n \to \infty} \left(\dfrac{1}{2} - \dfrac{1}{2n}\right) = \dfrac{1}{2}$
**3.** $A = \lim\limits_{n \to \infty} \left(\dfrac{2}{3} - \dfrac{1}{n} + \dfrac{1}{3n^2}\right) = \dfrac{2}{3}$
**5.** $A = \lim\limits_{n \to \infty} \left(\dfrac{4}{3} - \dfrac{1}{2n} + \dfrac{1}{6n^2}\right) = \dfrac{4}{3}$
**7.** $2$
**9.** $-3/4$
**11.** $83/6$
**13.** $1/2$
**15.** $157/6$
**17.** $8/12$
**19.** $52/3$
**21.** $4/33$
**23.** $3/2$
**25.** $2/3$
**27.** $15/4$
**29.** $16/3$
**31.** $1/4$
**33.** $1.2$ mC
**35.** $5.2L \times 10^6$ ft·lb

**Exercise 21-3**  *(Page 768)*
**1.** $2$ square units
**3.** $4/3$ square units
**5.** $4/3$ square units
**7.** $1/6$ square unit
**9.** $1/2$ square unit
**11.** $1/12$ square unit
**13.** $4/3$ square units
**15.** $1/6$ square unit

**17.** 1/8 square unit
**19.** 2.44 square units
**21.** 7.75 square units
**23.** 0.56 C
**25.** 0.06 ft·lb
**27.** 4/3 square units

**Exercise 21-4**   *(Page 773)*
 **1.** 0.83 ft·lb
 **3.** 10 m·kg$_f$
 **5.** 100,000 ft·lb
 **7.** $9.9 \times 10^{-7}$ J
 **9.** 2.0 mC
**11.** 5.8 mC
**13.** 0.22 J
**15.** $2\sqrt{2}/3$
**17.** 3.25 A
**19.** 21 W
**21.** 0.4 V
**23.** *(a)* $2.2 \times 10^8$ L
    *(b)* $3.7 \times 10^6$ L/min
**25.** $1.16 \times 10^6$ ft·lb
**27.** 4 s

**Exercise 21-5**   *(Page 782)*
 **1.** $\pi/5$ or 0.628 cubic unit
 **3.** $\pi/7$ or 0.449 cubic unit
 **5.** $4\pi$ or 12.6 cubic units
 **7.** $3\pi/2$ or 4.71 cubic units
 **9.** $128\pi/5$ or 80.4 cubic units
**11.** $8\pi$ or 25.1 cubic units
**13.** $3\pi/2$ or 4.71 cubic units
**15.** $\pi/6$ or 0.524 cubic unit
**17.** $32\pi/5$ or 20.1 cubic units
**19.** $\pi/30$ or 0.105 cubic unit
**21.** $\dfrac{2\pi}{3}$ or 2.09 cubic units
**23.** 39.2 cubic units
**25.** $\dfrac{8\pi}{3}$ cubic units
**27.** $2\pi$ cubic units
**29.** 754 m$^3$

**Exercise 21-6**   *(Page 791)*
 **1.** (2.9, 0)
 **3.** (1.1, 0)
 **5.** (3.0, 1.0)
 **7.** (0.78, 1.6)

 **9.** (0.75, 0.30)
**11.** (0.54, 1.2)
**13.** 1/3 unit from leg 2, 2/3 unit
    from leg 1
**15.** (5/6, 0)
**17.** (0, 4/3)
**19.** (1/4, 0) or (0, 1/4)
**21.** (1/2, 1/10)

**Review Questions 21-7**
*(Page 794)*
 **1.** $\dfrac{5x^4}{4} + C$
 **3.** $\dfrac{(2x + 1)^4}{4} + C$
 **5.** $\dfrac{(6x + 1)^4}{8} + C$
 **7.** $-22/5$
 **9.** 38/3
**11.** $-8/7$
**13.** 1/6 square unit
**15.** 16/3 square units
**17.** 81/2 square units
**19.** $\pi/20$ or 0.16 cubic unit
**21.** 14.2 cubic units
**23.** $\pi/2$ or 1.57 cubic units
**25.** (1.6, 0.76)
**27.** (1.7, 0)
**29.** *(a)* $\theta = 33.6t^2 - 0.70t^6$
    *(b)* 32.9 rad/s
**31.** 1000 ft·lb
**33.** 57,445 lb

**Chapter 22**

**Exercise 22-1**   *(Page 805)*
 **1.** $3 \cos x$
 **3.** $3 \cos 3x$
 **5.** $-6 \sin 2x$
 **7.** $-\sin x \cos x$
 **9.** $2 \cos 2x - 2 \sin x$
**11.** $\cos x - x \sin x$
**13.** $\dfrac{x \cos x - \sin x}{x^2}$
**15.** $2 \cos 2t$

**17.** $\dfrac{-\sin x \cos x}{\sqrt{1 + \cos^2 x}}$
**19.** $60\pi(\cos 60\pi t - \sin 60\pi t)$
**21.** $8 \sec^2 2x$
**23.** $3 \csc^2 (\pi - x)$
**25.** $2 \sec (x + 2) \tan (x + 2)$
**27.** $2 \tan x \sec^2 x$
**29.** $-3 \csc^3 t \cot t$
**31.** $3(\tan x + x \sec^2 x)$
**33.** $\dfrac{-2x \csc^2 2x - \cot 2x}{x^2}$
**35.** $\sin 2x \sec^2 x + 2 \tan x \cos 2x$
**37.** $\dfrac{\cos^3 x - 1}{(\sin^2 x)(1 - \cos x)^2}$
**39.** $ab \cos (bx + \phi)$
**41.** $4 \sin 6x\, dx$
**43.** $-\cot x(\csc x + \sin x)\, dx$
**45.** $y' = A\sqrt{k} \cos (t\sqrt{k})$;
    $y'' = -Ak \sin (t\sqrt{k}) = -ky$
**47.** $-6$
**49.** 1.732 ft/s
**51.** $V = -12 \sin 60t$
**53.** *(a)* $v = 2.0 \cos 0.5t$,
    $a = \sin 0.5t$
    *(b)* $-1$ cm/s$^2$

**55.**
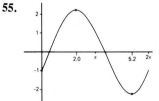

**57.** *(a)* Max: 2.24; min: $-2.24$
    *(b)*
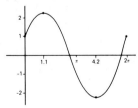

**59.** 1.1 mi/s
**61.** $2400 \sin 30t \cos 30t =$
    $1200 \sin 60t$
**63.** 90°

**65.** $y' = \dfrac{(-\sec^2 u)(u')}{\tan^2 u} =$

$(-\csc^2 u)(u')$

**Exercise 22-2** *(Page 812)*
1. $-\cos 2x + C$
3. $\dfrac{\sin (2x + 1)}{2} + C$
5. $\dfrac{\tan 2x}{2} + C$
7. $\dfrac{-\csc x^2}{2} + C$
9. $\tan x - x + C$
11. $\dfrac{\sin 2x}{2} + C$
13. $\dfrac{x}{2} + \dfrac{\sin 2x}{4} + C$
15. $\dfrac{-\cot 2x}{2} - x + C$
17. $\dfrac{\sin^2 x}{2} + C$
19. $\dfrac{\sin^3 3t}{9} + C$
21. $\dfrac{\tan^3 x}{3} + C$
23. $\sec x + C$
25. $\tan x - \sec x + C$
27. $\dfrac{-3 \cos 2\pi t}{2\pi} + C$
29. 5
31. 1/2
33. 2
35. $\sqrt{2} - 1$ or 0.414
37. $5\sqrt{2}/2 = 3.5$
39. $I_{rms} =$

$\sqrt{f \displaystyle\int_0^{1/f} (I_{max} \sin 2\pi f t)^2 \, dt} =$

$I_{max} \sqrt{\dfrac{f}{2} \displaystyle\int_0^{1/f} (1 - \cos 4\pi f t) \, dt} =$

$I_{max} \sqrt{\dfrac{f}{2}\left(\dfrac{1}{f}\right)} = \dfrac{I_{max}}{\sqrt{2}}$

41. $\dfrac{-A \sin 2\omega t}{2} + C$

**43.** (a) $-12 \sin 4t$
(b) $3 \cos 4t$

**Exercise 22-3** *(Page 818)*
1. $\dfrac{2}{\sqrt{1 - 4x^2}}$
3. $\dfrac{1}{x^2 + 9}$
5. $\dfrac{-1}{\sqrt{4 - x^2}}$
7. $\dfrac{4}{x\sqrt{16x^2 - 1}}$
9. $\dfrac{x + 1}{\sqrt{1 - x^2}}$
11. $\dfrac{1}{t^2 + 1}$
13. $\dfrac{-1}{(x + 1)\sqrt{(x + 1)^2 - 16}}$
15. $\dfrac{2 \tan^{-1} x}{x^2 + 1}$
17. $5 \sin^{-1} x + C$
19. $\dfrac{1}{2} \tan^{-1} \dfrac{x}{2} + C$
21. $\sec^{-1} \dfrac{x}{3} + C$
23. $\dfrac{1}{10} \tan^{-1} \dfrac{5x}{2} + C$
25. $2 \sin^{-1} \dfrac{2x}{5} + C$
27. $\sec^{-1} 4x + C$
29. $\tan^{-1} (x - 1) + C$
31. $\sin^{-1} (x - 1) + C$
33. $\pi$ or 3.14
35. $\pi/6$ or 0.524
37. $u = \cos y$
$u' = -\sin y \, (y')$
$y' = \dfrac{-u'}{\sin y}$,
$\sin y = \sqrt{1 - \cos^2 y}$
$= \sqrt{1 - u^2}$;
$\therefore \; y' = \dfrac{-u'}{\sqrt{1 - u^2}}$
39. $\dfrac{dp}{dt} = \dfrac{-t}{\sqrt{1 - t^2}}$

**41.** $y' = \dfrac{1/a \cdot u'}{\sqrt{1 - (u/a)^2}} =$

$\dfrac{u'}{a\sqrt{1 - u^2/a^2}} = \dfrac{u'}{\sqrt{a^2 - u^2}}$

43. $\pi/4$
45. $\tan^{-1} 2.0t$

**Exercise 22-4** *(Page 827)*
1. $\dfrac{3 \log_2 e}{x}$
3. $\dfrac{1}{x} - 1$
5. $\dfrac{\log e}{2x}$
7. $\ln x + 1$
9. $\cot x$
11. $\dfrac{0.75}{x\sqrt{\ln x}}$
13. $\dfrac{2\sqrt{p} - 1}{2p(\sqrt{p} - 1)}$
15. $\dfrac{1 - x^2}{x(x^2 + 1)}$
17. $\dfrac{\cos x - x \sin x}{x \cos x}$
19. $6e^{3x}$
21. $3^{1+x} \ln 3$
23. $e^{2x}(2x + 1)$
25. $\dfrac{xe^x}{(x + 1)^2}$
27. $e^x(\cos x + \sin x)$
29. $\dfrac{e^x}{(e^x + 1)^2}$
31. $\dfrac{4 \cos 2t - \sin 2t}{2e^{t/2}}$
33. $\dfrac{e^x + 1}{e^x + x}$
35. $e^x \cot e^x$
37.

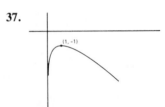

(1, -1)

**39.**

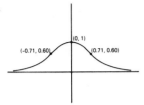

**41.**

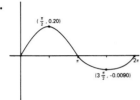

**43.** (a) $-kt(1 + 2 \ln t)$
(b) $k/2e$
**45.** $e^{-at}[(Ab - a)$
$\cos bt - (Aa + b)\sin bt]$
**47.** $10^2 e^{0.01t}$
**49.** $\dfrac{dq}{dt} = \dfrac{Ve^{-t/(RC)}}{R}$;

$R\dfrac{Ve^{-t/(RC)}}{R} +$

$\dfrac{CV(1 - e^{-t/(RC)})}{C} =$

$Ve^{-t/(RC)} + V - Ve^{-t/(RC)} = V$
**51.** $\ln y = x \ln x,\ \dfrac{y'}{y} =$
$1 + \ln x,\ y' = y(1 + \ln x) =$
$x^x(1 + \ln x)$

**Exercise 22-5**　(*Page 834*)
**1.** $\ln |2x + 1| + C$
**3.** $\dfrac{1}{2} \ln |2x^2 + 5| + C$
**5.** $\ln |\sin x| + C$
**7.** $\ln |e^x + e^{-x}| + C$
**9.** $\dfrac{x^2}{2} + \ln |x| + C$
**11.** $\dfrac{x^2}{2} + 2x - \ln |x + 1| + C$
**13.** $\ln |\sin^2 t + 1| + C$
**15.** $3e^{2x} + C$
**17.** $3e^{x/3} + C$

**19.** $\dfrac{e^{x^2}}{2} + C$
**21.** $\dfrac{e^{2x}}{2} + 2e^x + x + C$
**23.** $\dfrac{2e^{3t/2}}{3} + C$
**25.** $e^{\sin t} + C$
**27.** $-e^{1/x} + C$
**29.** $\dfrac{kt + e^{-kt}}{k} + C$
**31.** $\ln 4$
**33.** $\dfrac{e^2 - 1}{e}$
**35.** $\ln 2$
**37.** $e - 1$
**39.** $\dfrac{-\ln |E - IR|}{R} + C$
**41.** $1099$
**43.** $0.43$ C
**45.** $m = 10.0e^{-0.103t}$
**47.** $s = \dfrac{\ln |t^2 + 1|}{2}$

**Review Questions 22-6**
(*Page 836*)
**1.** $\dfrac{3}{2} \cos \dfrac{x}{2}$
**3.** $x(2 \cos x - x \sin x)$
**5.** $\sec^2 3x$
**7.** $\dfrac{\sqrt{\sec t}}{2}(t \tan t + 2)$
**9.** $\dfrac{6}{\sqrt{1 - 9x^2}}$
**11.** $\dfrac{1}{\sqrt{2x}(1 + 2x)}$
**13.** $\dfrac{20 \log e}{4x - 3}$
**15.** $-\tan t$
**17.** $e^{2x}$
**19.** $\dfrac{3\sqrt{e^{3x}}}{2}$
**21.** $\dfrac{3 \sin 2x}{2} + C$
**23.** $\dfrac{\tan x}{2} + C$

**25.** $x + \dfrac{\sin 2x}{2} + C$
**27.** $\dfrac{\tan^2 x}{2} + C$
**29.** $-\sqrt{4 - x^2} + C$
**31.** $2 \tan^{-1} 3x + C$
**33.** $2 \ln |x - 2| + C$
**35.** $-\ln |\cos 2x| + C$
**37.** $\dfrac{e^{2x} - e^{-2x}}{2} + C$
**39.** $6e^{t/2} + C$
**41.** $3(\sqrt{3} - 1)$
**43.** $\dfrac{\ln 2}{2}$

**45.**

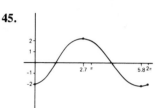

**47.** $2$
**49.** $\ln 4$
**51.** (a) $v = 3 \cos t - 5 \sin t,$
$a = -3 \sin t - 5 \cos t$
(b) $5.83$ cm/s$^2$, $-5.83$ cm/s$^2$
**53.** $-5.08$ W/s
**55.** $V = 6.6 \cos 120\pi t - 6.6$
**57.** $\rho/(2\pi r)$
**59.** $0.75$ J

**Chapter 23**

**Exercise 23-1**　(*Page 845*)
**1.** $-\ln |\cos 2x| + C$
**3.** $2 \ln |\sec 3x + \tan 3x| + C$
**5.** $\frac{1}{2} \ln |\sin (2t - 1)| + C$
**7.** $\theta - 2 \ln |\csc \theta - \cot \theta| -$
$\cot \theta + C$
**9.** $\tan x - 2 \ln |\cos x| + C$
**11.** $\ln |1 - \cos x| + C$
**13.** $\frac{1}{4} \ln |1 + \sin 2t| + C$
**15.** $\ln |\csc x - \cot x| -$
$\ln |\cos x| + C$

**17.** $\dfrac{-\cos^3 2x}{6} + C$

**19.** $\dfrac{\cos^5 x}{5} - \dfrac{\cos^3 x}{3} + C$

**21.** $\dfrac{3}{2}\left(\sin 2t - \dfrac{\sin^3 2t}{3}\right) + C$

**23.** $\dfrac{x}{2} - \dfrac{\sin 4x}{8} + C$

**25.** $\dfrac{3x}{8} + \dfrac{\sin 2x}{4} + \dfrac{\sin 4x}{32} + C$

**27.** $\dfrac{\tan^4 x}{4} + C$

**29.** $\dfrac{-2\cot^3 2x}{3} - 2\cot 2x + C$

**31.** $\dfrac{\tan^2 x}{2} - \ln|\cos x| - x + C$

**33.** $\dfrac{\tan^3 x}{3} + \tan x + C$

**35.** 1.193

**37.** 2/3

**39.** 0.347

**41.** 2/15

**43.** 1.67 C

**45.** $\sqrt{2}$

**47.** $\displaystyle\int \dfrac{\cos u}{\sin u}\,du = \ln|\sin u| + C$

**Exercise 23-2** *(Page 849)*

**1.** $\sin^{-1} x + C$

**3.** $\dfrac{-\sqrt{9 + x^2}}{9x} + C$

**5.** $\sqrt{x^2 - 1} - \sec^{-1} x + C$

**7.** $8\sin^{-1}\dfrac{x}{4} + \dfrac{x\sqrt{16 - x^2}}{2} + C$

**9.** $\dfrac{(3x^2 - 8)\sqrt{(x^2 + 4)^3}}{3} + C$

**11.** $\dfrac{1}{2}\left(\sec^{-1} x + \dfrac{\sqrt{x^2 - 1}}{x^2}\right) + C$

**13.** $\ln|3t + \sqrt{9t^2 - 1}| + C$

**15.** $\dfrac{x^2\sqrt{x^2 - r^2}}{3} + C$

**17.** 0.881

**19.** 0.447

**Exercise 23-3** *(Page 853)*

**1.** $-x\cos x + \sin x + C$

**3.** $\dfrac{e^{2x}}{2}(2x - 1) + C$

**5.** $t\ln\sqrt{t} - \frac{1}{2}t + C$

**7.** $x\sin^{-1} x + \sqrt{1 - x^2} + C$

**9.** $\dfrac{x^2}{4} - \dfrac{x\sin 2x}{4} - \dfrac{\cos 2x}{8} + C$

**11.** $e^x(x^2 - 2x + 2) + C$

**13.** $(t + 1)\ln(t + 1) - t + C$

**15.** $\frac{1}{2}e^x(\sin x - \cos x) + C$

**17.** $\frac{2}{3}(x + 1)^{3/2}(3x - 2) + C$

**19.** $\pi/4 - 1/2$

**21.** $\pi/4 - \ln\sqrt{2} = 0.439$

**23.** 1

**25.** $q = 2e^{-t}(\sin t - \cos t) + 2$

**27.** 8/3

**Exercise 23-4** *(Page 856)*

**1.** $-\dfrac{1}{2}\ln\left|\dfrac{2 + 5x}{x}\right| + C$

**3.** $\dfrac{2(9x - 8)}{45}(4 + 3x)^{3/2} + C$

**5.** $\frac{1}{2}\sin^{-1} 2x + C$

**7.** $\dfrac{\sin 2t}{2} - \dfrac{\sin^3 2t}{6} + C$

**9.** $x\tan^{-1} x - \frac{1}{2}\ln|1 + x^2| + C$

**11.** $\dfrac{x}{\sqrt{x^2 + 1}} + C$

**13.** $x^3\left(\dfrac{\ln 3x}{3} - \dfrac{1}{9}\right) + C$

**15.** $\ln\left|\dfrac{4 + x}{4 - x}\right| + C$

**17.** $\ln|x + 1| + \dfrac{1}{x + 1} + C$

**19.** $\dfrac{\cos x}{2} - \dfrac{\cos 5x}{10} + C$

**21.** $2 + x - \ln|x + 2| + C$

**23.** $\sqrt{t^2 + 0.04} - 0.20\ln\left|\dfrac{0.2 + \sqrt{t^2 + 0.04}}{t}\right| + C$

**25.** $\dfrac{3x}{8} - \dfrac{\sin 2x}{4} + \dfrac{\sin 4x}{32} + C$

**27.** $\dfrac{x}{4}(x^2 - 10)\sqrt{x^2 - 4} + 6\ln|x + \sqrt{x^2 - 4}| + C$

**29.** $\dfrac{e^x}{2}(x\cos x + x\sin x - \sin x) + C$

**31.** $1 - 2\ln 2 = -0.386$

**33.** $\frac{1}{4}$

**35.** $e - 2 = 0.718$

**37.** 0.489 C

**Exercise 23-5** *(Page 861)*

**1.** (a) 0.3438
   (b) 0.3333

**3.** (a) 1.218
   (b) 1.219

**5.** (a) 1.389
   (b) 1.386

**7.** 1.500

**9.** 0.3860

**11.** 0.4346

**13.** 0.2689

**15.** 0.8818

**Review Questions 23-6**
*(Page 862)*

**1.** $-\frac{1}{3}\ln|\cos 3x| + C$

**3.** $4\ln|\sin\theta| - 4\cot\theta - 3\theta + C$

**5.** $\dfrac{-\cos^4 x}{4} + C$

**7.** $\dfrac{\tan^5 x}{5} + \dfrac{\tan^3 x}{3} + C$

**9.** $\dfrac{-\sqrt{1 - x^2}}{x} + C$

**11.** $\dfrac{(x^2 - 9)^{3/2}}{3} + C$

**13.** $x\tan x + \ln|\cos x| + C$

**15.** $-x^2\cos 2x + x\sin 2x + \frac{1}{2}\cos 2x + C$

**17.** $5 + 2x - 5\ln|5 + 2x| + C$

**19.** $3\left(\sqrt{9x^2 - 4} - 2\sec^{-1}\dfrac{3x}{2}\right) + C$

**21.** $\dfrac{e^{3t}}{13}(3\cos 2t + 2\sin 2t) + C$

**23.** $\ln\frac{1}{2} = -0.6931$

**25.** $\frac{2}{3}$

**27.** 1.094

**29.** $\pi/2 - \ln(1 + \sqrt{2}) = 0.6894$

**31.** 4.93 C

**33.** $\frac{2}{3}[(t - 8)\sqrt{t + 4} + 16]$

## Chapter 24

### Exercise 24-1 (Page 868)

**1.** $1 + 2x + \dfrac{4x^2}{2!} + \dfrac{8x^3}{3!} + \cdots$

**3.** $1 - 2x + 3x^2 - 4x^3 + \cdots$

**5.** $x - \dfrac{x^3}{3!} + \dfrac{x^5}{5!} - \dfrac{x^7}{7!} + \cdots$

**7.** $-x - \dfrac{x^2}{2} - \dfrac{x^3}{3} - \dfrac{x^4}{4} - \cdots$

**9.** $x - \dfrac{x^3}{3} + \dfrac{x^5}{5} - \dfrac{x^7}{7} \cdots$

**11.** $x + \dfrac{x^3}{3!} + \dfrac{x^5}{5!} + \dfrac{x^7}{7!} \cdots$

**13.** $x + x^2 + \dfrac{x^3}{2!} + \dfrac{x^4}{3!} + \cdots$

**15.** $\dfrac{-x^2}{2} - \dfrac{x^4}{12} - \dfrac{x^6}{45} -$
$\dfrac{17x^8}{2520} - \cdots$

**17.** $1 + x + \dfrac{x^2}{2!} - \dfrac{3x^4}{4!} - \cdots$

**19.** $\dfrac{633}{384} = 1.6484$

**21.** 0.1564

**23.** Derivatives are undefined at $x = 0$

**25.** $1 - x + x^2 - x^3 + \cdots$

### Exercise 24-2 (Page 873)

**1.** $1 - \dfrac{x^2}{2^2 \cdot 2!} +$
$\dfrac{x^4}{2^4 \cdot 4!} - \dfrac{x^6}{2^6 \cdot 6!} + \cdots$

**3.** $1 + 4x + 8x^2 + \dfrac{32x^3}{3} + \cdots$

**5.** $2x - 2x^2 + \dfrac{8x^3}{3} - 4x^4 + \cdots$

**7.** $x^2 - \dfrac{x^4}{3!} + \dfrac{x^6}{5!} - \dfrac{x^8}{7!} + \cdots$

**9.** $1 + x - \dfrac{x^3}{3} - \dfrac{x^4}{6} - \cdots$

**11.** $x^2 - \dfrac{x^4}{3} + \dfrac{2x^6}{45} - \dfrac{x^8}{315} + \cdots$

**13.** 1.371

**15.** 0.1756

**17.** $-0.0102$

**19.** $1 + \dfrac{3x^2}{3!} + \dfrac{5x^4}{5!} - \dfrac{7x^6}{7!} + \cdots$

**21.** $\displaystyle\int e^x \, dx = x + \dfrac{x^2}{2} + \dfrac{x^3}{3 \cdot 2!} +$
$\dfrac{x^4}{4 \cdot 3!} + \cdots + C, C = 1$

**23.** $0.3 + 0.09t + 0.0135t^2 + 0.00135t^3 + \cdots$

**25.** $-0.0344$ mC

**27.** (a) $x + \dfrac{x^3}{3!} + \dfrac{x^5}{5!} + \dfrac{x^7}{7!} + \cdots$

(b) $1 + \dfrac{x^2}{2!} + \dfrac{x^4}{4!} + \dfrac{x^6}{6!} + \cdots$

### Exercise 24-3 (Page 876)

**1.** $1 - (x - 1) + (x - 1)^2 - \cdots$

**3.** $e + e(x - 1) +$
$\dfrac{e(x - 1)^2}{2!} + \cdots$

**5.** $\dfrac{1}{2} - \dfrac{\sqrt{3}(x - \pi/3)}{2} -$
$\dfrac{(x - \pi/3)^2}{4} + \cdots$

**7.** $\ln 2 + \dfrac{(x - 2)}{2} -$
$\dfrac{(x - 2)^2}{8} + \cdots$

**9.** 0.4118

**11.** 2.460

**13.** 0.5299

**15.** 0.8572

**17.** (a) $\cos a - (\sin a)(x - a) -$
$\dfrac{(\cos a)(x - a)^2}{2!} +$
$\dfrac{(\sin a)(x - a)^3}{3!} + \cdots$

(b) 0.6820

### Exercise 24-4 (Page 881)

**1.**

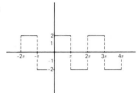

$\dfrac{8}{\pi}\left(\sin x + \dfrac{\sin 3x}{3} + \dfrac{\sin 5x}{5} + \cdots\right)$

**3.**

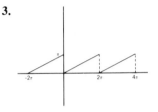

$\dfrac{\pi}{2} - \left(\sin x + \dfrac{\sin 2x}{2} + \dfrac{\sin 3x}{3} + \cdots\right)$

**5.**

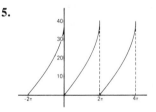

$\dfrac{4\pi^2}{3} + 4\left(\cos x + \dfrac{\cos 2x}{4} + \dfrac{\cos 3x}{9} + \cdots\right) - 4\pi\left(\sin x + \dfrac{\sin 2x}{2} + \dfrac{\sin 3x}{3} + \cdots\right)$

**7.**

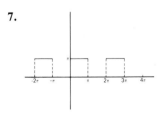

$$\frac{\pi}{2}, \; 2\sin t, \; 2\left(\frac{\sin 3t}{3} + \right.$$

$$\left. \frac{\sin 5t}{5} + \frac{\sin 7t}{7} + \cdots\right)$$

**9.**

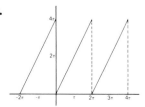

$$2\pi, \; -4\sin t, \; -4\left(\frac{\sin 2t}{2} + \right.$$

$$\left. \frac{\sin 3t}{3} + \frac{\sin 4t}{4} + \cdots\right)$$

**11.**

$$\frac{1}{\pi}, \; \frac{1}{2}\sin t, \; \frac{-2}{\pi}\left(\frac{\cos 2t}{3} + \right.$$

$$\left. \frac{\cos 4t}{15} + \frac{\cos 6t}{35} + \cdots\right)$$

**Review Questions 24-5**
*(Page 882)*

**1.** $1 + 3x + \dfrac{9x^2}{2!} + \dfrac{27x^3}{3!} + \cdots$

**3.** $3x - \dfrac{3^3 x^3}{3!} + \dfrac{3^5 x^5}{5!} -$

$\dfrac{3^7 x^7}{7!} + \cdots$

**5.** $x^{1/2} - \dfrac{x}{2} + \dfrac{x^{3/2}}{3} - \dfrac{x^2}{4} + \cdots$

**7.** $1 - x + \dfrac{x^3}{3} - \dfrac{x^4}{6} + \cdots$

**9.** $-\ln 2 + 2(x - \tfrac{1}{2}) -$

$\dfrac{4(x - \tfrac{1}{2})^2}{2} + \cdots$

**11.**

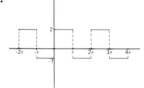

$$\frac{1}{2} + \frac{6}{\pi}\left(\sin x + \right.$$

$$\left. \frac{\sin 3x}{3} + \frac{\sin 5x}{5} + \cdots\right)$$

**13.** $-0.1053$

**15.** $0.5706$

**17.** *(a)* $V = 2t^2 + \dfrac{4t^3}{2!} +$

$\dfrac{8t^4}{3!} + \dfrac{16t^5}{4!} + \cdots$

*(b)* $0.8542$

# Chapter 25

**Exercise 25-1** *(Page 886)*
(Answers in this section show
first step of substituting functions
before simplification.)

**1.** $2 = 2$, part.

**3.** $\dfrac{-c}{x^2} = \dfrac{-(c/x)}{x}$, gen.

**5.** $e^{x+c} - e^{x+c} = 0$,     gen.

**7.** $-\cos x + \cos x = 0$, part.

**9.** $x(2c_1) = 2c_1 x$, gen.

**11.** $-xe^{-x} + (xe^{-x} + e^{-x}) = e^{-x}$, part.

**13.** $1 - \dfrac{c}{x^2} + \dfrac{x + c/x}{x} = 2$, gen.

**15.** $y(-x/y) = -x$, part.

**17.** $x(-1/x^2) + (1/x) = 0$, part.

**19.** $-4c_1 \sin 2x - 4c_2 \cos 2x +$
$4(c_1 \sin 2x + c_2 \cos 2x) = 0$,
gen.

**21.** $L\left(\dfrac{-AR}{L} e^{-Rt/L}\right) +$

$R\left(\dfrac{V}{R} + Ae^{-Rt/L}\right) = V$

**Exercise 25-2** *(Page 892)*

**1.** $y = ce^{x^2/2}$

**3.** $y = 1/x + c$

**5.** $y = 2/x^2 + c$

**7.** $y^2 = ce^{\tan^{-1}(x/2)}$

**9.** $y^2 = 4e^t + c$

**11.** $e^{-y^2} = c(x^2 + 1)$

**13.** $\ln y + e^x = c$

**15.** $y = -\tan x + c$

**17.** $\sqrt{1 + y^2} + \ln x = c$

**19.** $x^2 + (\ln y)^2 = c$

**21.** $y = 2x$

**23.** $y = \sin^{-1} x + 3$

**25.** *(a)* $v = 10(F - Ae^{-t/10})$

      *(b)* $v = 10F(1 - e^{-t/10})$

**27.** $y = e^{x^3 - 1}$

**29.** $q = Et/R + c$

**Exercise 25-3** *(Page 897)*

**1.** $y = 1 + ce^{-x}$

**3.** $y = 2xe^x + ce^x$

**5.** $y = x^2 + c/x$

**7.** $y = e^x - e^x/x + c/x$

**9.** $y = \dfrac{x^3}{10} + \dfrac{c}{x^2}$

**11.** $y = \dfrac{-\cos x}{x} + \dfrac{c}{x}$

**13.** $y = \dfrac{\sin t - \cos t}{2} + ce^{-t}$

**15.** $y = (x^2/2 + c)e^{2x}$

**17.** $y = -\tfrac{1}{2}\csc x + c\sin x$

**19.** $y = e^x + ce^{-x^3/3}$

**21.** $y = (x + 1)e^x$

**23.** $y = \dfrac{1}{x}(\tan x - 1)$

**25.** *(a)* $qe^{t/RC} =$

$\dfrac{1}{R}\displaystyle\int ve^{t/RC}\,dt + A$

      *(b)* $q = CV + Ae^{-t/RC}$

**27.** $v = \dfrac{1 + kt}{k^2} + ce^{kt}$

**29.** *(a)* $i = (V/R)(1 - e^{-Rt/L})$

      *(b)* As $t \to \infty$,

$e^{-Rt/L} \to 0$ and

$\displaystyle\lim_{t \to \infty} i = V/R$

**Exercise 25-4** *(Page 904)*

1. $y = c_1 e^x + c_2 e^{2x}$
3. $y = c_1 e^{2x} + c_2 e^{-3x}$
5. $y = c_1 + c_2 e^{-x}$
7. $y = c_1 e^{2x} + c_2 e^{-2x}$
9. $y = c_1 e^x + c_2 e^{-x/3}$
11. $y = c_1 e^{1.2x} + c_2 e^{-1.2x}$
13. $y = (c_1 + c_2 x) e^x$
15. $y = (c_1 + c_2 x) e^{-x/2}$
17. $y = c_1 \sin 2x + c_2 \cos 2x$
19. $y = e^{-x}(c_1 \sin 2x + c_2 \cos 2x)$
21. $y = e^x + e^{3x}$

23. $y = x e^{-2x}$
25. $x = c_1 \sin \omega t + c_2 \cos \omega t$
27. $i = c_1 + c_2 e^{-Rt/L}$
29. $i = (c_1 + c_2 t) e^{-t/\sqrt{LC}}$

**Review Questions 25-5**
*(Page 905)*

1. $y = e^{2x^3/3} + c$
3. $y^2 - e^{-2x} = c$
5. $y = \cot x + c$
7. $q = 1 + c e^{-2t}$
9. $y = \dfrac{3x}{2} + \dfrac{c}{x}$

11. $y = (1/x)(\sin x + c)$
13. $y = c_1 e^{-x} + c_2 e^{-3x}$
15. $y = c_1 + c_2 e^x$
17. $y = (c_1 + c_2 x) e^{-x}$
19. $y = c_1 \sin (x/2) + c_2 \cos (x/2)$
21. $y = e^{x^2/2 - 2}$
23. $y = \dfrac{1}{x} - x^3$
25. $m = m_0 e^{kt}$
27. $i = \dfrac{2L}{R^2} (Rt - L) + A e^{-Rt/L}$

# Partial Fractions Supplement

In calculus and other advanced courses it is necessary to do the opposite of combining fractions, that is, break up or resolve a fraction into *partial fractions* such as:

$$\frac{7x + 1}{(x - 2)(x + 3)} = \frac{3}{x - 2} + \frac{4}{x + 3}$$

The procedure for resolving a fraction into partial fractions is best understood by studying examples. The first two examples show the case of resolving a fraction whose denominator contains linear fractions that do not repeat.

### Example I
Resolve into partial fractions:

$$\frac{7x + 1}{(x - 2)(x + 3)}$$

### Solution
This is the example shown at the top. *A fraction whose denominator contains nonrepeating linear factors can be expressed as the sum of fractions whose denominators are the factors and whose numerators are constants:*

$$\frac{7x + 1}{(x - 2)(x + 3)} = \frac{A}{x - 2} + \frac{B}{x + 3}$$

The problem is to find the constants $A$ and $B$. Multiply the above equation by the lowest common denominator $(x - 2)(x + 3)$ to obtain:

$$7x + 1 = A(x + 3) + B(x - 2)$$

Let $x = 2$ to make $(x - 2)$ zero and solve for $A$:

$$7(2) + 1 = A(2 + 3) + B(2 - 2)$$
$$15 = A(5) + B(0) = 5A$$
$$A = 15/5$$
$$= 3$$

Let $x = -3$ to make $(x + 3)$ zero and solve for $B$:

$$7(-3) + 1 = A(-3 + 3) + B(-3 - 2)$$
$$-20 = A(0) + B(-5) = -5B$$
$$B = -20/-5$$
$$= 4$$

Therefore, the solution is as shown above. It can be checked by combining the two fractions:

$$\frac{3}{x - 2} + \frac{4}{x + 3} = \frac{3(x + 3) + 4(x - 2)}{(x - 2)(x + 3)}$$
$$= \frac{3x + 9 + 4x - 8}{(x - 2)(x + 3)}$$
$$= \frac{7x + 1}{(x - 2)(x + 3)}$$

## Example II
Resolve into partial fractions:

$$\frac{x + 8}{2x^2 - 3x - 2}$$

## Solution
Factor the denominator and express as the sum of fractions whose denominators are the factors and whose numerators are constants:

$$\frac{x + 8}{(x - 2)(2x + 1)} = \frac{A}{x - 2} + \frac{B}{2x + 1}$$

Multiply by the LCD $(x - 2)(2x + 1)$:

$$x + 8 = A(2x + 1) + B(x - 2)$$

Let $x = 2$ to make $(x - 2)$ zero and solve for $A$:

$$2 + 8 = A[2(2) + 1] + B(0) = 5A$$
$$A = 2$$

Let $x = -1/2$ to make $(2x + 1)$ zero and solve for $B$:

$$-1/2 + 8 = A(0) + B(-1/2 - 2)$$
$$15/2 = (-5/2)B$$
$$B = -3$$

Therefore the solution is

$$\frac{x + 8}{2x^2 - 3x - 2} = \frac{2}{x - 2} - \frac{3}{2x + 1}$$

The next example shows the case of a fraction whose denominator contains the square of a linear factor.

### Example III
Resolve into partial fractions:

$$\frac{x^2 - 6x - 10}{(x - 2)(x + 1)^2}$$

### Solution
*A fraction whose denominator contains the square of a linear factor can be expressed as the sum of fractions whose denominators are the linear factors and whose numerators are constants plus a fraction whose denominator is the square of the linear factor and whose numerator is a constant:*

$$\frac{x^2 - 6x - 10}{(x - 2)(x + 1)^2} = \frac{A}{x - 2} + \frac{B}{x + 1} + \frac{C}{(x + 1)^2}$$

Multiply by the LCD $(x - 2)(x + 1)^2$:

$$x^2 - 6x - 10 = A(x + 1)^2 + B(x + 1)(x - 2) + C(x - 2)$$
(Equation 1)

Let $x = 2$ to make $(x - 2)$ zero and solve for $A$:

$$(2)^2 - 6(2) - 10 = A(2 + 1)^2 + B(0) + C(0)$$
$$-18 = 9A$$
$$A = -2$$

Let $x = -1$ to make $(x + 1)$ zero and solve for $C$:

$$(-1)^2 - 6(-1) - 10 = A(0) + B(0) + C(-1 - 2)$$
$$C = 1$$

*To find the third constant B let* $x = 0$ and substitute the values for $A$ and $C$:

$$0^2 - 6(0) - 10 = -2(0 + 1)^2 + B(0 + 1)(0 - 2) + 1\,(0 - 2)$$
$$-10 = -2 - 2B - 2$$
$$B = 3$$

Therefore the solution is

$$\frac{x^2 - 6x - 10}{(x - 2)(x + 1)^2} = -\frac{2}{x - 2} + \frac{3}{x + 1} + \frac{1}{(x + 1)^2}$$

*The above example can also be solved by equating coefficients as follows. Multiply out the left side of Equation 1:*

$$x^2 - 6x - 10 = Ax^2 + 2Ax + A + Bx^2 - Bx - 2B + Cx - 2C$$

Group like terms:

$$x^2 - 6x - 10 = (A + B)x^2 + (2A - B + C)x + (A - 2B - 2C)$$

Equate coefficients to form three equations in three unknowns:

$$A + B = 1$$
$$2A - B + C = -6$$
$$A - 2B - 2C = -10$$

Solve this linear system. The answers are the same as above:

$$A = -2, \; B = 3, \; C = 1$$

The last example shows the case of a fraction that contains a quadratic factor that cannot be factored.

## Example IV
Resolve into partial fractions:

$$\frac{2x - 2}{x(x^2 + x + 2)}$$

## Solution
The quadratic factor $(x^2 + x + 2)$ cannot be factored. *A fraction that contains a linear factor and a quadratic factor that cannot be factored can be expressed as the sum of two fractions, one whose denominator is the quadratic factor and whose numerator is a linear factor of the form $Ax + B$, and the other whose denominator is the linear factor and whose numerator is a constant:*

$$\frac{2x - 2}{x(x^2 + x + 2)} = \frac{Ax + B}{x^2 + x + 2} + \frac{C}{x}$$

Multiply by the LCD $x(x^2 + x + 2)$:

$$2x - 2 = (Ax + B)(x) + C(x^2 + x + 2)$$

Multiply out and group terms:

$$2x - 2 = (A + C)x^2 + (B + C)x + (2C)$$

Equate coefficients:

$$A + C = 0$$
$$B + C = 2$$
$$2C = -2$$

Solve this linear system to obtain: $A = 1$, $B = 3$, $C = -1$. The solution is therefore:

$$\frac{2x - 2}{x(x^2 + x + 2)} = \frac{x + 3}{x^2 + x + 2} - \frac{1}{x}$$

**Exercises**

Resolve into partial fractions and check.

1. $\dfrac{3x + 5}{(x + 1)(x + 2)}$     2. $\dfrac{x - 8}{(x + 1)(x - 2)}$

3. $\dfrac{2x - 7}{(x - 2)(x - 3)}$     4. $\dfrac{6x + 4}{(2x - 1)(x + 3)}$

5. $\dfrac{7x + 5}{x^2 + x - 2}$     6. $\dfrac{3x + 13}{x^2 + 4x + 3}$

7. $\dfrac{4x - 7}{2x^2 - x - 1}$     8. $\dfrac{12x + 6}{3x^2 + 4x + 1}$

9. $\dfrac{3x^2 - 3x - 9}{(x - 1)^2(x + 2)}$     10. $\dfrac{2x^2 + 5x + 4}{(x + 1)(x + 2)^2}$

11. $\dfrac{5x^2 + 14x + 13}{(x + 1)^2(x + 3)}$     12. $\dfrac{-5x + 13}{(x + 1)(x - 2)^2}$

13. $\dfrac{19}{(x + 2)(x - 3)^2}$     14. $\dfrac{5x^2 + 24x + 32}{x(x + 4)^2}$

15. $\dfrac{3x^2 + 4}{(x^2 + 4x + 4)(x - 2)}$     16. $\dfrac{7x - 3}{(x^2 - 1)(x + 1)}$

17. $\dfrac{2x^2 - 1}{x(x^2 + x - 1)}$     18. $\dfrac{3x^2 - x}{(x + 1)(x^2 - 2x - 1)}$

19. $\dfrac{x - 5}{(x + 1)(x^2 + x + 2)}$     20. $\dfrac{x^2 - 2x - 3}{(x^2 + 1)(x - 1)}$

# Index